WITHDRAWN
FROM
COLLECTION

FORDHAM
UNIVERSITY
LIBRARIES

Overskill

The
Decline of
Technology
in Modern
Civilization

Overskill

by Eugene S. Schwartz

Chicago: **Quadrangle Books**, 1971

OVERSKILL. Copyright © 1971 by Eugene S. Schwartz. All rights reserved, including the right to reproduce this book or portions thereof in any form. For information, address: Quadrangle Books, Inc., 12 East Delaware Place, Chicago 60611. Manufactured in the United States of America. Published simultaneously in Canada by Burns and MacEachern Ltd., Toronto.

Library of Congress Catalog Card Number: 73–143575

SBN 8129–0180–0

To Amitai and Elana
and all the young of the earth
who seek a future

Preface

THIS BOOK will probably be indexed under science and technology. It could also be indexed under autobiography, for it is the story of my search for order and truth in a restless universe and a fast-changing and fearful world. There is something challenging about an ordered set of ideas, a philosophy, a system, a credo, a religion—and something comforting in its acceptance. The ideas provide landmarks, a mooring, a home to return to. They can appear to provide a stable base for one's journey in life.

The ordered set of beliefs, however, can become a false sanctuary and an escape from reality. The order is always man-made. Wherever he goes man encounters only himself. Whenever he attempts to understand things outside himself he finds answers only to the questions he formulates; the patterns he finds are those he created; the order he discovers is one he establishes by distilling away an accompanying disorder. The beliefs man finds are ultimately those he seeks.

I have sought for order: in evolution, in psychoanalysis, in philosophy, in Marxism, in science. Each set of beliefs defined order in a different way and distilled away different patterns of disorder.

They all perceived the elephant in a unique way. While they were temporarily comforting, they were all remiss in providing only a partial vision.

Science, I had once thought, was the ultimate answer to understanding and freedom. Science did away with metaphysics and philosophy. It established standards for understanding a fact as a fact; it afforded a universal language in mathematics for communication. Its methods and conclusions were open for revision, and its truths were assured by the error-correcting capability of an international community of scholars. Cybernetics and the automatic factory, intelligence amplifiers and world-wide economic planning, social feedback mechanisms and technological innovations—these appeared to be the ingredients from which an ordered society could be constructed. Homeostasis, the self-regulating mechanism of biological organisms, could serve as a model of the stable society.

I found, however, that homeostasis is a static concept. It cannot account for change. It cannot deal with conflict nor can it comprehend dialectics. The homeostatic society is a society for robots. Cybernetics is for automata, and the planned society is a prelude to the universal concentration camp. As I probed deeper into the meaning of science I saw that the euphoric vision of science and technology is a myth—now become dangerous and threatening. The set of beliefs that constitute science have no greater validity than any other set ever propounded by man in his long quest for meaning. Although they have been proclaimed as being free of subjectivity and metaphysics, they are both subjective and metaphysical. Science is a twentieth-century religion and, to paraphrase Marx, it has become a new opiate of the people. The opiate in the short run may help to delude people to think all is well. In the long run I believe the denouement will be along the lines suggested in the concluding chapters of this book.

The excesses of technological civilization threaten all present and future life on earth. Ignorance, hubris, deceit can no longer be accepted as erroneous policies, mistakes by mortal men. A cloak of expertise and scientific jargon cannot mask or dissipate the dangers. It is time to consider the policies that lead to destruction of the fabric of life as crimes. Unintended crimes, perhaps, but crimes no less. These crimes against man and nature can no longer be tolerated as being the inevitable consequence of an inexorable progress. They cannot be accepted as the product of the "invisible hand" of science and technology that eschews values and ethics. The "laws" of science and the myth of progress must not be allowed to continue to provide a cover for irresponsibility.

The earth belongs to all men who perceive it in infinite ways. Tomorrow can be an infinite of futures if we do not foreclose them. To help to keep our present alive and our futures open to promise is the purpose of this book.

The thesis of the book is suggested in the introductory chapter: the crises that threaten human survival are inherent in science and technology and are not amenable to rectification by more science and technology. Chapter 2 develops this theme and traces the rise of science especially during the past three hundred years. Science is shown to be based upon a set of inconsistent axioms that are accepted as an act of faith. The scientific religion is canonized in a false progress that is a pillar of the new religion. The relations of technology to science and the technological process are described in Chapter 3. The concept of the quasi-solution is introduced in Chapter 4, where it is shown that all technological solutions of problems leave a residue of unsolved problems that proliferate faster than their solutions. The euphoric technicism engendered by man's landing on the moon is questioned in Chapter 5, where the basic limitations to what man can know and what man can do are reviewed.

The next four chapters consider significant aspects of the major crises that confront mankind in terms of technological realism, a dialectic approach to the proliferating and interrelating residue problems arising from quasi-solutions to problems of land, food, population, energy, autos, waste, and war. These chapters are not another "rape of the earth" exposé, nor an appeal for an ecological conscience. They illustrate the increasing difficulty of extricating ourselves from problems by increasing usage of the techniques that helped to foster the problems.

The vast experiment to create an artificial environment that may be antithetical to the genetic program of mankind is considered in the "Laboratory in Vivo" in Chapter 10. The experiment seeks to test the limits of the earth on all living species, including man, with consequences that bear significantly on present and future generations. The decline of technological civilization is set forth in Chapter 11 as dialectical processes undermine the driving forces of science and technology. Law-and-order is shown as the shibboleth of technology in decline, as counter-technology will further sap the wealth of the world, a wealth that is a cover for a vast accumulation of debts that have impoverished the earth.

Is there hope? Perhaps—if we can achieve an inefficient society that is man-oriented, not machine-oriented. The philosophical underpinnings of such a society are suggested in the concluding chapter.

Contents

Preface vii
1 Introduction 3
2 The Faith That Dethroned Faith 14
3 The Technological Process 41
4 Quasi-Solutions and Residue Problems 62
5 Limits, Impotence, and the Unsolvable 78
6 Technological Realism 1: People, Land, and Food 112
7 Technological Realism 2: Energy 147
8 Technological Realism 3: Autos and Waste 180
9 Technological Realism 4: War 202
10 Laboratory in Vivo 219
11 The Decline of Technology 246
12 Toward the Inefficient Society 281
Suggestions for Further Reading 310
Index 325

Overskill

1 *Introduction*

A WALL OF WATER gushes from a fountain on the campus of the Illinois Institute of Technology in Chicago in a courtyard bounded by the austere, rigidly functional buildings of Mies van der Rohe. The water is forced four feet into the air in the shape of an inverted cone to form a chrysanthemum pattern before falling back into a pool below. Pressures, temperatures, and flow rates have been calculated to spew water from a specially designed circular spout that circulates it throughout the year. A passerby in subzero winter can witness the strange sight of water spouting and falling back on ice and snow encrusting the fountain.

No one lingers about the fountain in the winter, nor does it attract watchers in the warmer seasons. Students do not tarry to gaze at the fountain because its solid wall of rising water connotes power, not contemplation. No gentle spray rises from the spout to catch sunbeams and convert them into rainbows. There is no trickling or dripping or gurgling, only a ponderous rising and falling back. The fountain is too small to imitate the beauty of Chicago's Buckingham Fountain, several miles distant on the lakefront; it is too gross to capture the mood of the nearby Dove Girl and Turtle Boy fountains.

The fountain spouts forth in summer heat and winter cold, unattended, unwatched—a symbol of a technology unrelated to human purpose and human aspirations.

Such mindless technology is inundating the world, sweeping all before it. Although the graphs of "progress" continue to rise, man is only now beginning to perceive the price that is exacted to enable the fountains of technology to continue to spew forth their promise. Mingled with the "Eureka" cries that accompany the scaling of new heights of scientific achievement are the rising wails of mankind who suffer from these same achievements. An adventurous and truth-seeking science is becoming an appendage of the technology it spawned; truth is becoming that which politics and economics in their parochial national manifestations demand; freedom of inquiry and revolt against authority are being transformed into a new orthodoxy administered and controlled by a new elite. Science and technology have become established myths of the descendants of those forebears who sought to abolish myths in the name of reason. Science has become a secular religion; technology is its temple, efficiency its dogma.

What if the religion of technology is not a truth but a falsification of man and nature and society? What if this religion will bring about a homogenization of all cultures so as to preclude multiple responses to the dangers of the present and the challenges of the future? What if the elaborate mythology erected over the past three hundred years and reaching maturity in our era is found to be spun from the same figments of self-deception, unrequited wishes, and dreams that have characterized men's myths from the dawn of human life on earth?

It does not matter that the new mythology is shrouded in the cloak of science with its purity and vision attended by an esoteric priesthood of savants and corporate technicians. Their Pythagorean mystique is enshrined in huge temples built to worship the atom and the machine. They are the guardians of the cabbala that guides space ships, plots the trajectories of hydrogen warheads, and directs the "efficient" operation of the industrial economy. They communicate with the gods of automation who dwell in the wired intricacies of computers. They count and recite their symbolic littanies inherited from the notched sticks and abacus of the Orient, Napier's bones, and the calculators of Pascal, Leibniz, and Babbage.

Man treads upon the moon and on earth asks the age-old question: Is there a God? Man transplants human organs and ponders another timeless question: Is man immortal? Has modern man discovered the truths that ancient man had sought in vain, or are our answers self-satisfying merely because they are phrased in a language we have embellished with new symbolisms that reflect our uncertainties? Does our fear of the unknown differ from that of the ancients, whose fear was a reflection of the simple mystery of the world which surrounded them, whereas ours is a symptom of the overwhelming knowledge that still points to inexplicable mysteries?

Civilization, Oswald Spengler wrote, is the conclusion of a culture, "death following life, rigidity following expansion, . . . petrifying world-city following mother-earth." Rejecting the concept of eternal progress, Spengler believed that a civilization dies when it has exhausted its possibilities. "Pure civilization, as a historical process, consists in a progressive taking down of forms that have become inorganic or dead."[1]

While denying the Spenglerian metaphor that conceived of society as an organism that passes through successive ages, Arnold Toynbee, too, found a consistent pattern of breakdown and disintegration of all historical civilizations. He rejected the theory of the running down of the clockwork of the universe, or decline due to the deterioration of the quality of individuals, or, finally, the cyclical theory of history that appears as a theme in many ancient classics, in favor of his own tripartite explanation of the rise and fall of civilizations. The breakdown, he wrote in *A Study of History*, "arises from a failure of creative power in the minority, an answering withdrawal of mimesis on the part of the majority, and the consequent loss of social unity in society as a whole."

The Age of Reason inaugurated by the Enlightenment and por-

[1] Oswald Spengler, *The Decline of the West*, I (New York: Knopf, 1961), 31–32.

tended by Galileo is, in turn, a denial of all the theories of history that forecast the decline and dissolution of civilizations. Western scientific civilization was to be immune to the fatal flaws that undermined previous civilizations. "Progress" was the new philosophical armament that was to ward off decline and carry mankind onward and upward in a never-ceasing spiral of growth, expansion, and development. The touchstone of Progress was science, and the means to an earthly utopia was technology. Whereas other civilizations recognized their limitations and mortality, Western scientific civilization formulated a doctrine of unlimited possibilities to development, with ever-present new possibilities extending endlessly into the future.

The philosophy of "progress," although ostensibly based on the "reason" of science, was an act of faith, a belief in a secular religion. It supplanted earlier mythologies embedded in magic from which science sprang. Science triumphed over a less reasonable magic, because, pragmatically, science proved to be more efficient than magic. After all, were not the Whitunside festivals and sexual couplings in the fields of Peru and Java and West Africa to promote fertility of the soil and bountiful crops dispensed with because the application of trisodium phosphate to the soil resulted in increased yields of crops?

Technology became the servant of science in translating scientific "truths" into better life for the peoples of the earth. The secular religion was hailed as the new Messiah that would bring man surcease from toil and freedom from want. It would liberate the enslaved peoples of the earth; it would feed the hungry and cure the sick. It would carry man forward to progress, satiety, power.

The steel mill, the atomic reactor, one's country's name emblazoned on an orbiting satellite—these were the symbols in the dreams of leaders of Third World nations.* An infrastructure of

* The "Third World" refers, generally, to the pre-technological societies located in Central and South America and in Africa and Asia. The "First" and "Second" worlds (we will leave it to future historians to ascertain the order) refer, generally, to the technological societies of the capitalist and communist countries. "Third World" is used instead of "underdeveloped" or "developing" nations because the latter terms are normative in the sense that the technologically developed nations are considered to be the desired norm to which all nations must be compared. This

roads, communications, and factories—this was the promise of the "haves" to the "have-nots." Scientists and technicians invaded the developing nations, proselytizing for the new religion.

The religion of progress through science and technology combined elements of exaltation and grandeur with the arrogance of established religions. Mortal man has escaped earth's gravitation and trod upon the moon. Through biological alchemy, the promise of immortality was held to be within reach. Scientific man has no need for that other God when the universe is opened by rockets and the promise of immortality is made by genetics and molecular biology. The fire and brimstone of hell can no longer terrorize man who has experienced Dachau and Hiroshima. God, as a cultural anachronism, appears quaintly in an astronaut's prayers and is invoked to bless bombs and soldiers, but science and technology direct and power man's restless spirit.

In the historically short time span of three hundred years, science and technology have transformed the earth and its inhabitants. The earth pulses and labors. The electron and the atom have become genii to reshape the earth. Time is condensed, space telescoped. The world approaches a technological unity. The Messiah is nigh.

Alas! Even as the vision becomes intensified, the reality recedes. "The Road we have been travelling is deceptively easy, a smooth superhighway on which we progress with great speed, but at its end lies disaster," Rachel Carson warned in *Silent Spring*.[2] Mankind, meanwhile, ponders: We have sowed our fields with trisodium phosphate and there are more hungry people in the world than ever before. We have built hospitals and clinics and there are more sick people than ever before. We have built schools and illiteracy flourishes. We have built factories and filled them with machines and find that we are slaves to the machines. We have cut down our forests, depleted our natural resources, and overmined the earth. We have poisoned our lakes and streams, polluted

---

dichotomization is part of the arrogance of the technological societies. In a humanistic sense, the developed nations are becoming increasingly underdeveloped.

2 Rachel Carson, *Silent Spring* (Boston: Houghton Mifflin, 1962), p. 277.

the air we breathe, and transformed the face of the earth into a labyrinth of concrete ribbons, scarred hills, and monstrous slag heaps.

Living space is compressed on a crowded planet. We have cleared the jungles to build cities and have transformed our cities into jungles. We have erected skyscrapers to house millions, and we pass each other in the corridors as strangers. We have spewed out automobiles by the millions, and they crawl on the congested highways at a pace slower than a horse-drawn carriage. Speedy jet planes carry man back and forth across the earth as he seeks to escape problems at home and finds that they have taken root in foreign lands. The rich grow richer, the poor poorer. The satiated grow sick from overconsumption; the hungry sicken from starvation. The technological machine continues to disgorge products man neither needs nor wants.

The despoliation of the earth and the poisoning of man's habitat go hand in hand with the increasing anarchy between nations, the increasing breakdown of social institutions, and the growing alienation of individual men. Civilization is increasingly becoming the neurosis about which Freud wrote with great pessimism.

The period prior to the World Wars may well have marked the apogee of Western civilization. We may now be in what Toynbee termed a "time of trouble," if, indeed, we have not already entered the "interregnum." Warfare has become endemic to all nations. The military machines that devour an ever-increasing portion of the world's resources and manpower have paced the growth of technology and today threaten all mankind with annihilation, posing a threat to the continuation of biological life on this planet.

The confluence of overpopulation and lagging food production leads to predictions of mass famines with deaths numbering in the tens of millions within a decade. The pollution of our habitat provokes warnings that life on earth is endangered. Peace through a balance of terror becomes more questionable with each passing year as the arms race mounts and the probability of the ultimate holocaust grows. D. C. Somervell's resumé of Toynbee poses the question coldly before us as the torrent of problems converges to a frightening denouement: "A society does not ever die 'from natural causes,' but always dies from suicide or murder—and nearly always from the former."³

While each nation has sought to stave off its murder by preparing to murder other nations, collectively the nations have brought themselves to the verge of mutual suicide, a suicide not only by weapons but as a result of three hundred years of "progress" which may conclude in the rape of the earth. Man has posed an immediate threat to his own existence and, paradoxically, has menaced that posterity upon which the millennial faith of progress has been nurtured.

The issue is not one of technological optimism or technological pessimism. Posing the problem in this form begs the question. The problem is one of human survival. The role technology can play in the struggle to retain human life on the planet is one of the problems mankind now faces. "It seems possible that the new amount of technological power let loose in an overcrowded world may overload any system we might devise for its control; the possibility of a complete and apocalyptic end of civilization cannot be dismissed as a morbid fantasy."⁴ This statement by Don K. Price, former president of the American Association for the Advancement of Science, calls for a new technological realism. The drama now being played out on planet earth does not brook any facile simplifications of a Janus-faced technology that has the potential for both good and evil. Wise men will, it is implied, accentuate the good; venal or ignorant men will choose the evil. Technology as a human-devised means to solve problems is itself in question.

Belatedly, Western man is beginning to realize what he has wrought on his limited habitat. The earth is a finite territory with finite resources. In wasting and despoiling this bounty, man is faced with limits—limits in space, in food, in raw materials, in pure water, in fresh air. The ravaged earth does not fight back but presents its stricken face to a civilization that only now is dimly becoming aware of the limits on the possible and can acknowledge its hubris in nothing more than contrite despair and fear.

³ Note by D. C. Somervell in Arnold J. Toynbee, *A Study of History* (abridgment of Vols. I–VI) (New York and London: Oxford University Press, 1958), p. 273.
⁴ Don K. Price, "Purists and Politicians," *Science,* CLXIII (January 3, 1969), 25–31.

The violence against the earth and man that has been the mainspring of the Industrial Revolution has run amok as the possibilities for future civilizations begin to be foreclosed. Technology has hastened the process of foreclosure by homogenizing the world and further reducing alternative futures before the peoples of the earth. With its shibboleth of efficiency, "progress" has jeopardized human survival, for nature through countless millennia of evolution has never been efficient. Survival has been achieved through safety, through maintenance of reserves, through following myriads of paths, through exploration of many potentials, through proliferation of species, through unpredictable divergences and mutations, both biological and social.

What has gone wrong with the "progress" that has led man in but a short three hundred years to the edge of disaster? Has frail man with a propensity for both good and evil misused "reason"? Have the "relations of production" in private hands subverted the utopia that might have been? Has science for too long been influenced by a reductionist technique whereby it is assumed that breaking everything down into its smallest parts will reveal the whole, whereas a holistic science of interdisciplinary endeavors will make the necessary adjustments? Has piecemeal "progress" been too fragmented, whereas a cybernetic model of an integrated "system" will lead to more efficient planning and control? If man has unlocked the secret of the atom and trod upon the moon, cannot similar crash programs, through the mobilization of money and scientist-technicians, solve the problems that endanger man?

It is the thesis of this book that the answer to the above questions is uniformly "no." Man has not misused the "reason" of science and technology to bring us to our present state. Instead, the tragedy of the present was inherent in the basic premises of science from its early formulations of the modern age, beginning with Galileo Galilei, Francis Bacon, and René Descartes, and no reformulation of the questions with which science deals will alter these philosophical defects.

What then of technology? There is a school of thought that says all technology is bad. It is a destroyer of all human values; it is autonomous and has become an end in itself. Another school, the predominant one, states that technology is a great blessing, powering "progress" and advancing humanity while improving man's condition as predicted by the Enlightenment prophets. A third school maintains that technology is but the continuation of the advances that mankind has registered throughout the ages. It is nothing new, marks no revolutionary break with former practices, and is well recognized as a factor in social change.[5]

An increasing number of technicians, administrators, and scholars, however, confronted with the crises to which technology has contributed, are beginning to ask questions about the role of technology in society. Out of this questioning has come a range of suggested policies. There are those who speak of controlling science through law. The United States Constitution, for example, would add the right to a wholesome environment to the Declaration's rights "to life, liberty, and the pursuit of happiness." The futurists contend that by predicting possible futures on the basis of present and foreseen developments, the future can be controlled and policies, both private and governmental, can be undertaken to realize the most desirable futures. Technology assessment, its proponents claim, can analyze in advance the benefits and risks of exploiting new technological capabilities. Decisions would then be made by political and economic institutions to select those technologies that promise positive results, while those with deleterious results would be withheld. This latter policy is consistent with the views of those who urge technological renunciation or technological disarmament. Man's finger, in this view, is on an environmental trigger that can destroy man as surely as can nuclear weapons. The way to control is through abjuring those technological developments that may have short-range benefits but long-range defects.

The most vocal and widespread policy advocated to meet the crisis engendered by the revolutionary transformation of the earth and society by technology is—more technology. Writing in *Science*, the mouthpiece of the American scientific establishment, Professor Harvey Brooks of Harvard states the case for more technology to cure the ills technology has brought in its wake:

[5] Emmanuel G. Mesthene, *Technological Change: Its Impact on Man and Society* (New York: New American Library, 1970), pp. 15–19.

With respect to the great modern problems—what I call the four P's of population, pollution, peace, and poverty—it may be that articulating these is the most important part of the problem—that once these needs are formulated in the right way, the technological solutions will become obvious, or will fall into place.[6]

Can the reorientation of technology solve the problems technology has created? Can an extension of the scientific genius undo what the genius has done? Can more, larger, and more efficient techniques retrieve man from the catastrophes that threaten him and that arose from fewer, smaller, less efficient techniques?

Again the response is negative. New and more powerful techniques cannot solve the problems that technology has engendered because technology is a dialectical process arising from the relationships of man's interaction with nature. Technology is subject to change and conflict. It has limitations and constraints. It can also be self-destructive—and destructive in a way that is not derived from unreason or inefficiency but from the process itself.

The dialectical process of technology is not only self-destructive; it also undermines the science which supports it. The scientific enterprise and an expanded technology are likely to negate themselves and to extirpate mankind in the process.

Ours is an exhaustible world. We have lost much of our freedom to experiment and to choose because of the pressing urgency of the converging crises brought on, in the main, through the agency of the scientific and technological revolutions. We have lost space by shrinking the earth, and we have compressed time to the extent that a child grows up in a world that changes faster than he can adapt to it. If we rush pell-mell into a troubled and dangerous future with the same slogans and practices that have brought us to this situation, can we expect other than disaster?

Man's massive and total assault against nature has been characterized by abysmal ignorance and monumental stupidity. The assault has been led by a political economy that has elevated greed, selfishness, and acquisitiveness to holy virtues and has been abetted by a science and technology that excluded all morality and

---

[6] Harvey Brooks, "Applied Science and Technological Progress," *Science*, CLVI (June 30, 1967), 1712.

ethics from its practices. No individual or group bore responsibility for the uses to which knowledge and praxis were put. It is as if the "invisible hand" of political economy were to be matched with an "invisible hand" that would mesh the disparate and diverging sciences and technologies into a human enterprise that would promote human welfare.

The "invisible hands" have brought unimagined wealth and comfort to a small fraction of the earth's population in the short space of three hundred years. To achieve this ephemeral end, the earth has been stripped of its resources and the human habitat has been made nearly uninhabitable. Faced with converging crises and the failure of the "invisible hands" on the economic and the ecological fronts, we are now told that the answer to our problems is "more of the same." More science, more technology, more research. The hubris that has brought man to the brink of catastrophe propels man on to accelerate that catastrophe.

Man, at the peak of a nebulous "progress," is threatened on one hand by a suffocating death and on the other hand by annihilating disintegration. We must now re-examine the tenets that brought us to this state. The reason that is a faith must be re-evaluated. The science that reached for the moon as mankind began to lose the earth must be questioned. The tools that harnessed nature but destroyed her in the process must be recast.

2 *The Faith That Dethroned Faith*

*H*ISTORIANS are prone to paint time and events with a broad brush that accentuates some details and places others in shadow in order to depict the essence of a period. The distilled essence, as well as details and shadows, are always distorted by the vantage point of the historian who edits the past with the hindsight secured by living in what is the future of the historical object under study. As futures change and vantage points shift, re-analysis and re-evaluation are directed to repainting the canvas of history in new colors and patterns. What once were shadows are now highlighted; obscure details are made prominent; former major effects are blocked out.

It is time now to reassess the history of the past three hundred years in the light of the plight of the earth rather than from the euphoric vantage point of what is called "progress." The problems confronting mankind are not newly born but rather have evolved from a concept of man and nature that developed during the course of the past three centuries. Mankind finds itself in a dilemma not in spite of but because of the concepts of science and the developments of the technological process during this period. Our present troubles were inherent in the basic tenets of science. The continuous compounding of our difficulties is traceable, in part, to the practices of technology.

The history of modern science is comprised not only of the triumphs of scientific discovery and their intellectual implications but also of the development of a body of beliefs that constitute the philosophic foundations of what has become a secular religion. The edifice of modern science has been erected on a set of axioms whose truth can only be accepted on faith. "There is no science that is without unconscious presuppositions," according to Spengler, and "There is no Natural Science without a precedent religion. . . . Every critical science, like every myth and every religious belief, rests upon an inner certitude."[1] Science "has remained an antirationalistic movement based upon a naive faith," Whitehead observed. It has repudiated philosophy and "has never cared to justify its faith or to explain its meanings."[2] Bertrand Russell, in discussing the limitations of the scientific method, noted that "science started with a large amount of what Santayana calls 'animal faith,' which is, in fact, thought dominated by the principle of the conditioned reflex."[3]

Let us, then, examine those unstated axioms of science whose truths are accepted as an act of faith. Twelve of them can be identified:

1. Man is not naturally depraved.
2. The "good" life on earth is not only definable but attainable.
3. Reason is the supreme tool of man.
4. Knowledge will free man from ignorance, superstition, and social ills.
5. The universe is orderly.
6. This order can be discovered by man and expressed in mathematical quantities and relations.

[1] Spengler, *op. cit.*, pp. 380–381.
[2] Alfred North Whitehead, *Science and the Modern World* (New York: New American Library, 1953), p. 17.
[3] Bertrand Russell, "Limitations of Scientific Method," in *The Basic Writings of Bertrand Russell*, ed. Robert E. Egner and Lester E. Dennon (New York: Simon and Schuster, 1961), p. 624.

7. Although there may be many ways of perceiving nature—e.g., art, poetry, music, etc.—only science can achieve "truth" that will enable man to master nature.

8. Observation and experimentation are the only valid means of discovering nature's order.

9. Observed "facts" are independent of the observer.

10. Secondary qualities are not measurable and hence not real.

11. All things on earth are for use by man.

12. Science is neutral, value-free, and independent of morality and ethics.

These axioms are the culmination of a distinctive Western philosophy grounded in the Judeo-Christian view of man and nature. Taken together they constitute a philosophy based upon metaphysical presuppositions, although both the philosophy and metaphysics are denied by scientists. The belief in the redemption of mankind led by reason through science (axioms 1, 2, 3, and 4) has its origins in the millennial viewpoint of a new utopia that was to be attained on this earth. The redemption would come through faith in the historical process as influenced by science. All mankind and especially posterity, which to Diderot was a secularized version of the Christian after world, was to be redeemed by this new faith. The orderliness of the universe (axiom 5) is a concept derived from the religious belief in the rationality of God who set into motion a perfect world to demonstrate his omnipotence. This faith, Whitehead notes, "is a particular example of a deeper faith. This faith cannot be justified by any inductive generalization."[4] The belief in the reduction of this order to mathematical formulations (axiom 6) is derived from the Pythagorean view that the mystery of the universe is revealed through numbers. That the axioms are not self-consistent—the method of axiom 8 does not support the other eleven axioms—is a glaring inconsistency that is bypassed through faith in the remaining axioms.

That these axioms are derived from a long historical heritage is not in itself a reason to challenge them. Indeed, they would be more suspect if they emerged *de novo*. That the axioms appear to

---

[4] Whitehead, *op. cit.*, p. 20.

contradict the claims of scientists to have banished all metaphysics from the body of knowledge they have laboriously erected is a challenging philosophical problem, but again not a sufficient cause to reject the axioms. That science has rendered religion irrelevant only to erect a new religion in its stead and has substituted the inexorable laws of nature for the inexorable laws of Providence is an intellectual problem, but still not a cause for challenging the axioms.

The axioms can and must be challenged, however, especially axioms 7 through 12, because of their consequences. These consequences have led to an overweening pride on the part of the scientific community, a casting down of mankind, a divorce of men from the natural world which they inhabit, a greedy exploitation and despoliation of the earth and its resources, a freeing of man from moral and ethical responsibility, and, within the short space of three hundred years, bringing all of man and his habitat close to total annihilation.

How has this magnificent and grand adventure of man (but note, only Western man) that began with the discovery of language and fire and simple tools in prehistoric times and has seen man land on the moon come to such foreboding straits? What has led from the exultation of Johannes Kepler as he helped to lay the foundations of the new science ("I will endulge my sacred ecstasy"), to the remorseful confession of J. Robert Oppenheimer in the aftermath of Hiroshima ("The physicists have known sin")? How have the desires and hopes of science to free man come so close to enslaving him? In the exploration of this question, let me briefly note the major stages in the formulation of science's principal axioms.

Out of prehistory, the sacred texts of ancient peoples, and the myths of primitive tribes has come the story of man's early confrontation with the cosmos. It is a story intimately connected with man's search for meaning in a world about which he knew little. It is the story of man's quest for truth and certainty in a world in which no truth is self-evident and in which life amidst nature is uncertain. As early man developed in the nascent civilizations of fertile river valleys, he sought to structure the chaos of experience to bring order and meaning into life. Speaking the language of myth, his speculations on man and nature were conceived in terms of human experience viewed in the light of cosmic manifestations. Man and society were viewed as part of nature, and early cosmologies developed from this relationship. The daily rhythm of the rising sun and the annual flooding of the waters of the Nile inspired the Egyptian cosmology. Mesopotamian life developing along the banks of the Tigris and Euphrates rivers also noted the order of the great cosmic rhythms, although the vagaries of these rivers, unlike those of the Nile, introduced elements of force and violence and uncertainty.

In seeking to comprehend this prehistoric world man conceived of order arising from nonorder. For in the beginning was chaos, "And the earth was without form and void; and darkness was upon the face of the deep." According to an early Mesopotamian cosmology, in the beginning there were Tiamat, the primordial Substance, and Apsu, Matter, the mother and father pair that begat Mummu, Chaos. Among the many gods that followed in the course of eons was Marduk who slew Tiamat, the mother of the gods, and slit her in two, fashioning the heavens from one half and earth from the other.

Egyptian cosmology conceived this chaos as the Ogdoad, consisting of eight creatures, four frogs and four snakes. Chaos consisted of Nun and Naunet, the formless ocean and the primeval matter, of Huh, the Illimitable, and of Hauhet, the Boundless. Kuk and Kauket, Darkness and Obscurity, together with Amon and Amaunet, the Hidden and Concealed ones, completed the roster of chaos gods. From Chaos there came Atum, the Creator whose children were Shu, the Air, and Tefnut, Moisture; and from them issued the children, Geb, the Earth, and Nut, the Sky.

The ancients connected the order of their civilizations with the cosmic order, the manifestations of a hierarchy of different powers observed in nature. Ancient man sought an answer to the problem of telos in these manifestations, in the belief that the divine was immanent in nature and that nature, in turn, was intimately related to society. It was believed that nature's rhythms—the day following night, the passage of the seasons, the ebb and flow of the tides, the cycle of the regeneration of life and its decay in death—together with other cyclical and ordered manifestations of the universe, revealed that there was purpose and design in this order within which man and nature operate.

Man adjusted to nature by creating answers and explanations in myth and by magic sought to control the unknown forces. The control of nature for practical events by rites and spells was based on man's confidence that he could dominate nature directly if only he knew the proper methods. Discovering the limitations of his abilities to control nature, he made an appeal to supernatural beings—demons, ancestor spirits, and gods—for the primitive mind sought an answer in "who" rather than in "how." The deification of natural phenomena expressed in mythopoeic form led to the worship of a pantheon of natural gods who accounted for the phenomena. The gods were not created *ab initio,* as we have seen with the Egyptian and Mesopotamian gods, but themselves arose from a primeval substance in accordance with the particular theogony of the worshipers. Nor were the gods sovereign inasmuch as they had emerged from a pre-existent substance and were subject to decrees of fate as were mortal men. Magic transcended the gods who utilized its power for their purposes.

Agriculture was a creative act of union between man who sowed the seed and the earth that brought forth the plant. The earth was perceived as a primal Mother, and the act of plowing was regarded by some ancient tribes as inflicting a wound upon the earth. The Baiga of India used a wooden hoe for plowing because they believed an iron plow would tear the earth's breasts and break her belly. Ancient fertility rites were compounded of a blend of magical arts and confrontation with a vast and mysterious unknown wherein a connection was apprehended between the growth and decay of grains and fruits and the birth and death of human beings.

Each civilization modeled its viewpoint according to the state of its material achievements and its political characteristics. The ancients conceived of man as a part of society and "imbedded in nature and dependent upon cosmic forces." For them, Henri Frankfort relates, "nature and man did not stand in opposition and did not, therefore, have to be apprehended by different modes of cognition."[5] The Babylonian and Egyptian civilizations blended praxis and ritual in their material achievements. The gods were a part of nature which was a manifestation of the divine.

With the ancient Hebrews there arose a profound break with the then extant view of nature. Yahweh was declared to be a transcendent God who was not in nature. Theogony was dispensed with, for Yahweh's will was transcendent and sovereign over all. Natural phenomena were no longer manifestations of the divine but reflections of God's omnipotence. God transcended all phenomena, which became merely reflections of his will. All values were ultimately attributes of God, and man was to serve as his interpreter and servant. The Hebrew view on God and nature was well expounded in the Nineteenth Psalm: "The heavens declare the glory of God; and the firmament showeth his handiwork." Magic rites and incantations had no place in the new Hebrew religion; praxis became an act of understanding God's handiwork.

The transfer of the problems of man from the realm of faith and myth to the intellectual sphere was accomplished in the main by the Greeks. Although older civilizations had bequeathed to the Greeks a considerable body of scientific knowledge and acquaintance with techniques, it was Greek civilization that transformed this empirical knowledge into the beginnings of a theoretical science. In the period 600–400 B.C. a scientific outlook on the world and society was formulated for the first time in history. From Thales to Aristotle, the ancient Greeks speculated on the nature of reality in the light of everyday experience without recourse to ancient myths. From the tetrad, earth-air-fire-water of Anaximander to the "ceaseless flux" of Heraclitus and the "atoms" of Democritus, the Greek philosophers formulated systems to explain existence, the universe, man, and society. By the rejection of magical intervention, they laid foundations for the scientific approach to the interpretation of nature. The observation of technical and natural processes together with rudimentary experiments made possi-

---

[5] Henri Frankfort, *Before Philosophy* (Harmondsworth: Penguin, 1954), p. 12.

ble the reduction of these observations into a system of logical coherence.

The arrest of Greek science was not in the main due to errors of theory and methodology, for the germs of many of our currently held theories appear in the works of the Greek scholars. It was rather due to the divorce of technique from theory and the technician from the thinker. Philosophy became the province of a leisure class maintained by a slave society, and natural laws were once more made subordinate to divine laws. Science under Aristotle and Plato turned from *doing* something to *knowing* something, a reorientation arising from the changing character of Greek society. The Aristotelian dichotomies of mind and matter, body and soul, and the Platonic Idea mirrored the society in which these philosophies developed. Nature, conceived as infinitely various and inevitable in the fruition of its laws, was the conception of technicians who sought to control matter, while Nature as a power moving toward fixed goals was the conception of a master who governs slaves. While technicians were interested in the manipulation of matter, the priestly castes and the ruling strata of society were interested in the manipulation of man. Thus does Benjamin Farrington explain this episode of intellectual rise and decline. The *Republic* and *Laws* of Plato and the *Metaphysics* of Aristotle, Farrington points out, served as instructions by which the ruling classes were to help nature in achieving its teleological ends.[6]

The "Great Adventure," as Charles Singer called this phase of history, failed as science became alienated from philosophy and was separated from it in the Alexandrian period from 300–200 B.C. "Science was a way of looking at the world," Singer observes, "rather than a way of dealing with the world."[7] Whereas ancient man had sought after the "who" of nature instead of inquiring "how," beginning with the Greek philosophers and extending to the time of Galileo, the "why" of nature was substituted for the

[6] Benjamin Farrington, *Greek Science* (Harmondsworth: Penguin, 1953), pp. 133, 147ff.
[7] Charles Singer, *A Short History of Scientific Ideas to 1900* (New York and London: Oxford University Press, 1959), p. 132.

"who" with equal neglect of the "how." Nature, accordingly, was variably interpreted in relation to philosophy rather than in relation to praxis.

Both Stoicism and Neoplatonism derived analogies between man and the universe, the former believing man had been made for the universe, the latter stating that *the universe had been made for man.* This Neoplatonic doctrine had already been foreshadowed in Pliny's *Natural History:* "Nature and the earth fill us with admiration . . . as we contemplate the great variety of plants and find that they are created for the wants or enjoyments of mankind."[8]

With Saint Augustine, Neoplatonism passed over into Christianity. Wedded to the Hebrew creed of God's transcendent rule over man and nature, nature was now conceived as a manifestation of God. Natural theology in Western Christianity became a quest to understand and appreciate God through discovery of the workings of nature as established by the divine will. And man who was created by God in his image was instructed, according to the Book of Genesis, to "fill the earth and master it; and rule the fish of the sea, the birds of the sky, and all the living things that creep on earth . . . I give you every seed-bearing plant that is upon all the earth, and every tree that has seed-bearing fruit; they shall be yours for food."

Nature was God's, and God had commanded man whom he created to master it. The unfolding history of science in the three centuries following Galileo closely adheres to this basic tenet. The giants of science saw God's glory and infinite wisdom revealed through their discoveries. The *philosophes* who followed overthrew religion in the name of a secular science, even as the underlying assumptions of this tenet permeated their writings. The positivists, empiricists, and behaviorists of our own time, in seeking to exorcise all taint of metaphysics from their work, have unwittingly erected a new metaphysics based on truths that have been revealed to them. And, lo, the new metaphysics bears similarity to the old. The refrain that echoes through these raucous centuries in a frenzy

[8] Quoted in *ibid.,* p. 107.

of exploration, discovery, and constant change is the biblical command: "Fill the earth and master it." Like an echo that reverberates in the constricted space of a troubled earth, although he no longer addresses God but his own imperatives, man has responded with another biblical statement: "We shall listen and we shall obey." So once again is set the tragedy of an earthly paradise from which man faces permanent expulsion.

An entirely different cultural development unfolded in the Orient where, in the early periods of civilization, the technology of India and China was considerably more advanced than in the West. Production of cast iron, use of natural gas, windmills, silk, and paper were in evidence in the East well before their advent in the West. Such scientific instruments as the seismograph and the mariner's compass also originated in the East. Eastern civilizations, however, did not accept the Western urge to dominate nature. Their philosophies and religions turned man inward not outward toward nature, and their ruling classes were more versed in their classical literatures than in international commerce and exploitation of resources.

Medieval Western man had sought to develop a harmony of understanding in relation to man, the universe, and God. The order was conceived to be hierarchical, with God at the summit, then man, and finally the natural world. The task of medieval science was to elaborate a complete and comprehensive scheme of things in an all-embracing philosophy that was intimately connected to the moral world. But at the same time man was a part of nature, and his knowledge of it derived, as has been noted, by the same method by which he perceived and acknowledged the transcendence and omnipotence of God.

Francis Bacon and the other founders of modern science were the products of this medieval philosophy. Their work slowly altered the medieval outlook, substituting new concepts in its place. The foundations of these concepts and the new methods enunciated for obtaining knowledge gave rise to the axioms of contemporary science cited earlier.

The principal aim of knowledge for Francis Bacon, as for most of the other precursors of the school of scientific progress, was the utilitarian goal of improving man's condition by lessening his suffering and increasing his happiness. Casting aside the viewpoint of the Greek philosophers that knowledge affords its own intellectual satisfaction, Bacon defined the true goal of the sciences to be "the endowment of human life with new inventions and riches." He called on man "*to establish and extend the power and dominion of the human race itself over the universe.*" The means for erecting "the *empire of man over things* depends wholly on the arts and sciences" (italics added).[9] Thus was enunciated the primary goal of "riches" attached to the biblical injunction of dominion over nature. With Bacon, the primitive and medieval queries as to the "who" or "why" of nature were repointed in the direction of "how." The establishment of the "empire of man" for utilitarian purposes had begun.

Copernicus and Kepler, in their astronomical studies, led the revival of Pythagoreanism: the concept that the universe is inherently a mathematical universe and that, consequently, all human knowledge must be mathematical knowledge. They followed the Platonic viewpoint that postulated a universal mathematics of nature as against the Aristotelian viewpoint which, recognizing the qualitative aspects of nature, saw in logic the key to knowledge. To Kepler, the only real qualities in nature were those that could be expressed mathematically. The human mind was made to understand only quantity, and it was through the investigation of quantities that the harmony of the universe could be revealed. The *a priori* assumptions that the world was orderly and knowable only through mathematics were tacitly assumed by Kepler, who laid the basis for the complete distinction between primary and secondary qualities which Galileo was to make.

Galileo did for terrestrial dynamics what Copernicus had done for celestial dynamics but, more significantly for our discussion, he established, albeit not fully, the methodology of the new science. The maturation of this methodology was to come with Newton,

[9] Francis Bacon, "Aphorisms Concerning the Interpretation of Nature and the Kingdom of Man," *The Philosophical Works of Francis Bacon*, I; quoted in Lewis S. Feuer, *The Scientific Intellectual: The Psychological and Sociological Origins of Modern Science* (New York: Basic Books, 1963), p. 14.

building upon the foundations laid by Galileo. With Galileo the doctrine of uniformity and orderliness in the working of the universe became a fundamental tenet. The physical world began to be perceived as a giant machine whose actions could be calculated. These actions were distinct and resolvable. The limits imposed by our senses in apprehending the physical world necessitated that man should not attempt to solve ultimate problems but have only limited objectives. This leads to the need for specialization and, as a corollary, the fragmentation of the sciences.

Knowledge to Galileo was mathematical in character. "The book of nature is written in mathematical characters," he wrote in *Il Saggiatore*. Inasmuch as quality cannot be expressed mathematically, Galileo dispensed with qualitative aspects of the world, with repercussions that are evident today in the crisis that confronts science and man. He distinguished between primary and secondary qualities, the primary ones being length, width, weight, and shape. The secondary qualities—color, taste, odor, texture, etc.—he viewed as only modes of perception in man rather than real characteristics of matter. In another famous passage of *Il Saggiatore*, Galileo expresses this thought clearly: "To excite in us tastes, odours, and sounds I believe that nothing is required in external bodies except shapes, numbers, and slow or rapid movements. I think that if ears, tongues, and noses were removed, shapes and numbers and motions would remain, but not odours or tastes or sounds. The latter, I believe, are nothing more than names when separated from living beings."

Thus began the trend in science to disallow that which could not be measured and to accept as real only that which is quantifiable. But of even more far-reaching significance was the conclusion that if the universe were orderly and knowable through mathematics, and if the actions of its parts were calculable, then ends as teleology could be dispensed with and *the physical world had no connection with the moral world*. With Galileo a clean break was established between natural and moral philosophy.

Man, too, is reduced to the status of a bystander as the universe acts "through immutable laws which she never transgresses," for nature is "inexorable" and cares "nothing whether her reasons and methods of operating be or be not understandable by men."[10] With the reading of man out of nature because man is characterized by qualitative features that are not quantifiable, and because the universal mechanism does not require man for its functioning, there began what Charles Coulston Gillispie has termed the "fatal estrangement between science and ethics, which has left us in a world in which we have no place. There we drift toward a state of nihilism where ignorant technicians clash by night. For scientists have absolved themselves of moral responsibility and eschewed understanding in favor of measurement."[11]

This elimination of man from a role in the universe coupled with the complete divorce between the observer and the observed "facts" leads to what is perhaps the most fateful consequence of the scientific axioms. The "fatal estrangement" led not only to a break between action and morality, it began a "separation between phenomena which could be observed without calling on the resources of man's inner experience for final judgments and phenomena which are explicable, if at all, only in terms of the human personality, which cannot be understood by modern scientific methods." This separation, John U. Nef points out, separates the human personality from that which it observes and seeks to explain.[12] All objects and all phenomena become merely external "facts" to be observed and measured—including man. Moreover, all the subjective qualities that are the essence of an individual and that distinguish one individual from another are excised from the scheme of knowledge because they are tainted with metaphysics —or is it with a touch of humanity itself?

René Descartes affirmed the supremacy of reason and the invariability of the "laws" of nature. By this twofold affirmation he undermined the doctrine of Providence and the illusion of finality in the fate of man. In his *Discours*, Descartes saw in medicine a

---

[10] Letter from Galileo to the Grand Duchess Christina, 1615, quoted in Edwin Arthur Burtt, *The Metaphysical Foundations of Modern Physical Science* (Garden City: Doubleday Anchor, 1954), p. 75.

[11] Charles Coulston Gillispie, *The Edge of Objectivity: An Essay in the History of Scientific Ideas* (Princeton: Princeton University Press, 1961), p. 44.

[12] John U. Nef, *Cultural Foundations of Industrial Civilization* (New York: Harper and Row, 1960), p. 28.

"way to make men wiser and more clever." His studies of physics satisfied him that "it is possible to reach knowledge that will be of much utility in this life." With this knowledge it is possible to "make ourselves *masters and possessors of nature*" (italics added).[13]

Newton capped the new science with his exposition of the rules of observation and experiment. The universe had become a huge clock which, once set into motion by God, continued its movement. Newton rejected hypotheses although his work was intrinsically based on the hypothesis that measurement and observation could define the mathematical behavior of phenomena. He accepted the *a priori* assumptions of his predecessors that the universe was orderly and could be comprehended only mathematically. He held on to their distinction between primary and secondary qualities.

With the Enlightenment the reign of reason was proclaimed, the banishment of ignorance and superstition was advanced, and the spirit of the new science was set forth to lead mankind to a new millennium. The growth of "universal human reason" was seen as a cure for social and physical ills as knowledge supplanted ignorance and understanding replaced prejudice. Condorcet affirmed that errors in ethics and politics arose from false ideas in physics and ignorance of the laws of nature, and Voltaire popularized the work of Newton to spread the new ideas that were to free man.

While breaking new ground, the *philosophes* also based much of their position on the scientific outlook of their predecessors, including their unstated but implicit presuppositions. The *philosophes*, Peter Gay relates, were torn between the profuse and comforting metaphors about nature and the demands of the scientific view. They continued to "treat nature rhetorically as a bountiful measure, a treasure house lying open to be raided, a servant waiting for orders, a treacherous opponent requiring constant vigilance. . . . Optimists and pessimists among the *philosophes* debated just how ready nature was to be dominated, how shrewd or vicious its resistance. . . ."[14]

---

[13] René Descartes, *Discourse on Method and Meditations* (Indianapolis: Bobbs-Merrill, 1960), p. 45.
[14] Peter Gay, *The Enlightenment: An Interpretation*, Vol. II of *The Science of Freedom* (New York: Knopf, 1969), pp. 10, 160.

Diderot, the rationalist, was urging that both practical and theoretical thinkers must unite against *"the resistance of nature,"* as if following the ancient biblical injunction. D' Alembert carried Galileo's limited objectives to an extreme and renounced the possibility of complete knowledge of the nature of things outside of isolated observable facts.

The primal rule of mathematics was popularized by Voltaire, who stated, "When we cannot utilize the compass of mathematics or the torch of experience and physics, it is certain that we cannot take a single step forward."[15] The rule was carried to its ultimate extreme in the writings of David Hume: "When we run over libraries, persuaded of these principles, what havoc must we make? If we take in our hand any volume—of divinity or school metaphysics, for instance—let us ask, *Does it contain any abstract reasoning concerning quantity or number?* No. *Does it contain any experimental reasoning concerning matter of fact or existence?* No. Commit it then to the flames, for it can contain nothing but sophistry and illusion" (Hume's italics).[16] Closer to our own times, the mysticism of numbers was again hailed by Lord Kelvin, who affirmed the legacy inherited from Kepler, Galileo, Descartes, Newton, and Hume: "I often say that when you can measure what you are speaking about and express it in numbers, you know something about it; but when you cannot express it in numbers, your knowledge is of a meagre and unsatisfactory kind; it may be the beginning of knowledge, but you have scarcely, in your thoughts, advanced to the state of science whatever the matter may be."[17]

The idea of progress received its greatest impetus from the Enlightenment when reason was thought to be enthroned, and it has since been nurtured by an industrial civilization that sees progress both as the rationale for its existence and as a source of hope for human betterment. The idea of progress first arose to fulfill human

---

[15] Voltaire, *Traité de Métaphysique*, Chs. 3, 5; quoted in Ernst Cassirer, *The Philosophy of the Enlightenment* (Princeton: Princeton University Press, 1951), p. 12.

[16] David Hume, *An Inquiry Concerning Human Understanding* (New York: Liberal Arts Press, 1955), p. 173.

[17] Lord Kelvin (William Thomson), *Popular Lectures and Addresses*, 2d ed. (London: Macmillan, 1891), I, 80–81.

needs that were pragmatic and materialistic—at the same time that they were spiritual and moral. It evolved out of man's changing image of himself and his relation to the external environment he sought to control. By providing a rationale, progress explains and justifies the materialistic pursuits of industrial society. By offering hope for human betterment, progress seeks to condone the excesses of industrial society while giving assurance of their temporary existence. Implicit in this hope is a promise of fulfillment, and progress offers a belief in a messianic pathway to a Promised Land. Progress serves a pragmatic purpose in seemingly providing a basis for implicit measurement: the development of societies can be measured, as it were, on a scale of progress. At any point in time, man can pause, take stock, and mark his position on the escalator of advancement. Progress, as J. B. Bury has written, is conceived of as a slowly advancing movement in a definite direction toward desirable ends. This movement will supposedly continue indefinitely and is not subject to any external will.[18] But progress, with its implied doctrines of inevitability and constant advance toward desirable goals, has not always been an accepted tenet of belief. The Greeks perceived change as degradation and the passage of time, as in human life, as bringing on old age and social degeneration. Life was then renewed in continuing cycles. Medieval theory portrayed history as events that were ordered by divine intervention, though the concept of original sin prevented a theory of moral advance. Christianity, with its concern for salvaging souls for the next world, was neither capable of nor interested in developing a theory that guaranteed man's advancement in this world and was, moreover, subject to man's direction and intervention.

Three preconditions for the development of the idea of progress are cited by Bury. First, it was necessary to dethrone Greece and Rome as the apogee of human civilizations, for this status rendered all subsequent civilizations as inferior imitations and thus negated any idea of human advancement. Second, it was necessary to establish the value and desirability of life in this world as opposed to the belief that life on earth was but so much travail that served

---

[18] J. B. Bury, *The Idea of Progress* (New York: Dover, 1955).

only as preparation for the world after death. Finally, it was necessary to erect a scientific foundation for knowledge.

The first two preconditions were essentially satisfied between the fourteenth and seventeenth centuries, the period that marked the height of the Renaissance and the early transition to the Enlightenment. During this time, the intellectual soil that would nurture a theory of progress was prepared. The scientific revolution, beginning with Francis Bacon and Descartes and culminating in Darwin's theory of evolution, provided the foundations for a theory of progress. With the scientific revolution, the idea of progress became indistinguishable from the idea of science.

The advancement of science was accompanied by a growth of knowledge that purported to be moving forward from "lower" to "higher" levels, onward and upward toward the improvement and betterment of mankind. Advancement lay in man's mastery over nature, his control over and freedom from the environment, and his understanding of the "laws" of nature so that he could dominate and exploit it. Reason, science, and progress were a triad that at once became the ends and means of modern scientific civilization.

There were those who were in opposition to the simplistic grandeur of the triadic rhapsody, notably Rousseau. Acknowledging the splendor of the "age of reason," he stated two hundred years ago what is becoming ever clearer to our own generation: "Our minds have been corrupted in proportion as the arts and sciences have improved."[19] Rousseau was perceptive enough to challenge the elimination of finality in the human state. "The body politic, as well as the human body, begins to die as soon as it is born, and carries in itself the causes of its destruction."[20] Although alluding to the death of the human body, Rousseau suggested that the development of society is a dialectical process, not only a process of aging, and that self-destruction is inherent in construction.

Nineteenth-century science was inextricably bound up with the concept of progress. Auguste Comte, the father of sociology, con-

[19] J. J. Rousseau, "Discourse on the Moral Effects of the Arts and Sciences," in *The Social Contract and the Discourses,* trans. G. D. H. Cole (New York: Dutton, 1950), p. 150.
[20] *Ibid.,* p. 88.

sidered progress to be the soul of his system of positive philosophy, and in his law of the three stages he sought to enunciate the laws of progress. In passing through the theological stage, the mind invents ideas; these ideas are abstracted in the metaphysical stage; submission to positive, or scientific, facts takes place in the highest, or positive stage. It is in this highest stage that society will be organized on the basis of scientific sociology.

Herbert Spencer, the prophet of Darwinism, proclaimed in his *Social Statics:* "Progress, therefore, is not an accident, but a necessity. . . . It is a part of nature." Substituting Nature for Providence, Spencer saw the chaos of phenomena reduced to "a gigantic plan" which did not depend on accidents or chance but tended everywhere to order and completeness. "Always towards perfection is the mighty movement—towards a complete development and a more unmixed good."

Spencer was not only the spokesman of Darwinism; he was also the child of a rampant industrialization and a burgeoning technology. As Richard Hofstadter has demonstrated in *Social Darwinism in American Thought,* Spencer's system was conceived in and dedicated to an age of steel and steam engines, competition, exploitation, and struggle. Spencer believed in laissez-faire economics which he supported by analogies to the survival of the fittest and competitive struggle for life—arguments that in themselves were vulgarizations of evolutionary theory.

The twentieth-century version of progress turns out to be a blindly hurtling technology that has carried man to the moon, split the atom, created a cornucopia of commodities for a privileged few of the earth, and holds out a promise to carry along with it the remainder of mankind. Whereas flaws and dangers inherent in progress were becoming more apparent, in the twentieth century the "laws" of progress were becoming ever more elusive. The difficulties that beset the formulation of a theory of progress arose from the multifaceted components that had to be incorporated in the theory. Specifically, the theory must identify and describe the following:

1. Element: what progresses?
2. Motive force: what are the causal factors?

3. Process: how do the elements change with time?
4. Path: what is the route of change?
5. Ends: toward which goals are the elements moving?
6. Criteria: how are the goals measured?

In the past two hundred years many attempts have been made to complete the edifice of the theory of progress, and on numerous occasions claims have been made that the elusive, universal law of progress had been discovered. But the "law" of progress is still undiscovered. Nor is it likely to be discovered, for in fact progress is a state of mind based upon faith rather than an element of nature.

The observations of Niccolò Machiavelli, who wrote 200 years before the Enlightenment *philosophes* and 350 years before Spencer, have more validity than their millennial prognostications: "As a whole, the world remains very much in the same condition, and the good in it always balances the evil." Machiavelli, however, lived before the era of modern science, and were he alive today he might conclude that the evil now overbalances the good. As a supreme irony, the scientific progress which was to have eliminated the "illusion of finality" from a fated future now has brought mankind face to face with the prospect of complete finality.

In the three centuries of the modern scientific era, the axioms presupposed by scientists sought to reconcile facts about the universe with truths that were already revealed to man. "The underlying preconceptions of eighteenth-century thought," Carl L. Becker notes in *The Heavenly City of the Eighteenth-Century Philosophers,* "were still, allowance made for certain important alterations in the bias, essentially the same as those of the thirteenth century ... the *Philosophes* demolished the Heavenly City of St. Augustine only to rebuild it with more up-to-date materials."[21]

The principles the *philosophes* sought were the ones they started out with. Reason is supreme because it will lead to a good life free from constraints which can be reached by virtue of that reason. The good life on earth is that life reached through the exercise of reason which reveals to us the good life. These tautologies are then, not exercises in logic, but exercises in wish-fulfillment and as such are a study of the diverse forms of human aspirations. The progress

[21] Becker, *op. cit.* (New Haven: Yale University Press, 1961), p. 31.

that was deductively derived from the axioms, according to Bury, "belongs to the same order of ideas as Providence or personal immortality. It is true, or it is false, and like them it cannot be proved either true or false. Belief in it is an act of faith."[22]

If science were merely a question of faith, the matter could be allowed to rest, with some individuals believing and propounding and some unbelieving and denying. But science and the scientific attitude today are the dominant powers in the world, and they are driving the world toward destruction on the self-revealed authority of questionable axioms and a scientific philosophy that endangers the very ends the philosophy purports to attain.

It was a great misfortune that at the very time that science was achieving its triumphs and establishing a basis for harnessing and transforming the earth, restraints on its excesses were removed by the very axioms that made the triumphs possible. During the years of the Industrial Revolution and the urbanization of Western society, the divorce between science and ethics and morality made the exploitation of both man and nature cruel and brutal. Not that science is fully to blame for the excesses that human beings perpetrated, but it had provided powerful new tools for "the endowment of human life with new inventions and riches" while at the same time undermining existing faiths with their moral and ethical values and denying any value judgments to be associated with the new faith. The rejection of qualitative values had its effect on aesthetics, for the exploitation of the earth under the banners of scientific progress scarred and defaced the natural world and agglomerated millions of people in ugly and squalid urban centers.

"Just when the urbanization of the Western world was entering upon its state of rapid development," Whitehead writes, "and when the most delicate, anxious consideration of the aesthetic qualities of the new material environment was requisite, the doctrine of the irrelevance of such ideas was at its height."[23] In a period of expanding capitalism, Lewis Mumford observes, "the change from organized urban handicraft to large-scale factory production transformed the industrial towns into dark hives busily puffing, clanking,

[22] Bury, *op. cit.*, p. 4.
[23] Whitehead, *op. cit.*, p. 195.

screeching, smoking for twelve and fourteen hours a day. . . . Industrialism, the main creative force of the nineteenth century, produced the most degraded urban environment the world had yet seen. . . ."[24]

The cruelty and ugliness that were concomitant developments of the rise of science and technology were made possible not only by the axioms of science but also by an implicit understanding between the practitioners of science and the practitioners of state, church, and industry. The scientists had entered into a pact with society which stated, in effect, let us make our observations and measurements without your interference and you can use the results of our studies without our interference. Such a pact was openly avowed by Robert Hooke in the mid-seventeenth century in describing the activities of the British Royal Society: "The business and design of the Royal Society is—to improve the knowledge of natural things, and all useful Arts, Manufactures, Mechanik practices, Engynes and Inventions by Experiments—(not meddling with Divinity, Metaphysics, Moralls, Politicks, Grammar, Rhetorick or Logick)."[25]

In modern times, a major portion of science is directed and paid for by governments with the same tacit understanding. Science has become a commodity as much as an intellectual pursuit and is available to the highest bidder to be used as the bidder sees fit. Inasmuch as the goal of science is the determination of the "how" of things, the "why" of things is extraneous to science. Unfortunately, only with the "why" are values and ethics associated—and understanding.

It is not surprising that among the great driving forces of science are the assault upon nature, the desire for profit, and, above all, the pursuit of war. The affinity between science and violence has been evident throughout history and today is the prime mover of the scientific establishment. Archimedes contrived war engines for the defence of Syracuse against the Romans. Plutarch wrote that me-

---

[24] Lewis Mumford, *The City in History* (New York: Harcourt, Brace and World, 1961), pp. 446–447.

[25] Quoted in C. F. Waddington, *The Scientific Attitude* (West Drayton: Pelican, 1948), p. 75.

chanics was made distinct from geometry and came to be regarded as one of the military arts. And although machinery played a small part in Greek and Roman antiquity, Friedrich Klemm notes, this did not apply to machines of war. Vitruvius stated that the bulk of experiments and researches were conducted to construct artillery, and Pliny wrote of the use of iron in the service of war.[26] Leonardo da Vinci, the epitome of Renaissance man, was equally at home in the design of war machines or in painting. In offering his services to the Duke of Milan he cited a number of military devices he could make and added, almost as an afterthought, that he could also paint.[27] To his credit, he suppressed the design of a submarine, writing: "This I do not . . . divulge on account of the nature of men, who would practice assassinations at the bottom of the seas, by breaking the ships in their lowest parts and sinking them together with the crews who are in them."[28] And Galileo, seeking protection and financial aid, wrote to his friend Belisario Vinta: "I have many and most admirable plans and devices; but they could only be put to work by princes, because it is they who are able to carry on war, build and defend fortresses, and for their regal sport make most splendid expenditure, and not I or any private gentleman."[29]

This brief survey of the underlying metaphysics of modern science cannot bypass logical empiricism, the dominant school of scientific thought and the one that carries the axioms of science to their furthest extreme. Logical empiricism seeks to establish science as a self-sufficient activity that is the sole method of apprehending the world intellectually. In its view, knowledge is nought but biological behavior, and truth in any transcendental sense is denied. Induction is a form of conditioned reflex, and "science is an extension of animal experience and has no other meaning than the totality of experiences on which it is based." There is no necessary knowledge, and the meaning of a statement is the method of its

---

[26] Friedrich Klemm, *A History of Western Technology* (Cambridge, Mass.: M.I.T. Press, 1964), pp. 30–51.

[27] Cited in J. D. Bernal, *Science in History* (London: Watts, 1954), p. 269.

[28] Quoted in O. T. Benfey, "The Scientific Conscience: Historical Considerations," *Bulletin of the Atomic Scientists*, XII (May 1956), 177–178.

[29] Quoted in Georgio de Santillana, *The Crime of Galileo* (Chicago: University of Chicago Press, 1955), p. 6.

verification. Since experience does not reveal a world of values, value judgments are meaningless. Thus logical empiricism and the effort to convert to physics all areas of knowledge are fitting attributes of a technological society. "It is a technocratic ideology in the mystifying guise of an anti-ideological, scientific view purged of value judgments . . . it rejects by definition the possibility of insoluble problems. . . . It is an act of emancipation from troublesome philosophical questions, which it denounces in advance as fictitious."[30]

The movement of science into the biological and especially the behavioral fields can only be regarded with great trepidation. Science, perhaps unwittingly, has accentuated the mind-body dichotomy and has established the intellect, the supreme reasoner, as the predominant force in man. While admitting the body as an area of scientific inquiry, it is conceived of as a quasi-machine, controlled by the laws of chemistry and physics, in which instincts, emotions, and the affective qualities are ruled out. It was not by accident that psychology, as a scientific discipline, began only after the physio-chemical foundations for nature, in general, were laid down. It was but an extension of these disciplines to problems of the human mind, as differentiated from the body. Nor was it strange that psychology began as a quasi-mechanistic explanation of the working of the human psyche. Psychology in the first half of the twentieth century, according to Ludwig von Bertalanffy, was dominated by a "positivistic mechanistic-reductionist approach which can be epitomized as the *robot model of man*" (author's italics). This robot model is based upon the principles of stimulus-response, equilibrium theory, and utilitarianism.[31] No significant changes in these basic concepts have occurred to change the model.

Sociology, we have noted, was born as a scientific discipline in the positivist doctrines of Auguste Comte, who explained all human development in his psychological law of the three stages, of which social science was declared to be the highest in the hierarchy of sciences. The social sciences and psychology have lagged behind

[30] Leszek Kolakowski, *The Alienation of Reason: A History of Positivist Thought*, trans. Norbert Guterman (Garden City: Doubleday Anchor, 1969), Chs. 8, 9.
[31] Ludwig von Bertalanffy, *Robots, Men and Minds* (New York: Braziller, 1967), pp. 7–10.

the natural sciences because their subject matter is more unruly than nature, and man and his societies are less amenable to experiment and measurement.

The order that science has contended to exist in nature is part of the axiomatic fallacy. While partially orderly, many facets of nature and perhaps most facets of human behavior are not orderly in a classificatory or predictable sense. This realization has been admitted, in part, by the quantum and relativistic reformulations of scientific theory wherein the act of measurement is known to perturb the subject under measurement so that deterministic prediction is no longer possible. (This is discussed at length in Chapter 5.)

The social and behavioral sciences have resorted to three approaches to establish themselves in the image of physics. Bound as they are by the axioms that permit only observation and measurement, they have found it necessary to eliminate unruly man as an object of study. They have sought instead to isolate the mathematical "reality" that underlies the processes of life and human behavior. Thus: (1) What the social sciences cannot measure they remove from the area of science, e.g., quality of life, emotion, art, poetry, music, etc. Their successes in counting ought to be tempered by Whitehead's astute observation: "If only you ignored everything which refused to come into line, your powers of explanation were unlimited."[32] (2) For what cannot be measured but should be measured, they either make a simplifying assumption to render the complex measurable, or they seek to mold the recalcitrant statistic into a measurable pattern. (3) When faced with the inability to describe or predict individual behavior, they aggregate individuals into macrostatistics to derive a probability measure of predictable behavior.

An experimental scientist, denying any interest in the identity of experimental animals, has explained that "what experimental psychologists are interested in is not organisms, but data . . . the data generator is at best a nuisance chosen because it is clean, docile, and traditional."[33] The data generators, in this case, were rats

[32] Whitehead, *op. cit.*, p. 103.
[33] Merrill E. Sarty in *Science*, CLVIII (October 13, 1967), 205.

—called rats even in scientific papers. Compare this with the elimination of all personality when the generator is a human and is referred to only as "S" for subject in scientific papers.

The thrust of the behavioral sciences is to stamp out disorder because the sciences cannot deal with it; to create situations that are conducive to control and hence to prediction; to make the complex simple because otherwise it cannot be comprehended. More serious is that science declared it a *sine qua non* that it must be objective by eliminating all subjective values and judgments, although, as has been shown earlier, subjective judgments were the bases of the axioms upon which science itself was built. Science, it is claimed, is thus neither good nor bad, but neutral. But if science is ahuman in its neutrality by banishing all human qualities, is it not but a short step to become inhuman?

Efficiency is a scientific concept, objective and rational. The modern factory, with its degradation of the human spirit harnessed to the machine, is efficient in this objective sense but inhuman in practice. The Nazi crematoria were efficient devices for committing human genocide, but inhuman establishments. The atomic bombs over Hiroshima and Nagasaki were efficient devices for carrying out the objectives of war, but inhuman devices in exterminating masses of people. The flight of Apollo 11 was a scientific achievement that demonstrated great technical efficiency, but may be judged as inhuman in squandering resources that might have alleviated hunger.

The transition from objective to ahuman to inhuman is not totally unexpected. Science, as the writings of the early scientists reveal, was viewed as an assault upon nature, to "extend the power and dominion," "the empire of man over things," and to "become the masters and proprietors of nature." Violence and exploitation are corollaries of this viewpoint. The rape of the earth, which we are now forced to acknowledge, was inherent in the scientific philosophy that sought to dominate the earth. From domination of the earth to domination of man was a logical step in scientific reason, for man had become an object in a mechanical world. The separation of moral and natural philosophy led in the Industrial Revolution to the carrying over of the assault upon nature to an assault upon man. The emotional and aesthetic qualities of man which eluded scientific inquiry were sacrificed to labor power and mechanical skills which could be quantified.

The dehumanization of man in the technological process was a denial of man by virtue of scientific reason. It is a cultural phenomenon that has been described by Norman O. Brown as the "dominion of death-in-life." The mentality that reduced a vibrant earth and man himself to mechanical objects "is an awe-inspiring attack on the universe." By virtue of the divorce from the subjective sources of man, pure intelligence becomes a stultifying and deadening insensitivity, a product of dying.[34]

Science purports to be value-free and yet has established itself as the supreme value of Western civilization. A strong scientific establishment is seen as the basis for all power: military, industrial, intellectual. The schools and universities are harnessed to the chariots of science. Business and economics are becoming scientific disciplines in the quest for orderliness and efficiency. All mankind is oriented to techniques in the image of science.

Yet science cannot establish any values and, by its own definition, is not qualified to provide a scientific ethics. Science cannot furnish goals, according to Einstein, nor does it contain any ethical values, according to Schrödinger. Further, it cannot in any manner sustain its own establishment as a final good.

A new orthodoxy has replaced the old faiths. By being alienated from nature and absolved from moral participation in the cosmic process, man has been left in a world without humanity. "After Galileo," Gillispie writes, "science could no longer be humane...."[35] This "fatal estrangement" reached a climax on July 16, 1945, when the S-1 gadget of the secret Manhattan Project exploded over the desert at Alamogordo, New Mexico. In that blinding instant of history, the lines of the Bhagavad-Gita flashed through the mind of the bomb's creator, J. Robert Oppenheimer:

> I am become Death,
> The shatterer of worlds.

[34] Norman O. Brown, *Life Against Death: The Psychoanalytical Meaning of History* (New York: Vintage, 1959), p. 316.
[35] Gillispie, *op. cit.*, p. 16.

"The physicists have known sin," Oppenheimer remarked in a lecture two years later, "and this is a knowledge which they cannot lose."[36]

[36] Lecture at Massachusetts Institute of Technology, November 25, 1947.

3 *The Technological Process*

*A*LTHOUGH other scientists as well as the physicists have known sin, and both man and nature have been sinned against, men continue to look to technology to save them from the situation that science, as applied through technology, has created. Earlier, I suggested that technology cannot be the savior. Indeed, more technology may well worsen the situation, hastening the destruction of technology itself along with the earth and its inhabitants. In developing this theme I shall briefly review the relations of science and technology in history, describe the technological process, and investigate several of the major methods that this process employs.

Technology is the process of applying knowledge for practical purposes. Three elements are involved in the process: the actor, who initiates and carries out the process; the mechanism, the particular technique or tool that is employed; and the acted upon, the material or persons to whom the mechanism is applied. The relations between these elements fairly well define the state of technology at any given period.

The English language, unfortunately, is poor in making subtle distinctions that differentiate the various stages of technology.

Many primitive languages are much richer in expressing the nuances observed in growth and process. The Trobriand Islanders have at least nine distinct names for the taytu or yam, depending on its size, shape, ripeness, time of harvesting, etc. The Arabs have more than six thousand words to describe a camel. These languages, on the other hand, are poorer in their ability to express abstractions and more complex concepts.

Since there is no spectrum of words to capture the essential characteristics that differentiate between the technical behavior of Paleolithic, Neolithic, ancient, medieval, and modern man, I shall use the term "technics" to designate all the earlier forms and "technology" to characterize modern practice.

Technics were a part of human culture from the dawn of civilization. These technics were manifested in the unitary systems of myth, magic, and praxis that were described in the preceding chapter. Magic and art were integrated with technics in the intimate relationship that existed between early man and his environment. The basic forms of the first human societies arose out of man's relations to the physical and social environment and consisted of small, loosely organized structural units based upon blood relationships within a spatial setting. Life was lived close to nature. Early man had neither dwellings nor tools. He roamed the countryside in search of food that he hunted. He migrated when food was scarce or when the climate changed.

To Paleolithic man we owe much that is fundamental to our own technical culture. All the major methods of handling and shaping materials, including the use of fire, were developed by these early huntsmen. Knowledge concerning the habits of many animals and plants arose in connection with their food-gathering activities. The monumental tasks of developing language, ritual, and painting were accomplished by them.

Neolithic culture, with its agriculture, domestication of animals, and development of tools and pottery, marked an advance in man's control over the environment, a control which was accompanied by more complex forms of social organization. Greater control over the sources of food and increased use of implements and primitive tools along with the development of language and arts brought about an increased population, which led to further structuring of the social pattern. Neolithic village culture left a legacy of agriculture, weaving, pottery, and the social inventions of pictorial symbolism and organized religion.

During the Bronze and Iron Ages most of the remaining basic technics of civilization were developed, including the use of metals, architecture, the wheel, and simple machines. Use of the alphabet, numbers, writing, commerce, navigation, and organization of the city and government were the legacy of these early peoples.

Fascinated as we are by the seeming marvels of present-day technology, it is easy to take for granted the giant achievements of the forerunners of civilization. "For the greater part of our lives," J. D. Bernal has reminded us, "we are surrounded with and use equipment evolved at that time and scarcely altered in the intervening 5,000 years."[1] We could manage to live without many of the modern pseudo-necessities like electricity, TV, telephones, automobiles, etc., but we would find life nearly intolerable without the foods, fire, simple tools, agriculture, language, and other major social inventions early man bequeathed to us.

Actor, mechanism, and acted upon stood in a unitary relationship in early technics as they do even today among primitive tribes. The act of agriculture was at one and the same time a social event, a religious affair, a magical rite, and a productive art. The construction of a canoe blended the life of the tree with the act of man in felling it, hollowing it out, and consigning it to the water. Art and decoration, ritual and incantation, work and play blended in the creative act. Johan Huizinga, indeed, contends that "civilization arises and unfolds in and as play." Myth and ritual give rise to all the great instinctive forces of civilized life which are all "rooted in the primeval soil of play." Man has created a "second, poetic world alongside the world of nature," and plays the order of nature as imprinted on his consciousness.[2]

This play, or unitary act of creativity, expressed in early technics was, according to Lewis Mumford, aimed more at utilizing man's

[1] Bernal, *op. cit.*, p. 91.
[2] Johan Huizinga, *Homo Ludens: A Study of the Play Element in Culture* (Boston: Beacon Press, 1955), pp. ix, 4, 5, 15.

capabilities than at the direct acquisition of food or control of nature. The Greek term "tekhne," Mumford notes, made no clear distinction between industrial production and symbolic art, and for the greater part of human history these two aspects of human creation were intimately associated. Technics "was broadly life-centered, not work-centered or power-centered."[3]

Technics predated science, which began to develop only after the major technics of civilization had been discovered by early man. In its early days, science was distinct from technics, being based upon speculation and philosophy rather than upon craftsmanship. This aloofness, as has been noted, led to the decline of Greek science. The crafts were regarded as proper for slaves but not for the citizens of the Greek city-states. The Romans, too, displayed a contempt for manual labor which, Seneca wrote, "involves bowed shoulders and earthward gaze." To Archimedes, technics was "mechanaomai," a Greek term meaning "I contrive a deception." The objective of mechanical technics was to outwit nature by overcoming the difficulties it posed.[4]

It was not until the seventeenth century that technics began to be used for the further development of science. The tools for systematic observation and experimentation were provided by technics, and Galileo, a superb technician as well as scientist, was the first to bring together the skills of both science and technics in the same person. During the seventeenth and eighteenth centuries, science learned a great deal from technics and offered it little in return. This situation arose because the practical arts had become highly developed on an empirical basis throughout the medieval period. The *Pyrotechnica* of Biringuccio (1480–1539) contained detailed descriptions of metal- and glass-working and the chemical industry. *De Re Metallica* of George Bauer (1490–1555) was a comprehensive treatise on minerals, metals, and mining.

In the Middle Ages, the mariners' compass stimulated inquiries into magnetic phenomena. Up to and through the Industrial Revolution this flow from technics to science continued. Science played

[3] Lewis Mumford, *The Myth of the Machine: Technics and Human Development* (New York: Harcourt, Brace and World, 1966), p. 9.
[4] Klemm, *op. cit.*, p. 25.

a minor role in the major innovations of that revolution which was carried out, in the main, by tinkerers and artisans who had no knowledge of science. The instrument maker, Watt, the jeweler's apprentice, Fulton, the millwright, Rennie, the brakesman, Stephenson, and hundreds of others, many of whose names have been forgotten, were the creators of this new industrial civilization.

The invention of the steam engine preceded the study of thermodynamics. Waterwheels and windmills antedated the study of hydraulics. Technics merged into technology with the advent of the Industrial Revolution and, henceforth, technology was to become a cultural force in its own right—a shaper (and destroyer) of societies and the cornerstone of industrial civilization.

With the coming of electricity and new energy sources in the mid-nineteenth century, technology came to be more closely associated with science. But even now, technology is not the exclusive servant of science. Both science and technology are related—now one leads, now the other. "A technological device can lead to both a scientific advance and a new technological device, and a scientific advance can lead to both a new scientific advance and a new technological device," Melvin Kranzberg explains, "and the potential in the association of technology and science is a chain reaction of scientific discovery and technological invention."[5] It may be appropriate to refer to both scientific technology and applied science.

The marriage of science and technology, Kranzberg adds, went through a long and difficult courtship. It cannot be gainsaid that the medium that helped bring the two together most often was, and continues to be, the pursuit of war. We have already seen how war was one of the principal motivations of science throughout its history, and we have noted the association of great scientific figures with the institution of war. Mumford observes that until the thirteenth century, mechanical invention owed a greater debt to warfare than to the peaceful arts. Metallurgical and chemical advances up to the present time were obtained in the service of war.[6] The

[5] Melvin Kranzberg, "The Unity of Science-Technology," *American Scientist*, LV (March 1967), 48–66.
[6] Mumford, *The Myth of the Machine*, p. 227.

research and development effort in the United States is a $25 billion a year program, of which nearly two-thirds is funded by the government; and of this, approximately three-fourths of the government's research and development effort is expended on military, space, and atomic programs, the latter two being closely associated with military programs.

Galilean and Newtonian physics were well fitted to the capitalism that was developing in Western Europe and to the individualistic, competitive, laissez-faire doctrines being promulgated by Adam Smith. The new science was helping to overthrow the heavy hand of the past that acted as a brake on commerce and manufacture. It also opened the door to that exploitation of the earth which Bacon had so strongly urged. It was the lot of technology, however, to be the mainspring of the grand assault upon nature. Because of its piecemeal character and its amenability to individual enterprise, technology became one of the prime motive forces of the developing capitalistic economics.

Already in 1667, Thomas Sprat, in *The History of the Royal Society*, was complaining that "Knowledge still degenerates to consult profit too soon; it makes an unhappy disproportion in their increase; which not the best, but the most gainfull of them flourish." The haste to profit, Sprat added, really diminished profit, for it allowed the grand prize, the Treasures of Nature, to slip from the hands of the leisurely gentlemen of the Royal Society.[7]

From the Industrial Revolution to the present day, technology has been a fomenter of revolution in the way people live and work, in the way they perceive the world about them, in the way they fight their wars, in the way they exploit the riches of nature for personal and social gain. The engineer, the descendant of the craftsmen of the Industrial Revolution, emerged during the latter stages of that revolution under the impetus of coal and iron, transport and communications. Even more directly than the scientist, the engineer saw as his purpose, as Thomas Tredgold wrote of civil engineering in 1827, "the art of directing the great sources of power in nature for the use and convenience of man." This purpose reiterates the Baconian credo with its basis in Judeo-Christian

[7] Quoted in Feuer, *op. cit.*, pp. 60–61.

metaphysics. Only the "great sources of power" have been added to the anthropocentric motif.

Technology is intrinsically utilitarian. It aspires to no cosmic theories of the universe or systematic theories of knowledge. Elements of "why" do not bedevil its workings as it proceeds to build upon the "how" of things. But technology has been affected by the metaphysical assumptions of science and the exploitation of the Treasures of Nature for greed, profit, and war has proceeded at an accelerating rate with barely a ripple of doubt or thought about consequences, effects, or morality. Pragmatism is the key concept of the engineer and the technician: a thing is successful if it accomplishes that for which it was intended. As a corollary, if it is possible to do something, it will be done. The end is the "thing" that was to be accomplished whether the "thing" be a device, a process, an institution, or a social creation. There exists no end beyond this immediate end. Values, morality, ethics, good and evil have no place in the workings of the technological process. The elements of play and craftsmanship arising out of a unitary relationship between the actor, the mechanics, and the acted upon have long since been sacrificed to the idols that modern man worships.

Let us examine in detail the workings of the technological process. I have called it a process because it is social in character; it combines the three elements in many ramifications; it is dynamic and changing, rather than static; it is a blend of science and technique; it generates its own motive force, in part, for it is continually adapting or destroying the old and generating the new; and, finally, it is subject to the dialectical laws of change.

These characteristics of the technological process militate against technology's being an autonomous force that is self-propelling and all-pervasive, as Jacques Ellul contends. Ellul uses the term "technique" rather than "technological process," and defines it as "the totality of methods rationally arrived at and having absolute efficiency (for a given stage of development) in every field of human activity."[8] "Rationally arrived at" and "absolute efficiency" are, for all their seeming clarity, metaphysical concepts that can be

---

[8] Jacques Ellul, *The Technological Society*, trans. John Wilkinson (New York: Knopf, 1965), p. xxv.

interpreted according to the faith of the interpreter. But more important, as I shall attempt to show, because of the social and dialectical character of the technological process, its triumph over man is by no means necessarily assured. Lewis Mumford's view in one of his early books, *Technics and Civilization,* is as legitimate as Ellul's: "Technics and civilization as a whole are the result of human choices and aptitudes and strivings, deliberate as well as unconscious, often irrational when they apparently are most objective and scientific: but even when they are uncontrollable they are not external. . . . No matter how completely technics relies upon the objective procedures of the sciences, it does not form an independent system, like the universe: it exists as an element in human culture. . . . The machine itself makes no demands and holds out no promises: it is the human spirit that makes demands and keeps promises."[9]

The remainder of this book will discuss the problems that technology is beginning to encounter, the limitations on what man can know and do, and the dialectical changes that militate against Ellul's thesis that "we can be confident that the final result will assimilate everything to the machine."[10] The final result may indeed be as dismal as now portends, but the causes will not be due primarily to the conquest of man by the machine.

The technological process entails six general steps:

1. *Formulation of a problem.* The problem may be either external or internal to the technological process. An externally formulated problem is one posed by a social need (whether real or not is irrelevant) such as production, communication, transport, energy, etc., or one posed by science to enable it to carry out some aspect of inquiry. An internal problem is one generated by the technological process itself. The incompleteness of every technological solution, which will be shown in the following chapter, is a fertile source for the continual formulation of new problems. The

---

[9] Lewis Mumford, *Technics and Civilization* (New York: Harcourt Brace, 1934), p. 6.
[10] Ellul, *op. cit.,* p. 12.

process is, therefore, doubly driven—from within as well as from without—but it is only the internally generated problems that grant any autonomy to the technological process *qua* process.

2. *Analysis of the problem.* The analysis of a technological problem involves at least four substeps: (a) a survey of possible solutions based on an evaluation of the state of the art—that is, a survey of what current technology, in related fields, has made possible, and on the statement of the scientific laws that are involved. Inasmuch as there is no *a priori* "best" solution, the technological process is rich in opportunity, for many of these possible solutions may be explored. At this stage, also, a deficiency in a particular branch of scientific knowledge may be noticed with the stimulation of scientific investigation leading to an interplay of science and technology as noted previously. (b) A definition of parameters: size, shape, weight, quantity, etc. is the second substep of the analysis. To a large extent, these parameters are socially determined—they are determined external to the technological process within the context of state-of-the-art possibilities and subject to the constraints listed in substep (c). (c) The constraints under which the solution is to be effected are gauged. These constraints include cost, profit, effectiveness, political and social factors, etc. In any economy, economic guidelines impose constraints upon the technological process. The solution must be viable in a given sociopolitical setting; it must have a certain effectiveness to warrant its development in terms of both cost and the degree to which it solves the designated problem. (d) The solution in light of the analysis performed is specified in regard to the formulated problem. In this step, one of the possible solutions is chosen subject to available science and technology and within the totality of constraints.

To a very large extent, then, the technological process is a social process dependent on other aspects of science and technology related to any given problem and subject to constraints, most of which are socially determined. An attempt can be made to formulate these constraints through other technological processes; yet the ultimate criteria of these formulations are human and social, not technological. Since these criteria are based upon metaphysical considerations, it is poor metaphysics, rather than an autonomous technology, that is to blame for excesses of the technological process.

3. *Initial plan or design.* The development of an initial plan or design is the third step in the technological process. The design may be that of a space ship or a new detergent, a new highway network or the reconstruction of a city. Whatever the design, it is tentative and its subsequent development is subject to scientific, technological, and social considerations.

4. *Development and testing.* The design or plan is implemented and tested to determine its conformity with the specifications.

5. *Modification and improvement.* The design is modified in accordance with the results of the tests and the new insights and knowledge developed in the course of the entire technological process. Attempts are made in this stage to compensate for deviations from the specifications, to diminish unforeseen secondary effects, to improve effectiveness, and to further reduce costs.

6. *Retest.* The modified and redesigned solution is retested.

These six steps are applicable to all aspects of the technological process, though specific problems may alter some of the steps. Each step is a complicated process entailing many substeps. Several of the steps, especially the last three, may be repeated many times before a "satisfactory" solution—that is, a plan, a design, or a product—is achieved.

In some respects, the technological process assumes the characteristics of a perpetual-motion machine: the end result of all development is more development and of all research is more research. Every "successful" achievement of the process makes obsolete much that preceded the latest development, and the uncovering of other techniques and other possibilities, together with unforeseen secondary problems and effects, opens up new vistas. In the development of solutions to many large-scale problems, it is often perceived that there is a "better" way to accomplish an objective, perhaps even through an alternative solution, at some point in the design or development steps. If costs and scheduling preclude backtracking, the original design may have to be implemented, even though completion may guarantee that a potentially obsolete solution is being effected. Obsolescence is thus not primarily a cultural phenomenon but an inherent result of the working of the technological process. Metaphysical criteria of profit, social need, and effectiveness can accelerate obsolescence, but they are not its primary generators.

The technological process by definition has limited objectives and is therefore fragmented and specialized. Those who develop resources have no interest or concern with the uses to which the resources will be put. Not only do the technologists who utilize a resource have no concern with how their product will be used, but the continuing availability of that resource is also none of their concern. The depletion of the resource appears as a challenge to develop a substitute and serves the continuing quest of the technological process. Once a product is produced, it becomes a challenge to produce a new product. After the new product is in its turn developed, a new challenge arises to improve or modify it. This thrust of the technological process, however, is socially conditioned by an economy that strives for a *perpetuo moto* characterized by continual growth and expansion.

We have seen in the axioms of science the glorification of numbers and observable "facts" with a consequent degradation of man. How does man fare in the technological process inasmuch as he cannot be excised from it completely?

Frederick Winslow Taylor introduced the principles of scientific management early in the twentieth century in response to Theodore Roosevelt's call for "national efficiency." "In the past man has been first," Taylor wrote in 1911, "in the future the system must be first." Declaring that his principles of management "are applicable to all kinds of human activities," Taylor sought to bring "the development of each man to his state of maximum efficiency." His aim was to replace the method of initiative and incentive under which workmen were left with responsibility for doing their jobs with a science that was beyond the understanding of the worker. The mechanic arts were to be studied by management, and the worker was to be assigned his scientific tasks as part of the system.¹¹ The Pavlovian conditioned reflex, which had already won a Nobel prize for its discoverer, offered a scientific basis for industrial management.

As technology "progressed" into the era of cybernetics and automation, new difficulties arose over man's place in the technological process. Bigger systems—that is, complex, interrelated multi-processes—incorporated man as a subsystem. As a subsystem, however, "man leaves much to be desired. What other system has no significant prospect of miniaturization or ruggedization, can work at full capacity only one-quarter of the time, must be treated as nonexpendable, requires a critical psychological and physical environment, cannot be decontaminated, and is so unpredictable?"¹² And man is a "bit of a nuisance to the system. Into the exquisitely refined calculations for a man-machine system, man's actions and reactions bring on unpredictable variability. . . . At some uncertain point, man becomes overloaded by environmental stress. No longer able to adjust, he becomes a liability to the system."¹³ "From the engineer's viewpoint, then, the mechanism of failure in man is an overload on certain built-in limitations. Thus, as a biological heat engine, man has limited fuel storage and practically no reserve of oxidizer. As a fluid-filled system with a built-in leak, he requires a continuous supply of fluids to maintain his structural integrity. . . . His internal pumping systems, fluid lines, communication systems, and various subcomponents are susceptible to wear, and the erosion of disease."¹⁴ To the computer, the foundation of a cybernetic system, "the human being is an imperfect mechanism, prone to mistakes and with a slow reaction time. His memory capacity is huge, but unfortunately he is unable to organize this capacity as efficiently as a computer. The net result is that the human can interact with a central processor only through devices which reduce his tasks to simple ones. . . ."¹⁵

[11] Frederick Winslow Taylor, *The Principles of Scientific Management* (New York: Harper, 1929), pp. 7–26.
[12] C. C. Cutler, *IEEE Spectrum*, IV (August 1967), 69.
[13] Editor's introduction to "Men Under Stress"; see following note.
[14] T. Morris Fraser, "Men Under Stress," *Science and Technology*, January 1968, pp. 38–44.
[15] "Remote Input/Output Data Devices," Rohm-Wheatley Technical Report, August 1966.

The engineering psychologist who aids in the design of a man-machine control system is scientific in stating that, speaking mathematically, "he [man] is best when doing least. . . . The more complex the human task, the less precise and the more variable becomes the man. . . . Human control behavior, it is asserted, reaches the optimum when the man becomes the analog of a simple amplifier."[16] The transfer function of man's behavior is constantly modified through a learning process in which a trial and error system acts to minimize the average error. But the machines which in every department of nonemotional thinking are superior to workers on the job can be employed as better amplifiers. The replacement of the human amplifier becomes not only feasible but desirable in those circumstances where an electronic system performs easily where man fails. The replacement of the fifty-pound pull afforded by man's muscular system with the gigantic force potential of electro-mechanical systems can now be accompanied by the replacement of his physical control over the machine system by speedier and more accurate techniques.

Thus have the Galilean concepts reduced man to the status of a deficient but tolerated subsystem in the imposing edifice of technology. Such conceit and hubris, born of value-free manipulations of nature and of man, were well characterized by Karel Capek in his play R.U.R. By chemical synthesis, Rossum had been able to imitate protoplasmic material. His son, proceeding as an engineer, perfected the process and describes the "facts":

—A gasoline motor must not have tassels, Miss Glory. And to manufacture artificial workers is the same thing as to manufacture gasoline motors. The process must be of the simplest, and the product of the best from a practical point of view.
—What sort of worker do you think is the best from a practical point of view?
—Perhaps the one who is most honest and hard-working.
—No; the one that is cheapest. The one whose requirements are the smallest. Young Rossum invented a worker with the minimum amount of requirements. He had to simplify them. He rejected everything that did not contribute directly to the progress of work—

[16] H. P. Birmingham and F. V. Taylor, "A Design Philosophy for Man-Machine Control Systems," *Proceedings of the Institute of Radio Engineers*, XLII (December 1954), 1748–1758.

everything that makes man more expensive. In fact, he rejected man and made the Robot.

—. . . Really a beautiful piece of work. Not much in it, but everything in flawless order. The product of an engineer is technically at a higher pitch of perfection than a product of nature.[17]

It comes as a great surprise, however, to discover the paradox that man serves best as a robot when he is more the man. The famous Hawthorne investigations revealed that productivity, with a given physical plant, is a function of the human relations that exist in the factory system as well as of the physical conditions. The social process, by means of which integration and identification of the individual with the collective work group takes place, is of great import in determining productivity levels. Work behavior is as much the function of social control as production techniques and wage-incentive plans. Group norms regulate output levels even more than the physical capacity to produce. Management's goal lay in "maintaining the equilibrium of the social organization so that individuals through contributing their services to this common purpose obtain personal satisfactions that make them willing to cooperate."[18] This common purpose was the furthering of the economic purpose of the organization.

In seeking to further production and economic profit, it was found necessary to study man in his role as a worker. As the modern industrial organization developed technologically from the engineering applications of scientific knowledge, it was also found necessary to apply scientific method to the human factor in the organization. Industrial engineering which began, in the main, as a technique of time-and-motion studies designed to increase productivity, developed into industrial sociology, where the social behavior of the worker was viewed as as important a factor as the physical behavior of the machines and raw materials. The field has developed into the "sciences" of industrial psychology and sociology, personnel management, management structure and relations, and communication and feedback techniques in huge industrial sys-

[17] Karel Capek, "R.U.R.," in *Best Plays of 1922–23*, ed. B. Mantle (New York: Dodd, Mead, 1923).
[18] F. J. Roethlisberger and H. A. Wright, *Management and the Worker* (Cambridge, Mass.: Harvard University Press, 1939), p. 569.

tems. Performance of the physical processes required the social organization of the industrial system. These studies are best characterized by their attempt to transform their associated "sciences" into the model established by the mathematics of physics. Thus man "ordered" the exploitation of nature, established a social organization around the quantifiable requirements of science and technology, and sought to incorporate ineffective, unpredictable, and frail man into this organization as a "subsystem."

But what is the system of which man is a subsystem? The technological process operates in a complex and dynamic world where literally *all is connected to all*. The cutting of a single tree in one country changes the balance of nature by altering the local water cycle, affecting local soil conditions, influencing local fauna, altering the local environment. By themselves, these effects are unnoticed or unfelt. Multiplied by thousands and millions, the denuding of acres and acres of land leads to changes in agriculture, erosion, problems of flood control, and so on. The exhaust of one automobile adds a few cubic centimeters of carbon monoxide to the air. The exhaust of millions produces an excess of carbon monoxide that threatens to alter the climatic patterns of the earth. A pound of phosphate on a field will help a larger crop to grow, and the residue that washes into a stream flows away. But the accumulation of phosphate from hundreds of acres in streams and lakes establishes new chemical conditions which can destroy all life in those waters.

An auto requires rubber that grows in Malaya and Indochina; autos require highways that cut through cities, woods, and fields; highways require sand and stone and concrete that are gouged from huge holes in the earth; and autos give off waste products that create deathly smogs. Autos consume gasoline that creates a demand for oil, that in turn creates a need to explore for oil beneath the surface of the earth's lands and waters. Tankers and pipelines are built, and oil spills occur destroying birds and seals that are far removed from autos.

All is connected to all in a finite world that represents a closed system. A closed system is one in which *all connectedness is within the system; there are no external connections.* What is input to one part of the system is output to another and each output, in its turn, is input to another part. Nothing can be thrown away because there is no automatic disposal. The refuse accumulates until it competes with other uses for the refuse pile. The polluted air stifles breathing and adds its contaminants to the atmosphere as diffusion takes place. Man can dig a hole in the earth, but a place is required to hold the removed earth. Man can burn his waste products, but the gases and particulate matter remain. A man can "solve" his problems by creating new problems for his neighbor. One nation can "solve" its problems by creating new problems for other nations. The accumulated problems of the world, however, cannot be transferred.

Similarly, *there can be no "external" costs in a closed system*. The wastes of an industry, calculated in the context of its private profit-or-loss statement, may be disregarded; the peoples of the earth cannot disregard them. The sewage discharged into a river by one community may in isolation be a cheap method of disposal, but it is exorbitantly expensive for the river and other communities. Starvation of a people far away may cost remote nations very little, but the expense appears in the defense budgets of the "unaffected" nations. The depletion of a resource may enrich the depleters and provide short-term affluence for those who expend the resource, but the earth is poorer. As gross domestic product curves of the nations soar and wealth accumulates, the earth is becoming progressively impoverished with the balance sheet of nature showing an ever-increasing deficit.

Can the disjointed and fragmented technological process be unified or controlled in some manner that will integrate it into broad social objectives? According to the cyberneticians, information specialists, system analysts, and planners, the technological processes that are now available will provide the unifying procedures that are required.

But, in fact, are not these procedures part of the technological process itself? Are they not subject to the same limitations of the process: short-range goals, no concern for past or future, no value judgments, no morality or ethics? True, technocrats might answer, they are subject to these limitations, but in a symbiotic system between man and machine these limitations can be overcome. What of values? These, it is claimed, are to be fed into the machines in the manner that they are developed by humans.

Can technology assimilate and unify a diverse world of multiple inputs and outputs that operate within one closed system that is man's habitat, workshop, and play site, his heaven and his hell? Cybernetics was defined by Norbert Wiener as "the science of control and communication in the animal and the machine."[19] Wiener derived the name of the new science from the Greek word *kubernētēs,* or steersman, although he later learned that André Marie Ampère had previously used this term in reference to political science. Cybernetics as a science is not concerned with what a thing is but with what it does, with the "how" of its behavior. Its range of inquiry is the system, large or small. Inasmuch as a thing consists of an infinite number of variables, a system is the collection of all the variables. Cybernetics, according to W. Ross Ashby, is the consideration of all the possible behaviors of the system considered as a machine, for it is "essentially functional and behaviouristic." It is obviously impossible to study "all" possible behaviors, and it therefore becomes necessary "that we should pick out and study the facts that are relevant to some main interest that is already given." The scientist's task is to continue to rearrange the variables, taking other variables into account, "until he finds a set of variables that gives the required singleness." This can be done for "every determinate dynamic system," Ashby states, ". . . simply because science refuses to study the other types . . . dismissing them as 'chaotic' or 'non-sensical.' "[20]

Ashby's discussion follows the scientific axioms religiously. Determinate machines can be controlled and their behavior predicted —with important reservations about unreliability of parts, complexity, and the association with fallible humans. Communication with machines can be precise and error-free—up to a point, for noise is a natural phenomenon and the atmosphere and lightning and thermal fluctuations of matter impose a ceiling on fidelity in communications. Coupled with the computer and sophisticated

---

[19] Norbert Wiener, *Cybernetics* (Cambridge, Mass.: M.I.T. Press, 1948).
[20] W. Ross Ashby, *An Introduction to Cybernetics* (New York: Wiley, 1958), Chs. 1–3.

feedback mechanisms, cybernetics can lead to a certain level of automation—although many activities and processes are not amenable to being automated—and thus increase productivity.

Difficulties arise when cybernetics is applied to society as a social system, to economics as an economic system, to politics as a political system, to the city as an urban system, to war as a cultural system, and to man as a biological system. For all these systems are either "chaotic" or "non-sensical" in the scientific sense. They can become determinate systems only through control and communications that establish the "required singleness." This is the thrust of the technological society that Mumford and Ellul and other critics have decried and against which Norbert Wiener warned in *The Human Use of Human Beings* and in *God and Golem*.

Marshall McLuhan's thesis that the "medium is the message" is a cybernetic approach to "singleness." McLuhan solves the "chaos" of language and thought by adopting the premises of information theory which exclude the meaning of information from communications. It is a simpler matter to control the medium—within limits—than to control the meaning of information conveyed through the medium.

The richness and vitality of human language have prevented its reduction to "singleness" even by the computer. The compact but untransistorized human brain continues to demonstrate intellectual superiority over the computer in all areas of human communication. After decades of intensive research on machine translation of languages, the Automatic Language Processing Advisory Committee of the National Academy of Sciences–National Research Council concluded that "There has been no machine translation of general scientific text and none is in immediate prospect." Unedited machine output was "decipherable for the most part" but "sometimes misleading and sometimes wrong," and "slow and painful reading."[21] Note, too, that the reference was only to scientific text, not literary text.

Although computers can sort and count words faster than hu-

---

[21] "Language and Machines: Computers in Translation and Linguistics," Publication No. 1416 (Washington, D.C.: National Academy of Sciences, 1967), summarized in *Science,* CLV (January 6, 1967), 58–59.

mans, they are unable to do much more as is evidenced by attempts to use computers for automatic indexing. I. A. Warheit reports: "Attempts to date to do machine indexing have only been partially successful. Both statistical methods and sentence analysis, which attempt to imitate human logic, are being tried. The present prognosis that these approaches will succeed is very poor. Machine indexing seems to be doomed to the same fate as machine translation. Human indexing, for the time being at least, remains a component of the total information system."[22]

The hubris of the cyberneticians who would control man is unlimited. Dr. Alfred C. Ingersoll, dean of the school of engineering at the University of California at Los Angeles, contends that "development of electronic brain cells that can genuinely think is under way. This is the first step in producing a computer that ultimately may reproduce itself. Man had better be watchful and keep the key to the switches in his pocket if he wants to survive."[23] Not to be outdone, a Soviet scientist, Dr. Nicholai M. Amosov, head of the Biological Cybernetics Department of the Ukranian Academy of Sciences, reported to the First Annual Conference of the American Society for Cybernetics that an artificial intelligence with all the attributes of human personality can be created. The Russians, he stated, have already built a model of purposive behavior which verified their original hypotheses about mechanisms of human thinking and behavior.[24]

It is these "original hypotheses" that are to be feared, for an indeterminate mechanism cannot behave in a mechanistic manner. Control, even in a probabilistic sense, implying prediction of a "singleness" of behavior, cannot be accomplished without transforming the object of study, including man, into a determinate mechanism.

This fallacy confronts all the practitioners of system analysis and planning, operations research, planning-programming-budgeting, simulation, and modeling as they apply their respective techniques to the solution of large man-society systems. The essence of these

[22] I. A. Warheit, "Current Developments in Library Mechanization," *Special Libraries*, July–August 1967, p. 425.
[23] *Computer Digest*, August 1968.
[24] *Computer Digest*, December 1967.

techniques is to consider the significant variables of the system in their many combinations and permutations to arrive at a "singleness" or a spectrum of "singlenesses" according to the criterion of a measure of effectiveness, be it optimization or minimization of some parameter or a composite measure. In choosing among a very large number of possible alternatives, the designer-planner seeks the "best" possible solution, preferably expressed in mathematical terms.

The elements of system analysis are similar to those we have already described for the technological process: (1) to define and describe the variables of the system; (2) to describe how the system works, showing the interactions of its subsystems; (3) to establish the measure of effectiveness; and (4) to examine alternative configurations of the system's variables and determine the effectiveness of these alternatives. The system analysis may employ simulation, which is an imitation of the behavior of the real system by means of a model constructed to correspond, to the extent possible, to the real system.

The axiomatic fallacy of the scientific method is apparent at every step of system analysis. In step (1), from the infinite number of possible variables to be selected and defined, only those are selected that (a) appear to be significant to the analyst, (b) are amenable to quantification, and (c) are restricted in number according to the means for their consideration. Factors (a) and (b) are value judgments that are called upon in ostensibly purely technical problems and are amplified in social problems. John R. Seeley emphasizes this point in describing the reality of the social scientist: "There is in each case a literal infinity of nonfalse representations that can be made. *Which* will be made in actuality is as poetic in motive and as political in effect as any other essentially expressive action ... a ritualized acting out of an internal choice or necessity. ..."[25] Factor (c) is a serious limitation that even the computer cannot bridge, and forces the analyst to limit further, if necessary, the number of variables originally selected according to his value judgments.

[25] John R. Seeley, *The Americanization of the Unconscious* (New York: International Science Press, 1967), p. 140.

Step (2) is also one of human judgment in defining subsystems and establishing functional relationships between them. Step (3), the measure of effectiveness, is almost exclusively based on human value judgments and thus constrains the "best" solution. In step (4) the trade-offs between different subsystem variables is similarly based on value judgments, as are the weights assigned to each consequence and the method of aggregation used to arrive at a composite measure of effectiveness. In short, every step of "scientific" system analysis, planning, budgeting, and simulating, even though it uses the most sophisticated computers and programming techniques, is but an exercise in human judgment and neither a picture of nor a prediction concerning the real world.

The search for artificial intelligence, the attempts to measure that which is incommensurable, and to control that which is free and unordered, will be continued by science and technology. They are not autonomous forces, however, and the dialectical laws of change will have their effect as they transform science and technology and the society in which they operate. In the vast reaches of the technological process, actors interact with actors, mechanisms with mechanisms, acted upon with acted upon, and all interact in innumerable permutations and combinations whose circular causal chains cannot be disentangled or foreseen.

> A wild confusion of order is clawing through
> a broken system of our most reliable wires.
> . . . . .
> In the back country a random raindrop
> has broken a dam.[26]

[26] William Stofford, "Report from an Unappointed Committee," Peace Calendar (New York: War Resisters League, 1968).

4 Quasi-Solutions and Residue Problems

*A*T FIRST GLANCE it appears that most problems that have been dealt with in the past three hundred years have been amenable to technological solution. Yet the success of science and technology was predetermined, in a sense, by selecting for solution only those problems that could be solved. Unsolvable problems were not accepted as real problems. Failures did not receive public notice. With the players themselves, scientists and technologists, establishing the rules and calling the score, success was inevitable—especially when the players were also allowed to define success.

Most of the industrial developments of modern civilization are technological solutions within the framework of limited objectives based upon criteria established for the technological process. Each problem that is amenable to solution has been considered an isolated problem within an open system. When the definition of a technological solution is broadened to include the myriads of interrelationships with other processes and materials within a closed system, as is required for a realistic approach to the technological process, then the solution is no longer a complete solution. The solution to the specific problem is almost always found to have engendered a set of new problems arising from the interrelationships and the finite characteristics of the closed system. Most such solutions, in fact, turn out to be *quasi-solutions*.

A technological solution is always a quasi-solution because it gives rise to a residue of unsolved problems. The residue arises from three sources. The first source is the *incompleteness of the technological solution*. A machine or a process, for example, is never 100 per cent efficient, so the problem of energy conversion is never completely solved. The process utilizes new materials in some form. A reduction in the amount of material—its size, shape, or weight—may be realizable and, hence, the solution is again incomplete inasmuch as further improvements are possible. The equipment required by the process is a given size and weight. It operates in certain specified ways. Reductions or changes in size, weight, or method of operation may bring about improvements in efficiency, quantities of materials used, or amount of energy required for operation. This has to do with incompleteness in the input and operating stages of a machine or process. But incompleteness is also involved in the output product. Considerations of size, speed, weight, efficiency, among other characteristics, apply to the output in a manner similar to that of the input.

The second source of residue problems is the *augmentation of the original problem*. In distinction to the problems arising from incompleteness, the problems of augmentation arise when a higher-level problem is engendered by the completion of a solution to an original problem to the extent possible, at which time the general problem demands a different type of solution. The new solution often brings in its wake a completely new technology, for, dialectically, transcending one level of solution poses a different set of problems. An increase in computer speed, for example, was obtained by using semi-conductor components in place of electromechanical relays. Semi-conductor technology was completely new and today is being further augmented by integrated circuits and large-scale integration of scores of logic modules in subminiature arrays.

Again, piston-powered aircraft were limited in speed by the number and size of the engines required to lift a load, of which the engines themselves were an appreciable portion. The problems of piston aircraft were transcended by the jet engine, which initiated new problems associated with jet propulsion. Higher speeds necessitated the development of new metals such as titanium. The processing of titanium required a new technology of titanium recovery from ore. The machining of titanium gave rise to new problems in metallurgy and machining. Electrical discharge machining, electro-chemical milling, chemical milling, sonic abrasion, sonic impact, plasma arc, and lasers are among the new machining techniques that were engendered by the use of exotic heat-resistant alloys and new refractory metal alloys from tungsten, molybdenum, columbium, and hafnium whose use, in turn, was necessitated by space and military rockets and missiles.

The theoretical knowledge of atomic fusion was utilized in the development of the hydrogen bomb. The drive for completion of the solution to bomb development occurred in devising "cleaner" bombs which had less nuclear fallout potential than "dirty" bombs, developing more powerful bombs, and developing improved techniques for fabricating the bombs. When efforts were directed to harnessing fusion for civilian-oriented energy conversion, the problem was augmented to finding a container that would encompass the fusion process at a temperature greater than 100 million degrees. Experiments with magnetic plasmas have dominated this search for more than a decade.

The third source of residue problems is *secondary effects*. These arise when, in the process of solving an original problem, a number of secondary effects are initiated—some foreseen, others unexpected, some recognized, others unrecognized.

The use of coal is one quasi-solution to the problem of energy conversion. The mining of coal has secondary effects on the development of mining machinery, transportation, mine safety, gas-detection devices, and mine-associated diseases. The debris removed from mines creates secondary problems of slag heaps and the destruction of the countryside. The housing and health of the mining community is another aspect of the mining problem. The burning of coal is similarly associated with a number of secondary effects: the release of particulate ash, sulfur, and other gases to the atmosphere.

A quasi-solution is thus seen to generate three new sets of problems arising from incompleteness, augmentation, and secondary effects of the original solution to a problem. Whereas the problems rising from incompleteness are mainly extensions of the original problem, the problems engendered by augmentation and secondary effects are essentially new problems. Invariably, the combined problems that exist after the initial solution are greater in number than the original problem.

It is this compounding of residue problems arising from quasi-solutions that is one of the most powerful driving forces of the technological process. The drive for completeness pushes technology to current limits of developments, at which point augmentation takes effect and new problems are generated. Completion problems and augmentation problems both generate secondary effects which initiate a new set of problems that call for solutions. Each new solution, however, is a quasi-solution and it, in turn, gives rise to new sets of problems in each category of problem sources.

The proliferation of residue problems is shown diagramatically in Figure 1. A quasi-solution to an original problem (1) generates three sets of residue problems in the first generation. As explained above, each newly generated problem can generate another set of residue problems when a quasi-solution is applied to the problem. The figure illustrates a quasi-solution applied to problem 2, a secondary-effect problem, in the first-generation residue. This quasi-solution, in turn, generates a second generation of residue problems, of which problem 3 arises in augmenting a quasi-solution to problem 2. The quasi-solution of problem 3 generates a third generation of residue problems, of which problem 4 is one of the secondary effects. A quasi-solution of problem 4, in turn, again generates a new set of residue problems, and the process continues to proliferate.

The varying size circles in Figure 1 indicate varying intensities or complexities of the problems. Some residue problems may be

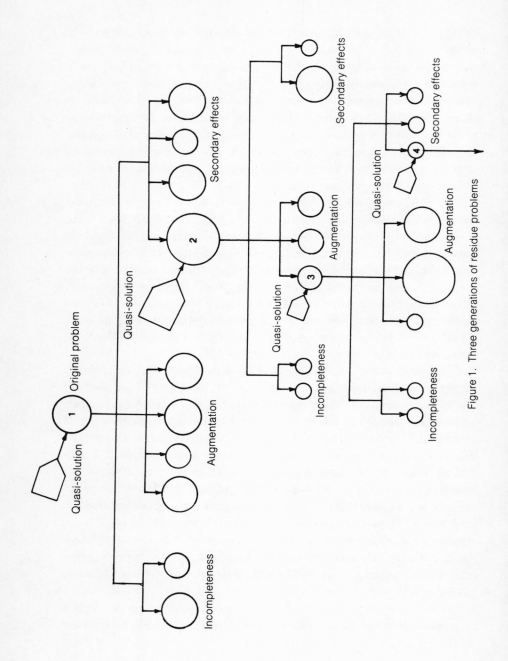

Figure 1. Three generations of residue problems

ignored, because (1) they are not recognized, (2) no immediate quasi-solution is at hand, or (3) other problems appear more pressing or more promising of quasi-solution. The longevity of the problems varies, and generations of problems can overlap.

The proliferation of residue problems gives rise to a residue chain as shown in Figure 2. Each distinct problem, 1, 2, and 3, generates a set of residue problems. A quasi-solution is shown for only one problem in each generated set. The quasi-solution again generates a set of residue problems, for which another quasi-solution is shown for only one problem in each set. By the fourth generation of problems, a complex network is seen to exist even with the sparse application of quasi-solutions.

A quasi-solution to problem 1, for example, generates a set of residue problems $1A_1$, $1A_2$, and $1A_3$, among others in the first generation. A quasi-solution to problem $1A_1$ generates a second generation of residue problems, including $1B_1$ and $1B_2$. A quasi-solution to problem $1B_2$, in turn, generates a third generation of residue problems, including $1C_1$ and $1C_2$. Because of varying longevity of the problems, generational overlap can occur and relationships that *cannot be foreseen* can be established. In the figure, $1A_2$ is related to $1B_1$, which is in turn related to $1C_2$. More significantly, relationships can appear between problems that are generated in the residue of quasi-solutions to completely distinct and unrelated problems. Problem $1A_3$ in Figure 2, for example, is shown related to $2B_2$ and $3B$. Other relationships are illustrated between diverse sets of residue problems.

The relationships between problems can affect the problems in two ways. (1) A residue problem of one quasi-solution can cancel or negate the quasi-solution of another problem, or (2) a residue problem of one quasi-solution can reinforce a residue problem of another quasi-solution and make a quasi-solution to one or both problems more difficult or impossible. The cancellation or reinforcing effects may be slow and cumulative so that many generations of residue problems may be generated before these effects are noted.

A classic example of cancelling effects is the saving of lives and increased longevity as a secondary effect of solutions to medical

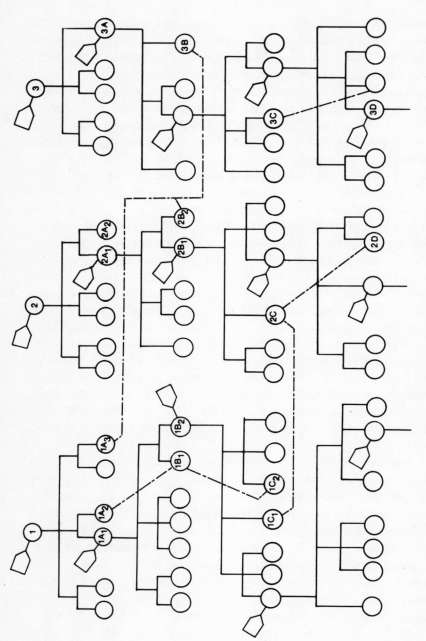

Figure 2. Residue chains with linking problems

problems of disease. The resulting expansion of population in a Third World country negates gains in an improved standard of living that is sought in economic development. An example of reinforcing effects is the presence of pollutants from the exhausts of autos that were developed to solve a transportation problem, combining with the pollutants that are emitted to the atmosphere from factories that were built to solve problems of manufacture. These reinforcing effects were slow and cumulative and not perceived as a menace until a dialectical change from a quantity of pollutants that was barely measurable to a qualitative change in the unhealthful characteristics of the atmosphere had occurred.

The relationships of Figure 2 can be illustrated by considering problem 1 to be that of developing an improved fuel for higher-performance autos. An improved fuel is developed by a quasi-solution to the problem. The quasi-solution is augmented in $1A_1$ by increasing the octane rating of the fuel, and is augmented in $1A_2$ by adding tetraethyl lead to eliminate pinging and improve engine efficiency. As a secondary effect of the higher octane rating, in $1B_1$ there arises an exhaust consisting of unburned hydrocarbons and, in $1B_2$, carbon monoxide. A quasi-solution to the reduction of carbon monoxide and the hydrocarbons is the introduction of antipollution devices. A secondary effect of these devices is to increase nitrogen oxide emissions. Lead that results from the burning of the fuel containing tetraethyl lead is related to the carbon monoxide and to the nitrogen oxide by being constituents of polluted air.

A secondary effect of the development of improved fuel, shown in $1A_3$, is the heat dissipated to the atmosphere by the burning of the fuel in millions of autos. Problem 2 is that of generating energy, and a quasi-solution is to burn fossil fuel. One secondary effect, shown in $2A_1$, is the depletion of fossil fuels. A quasi-solution to the depletion is the development of atomic energy. A secondary effect of energy generation by means of atomic fission is thermal pollution, shown in $2B_2$. Thermal pollution may raise the temperature of streams and lakes that are used as coolants in energy generation; in order to limit this pollution, work will be required that will produce more heat that must be dissipated to the atmosphere. The heat from $1A_3$ and $2B_2$ combine in the atmosphere to raise the ambient temperature. Problem 3B is the carbon dioxide that is generated as a secondary effect from a quasi-solution to energy generation in manufacturing problem 3. Carbon dioxide in the air tends to raise the temperature of the atmosphere because this gas absorbs and traps heat and prevents some of the heat from radiating back into space. The heat effects of $1A_3$, $2B_2$, and $3B$ are reinforcing, although they arise from unrelated technological processes.

In a technological society there arises the paradox that the more problems are solved, the more problems await solution. If the residue problems were only those arising from incompleteness, there would exist a possibility that, *in toto*, solutions would be achieved at a rate equal to or greater than the rate of appearance of the residue of incompleteness. This possibility would exist because, in general, the initial solution would be a major solution to a problem, and the problems arising from incompleteness would be subordinate and less important. The presence of augmentation and secondary-effect problems, however, precludes a balance between problems and solutions, and establishes a preponderance of unresolved problems. *Problems proliferate faster than solutions can be found to meet them.* The residue of problems arising from quasi-solutions accumulates faster than solutions can reduce the residue.

Is there not, it may be asked, a calculus of problems such that one solution can effectively solve many problems? Is there not in the alchemy of the technological "breakthrough" a way out of problem accumulation? A scientific or technological "breakthrough," it may be claimed, wipes the slate clean, and restores a balance between problems and solutions. I have called this approach alchemy because even a "breakthrough" (however technologists may wish to define the term) cannot be a universal solution to all problems, and a rapid advance in one problem area does not solve the proliferating problems in other areas. Further, the "breakthrough" solution immediately establishes a new set of residue problems, etc., etc.

The concept of a breakthrough is associated primarily with the set of augmentation problems where quasi-solutions of completeness have imposed a limit or bound upon the problem, a limit which must be transcended, or the problem must be re-established in another framework that is free of the former limit. The quasi-solutions of completeness and secondary-effects problems remain, however, and new residue problems follow in their wake.

Up to this point problems and solutions have been discussed in a general way. The generality has been intentional, for there is no distinctly technological problem and no distinctly technological solution independent of social enterprise, response, and change. The technological process is irrevocably bound up with the socio-economic, political, and cultural affairs of mankind. Every technological solution is effected by people, on one hand, and for people, on the other. Even in highly developed technological societies machines do not control other machines in the interest of machines. Science is transformed into utility for people by people.

Like science, technology is subjective and conditioned by prevailing social and psychological attitudes. The set of problems for which technology seeks solutions arises from economic necessity. Technology is socially determined. "Technique itself," Ellul has written, "is a sociological phenomenon."[1]

Given the close relationship between society and technology, no problem is truly technological but rather all are techno-social problems. Similarly, there cannot be a truly technological solution to a techno-social problem but only techno-social solutions. It is the social character of technology that establishes the secondary effects of quasi-solutions as a primary source of the residue of unsolved problems.

Not only do techno-social problems proliferate faster than techno-social solutions can be found to meet them, but the quantity of problems changes their qualitative character. *Each successive set of new or residue problems is more difficult to solve than predecessor problems.* The increased difficulty stems from seven factors: (1) dynamics of technology, (2) increased complexity, (3) increased cost, (4) decreased resources, (5) growth and expansion, (6) requirements for greater control, and (7) inertia of social institutions.

Each of the first six factors has the tendency to impede finding solutions and results in expanding the residue of unsolved problems

[1] Ellul, *op. cit.*, p. xxvi.

arising from each new quasi-solution. These factors create a technological squirrel cage wherein technicians must run faster and faster to remain in the same place. Unlike the squirrel cage, however, the faster they run, the further they fall behind. Each quasi-solution has a multiplier effect on the residue of problems.

(1) The belief that the discovery of solutions can keep pace with problems is essentially derived from the viewpoint of statics. In statics, if all things remain constant, then given a problem, it is believed that a solution can be found to meet the problem which once solved remains solved. The approach through statics plus a "breakthrough" outlook has led to the self-deception of scientists and technologists that has given rise to the euphoria of "all things are possible, and if possible they will be accomplished."

Neither the physical nor the biological worlds, of course, are static but highly dynamic. Technology and science, indeed, have demonstrated the dynamic characteristics of the earth and the universe and have accentuated the dynamism inherent in every field of human life and activity. This dynamism negates the possibility of a problem's remaining solved for long. Moreover, the set of residue problems, especially the secondary effects, effectively changes the conditions which elicited the problem, altering the extent to which the solution to the original conditions is a solution.

(2) Each quasi-solution uses up a range of alternative solutions. The set of alternatives available for solution of a residue problem is of a different kind and necessitates a search for a new set of solutions. The new set is more complex because the initial conditions have been transformed, simpler solutions have already been attempted, and a residue of problems from the initial quasi-solution has been created, some of which impose constraints on the new set of solutions.

The mining of coal, for example, is a process of continually growing complexity. As upper strata of coal are depleted, it becomes necessary to burrow deeper into the earth. The deeper the mine, the greater the depth of the mine shaft, the more extensive the mine tunnels, the greater the need for supplying fresh air and exhausting harmful gases. The more extensive the mine complex, the greater the need for safety devices, for alarm and control devices, for rescue equipment. The increasing depth of the mines and the inferior grades of coal encountered, in addition to the factor of human safety, engendered the transition to automatic mine equipment, which in turn has engendered new problems.

Even an innovation or invention is a quasi-solution. Each brings in its wake a residue of new techno-social problems, the complexity of which are greater than the initial innovation. Researchers who traced a group of innovations back historically to the theoretical concepts which led, unknowingly, to the later innovation have noted: "Because of the increasing comprehensiveness and sophistication of innovation, it is almost certain that the diversity of knowledge required for tomorrow's innovation will be even greater."[2] Advances in new technologies are especially prone to the growing complexity that accompanies advance. Thus the computer, automation, and synthetic chemical industries, for example, are opening up new problems of growing complexity with each new development of their respective industries.

(3) With increasing complexity comes increasing cost. In general, the cost of solving a set of residue problems is greater than the cost of the initial quasi-solution. Chemical fertilizers cost more than manure; tractors cost more than animal-pulled plows. Harvesting machines cost more than scythes. The increase in cost per unit of output is held down by increased yield and increased productivity. But increasing costs associated with factors that do not increase productivity are becoming more prevalent in relation to quasi-solutions for secondary-effect problems. The introduction of seat belts, collapsible steering shafts, afterburners, etc., have not improved the efficiency of the automobile, for example, but have been necessitated by the campaign for auto safety to reduce highway slaughter. The costs of attempting to control pollution, reclaim land and water, and re-create a livable earth may well be astronomical.

(4) Every techno-social quasi-solution to a techno-social problem consumes resources. Since resources are finite—as, to our

---

[2] C. A. Stone, *et al.*, "Technology in Retrospect and Critical Events in Science," I (Chicago: IIT Research Institute, December 15, 1968). This report was prepared for the National Science Foundation.

alarm, we seem only recently to have come to realize—the solution of the residue of problems arising from the quasi-solution must proceed against a background of depleted resources. Every technological process requires energy and consumes fuel. Hence the supply of fuels must always decrease. The requirements of steel production led to the depletion of high-grade iron ores from the Mesabi range. The development of the taconite process was dictated by the depletion process. Every highway that is laid to provide a route for an increasing number of autos decreases the amount of land available for further highways. Every home that is built removes a plot of land from further building or from agricultural production.

With large reserves of resources, the incremental depletion goes unnoticed but is always present and is cumulative. When depletion is evident in one resource, the problem can be transferred to another. Thus the move from coal to oil to gas to atomic fission for energy conversion. Other factors then enter to impede solutions: among them, greater complexity and greater cost until depletion of the new resource sets in.

(5) Technology fosters growth and expansion. The expansion arises from the impetus of economy of scale, since the cost per unit decreases with greater output to a certain point. Growth is a concomitant of technological advances, for its effects provide larger food supplies (within limited geographical boundaries) to feed growing populations. Medical advances lower death rates and contribute to growing populations. Increasing populations require more homes, clothing, autos, foods, etc., etc.

All this growth and expansion, however, tends to increase the complexity of solutions, increase the overall cost, and consume more resources. Further, since most growth in technological societies is not linear but exponential, the proliferation of problems at a faster rate than solutions to them is assured. Dialectically, quantity becomes transformed into quality, and previous quasi-solutions to problems under one set of conditions are negated by the qualitative change.

Highway engineers have discovered, for example, that many expressways they build are congested at completion time, for the number of autos available to use the expressway has increased during the construction period at a greater rate than the provided capacity. Small incremental gains achieved by decreasing pollution from autos are negated by the increasing total amounts of exhaust from more autos. Anti-pollution controls in smaller jets are made more difficult to transfer to jumbo jets that consume 30 to 40 per cent more fuel.

Bertrand Russell has pointed out that each improvement in locomotion has increased the area over which people are compelled to move. Before the auto age, a worker, for example, might have lived a half-hour's walking distance from his work site. Today with the auto the worker can live further away but is fortunate if he can drive to work in a half-hour. The classic example of growth that negates quasi-solutions is the increasing immobility of autos in the urban environment.

(6) Each successive techno-social solution to a techno-social problem requires greater and increasing control over (a) resources, (b) processes, (c) science, (d) technology, (e) individuals, and (f) institutions.

Control over resources will be dictated by the scarcity or disappearance of the resources in question. That which can be done will be established by what resources are available. The cost of developing substitutes will demand a system of rigid priorities. Some processes will be too costly to implement; others will be banned because of deleterious secondary effects. The proliferation of complexity will attend all developments. Science will accelerate its development as a planned activity that must justify itself in the economic marketplace. Complexity will increase the cost of probing the unknown, and priorities will be established for research and study. Technology, too, will be constrained by an economic calculus as waste is forbidden and recycling of resources becomes mandatory.

The requirements of increasingly technological civilization will demand further developments in technology to sustain this civilization. The development of new and increased supplies of food will call for the creation of new food-producing and food-processing industries and techniques. Harnessing the atom, the wind, and the sun will become imperative to furnish the energy to provide for a growing population and to enable society to continue its technological growth. Vast new quantities of minerals, ores, and fuels will be required to furnish raw materials to fill growing needs and demands.

The demands for resources will accelerate as our natural resources are depleted. The depletion of fossil fuels can be calculated. Mineral wealth has been reduced to a point that demands boring into the bowels of the earth and processing ores with but small concentrations of metal. Agriculture requires ever-increasing amounts of fertilizer to maintain soil fertility. Water requirements demand large-scale irrigation systems and intricate complexes of dams and aqueducts. Deserts and mountains are surveyed for possible extension of food-producing areas.

(7) The inertia of social institutions makes the solutions to residue problems more difficult because (a) the institutions developed to take advantage of one set of solutions may interfere with the development of solutions to the set of residue problems, and (b) because of the time lag between the generation of a problem and the development of an institution that is amenable to the solution.

As a consequence of the proliferation of residue problems at a faster rate than solutions can be found for them, and because each successive set of residue problems is more difficult to solve than its predecessors, *the residue of unsolved problems arising from a multiplicity of quasi-solutions converges to a point where no techno-social solution is possible*. Thus a situation arises wherein a solution may exist for an individual problem while no solution exists for the totality of problems.

The dialectical process whereby a solution to one problem generates sets of new problems that eventually preclude solutions is summarized in the five steps of techno-social development.

1. Because of the interrelationships and limitations existing within a closed system, a techno-social solution is never complete and hence is a quasi-solution.

2. Each quasi-solution generates a residue of new techno-social problems arising from: (a) incompleteness, (b) augmentation, and (c) secondary effects.

3. The new problems proliferate at a faster rate than solutions can be found to meet them.

4. Each successive set of residue problems is more difficult to solve than predecessor problems because of seven factors: (a) dynamics of technology, (b) increased complexity, (c) increased cost, (d) decreased resources, (e) growth and expansion, (f) requirements for greater control, and (g) inertia of social institutions.

5. The residue of unsolved techno-social problems converge in an advanced technological society to a point where techno-social solutions are no longer possible.

5   *Limits, Impotence, and the Unsolvable*

ALEXANDER the Great, Plutarch relates, was reputed to have wept when he was told by his friend Anaxarchas that the number of worlds was infinite and he realized that he had not yet conquered one. His later conquest of a large portion of the then-known world was but a small part of the world we now know—and even that conquest was short-lived.

The history of science and technology is akin to Alexander's impatience and his ultimate denouement. There appear to be endless worlds, and worlds within worlds, still to conquer—yet every conquest of the known expands the unknown, and that which is conquered once appears anew to be reconquered. As discovery continues, some limits are approached while others are expanded in a dialectical process that is continually enlarging and contracting the possibilities of further discovery.

Is science an "endless quest," or are there limits on what science can discover? It is a relatively simpler task to indicate limits on technology. Some of the limits on technology are imposed by science, others by the materials and processes through which technology expresses itself, still others by the dialectics of the technological process.

It would be rash to attempt to predict how these limits will affect science and technology in specific cases. It is sobering, however, to reflect seriously on a number of factors that impose various constraints on what man can do. Such reflection is especially important to counter the euphoric optimism exemplified by the moon syndrome. After two United States astronauts set foot on the moon, it has become commonplace to respond to every problem with, "If man can go to the moon, he can do anything."

The moon landing, as with much of the space program, was an exceedingly costly engineering enterprise that stemmed from no basic scientific need and certainly no social need. The decision to go to the moon was a technological impulse, rather than a reasoned response to a techno-social problem, which arose from national pique at the shame of Sputnik and the humiliating defeat at the Bay of Pigs in Cuba. It was a Cold War decision with an eye on the military potential of space, and was made, as Leonard Mandelbaum put it, "without reference to any comprehensive and integrated national policy designed to maximize the use of scientific and technological resources for social objectives."[1]

Perhaps the greatest motivation behind the moon landing, albeit not a fully conscious one, was to demonstrate that science and technology could solve complex problems in the face of intractable techno-social problems that had already begun to proliferate and converge to the point where they had become unsolvable. Historians will wonder, with justification, that a nation invested the effort and the amount of resources on the moon project in the decade that preceded the decline of technology and the realization that life on earth was not merely endangered but would be faced with possible disaster within the short space of one or two generations. They will explain that, like wars which generate collective enthusiasm and distract attention from other problems, the moon project was a national mobilization to prove that a powerful nation could accomplish the "impossible" by landing men on the moon when it could not solve the many more demanding problems on earth. Like other techno-social solutions, however, the moon land-

[1] Leonard Mandelbaum, "Apollo: How the United States Decided to Go to the Moon," *Science*, CLXIII (February 14, 1969), 649–654.

ing has raised a host of new problems whose solutions will follow the path of proliferation described earlier.

There are eight types of limits that impose constraints upon man in his quest to accomplish the "impossible." These limits, singly and in combination, have already begun to impose constraints on the development of science and technology and will play an increasingly larger role in the future. Such constraints include: (1) Finitude—the physical limits on the earth and its resources, water, and air; (2) the narrow limits within which life is maintained in man; (3) unsolvable problems, or classes of problems for which there can be no solutions; (4) impotence principles that define the impossible; (5) perimeters that define the limits of the possible; (6) limitations on observation and measurement; (7) limitations on knowledge arising from complexity; (8) epistemological constraints.

Finitude

Modern cosmologists contemplate the universe as did the ancients, and more than four hundred years after the publication of Copernicus's *De Revolutionibus Orbium Coelestium* their theories on the origin of the universe are still speculative. The "big bang" theory, in which a primeval atom is believed to have exploded and sent material outward in all directions, vies with the "steady state" theory, according to which the universe has always existed and will continue to exist. Theoretical problems arising from each theory have led to others which hypothesize an oscillatory process of expansion and contraction, or a series of little bangs.

The cosmological mystery has impressed upon man, within the framework of human dimensions, the concepts of timelessness and infinitude. Space and time stretch endlessly out of the past and endlessly into the future. The light year, approximately one trillion miles, based upon the speed of light of 186,000 miles per second, is the measurement base of the universe. The landings on the moon are construed to initiate the exploration of the infinite.

The infinitude of the earthly habitation, too, has been intimated to man by the physical scientists who have harnessed the energy of the atom through fission and who promise unbounded energy when the problem of nuclear fusion is solved. The deuterium of the seas, it is suggested, will provide energy resources for untold time.

The belief in infinitude has been a corollary of the triadic concept of unbounded reason, continuing progress, and knowledgeable science. Homo sapiens, puny physical creature that he is, was mastering, albeit slowly, nature and the universe in the painstaking and boundless exploration of infinite paths. The earth was vast, endless stretches of land, countless waves upon the seas, the atmosphere stretching upward into the firmament.

The euphoria of boundless, limitless infinitude nurtured through more than ten thousand years of human civilizations has been pierced in this generation. Today we are face to face with the limits of the earth. The earth is a finite living space for man and all the living creatures and vegetation with which he shares a common habitat. Except for the occasional impact of a meteorite on the earth's surface, or the return to earth of rocks from the surface of the moon, the dimensions of the earth and its resources are relatively fixed. Life is sustained on a small portion of the approximately 200 million square miles of the earth's surface. More than two-thirds of this total surface is water, and of the land surface, high mountain peaks, glacial areas, and deserts are unfit for human habitation. Less than one-eighth of the total area remains for regular habitation. Upon this finite area man must dwell, produce his food, and maintain the conditions necessary for his survival.

The total space of the earth is a fixed space. More land can be obtained by filling in water or draining water. But land fills and land reclamations only reshuffle the ratio between land and water; they do not create more total area. The resources of this area are also finite. They were laid down in that unknown past when the earth was formed. They can be transformed or transmuted but new mass-energy cannot be created *ab initio*. Some of the transformations are natural processes like the formation of soil, and coal and oil. These natural processes have occurred slowly, requiring thousands or millions of years to come to fruition. Under natural conditions, it requires approximately one thousand years to form one inch of humus. The laying down of the great coal beds which provide the bulk of man's energy sources occurred many millions of years ago in the Carboniferous period when warmth and giant plants and water abounded. The formation of coal occurred over a period of millions of years in a particular stage of geological change that can never recur.

Minerals and other chemical substances are situated at scattered places on the surface of the earth, on the land and in the waters. Sometimes they appear in pure form; usually they appear mixed with other substances in the form of ores. Pure substances can be retrieved, in principle, from any concentration of ore. Substances can be transformed, mixed, synthesized, and reduced—but the totality of substances cannot be changed.

The resources of the earth, organic and inorganic, fall into two broad categories: the renewable and the nonrenewable. A renewable resource is one which has an inherent capacity to maintain itself. Most organic resources are renewable, *up to given limitations*. Thus foods in the form of grains, fruits, and meat are essentially renewable, as are forests. Renewal, however, is itself subject to biological limitations. The conditions under which the processes of photosynthesis in the case of plants and metabolism in the case of animals take place must be met. The renewal of the latter is dependent upon photosynthesis in the former for, as Raymond Bouillenne has truly remarked, "all flesh is grass."[2] Photosynthesis is dependent on complex processes by which green leaves synthesize carbon dioxide and water into chlorophyll and sugars that effect a transition from inorganic to organic matter. The process of photosynthesis is dependent upon an intricate balance of sunlight, temperature, carbon dioxide, and water. Interference with any of the requirements of the process has the effect of transforming once renewable plant resources into an unrenewable resource, with similar effects upon other heretofore renewable resources.

Most metals and fossil fuels will be exhausted within decades or, as in the case of coal, within several centuries depending on their continued utilization as an energy resource. Tin, mercury,

[2] Raymond Bouillenne, "Man, the Destroying Biotype," *Science*, CXXXV (March 2, 1962), 706–712.

helium, and tungsten are now rapidly approaching exhaustion. Most of the oil and gas will be used up within from fifty to sixty-five years. Water and wood resources are already becoming limited in portions of the earth.

A renewable resource can become unrenewable. Its unrenewability can be temporary or permanent, depending on whether the process that induced the change is reversible or irreversible. A depleted soil may establish conditions under which a plant becomes nonrenewable. Replenishment of the soil may re-establish the requisite environmental conditions to re-establish a renewable plant. But the extinction of the dodo and the passenger pigeon, both of which were renewable in their time, was irreversible. Now they cannot be reconstituted. During the twentieth century an average of one species per year has become extinct, and more than five hundred additional species of birds and animals are on the endangered list.

Inorganic resources are invariably nonrenewable, for these resources cannot maintain themselves. Coal is nonrenewable because through combustion, heat and waste products are produced and the coal is destroyed. New coal is not being deposited because the plant and climatic conditions that prevailed during the Carboniferous period can no longer be repeated. The same is true for oil, natural gas, and other resources. Nonrenewable resources, by definition, became depleted as they are used since their quantity is finite and they cannot be reconstituted.

Man-made resources like steel, autos, synthetic compounds, and the myriads of commodities produced by a technological society are artificially renewable within the limits of the availability of finite basic resources and the availability of energy to effect the necessary transformations. As depletion continues in an irreversible process, limits on production will become necessary while the energy required for transformation will increase in proportion to the depletion.

The renewable resources as well as the transformed products and waste products must all compete with man for the finite space available on earth. An area of earth cannot at the same time serve as a field of corn, a homesite, a highway, a factory location, a mine, and a garbage dump. With a sparse population on earth, conflict did not arise. With a burgeoning population in all areas of the earth, conflicting claims on land use become a source of contention which also impose limits on the utilization of the space.

Plants and animals as biological organisms can grow individually and increase in numbers. Their biomass, however, is part of the totality of the available mass-energy on the earth. Some plants and animals, by serving as food for other organisms, are a form of resource upon which man is dependent.

Limits of Maintaining Life

The air, the sunlight, the soil, plants, living and dead organisms, are all intimately linked in the biosphere. At some point in geological time, the totality of conditions on earth made life possible, and living organisms evolved through biological time in the process of evolution. At each stage of evolution life was made possible by surviving organisms meeting the minimum requirements for survival in the available environment. These conditions constitute the ecology of an ecosystem. Ecology is the study of the mutual relations of organisms with their environment and with one another. Man's survival in the ecosystem depends on very narrow internal limits and equally narrow external limits.

Man as a biological organism can counteract the tendency of nature to disorganize and can grow and develop. By means of the flow of energy and information across the boundaries between the living organism and its external environment disintegration and decay are prevented. "What an organism feeds upon is negative entropy," Schrödinger has explained. "Or, to put it less paradoxically, the essential thing in metabolism is that the organism succeeds in freeing itself from all the entropy it cannot help producing while alive."[3] When the organism can no longer counteract the forces of disorganization and disorder because of disease, malfunction of organ or tissue, or old age, death ensues. Death is thus the state of static stability of an organism.

[3] E. Schrödinger, *"What Is Life?" and Other Essays* (Garden City: Doubleday Anchor, 1956), p. 71.

A biological organism is a hierarchical system of processes in which the integrated organism represents a balance between the ordering and the disordering processes of life. Each cell is related to other cells—sensory cell with motor cell, organ with organ, tissue with tissue in a dynamic self-regulating system that tends to maintain itself. Life and well-being depend upon a delicate balance maintained by the integrated differentiation that is man. The complex chemistry of the system is maintained by sensitive sensors in blood and gland, cell and organ. The fluid matrix preserves the constancy of the internal environment. Thirst, perspiration, blood clotting, and respiration help to maintain this constancy. An equilibrium level of protein, fat, salt contents of the blood, the blood sugar, and other chemicals is maintained. Blood pressure, oxygen supply, waste disposal, and other processes are automatically regulated.

The conditions of life lie within a narrow range of parameters. Body temperature, chemical composition of the blood, osmotic pressure, hydrogen-ion concentration, blood pressure, the endocrine balance, etc.—all must be maintained within narrow limits. The malfunction of any part of the organism is reflected through complex interrelationships in other parts of the organism, resulting in the instability of the organism itself.

Against a background of relatively constant factors there are other factors that allow wide variability. This variability enables the organism to modify its behavior according to the external environment and to adjust to the new situation. A relative constancy is maintained by the variation of one aspect of the balanced living system that induces compensating changes in other aspects. The structural integration of the organism is thus maintained because it is modifiable.

The constancy of the internal environment required for survival, with its narrow limits, is highly dependent on the availability and maintenance of requisite conditions in the environment. The body must have access to those resources which make the creation of negative entropy possible: pure air, pure water, sufficient and proper food, and communication between the internal and external environments. The self-maintaining functions of the environment require proper balance between gas exchange, water purification, nutrient cycling, and the other processes that maintain life.

Unsolvable Problems

The concept that there can exist problems for which there are no solutions is as difficult for modern man to comprehend as the concept of the finitude of food, land, water, air, and other resources. Yet as there are limits to these resources, so are there limits to problem-solving. For some classes of problems there are no solutions. For other classes of problems there may or may not be solutions. The inability to solve certain classes of problems does not arise from lack of knowledge but from difficulties inherent in the problem itself.

The history of mathematics is a twofold story involving breaking ground in new areas of mathematics and the development of proofs for previously unsolved problems. At present there is still a considerable body of unsolved problems—among them, for example, Goldbach's conjecture and the problem in diophantine analysis known as Fermat's Last Theorem. In 1931, Kurt Goedel introduced to mathematics a concept that is analogous to the impotence principle of science: namely, that the axiomatic method contains certain inherent limitations when applied to complex systems, and that it is impossible to demonstrate the internal consistency of such systems without resort to principles of inference outside the system. These principles, in turn, contain the same inherent limitations in their own logical consistency. Thus there arises a class of problems, or propositions, which are formally undecidable. It should be noted that these propositions cannot be proved, not because mathematicians are not sufficiently clever or do not have available powerful enough tools, even the largest computers; they cannot be proved because of inherent logical limitations. Further, any system of arithmetic is incomplete because, given any consistent set of axioms, there are true arithmetical statements that cannot be derived from the axioms.

On a more prosaic level, everyone is familiar with games of chess and tic-tac-toe in which, under certain circumstances, the victory of one player over another, which might be considered a form of solution of the problem posed by the game's rules, cannot be accomplished. In these circumstances a stalemate results and neither opponent wins. A solution or victory in these games may therefore be either possible or not possible.

Impotence Principles

A class of problems for which there can be no solutions arises from fundamental restrictions imposed by the laws of nature. It has been pointed out that the axioms of science are not derived from a philosophical basis or world-view. Indeed, an unstated assumption of the scientific viewpoint specifically excludes a worldview. Science was to be free and unfettered. Its hypotheses were to be objectively formulated, their testing through experiment by measurement and observation, their modification when necessary, and their validation—all were to be totally objective, free from individual bias, untrammeled by allegiance to any pre-existing constraints. The speculations of the Greeks were upset by the onward push of science as was the rationalism of the scholastics overturned. The Ptolemaic epicycles and body humors, phlogiston and ether, successively gave way to scientific views on cosmology, medicine and physiology, chemistry and relativity. Man was emancipated from philosophy; the struggle against nature continued remorselessly. Power was to be grasped through understanding, which was to make man free. The success of theories was measured solely by practical results. The method of science was empirical; the evaluation of science was pragmatic.

Just as science dethroned a faith only to establish a new faith in its stead, so too did science, although inveighing against the doctrine that nature was but the inexorable workings of a Providence, establish instead the inexorable working of natural "laws." If an ominiscient God had guided the planets and sun in Ptolemy's cosmology, the laws of gravitation were the guiding force in Newton's cosmology. Newton, however, did not make a complete break with Providence, for he still ascribed to God the laws of motion on earth and those in the heavens. And "perhaps the whole frame of nature," he wrote, was established "at first by the immediate hand of the Creator; and ever since by the power of nature." God had set the primal machine in motion and kept it running. Although evolution has made it possible to dispense with the need for initial creation, the mechanism for starting the primal machine has not yet been discovered—if, indeed, it is discoverable.

But the machine was orderly and obeyed laws. These laws were at once first cause, process, and end. The laws could be modified, some could even be repealed, but old laws and new laws merged into the disciplines of science and established what could be and what could not be. The order of nature, Whitehead said, took on the attributes of the classical Greek Fate. "This remorseless inevitableness is what pervades scientific thought. The laws of physics are the laws of fate."[4]

Edmund T. Whittaker expressed a similar thought in 1947 when he said that "the whole of physics can be derived from certain principles of impotence." These principles state what cannot be done by defining the impossibilities of the world. They are not negative in the sense of forbidding an action or event, but impose limits on that which can be accomplished in a positive sense. The impotence principles are vital to science for, as Garrett Hardin notes, "Only if some things are impossible can other things be."[5]

The classic examples of principles of impotence are the first and second laws of thermodynamics and the quest for a perpetual-motion machine which has intrigued man as far back as the time of Archimedes. The first law states that energy cannot be created or destroyed, from which is drawn the corollary that the total energy of the universe is constant. Joule's law relating heat and work is a further extension of the first law. The ingenious contrivances developed in medieval Europe, built largely about the principle of the waterwheel, attest to the fascination of man with the idea of perpetual motion. Although Hermann von Helmholtz stated that a machine of this type was axiomatically impossible in a paper before the Berlin Physical Society in 1847, many attempts were

[4] Whitehead, *op. cit.*, p. 11.
[5] Garrett Hardin, *Nature and Man's Fate* (New York: Rinehart, 1959), p. 307.

made in subsequent years to develop machines that, once started up, would continue endlessly without need of new energy.

Machines whose design was relatively free of mechanical or electrical friction were still doomed to failure by the second law of thermodynamics, which states that entropy always increases, or more simply, that heat must be transferred from a higher to a lower temperature to do work. In a steam engine, cooling and condensation, for example, involve heat losses.[6] The laws of thermodynamics establish limits on heat conversion and energy production. Total efficiency can never be achieved in any physical process; there will always be losses. These inefficiencies are not due to ignorance on the part of designers or poor craftsmanship in building machines. They are due to the "remorseless inevitableness" of the laws of thermodynamics.

Hardin, a biologist, considers the accumulation of impotence principles in the field of biology as a sign of maturation. He cites, for example, five impotence principles that relate to evolution: (1) Weismann's anti-Lamarckian principle of the separation of soma and germ plasma, a principle that disallows the inheritance of acquired characters. Even the full authority of the Soviet state could not for long sustain Lysenkoism, which sought to assert the opposite. (2) The competitive exclusion principle which states that no two organisms that compete in every activity can coexist indefinitely in the same environment. (3) Waste is inevitable: "There is no heredity without its tax of mutation; most mutations are bad; their production and elimination is a kind of waste." (4) The Haldane-Muller principle that in a state of nature, each lethal mutation causes on the average of one "genetic death." (5) In a state of nature, all bad mutations are, in their cumulative and ultimate effects, equally bad.[7]

The development of science can be considered as a search for impotence principles, with each discipline laboriously formulating its own set of such principles. These principles, it is true, are based

---

[6] Stanley W. Angrist, "Perpetual Motion Machines," *Scientific American*, CCXVIII (January 1968), 114–122.

[7] Hardin, *op. cit.*, pp. 308–309.

on negative evidence. They are necessary but not sufficient conditions and, hence, cannot be completely deterministic. A stream of water has never been found that runs uphill; heat has never been noted to flow from a colder to a warmer object; no instance of the inheritance of acquired characters has ever been cited. Whereas a large number of positive confirmations are required to validate a physical law, only one negative case is required to refute it. Thus, while every positive case increases the odds that a law is valid, the finite probability of a negative case cannot be ruled out on theoretical grounds. This principle is a heritage from Newton, who forcefully enunciated his rejection of hypotheses in his fourth rule of reasoning in experimental philosophy: "We are to look upon propositions collected by general induction from phenomena as accurately or very nearly true, notwithstanding any contrary hypotheses that may be imagined, till such time as other phenomena occur, by which they may either be made more accurate or liable to exceptions. This rule we must follow, that the argument of induction may not be evaded by hypotheses."[8]

Perimeters of the Possible

The laws of nature impose bounds upon man in another way. The impotence principles define the impossible; other principles set definite limits on the possible. Given that the maximum speed possible is that of the speed of light, and that electricity travels at this speed, it follows that once electrical communication has been set up to circumscribe the earth, no further improvement in speed of communication can be achieved. This establishes what Roderick Seidenberg has called a "perimeter of the future."[9] Once a limit has been reached there is no place further to go.

The limitation on the speed of electricity became a practical concern when communication satellites were placed in stationary orbits 22,000 miles above the earth. The transit time of signals to the satellite and back to earth caused a time delay in the transmis-

[8] Isaac Newton, *Mathematical Principles of Natural Philosophy*, II; quoted in Burtt, *op. cit.*, p. 219.
[9] Roderick Seidenberg, "Justice for All, Freedom for None," *Center Diary* (Center for the Study of Democratic Institutions), XVII (March–April 1967), 33.

sion of speech which caused it to sound garbled. It became necessary, therefore, to develop compensating circuits to overcome the effects of the time delay. Transmission time of the Mariner satellites that flew past Mars was measured in minutes. More distant space travel will require ever-increasing transmission times.

At the other extreme of time duration of electric signals is the digital computer, in which signal time is measured in nanoseconds, or billionths of a second. Since it takes about three nanoseconds for electricity to travel one meter, the length of interconnections becomes a limiting factor on processing speed. The limit imposed by the speed of light may prevent man from discovering the secrets of the universe. In cosmic time, where distances are measured in light years, man is barred from knowledge of the present because the light signals he receives on earth have all been emitted in the distant past.

Another limit is absolute zero, or 273 degrees Centigrade below zero. At this temperature molecular activity ceases and the phenomenon of superconductivity appears. This phenomenon has been used in developing a cryogenic computer. A better computer cannot be built, however, by exceeding absolute zero. Absolute zero is a physical perimeter.

Sound travels at a rate of 1,090 feet per second, or 746 miles per hour in the air at sea level. It cannot be made to go faster under the same conditions. A jet plane that flies faster than the speed of sound creates a sonic boom as its speed exceeds that of sound. The sonic boom cannot be prevented by changing the speed of sound.

Other perimeters exist and establish what engineers call theoretical performance limitations. These limitations are upper bounds on what can be achieved. They differ from "state-of-the-art" limitations in that the latter imply that the limits are extensible and await further knowledge. "State-of-the-art" is a deceiving concept because it abets the illusion of endless extension of a technology without limits.

In rapidly expanding technologies which have a short time span between machine or component design and end product, the ability to predict technological expansion assumes a significant role in the design process. The degree of expansion is in part dependent upon the proximity of achieved results to the theoretical limits of the application. If a technology is close to the limit, then progress in design improvements will tend to be slow and obsolescence will not occur quickly. If the converse holds, then progress may be explosive with obsolescence being exceedingly rapid.

Limits exist in the biological sciences, too, as, for example, in the size and form of living organisms. J. B. S. Haldane, in his entertaining essay "On Being the Right Size," demonstrates that for every type of animal there is a most convenient size and that a significant change in size necessitates a change in form.[10] This same point was made by Galileo who noted that there was a growth limit for each species according to its physiological nature. Thus, to become larger than its present size, a gazelle would have to make its legs short and thick and would become rhinoceros-like, so that every additional pound of weight would have the same bone area for support. On the other hand, it could stretch its legs and compress its body, but then it would resemble a giraffe.

An insect dropped from a height would not be affected by the fall because its weight is a function of its volume and the resistance to the fall is proportional to the surface area, whereas a larger animal would be killed. In respect to surface tension, however, the roles are reversed: an insect may be trapped by surface tension whereas a pound of water in a film a fiftieth of an inch in thickness on a wet man will have little effect. The possibility of giant insects is limited by their method of respiration. Crustacea are limited by moulting, land animals by their skeletons, sea animals by their food-gathering capacities. From the microscopic to the giant organisms, size brings advantages and disadvantages until, according to Julian Huxley, "life finally comes up against a limit of size where advantage and disadvantage balance."[11]

Similarly, there appears to be a natural limit on the longevity of organisms. Years of selective breeding have not appreciably altered the life span of domesticated animals, nor has evolution expanded the life span of given species. It is believed, too, that the life span

[10] In *The World of Mathematics,* ed. James R. Newman (New York: Simon and Schuster, 1956), II, 952–957.
[11] Julian Huxley, "The Size of Living Things," in *Man in the Modern World* (New York: New American Library, 1955), p. 88.

of man has not been lengthened by advances in public health and medicine—instead such advances have made it possible for more people to survive to live out their potential life span.

The increasing rate of survival and the proliferation of population has begun, dialectically, to affect man in a negative manner. Thus medical progress is beginning to be outdistanced by an increase in the incidence of chronic and degenerative diseases that characterize civilized societies. Life expectancy has not increased significantly in the last several decades, and the prolongation of life has in effect come to a halt. Indeed, given the increase in accidents in industry, by autos and airplanes, and the countless other injuries that arise in a complex technological society, man may be on the threshold of a period of deteriorating health and declining longevity. There are also indications that the limit of effectiveness of current methods of disease control and prevention may be approaching.[12]

The most significant limitation on man arises from the limits to the range of his adaptive capabilities. The accelerating rate of change in technological development is taking place while man's biological development remains practically the same. In primitive society the long period of childhood served to prepare the growing child for taking his place in an essentially static framework. His life was patterned on the traditions and mores of his ancestors, and his environment was often little changed from that of his forebears. But in our rapidly developing civilization of technological and social change, the lessons of childhood often have to be unlearned by the adult, for these lessons may well become outdated between childhood and maturity.

Although attempts are made to retain the ways of the past, the changed conditions of the present and the near future make these ways antiquated. "Technical change disrupts old habits....," Margaret Mead has said. "Technical change involves new learning, after adulthood, and changing types of behavior."[13] Even in the most advanced countries of the world, a parent born at the begin-

[12] Amasa B. Ford, "Casualties of Our Time," *Science*, CLXVII (January 16, 1970), 256–263.
[13] Margaret Mead, *Cultural Patterns and Technical Change* (New York: New American Library, 1955), p. 269.

ning of the twentieth century did not know the automobile, the airplane, radio, television, and the innumerable other inventions of our century's technology. To their children, these innovations are taken for granted as part of their cultural milieu while radar, automation, computers, guided missiles, and solar and nuclear energy in turn become new to them. The grandchildren may well live in a world that is utterly strange to the grandparents and unfamiliar to the parents, not only from the point of view of gadgets, but from the changing social patterns which technological change brings in its wake. The social complex of childhood may be vastly changed by maturity as the pace of technological development increases. A child must be educated for a world whose future outlines are unknown to his teachers.

"The usefulness of all technological developments," René Dubos has warned, "has limitations that are inherent in the unchangeable aspects of man's biological nature." Man is destroying all aspects of the environment under which he evolved as a species and developed into man.[14] Hugh H. Iltis, a botanist, has emphasized this same limitation in reminding us that man evolved in a lengthy process of over 100 million years as a mammal, 45 million years as a primate, and over 15 million years as an ape. "As unique as we think we are, we are nevertheless programmed genetically to clean air and sunshine, to a green landscape and unpolluted water, to natural animal and vegetable foods. To be healthy, which after all has nothing to do with I.Q. or culture, means simply allowing our bodies to react in the way the ten millions of years of evolution in tropical or subtropical nature have equipped us to do."[15]

We have seen that the body mechanisms that maintain life operate within very narrow ranges and are highly susceptible to rapid and excessive fluctuations in the environment. Man can don a gas mask to filter out air pollution, but he cannot change his method of respiration nor his dependency on oxygen. Man can consume

---

[14] René Dubos, "Man's Unchanging Biology and Evolving Psyche," *Center Diary*, XVII (March–April 1967), 38–43.

[15] Hugh H. Iltis, "Ecological-Evolutionary Relationships, Part I," Newsletter of the Society for Social Responsibility in Science, No. 189, May 1968.

food pills in lieu of plants and meat, but his metabolic processes remain unchanged and he continues to require light, air, and a host of chemicals in intricate combinations.

Measurement Limitations

Scientists in the eighteenth century believed that the great forces of nature were determined by configurations of mass. The future path of a moving body could be predicted and its path known if its present condition and the forces acting upon it were known. All phenomena were considered the result of the action of attraction and repulsion forces as a function of the distance between unchanging particles. Newton postulated the existence of an "absolute, true, and mathematical time, which of itself flows equably without regard to anything external." Science conceived of atoms as fundamental, indivisible building blocks of nature. The laws of conservation of energy and electromagnetic theory involving electric and magnetic fluids were formulated. By the end of the nineteenth century, Maxwell had unified optics and electrodynamics in his four famous equations. The kinetic theory of matter had united thermodynamics and mechanics.

In these endeavors it appeared that physical science had succeeded in discovering the fundamental and absolute laws of the universe, and had learned to express these laws in the precise language of mathematics. The laws were derived from precise, repeatable experiments. They met the test of objective criteria. They were verifiable. A new truth had indeed been created, and this time, so it appeared, man had tested his knowledge and knew that he knew.

But even in the scientific "absolute" a crisis was brewing, and the Michelson-Morley experiment of 1881, designed to measure the ether drift, ushered in a new revolution in science that is still continuing. The contradiction revealed in this scientific crisis, as in others of the past and the future, lay not in the realm of the objective world but only in man's formulation of the objective world. Once again it was confirmed that even in science there are no eternal theories. "Every advance in science," Einstein noted, "arises from a crisis in the old theory through an endeavor to find a way out of the difficulties."

"Space," "time," "mass," and "length" had become objectivized in classical physics and semantically had become symbols of real things, of universals. It had been believed that there was a unique meaning given to "space" and a unique meaning to "time." It had been believed that there was such a thing as simultaneity—that an event on the earth observed by an observer on earth occurred at the same time when viewed by an observer in space. The Michelson-Morley experiment, however, revealed that the velocity of light is constant irrespective of the motion of the observer. To explain this experiment, Einstein had to rise above the current abstractions of scientific thought and challenge fundamental axioms. He operationally described the methods of determining space and time, mass, and length. By showing that the velocity of light did not vary with any observer regardless of his relative motion, Einstein concluded that space and time vary with different observers. The rigid *a priori* metrical structure of space and time was surrendered for a structure which was influenced by its specific physical content. Space and time, it was shown, were not revelations of the Platonic world of Ideas or a Kantian vision grown from pure reason.

"Length," "mass," "time," and "space" are not absolute facts about objects, but instead describe a relationship between objects and observer. "Length" and "time" become functions of relative velocities; and as the velocity of an observer approaches that of light, the length of an object in the line of motion would approach zero. A three-dimensional figure would appear as two-dimensional, and a sphere would appear ovate. The equivalence of mass and energy as expressed in the relationship $E = mc^2$ and the bending of light rays follow from relativity theory. Hermann Minkowski, by combining geometry with physics, showed that in our experience space and time can never be separated, that a datum is not a place and a point of time, but an event, a place at a definite date. The four-dimensional "space-time" continuum relates the space-time of our experience, as symbolized in words, to the structural space-time of reality, and thus does away with the duality which had been thought to exist.

The revolution in physics as revealed by quantum theory is also instructive in its conception of reality. In quantum mechanics the explanation of nature and matter became more difficult to determine and more difficult to conceptualize. Immutable elements gave way to atoms, which were discovered to consist of a complex structure made up of many types of particles. The submicroscopic size and astronomical velocities involved precluded normal measurement techniques and analysis. As a result there has emerged a fascinating and challenging picture of the nature of reality.

Classical physics had assumed that each physical event could be localized independently of all the dynamic processes around it. It was thought that greater refinement of measuring techniques would lead to greater accuracy of observation and approximation. But the existence of quanta, now considered as basic, discrete quantities of energy, imposes a lower limit on the measurements which the physicist can make on a system under observation. Due to the submicroscopic character of the quantum under study, there results an indeterminacy caused by the interaction between the system under study and the measuring device. If an attempt is made to measure the velocity of an electron, its displacement must be observed. A light must either be focused upon it or it must emit a light. The time required for a light to emerge from its source, its birth period, is 10^{-10} seconds. An electron revolves in its orbit 10^{16} times per second. A light on the electron would therefore reveal only a circular smear since an electron would have already revolved one million times. If an individual electron were illuminated, the light rays would disturb the momentum of the particle. This change of momentum would be greater the shorter the wavelength of the light employed. But it is the shorter wavelength that will most accurately determine position.[16]

Quantum theory has demonstrated that other paired quantities that appear simultaneously, such as time and energy, also cannot

[16] Henry Margenau, *The Nature of Physical Reality* (New York: McGraw-Hill, 1950), p. 41.

be measured in pairs simultaneously with any degree of accuracy. The conditions of observation of conjugate quantities are thus incompatible. Physical science found it necessary to express latent observables in mathematical terms. A collection of observables became a state as a function of space coordinates or probability amplitudes. In Newtonian physics, it is recalled, the sequence of physical phenomena was completely determined by its past and, in particular, by the determination of all positions and moments at any given time. It was now discovered that the whole past of an individual system does not determine the future of that system in an absolute way, but indicates the distribution of possible futures of that system. There was no longer a strict causal bond between successive measurements but a bond of probabilities. Such a system was characterized by its statistical relations and was analyzed by probability calculus. Instead of a system being characterized by the individual nature of its parts, groups of individuals were dealt with in accordance with the probabilities of their collective behavior.

The Heisenberg principle of uncertainty established a limit on experimental observations and measurements by showing that the measurement itself perturbs an observed system in such a way that corrections cannot be made for the disturbing effect of the instruments. This limit does not arise from experimental error in a mathematical sense, from the design of the experiment, or from the lack of precision of instruments. At the quantum level, the limit is erected by physical laws.

Niels Bohr pointed out the significance of this limitation in regard to living organisms in 1933 when he showed that measurements at the microscopic level would alter a living system so radically that the act of measurement would lead to the sickening and eventual death of the organism. Modern biologists, chafing at the possible metaphysical taint of vitalism in dealing with life, are seeking to follow the path of physics and reduce life to a statistical ensemble—a probabilistic concept that makes mathematics a primary representation of objects. To exorcise metaphysics from biology, Walter M. Elsasser, for example, has called for the utilization of probabilistic induction along the lines of the physicists to explain the problems of organism. "Instead of looking for causal connections and de-evaluating statistics," he contends, "we represent objects of experience in such a manner as to get the maximum benefit from statistical indeterminacies."[17]

Even as the act of measurement perturbs the object of measurement, speed limitations and the uncertainty principle also affect the act of measurement itself. Wave-mechanical restrictions on the limit of speed of a measurement arise because of a time uncertainty which becomes comparable to the measurement time at extremely high speeds. The information obtained from a measurement brings about an increase in the entropy of the measuring equipment. In a closed system that includes the measured object and the measuring apparatus, the act of obtaining information reduces the entropy of the system by establishing order—that is, selecting a specific measurement from a diversified set of probable measurements. In accordance with the second law of thermodynamics, an impotence principle discussed earlier, there must be an increase of entropy elsewhere. This increase in entropy is related to an amount of energy that is degraded during the measurement.

The implications of this increase in entropy are significant in relation to the task of measuring and sensing the state of high-speed devices or circuits, such as are used in electronic computers. The amount of information gained depends upon the expenditure of power—up to a limit. To detect a signal in zero time would require an infinite amount of power. H. J. Bremerman has shown that there is a finite limit on the number of configurations that a computer can assume, a number he calls transcomputational.[18]

Rolf Landauer has extended this concept one step further by stating that "the tools of mathematics are physical, and therefore that mathematics is limited by nature." Because of thermal agitations and the uncertainty principle, Landauer argues, "the mathematical processes themselves are not really capable of being car-

[17] Walter M. Elsasser, "Acausal Phenomena in Physics and Biology: A Case for Reconstruction," *American Scientist*, LVII (Winter 1969), 502–516.
[18] H. J. Bremerman, *Proceedings of the 5th Berkeley Symposium on Mathematical Statistics and Probability* (Berkeley: University of California Press, 1967), IV, 15.

ried out, and . . . we should not (even in pure mathematics) invoke processes that nature will not permit."[19]

Limitations on Knowledge

Is man's knowledge infinitely extendable, or are there natural limits on human comprehension? Are there laws of human knowledge analogous to the laws of thermodynamics in physics—that is, that the amount of information will tend to decrease and that greater and greater amounts of energy (human and physical) will have to be expended to obtain ever-increasing amounts of information? Will knowledge eventually overwhelm man in that no individual or group will be able to integrate all the facets of information and mold them into a comprehendable entity?

We can only speculate at this point in time, but that there are some limitations on human knowledge is already evident. We have seen how modern physical theory has established limits on observation and measurement. We have also seen that impotence principles limit the very act of measurement. These limitations occur at the quantum level. Are there limitations on the observable at the astronomical level as well?

Cosmologists, John A. Wheeler reports, have consolidated the quantum principle and the theory of relativity into what is called geometrodynamics. This theory of quantized general relativity postulates a superspace in which a probability wave propagates. But the configurations of space that occur with appreciable probability are so numerous that they cannot be nested together and accommodated in any one space-time. Hence, space-time is eliminated, and with it time itself is gone. With the disappearance of time, "before" and "after" lose their meaning. The question, "What happens next?" has no well-defined significance in this context. The geometry of space at small distances, too, is not well defined. "If there is no such thing as the geometry of space at small distances,

[19] Rolf Landauer, "Wanted: A Physically Possible Theory of Physics," *IEEE Spectrum*, IV (September 1967), 105–109. See also Panos A. Ligomenides, "Wave-Mechanical Uncertainty and Speed Limitations," *IEEE Spectrum*, IV (February 1967), 65–68; Louis Brillouin, "The Negentropy Principle of Information," *Journal of Applied Physics*, XXIV (September 1953), 1152–1163.

then it is also true," Wheeler continues, "that there is no such thing as 'the' universe at large distances . . . there is no unique history that one can ascribe to the universe." There are probabilities of different histories.[20]

Will one of these histories be that of an expanding universe or an oscillating condition of expansion and contraction? According to Hubble, more remote galaxies are receding faster than closer ones, and their speed increases the further they are from us. If, indeed, this relationship is valid until a point is reached at which velocities approach the speed of light, the galaxies will be effectively unobservable.[21]

If the universe is expanding and may be infinite, is infinity one-sided, extending only to the macrocosm but not the microcosm? The atom, once conceived of as an indivisible unit of matter, has long since been broken into hundreds of smaller particles. With the increasing power of newly built atom-smashers, new particles are continuing to be found. So bewildering is the proliferation of particles that a Nobel prize was awarded to Murray Gell-Mann in 1969 for his contributions to the theory of classification of particles, including the "theory of strangeness" and the "eightfold way" which sought to establish a theoretical order to the complex universe of the atom. Will Gell-Mann's theory that the quark is the fundamental particle from which all others are made outlive the next generation of atom-smashers?

There is yet another problem that physical theory must face. In biology, Darwin made it possible to dispense with a Creator who breathed life into living creatures. Although the origin of life is not known, it is believed theoretically possible that life could have been begun by the action of heat and light on inorganic substances at a particular period in geological time when all the conditions for the transition from inorganic to organic substance were present.[22] The probability that such a transition could have taken place is exceedingly small, but in the time span of 4.5 billion years it was

[20] John Archibald Wheeler, "Our Universe: The Known and the Unknown," *American Scientist*, LVI (Spring 1968), 1–20.
[21] Patrick Moore, "The Queer Cosmos," *Science and Technology*, November 1967, pp. 28–37.
[22] See, for example, A. I. Oparin, *Origin of Life*, 2d ed. (New York: Dover, 1953).

only necessary that this event occur once. In the geological time scale, George Wald notes, the impossible became possible, the possible became probable, and the probable became certain.[23]

In physics, too, there is a question of "in the beginning." If, following Wheeler again, Einstein's views on nature are correct, a particle, "a quantum state of excitation of a dynamic geometry," will be "constructed out of space." To construct a particle from "empty" space there is no place in pure quantum geometrodynamics for the necessary factors to come from except from "initial conditions." These "initial conditions" are the initial shape of space and its initial speed of expansion in this cycle of the universe. But physics does not have anything to say about "initial conditions." From Newton's first law of motion to geometrodynamics, science is silent. Given "initial conditions," it is possible to state physical laws, but what is the Prime Mover? Does physics still await its Darwin, or is the possibility of a Darwin excluded by the galaxies receding at the speed of light and the possible divisibility of the hypothetical quark into nonmeasurable particles?

Thus, at both the quantum and cosmic levels of the physical universe there loom questions that grow in complexity and whose answers become ever more elusive. "We have sought for firm ground and found none," Max Born wrote of the quantum universe. "The deeper we penetrate, the more restless becomes the universe, and the vaguer and cloudier.... There is no fixed place in the universe; all is rushing about and vibrating in a wild dance."[24] Born is echoed at the astronomical level by Wheeler, who describes the geometry of space at small distances as fluctuating between one configuration and another like a carpet of foam "made up of millions of tiny bubbles . . . continually bursting and new ones being formed."[25] And this "restless universe" and "millions of tiny bubbles" of space must be comprehended by a human organ that itself may be uncomprehendable for, as Charles Sherrington has described the waking brain, "It is as if the Milky Way entered upon some cosmic dance. Swiftly the head-mass becomes an enchanted loom where

[23] George Wald, "Origin of Life," *Scientific American*, CXCI (August 1954), 45–53.
[24] Max Born, *The Restless Universe* (New York: Harper, 1936), p. 277.
[25] Wheeler, *op. cit.*

millions of flashing shuttles weave a dissolving pattern, always a meaningful pattern though never an abiding one."[26]

Ideally, the language of science was to have been precise. Phenomena were to be observed and measured. Definitions were to be made in terms of these observations and measurements. Subjectivity and human values were excluded. But the language has become many languages, and each language has become an esoteric jargon that is foreign to the ears of a scientist in another field. The physical world and man have become as elephants described in different ways according to specialized viewpoints.

"Nature must be considered as a whole if she is to be understood in detail," A. S. Lotka observed in 1925. Can the jargons be reconciled to unify the picture of man and earth that has been sundered? If they can be, is it possible for man to comprehend the immensity of nature in a unified way? Perhaps Shakespeare was prescient when he wrote
>In nature's infinite book of secrecy
>A little can I read.

Eugene Wigner has discussed some of these problems in a paper entitled "The Limits of Science."[27] As science grows and its literature expands there are limits, he believes, to the capabilities of the human intellect to communicate the mass of accumulated knowledge. In some areas the background information is so great that the older parts are forgotten. Information can be condensed up to a certain point, but beyond that point condensation degrades information. One consequence is that science tends to shift its direction toward fields where less information is available, and hence more information remains to be collected. But science cannot shift its growth indefinitely, because a new discipline is more complex than an older discipline while it also encompasses much of the old. These shifts always involve "digging one layer deeper into the 'secrets of nature.'" The uncovering of each subsequent layer requires more elaborate and longer studies; the interest to burrow may decrease with deeper layers.

[26] Charles Sherrington, *Man on His Nature,* 2d ed. (Garden City: Doubleday Anchor, 1951), p. 184.
[27] Eugene Wigner, "The Limits of Science," *Proceedings of the American Philosophical Society,* XCIV (October 1950), 422–427.

"The absence of interest and the weakness of the human intellect," Wigner observes, "may easily combine to postpone indefinitely the determination of the full adequacy of the nth layer of concepts." The subsequent shift to other areas may mean "the acknowledgment that we are unable to arrive at the full understanding of even the inanimate world...."

It is interesting to inquire into the possible depth of the layers of concepts which Wigner discussed. As one digs deeper, will the bedrock be exposed finally to view, or can burrowing extend endlessly? John R. Seeley has suggested that the latter view is more likely. Describing his "inexhaustibility theorem," Seeley remarks that "the subject matter of something cannot be exhausted if the first description both alters, and, in any case, increases the subject matter to be described."[28] Hence, further study expands rather than contracts the field of interest and requires further and more intensive study to continue an exploration.

The inexhaustibility theorem and layers of concepts are well illustrated by neurophysiological research. After thousands of man-years of research and study, the understanding of the nervous system is restricted in the main to events at or near the periphery of sensory and motor systems, leaving comprehension of the language and mechanism of nervous action almost nonexistent. The scientists, Leon Harmon of Bell Telephone Laboratories has explained, have developed an analogy between the computer and the brain and assumed that electrical signals are the language of information. The analogy may, however, be misleading, having arisen from a historical coincidence that made electrical measurements readily available to neurophysiologists. The electrical events, Harmon has suggested, may be second order, or perhaps even artifacts.[29]

The enormous complexity of the human brain so far exceeds any system that man has studied that attempts to understand every detail of the brain's functioning are regarded as a hopeless task—for the foreseeable future. Attempts to transform the brain, as with

---

[28] Seeley, *op. cit.*, p. 153.
[29] Nilo Lindgren, "To Understand Brains," *IEEE Spectrum*, V (September 1968), 52–58.

physics and biology, into pure mathematics, will perhaps create new artifacts, but neither thought nor life.

And what of the burgeoning literature in which the burrowing is to take place? The present 35,000 scientific journals published throughout the world are increasing in number at a rate of over 3 per cent each year, which results in their number doubling every twenty years. In 1960 it was estimated that approximately 7 million books and pamphlets were added to the world's libraries. The total world's holdings was estimated to range from between 2.2 to 7.7 hundred million books and pamphlets, depending on the degree of overlap assumed between libraries.[30]

The burden upon the scientist to keep abreast of this flood is continuing to grow, as scientific societies and journals extend their efforts to increase the total output. An International Congress of Biochemistry published its own proceedings in fifteen volumes. The 2,100 papers submitted to the Second Geneva Conference on the Peaceful Uses of Atomic Energy required thirty-three volumes for publication. The American Institute of Physics published 65,887 editorial pages in twenty-one journals in 1968 and another 26,065 pages in translated journals. The American Chemical Society published 40,225 pages in twenty journals, and the Institute of Electrical and Electronics Engineers published 23,759 pages in thirty-six journals and 9,979 additional pages in seven translated journals.

To assist the scientist in his labors, abstracting journals have been established to review the primary literature and condense it for faster perusal. The Medical Literature Analysis and Retrieval System (MEDLARS) of the National Library of Medicine reviewed 6,000 journals in 1963 and printed more than 500 pages monthly in the *Index Medicus* with nearly 12,000 abstracts monthly. In 1969 the *Index* reviewed nearly 25,000 journals and estimated printing of 250,000 abstracts for the year.

Chemical Abstracts Service published more than 131,000 pages in its abstracting publications in 1967 and reported on nearly 200,-

---

[30] John W. Senders, "Information Storage Requirements for the Contents of the World's Libraries," *Science*, CXLI (September 13, 1963), 1067–1068.

000 papers and reports and 37,000 patents. It scans nearly 1,500,000 articles a year in 12,000 journals and abstracted 240,000 in 1969. The five-year Index (1957–1961) contained 5 million separate entries on 22,000 pages in fifteen volumes. The 4 million published abstracts since 1907 cover 350,000 pages in sixty-eight volumes. Chemical literature has been growing at a rate of 9 per cent for the past twenty years, and deals with about 4 million known compounds, which themselves are increasing at a rate of more than 100,000 new compounds per year.[31]

When we consider that the information capacity of the human brain has been estimated to be about twenty binary digits per second, the rate of increase of new literature of nearly 2 million binary digits per second points up the enormous gap between information available and information that can be assimilated. It is not in a light vein, therefore, that it has been said that a physiologist would have to read four hundred papers a day to keep abreast of current work in his field. Similar situations exist in the other scientific disciplines.

George Gaylord Simpson, in *The Meaning of Evolution,* has summarized a lifetime of research in a manner somewhat similar to Wigner. "The ultimate mystery is beyond the reach of scientific investigation, and probably of the human mind. There is neither need nor excuse for postulation of nonmaterial intervention in the origin of life, the rise of man, or any other part of the long history of the material cosmos. Yet the origin of that cosmos and the causal principles of its history remain unexplained and inaccessible to science. Here is hidden the First Cause sought by theology and philosophy. The First Cause is not known and I suspect that it never will be known to living man. We may, if we are so inclined, worship it in our own ways, but we certainly do not comprehend it."[32]

These may seem like strange observations from men of science, from whom arrogance rather than humility might have been expected. In the days of the computer and Big Science and system

[31] *Chemistry,* July–August 1968.
[32] George Gaylord Simpson, *The Meaning of Evolution* (New Haven: Yale University Press, 1949), p. 278.

analysis and travel to the moon, are not these trepidations out of date? Haven't limits to man's capacities and capabilities been greatly expanded and perhaps eliminated? It is well, then, to consider the words of Norbert Wiener, whose study of cybernetics is the underpinning of these new sciences and technologies: "The world of the future will be an ever demanding struggle against the limitations of our intelligence, not a comfortable hammock in which we can lie down to be waited upon by our robot slaves."[33]

Problems of Epistemology

A limitation on our intelligence that cannot be overlooked stems from epistemological problems. As shown in the discussion of the basic scientific axioms, science is based intrinsically upon a number of *a priori* assumptions whose acceptance is dependent on faith. Among these assumptions is the concept that nature is orderly, and that through the exercise of reason this order can be discovered and employed to harness nature in the service of man.

The concept of harnessing nature through conquest was in error because it failed to recognize that man was a part of nature and that what happened to nature would in turn redound upon man. The concept of reason was, in fact, anti-rational because it eschewed an integrative philosophy that united disparate phenomena, for science dealt only with facts not beliefs. The enthronement of "facts" was self-serving, for "facts" were the only reality that science could manipulate after it had excised philosophy, morality, ethics, and all considerations of quality from its ken of observation. "Facts" were selected for scientific scrutiny only if they were amenable to observation and measurement, if they could be placed within an orderly structure.

In this scheme, knowledge became, as Norbert Wiener has wisely observed, the interpretation for man's convenience of a system that had not been designed for man's convenience. In short, science selected as "facts" those phenomena that accorded with the intuitive faith in natural order and were reducible to what Whitehead has termed the tenets of scientific materialism. In a

[33] Norbert Wiener, *God and Golem* (Cambridge, Mass.: M.I.T. Press, 1964).

reversal of Platonic dualism, science termed phenomena real and ideas and qualities unreal. Because of its narrow selection of "facts," the scheme was successful because just those areas were selected which were amenable to, and appeared to require, investigation.

The laws of nature that were promulgated may be not so much properties of nature as properties of man who describes and classifies phenomena. Nature, after all, does not have coordinates, and many of our basic concepts are man-imposed—inspired by man's conceit and the psychological craving for order. "The truths of mathematics and the mathematical formulations of the principles of science," Eric Temple Bell observes, "are of purely human origin; they are not eternal necessities but matters of human convenience."[34] Does time exist in the universe, or is it a convenience by which man seeks to order his affairs? Does distance exist, or is it a creation of man's yardsticks? The restless universe of the quantum and the mystery of the heavens is forcing man to redefine his concepts and reconsider his approach to the comprehension of nature. The heavens have not undergone any fundamental change since man has been observing the firmament. Space and time, as reifications, have been changing.

The incantations of science are becoming less precise as its vision of reality recedes further from the world of "common sense" and creates instead a highly artificial world of constructs cemented together by abstract mathematical symbols. The selection of constants and variables in the real world is becoming more arbitrary as the outlying point, the unconforming phenomenon, the unpredictable atom and human, and a disordered universe rebuff the scientists' efforts to snare them in his idealized, and artificial, Procrustean framework of quantification and order.

Just as it is difficult to select a "fact" from a universe before comprehending that it is a "fact," we must also consider the tool of that comprehension, human language. The hypothesis of Benjamin Lee Whorf, that the structure of a language influences the manner in which reality is understood and determines how man behaves toward that reality, is significant. This hypothesis suggests that different pictures of the universe may be held by observers of the

[34] Eric Temple Bell, *Magic of Numbers* (New York: McGraw-Hill, 1946), p. 355.

same physical evidence if their languages are fundamentally different. Thus, the picture of reality is not only a function of philosophical or religious background, but also of linguistic background.

The Hopi language, for example, contains no reference to time, though all observable phenomena of the universe can be explained by this language. Newtonian space, too, is not part of the Hopi language and, therefore, is not used to explain observed facts. Yet the Hopi introduce new concepts and abstractions that make their world picture as comprehensive as that of the scientists. If these Hopian postulates appear to be tinged with metaphysics, Whorf adds succinctly, they differ only in quality from the metaphysical basis of the scientific enterprise.[35]

The crystallization in language of a culture's philosophy is exemplified by the ancient Oriental world-views which saw the quandaries inaugurated by the quark and the geometrodynamics of space as irrelevant to their frame of reference. Since Being is timeless and flowing, there is no beginning and no end. The universe grows from the Imperishable and swells out and is drawn back in. Life continues endlessly through countless incarnations. Since the perceived reality is different, the observed facts are different. Which is the reality and which the fact? Does objective experience precede subjective experience, or is it the converse?

Perhaps there is no one answer because the wording of the question is itself so closely tied to the cultural and philosophical viewpoints of the questioner and the respondent. The positivist philosophy, as we have seen, denies those questions for which answers cannot be found directly through measurement and observation. Ludwig Wittgenstein summarized his position when he stated, "What we cannot speak of we must be silent about." And in his *Tractatus Logico-Philosophicus*, he cut the ground from beneath his own feet in the book's final words: "My propositions are elucidatory in this way: he who understands me finally recognizes them as senseless...."

We finally come to the argument of David Hume, who questioned the very principles upon which science erected its laws.

[35] *Language, Thought and Reality: Selected Writings of Benjamin Lee Whorf*, ed. John B. Carroll (Cambridge, Mass.: M.I.T. Press, 1966), pp. 57–85, *passim*.

Hume stated as a general proposition that knowledge of cause and effect cannot be attained by *a priori* reasoning but only by experience. If one were confronted by an unfamiliar object, and required to state the effect expected from the object, it would be necessary to invent an arbitrary effect, for the effect is completely different from the cause and can never be discovered in it. Reason, therefore, cannot assign the ultimate cause of any natural operation. "But as to the causes of these general causes, we should in vain attempt their discovery, nor shall we ever be able to satisfy ourselves by any particular explication of them. These ultimate springs and principles are totally shut up from human curiosity and inquiry.... The most perfect philosophy of the natural kind only staves off our ignorance a little longer.... Thus the observation of human blindness and weakness is the result of all philosophy, and meets us, at every turn, in spite of our endeavors to elude or avoid it."[36]

Writing nearly 175 years after Hume, Bertrand Russell reaffirms this basic problem in science and asserts that "the limitations of scientific method have become much more evident in recent years than they ever were before." Less and less is found to be datum and more and more is found to be inference, although the greater part of the inference is unconscious. Being unconscious, however, Russell warns, does not grant validity to the inferences. There remains a "doubt as to the validity of induction," which "remains an unsolved problem of logic."[37]

We have now come full circle. Whereas the restless universe appears to preclude our attaining the ultimate cause by virtue of the complexity of the effects and the limitations of both measuring devices and human comprehension, Hume and others affirm that this ultimate cause cannot be apprehended because causations cannot be inferred from any effects we may observe. To paraphrase Simpson, we may, if we are so inclined, establish *a priori* causes and embellish them as a faith, each in his own way, but we certainly do not comprehend them.

[36] Hume, *op. cit.*, p. 45.
[37] Bertrand Russell, "Limitations of Scientific Method," in *The Basic Writings of Bertrand Russell*, pp. 621–624.

Limits, Impotence, and the Unsolvable

Finitude, unsolvability, impotence principles, perimeters of the possible, limitations on observation and measurement, knowledge and comprehension—all combine to impose constraints on what man can know and do. While many things are possible, all things are not possible and some are impossible. Whereas some types of problems can be solved, others cannot. In the next four chapters we will consider how the proliferating residue problems of numerous quasi-solutions are beginning to converge to generate unsolvable problems.

6 *Technological Realism 1: People, Land, and Food*

AT THE TIME of the 1967 summer solstice, silvery islands began to appear in the southern reaches of Lake Michigan. As viewed from the air, the islands shimmered and glistened, reflecting the sun as they bobbled in the undulations of the waves. Hundreds of yards in width and sometimes stretching out for fifty miles, the silvery streaks moved with the wind and waves. The islands were ephemeral and shifting, disappearing and coalescing like huge, shiny oil slicks.

Easterly winds directed the islands to the shores at the eastern tip of the lake, and residents of Chicago awoke one morning to find the beaches inundated with the bodies of dead alewives, a species of saltwater fish that had once lived in the oceans but now made their homes in the Great Lakes. The rotting bodies were piled high on the beaches, forming a platform more than a foot in depth in the harbors and along the shores. It was estimated that more than twenty billion dead fish lay putrefying on the white sands, contributing an odor that competed with the effluvia of the refinery and steel mills. Huge bulldozers scooped deep trenches in the sand to bury the fish in a race against the oncoming waves bringing in new cargos.

The alewives had starved to death, some scientists claimed, but others pointed out that examination of their stomachs did not indicate starvation as the cause of death. Some scientists believed the deaths resulted from a normal life span of three years which, together with the lack of predators in Lake Michigan, resulted in a huge wave of dying when the life cycle was completed. Discussions relating to the alewives' adaptation to fresh water, the seasonal depletion of plankton on which the alewives fed, and other diverse causal explanations were offered for the dead fish.

The billions of dead alewives were nature's response to a converging chain of residue problems generated by a series of quasi-solutions to totally unrelated problems. When a continuous waterway was opened up from the mouth of the Gulf of St. Lawrence to the Great Lakes, making ocean ports of the inland cities of Cleveland, Detroit, Milwaukee, Chicago, and Duluth, the engineers did not consider the vast ecological changes that would be brought about as the biological equilibria of the Great Lakes and their interconnecting waterways were altered. It was not foreseen that the locks and canals that allowed passage of oceangoing vessels would also allow passage to the lampreys and the alewives. The lampreys came first and devastated the breeds of game fish that had made their homes in the fresh waters of the Lakes. Trout, perch, and other edible fish soon disappeared, and once-thriving fishing enterprises were destroyed. The economic charts that hail the progress of the waterway did not include in their profit-and-loss columns the destruction of the fishing industry, nor the cost of the research on lamprey control.

The alewives, originally saltwater fish, entered the lakes along the same route as the lampreys. The larger game fish, which would normally have preyed upon the alewives, had already been destroyed by the lampreys, so the alewives flourished—at least in some of the Great Lakes. Lake Erie had already been turned into a cesspool for the Cleveland megalopolis, and most marine life had been destroyed. Perhaps, after all, a dead Lake Erie was the most effective barrier man could create to prevent a further inundation of the other Lakes by strange species of water creatures, although mindless technology, not any deliberate purpose, had destroyed life in Lake Erie.

The alewives tale reveals many things about the fish and about humans. (1) The development of the St. Lawrence Seaway was treated as a problem of engineering subject to economic constraints. The effects of the Seaway on the totality of the human-nature matrix were neglected, if understood at all, as in so many similar undertakings. (2) Faced with unforeseen consequences, solutions were sought which were piecemeal and which brought in their train new, unforeseen consequences. (3) The attempted solutions were based upon engineering techniques subject to economic constraints, even though this shortsighted approach brought about the original problems. (4) Nature has its own logic which is not dependent upon man's thinking or unthinking response.

The alewives problem also illustrates the dialectical characteristic of technology. It is not an autonomous mechanism that operates independently of man and society, but instead within the social framework it is both good and bad, and quite independent of man's hopes and predictions. The good is transformed into the bad, and the bad is transformed into the good; the optimistic prediction is undermined by pessimistic prognoses, and pessimistic prognoses are belied by optimistic developments. In the long run, technology will run its course and will evolve into a new thesis—if a society remains in which a post-technological period can come into being.

As with the alewives problem, so there have arisen a number of other techno-social areas in which residue chains are proliferating in ever more complex linkages. The sets of residue problems are becoming interrelated in a heretofore totally unexpected manner. In one area the gross effect is the cancellation of another quasi-solution by the aggregation of residue problems in other areas. In another area, the effects of unsolved problems accumulate to make new quasi-solutions difficult. In still other areas, the residue problems reinforce each other as they converge and foreclose quasi-solutions to the convergent sets of problems.

Before investigating such illustrative residue chains, it is worth taking a short detour to an area of mathematics that has been pre-empted by the theoreticians of Progress—but only in part. The theoreticians bask in the growth curves that appear to confirm their prognoses of continual advances—on all fronts—but they have failed to reckon with *asymptotes*,* the limiting position which most of these curves must approach with the passage of time in a finite world bounded by perimeters and impotence principles, and constrained by the depletion of resources and the growing demands upon these limited resources.

Although a growth curve can take many forms, two common general forms are the linear and exponential.† In linear growth, the increment in each unit of time is directly proportional to the first power of a variable, that is, growth occurs by arithmetic progression. In an exponential curve growth is a function of a power or exponent that is the variable. The curves are illustrated in Figure 3.

If two related events both have independent growth curves, and the relationship requires that the distance between the two curves be fixed or in a specific ratio, a problem will arise when the growth rates diverge to alter the fixed distance or the given ratio, as in the Malthusian curves of food and population. According to Malthus, the food supply grows linearly while population grows in a geometric progression. Since a given amount of food per unit of population is required to sustain life, Malthus stated, the population growth must always run far ahead of the available food supply, and with the tragic consequences he foretold.

Obviously, an unchecked growth curve will eventually approach infinity. But inasmuch as in a finite world it is impossible to reach infinity, the curve must be slowed in its rate of growth at some point, and it must become asymptotic to a given limit, or else it must actually decrease.

A growth curve of a selected group of boys is shown in Figure

* An asymptote is a line that is the limiting position which the tangent to a curve approaches as the point of contact recedes infinitely along an infinite branch of the curve. It is familiar from geometry where a hyperbola is drawn about the foci and the approaching diagonal lines drawn through the vertices of a symmetrical rectangle that is drawn parallel to the transverse and conjugate axes. The diagonal lines are the asymptotes of the hyperbola.

† A linear growth curve is of the form $y = ax$, where a is a constant and x is a variable. An exponential curve is of the form $y = ab^x$, where a and b are constants and the exponent x is the variable.

116 *Overskill*

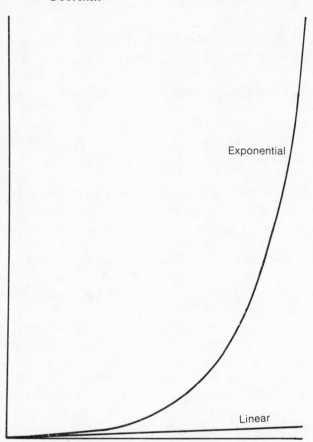

Figure 3. Linear and exponential curves

4. At birth, the average height (length) is approximately twenty-one inches; an average growth of eight inches is expected in the first year. If growth were to continue at the same rate in subsequent years, the A curve would be followed—with disastrous consequences. During the second year, although a slowing of the growth rate is experienced, if the new growth rate were continued, curve B would be followed, again with tragic consequences. The spurt in growth at adolescence, curve C, would once again carry the child beyond human limits if the growth were not braked. At age eighteen full growth is nearly attained, with only small incre-

Technological Realism 1: People, Land, and Food 117

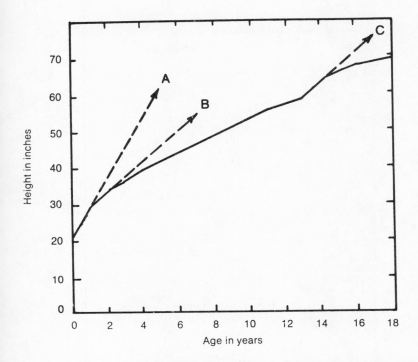

Figure 4. Growth curve for a selected group of boys (Source: J. M. Stephens, *Educational Psychology*, revised edition, New York, Holt, Rinehart and Winston, 1960, p. 77.)

ments experienced until the age of twenty when full growth is attained.

The process of human growth is much more complex than the simple curve of Figure 4 indicates. Each general system and specific organ must grow, and its growth must be coordinated in the general growth pattern. Figure 5 illustrates growth curves for four different parts and tissues of the body. The general growth curve that includes the body (except the head) and the respiratory and digestive organs and musculature follows the general curve of Figure 4. The growth curve of the reproductive organs is rather quiescent until the age of puberty, when it shoots rapidly upward. The brain and head growth curve rises rapidly in the first four years of

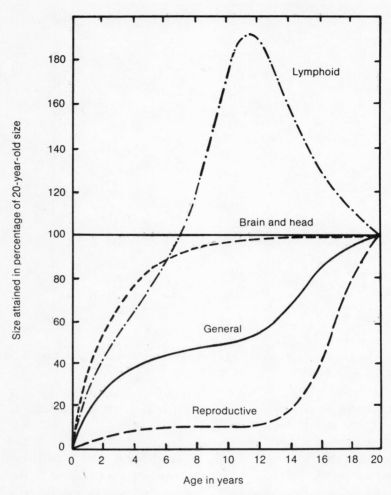

Figure 5. Growth curves of different parts and tissues of body (Source: P. H. Mussen, J. J. Conger, and J. Kagan, *Child Development and Personality,* second edition, New York, Harper and Row, 1963, p. 511.)

life, then tapers off gradually to reach its natural asymptote shortly after puberty. The curve of lymphoid growth grows to nearly double the normal size at around age eleven, and then diminishes to normal at maturity. For each curve there is a normal limit which is established by the hormonal regulators of organic development. Slow growth of certain organs is accelerated at given ages, rapid growth is decelerated, all is brought into a balanced unity.

Almost all natural processes are characterized by growth curves that approach definite limits, as in the growth of the human body. In the wild state, foliage will grow in height, weight, and density according to limits imposed by space, temperature, light, and nutrient materials. In an environment of prey and predators, a balance will be established, for if a predator devours all the prey, the predator will starve. Expansion in the population of the prey will permit a like expansion in the predator population.

By intervening massively in the cycles of nature, man has artificially destroyed the asymptotic limits that maintain the delicate balance that supports life. Through arrogance and inordinate egotism, modern man has decreed that asymptotes denote stagnation and regression. The curves that limit must become limitless. Growth becomes the final end. Population, gross national product, all indexes must continuously rise, for it is as if failure were writ large on stationary or descending (negative) growth: failure of intelligence, failure of will, failure of science and technology, failure of an economic system.

Limitless growth, however, is impossible. Heinz von Foerster and his associates at the University of Illinois have estimated population growth, for example, by considering all extant estimates of human population from the beginning of man and making certain assumptions regarding fertility, mortality, and environmental influences. With these minimal constraints, they have calculated that the population would reach infinity on "Doomsday: Friday, 13 November, A.D. 2026."[1] "Our great-great grandchildren," von Foerster assured his readers, "will not starve to death. They will be squeezed to death."

Von Foerster's Doomsday will mark the passage of man from the earth in but the short span of 6,030 years, if we consider the finding of an earlier scholar. Dr. John Lightfoot, vice-chancellor of the University of Cambridge, declared in the seventeenth century that

---

[1] The title of a paper in *Science*, CXXXII (November 4, 1960), 1291–1295.

as a result of his most profound and exhaustive study of the Scriptures, "heaven and earth, centre and circumference, were created all together, in the same instant," and that "this work took place and man was created by the Trinity on October 23, 4004 B.C., and at nine o'clock in the morning."[2]

Lightfoot's date has long since been shown to have been in error. Perhaps von Foerster's date, too, will prove to be in error, but his general conclusion is essentially valid, for the exponential curve of population growth will, if unchecked, reach a saturation level at some point in time.

Scientists since World War II have glowed at the rising funds allocated for research and development in the United States. Between 1959 and 1965 allotted funds rose from $13 billion to $21 billion, an average growth rate of 9.5 per cent. With a growth rate exceeding that of the Gross National Product (GNP), it can be shown that in a stated period of years research and development funds would consume the total GNP before it too would reach infinity.

Similarly, it can be shown, following the logic of the mathematical curves, that (1) the earth will eventually be totally paved with asphalt and concrete; (2) a number of vehicles exactly equal to the area of the highways will rest upon the paved land mass; (3) human bodies will simultaneously take up every square foot of land mass; (4) dwellings, factories, and other buildings will simultaneously cover every square foot of land mass; (5) the produce of mills and factories will simultaneously cover the total land mass; (6) solid wastes, mine tailings, slag heaps, garbage, and industrial wastes will cover the entire land mass; and (7) all natural resources will be reduced to zero in a negative exponential curve at a point in time antedating all the other developments.

The extrapolation of curves is always a risky undertaking, as Mark Twain pointedly noted in *Life on the Mississippi*. Remarking that "There is something fascinating about science," Twain perceived the "wholesale returns of conjecture out of such a trifling investment of fact," as he conjectured on the future of his beloved

---

[2] Quoted in Andrew D. White, *A History of the Warfare of Science with Theology in Christendom* (New York: Braziller, 1955), p. 9.

river. "In the space of one hundred and seventy-six years the lower Mississippi has shortened itself two hundred and forty-two miles. That is an average of a trifle over one mile and a third per year. Therefore, any calm person who is not blind or idiotic, can see that in the old Oolitic Silurian period, just a million years ago next November, the Lower Mississippi was upwards of one million three hundred thousand miles long, and stuck out over the Gulf of Mexico like a fishing-rod. And by the same token any person can see that seven hundred and forty-two years from now the Lower Mississippi will be only a mile and three quarters long, and Cairo and New Orleans will have joined their streets together. . . ."[3]

All extrapolations cannot be reduced so easily to a *reductio ad absurdum* as the moving of Cairo and New Orleans together. Given existing trends, and all other things being equal, growth will follow the inexorable paths of the mathematical formulations of the scientists and statisticians. This is the direction in which mankind is racing. To hasten the growth in every field, and all at the same time, nations are investing heavily in science and technology and relying upon them more and more to solve the problems that are engendered in increasing numbers as the exponential curves shoot skyward. The residue chains are themselves proliferating on their own exponential curves and are faced with the same denouement as the other progressive achievements of growth and expansion.

Let us, then, examine several techno-social problems from the viewpoint of residues and quasi-solutions. These studies are not intended to be essays on ecology, although the holistic approach required by technological realism essentially involves an ecological approach. Nor are the studies intended to be technological investigations of specific problem areas. The examples discussed illustrate the viewpoint advanced in the previous chapter against the background of finitude, unsolvability, impotence principles, and constraints on what man can know and do. The case studies all illustrate that for every exponential curve an asymptote will be established. Either man will voluntarily set the limit or nature will establish it for him. The longer the decision to establish the nec-

[3] Samuel L. Clemens, *Life on the Mississippi* (New York and London: Oxford University Press, 1962), p. 129.

essary limit is delayed, the more difficult and impossible will be any intelligent response. The alewives kill-off does not wait on man's readiness nor, it will be shown, do any other of the pressing problems that confront technological societies.

The biblical injunction "Be fruitful and multiply" was written by one who must have been familiar with the fecundity of nature. For nature is fecund, not in a profligate manner but conservatively, designed to aid and ensure survival. A tapeworm can lay up to 120,000 eggs a day; a frog can lay 10,000 eggs in a single spawning, and a cod can lay four million eggs in a year. Excessive numbers in species other than man have been kept in check by interspecies competition, by disease, or by limited food supplies.

The human ovary is estimated to contain about 200,000 ova at puberty, while an ejaculation of male semen can contain more than 200 million spermatozoa. Fortunately, human reproduction is usually effected by the union of one ova and one sperm; a gestation period of nine months places a limit on the number of children a woman can bear during her fertile years. Nevertheless, because of the lengthening life span of humans, and because of man's massive intervention in the death cycle, accompanied by his technological assault upon nature, man threatens to inundate the earth with his progeny—a feat the tapeworm, the frog, and the cod were never able to accomplish.

The dialectical process of technology is well exemplified by the disintegrative and destructive forces set into motion by the creative act of saving life through medical science. Brought into being by technological developments that sought to strip the earth of its resources, technological civilization has brought about a demographic revolution. By changing survival patterns, science has transformed birth rates that once merely balanced deaths to a now potentially destructive force. Life-giving medicines and medical techniques are helping to expand the world's population at a time when the majority of mankind is still the victim of biological slavery, hunger. The science that heals and saves life, paradoxically, is relegating millions of lives to continuing or increasing suffering and deprivation.

The exponential curve of population growth shown in Figure

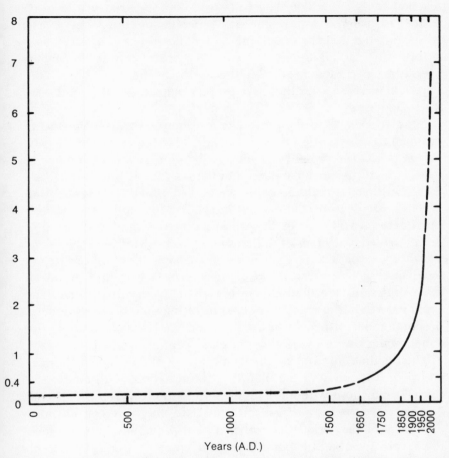

Figure 6. Estimated population of the world, A.D. 1 to A.D. 2000 (Source: Harold F. Dorn, "World Population Growth: An International Dilemma," *Science*, January 26 1962, p. 284.)

6 is a result of the scientific-technological revolution of the past three hundred years. The dashed line up to the year 1650 is an estimate of human world population to that date in periods during which there were no censuses. The approximately 470 million inhabitants of the earth in 1650 were the product of more than 100,000 years of human evolution and increase. By the year 1800

this figure had nearly doubled, with a population of one billion reached by 1830. A doubling to two billion occurred in the next hundred years; the third billion was added in only thirty-one years. It is estimated that the fourth billion will be added in fifteen years and the fifth billion in only ten years. Approximately seven billion people are expected to inhabit the earth by the year 2000. The potential increase after that date propels man on toward von Foerster's Doomsday.

Population may well be the prime example of a techno-social problem that is fast becoming unsolvable, for the reality of the situation is that the present human population may already be much greater than the earth can adequately maintain as free human beings, and there is no foreseeable method in which the exponential growth curve can be diminished in time to forestall disaster. Approximately two-thirds of the world's present population are undernourished, with tens of millions eking out existence at the near-starvation level. The largest increases in population are occurring in precisely those areas where hunger is most endemic. Some of the populations are doubling in less than twenty-five years. The rapid growth in these populations is resulting in a disproportionate preponderance of people in the younger age brackets, and an accompanying rise in the number of child-bearing women.

The population problem affects all the peoples of the earth, rich and poor, the few and the plentiful, those nations with ample resources and those without. The wealthy nations or those that can commandeer the resources of other nations may stave off Doomsday for a longer period than other nations, but they, too, are facing severe social and economic problems arising from population growth. The situation is such that a policy of zero population growth is mandatory for the entire world, and such a policy must be effected in a very short time.

Zero population growth means a birth rate that will do no more than replace the present population; it is estimated at a rate of fourteen births per thousand, and is the equivalent of about two children per family.

Obviously the exponential growth in population gives rise to serious problems in food, resources, land, and the quality of life.

Failure to solve any of these problems will lead to catastrophe, for species whose populations exceed or press upon the carrying capacity of resources in a finite space will undergo reduction. Either the birth rate must fall or the death rate must rise. The specter of Malthus cannot be wished away by churchmen, statesmen, or scientists.

Currently, one-third of the world's population is satiated by consuming three-quarters of the earth's harvested crops. Ironically, the greatest increases in food production in the past decades have occurred in the satiated countries, not the hungry ones. Agricultural science and technology flourish among the well-fed and languish among the hungry. As a consequence, the satiated are threatened with obesity, the hungry with starvation. In the United States, farmers are paid not to grow some crops so as to artificially maintain a price structure, which has precedence over feeding hungry people, including millions in the United States itself. Some surplus foods are prohibited from being shipped to hungry countries because of America's ideological differences with their ruling elites.

How many people can the earth feed? The Committee on Resources and Man of the National Academy of Sciences has estimated the foreseeable increases in food supplies over the long term at about nine times the present available capacity. This optimistic increase would be accomplished by quadrupling present land production of food and increasing the ocean yield two and one-half times. To attain this goal would require maximum increases in productivity of existing lands; cultivation of all potentially arable lands; development of new crops; use of more vegetable and less animal protein; continued use of pesticides and herbicides; increased use of fertilizers; chemical or microbiological synthesis of foods; and other innovations.

Even with this ninefold increase in food production, the ultimate limit of world population would be about thirty billion people "at a level of chronic near-starvation for the great majority and with massive immigration to the now less densely populated lands."[4]

---

[4] National Academy of Sciences–National Research Council, "Resources and Man," Introduction and Recommendations in U.S. Congress, House, Committee on Government Operations, Conservation and Natural Resources Subcommittee, *Hear-*

At present rates of population growth, this population figure would be reached by the year 2075. Harrison Brown has optimistically estimated a maximum of fifty billion people living on largely synthetic foods, nuclear energy derived from the deuterium in the sea, and resources derived from sand and rock. This population would be reached, at current rates of increase, before 2100.

Given effective controls, it is hoped that a leveling of population near a maximum of ten billion people could be achieved by the year 2050, although it is suggested that a leveling off at present population (3.5 billion) "would offer the best hope for comfortable living for our descendants, long duration for the species, and the preservation of environmental quality."[5]

What is not yet fully realized by the peoples of the industrialized nations of the West is that their high standards of living have been —and, in part, are continuing to be—achieved by a massive exploitation of the world's total resources and a period of capital accumulation that was hardly noted for civilized practices. It is significant to recall that the first fruits of man's increasing control of nature in England, the birthplace of the Industrial Revolution, were, in the words of C. Wingfield-Stratford, "a misery and degradation such as it would be hard to parallel at any time in our [English] history." The Industrial Revolution in England, which served as the model of capitalism in the work of Karl Marx was, to Marx, a violent encroachment upon the lives of the English. All bounds of morals and nature were swept away by the insatiable demands of the burgeoning factory system and machine production. Violently torn from their yeoman holdings by the Enclosure Acts, the dispossessed Englishman was driven from his land into the slavery of the factory system, where a comparatively free and prosperous peasantry were reduced to pauperized laborers. Generations of women and children and displaced peasantry were blighted by the unplanned and unforeseen consequences of this early revolution brought about by science and technology.

---

ings, *Effects of Population Growth on Natural Resources and the Environment*, 91st Congress, 1st Session (Washington, D.C.: Government Printing Office, 1969), Appendix 2, p. 122.

[5] *Ibid.*, p. 123.

Not content with plundering its own land and people, the rising industrial class of England also set out to exact tribute from its conquered possessions across the seas in that mighty empire upon which the sun was supposedly never to set. Capital accumulation in the other industrial nations of Europe was similarly achieved by internal exploitation and external thievery. The record of the Spanish conquistadores, the colonization of Africa and Asia, and the conquest of the ocean islands play important roles in the "progress" of mankind through science and technology.

The effects of the Industrial Revolution in other nations of the West, though different in magnitude, were comparable in their ruthless exploitation of both natural and human resources. In the United States, child labor, sweatshops, and hazardous working conditions characterized the industrialized North, while a system of human slavery flourished in the South. As many of the most illustrious of English families had their beginnings in the expropriation of church estates under Henry VIII and the exploitation of the English people, so in America many of the illustrious names of the builders of American industry found their beginning in the era of the Robber Barons.

In their "great leap forward," the Third World nations of our generation have before them the historical models of the affluent societies which rose to wealth and world power by violence against their own peoples, followed by exploitation of the peoples and spoliation of the resources of the world. The tensions arising in the awakening nations of the world are not the result primarily of Cold War and conflicting ideologies, though the compounding of these tensions with other conflicts serves to exacerbate problems. The revolt of the domesticated natives of once paternalistically ruled colonies is not a subversive ideology, but a quest for that freedom which has marked the rise of every people.

But with the independence of every new nation one more contender exists to press the world's diminishing stock of resources. The peoples of these new nations seek to free themselves from exploitation and to enter the arena of competition for world resources and markets. The economies of Germany and Japan, arisen from the ashes of World War II, are again entering the international marketplace with their wares and products. And looming ever larger is the vast industrial potential of a giant unchained, China, with its more than 800 million people and a contending social system.

The people of the United States, with 5.8 per cent of the world's population living on 6 per cent of the earth's land surface, are consuming nearly 50 per cent of the world's resources. Harrison Brown and his associates have estimated that to sustain the American standard of living the following quantities of materials are required for every individual in the United States annually:

> 1,300 pounds of steel
> 23 pounds of copper
> 16 pounds of lead
> 3.5 tons of stone, sand, and gravel
> 500 pounds of cement
> 400 pounds of clay
> 200 pounds of salt
> 100 pounds of phosphate rock

A total of twenty tons of raw materials must be dug from the earth and processed to produce these quantities.[6]

Philip M. Hauser has estimated that in 1950, with a world population of 2.5 billion people, the total world production of goods and services would have supported only 500 million people at the United States level and 1.5 billion at the European level. If all the world's present population were brought to the United States level, Brown and his associates have estimated the following quantities of resources would be required:

> 18 billion tons of iron
> 300 million tons of copper
> 300 million tons of lead
> 200 million tons of zinc
> 30 million tons of tin
> etc.

"*These totals,*" Brown notes, "*are well over one hundred times the present world annual rate of production.*" Furthermore, the required quantities of copper, lead, zinc, and tin "*are considerably*

[6] Harrison Brown, James Bonner, and John Weir, *The Next Hundred Years* (New York: Viking, 1956), pp. 18–19.

greater than could be removed from all measured, indicated, and inferred world reserves of ores of these metals" (italics added).[7] This alarming situation, which dooms the "revolution of rising expectations," already exists when the world population is only 3.5 billion people.

The population of the United States is also growing, at a rate approaching 1 per cent. This growth rate will add 100 million people to the United States' population by the year 2000. If the growth in population is accompanied by a further projected increase in the standard of living, "Americans will be confronted with the hard choice of foregoing some of their affluence or continuing to import, *at increasing rates, the raw materials on which the underdeveloped countries might base their own industrial growth*" (italics added).[8] The supplying of materials for our burgeoning population, Congressman Henry S. Reuss observes, may be attempted "only at the risk of raiding the rest of the world."[9]

This raiding, however, is not merely a possibility in the near future; it is a reality in the present. In 1965 almost half of the world's population shared 11.5 per cent of the world's total gross domestic product, while the one-fifth of the population in the advanced capitalist nations shared about 60 per cent. This relative economic backwardness of the Third World nations compared with the advanced capitalist countries has grown worse in the recent past and continues to deteriorate. One of the reasons for this state of affairs is the raiding that is constantly accelerating as agricultural and mineral products are taken in increasing quantities from the Third World nations to maintain the affluence of the developed countries. In the Third World countries, agricultural products destined for export, for example, increased at more than twice the rate of products destined for internal consumption. While famine is foreseen in some of these countries within the next decade, agricultural development is concentrated on exportable products. "The Third World tightens its belt," Pierre Jalée has written, "and reaps for others." In 1965 the Third World countries supplied about 28

[7] *Ibid.*, p. 33.
[8] Statement of Preston E. Cloud, Jr., in *Effects of Population Growth on National Resources and the Environment*, p. 6.
[9] *Ibid.*, p. 1.

per cent of the world's production from the extractive industries, the major proportion of which was exported to the developed capitalist countries.[10]

During the period 1900 to 1961, average mineral imports increased tenfold in the United States—from $323 million to $3.6 billion—while imports as a percentage of consumption went from 1.5 per cent in favor of exports to 14 per cent in favor of imports. Iron ore, copper, lead, zinc, bauxite, petroleum, columbium, chromium, cobalt, are being imported in increasing quantities, with 100 per cent of the latter three minerals imported. As early as 1958, of sixty-two strategic materials stockpiled by the United States, between 80 and 100 per cent of thirty-eight of these materials was imported.[11]

The reason for the continuing raiding of resources by the wealthy nations is clearly seen in the bar graphs of Figure 7. The lifetime of estimated recoverable reserves of mineral resources calculated at *current* rates of recovery at *current* population levels of 200 million people for the United States and 3.3 billion for the world (1966 figures) is shown in the graph. Only four or five mineral resources in the United States have assured lifetimes beyond 1984; fourteen commodities have world reserves beyond 1984. By the year 2000, only three commodities will be present in the United States and ten in the world. By the year 2038, only eight commodities will still have reserves in the world. As Preston E. Cloud notes, new reserves may be found, the estimates may be high or low, but the total picture indicates *a rapid and soon-to-be-attained total depletion of the earth's primary resources.*[12]

The "have-not" position of the United States will be shared by the other developed countries. France, for example, which imported 40 per cent of its minerals in 1961, is expected to import 80 per cent by 1985. A report prepared by the French government in

---

[10] Pierre Jalée, *The Third World in World Economy* (New York: Monthly Review Press, 1969), pp. 7, 21, Chs. 2–3.

[11] Harry Magdoff, "The Age of Imperialism," *Monthly Review*, XX (June 1968), 11–54.

[12] Preston E. Cloud, Jr., "Realities in Mineral Distribution," *Texas Quarterly*, Summer 1968. Reprinted in *Effects of Population Growth on National Resources and the Environment*, Appendix 11, pp. 219–242.

Technological Realism 1: People, Land, and Food

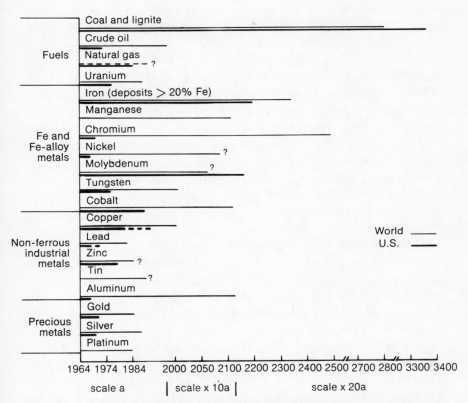

Lifetimes of estimated recoverable reserves of mineral resources at current minable grades and rates of consumption (no allowance made for increasing populations and rates of consumption, or for submerged or otherwise concealed deposits, use of now submarginal grades, or imports).

Figure 7. Lifetimes of estimated recoverable mineral resources (Source: Preston E. Cloud, Jr., "Realities in Mineral Distribution," *Texas Quarterly*, Summer 1968. Reprinted in *Effects of Population Growth on Natural Resources and the Environment*, op. cit., p. 239. Data from Peter Flawn, *Mineral Resources*, Rand McNally, 1966.)

1964 points out that "the economy of Europe as a whole will find itself in a similar position."[13]

Thus the "have" nations of the world are on the way to becoming

[13] Jalée, op. cit., pp. 88–89.

"have-not" nations. They can attempt to maintain their favored position with regard to the agricultural and mineral wealth of the world by continuing to exploit the Third World countries—but only up to a point. The affluent nations will have to cease their plunder because of the depletion of the Third World's resources, if indeed this eventuality is not antedated by revolution. The very term "have-not" is an imperialistic euphemism that seeks to mask this plundering, and from a resource standpoint is a falsehood. But with growing populations and growing internal rates of resource utilization, the depletion rates can only move forward in time.

Together with food and resources, people need land. Competition for land ensues between requirements for food production, dwelling, manufacture, health, education, and recreation. In 1900, with a population of over 75 million people, the density per square mile of land area in the United States was 25.6 people. By 1960 the density had more than doubled to 60.1, and it continues to grow with population. Although these densities are far below equivalent densities of some European and Asiatic countries, the United States is becoming increasingly more urbanized and, consequently, 53 per cent of the people live in only 13 per cent of the land area. In the seventeen states of largest urban concentration, density is 201 per square mile—and increasing.

In 1969, assuming the population could have been spread evenly over the 1.9 billion acres of land in the United States, approximately ten acres would have been available for every individual. Of these ten acres, one and one-half acres are required for food production, two and one-half acres to supply wood, and another three and one-half for cattle and sheep grazing. Other requirements of an urban, technological society consume nearly an acre for cities, two-tenths of an acre for highways, railroad rights-of-way, and airports; one-tenth of an acre for military camps; two-tenths of an acre for parks and monuments; and one acre for recreation. The remaining portion of the per capita land area is desert, mountain, or wasteland. By the year 2000, the per capita area will be in the vicinity of six acres—and, as always, continuing to decrease as population grows. Because land is continuing to be absorbed by concrete and asphalt, dwellings and factories, waste heaps and quarry pits, some scientists estimate that there will be a land deficit by the year 2000—that is, a deficiency of land to provide the standard of living currently available in a semi-humane society. Each exploiter of land for nonfood purposes is assuming that someone else will use other land for food production. Thus, California is destroying its rich agricultural lands at a prodigious rate, but so also are Florida, Texas, the Midwest, and indeed the entire nation.

The United States Soil Conservation Service estimates that about 1.5 million acres of prime farm land are lost to city growth and highways every year. Since 1958, more than ten million acres have been taken from croplands, 3.5 million acres from pastures, and nearly eleven million acres from all other uses. During this same period there was a net loss of fifteen million acres of forest lands.

China was once a heavily forested land, but its civilizations were built at the expense of its forests. Today it is estimated that only 9 per cent of China's land remains forested. The United States was originally covered by 900 million acres of forest, but its rate of deforestation exceeds any that China has experienced. About one-third of these forests have been permanently destroyed, and only about forty million acres remain in virgin timber.[14]

Land, clearly, is a finite, nonrenewable resource. As man seeks to intensify exploitation of the land for mining and food production, he generates a complex chain of residue problems that are proliferating with unforeseen consequences. Cultivation in many areas has completely replaced natural conditions, and virgin regions are fast disappearing in the world. The reduction of forest areas has brought in its wake a series of ecological disasters exemplified by floods, drought, erosion, depletion of soils, lowering of the water table, dust storms, and climatic changes.

Archaeologists are uncovering the artifacts of long-buried ancient civilizations that once flourished in what are now semi-arid or desert areas. These ancient remains are mute testimony to the

[14] Raymond F. Dasmann, *The Last Horizon* (New York: Macmillan, 1963), pp. 129–130.

fate of those who treat with other than respect the eight to twelve inches of the earth's topsoil that sustain plant and human life.

By the Golden Age of Greece the oak and pine cover of the Mediterranean area had already been stripped. Plato wrote about the harmful effects of deforestation and grazing that caused water to "run from the naked earth into the sea," leaving lands that "resembled the bones of a diseased body." Raymond Bouillenne describes the apparently irreversible reduction of the surface of cultivatable lands that causes an irretrievable loss of twenty million acres a year:

> Egypt, which now occupies only the Nile Valley, like China, which clings to the valleys of the Blue River and the Yellow River, formerly covered the now desert spaces with its fields.
>
> Cyrenaica, at the time of Imperial Rome, possessed the famous gardens of Berenice. The Libyan Desert hides the ruins of great towns such as Thysdrus (El Djem), whose stadium was built to hold 60,000. The French explorer Auguste Chevalier discovered in the Sahara, beneath thick layers of sand, traces of dense forests which, less than 2,000 years ago, made a rich colony of what is now synonymous with barrenness and aridity. And the list does not end here: Arabia, Babylon, and Tibet can be added. In Morocco, since the Roman period, 12.5 million acres of forest have disappeared as a result of fire and overgrazing by sheep and goats.
>
> In the great forest massif of Central Africa, where the state of regression is rapid, the present state of affairs is becoming disastrous. . . . This forest area is receding with alarming speed. On all sides the deserts are advancing . . . in French Equatorial Africa alone the loss of fertilizing matter during the present generation has amounted to half a billion tons. . . .[15]

In Europe, in the Americas, and now in Asia and Africa, onrushing hordes of people are transforming a natural habitat into manmade reservations of settlements, farms, and factories. By 1950 it was estimated that 40 per cent of the rich rain-forest zone of Africa had been cleared. In some countries, as for example Nigeria, nearly three-fourths of the forests had been cleared. The last great reservoir of rain forest left in the world is in the Amazon basin, but by 1955, 40 per cent of this, too, had been cleared.[16]

[15] Bouillenne, *op. cit.*, pp. 706–712.
[16] Dasmann, *op. cit.*, pp. 124–125.

The very act of cultivation is partially to blame for some of the ecological disasters that follow civilization. The rate of erosion is increased by man's occupancy of the land, while intense cultivation causes an increase in erosion that is one or more orders of magnitude greater than when the land is under its natural vegetative cover. The amount of sediment carried away by large river systems is estimated to have increased two and one-half times since large-scale human intervention.[17]

Surface mining in the United States is destroying 150,000 acres of land annually and has already defaced 3.2 million acres over the years. Sand and gravel, gold dredging, rock, and the stone and clay industries are responsible for almost half of this devastation and defacement. At present about twenty thousand surface and stripmines are in operation; and by 1980 an estimated additional five million acres will have been defaced by these operations or from their waste.[18]

In Soviet Russia, Boris N. Bogdanov, head of the Ministry of Agriculture's department for the protection of nature preserves and hunting grounds, reports that much good farm land is lost "through its indiscriminate allotment to nonagricultural purposes." Additional Russian land is wasted through allocation to industry. "Hundreds of thousands of acres of land in the Soviet Union have been lost to the mining industry. Open-cut mining alone causes some 75,000 acres of land to be wasted annually," Bogdanov notes. More than 45 million acres of forest have been cut and burned in the last twenty-five years without being restored, as forest depletion continues to exceed reforestation.[19] These admissions by a Soviet official lend support to the thesis that depletion, pollution, and waste are an integral part of technological and industrial society quite independent of the "relations of production." The significance of this point, with its indication of the need to restructure

[17] Sheldon Judson, "Erosion of the Land—Or What's Happening to Our Continents," *American Scientist,* LVI (Winter 1968), 356–374.
[18] *Solid Waste Management: A Comprehensive Assessment of Solid Waste Problems, Practices and Needs,* Prepared by Ad Hoc Group for Office of Science and Technology, Executive Office of the President (Washington, D.C.: Government Printing Office, 1969), p. 37.
[19] Article in *Ekonomika Selsoko Khozyaistva* [Agricultural economy], reported in *New York Times,* April 9, 1970.

the "forces of production" in both capitalistic and communistic societies as a requirement for a humanistic society, will be taken up later.

The coasts of the oceans and the seashores are also under constant assault as industrialists seek water outlets near which to build their plants. Meanwhile, hotels and private dwellings also inundate these previously untouched areas. The United States Conservation and Natural Resources Subcommittee of the House of Representatives has summarized this development in a report to the Congress: "The natural environments of our Nation's bays, estuaries, and other water bodies are being destroyed or threatened with destruction by water pollution, alteration of river courses, land-filling of the shallow and marshland areas, sedimentation, dredging, construction of piers and bulkheads, and other manmade changes."[20]

Indeed, such is the growing need for land that the practice of burying the dead in land set aside for that purpose is competing with the needs of the living for the same land. In Tokyo the metropolitan government has announced a shortage of burial sites. In large urban areas, cemetery space is becoming increasingly scarce as cemeteries vie with housing projects and industrial sites for expansion to new locations.

To counteract the ravages of yesterday, man seeks through technology to reassert his mastery over the blighted lands. Thus the bulldozer, the turbine, fertilizer, and biocide are called into play. But each turns out to be a quasi-solution engendering a chain of proliferating residue problems.

Consider the Aswan Dam on the upper Nile. It was built to prevent floods, to provide a regulated supply of water for irrigation, and to generate electrical energy. Though the dam was conceived as an answer to Egypt's pressing population, during the time it took to build it, the population had increased to a point where any gains in agricultural productivity would be fully absorbed by

---

[20] U.S. Congress, House, Committee on Government Operations, Conservation and Natural Resources Subcommittee, 21st Report, *Our Waters and Wetlands: How the Corps of Engineers Can Help Prevent Their Destruction and Pollution,* 91st Congress, 2d Session (Washington, D.C.: Government Printing Office, 1970), p. 1.

the incremental population. With its very inauguration, unforeseen consequences appeared. The rich nutrients that were annually carried to the sea by the flooding Nile were cut off with a catastrophic destruction of phytoplankton that had long been a basic food for marine life. The sardine fisheries that formerly produced eighteen thousand tons per year are now producing about five hundred tons. The formation of a stable lake has permitted aquatic snails to maintain large populations that had formerly been kept under control by decimation during the dry seasons. Population has increased near the irrigation channels, and people are more prone to the disease schistosomiasis which is carried by a river fluke that uses the snails as a host: the snails harbor a more virulent species of fluke than the species that formerly abounded in running water, so the Dam has increased both the virulence and the incidence of the disease.

The Kuriba Dam on the Zambesi River between Zambia and Rhodesia has been beset by a series of disasters resulting from unforeseen changes in the water level and water table. The growth of weeds in the new lake held the catch of fish to 2,000 tons instead of an anticipated 20,000 tons. The Dam displaced more than 29,000 farmers, and now serves instead as a home for the dread tsetse fly.

In India, irrigation systems on the Indus and Ganges rivers are resulting in a net annual loss of arable land arising from the salinization process. Continued irrigation of flat plain lands brings to the surface of the soil salt from lower layers of the earth which contaminates the topsoil as it is carried upward by a rising water table; this is a twice-told tale, first told long ago by overirrigation in ancient Ur and Nineveh. In Arizona and the plains of Texas, disastrous drops in water tables resulted from excessive pumping to irrigate cotton acreage—a commodity that was already produced in surplus. In the Soviet Union the construction of the Rybinsk reservoir flooded 432,000 acres of good farm land that provided fodder for a large livestock farming region. In the United States, the Corps of Engineers has built dams and reservoirs that cover more than four million acres and have a total shoreline longer than the mainland. Plans for the flooding of millions of additional acres are on the drawing boards. In the process of construction, with prime goals being flood control and irrigation on a cost-effectiveness basis (which is, by definition, shortsighted because only engineering efficiency and obvious economic factors are considered), valuable natural areas are destroyed, rivers and streams disappear, and the ecology of entire regions is transformed.

Little wonder, then, that talk by scientists to divert such large rivers as the Yukon and Fraser in Canada to irrigate the deserts of the southwest United States, or the Ob and Yenisei rivers in Russia from the Arctic Ocean to irrigate the plains of Turkestan, raises serious alarm. The augmentation problems and secondary effects that will arise can only generate a new set of problems whose consequences cannot be foreseen. Even the immediate consequences involve many assumptions and guesses.

The effort to reclaim land is constantly acompanied by actions that in fact turn out to destroy the land. Even acts of reclamation designed to increase food production for a pressing population often have an opposite effect. Barry Commoner has suggested that the increasing use of nitrate fertilizers has resulted in an unforeseen vicious cycle. The continuing use of artificial nitrates tends to reduce the natural production of nitrogen compounds in the soil. Consequently, the more artificial fertilizer that is used, the more is required.

The "mining" of food on large land "factories" under greater and greater artificial conditions has been made possible by agricultural processes that are scientifically conceived and technologically developed. The agricultural revolution in the industrialized nations is as much a triumph of modern chemistry as of agricultural knowledge. Yet chemical agriculture in the form of fertilizers, herbicides, and pesticides has its own dialectical logic—a logic revealed in increasingly distorted and unbalanced ecosystems.

Another serious secondary effect of the use of fertilizers is the runoff of nitrates and phosphates from fertilized fields into streams and rivers, where the chemicals are carried into lakes and contribute to the problem of eutrophication, the aging process by which lakes mature, age, and die. Normally this process requires thousands of years. Since 1940 in most lakes in the world that have remained clear and pure since the glacial period, the eutrophication process has accelerated to the point where some of them are dying. The prime contributor to this accelerating aging process is the phosphorus contained in agricultural runoff, human waste, and detergents. Because phosphorus is a fertilizer, it stimulates the growth of aquatic plants. With overdoses of fertilizer, the plants grow in excess quantities and die off rapidly. In the process of decay they use up the dissolved oxygen of the water, thus depriving fish and other aquatic species of this oxygen. The decay of the organic plants releases inorganic phosphorus to the waters, which initiates a new cycle of growth. The water is gradually filled with the accumulated masses of rotting vegetation and decay products. The lake becomes a cesspool, then a bog, and finally dry land.

The surface waters of the United States receive between five and eight billion pounds of nitrogen and between 0.9 and 1.7 billion pounds of phosphates annually, of which more than 75 per cent is from man-made sources. Half of the nitrogen runoff is from cultivated lands and another one-third from sewage.[21]

The once clear waters of Lake Erie have been transformed into a vast monument to man's arrogant disregard of natural equilibria. The lake has become a cesspool; trash and oil slicks float on a bed of putrefying algae as forty billion tons of raw sewage are dumped annually into the lake. Daily 9.6 billion gallons of industrial waste are poured into the lake together with a daily dosage of 150,000 pounds of phosphates. A process that would normally require fifteen thousand years has, in the case of Lake Erie, been accomplished in less than fifty years as eutrophication engulfs the lake.

The filth of the lake, ironically, can be traced in part to man's penchant for cleanliness, because one of the major sources of phosphates in the cesspool comes from detergents. Between 35 and 50 per cent of the phosphorus input to the lake is attributed to detergents.[22] Nationwide, over 5.3 billion pounds of detergents are consumed in the United States in a year—an amount that has in-

[21] F. Alan Ferguson, "A Nonmyopic Approach to the Problem of Excess Algae Growth," *Environmental Science and Technology*, II (March 1968), 188–193.

[22] U.S. Congress, House, Committee on Government Operations, Conservation and Natural Resources Subcommittee, 23d Report, *Phosphates in Detergents and the Eutrophication of America's Waters*, 91st Congress, 2d Session (Washington, D.C.: Government Printing Office, 1970), p. 12.

creased more than tenfold since 1947. About 1.7 billion pounds of the total detergent is sodium tripolyphosphate, the builder of the detergent, of which slightly more than one-fourth is elemental phosphorus. Together with other phosphorus compounds, a total of 370 million pounds of elemental phosphorus enter the surface waters of the country.[23]

Although the effects of accelerated eutrophication have been noted for many years, the percentage of phosphorus in detergents has risen from 7.7 per cent in 1958 to 9.4 per cent in 1967. The per capita consumption of detergents has risen in the United States from 2.8 pounds in 1947 to 26.6 pounds in 1968.[24] Nearly a billion dollars a year is expended by the detergent manufacturers to increase sales. The industry was forced to change one of the major components of detergents, the surfactant, in 1965 because it was not biodegradable and was converting the nation's waters into foaming bubble baths. The builder component, the phosphate, was not changed, however. Now the industry is faced with a new problem that threatens the lakes of the nation and indeed the world, for the fate of Lake Erie is but a harbinger of that which threatens rivers and inland lakes everywhere.

The destruction of Lake Michigan and the other Great Lakes is already under way, as is that of Lake Baikal, in the Soviet Union, the world's deepest freshwater lake. The rivers that feed the lakes are becoming sewer mains channelling sewage, chemicals, and other effluent into the lakes. The Rhine River annually carries more than fifteen million cubic yards of industrial waste and sewage into the North Sea, where it has already led to the extinction of trout, salmon, and other marine species. The Ural River in the central Soviet Union is similarly polluted. The Sumida River in Tokyo is an open sewer. The River Seine carries a high level of detergent wastes, as does the Tama River on Honshu Island in Japan. The Meuse and Schelde rivers in Belgium are considered heavily polluted. Lake d'Annecy in France, Schliersee and Tegernsee in Ger-

[23] *Ibid.*, p. 15.
[24] U.S. Congress, House, Committee on Government Operations, Conservation and Natural Resources Subcommittee, *Hearings, Phosphates in Detergents and the Eutrophication of America's Waters*, 91st Congress, 1st Session (Washington, D.C.: Government Printing Office, 1970), p. 261.

many, Zurichsee in Switzerland, and a host of other lakes in the United States are all undergoing rapid eutrophication.

Agricultural runoff, sewage, and detergents are all secondary effects of technological processes. The processes are quasi-solutions to problems of technological society. Attempts to deal with the secondary effects face all the factors outlined earlier that make solution of residue problems more difficult, with each new solution, in turn, becoming a new quasi-solution generating new problems.

Finding a new substitute for the phosphate in detergents is a difficult and critical problem. The effects of sodium nitrilotriacetate (NTA), for example, are unknown, and it must be remembered that more than five billion pounds of detergents are used annually. Because NTA can combine with metals, it may result in heavy metals getting into public waters. In addition, NTA may be corrosive. Experimental detergents in the United States that employ NTA still use phosphates in considerable quantities.

Sewage treatment, on another branch of techno-social problems, also contributes to eutrophication by breaking down sewage into nitrogen and phosphorus compounds. These compounds are discharged into streams and rivers and enter the lakes where the eutrophication process re-creates the same noxious organic matter in decaying algae that the sewage treatment was designed to control. The removal of the phosphorus from the sewage at the sewage disposal plants will require a new concept of sewage management in which the nutrients must be separated out and the resultant liquid, sludge, or solids disposed of. The first process will entail a large quantity of other chemicals, which will in turn create other pollution problems. Professor John R. Sheaffer has described a process of precipitating phosphates from waste water by the use of ferric chloride. Incineration of the resulting sludge would release the phosphorus into the atmosphere, whence it could return back to the land in rainfall. Disposal can also be by means of subsurface injection, or dumping on the land or in the oceans, but this method merely shifts the point of pollution. Disposal can be preceded by conversion and product recovery with disposal of the remainder, but new processes will be required which will necessitate the use of additional energy and the creation of other by-products. The establishment of a closed system for nearly 100 per cent removal of the nutrients would require 130 acres to treat one million gallons of waste water daily. For the entire nation, 3.7 million acres would be required at enormous cost and without consideration of the growth of population, expanding water usage, and increasing quantities of sewage and industrial waste.[25]

Population growth, as has been noted, is closely associated with the intensification of agricultural "mining" with its reliance on chemical fertilizers, herbicides, and pesticides. The latter two classes of chemicals are quasi-solutions to problems of undesirable plants and harmful insects. Like the other quasi-solutions we have investigated, these, too, are generating a proliferating chain of residue problems that are combining with other problems of pollution that now threaten man's ability to survive. And, as in so many other areas of technology, the effects of the quasi-solutions are counterproductive, have effects contrary to their intended effects while generating unforeseen and unpredicted side effects.

Because the immediate short-term effect of pesticides led to an increase in food production, their use multiplied in the years following World War II with statistics of production of a wide variety of chemical compounds rising on a familiar exponential curve. Pesticide production is increasing at an annual rate of 15 per cent; it is expected that insecticides will double by 1975 with herbicides increasing even faster.[26]

The dangers of pesticides have been well documented by Rachel Carson. As the "Silent Spring" spreads to more streams and fish kills, more birds and animal species, conservationists and ecologists continued to point out the growing dangers even as pesticide production continued to climb the exponential curves. It was not until it was clearly demonstrated that residues of major pesticides are stored in the fatty tissues of fish, bird, and animal species, are eventually ingested by humans, and are potentially harmful, that belated action has been taken.

[25] John R. Sheaffer, in *ibid.*, pp. 211–213, and p. 47 of report cited in note 22, above.
[26] U.S. Department of Health, Education and Welfare, *Report of the Secretary's Commission on Pesticides and Their Relationship to Environmental Health*, Pts. I and II (Washington, D.C.: Government Printing Office, 1969), p. 21.

The wonder pesticide, DDT, hailed in the late 1940's and 1950's as one of mankind's greatest blessings, became the "uninvited additive" in the late 1960's. By 1969 the beginning of phasing out its "nonessential" use was initiated. DDT was acclaimed as the answer to the malarial-carrying anopheles mosquito; a world-wide campaign to eradicate malaria by means of DDT was launched by the World Health Organization, and tons of DDT were spread in eighty-five countries and territories. The spraying of DDT upon the earth was the beginning of a new chemical revolution. Even as the United States begins to reduce its internal consumption, the export of DDT to other countries is increasing, especially under foreign aid programs.

As is now well known, DDT and the other chemicals unleashed an assault upon the earth that in many places completely upset the balance of nature. Most of the pesticides were broad-spectrum in effect, killing not only target insects but harmless and beneficial species as well. DDT used to kill apple pests, for example, also killed ladybird beetles; carbaryl was very toxic for bees. Deleterious effects were also noted on fish, birds, and animals that ate pesticide-contaminated species or absorbed the pesticides in streams as runoff carried the chemicals to the waters. Traces of DDT have been found in seals and penguins in Antarctica.

A sampling of fish in 1969, for example, indicated that nearly 100 per cent of the fish tested contained traces of DDT, 75 per cent contained dieldrin, 32 per cent heptochlor, and 22 per cent chlordane. Some fish contained more than nine times what was considered a "safe" limit of DDT.[27] The amount of insecticides in Lake Michigan was considered to pose a threat to the reproduction of coho salmon, lake trout, alewives, perch, smelt, and other species of fish. The salmon and lake trout, ironically, are being introduced into Lake Michigan to feed upon the alewives. Again, in the estuarine areas of the United States, where salt and fresh water meet and from which two-thirds of the seafood of the country is obtained, the catch of fourteen species decreased by 50 per cent during 1960–1965 as a result of pollution.[28]

[27] *Environmental Science and Technology*, III (July 1969), 613.
[28] *Environmental Science and Technology*, II (April 1968), 241.

Many of the pesticides, including DDT, are persistent and do not break down easily. Furthermore, they are essentially nonbiodegradable, and hence not readily metabolized in the human or other animal species. They accumulate in an organism, and the traces are aggregated as one proceeds higher up the food chain.

A dialectical side effect is the resurgence and multiplication of target species of insects in the wake of developing resistance to pesticides. About 224 species of insects in various parts of the world have developed resistance to one or more groups of insecticides, including eighty-nine species that have become resistant to DDT. Several species have become resistant to two or three insecticides. In a survey of twenty years' use of pesticides in an area of Louisiana, it was found that cotton pests are becoming resistant faster than substitute pesticides can be found. The rice water weevil became resistant to aldrin in five seasons. In some areas, certain species of insects could no longer be controlled with any available pesticide. While chlorinated hydrocarbons were effective against stinkbugs in rice fields, they were harmful to fish and birds. Organophosphorus or carbamate pesticides were also harmful to crawfish, fish, and fowl. The permanent types of pesticides containing mercury, arsenic, or lead have accumulated in the soil in orchards, making it toxic and shortening the life of trees the sprays were intended to protect.

The decimation of certain species, by upsetting the natural balance, also gives rise to invasions of other insects, which in normal numbers are not necessarily harmful. Birds are also destroyed by the influence of pesticides on their reproductive systems, the thinning of egg shells, and even possible egg-eating by parents. The destruction of birds as a secondary effect of pesticide use also helps to upset a natural balance and to develop new insect pests.

The development of pesticide-resistant species of insects is a reminder that some problems do not remain solved. Resistance is an evolutionary process by which species come into being through a process of natural selection. Similar processes have been found among bacteria that develop resistance or immunity to chemical drugs. Staphylococci in hospitals and Plasmodium falciparum, the protozoan that causes malignant malaria, are examples of this evolutionary process. In some instances, George Gaylord Simpson observes, resistant strains thus evolved can pass on this resistance to micro-organisms that have not been exposed to selective action of the drug. "The attempt to solve one problem of applied biology," he states, "has caused numerous new problems."[29]

As with the fertilizers, the more pesticides are used, the more the dosage must be increased for comparable results; eventually, the pesticides can lose all effect on target species. Their persistence in the soil and their runoff into streams, however, has continuing ecological effects. The precious topsoil is also host to a myriad of tiny invertebrates. An average acre of soil in the temperate zone can contain more than one thousand pounds of earthworms; a square yard can contain as many as a million arthropods and even more nematodes and protozoans. Pesticides and herbicides are mixed with the soil during cultivation and can have a lethal effect, on a selective basis, on many species of these fauna. The more efficient the cultivation, the greater is the effect on the fauna upon whose continued existence the fertility and structure of the soil in part rest. The earthworms that are generally not susceptible to many insecticides, ingest them and pass the traces on to birds that eat the worms.[30]

Agricultural "mining" has yet another dialectical effect, for the agricultural operations that are required for efficient crop production favor many plant diseases and create new diseases. The number of recognized plant diseases nearly tripled between 1926 and 1960. Although some of these diseases were probably overlooked in the past, most of them are attributable to man's activities. The principal mechanism by which man introduces disease are by plant introduction, vegetative propagation, monoculture, tillage, harvesting, storage, fertilization, irrigation, herbicides, plant breeding, site location, and release of disease-producing chemicals.[31]

Thus has the life-giving earth been despoiled and poisoned, with

---

[29] George Gaylord Simpson, "Biology and the Public Good," *American Scientist*, LV (June 1967), 161–175.

[30] Clive A. Edwards, "Soil Pollutants and Soil Animals," *Scientific American*, CCXX (April 1969), 88–89.

[31] C. E. Yarwood, "Man-Made Plant Diseases," *Science*, CLXVIII (April 10, 1970), 218–220.

man imbibing his own poisons. The gentle rains that once cleansed the air and purified the soil now carry poisons into the earth's crevices and streams. The streams carry their loads of filth and destructive chemicals into the lakes and the oceans. And the scientists and technicians who direct this mighty torrent of abuse and destruction, albeit in the name of creativity, progress, and human welfare, cry out for more and more science and more complex technology. Wherever they turn, however, the proliferating chains of residue problems generated by an exponential rise in population and an exponential increase in nearly all exploitative and production categories act and react in dialectical processes that generate new problems in attempts to solve old problems.

As growing populations require ever more land, the requirements of the population are destroying the land. The demand for more food calls for more fertilizer and more irrigation, but, dialectically, overfertilization destroys the soil and overirrigation renders it saline. Overcultivation increases soil erosion; the destruction of forests and grasslands affects the natural balance of an entire region. Urbanization destroys farm land; deforestation and concrete-asphalt highways change drainage patterns and alter water tables. The increasing use of herbicides and pesticides damages the ecology of the entire earth, adds to the pollution burden of the environment, destroys harmless and beneficial plants, insects, and birds and animal species, and, finally, enters the food chain to cause known harm to many species while posing unknown hazards to man.

Since land cannot be created and a growing population places increasingly competitive demands on the land, further increases in population can only add to the crises of food deficiency and depletion of scarce resources. Efforts to cope with the crises through technology only serve to further inflame the crises by creating proliferating chains of residue problems that make any ultimate solution less possible. The Third World already has a date with widespread famine, estimated sometime around 1975–1985.[32] The developed world is fast rushing to the same appointment.

[32] See, for example, William and Paul Paddock, *Famine 1975! America's Decision: Who Will Survive?* (Boston: Little, Brown, 1967); Paul Ehrlich, *The Population Bomb* (New York: Ballantine, 1968).

7 Technological Realism 2: Energy

"*AS FREE AS THE AIR*" is a folk proverb that has been laid to rest by technological society. The air is no longer free and has joined other seemingly inexhaustible resources in becoming a finite and much sought-after resource. Purveying announcements of air scrubbers, the advertising industry has jumped on the anti-pollution bandwagon by displaying in multicolored advertisements: "Caution . . . breathing may be hazardous to your health." Fly-ash precipitators are introduced with the caption, "Does your plant have bad breath?" Other technological devices are shown against a background of a person protesting, "My lungs aren't meant to be dust collectors."

This growing concern with the air contrasts strongly with that shown by Charles II of England, who is said to have laughed at scientists of the Royal Society who were reported to be weighing air. Thomas Shadwell, in a play written three hundred years ago, satirized the scientific study of air by having Sir Nicholas Gimcrack weigh and bottle air like champagne. The bottled air was stored in his cellar, and he opened a bottle in his room when he desired a change of climate. Shadwell's satiric solution might become a necessity as the pollution of the air joins the poisoned waters and blighted land as a monument to the science and technology that were to have freed mankind from subservience to his environment.

As the assault upon nature continues under ever more difficult conditions, the exponential curves of "progress" continue to reach for infinity, leaving in their wake more proliferating chains of residue problems arising from quasi-solutions to problems of energy production. These residue problems, like those discussed earlier, intertwine with other residue chains in strange and unexpected patterns whose consequences are unforeseen.

While the rendezvous with famine approaches, growing populations and expanding technological societies create ever larger demands for energy, the motive force of industrial civilization. The graphs of energy production in the United States and the world exhibit familiar exponential growth curves, even as the final depletion of the fossil fuels that power most energy-generating facilities draws near. Atomic energy, hailed as a successor source to the fossil fuels, is already being generated in growing quantities. The shift to atomic energy, we are told, is a natural development which has seen energy sources continually changing from wood to charcoal to coal to oil and gas and, finally, to atomic fission and, hopefully, fusion. Each technology of power generation, however, is a quasi-solution, generating a new and more complex set of residue problems.

The growth of energy production obtained from mineral fuels and hydropower in the United States between 1850 and 1965 is shown in Figure 8. From 216 trillion British thermal units (Btu)* in 1850, production had increased 160 times in a century, and by 1965 the figure had reached nearly 50 quadrillion Btu's. Projected growth in 1975 will raise the total to approximately 75 quadrillion Btu's. The average annual growth rate between 1900 and 1955 was 3.1 per cent.[1]

Two-fifths of United States energy production in 1965 was accounted for by industry, one-fifth by transportation, of which half was consumed by the automobile, and one-fifth by households for

---

* A British thermal unit is the amount of heat required to raise the temperature of one pound of water one degree Fahrenheit.

[1] Sam H. Schurr, et al., *Energy in the American Economy 1850–1975* (Baltimore: Johns Hopkins Press, 1960), p. 145.

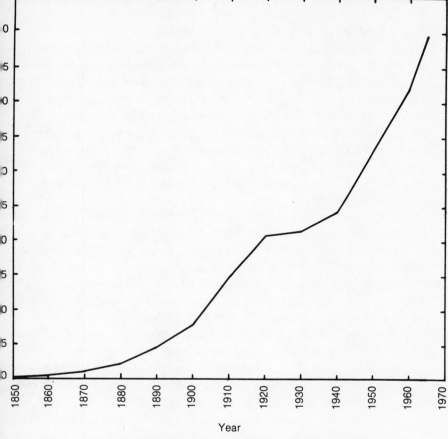

Figure 8. Energy production (mineral fuels and hydropower) in the United States, 1850–1965

heating and cooking. The remainder was used by government, commercial establishments, and agriculture. The source of energy has shifted from coal, which supplied nearly half the energy in 1947 but only one-fifth in 1965, a percentage that will hold relatively constant through 1980. Natural gas has increased as a source of energy supply from 14 per cent in 1947 to 31 per cent in 1965 and will decline to approximately 29 per cent in 1980 as depletion nears. Petroleum, which accounted for 34 per cent of the energy in 1947, rose to a peak of 43 per cent in 1965 and will decline to 41 per cent in 1980. Nuclear energy is expected to supply 5 to 9 per cent of the total need in 1980.[2]

Coal has until now been the major source of energy for the industrial countries. The availability of coal and iron ore were among the major factors that led England into the Industrial Revolution. The availability of a large coal supply also has been a significant factor in the rise of the United States as an industrial power. Coal production in the United States showed an exponential rise during the period 1850 to 1920, when annual tonnage rose from eight million tons to 658 million tons. Production dropped during the Depression years, rose again during World War II, and continued to drop through 1955.[3]

The production of natural gas in the United States in the period 1900 to 1965 is shown in Figure 9. From 128 billion cubic feet in 1900, production rose to sixteen trillion cubic feet in 1965, an increase of 128 times in sixty-five years. Production has nearly doubled in the past ten years, and is projected to rise to twenty trillion cubic feet in 1975. The coal equivalent of the nearly 9.5 trillion cubic feet produced in 1955 was 390 million tons of bituminous coal. Per capita consumption rose from 3,000 cubic feet in 1900 to 53,000 cubic feet in 1955, an increase of nearly eighteen times.[4]

The rapid rise in the production of crude oil in the United States is illustrated in Figure 10. Production of oil and oil products rose from 134 million barrels in 1905 to 2,849 million barrels in 1965, an increase of over eighteen times in sixty years. Output has more than doubled since 1940. In 1940 the United States began to import oil, the number of barrels rising from 41 million in 1940 to 449 million in 1965. The imports in the latter year amounted to 17 per cent of the domestic production.[5]

The world-wide production of oil has also shot upward as a con-

[2] William A. Vogely and Warren E. Morrison, "Patterns of U.S. Energy Consumption to 1980," *IEEE Spectrum*, IV (September 1967), 81–86.
[3] Schurr, *op. cit.*, Table 12, p. 63.
[4] *Ibid.*, p. 130.
[5] Statistics 1905–1955: *Historical Statistics of the United States, Colonial Times to 1957* (Washington, D.C.: Bureau of the Census, 1957), Tables M 133–137, p. 360; 1960–1965: *Statistical Abstract of the United States, 1968* (Washington, D.C.: Bureau of the Census, 1969), Table 1028, p. 675.

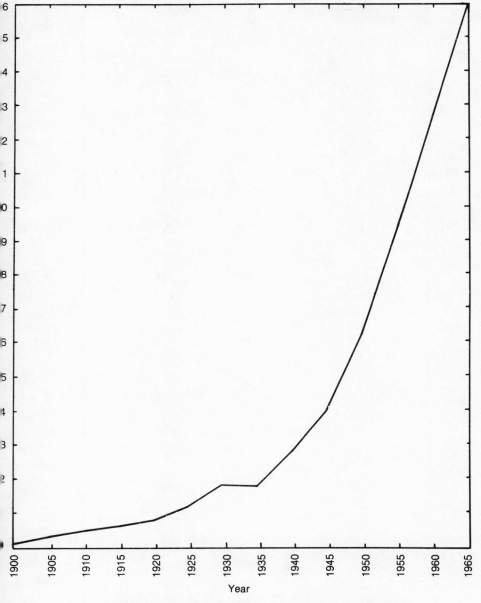

Figure 9. Production of natural gas in the United States, 1900–1965

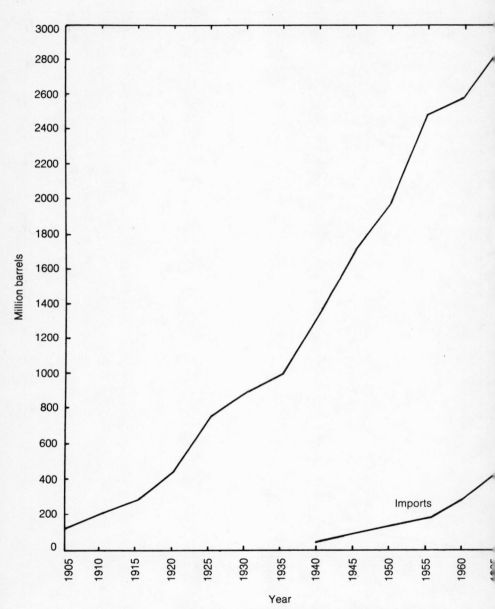

Figure 10. Production of crude petroleum in the United States, 1905–1965

sequence of world-wide industrialization and the rapid increase in the number of motor vehicles. A graph of world production of oil appears in Figure 11, which displays the all too familiar exponential growth curve. In 1860, 508,000 barrels of oil were produced in the world, of which the United States produced 98.4 per cent. By 1900 the annual world output was 149 million barrels. More than two billion barrels were produced in 1940, with the United States'

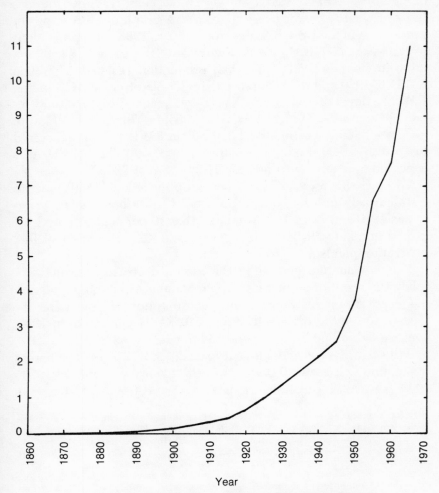

Figure 11. World production of crude oil 1860–1965

share dropping to 62.9 per cent. By 1965 the world total had increased more than five times the 1940 figure to reach more than eleven billion barrels, with a further decrease in the United States' share to 25.7 per cent.[6]

Whereas total energy production is estimated to grow at a rate of 3.3 per cent per year in the United States between 1965 and 1980, the percentage utilization of this energy by the generation of electricity will grow at a rate of 6.6 per cent, leading to a doubling of electric power every ten years. A graph of electric power production in the United States from 1902 to 1965, with projections to the year 2000, is shown in Figure 12. Starting with six billion kilowatt hours (KWH) in 1902, production reached 57 billion KWH in 1920 and by 1930 more than doubled to reach 116 billion KWH. A level of 182 billion KWH was reached in 1940, and increased another 3.5 times by 1955. In 1965, 1,057 billion KWH were generated, and an estimated 1,520 billion KWH will be generated in 1970. By 1990 production will soar to 5,830 billion KWH; it is expected to reach ten trillion KWH by the year 2000.[7]

To meet this increase, the electric utility industry will have to install nearly one million megawatts of additional capacity between 1970 and 1990. It is estimated that 40 per cent of power installed in 1990 will be nuclear power and 45 per cent fossil fuel generating plants.

The exponential increase in the use of electrical energy is not solely the result of an increase in population. Although population has been growing, the per capita consumption of power has increased at a more rapid rate. From 540 KWH per capita in 1920, average individual usage is expected to reach 7,950 KWH in 1970, and this figure is projected to soar to 22,200 KWH in 1990. Whereas population will grow 40 per cent in the twenty-year period 1970–1990, per capita use of electric power is projected to nearly triple.

---

[6] Statistics 1860–1955: *Petroleum Facts and Figures, Centennial Edition* (New York: American Petroleum Institute, 1959), p. 437; 1960–1965: *Petroleum Facts and Figures, 1967* (New York: American Petroleum Institute, 1967).

[7] Statistics 1902–1965: *Historical Statistics of the United States*, Table S–15, p. 506; 1960–1965: *Statistical Abstract of the United States, 1968*, Table 756, p. 513; projections 1970–2000: Bureau of Power, Federal Power Commission in U.S. Congress, Joint Committee on Atomic Energy, *Hearings, Environmental Effects of Producing Electric Power*, Pt. 1, 91st Congress, 1st Session (Washington, D.C.: Government Printing Office, 1969), p. 54.

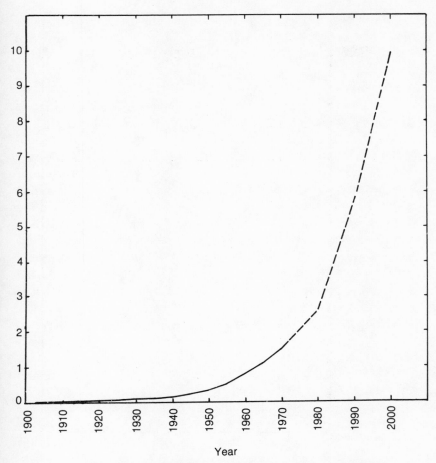

Figure 12. Production of electrical energy in the United States, 1902–1965, with projections to 2000

The familiar exponential curve of per capita use is shown in Figure 13.

The production of electric power represented only 13 per cent of total energy production in 1947. This grew to 21 per cent in 1965, and is expected to rise to 28 per cent in 1980 and 43 per cent by 2000.

The burning of 29.7 million tons of coal in 1968 accounted for nearly 52 per cent of electric power generation in that year. Natural gas in the amount of 3.1 trillion cubic feet accounted for 23 per

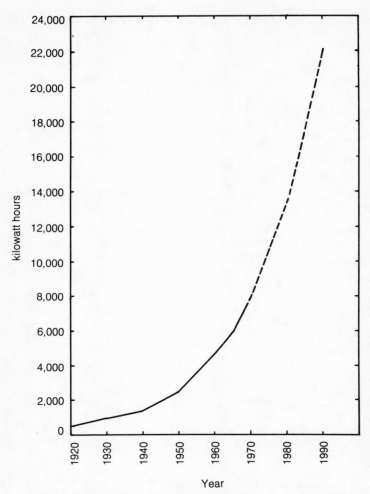

Figure 13. Per capita power consumption in the United States, 1920–1965, projected to 1990

cent; 187 million barrels of oil provided 7 per cent. Hydropower supplied 17.3 per cent, while nuclear power provided less than 1 per cent. By the year 2000, however, nuclear plants will generate nearly 53 per cent of the power, coal 30 per cent, hydropower 7 per cent, oil 5.5 per cent, and gas 4.8 per cent.

The profound changes that will be effected by the significant shift in energy resources, the continued increase in energy production and the per capita consumption, together with the growing percentage of the energy expended in the generation of electric power, exemplify quasi-solutions that are creating a proliferating host of residue problems. Among the problems are depletion of nonrenewable resources and attendant air, water, and land pollution.

According to Peter Flawn, oil reserves in the United States will be depleted by the end of the century.[8] Sam H. Schurr and his associates cite nine different estimates ranging from 140 to 300 billion barrels of oil reserves in the United States. With allowance for future discovery, the estimates range from 357 to 687 billion barrels.[9] Harrison Brown and his associates estimated known worldwide oil reserves in 1957 at 250 billion barrels of crude oil, with perhaps an additional 900 billion barrels awaiting discovery. The exponential curve of Figure 11, if extrapolated, will consume the total projected reserve in less than one hundred years. If electric power generation and fuel consumption by motor vehicles continue to increase at current rates, this period will be considerably shortened. As noted in *The Next Hundred Years,* "although the age of fossil fuels has barely begun, we can already see its end."[10]

The reserves of natural gas both in the United States and the world, according to Flawn, will be exhausted before the end of the century.[11] A National Academy of Sciences committee reported in 1962 that in the United States the peak in crude oil production was expected to occur by 1970 and that of natural gas by 1980.[12] Schurr and his associates estimated the United States reserves at 167 billion barrels, which would provide a total future supply of natural gas of 1.2 quadrillion cubic feet.[13] At a *constant* rate of 1955 consumption, these estimated reserves would be depleted in 120 years beyond the year of estimate, 1960. At *constant* 1975 projected rates, depletion would take place in sixty years. If the rate of consumption increases, depletion will come earlier. The American Gas Associa-

---

[8] See Figure 7.
[9] Schurr, *op. cit.,* Table 100, p. 358.
[10] Harrison Brown, *op. cit.,* p. 99.
[11] See Figure 7.
[12] *Energy Resources,* Publication 1000–D (Washington, D.C.: National Academy of Sciences, 1962).
[13] Schurr, *op. cit.,* p. 411.

tion reported in 1970 that proven gas reserves had declined for two years, during which time sales exceeded new sources. With proven reserves of 275 trillion cubic feet, at the 1969 rate of consumption of 20.7 trillion cubic feet, there would remain only little more than a thirteen-year supply.[14]

United States coal reserves were estimated in 1960 to be at least two trillion tons.[15] Flawn shows United States and world reserves sufficient to last beyond the year 2800 at *current* minable grades and rates of consumption. Brown and his associates also estimate that world coal reserves will last to the year 2800.[16]

Although coal is the most plentiful fossil fuel, and is used extensively in the generation of electrical power, it is also the fuel that gives rise to the largest quantities of polluting materials. The secondary effects of power generation are cumulative, combining with effects of other quasi-solutions, and today pose a serious threat to health and general well-being. These effects also threaten to interfere with natural climatic processes in potentially catastrophic proportions.

Electric power generating plants generated nearly 25 million tons of pollutants, approximately 80 per cent of which was in the form of fly ash, from burning 240 million tons of bituminous coal and lignite in 1965. The total residue in 1970 from coal-fired plants is estimated at thirty million tons, with a concomitant increase in fly ash.[17] The ash fouls the air and is eventually deposited on the earth, alighting on habitations and on foliage. This particulate matter is joined by ash from industrial plants, incinerators, the burning of leaves, etc., with the result that hundreds of tons are deposited per square mile in large urban areas.

In a step to eliminate the extent of the ash fallout in urban areas, some electric generating plants are being moved to the vicinity of the coal mines, where environmental protection is supposedly being designed into the plants. The Keystone plant in Pennsylvania, for one, has designed recirculating water systems for cooling and large precipitators for removing fly ash from the smokestacks. An

[14] *New York Times,* April 13, 1970.
[15] Schurr, *op. cit.,* p. 323.
[16] Harrison Brown, *op. cit.,* Figure 20, p. 109.
[17] *Solid Waste Management,* p. 12.

artificial lake 4.5 miles long and one mile wide has been created for cooling purposes. The precipitated fly ash is expected to amount to about one million tons in the early years of operation. "We are going to make the first man-made mountain of the material," an employee stated.[18] The effects of this mountain and the artificial lake on the ecology and aesthetics of the area are not considered when boasts are made of a 99.5 per cent ash-removal efficiency.

Sulfur dioxide is another pollutant resulting from the burning of fossil fuels. In 1967 about 26 million tons of this gas were released to the atmosphere in the United States. The estimate for 1970 is 37 million tons and a projected volume of 90 million tons is forecast for 1990. Nearly 60 per cent of the sulfur dioxide emissions comes from burning coal, with another 20 per cent from the burning of petroleum products. Although by the year 2000 coal will provide only 30 per cent of generated electric power, its total consumption will increase threefold, and more than 45 million tons of sulfur dioxide are expected to be emitted from the burning of coal in 1990.

In the atmosphere, sulfur dioxide is slowly converted to sulfuric acid which is a corrosive substance. It is irritating to the respiratory system and attacks materials: metals corrode, paints disintegrate, fibers weaken and fade, building materials discolor and disintegrate. Sulfur dioxide has also been found to suppress plant growth. Together with oxides of nitrogen and particulates, sulfur dioxide has been issuing from smokestacks around the world without scientists being fully aware of its effects upon man and the environment. Today, when the sheer tonnage of pollution can no longer be ignored, scientists can only speculate on its effects. Is there a threshold limit beyond which the gas is harmful on a cumulative and long-range basis? Is there an acceptable level of risk—for the individual, for society? The same questions arise in relation to treatment of sewage, detergents, and pesticides. In fact, the synergistic effects of sulfur-related compounds with these and other chemical pollutants are also not known.

What is known, however, is that air pollution can produce acute or chronic human effects. Besides the well-known air inversion in

[18] "Electric Utilities Include Built-in Safeguards for Environmental Protection," *Environmental Science and Technology*, III (June 1969), 523–525.

Donora, Pennsylvania, in October 1948, London experienced killer air pollution periods in 1952 and 1962, in which four thousand and six hundred excess deaths were reported, respectively. A similar period was reported in Belgium's Meuse Valley in 1960. In New York and other major American cities a rise in the number of deaths from pulmonary emphysema has been noted. Although no specific "air pollution disease" has been diagnosed, evidence indicates that air pollution can cause serious physiological damage and death. Of even greater danger are the as yet unknown long-range cumulative effects of living in environments that are becoming increasingly affected by a growing number of pollutants—some in major proportions, others in only minute traces. As noted earlier, longevity in the United States has leveled off or has begun to decrease as the cumulative effects of a technological society begin to act adversely upon populations.

As for the control of pollution, the secondary effect of the appearance of sulfur dioxide has called into operation a set of quasi-solutions to deal with the problem. Build higher smokestacks, one quasi-solution states, but the only effect of this is to disperse the pollution over a wider area. Another suggested method for control of the sulfurous gases is to add powdered limestone to the flue gas after combustion to convert the gases into solid form. The solids must then be extracted instead of the gases. Assuming that a precipitation process can be devised, this process would not be totally effective, so that some gases would still be discharged into the atmosphere. If the precipitated solids are waste, they must be disposed of, thus joining the general problem of solid waste disposal. If the precipitated solids are reclaimed, or even partially so, then an entire reclamation process and plant must be developed which, in turn, will be a quasi-solution to the reclamation problem from which will emerge still a new residue of problems, perhaps of a different and even more complex nature. Whether reclaimed or not, the precipitation process will require energy for operation. Hence, more fuel must be transformed into energy—which is the point at which the initial problem began.

Another possible solution to the waste gas problem is to convert the coal to gas, then remove the sulfur from the gas before it is burned. Here a new process is once again substituted for an older process. The new process, however, will generate a residue of unsolved problems in that waste disposal and energy generation are again required. At present "an acceptable technology for the control of sulfur dioxide emissions . . . does not exist . . . ," an official of the Bureau of Power of the Federal Power Commission has stated.[19]

Other emission products such as nitrogen oxides produced by coal-burning generating plants can be converted into nitric acid when inhaled by man and combined with the water in the body. The acid damages cell tissues, especially in the lungs. A million megawatt station can discharge about eighty tons of nitrogen oxides daily into the atmosphere, where they combine with the oxides discharged from autos and industrial plants.

One of the synergistic effects of the burning of fossil fuels is interference with the climate on a global scale. Our understanding of climate and our ability to control it are rudimentary. Our ability to meddle with the climate, unfortunately, is not commensurate with our knowledge.

The effective functioning of the carbon cycle, whereby carbon dioxide is given off to the atmosphere through the process of oxidation and is in turn converted into carbohydrates by the process of photosynthesis in plants, depends upon an intricate balance between oxidation and photosynthesis. Under normal conditions, the amounts of oxygen and carbon dioxide in the air are relatively constant. It has been estimated that 150 billion tons of carbon are utilized in photosynthesis annually, the tonnage about equally divided between terrestrial and marine plants. One-fifth of this quantity is present in atmospheric carbon dioxide, the remainder arising from oxidation of organic matter.

Every act of oxidation removes oxygen and puts carbon dioxide into the air. The amount of atmospheric carbon dioxide has increased about 8 per cent in the last seventy years, and is increasing more rapidly at present because of the increased burning of fossil fuels. At the same time, changes are taking place in the oxygen

---

[19] *Environmental Effects of Producing Electric Power*, p. 63.

content of the air. The atmosphere today contains about 0.032 per cent of carbon dioxide. The National Air Pollution Control Administration has predicted that carbon dioxide emissions in the United States will increase about 4 per cent per year through 1985. If the reserves of fossil fuels are used as projected, a 185 per cent increase in carbon dioxide in the air is estimated over the amount in 1950. The carbon dioxide in the air could then reach 0.092 per cent, and the oxygen could decrease by a small amount.[20]

The possible depletion of oxygen is a controversial problem. Wallace S. Broecker has estimated that burning fossil fuels at the projected rates will at most consume 0.2 per cent of available oxygen, and if all known fossil fuel reserves were burned, less than 3 per cent of available oxygen would be consumed.[21] These conclusions were based on global averages and do not consider such "hot spot" areas as the United States, where some scientists contend there is already an oxygen deficiency. Although a study of the oxygen in *clean* air measured over the oceans and continental shelf between 1967 and 1970 showed a statistical similarity with all reliable measurements since 1910,[22] similar measurements have not been made over highly industrialized areas. It is generally recognized that the oxygen of the earth's waters is being seriously depleted. We are assured, however, that man would succumb to pollution of the environment long before the oxygen supply would be depleted.

What is the effect of a change in the carbon dioxide content of the air? In 1965 the President's Science Advisory Committee Environmental Pollution Panel projected a 25 per cent increase in carbon dioxide between 1950 and 2000, and stated: "This may be sufficient to *produce measurable and marked changes in climate, and will almost certainly cause significant changes in the temperature and other properties of the stratosphere. At present, it is impossible to predict these effects qualitatively,* but recent advances in mathematical modeling of the atmosphere, using large com-

[20] Eugene K. Peterson, "Carbon Dioxide Affects Global Ecology," *Environmental Science and Technology*, III (November 1969), 1162–1169.

[21] Wallace S. Broecker, "Man's Oxygen Reserves," *Science*, CLXVIII (June 26, 1970), 1537–1538.

[22] L. Machta and E. Hughes, "Atmospheric Oxygen in 1967 to 1970," *ibid.*, pp. 1582–1584.

puters, may allow *useful* predictions within the next two or three years" (italics added).[23]

The "useful" predictions have not been forthcoming. One scientific study estimates a doubling of the carbon dioxide content of the air would increase the world-wide mean temperature by 6.5 degrees F, while halving the content would lower the temperature 6.8 degrees F. Another study estimated that a 25 per cent increase in carbon dioxide would raise the temperature from 1.1 to 7 degrees F, depending upon the amount of additional water vapor in the atmosphere. Assuming constant relative humidity, another report estimated a temperature rise of 4.25 degrees F if the carbon dioxide content were doubled and average cloudiness assumed.[24]

Together with carbon dioxide and other pollutants that are being poured into the atmosphere, the growing fleets of jet planes that criss-cross continents and oceans are contributing their effluvia directly into the upper reaches of the atmosphere. It has been estimated that cirrus cloud cover over the North Atlantic has increased almost 10 per cent in recent years, with even larger increases anticipated. A study group at the Massachusetts Institute of Technology has indicated that stratospheric flights of supersonic transport planes would tend to double global averages of water vapor in the stratosphere, which would lead to increased cloud formation and higher stratospheric temperatures with consequences that are unpredictable. On the ground a pall of highly polluted air extending for miles already hangs over major airports.

The "significant changes" that may occur are not well understood or predictable and indeed appear to be paradoxical. A warming of the earth's atmosphere would result in the gradual melting of the earth's ice caps with a consequent rise in sea level amounting to perhaps two feet every ten years. The melting, Eugene K. Peterson observes, could also bring about a tremendous redistribution of the weight and pressure exerted on the earth's crust resulting, in turn, in an increase in earthquakes and volcanic activity. The latter would further increase the carbon dioxide content of the atmosphere. Rising temperatures of surface ocean waters would

[23] Quoted in Peterson, *op. cit.*
[24] *Ibid.*

have the effect of absorbing less carbon dioxide, thus again increasing the carbon dioxide content of the atmosphere.

If population does not stabilize by the year 2020 and consumption of fossil fuels increases proportionately, temperatures could rise nine degrees F over 1950 levels. In that case, according to Peterson, some possible effects are:

- Most areas would get more rainfall, and snow would be rare in the contiguous 48 states, except on higher mountains.
- Ocean levels would rise four feet.
- There would be major increases in earthquakes and volcanic activity.
- The Arctic Ocean would be ice-free for at least six months each year, causing major shifts in weather patterns in the northern hemisphere.
- The present tropics would be hotter, more humid, and less habitable, but the present temperate latitude would be warmer and more habitable.[25]

Other scientists suggest that the above process, instead of triggering a grand melting, will increase the water content of the atmosphere and that the carbon dioxide, together with dust and other particulate pollutants, will effectively act as a shield from the sun's heat and cause the temperature of the earth to fall, thus leading to a new ice age.

While the actions of man's interference with nature may lead to either the melting of the ice caps or a new ice age, *man is continuing to intervene massively in the carbon cycle which provides him with air to breathe and which transforms grass into flesh, without understanding the processes by which he is affecting the balance or comprehending the effects of these changes.* Carbon dioxide and oxygen are essentially nonrenewable resources, for there is no known way to replenish the oxygen of the atmosphere or to restore the atmospheric or oceanic balance of carbon dioxide. The oxygen reservoir is maintained by the action of sulfate-reducing bacteria in anerobic environments and, clearly, destruction of the organisms or bacteria that carry on the reduction process would lead to crisis. The bacteria, diatoms, and plankton of the ocean also depend on the maintenance of a delicate chemical balance, as do the nitrogen-

[25] *Ibid.*

fixing bacteria of the soil. If the *Torrey Canyon*, which spilled thirty million gallons of oil into the ocean in 1967, had been carrying pesticides, it has been observed, the oceans might have been transformed into biological deserts. Similarly, destruction of any one of a half-dozen nitrogen-fixing bacteria could bring about an end to life on earth. Might this destruction be in its initial stages with massive doses of fertilizers being applied to soils, as Barry Commoner has noted?

The rising content of carbon dioxide in the atmosphere may well play another trick upon unsuspecting man. For man, the upper limit for prolonged exposure to carbon dioxide is only fifteen times the present level, whereas plants may thrive at levels four to six times greater than man's upper limit. Thus, dialectically, by permeating the atmosphere with carbon dioxide, food production may be increased—though man may not survive to partake of the food.

Given this complex of residue problems wherein energy requirements continue to increase while (1) the pollution arising from the burning of fossil fuels has reached an unacceptable level that is threatening human life and survival; (2) the present unacceptable levels of pollution are expected to increase greatly; and (3) the supply of fossil fuels is approaching depletion, atomic energy is being hailed as a new savior. Atomic energy will, it is averred, provide a practically inexhaustible supply of energy and prevent pollution.

The Atomic Energy Commission, under whose aegis the nuclear energy program is being implemented, epitomizes the character of much of science. On one hand, the AEC is the producer of bigger and deadlier hydrogen bombs, nuclear missiles, and the panoply of space-age horror weapons. On the other hand, the AEC is promoting the peaceful uses of nuclear energy, especially the production of electrical energy, under licensing arrangements with private utility companies.

As if to expiate the sin the physicists have known in unleashing the power locked in the atomic nucleus, "the shatterer of worlds" is now to be harnessed for man's benefit. The atonement for Hiroshima-Nagasaki and the mitigation of the threat that hangs over all the peoples of the world are unconsciously sought in offering to mankind the manna of infinite energy. It is as if from the rocks of the earth and the waters of the sea there will arise a peaceful genie to serve man in the final conquest of nature by harnessing the totality of the environment to man's ephemeral but overweaning gluttony.

Alas, atomic energy is but another quasi-solution to another techno-social problem, generating still another proliferating chain of residue problems. The promise of cheap and infinite energy, rather than expiating a prior sin, portends to unleash a new, possibly greater sin upon the peoples of the earth. Radioactive pollution of the earth, the air, and water threatens to add another deadly ingredient to the invisible web man is fashioning to render his habitat unfit for the propagation of life. Nuclear energy is a genie tainted with the memory of a mushroom cloud . . .

> And much of Madness, and more of Sin,
> And Horror the soul of the plot.

Atomic energy is estimated to provide 5 to 9 per cent of the total energy production of the United States by 1980, including 27 per cent of electrical energy. World-wide, Harrison Brown estimates that up to 65 per cent of the world's energy requirements would have to be met by nuclear energy by the year 2000. This energy would be provided essentially by current models of reactors that utilize uranium-235 obtained from cheap uranium ores. The reserves of these ores, however, are limited and represent only a small percentage of the energy content of the world's coal supply. Like other mineral resources, their depletion may be complete by the end of the century (see Figure 7). The AEC estimates the depletion of the cheap ores to take place by the mid-1980's, and lower-grade ores by the year 2000.

In 1962 the AEC projected a nuclear-generating capacity in the United States of forty million kilowatts by 1980. But as a result of its promotional activities and the increasing demand for electricity, as early as 1969 plants with more than 72 million kilowatts were already in operation, under construction, or in the planning stage. The Federal Power Commission now estimates requirements for 150 million kilowatts by 1980 and 509 million kilowatts by 1990. As a consequence, reactors are being built and more are planned to utilize a fuel that will soon be depleted. Attempts to import uranium will be faced with competition from producing countries, many of which are also planning nuclear generators that loom large in their development plans. Beginning in the early 1970's, uranium production will fall behind demand on a world-wide scale.

Technologists hold out two possible reprieves from the fated depletion of cheap uranium ores at the very time when atomic energy is looked to as a panacea. The first is the breeder reactor. This reactor breeds fissile uranium-233, or plutonium-239, and creates more fissile material than is burned, making every nucleus of uranium and thorium a source of energy. It is estimated that the energy potential is multiplied by a factor of 400, representing the ratio of the number of thorium and uranium-238 nuclei found in nature to the number in naturally occurring uranium-235 isotope, and again by a factor of 100 million, representing the ratio of the quantities of accessible uranium and thorium to the amount of cheap uranium ore.

An average piece of granite, according to Brown, contains only four parts per million of uranium and about twelve parts per million of thorium. The atomic energy equivalent in one ton of granite is about fifty tons of coal. Reducing this quantity by the energy required in mining and extraction would leave approximately the equivalent of fifteen tons of coal per ton of granite. The Conway granites of New Hampshire are estimated to contain thirty million tons of thorium at an average concentration of fifty parts per million which, at Brown's estimate, would suffice for thousands of years at the rate of forty tons of uranium or thorium daily. Granite, of course, would become an object of competing demands—e.g., it would still be needed for construction—as have other resources with multiple uses.

The breeder reactor is a research development funded by the AEC at over $2 billion in addition to large expenditures by private industry. It is a difficult and complex technological undertaking fraught with normal engineering problems as well as with many others stemming from the aspects of radioactivity. Experience with the breeder reactor at the Enrico Fermi power plant near Detroit indicates that successful operation of the reactor has not yet been achieved. The experience also points up the residue problems that will be created by this quasi-solution to nuclear power generation, most of which are shared by nuclear reactors in general. The obstacles that continue to develop in the long-range experiment to develop the breeder reactor raise serious questions about its success, even at the pure engineering level, quite apart from consideration of its side effects.[26]

The achievement of atomic fusion that would make possible the burning of the sea's deuterium would provide even larger reserves of energy. Yet the very difficult and complex problems associated with the process have not been resolved, if ever they can be. These problems have some of the qualities of developing a container for a universal solvent: a plasma of independently moving ions and electrons must be contained at a pressure of fifty atmospheres and a temperature of hundreds of millions of degrees by a magnetic field.

Thus the world is living on borrowed time, for the depletion of fossil fuels is fast approaching or, as in the case of coal, a level of pollution is created that is unacceptable. But harbingers of the deeper problems that will beset man with the introduction of atomic energy are already visible. Among the set of residue problems arising from atomic energy are the possibility of catastrophic accident, disposal of radioactive wastes, radioactive pollution of the soil, air, and water, and thermal pollution. Technologists, in general, belittle the dangers, as they have belittled previous caveats against mounting despoliation and pollution. But quasi-solutions have their own logic.

Present plans call for the storage of solid wastes, mainly spent fuel rods, at reactor sites for 150 days, where radioactivity will be reduced by a factor of thirty. Fuel-processing plants will then treat the waste and store it for ten years, after which the wastes will be

[26] Richard Curtis and Elizabeth Hogan, *Perils of the Peaceful Atom* (New York: Ballantine, 1969), pp. 1–18, 265–266.

buried. The AEC estimates that by the year 2000, about 800,000 cubic feet of solid waste will require seven hundred acres of abandoned salt mines for storage.[27]

Some of the wastes, with half-lives of seven hundred or more years, slowly dissipate their radioactive energy. Present storage is in large underground tanks that pose problems of eventual water contamination and entry into ecological cycles. The AEC is conducting research to reduce liquid waste to solid cakes that can be "stored forever" in salt and rock caverns. That there is a potential hazard from these stored wastes is attested to by the planned burial of some wastes at a remote site in Oregon, near Wagontire Mountain and Jackass Creek, by a company that was formed to "handle disposal of materials too hazardous to be dumped in inhabited regions."[28] Waste material would be mixed with cement, poured into steel drums, and buried deeply in labeled trenches that would be continuously monitored.

The AEC has announced the development of the Kansas Nuclear "Park" in an abandoned salt mine near Lyons, Kansas, as the nation's first underground radioactive waste repository. The salt mine was selected because it is *almost* impervious to water, *almost* independent of all water sources, and *generally* free of earthquake activity. Studies will continue over a period of several years to determine the significance of the "almosts" and "generally" to assure that the "Park" will cause no hazard. A deep well that will extend over one mile into a shale deposit in New York State is also under consideration. Meanwhile, these wastes are stored at four sites that have previously also been cited as hazard-free, although their safety has been called into question by another government agency.

Solved problems, as noted earlier, have a tendency to become unsolved. The "stored forever" problem is perhaps wishful thinking. Already in 1965, storage tank failures were found in several of the AEC storage depots, where it was found necessary to re-

[27] "Power Generation and Environmental Change," Symposium of American Association for the Advancement of Science, December 1969, reported in *Science*, CLXVII (January 9, 1970), 159–160.
[28] *New York Times*, January 5, 1970.

place leaking tanks after a period of only twenty years when the radioactive wastes they contained would be lethal for hundreds and thousands of years.

Billions of gallons of intermediate-level wastes have been consigned to underground storage, and by the end of the twentieth century billions more of high-level wastes will have to be stored. The AEC is planning to bury 200,000 gallons of water containing tritium in deep-well injection facilities. These deep-well facilities, however, are but another quasi-solution with accompanying residue problems. David M. Evans has demonstrated that over seven hundred minor earthquakes occurred in Denver, an earthquake-free region, since 1882, when the nearby U.S. Army Rocky Mountain Arsenal began injecting waste into a twelve-thousand-foot disposal well starting in 1962. Evans also cites instances in which cyanide appeared in the drinking water of Buffalo; the front yard of a family in Wichita split open and spilled tons of salt-saturated water; a Denver water well from which the pesticide 2,4-D can be drawn, and another well that produced anti-knock gasoline. Fourteen per cent of the disposal wells existing in 1968 were shut down or suffered failure.[29]

A 1966 report by the National Academy of Sciences criticized AEC storage practices, stating: "The current practices of disposing of intermediate and low-level liquid wastes and all manner of solid wastes directly into the ground or in the fresh water zones, *although temporarily safe, will lead in the long run to a serious fouling of man's environment*" (italics added). The report also criticized the general practice of ground storage of waste because of build-up problems. Earth materials under a storage site become saturated with the passage of time, and in the process become equivalent to the radioactive concentration of a lesser volume of high-level wastes. The Academy also criticized the present storage sites as "poor geological locations" because they could be subject to earthquakes.[30]

Up to 1960, a quantity of radioactive waste equivalent to more

---

[29] David M. Evans, "Pollution Goes Underground," *Nation*, CCIX (December 8, 1969), 632–635.
[30] *New York Times*, March 7, 1970.

than fourteen thousand curies had been dumped into the Pacific Ocean, and another eight thousand curies of waste had been dumped into the Atlantic. If problems arise from sealed tanks on land, the problems of tanks in the oceanic environment are many times more serious. A related problem, that of storing forever lethal gases prepared during World War II, is a reminder of the fallibility and ephemerality of quasi-solutions. Lethal mustard gas, part of eighteen thousand metric tons of material captured from the Germans and dumped in the Baltic Sea in 1945, began escaping from rusted containers and injured six fishermen who were handling contaminated nets and fish in 1969. Danish authorities feared that a large area of the Baltic would be contaminated.[31]

The unknown consequences of dumping wastes, chemicals, and radioactive materials in the oceans has not deterred the U.S. Army from continuing the practice. In mid-1970, 418 steel and concrete coffins containing sixty-six tons of deadly GB nerve gas were dumped in a deep trench of the ocean 280 miles off the coast of Florida, although the chairman of the Council on Environmental Quality admitted that "our knowledge of [the oceans] is still so limited that we cannot confidently predict the consequences of placing in them any dangerous material."[32]

Mixed in the coffins is a land mine containing the even more virulent gas, VX, manufactured from a captured Nazi formula, which was responsible for the death of 6,400 sheep in Skull Valley, Utah, in the spring of 1968. The agency, which had lied about the cause of the sheep kill, acknowledged that it did not know what would happen if the coffins were to rupture at great depth. Burial was imperative, however, because it was believed the gas rockets had begun to leak and a land explosion could lead to a major tragedy. Oceanographers have pointed out that life conditions in the ocean depths have been constant for millions of years, and the introduction of pollution into this stable environment could upset the marine food chain with severe, perhaps catastrophic implications.

[31] Letter in *London Sunday Times*, quoted in *Industrial Research*, November 1969, p. 13.
[32] Russell E. Train in testimony before the Subcommittee on Oceanography of the Senate Commerce Committee, reported in *New York Times*, August 6, 1970.

The probability of a nuclear accident increases with the increase in the number of nuclear plants, their growing capacity, the increasing complexity of nuclear energy conversion, and the growing trend to site the plants near large cities where electrical energy is to be used. Accidents can involve the processes that control fission, the discharge of effluents, the refueling process and removal of wastes, the transportation of radioactive materials, effluents, and isotopes, and the storage of fissionable material and wastes. Such accidents can be brought about through human error or mechanical, electrical, or chemical failures, not to speak of earthquakes, tornadoes, or other vagaries of nature. They can also be brought about by lack of knowledge about the complex interrelationships that take place in a process in which the basic chemical elements are transmuted into other elements and mass is transformed into energy.

Although the incidence of nuclear accidents has been low, some have occurred, causing fatalities and contaminating the environment. The Windscale Works in England suffered a serious accident in 1957 in which an area of four hundred square miles around the site was contaminated with radioactive materials. In 1961 three men died at the Idaho Falls experimental reactor. A number of serious accidents have occurred in other reactors that were contained before a hazard to the public became serious. The escape of radioactive material from hydrogen bombs accidentally dropped off the coast of Greenland and on farm land at Palomares in Spain indicates the great danger the world faces in augmenting the spread of nuclear material throughout the world, whether for war or peaceful purposes.

What would be the effect of a serious accident in a nuclear reactor plant? No one really knows. The Brookhaven Report of 1957 on the "Theoretical Possibilities and Consequences of Major Accidents in Large Nuclear Plants" estimated that as many as 3,400 people could be killed at distances up to fifteen miles away, and 43,000 injured up to distances of forty-five miles. Approximately 150,000 square miles could be radioactively contaminated, with property damages soaring into the billions of dollars. Since 1957 the size of the nuclear energy plants has not only greatly increased but they have been sited nearer to large cities. If the probability of a nuclear catastrophe is small, as the AEC contends, it should be remembered that, as in evolution, given a long time period, small probabilities become certainties.

An equally serious danger that is not slightly probable but is planned involves the necessary release of radioactive materials into the air and water as normal industrial wastes. The Dresden plant, fifty miles from Chicago, for example, discharges into its coolant canal the following radioactive materials: tritium, cobalt-58, cobalt-60, strontium-89, strontium-90, iodine-131, cerium-137, and barium-140. Radio-cobalt and strontium-90 have been found in river water where the city of Peoria draws its water supply. Traces of sulfur-35, cobalt-57, zinc-65, zirconium-95, niobium-95, ruthenium-106, and cerium-141 have also been detected in the Illinois River, into which the Dresden coolant canal drains. In the vicinity of the plant iodine-131 has been found in heifers, strontium-90 in snow, and cesium-137 in corn kernels. Released to the atmosphere by a boiling water reactor plant are: tritium, nitrogen-13, krypton-85, krypton-87, krypton-89, krypton-90, xenon-133, xenon-135, and xenon-138.

Tritium produced from lithium and boron combines with oxygen to form water. At the reactor site, tritium becomes part of the effluent; at the processing plant, it enters the atmosphere as water vapor. Tritium can enter the body through the skin and by way of the lungs. Animals and man can convert tritium-labeled hydrogen gas to tritiated water by oxidation in the tissues. Because tritium is an isotope of hydrogen, tritiated water behaves chemically and biologically like ordinary water. Although its radioactive half-life is 12.3 years, it is estimated that its biological half-life is approximately twelve days and, it is claimed, there is no build-up in the body. There is some evidence, however, of tissue-bound tritium which indicates that the level recommended as acceptable by the International Commission on Radiological Protection should be lowered. The biological cycle also indicates that tritium does not behave only as an element of tritiated water but can become concentrated as its distribution in the body is not as regular as was first believed.[33]

At the Hanford plutonium plant a large fraction of the produced tritium is released into the ground, where it moves toward the Columbia River. The maximum recommended concentration allowable in drinking water for continuous nonoccupational exposure has been found in wells located six miles from the disposal area. The prediction of the course of tritium dispersion in water, researchers have reported, cannot be determined by simple assumptions, and the magnitude of its spreading is almost impossible to predict *a priori*.[34]

The build-up of tritium from the nuclear power industry will continue in the atmosphere and will be superimposed on the amount present from natural sources and weapons testing. By 1985, atmospheric levels from nuclear wastes will reach natural levels. The tritium produced from weapons testing will predominate until 1995, at which time reactor products will be the predominant source.[35]

At reactor sites and reprocessing plants, krypton-85, a radioactive substance with a half-life of 10.8 years, is released into the atmosphere. Not separated by any existing waste-handling technique, krypton-85 is discharged directly into the atmosphere. It is chemically and biologically unreactive and can be inhaled, or its energy can be delivered to man by the surrounding air. This isotope subjects the whole body to exposure and is persistent. Its incidence in the atmosphere builds up as the number of nuclear energy plants increases, and its effects will be long-lived.

Edward P. Radford, testifying at a hearing of the Joint Committee on Atomic Energy in January 1970, said that tritium concentrations at the only commercal reprocessing plant in operation have on occasion been 100,000 times the natural level in rain water.[36] In March 1970 several ducks feeding in waste water trenches at

[33] *Environmental Effects of Producing Electric Power*, Appendix 13, p. 772.

[34] D. G. Jacobs, "Sources of Tritium and Its Behavior upon Release to the Environment," *ibid.*, pp. 568–569.

[35] H. T. Peterson, *et al.*, "Environmental Tritium Contamination from Increasing Utilization of Nuclear Energy Sources," *ibid.*, p. 765.

[36] *Chicago Sun-Times*, January 29, 1970.

the Hanford Works were found to be highly radioactive. The phosphorus-32 radioactivity would have given a person who might have eaten the fowl immediately after killing them five times the maximum permissible dosage of radiation.[37] Traces of cobalt-60 and manganese-54 have been found in the lower Hudson River, downstream of the nuclear plant at Indianpoint. The radioactive manganese has been found in acquatic plants. Blue crabs showed traces of radioactive cesium-137 and cerium-144. Freshwater clams that feed by filtering plankton and sediment showed traces of the radioactive manganese.[38] A Soviet scientist at an international conference in June 1970 reported that radioactive wastes from the Windscale nuclear plant had so contaminated the Irish Sea that the backbones of embryo fishes were being deformed.[39]

The radioactive contamination of the environment by nuclear energy plants is added to the contamination from nuclear weapons tests. Although the Nuclear Test Ban Treaty of 1963 banned the testing of nuclear weapons in the atmosphere, only those nations that are signatories to the treaty carry out its provisions. Thus while some of the fission products from pre-1963 tests made by the United States, the Soviet Union, and England still remain in the atmosphere and are slowly falling to earth, the radioactive debris of Chinese and French atmospheric bomb tests rains upon the earth.

Underground nuclear tests provide no real assurance that contamination will be controlled. In tests conducted by the United States radioactive material has been vented into the atmosphere from fissures in the containing earthen chambers. More serious is the possibility that the tritium produced in the underground explosions will enter the groundwater and be carried along to where it may be used by people.

The Colorado Committee for Environmental Information has alleged that the Rocky Flat atomic bomb manufacturing plant is releasing plutonium into the air, water, and soil. The AEC ac-

[37] *Chicago Tribune*, March 15, 1970.
[38] Gwyneth Parry Howells, Theo. J. Kneipe, and Merril Eisenbud, "Water Quality in Industrial Areas: Profile of a River," *Environmental Science and Technology*, IV (January 1970), 26–35.
[39] *New York Times*, July 1, 1970.

knowledges the presence of plutonium but minimizes its effects. One should be very cautious about official attempts to minimize pollution effects, for such effects have been denied or treated as inconsequential in the past, when ultimately they have proved of considerable consequence. Such an example, also found in Colorado, concerns initial official appraisals of radiation dangers from uranium mine tailings, the fine, sandy material that remains after radioactive ores have been extracted. In Grand Junction, a uranium-mining town, these tailings were given away to builders over a fifteen-year period to be used in place of sand for backfilling and under concrete slabs in home construction. Fifteen years later, eighty homes showed very high radioactive levels, and in the nearby town of Uravan it became necessary to vacate two homes. The highest readings in Grand Junction were 180 times in excess of levels set by the Colorado Health Department, and one Uravan house showed a level three to four times above the worst Grand Junction level. All these levels exceeded those allowed to miners working in a uranium mine.[40]

Another secondary effect of nuclear energy production is thermal pollution. Steam boilers lose almost 60 per cent of their energy to the atmosphere and contribute to its heat build-up. Gas turbine boilers lose nearly 50 per cent of their energy. Light water reactors are currently 30 per cent efficient. Breeder reactors are expected to be 40 per cent efficient, though their waste heat cannot be vented into the atmosphere. The reactors must be cooled by passing colder water through them to carry off the heat. The cooling water is warmed as a consequence, and returns to its source in a heated condition.

Because fish and most other marine species are cold-blooded and lack heat-regulating mechanisms, they are unable to control their body temperature under changing conditions. Their metabolism is in direct proportion to the temperature, and as temperature increases their rate of metabolism increases with an accompanying demand for more oxygen. It is estimated that temperatures over 90 degrees F are injurious to most fish. Raised temperatures that are below the lethal limit have been found to have adverse effects

[40] *New York Times,* February 17, 1970.

on fishes' reproductive systems as well as on their longevity. Since in the summer some streams naturally approach a temperature of 90 degrees, an artificial increase can be harmful.[41]

Given the known facts about increasing water temperatures, and keeping in mind that man's knowledge of the effects of thermal pollution are very meager, let us consider the magnitude of the secondary-effect problem. In 1968 the cooling of steam condensers in electric generating plants in the United States accounted for three-fourths of the sixty trillion gallons of water used for cooling. Whereas a thermal electric plant requires approximately three-fourths of a gallon of cooling water per minute for every kilowatt hour of energy, a nuclear energy plant will require a 67 per cent increase in cooling water. It is estimated that by 1980 one-sixth of freshwater runoff in the United States will be required to cool power plants, and that by the end of the century one-third will be required. But because runoff is not constant, being greater in the spring, even in 1980 half of the water is expected to be used during three-quarters of the year. Electric power plants that are cooled by a single stream can raise the temperature of the water ten degrees.

At the very time that the earth's rivers and the shrimp, crab, oyster, and fish beds lying off coastal waters are being looked to as a supply for a portion of increasing food demands, the exponential rise in power generation threatens to add another obstacle to the water's ability to support life. As a secondary effect of a quasi-solution, the increase in nuclear power plants will accentuate the problem.

The Florida Power and Light Company, for example, which operates a power plant at Turkey Point, twenty-five miles south of Miami, is building a six-mile-long canal to cool the ten thousand gallons of water that are discharged per second, at a temperature between 90 and 100 degrees, into Biscayne Bay. The building of two nuclear reactors to supplement the currently operating oil-fueled generators is expected to quadruple the heated discharge.

---

[41] John R. Clark, "Thermal Pollution and Aquatic Life," *Scientific American,* CCXX (March 1969), 18–27; see also statement of Donald Mount in *Environmental Effects of Producing Electric Power,* pp. 356–373.

By diluting the discharged water with cooler ocean water and running it through the canal, the company had hoped to decrease thermal pollution of the bay with its rich plant and fish life. The Federal Water Pollution Control Administration (FWPCA), however, has called upon the company to cease dumping heated water into the bay, contending that even at current rates of discharge serious effects are noted in marine life. It is claimed that the heated water has already damaged 670 acres of the bay.

Longer-range effects have been noted in the Thames River in England, into which the heat released rose from 555 million watts in 1930 to 3,700 million watts in 1950. The yearly average temperature rose from 53 to 60 degrees F during the twenty-year period; three-fourths of the increase was attributed to heat from fossil-fuel power plants. The oxygen deficit of the river was increased by about 4 per cent during the same period, with unknown biological effects.[42]

Unfortunately, heat cannot be dissipated or concentrated without doing work that creates more heat. Where should the heat be dissipated? In the case of the Bay of Biscayne, the FWPCA has recommended that a standard industrial cooling tower be used on which discharged water is passed over vanes releasing the excess heat into the air. The cooling towers also evaporate large quantities of water. The towers required for a 2.5 million watt power plant, for example, are estimated to produce more than twenty thousand gallons of evaporated water per minute. This quasi-solution, as has been noted, is already causing serious secondary effects by contributing to the unknown alteration of the climate.

Although nuclear power plants are being touted as an answer to both the exponential demand for electrical energy and the demand for pollution-free energy sources, these nuclear power plants present a greater overall burden on the environment than fossil fuel plants—both in the short run and in the long run. Radioactive pollution is essentially an irreversible phenomenon. If radioactive contamination is one day found to be excessive in the light of new knowledge and understanding, its effects cannot be reversed but will remain for centuries. Even if it were found necessary to stop

[42] Howells, *op. cit.*

all nuclear energy production, the radioactive isotopes that had been previously created would remain. The effects of the radioactivity can damage individuals in the short run or, more seriously, can affect unborn generations through genetic mutations. In effect, then, this generation is holding future generations hostage to care for the wastes and by-products of our atomic age.

8 *Technological Realism 3: Autos and Waste*

*T*HE SEARCH FOR perfect mobility with the internal combustion engine is leading to total immobility, while the by-products of the engine transform society and threaten its extermination through massive pollution of the air. When he developed the first model of his automobile, Henry Ford could not have foreseen that only sixty years after the first successful road test this situation would develop. The internal combustion engine appeared to be a perfect solution to the problem of human transportation. It was a highly individualized solution, when incorporated in a wheeled vehicle that provided a self-contained unit. The unit was amenable to mechanical fabrication. It carried its own supply of fuel. Its relatively simple operation was easily learned. Indeed, it appeared to be a near-perfect solution to a basic transportation problem—and one which further advances in technology could only improve.

The development of the automobile instead engendered a vast residue of problems that clearly revealed it to be a quasi-solution. Immediately, it gave rise to a number of extensive economic, social, and psychological problems. First, it was necessary to augment an industrial system that was based on mass production, the extensive use of the conveyor belt, and refinement of the concept of the division of labor. The factory became a micro-social system with attendant management problems. Each problem spawned its own set of solutions to isolated problems which in turn spawned a set of residue techno-social problems. As in other mass-production industries, the needs of the workers heightened the demand for industrial unionism, safety, workmen's compensation, unemployment compensation, and provisions for illness and old age—each of which was a quasi-solution to a specific problem that would later engender still another set of residue problems.

Among the residue problems were also the social effects of the automobile, which led to mass mobility, rootlessness, the agglomeration of popul ions centering about large urban megalopoli, and the devouring appetite for land demanded by urban sprawl. The residue problems generated by growth in the number of vehicles, their insatiable demand for road space, the development and quest for fuel, and the growing accumulation of deleterious by-products fully illustrate the concept of the quasi-solution.

Figure 14 illustrates the growth in auto vehicle registrations in the United States from 1900 to 1967, for all types of vehicles including buses and trucks. In 1967 nearly 80.5 million autos and 16.5 million trucks and buses were registered for a total of 97 million vehicles. It is estimated that by 1980 the total number of vehicles on the nation's highways will approach 150 million. The exponential growth of the number of registrations will lead to a rise of 50 per cent in the vehicle population in ten years—at a time when the human population is expected to increase by half, to 300 million in thirty years. The vehicle population is thus increasing three times as fast as the human population.

With production and sales of domestic and imported vehicles approximating ten million annually, the growth in the number of registered vehicles would increase even more precipitously than indicated in Figure 14 if an increasing number of vehicles were not scrapped each year. In 1950, 3,234,000 vehicles were junked, and in 1960 the figure rose to 4,783,000; by 1965 it jumped again to 6,841,000.[1]

[1] *Statistical Abstract of the United States, 1968*, Table 821, p. 551.

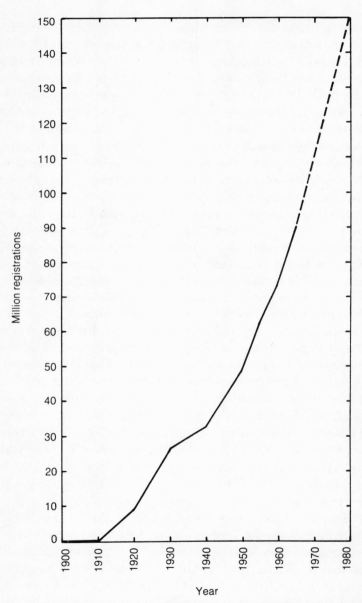

Figure 14. Motor vehicle registrations in the United States, 1900–1965 and projections to 1980

These scrapped vehicles are a brutal testimony to an economy of waste, for most of them are not reclaimed and their constituent components, made of increasingly rare substances, become irretrievably lost. The failure to reclaim the vehicles is primarily an economic one, for components can be produced from newly mined and fabricated resources cheaper than by reclaiming them from scrapped vehicles—at least for a time. There is no reason to believe that those whose extravagance is producing vehicles with very short life spans will not continue such production—for a time.

The scrapping of vehicles generates a residue of economic and technical problems ranging from landscape defilement to occupying needed land space, physical removal and disposal, and resource waste. The problem of abandoned vehicles in urban streets already clogged with moving vehicles becomes more serious. In New York City alone, 5,100 vehicles were abandoned in the streets in 1961—a figure that grew to 57,742 in 1969. Nation-wide, over half a million cars were abandoned in 1967.

The motor vehicle has a voracious appetite, not only for the many finite resources used in its fabrication but also for land and fuel. In 1930 there were slightly more than three million miles of rural and municipal roads and 193,000 miles of the federal-aid highway system. By 1966 the respective mileages were up to 3,698,000 and 911,000—for a total highway system of 4,609,000 miles.

The registration curve of Figure 14 and the highway figures reveal the cause of traffic congestion. In 1940 there were approximately ten vehicles per mile of highway. In 1950 the ratio had increased to 12.5 and in 1960 to 16.7. By 1966 it had risen to nearly twenty vehicles per mile of highway—a doubling of vehicle density in twenty-five years. By extrapolation, we can expect a density of thirty vehicles per highway mile by 1980. Unfortunately, the total number of vehicles is not spread out statistically along the total length of available highways but is highly concentrated in urban areas. And the great bulk of vehicles are constrained to use the same small portions of highways at the same times. The result is the familiar transformation of expressways to, in effect, gigantic parking lots, and the growing trend to immobility in transportation.

The highway planners and designers are familiar with these figures and trends. But their answer is more expressways—a quasi-solution doomed to fail and one that is generating residue problems that are helping to destroy the city and its neighborhoods. The logic and the mathematics of the planners are impeccable. In the city of Chicago, for example, it has been found that in an optimum expressway pattern, expressways should be located at approximately four-mile intervals. The planners have neglected, however, to examine their assumptions. If their assumptions follow the rise of automobile registration, and they seek to expand the highways to meet this exponential growth of vehicles, their mathematics will reduce the optimal distance between expressways from four to three miles, then to two miles, and so on, till the cities are eventually paved over with concrete.

As the motor vehicle has become a world-wide phenomenon, the world registration of vehicles has jumped from seventy million in 1950 to 190 million in 1966—an increase of 170 per cent.[2] In the same period, United States registrations increased at only half the world rate, or 86 per cent.

An important secondary effect of the development of the motor vehicle is the highway slaughter attendant upon the masses of vehicles competing for space in machines that, because of malfunction or carelessness, become weapons upon impact. In the United States, 34,763 persons were killed in auto accidents in 1950, while 1,800,000 were injured. By 1960 the number of dead had risen to 38,137 and the injured had increased to 3,078,000.[3] In 1969, 56,400 persons died and 4,600,000 were injured in more than twenty million accidents.

Although the traffic death rate per 100,000 population rose from 0.4 in 1904 to 26.2 in 1940, the exponential growth was halted in 1950 when the rate fell to 23.0 and then fell again in 1960 to 21.2.[4] The death rate per 10,000 autos stood at 7.1 in 1950, fell to 5.1 in 1960, and rose to 5.5 in 1966. Per 100 million vehicle miles

[2] *Ibid.*, Table 821, p. 552.
[3] *Ibid.*, Table 835, p. 560.
[4] Rates from 1904–1940 from Series B126, *Historical Statistics of the United States, Colonial Times to 1957*, p. 25; other rates from Table 836, *Statistical Abstract of the United States, 1968*, p. 560.

the respective rates were 7.6, 5.3, and 5.7.[5] In absolute figures the number of accidents, injuries, and deaths is increasing with the increase in vehicles, although at some level they may become asymptotic.

While consuming land and material resources, the motor vehicle is also devouring fossil fuels at an ever-increasing rate as shown in Table 1. In 1940 32.5 million registered vehicles were

TABLE 1
Vehicle Miles and Fuel Consumption in the United States

Year	Vehicle miles (millions)[1]	Motor fuel consumption (millions of gallons)[2]	Average gallons per vehicle[2]		Average mileage per gallon[2]	
			Passenger	All	Passenger	All
1940	302,188	24,404	594	678	15.29	13.79
1950	458,246	40,280	603	728	14.95	12.87
1960	718,845	63,714	661	777	14.28	12.42
1965	887,812	71,104	656	775	14.15	12.49

SOURCES: [1] U.S. Bureau of the Census, *Statistical Abstract of the United States, 1968* (Washington: Government Printing Office, 1969), Table 816, p. 550.
[2] *Ibid.*, Table 817, p. 550.

driven 302 billion miles. By 1965, 90.3 million vehicles were driven 888 billion miles. During this period, motor fuel consumption rose from 24.4 billion gallons to 71.1 billion gallons. The ratio of increase in each category was approximately three times, and the two ratios were essentially equal. The average gallons per vehicle, however, has tended to increase—rising from 678 to 775 gallons per vehicle in the twenty-five-year period. The resultant demand for gasoline has helped trigger the exponential growth curve of production of crude oil. In spite of warnings about the impending total depletion of oil resources, Table 1 shows a declining average mileage of all vehicles, and especially passenger vehicles. Thus, waste and profligateness are increasing as resources diminish.

On another branch of the chain of residue problems generated

[5] *Ibid.*, Table 836, p. 560.

by the development of the motor vehicle are the problems involved in the exploration, extraction, and transportation of oil. These involve not only technical problems but the complex techno-social problems of geopolitics, imperialism, and, ultimately, war. The search for oil continues man's assault against nature as the oceans of the earth are exploited—regardless of ecological effects upon the organisms of the water. Within the 1970's, 35 per cent of the world's increasing demands for oil will be met by offshore production. Oil has been found in the North Sea off of Norway, in the Arabian Gulf near Iran, and off the coasts of Nigeria and Angola. Oil is also believed to lie in sedimentary formations in the northern Bering Sea. Major oil companies are meanwhile searching the sea beds off a score of other countries in a frenzied quest for more oil and gas.

Offshore oil production poses a growing threat of pollution of the oceans at the very time that the world looks to the oceans as a growing source of food. The dangers of offshore oil production merge with those arising from the increasing transport of oil on the oceans as the industrialized nations increase their oil imports from Third World countries. With consumption of petroleum increasing at a rate of over 7 per cent per year, the transportation of crude oil is becoming a major industry. Tankers in World War II carried 16,000 tons, while 1960 tankers carried 100,000 tons. By 1968 a tanker with a capacity of 312,000 tons was built and another with a capacity of 370,000 tons was being planned in response to the growing demand for seaborne oil transport. Even as the incidence of oil spills was growing with concomitant increases in ocean and beach pollution, the major design and construction changes in the tankers were generally increasing the potential pollution hazards that would result from collision and stranding. Simplification of piping and valves provided less flexibility in controlling the cargo, while designing for increased drafts increased the probability of grounding. The elimination of wing bulkheads and scantlings resulted in less energy-absorbing structures and simplification of machinery; the elimination of stripping systems in favor of free-flow pumping systems reduced the capability to cope with emergency situations.⁶ These various quasi-solutions guaranteed innumerable residue problems.

The spilling of thirty million gallons of oil by the tanker *Torrey Canyon* off the coast of England in 1967 was the most notorious of a succession of oil spills that are becoming more common and more dangerous as oil traffic grows. The *Torrey Canyon* spill contaminated 140 miles of English coast as well as stretches of the Brittany coast one hundred miles from the spill. At Delaware Beach over one million gallons of heavy fuel oil were beached in 1968. Sixty thousand gallons of diesel oil were spilled into Humboldt Bay in California. The release of eighteen million gallons of oil from an offshore oil well near Santa Barbara, California, ruined miles of beaches and adversely affected marine biology. These spills were publicized disasters that tend to mask the nearly ten thousand spillings of polluted materials that are estimated to occur annually by the U.S. Coast Guard, one-third of which involve oil.

The growing menace of oil spills was evident in February 1970 when in a period of sixteen days four "accidents" occurred at various locations on the North American continent. On February 4 the tanker *Arrow*, bringing more than 3.8 million gallons of oil from Venezuela to a Nova Scotia pulp mill, ran aground and began spilling oil into remote Chedabucto Bay. On February 10 a fire started on an oil platform over the Gulf of Mexico about fifty miles southeast of New Orleans, and was fed by crude oil and gas spouting from wells drilled into the seabed. On February 13 the tanker *Delian Apolon* spilled more than ten thousand gallons of heavy oil into Tampa Bay when it ran aground; an oil slick covering more than one hundred square miles was created before it began washing ashore, blanketing beaches with a thick covering of oil and trapping thousands of pelicans, gulls, grebes, and other shore birds. And on February 20 a barge carrying 2.5 million gallons of gasoline and diesel fuel collided with a jetty in Humboldt Bay on the California coast and spilled 84,000 gallons into a flammable slick. Per-

---

⁶ *Oil Spillage Literature, Search and Critical Evaluation for Selection of Promising Techniques to Control and Prevent Damage* (Richland, Washington: Battelle Memorial Institute, 1967), pp. 2–3, 2–4.

haps the most ironical oil-gasoline incident of this kind is the reported calling in of a fire department to extinguish a blaze on the polluted Cuyahoga River near Cleveland, which carries its grimy effluvia into the cesspool that once was Lake Erie.

Most oil spills occur near land and are harmful to plant life, shellfish, fish, shore birds, and other animals. The British auk has dangerously declined in population because of oil spills; more than twenty thousand guillemots were estimated to have died in the *Torrey Canyon* disaster. More serious is the possibility that petroleum substances will enter into the oceanic food chain, whence eventually they will be passed on to humans who obtain food from the oceans. A hitherto unsuspected effect for which evidence is mounting is that blobs of oil can be carcinogenic and are entering the cycle of marine life.

The technological response to oil spills involves the familiar set of quasi-solutions with their residue problems and the continually proliferating residue chain. Some chemists believe, for example, that chemical dispersants merely hide oil spills and pose an additional danger. Some of the chemicals are believed to be more harmful to marine life than the oil they were intended to disperse. The breakup of the oil into tiny particles makes the oil more readily available to marine organisms, and so aids in the contamination of ocean food. Further, the application of dispersants to beaches makes the oil penetrate more deeply, with the result that the beaches become more vulnerable to erosion by wave action. Thus, in the *Torrey Canyon* spill some scientists have contended that the dispersants did more harm than the oil. To cap a humiliating situation in which man's ingenuity has been found wanting, it has been admitted that the most effective technique used to clean up the Santa Barbara oil spillage was to spread straw as an absorbent and to collect the straw with rakes and pitchforks.

The Gulf of Mexico and Chedabucto Bay accidents are powerful reminders of the helplessness of man in the face of adverse conditions and the irrelevance of much of his technology. The Gulf of Mexico accident unfolded like a Greek tragedy. Oil wells drilled into the seabed fifty miles southeast of New Orleans began burning from an undetermined cause on February 10. The raging fire was fed by oil and spouting gas. Acknowledging that once the fire was put out oil would gush from the wells until they were capped, oil company officials assured authorities that millions of dollars of equipment and technical ingenuity would contain the danger of oil contamination. A string of barges was established to create a boom to block the oil from endangering several wildlife refuges and the rich Gulf shrimp and oyster beds. A flotilla of boats with skimmers and vacuum machines were ready to scoop up spilled oil. Held in abeyance were twenty thousand bales of straw to soak up the oil in the improbable circumstance that it would be washed ashore.

After the fire had been extinguished with the help of dynamite on March 10, oil began to pour from the wells and to form a heavy oil slick. On the same day, the National Wildlife Refuge on Breton Island was menaced when an oil-collecting boom broke. The clean-up was reported to be "going well" as the boom of heavy mesh fence covered with vinyl was repaired—only to break again. On March 11 the vinyl and plywood dams collapsed in heavy seas and over 1,500 barrels of crude oil began to move toward the oyster beds. The skimmer boats could not operate because of wind and high seas. On March 12 the incident was officially termed a "disaster" as oil slicks covering fifty square miles of the Gulf neared the oyster beds. If necessary, it was planned to set off fireworks to startle a quarter-million geese to begin an earlier migration northward. On March 13 officials considered setting the oil on fire. An oil slick moved into the marshes of a wildlife refuge the next day while officials scanned the wind notices to determine the course of the oil slicks. A well head used to cap a spouting well blew off on March 15, and the escaping oil added to the fifty-two-square-mile slick.

Faced with a growing oil slick, the oil well's owners smothered the spouting wells with tons of mud and dynamite. They poured dispersant chemicals on the slicks, though the effects of these chemicals on the marine life threatened by the oil had not been established. It is known, for example, that dispersants act to sediment the oil, and sedimented oil acts as a partitioner and concentrates substances from the water. Thus DDT, which is highly soluble in oil, can become concentrated to reach insecticidal application levels.[7] The ecological effects of this offshore oil production are unrevealed, if they are known.

The Chedabucto Bay spill transformed the bay into a cold-water laboratory—with primitive measures taking precedence over scientific ones. Efforts were made to burn the spilled oil, but low sea temperatures frustrated ignition efforts with benzine, magnesium, and flame-throwers. Old tires filled with napalm burned doughnut-shaped holes in the congealed oil and sank to the bottom. Chemical dispersants were halted by the government as being harmful to marine life. As at Santa Barbara, sawdust and peat moss were used to soak up the oil on the beaches, and bulldozers scraped up the contamination.

While avarice and ignorance threaten the oceans and waterways of the world, the same forces now threaten the last remnants of the earth that till today have escaped exploitation by man. The oil find on the North Slope of Alaska and the oil prospecting now going on in the Canadian Arctic regions pose a threat to the frozen tundra and its inhabitants, both animals and men. Food chains are short and uncomplicated in these regions whose ecosystems are relatively simple. As a consequence, they are subject to fast and devastating disruption.

The plan to build roads through the permafrost region, and to move hot oil through pipes—some on raised platforms, some buried in the earth—has raised a storm of protest over this newest assault against nature. The hot pipeline, it is believed, could cause intense thawing, thus jeopardizing the stability of the pipeline. The movement of ice and the threat of earthquakes add to the fear of huge oil spills in an environment that would be more difficult to handle than that of Chedabucto Bay. The exploitation of this oil is a vast experiment which may mean profits for some in the short run and catastrophe for a region in the long run. Writing

---

[7] Rolf Hartung and Gwendolyn W. Klingler, "Concentration of DDT by Sedimented Polluting Oils," *Environmental Science and Technology,* IV (May 1970), 407–410.

about the oil rush in Alaska, a scientist, while calling for "rational exploitation without ruination," conceded that "when rape is inevitable, relax and enjoy it."[8]

Finally, in this sample of residue problems arising from the quasi-solution of transportation, we come to the problem of air pollution, perhaps transportation's major secondary effect. The emergence of the problem illustrates the dialectical transformation of quantity into quality, for pollutants were produced by the first automobiles, though of course in smaller quantities per auto than in today's autos. Until the pollution from tens of millions of vehicles accumulated to reach a given level, pollution was not perceived as a social and health problem. Having reached this level, pollution has become a problem in its own right.

Autos produce many waste products as a result of the combustion of motor fuel. Among them are hydrocarbons, carbon monoxide, nitrogen oxides, and lead. More than ninety million tons of pollutants are poured into the atmosphere each year by motor vehicles—a tonnage which increases with the growth in the number of vehicles. Carbon monoxide, the unseen and unsmelled killer in closed quarters where an auto engine operates, is the largest waste component. At incidence levels as low as ten parts per million (ppm), carbon monoxide begins to affect oxygen transport in the blood adversely by the formation of carboxyhemoglobin. At thirty ppm, the level of carboxyhemoglobin rises to over 5 per cent, resulting in a loss of energy and decline in people's mental and physical reactions. In cities, carbon monoxide levels of fifty ppm have been reached, and in traffic jams levels have risen to 140 ppm.

The lead from leaded gasoline, vented into the atmosphere as part of the waste product of vehicles, also enters the human bloodstream. With lead poisoning occurring at levels of from 800 to 1,000 ppm, measurements of 100 to 300 ppm have been found in urban residents.

Hydrocarbons mix with other pollutants in the air to form photochemical smog. Nitrogen dioxide is toxic to man and animals, and can cause death and chronic respiratory diseases in humans at levels less than five ppm. Consequently, pollution alerts in cities

[8] F. F. Wright, letter in *Science*, CLXVI (December 5, 1969), 1220–1221.

have become part of the urban picture in an ever-increasing number of cities, where the population is warned to exercise care and attention to their health, and temporary restrictions are placed on pollution-producing activities.

The convergence of these pollutants poses an immediate and direct threat to continued life in large cities, and some believe that mass deaths will be experienced in many cities throughout the world within the next decade. Already during pollution alerts people with respiratory and heart conditions are warned not to leave their homes.

These residue problems, like others, propel the technological machine all the faster, and thus technologists are hard at work to provide more quasi-solutions to them. Except in the case of the internal combustion engine that drives the motor vehicle, the convergence of the many proliferating chains of problems portends to foreclose a further quasi-solution.

Let us proceed backward up the residue chain of problems to view attempted solutions, starting with pollution. Anti-pollution devices were made mandatory on all autos manufactured after 1966, with increasing levels of control being required in subsequent years. These devices are usually attachments to either the crankcase or the tailpipe and are intended to produce more complete combustion which will filter out pollutants. Each solution, however, turns out to be a quasi-solution generating as many problems as it solves. The combustion of fuels at higher temperatures, for example, to reduce carbon monoxide and hydrocarbon wastes, increases nitrogen oxide emissions. The elimination of lead requires the development of a new type of gasoline and changes in the motor and carburetor systems. Exhaust control devices are designed to operate at moving speeds of fifty miles per hour—speeds unattainable in many cities where long idling periods and bumper-to-bumper traffic characterize what is termed transportation.

It would be necessary, in part, to undo many of the achievements of decades of automotive technology to reduce pollution, for over the past forty years nearly every improvement has contributed to pollution. Consequently, nearly every method for reducing pollution would also seriously impair performance and increase cost.

There is a further caveat to the problem of reducing auto pollution, arising from the cancellation effects of different sets of residue problems. Assume, for example, that a 1970 auto produces, on an average, one ton of pollutants each year. At added cost and diminished performance, let us further assume that anti-pollution devices attached to an auto decrease pollution by 50 per cent. An auto would then produce about one thousand pounds of pollutants per year. A doubling of the number of autos would restore total pollution to the original level. Further reductions in per auto pollution would again be wiped out by an increase in the number of autos.

The above illustration indicates another reason why technology is always expanding and refining its processes. The technology developed for one quantity is rendered obsolete by a larger quantity. The quantitative aspects of pollution are transformed qualitatively when a set of residue problems within the framework of sufficiently large quantities effectively negates technology to preclude a techno-social solution.

At a time when all pollution will have been eradicated, the motor vehicle as we know it would no longer exist; it would have succumbed to the impotence laws of physics. Long before this time, however, the exponential increase in the number of motor vehicles will have been brought to a halt by depleted resources, oil among them. The increase in number and the continued expansion of the highway network may bring about a decrease earlier, either through voluntary action or through unplanned chaos. In any case, there are no techno-social solutions to the proliferating chains of residue problems generated by the development of the privately driven motor vehicle.

The invasion of the auto is generating a familiar set of residue problems around the world: congestion leading to immobility, highways competing against other imperative land needs, rising death rates (currently running at 200,000 per year), increasing demands upon fuel reserves, and, finally, world-wide air and ocean pollution.

Waste and Waste and Waste

Every technological process produces waste as a necessary and normal product of its operations. Slag heaps and mine tailings are natural by-products of mining; manure is a natural by-product of meat and milk production; sugar cane trash and cereal straw and stubble are natural leftovers from agricultural operations; tree prunings are a corollary of fruit growing; metal shavings, chips, and powdered metal are necessary wastes produced by machining of metals; scrap materials and damaged or out-of-tolerance components are legitimate wastes produced by manufacturing processes; every chemical process produces waste by-products.

In pursuit of efficiency, technology always attempts to reduce waste products by improving the process. The improvement can be the reduction of scrap in a fabricating process, the increased use of raw materials, or the transformation of waste products into useful products. Some processes are not amenable to the first two types of improvement, which can be achieved only through transformations. Increased meat production will be accompanied by an increase in manure; increased yields of grains will result in increased quantities of straw and stubble. Increased mining will increase mine tailings. Their reduction through transformation into useful products, like the original processes, are quasi-solutions that generate new chains of residue problems.

Another type of waste is a by-product of man's living in social units. Garbage, sewage, discarded objects, broken and damaged materials, and all the other detritus of a technological civilization fall into this second category. These wastes are not amenable to efficiency calculations, and their production cannot be directly controlled by technological processes. Technology can only deal with their collection and disposal. Part of the disposal process can be the transformation of the waste for easier or less costly disposal, less harmful effects, or for reclamation and reuse. The transformation processes are again quasi-solutions that establish still other chains of residue problems.

In a highly developed technological society the accumulation of the two categories of waste becomes waste in a third sense of the word: unneeded and lavish. Waste becomes a squandering of scarce resources, a blindness to the needs of the entire community of man now and in the future, and a profligate approach to the natural world. Technological society becomes a junkheap society where planned obsolescence and a replace-and-throw-away attitude suffuse all aspects of the society, starting with production of commodities and ending with consumption of commodities. Production for waste assumes an ever-increasing role in maintaining an expanding economy in the technological society. The service industries serve as the primary vehicle for an economy based upon waste and become more and more parasitic on essential production processes. In 1966, for example, more than $16 billion worth of materials were used in packaging in the United States, of which 90 per cent was intended to be thrown away. The degradability and combustibility of waste have diminished as its volume has increased.

Waste products consume space. They can be scattered or centralized in specified areas on the land; they can be discharged into streams and oceans; they can be dispersed in the air. Technology can move them, disperse or concentrate them, or transform them, but cannot command them to disappear. In a closed system, every technological process that operates on augmentation or secondary-effect problems in connection with waste generates new sets of residue problems through quasi-solutions. We have discussed waste problems encountered in quasi-solutions of food production problems in connection with fertilizers and biocides. We have noted the effects of detergent wastes and the deleterious consequences of fossil fuel utilization in energy production and automobile vehicles. Some of the residue problems arising from an increasing production of nuclear energy have been reviewed. In this section, other residue problems arising from the wastes generated by a technological society are investigated.

Commercial and industrial wastes can be solid, liquid, or gas. Although pollution from wastes is most noticeable in its effects on man in liquid or gaseous form, solid wastes are fast becoming a problem in their own right due to sheer volume. Combining solid wastes from individuals and communities, it is estimated that almost 1.5 billion pounds of urban solid waste are produced in the United States each day. In 1967, considering all sources of solid waste, one hundred pounds were produced per capita daily for a total of 3.6 billion tons for the year. Of this total, 299 million tons were produced by domestic, municipal, and commercial sources, 110 million tons by industry, 2.1 billion tons by agriculture, and 1.1 billion tons by mineral operations. An increase to five billion tons is estimated by 1980. Of the ten pounds of urban and industrial wastes produced per capita daily, it is estimated that only half is collected, the other half being discarded as litter or burned to produce ash and gaseous pollutants.[9]

As the world seeks to increase food production, it is sobering to consider the wastes that are generated as by-products: animal manures, vineyard and orchard prunings, crop-harvesting residues, animal carcasses, greenhouse wastes, and pesticide containers. Ten thousand cattle can produce 260 tons of manure per day, and four hundred milk cows can produce about fourteen tons of solid waste daily. Five tons of waste are produced daily by 100,000 chickens. Over a billion tons of fecal wastes and over 400 million tons of liquid wastes were produced in 1967 by domestic animals in the United States alone. Used bedding and dead carcasses raised the total production of animal wastes to close to two billion tons.[10]

Major agricultural crop wastes in the United States in 1967 amounted to more than one-half billion tons, exclusive of an estimated 25 million tons of logging debris left in forests. Over fifteen pounds of agricultural waste were produced daily per capita. The leaves and stalks from 57 million acres of corn were estimated to total 256 million tons, or 4.5 tons per year for every individual in the country. Wheat stubble amounted to 65 million tons on 50 million acres. Nearly 4.5 million tons of potato waste were generated from 1.5 million acres. Over 10 million tons of vegetable waste were produced on 3.6 million acres.[11]

Waste in a technological society is generated in the process of mining or producing raw materials, and again in the various tech-

[9] *Solid Waste Management*, pp. 7, 15.
[10] *Ibid.*, p. 30 and Table 11, p. 33.
[11] *Ibid.*, p. 30 and Table 10, p. 32.

nological processes that use these materials. More than twenty billion tons of solid mineral wastes have been generated in the United States by the mineral and fossil fuel mining, milling, and processing industries. Current annual production is 1.1 billion tons —and is expected to rise to two billion tons by 1980. If ocean and oil shale mining become commercial enterprises, as is anticipated, the annual volume of waste from mining will double to four billion tons.

Eight major industries contribute 80 per cent of the total mineral waste. Nearly 100 million tons of coal wastes were produced in 1965, 233 million tons from iron and copper mining.[12] These and other wastes from other sources are all expected to increase annually—until depletion of the major sources.

What does a technological society do with these vast quantities of waste? A government report reveals the simple answer: "Prior to 1965, an estimated 5 million acres—equivalent to about 7,000 square miles—were either covered with unsightly material and solid fuel mine and processing waste, or the land was so devastated by current and past operations as to be not only useless and ugly, but in some cases a hazard to human life and property." The report describes the mineral wastes in the following manner: "Mineral mine wastes are, for the most part, barren overburden or submarginal grade ore from open pit or surface mining. . . . These mountains of waste, often hundreds of feet high and covering extensive land areas, accumulate over the landscape adjacent to the mining operations. Large tonnages of solid waste . . . have been disposed of in remote or sparsely populated areas. Equally large tonnages, however, . . . have been deposited in or near populated or frequently visited areas. . . . The possibility of another rock and mud slide from such towering piles, such as that which occurred in Aberfan, Wales, is an ever-present threat to life and property in the vicinity of these banks. Wind erosion of the banks contaminates the nearby areas with dust. When the banks are ignited, as frequently occurs, they pollute the air with noxious fumes which destroy the surrounding vegetation, corrode domestic and commercial structures, and cause accident-provoking smog."

[12] *Ibid.*, pp. 30, 34, and Table 12, p. 35.

The metallurgical and chemical processing of mineral concentrates also generates numerous types of solid wastes. "Some of the wastes," the report continues, "are toxic and present serious local air, water, and land pollution problems. For example, each year about 1.7 million tons of [red muds generated in preparing alumina from bauxite] are discharged directly into the Mississippi River by two plants in Louisiana, and an estimated 3.6 million more tons are stored in waste ponds in other parts of the nation."[13]

To these man-made mountains of mineral waste will soon be added the man-made mountains of fly ash and other pollutants that it is hoped will be removed at one point in the technological process, but which will appear as new problems at other points. What can be done with the scars and blemishes man has inflicted on the earth? These mountains contain no nutrients and have poor soil texture. Efforts to stabilize them are generally incompatible with efforts to cover them with vegetation. They can be moved—but to where in a world in which space is becoming more elusive? Iron ore tailings in the Lake Superior region have been dumped into the lake for the last thirteen years at the rate of sixty thousand tons daily. More than a third of the material is pulverized and dispersed in the lake; the solids have piled up in an artificial peninsula that currently extends a half-mile into the lake. The discarded waste contains an estimated ton of nickel, a ton of zinc, 3 tons of lead, 4.5 tons of chromium, 30 tons of phosphorus, and 372 tons of manganese daily.[14] Thus as one scarce resource is depleted, other equally scarce resources are squandered. The quasi-solution of ultimate recovery of all materials from the waste will lead to additional energy requirements which will in turn create different forms of pollution and waste. Technologically, material can be separated at concentrations of several parts per million, as has been demonstrated with uranium. The set of residue problems generated by this mode of recovery, however, will inevitably bring new sets of complications.

Wastes are produced by every industry at ever-increasing levels.

---

[13] *Ibid.*, pp. 37, 34, 36.
[14] Gladwin Hill, "Lake Superior, Private Dump," *Nation*, CCIX (June 23, 1969), 795–796.

In 1965 the total industrial waste in the United States was estimated at 115 million tons. The paper industry produced over 30 million tons; industrial scrap metal was produced at a rate of 15 million tons; 56,000 tons of waste were generated by chemical industries, and 1.5 million tons by the rubber industry.[15]

The food-processing industry, like its counterpart in agriculture, increases its output of waste together with its output of food. Of the total weight of the corn crop used in canning, for example, 50 per cent is estimated to be field waste, 30 per cent process waste, and less than 20 per cent is marketable product. Changes in the technology of mechanical harvesting are shifting the location of a greater proportion of the waste from the field to the cannery. Solid wastes produced as by-products from forest and fiber crops are more difficult to process than those from other crops.[16]

Nearly two thousand establishments produce more than seven thousand commercial chemical products in the chemical industry and generate uncounted solid, liquid, and gaseous wastes, all of which must be disposed of in some way. The places of disposal are limited again to the earth, the water, or the air. The melange of mixtures, soups, and gaseous effluvia that are produced by these wastes in combinations with wastes from other sources is extremely large.

In Sweden there is a growing concern about lakes that are being contaminated from mercury from crop seed coatings and the pulp industry. Canada has recently banned the sale of fish from one of its lakes for the same reason. Traces of fluorine have been detected from decomposition products of a billion hair spray cans that are used annually in Europe. In Tokyo deaths have been reported from mercury and cadmium poisoning traced to industrial wastes, and cyanogen, a cyanide, has been found in the city's rivers. As noted earlier, cyanide, 2,4-D, salt, and gasoline have appeared in drinking wells of the United States.

The wastes discussed thus far are by-products of technological processes in mining and industry. To these wastes must be added those consumer wastes produced by society in the form of garbage

---

[15] *Solid Waste Management*, pp. 19–25.
[16] *Ibid.*, pp. 25, 27.

and sewage. Total domestic consumer wastes amounted to 128 million tons in 1967, or 3.5 pounds per capita daily.[17] In a year, the average American discards 188 pounds of paper, 250 metal cans, 135 bottles and jars, 338 caps and crowns, plus a considerable amount of miscellaneous packaging.[18] The total quantity of packaging material discarded in 1966 was more than 46 million tons. Per individual this was 121 pounds greater than in 1958, and the quantity is expected to increase by 136 more pounds per capita by 1976. The production of bottles and cans grows exponentially, with 10 billion produced in the United States in 1958, 25 billion in 1966, and 58 billion estimated by 1976.[19]

Waste-disposal techniques are generally similar to those used by the ancient Romans: open dumps, land fills, incinerators or composting plants, or ocean dumping. Each technique is a quasi-solution that is becoming less feasible and less desirable. The open dumps are unsightly, unhygienic, and the land is becoming less available. Sites for land fill are becoming unavailable as competing demands make their inroads. A survey of six thousand existing land fills, for example, of which nearly 80 per cent were operated by public agencies, revealed that less than 6 per cent were truly "sanitary" land fills, since the land was not filled after dumping.[20] Incineration produces its own wastes in the form of ash and gaseous pollutants. The effect of the use of streams and lakes as disposal sites has already been noted. Used as the world's dump, the oceans have also been seriously affected by the thousands of chemical products that find their way to these bodies of waters.

The primary requirement of any disposal method is that a pollutant from one medium be prevented from causing pollution elsewhere. The realization of this requirement is prevented, however, by the impotence principle of the indestructibility of matter. Depositing iron ore tailings in Lake Superior solves the disposal problem at the mine by relocating the problem in the lake. Incin-

[17] *Ibid.*, p. 7.
[18] Charles A. Schweighauser, "The Garbage Explosion," *Nation*, CCIX (September 22, 1969), 282–284.
[19] Arsen J. Dornay, Jr., "Throwaway Packages—A Mixed Blessing," *Environmental Science and Technology*, III (April 1969), 328–333.
[20] *Solid Waste Management*, p. 59.

eration of garbage or sewage produces particulate ash. Wet scrubbers can remove some of the ash by producing water-ash slurry which, in turn, must be disposed of, or dry precipitators produce mountains of ash. Corn stubble can be burned but produces tons of carbon and ash. The stubble can be buried, but the danger of perpetuating plant diseases and insects remains. Pesticides and other chemicals can be used against bacteria and insects but these, too, have their own countereffects when they remain in the soil or run off into streams.

9 *Technological Realism 4: War*

*O*URS IS A generation that may not have a future. This is the stark fact the younger generation perceives and the older generation denies. Although warfare has been a constant of all human societies from prehistoric times to the present, only our generation has the capability to destroy all life on the planet. As Bertrand Russell has remarked, previous generations were prevented from destroying the world because of their ignorance and incompetence. We are not wise, but through science and technology we have become sufficiently competent to add man to the 99 per cent of all living creatures that have become extinct in the course of life on earth.

War represents the epitome of the scientific-technological genius laboring in a meaningless universe, unrestrained by values or morals. Death and destruction, massive waste of resources, irretrievable expenditure—these are technical objectives to which science and technology have lavishly applied themselves. The technology of modern weapons demonstrates, as has no other facet of technological society, the unsolvability of the major techno-social problem of our era—peace—by science and technology. Scientific research has undermined the very existence of the nation-state by destroying its protection in space and time. Military leaders can no longer plan on "successful" wars, and political leaders can no longer protect their citizens from mass destruction. National security, however, is still proclaimed as an absolute value, taking precedence over all other social values.

In their lucid moments, statesmen and generals bluntly acknowledge that there can be no victors in a future war. Douglas MacArthur stated that "global war has become a Frankenstein to destroy both sides. If you lose you are annihilated. If you win, you stand only to lose. No longer does it possess even the chance of the winner of a duel. It contains only the germs of double suicide."[1] President Dwight D. Eisenhower commented after the collapse of the ill-fated 1960 summit meeting: "All of us know that, whether started deliberately or accidentally, global war would leave civilization in a shambles. In a nuclear war there can be no victors—only losers."[2] Nikita Khrushchev, then Premier of the Union of Soviet Socialist Republics, showed his awareness of the consequences of nuclear war: "Only madmen and lunatics can now call for another world war. As for the men of sound mind—and they account for the majority even among the most deadly enemies of communism—they cannot but be aware of the fatal consequences of another war."[3] This truth was also evident to President John F. Kennedy, who delineated the alternatives clearly in his address to the Sixteenth General Assembly of the United Nations: "The weapons of war must be abolished before they abolish us."[4]

In spite of universal awareness of the consequences of large-scale war, the generals and statesmen of the world prepare for war and express belief in victory. Military planning, defense, deterrence, and the problems associated with push-button, missile-age warfare are evolved almost entirely on the basis of the technological alternatives offered by scientists and technicians. Political, economic, and social goals become secondary to the implications of technological developments. The nuclear-tipped intercontinental

[1] Commencement speech at Michigan State University, quoted in *Chicago Sun-Times*, June 12, 1961.
[2] *New York Times*, May 26, 1960.
[3] *New York Times*, June 22, 1960.
[4] *New York Times*, September 26, 1961.

missile, the earth-orbiting satellite, and the sea-lurking submarine today have a compulsive logic of their own which dictates a nation's policy toward war.

After President Lyndon B. Johnson, who campaigned in 1964 on a promise to keep American youth from fighting and dying on Asian soil, embroiled the nation in what has proved to be its longest and second most costly war in terms of casualties, technology was called upon to devise new techniques of counterinsurgency and guerrilla warfare. The war in Vietnam has pitted the technological prowess of the world's foremost technological society against the struggle of a small nation to achieve its freedom and domination from all the large powers. The technology developed for and tested on the people and land of Vietnam has provided a blueprint for the United States' attempt to retain hegemony over the nations whose raw materials are essential to the maintenance of an affluent and expanding American society.

The Vietnam War has also disclosed the full power of what President Eisenhower called the military-industrial complex, which has mobilized the nation's wealth and resources for war and enlisted the enthusiastic support of the universities and the nation's scientific establishment in the quest for a more sophisticated military technology. To conduct the war in Southeast Asia and be prepared to engage simultaneously in other wars, the United States is utilizing more than 25 per cent of the nation's scientists and engineers in war-related research, development, and production, and employing nearly 10 per cent of the labor force in war-related activity. The technological boom of the past three decades, and the growing prosperity of this island among nations, has been fueled, in large measure, by the wars that have been fought, are being fought, and are being planned for the future.

Although faced by diminishing resources, threats of mass famine, mass poverty, and the world-wide menace to the earth's environment, the nations of the world spent $200 billion in 1969 on wars and preparation for wars—an increase of 44 per cent since 1964. In the six-year period 1964–1969, more than $1 trillion was expended for arms and armed forces. This sum exceeds two years' income of ninety-three Third World countries with a combined population of more than 2.5 billion people. "Larger than any civilian programs financed by public funds, the world's military budget in this period took as much public money as was spent by all governments on all forms of public education and health care," the U.S. Arms Control and Disarmament Agency reports. The United States spent $82 billion in 1969 compared with $56 billion by the USSR. In the past two years the greatest rise in military expenditures has been made by those nations which could least afford an arms race, the Third World nations.[5]

While the world's standard of living in real terms, discounting inflation, has improved relatively little during the past six years, "the diversion of resources to military purposes has expanded in step with the world's capacity to produce," the Arms Control and Disarmament Agency observes. The Gross National Product in the Third World countries rose at an annual average of 2 per cent during this period, while arms expenditures rose an average of 7 per cent per year. As a consequence, "part of the growth dividend since 1964 has been dissipated in higher military expenditures, rather than contributing to the improvement of living standards."[6] Nearly 22 million men were estimated to be in the world's armed forces in 1967, with another 33 million engaged in military manufacturing and other supporting work.[7]

The Vietnam War is but a prelude to the ultimate drama of nuclear war that hangs over all civilization and biological life. "Overkill" and "megadeath" are terms that attempt to describe the technological capability to kill, several times, every man, woman, and child on earth. Both the United States and the USSR have over a thousand nuclear missiles in readiness and aimed at "enemy" targets. The United States has more than two thousand additional missiles in nuclear submarines and a fleet of over 1,800 long-range bombers. Although it can already destroy any possible "enemy," plans call for a doubling and tripling of this arsenal.

The paradox of maintaining peace by preparing for war is plau-

---

[5] U.S. Arms Control and Disarmament Agency, *World Military Expenditures 1969* (Washington, D.C.: Government Printing Office, 1969), pp. 1–2.
[6] *Ibid.*, p. 3.
[7] *Ibid.*, p. 8.

sible for one nation but degenerates into nonsense when practiced by two nations. Lewis F. Richardson explains the paradox: "Common sense might suggest, and bitter experience confirms, that governments which develop their armaments cannot expect their professions of peaceful intentions to be believed, however sincere they really may be. There is, therefore, a dilemma: preparations for victory in defense are incompatible with the continuance of peace."[8] Consequently, the more peaceful one nation attempts to become by acquiring new weapons, the more warlike it appears to its neighbor and, by a dreadful symmetry, the peaceful preparations of the neighbor are, in turn, viewed as hostile by the first nation. Thus weapons, counterweapons, and counter-counterweapons escalate in an insane arms race. Nations become victims of their own weapons systems and technology. The supreme irony is that, in a manner that completely destroys their freedom and independence, as Richardson notes, the actions of nominally sovereign and independent nations are really controlled by their neighbors.

The arms race is also stoked because "the motives of defense and attack easily cohabit in the same mind," Richardson observes.[9] This is well demonstrated in the advertisements of Doomsday that appear in technicolor in the nation's press as industry extols its scientific prowess. The systems manager of the Polaris missile proclaims in a two-page ad, "Today it's a Polaris world," and quotes President Eisenhower's characterization of the weapon as "a revolutionary and practically invulnerable ballistic missile system." But another company belittles the idea of the "ultimate weapon": "The ICBM is often called the 'ultimate weapon.' Yet, throughout history, there have been many so-called ultimate weapons. Men of science have always found a defense." One ad describes the operation of Polaris: "Concealed by ocean depths, these ever moving missile bases cannot be accounted for by an enemy planning a surprise attack." But another vista-colored ad informs the reader that a "radical new Passive Underwater System apprehends subma-

---

[8] Lewis F. Richardson, *Arms and Insecurity: A Mathematical Study of the Causes and Origins of War*, ed. N. Rashevsky and E. Trucco (Pittsburgh: Boxwood Press; Chicago: Quadrangle Books, 1960), p. 65.
[9] *Ibid.*, p. 28.

rine craft almost intuitively . . . with uncanny accuracy. . . ." Still another ad describes ASROC, the first long-range anti-submarine weapon: "From initial sonar detection through firing and ultimate target contact, ASROC's automatic system delivers an unerringly destructive missile from thousands of yards away—all in a matter of seconds."

We are assured by a manufacturer of pressurized dehydrator systems than an "anti-missile radar is kept on the alert," while another company proudly proclaims that its leadership in electronic warfare is "typified by its advances in countermeasures and counter-countermeasures against all known types of electro-magnetic radiation." But pity the poor manufacturer whose stock in trade is so secret he can't advertise it except in metaphors like: "Countermeasures and the myxine glutinosa. Formidable, indeed, is the horrendous hagfish. Though small in size, his capabilities are enormous. In a matter of hours he can eat his way through the skin of his victim and then proceed to devour everything except bone. Things are not always what they appear, a theme common in countermeasures."

And that's the way of science, as predicted in the ads. We build a missile-launching submarine that "cannot be accounted for" while at the same time devise a system that "apprehends submarine craft almost intuitively" and develop anti-submarine weapons that "deliver an unerringly destructive missile." If the Russians and Chinese and all our potential enemies have engineers as brilliant and creative as our own, then clearly their missiles are as invulnerable as ours and as hard to detect and as full of tricks. (Their ads—if they had them—would surely portray the "American hordes" in mortal fear before their ultimate weapons!) And if their creative genius has created an anti- for all our weapons, as ours has for theirs, then surely our weapons are not absolute! And what is *vice* is *versa*.[10]

As the deadly game continues, we are brought to the stage of

---

[10] The above three paragraphs are excerpted from "Advertising Doomsday" by Eugene S. Schwartz, which appeared in slightly different form in the *Nation*, March 18, 1961, and were reprinted in *Peace and Arms*, ed. Henry M. Christman (New York: Sheed and Ward, 1964).

the anti-ballistic missile (ABM) and the multiple independently-targeted re-entry vehicle (MIRV) with its cluster of independently steerable nuclear warheads. And *game* it is considered by a generation of academic militarists who have enlisted mathematics to create a science of destruction. The mathematics of destruction came to maturity in World War II, when operations research was developed by scientists and mathematicians who joined hands with the military in solving some of the complex problems that arose when one portion of mankind sought to annihilate another. With the advent of the hydrogen bomb, the diversification and multiplication of missiles, and the launching of space satellites, the technology and conduct of war have become so complex that a new type of "scholar," the game-strategist, has come into prominence. Utilizing intricate mathematical tools and assisted by giant computers, the game-strategist has harnessed the intellectual inheritance of science and mathematics in a new discipline and industry—the planning of modern war.

The strategists are organized in numerous government agencies attached to the military services to provide advice on all phases of military technology and policy. They are aided by for-profit companies whose sole product is ideas, by industries which support their own analysts to prove the efficacy of the weapons systems they develop, and by universities where scholarship is available for application to any noble and some ignoble purposes. Their work purports to be scientifically neutral and free of bias and emotion. All traces of humanity have been excluded from their idealized equations, statistical research, and logical deductions. Their work is a monument to scientific objectivity, but because it treats of man even as it consciously excludes him from consideration, it in fact becomes inhuman, immoral, and exhibits that virulent symptom of twentieth-century insanity: genius in the service of "that devil's madness—war."[11]

Conceived as a tool to be used "when decisions become difficult," war gaming is based upon the use of models, simulation, and probabilistic analysis. Because confidence must be attached to answers, "righteousness of means, although a factor, is not very mea-

[11] Robert William Service, "Michael."

surable and, as such, is relegated to someone higher up. We must ignore it or code around it in computer models. . . . Add to this such complications as defensive weapons, passive shelters, backfiring, and immoral acts (killing civilians, etc.) and you have some of the elements of a brilliant (and important) game."[12]

The "physics of war" concerns a comprehensible body of "natural phenomena that may be treated scientifically . . . in the specific, theoretical, and quantitative way so effectively pursued in the study of physical phenomena," an article in *Operations Research* asserts. The fundamental facts of the physics of war are:

1. The past and continuing existence of irreconcilable conflict in human society.

2. Man's past and continuing use of military force in the attempt to resolve such conflict decisively.

3. The primary and indispensable role of weapons in the exercise of such force.

4. The necessity generally of searching for and finding targets before weapons in warfare can be used effectively.[13]

It can readily be seen that if "facts" 1 and 2 are granted as invariants of human societies, the physics of war need not consider the alternative hypothesis of peaceful settlement of international disputes by negotiation and law. "Facts" 3 and 4 then lead logically to explorations of optimum employment of military technology for imposing one nation's will "decisively" on another.

Certain concepts are fundamental to this physics. First is the concept of *kill*. This is a neutral term that refers to the destruction of any object: a missile, a submarine, a tank, or a human being. War is made scientific by ignoring the distinction between weapon and human altogether. Recognizing the uncertainties of the world and war, *kill* is transformed into a probability, and the symbol p_k, the *kill probability*, is the purified essence of an intellectual approach to destruction. Use of the symbol in complex mathematical routines makes it possible to determine combat losses physically imposed at measurable rates. Only a little more mathematics is

[12] James P. Dix, "Game-Theoretic Applications," *IEEE Spectrum*, V (April 1968), 108–117.

[13] H. Brockney, "The Dynamics of Military Combat," *Operations Research*, VII (January–February 1959), 30–44.

required to calculate the *maximum possible kill* under a variety of circumstances.

When the tables are reversed and we become the objects for which the enemy calculates its p_k, our interest is to maximize our *survival worth*. In like manner, terms may be defined, equations formulated, and solutions found for the conduct of war.

In its pristine form, the physics of war can be formulated as a simple problem: "As two forces engage in combat without replacements, the number of survivors on the two sides will diminish. The force first reaching zero loses; the other wins."[14] It was perhaps to test this theory that the United States introduced the ghoulish policy of the body count in Vietnam to accurately measure its impending "victory." Such a policy is in accord with the war physics' requirement "to test the predictions of that theory to see if they agree with the fundamental, already established facts of warfare."[15] In view of the longevity of past wars, and especially the Vietnam War, the duration of the war in the model can vary as can the time interval between *kills*. Replacements can be considered, ammunition can be limited or unlimited, fighting units can be fixed or mobile.

Although formulations cannot be too realistic, by means of probabilistic analysis, such as the Monte Carlo technique, a simple curve is developed that "answers satisfactorily the question: How many people will die if I push this button? Complicated chains of different logic can be linked together by merely playing a game in which you simulate conditions by throwing switches on or off, depending on random numbers."[16]

These computational difficulties carry over into cost evaluation in regard to weapons systems in general and to the soldier in particular. A soldier can be considered as a weapons system and can be assigned a cost. The problem is to "decide whether a soldier is to be valued for his combat capabilities alone or for his socio-economic values also. . . . It is the *mathematical average of*

---

[14] Richard H. Brown, "Theory of Combat: The Probability of Winning," *Operations Research*, XI (May–June 1963), 418–424.

[15] Brockney, *op. cit.*

[16] Dix, *op. cit.*

an anonymous life for which we seek an acceptable value" (italics added).[17]

Other studies have shown that the scientific analysis of war appears to be applicable to the nuclear age. Solutions have been found for "optimum" deployment of both offensive and defensive weapons for a nuclear missile or aircraft attack against point targets. Even cities of moderate size can be regarded as point, rather than area, targets due to the wide area of *kill* of megaton missiles. Methods for "optimally" distributing weapons among large areas to maximize radiation casualties have also been developed; in such models, casualties produced are calculated as a function of the total tonnage delivered.

Early calculations derived from the models were revealed in a report on the *Biological and Environmental Effects of Nuclear War*, prepared by the Special Subcommittee on Radiation of the U.S. Congress in 1959. Using computers to determine the effect of 263 hydrogen bombs of 1,446 megatons delivered on 224 targets in the United States, experts estimated that nearly 20 million people would be killed the first day with another 22 million being fatally injured and still another 18 million being injured. Calculation of *surviving worth* would require subtracting the following losses, among others: nearly 12 million homes severely damaged, 8 million homes moderately damaged, and 1.5 million homes lightly damaged, for a total of 45 per cent of the homes in the United States.

Approximately half a million homes would be so radioactive that they would have to be evacuated for more than a year, while another two million homes would have to be evacuated for periods of several months to a year. The mathematical models do not take into consideration long-range effects, of which the congressional committee noted that casualties would be falling for generations as a consequence of genetic defects induced by radiation and fallout from a nuclear war.[18]

[17] Thomas L. Saaty, "A Model for the Control of Arms," *Operations Research*, XII (July–August 1964), 586–609.

[18] U.S. Congress, Joint Committee on Atomic Energy, Special Subcommittee on Radiation, *Hearings, Biological and Environmental Effects of Nuclear War*, 86th Congress, 1st Session (Washington, D.C.: Government Printing Office, 1959).

The above calculations have long since been superseded by progress in the technology of destruction resulting from the arms race. In his 1965 "military posture" report to Congress, the Secretary of Defense revised the catastrophic forecast. Assuming a population of 210 million Americans in 1970, calculations revealed that 149 million Americans would be killed in a nuclear attack if both cities and military targets were simultaneously attacked. Only 122 million might die if the attack on the cities were delayed long enough to allow retaliation against enemy missiles. The retaliation would exact a toll of more than 100 million Russians and destroy 80 per cent of that nation's industrial capacity.

A fallout shelter program costing $5 billion, the computers calculated, would reduce American casualties to 120 million in a simultaneous attack, 90 million if retaliation blunted the attack on the cities. Expenditure of $25 billion on shelters, including $17 billion for an anti-missile program, would cut the toll, it was estimated, to 78 million American dead in a simultaneous attack, 41 million if the attack on cities were delayed. The additional expenditure of $25 billion, it is noted, may hold the 1970 toll to that forecast in 1959, *given that the equations are realistic, anti-missiles are effective, and the bomb shelters do not become tombs.* Since the 1965 estimates, megatonnage, delivery systems, etc., have all been escalated. The calculable results are correspondingly more obscene as the weapons systems become scientifically more sophisticated.

Mathematics is versatile, and p_k can serve to solve other military problems in addition to those dealing with nuclear exchanges. It is no surprise to find studies being conducted on *casualty production models* for both chemical and biological warfare. The euphemistic term *munitions* is used as a generic term for: GB, of which a drop on the skin causes instant death; bacteria capable of producing deadly diseases in man and animal; and a Pandora's box of micro-organisms and pests capable of destroying man, animal, or plant.

With the positivistic necessity to verify predictions, the following statement was candidly reported at a symposium celebrating the first decade of military operations research: "We in the Operations Research Office, *purely from the view of progress in opera-*

tions research, considered the Korean War a very happy circumstance. Analysts had the opportunity to try to measure a shooting war and to get their feet muddy after the tradition of that Queen of Battles, the Infantry" (italics added).[19]

What do the analysts measure? Combat effectiveness based largely on firepower is the primary measure. But, the analysts write, "the measures of firepower themselves are often based on intuition." In searching for a measure of effectiveness that is acceptable "at least on common sense grounds," there are many obstacles. "First . . . there is not one but an infinity of functions that could serve as a measure of effectiveness. . . . Second, we have no recipe for constructing any of these functions. Third, the combat effectiveness measures arrived at in this way are too abstract; we would not know how to verify combat effectiveness ratings experimentally, even assuming that the required tests were practicable." After several elaborate formulae and much mathematical legerdemain, we are finally informed that combat effectiveness "is the ability to prevail in combat."[20]

In Vietnam the military and the scientists have had another opportunity to test a new generation of weapons and strategies, *in vivo*, on a peasant people of Oriental stock. Chemists, biologists, sociologists, and economists, a coterie from all segments of academia, have participated in the orgy of technological warfare. They helped to perfect napalm, anti-personnel pellets, and gases. They helped to develop an arsenal of chemical and biological weapons that were held in reserve. The behavioral scientists were meanwhile developing new techniques of pacification, techniques that would be used increasingly in the cities of America and throughout the Third World as revulsion mounted against a technology gone berserk. The anti-guerrilla weapons and strategies were studied as models of universal repression against those demanding a future.

The technologists, however, both those who fashion the weap-

---

[19] W. L. Whitson, "The Growth of the Operations Research Office in the U.S. Army," *Operations Research*, VIII (November–December 1960), 812.

[20] Philip Hayward, "The Measure of Combat Effectiveness," *Operations Research*, XVI (March–April 1968), 314–323.

ons and those who seek to fashion the mind, overlooked a prime ingredient in the "poor man's power"—guerrilla warfare. "You cannot destroy a political belief without killing, one by one, all the people who possess it," Vladimir Dedijer, the Yugoslav historian, reminds us. Scientific advances in weaponry can make the killing easier to accomplish, but they do not alter this basic situation.[21] The ghoulish "body count" and computer calculation of "pacification" are additional contributions by the United States to the physics of war in which a military technology seeks to destroy a political belief.

A perhaps inadvertent *reductio ad absurdum* of the new physics appeared in a study which posed the proverbial immovable object against an irresistible force. A system based upon four, admittedly oversimplified postulates was described:

1. Either opponent can provide absolute assurance that his own destruction will entail that of the other.
2. Either opponent can meet the application of limited destructive force with effectively equal or with greater force.
3. Neither opponent can trust the other absolutely in any agreement.
4. Neither opponent will surrender to the other, nor submit without redress to incommensurately large losses, as judged by each opponent.

From these postulates a number of deductions were made and proved à la Euclid, including the Theorem: "There is no absolute insurance against mutual destruction." The proof is that postulate 1 implies that there is no possibility of unilateral insurance against mutual destruction, except through surrender of one opponent to the other. The latter eventuality is excluded by postulate 4. Insurance through cooperation is ruled out by postulate 3, since such an agreement cannot be absolute. Hence, neither unilateral nor mutual insurance can be absolute.[22]

Q.E.D. Science and technology cannot wage a war successfully

---

[21] Vladimir Dedijer, "The Poor Man's Power," in *Unless Peace Comes*, ed. Nigel Calder (New York: Viking, 1968), pp. 18–19.

[22] Douglas L. Brooks, "Choice of Pay-Offs for Military Operations of the Future," *Operations Research*, VIII (March–April 1960), 159–168.

and cannot afford protection to the people who would be involved in a war. Every technological development increases the annihilating potential of a war and further restricts the possibility of achieving peace. Technology begets counter-technology which begets an arms spiral. The treasure and resources of the world are expended in ever-increasing amounts in a chimerical quest for that which the quest destroys.

The futility of the quest is repeated over and over again: "Most informed people in this field [weapons technology] believe that it is highly improbable that either side or any nation can invent their way out of trouble by developing a new decisive weapons advantage through science and technology," Trevor Gardner, a former Assistant Air Secretary for Research and Development, stated in 1960.[23] Anatol Rapoport wrote a year later: "I believe that strategic thinking is powerless to resolve the problem of threatening total war. Deterrence is self-contradictory, since to be effective, it must be ready to be used, and the deterred must believe it is going to be used. If he does not believe it, deterrence fails, and we have total war. If he does believe it, it is only a short and inevitable step to a conviction that what poses as deterrence is a potential premeditated attack, and this conviction, in turn, leads to the contemplation of a preventive attack, which, in turn . . ."[24] In 1964, Joseph B. Wiesner and Herbert F. York stated: "It is our considered professional judgment that this dilemma has no technical solution. If the great powers continue to look for solutions in the area of science and technology only, the result will be to worsen the situation."[25]

Referring to the Vietnam War, which was being escalated at the time, John S. Foster, Jr., then director of Defense Research and Engineering, summarized the same point applied to a new situation: "If there is one indisputable feature of the Vietnam War, it is that a technology fix alone will not solve our problems. The hardcore problems are essentially political, social, and economic. The

---

[23] Quoted in *Electronic Design*, May 25, 1960.
[24] Remarks on "National Security Dilemma: Challenge to Management Scientists," *Operations Research*, VII (1961), 223.
[25] *Scientific American*, CCXI (1964), 27.

solutions to these problems will not be found in the products of research and development. Nor will it help to invoke any mythology about the potential of research and development."[26] Commenting on the spiraling arms race, George W. Rathjens observed: "Moreover, it appears that at the end of all this effort and all this spending neither nation will have significantly advanced its own security. On the contrary, it seems likely that another upward spiral in the arms race would simply make a nuclear exchange more probable, more damaging, or both."[27] Herbert F. York has repeated his conclusion of five years earlier that "there is no technical solution to the steady decrease in our national security that has for more than 20 years accompanied the steady increase in our military power."[28]

The unsolvability of the quest for peace through arms is a product of science and technology. The sets of proliferating residue problems generated by the quest have brought about a merging of the threats of nuclear annihilation with those of starvation, poverty, depletion of resources, and pollution of the environment. War and preparation for war contribute to and aggravate these other techno-social problems.

Wars and preparation for wars are currently consuming 7 per cent of the world's total output of goods and service. This sum is equivalent to the total Gross National Product of 52 per cent of the world's population. U.S. military expenditures amount to approximately 8 per cent of its Gross National Product. Against the background of the immanent depletion of its own natural resources, and the rising and competing demands of other nations for their own resources, the depletion represented by the inherently wasteful weapons and munitions with their high rate of obsolescence in peacetime and their total obliteration in wartime is unconscionable. In the area in which 75 per cent of the world's population produce one-sixth of the world's Gross National Prod-

[26] Quoted in *Electro-Technology*, XXI (September 1967).
[27] "The Dynamics of the Arms Race," *Scientific American*, CCXX (April 1969), 15–25.
[28] "Military Technology and National Security," *Scientific American*, CCXXI (August 1969), 17–29.

uct, the combined expenditure of $22 billion a year on wars and preparation for war is a tragedy.

Every resource, especially a developed resource, is a treasure that the peoples of the world cannot afford to squander. The maxim that a modern army is dependent on oil has increased the demand for this resource and accelerates the potential for accidents involving tankers with resulting oil spills and the other consequences that have been discussed previously. The potential for more serious and more harmful accidents of all types continues to mount together with the arms race. The derailment or explosion of munitions and chemicals is becoming more common as railroad accidents continue to increase. The routing of these dangerous cargos through and near large urban centers sets the stage for major catastrophes.

One of the dangers lies in the manufacture, testing, storage, and deployment of nuclear weapons. The manufacture of plutonium is accompanied by the creation of fission products that must be disposed of through burial, as was noted in Chapter 7. Contamination of the air and the water used for cooling has also been noted. Thermal pollution is a necessary by-product of the manufacturing process. The accidental dropping of bombs near populated areas because of crashes or equipment malfunction have occurred on a number of occasions. Nuclear-tipped missiles have gone astray in firing tests.

Mankind is moving into space-age warfare, with satellite spies in orbit and manned space platforms now being designed. Efforts are being made to militarize the oceans as well as the skies. Armies of robots in free-moving, self-regulating vehicles, walking tanks, and unmanned suicide bombers are under development. Scientists are seeking ways to modify the weather, produce earthquakes, and render the land barren. They are experimenting with brain waves to incapacitate entire nations, and with chemicals to render populations senseless.

But the fighting machines and clanking robots of tomorrow's wars have the same goal as all wars—human destruction. To prevail in combat requires that the conditions that sustain life be destroyed. The assault upon nature reaches its logical denouement in the assault upon man. In military technology science comes full circle and generates a proliferating chain of problems that defy containment.

To scientists and statesmen whose "games" pose a threat to the lives of all people on earth, pollution, waste, and upsetting the balance of nature through preparations for war are minuscule problems. To a tortured humanity and a despoiled earth, these problems are a serious threat. Any possible solution recedes as wars and preparation for new wars continue.

10 *Laboratory in Vivo*

NATURE is wild, random, disorderly. Plants and animals find niches in the complex disorder and establish life cycles that mesh with the myriads of cycles that sculpture the earth, nourish life, and replenish and cleanse the biosphere. Over the teeming, variegated, and multipatterned disorder, a natural order evolves that maintains the disorder through cyclical processes. All is connected to all, as life thrives and ebbs in the niche-filled successions of organisms.

The imposition of order on any portion of the "allness" of nature creates a disturbance that reverberates throughout the disordered pattern, abolishing sheltered niches and interfering with cyclical processes. The cycles that maintain disorder have tremendous healing properties, and a small disturbance can be absorbed and integrated into a new pattern. Thus is disorder preserved in the face of natural perturbations and disturbances.

Man's order, however, begets order. Once the disordered pattern is broken and the natural cycles are disturbed, only the imposition of more order can circumvent the breakdown of the cycles and the jeopardizing of life that is sustained by these cycles. But order, too, has its limits, and at some point the imposed order overwhelms the natural cycles and threatens to destroy them. At this point the natural cycles may have been upset and changed so that the ordering pattern can no longer be sustained. Chaos, breakdown, and disintegration follow.

Such has been the trend of all civilizations since man's prehistoric predecessors first made their appearance on earth. With the advent of the first primitive tool, the building of the first fire, the clearing of the first forest, the taming of the first wild animal, man has sought to impose order on the environment in which he found himself. This order was the establishment of an artificial environmental enclave in the midst of the natural environment. Every expansion of the enclave was gained at the expense of the natural environment and by extending the artificiality of the environment with which man surrounded himself. Expansion led to greater artificiality, and greater artificiality led to further expansion in a new cycle whose aim was to impose a man-made order upon the natural order. More significant, every artifice that altered the environment affected the natural "allness" and required further artifices to sustain the initial artifice.

Civilization is the progressive extension of an artificial environment at the expense of the natural environment, accompanied by a growing interference with the natural cycles and processes that have fashioned and sustained life. The successive stages of civilization have been chronicled by archaeologists and historians who have viewed its rise in terms of tools, ideas, and social forms. Civilizations can also be viewed in terms of the artificiality of the environment.

Three general stages can be delineated by this measure. The first stage extended from the dawn of mankind to the Neolithic period and was one in which man lived almost wholly within the natural environment. His few numbers and lack of skills and tools presented minor perturbations to the natural environment, and these were readily absorbed and dissipated by natural recuperative processes. In the second stage, which dates from the Neolithic period to the Industrial Revolution, man lived in relative symbiosis with the natural environment. The development of technics and an increase in numbers often led to an excess of artifices that overwhelmed nature's recuperative powers on a local scale, but the earth was large enough to prevent world-wide disturbances.

The third stage extends from the Industrial Revolution to the present. It is a period of civilization in which the natural environment is being replaced by a totally artificial environment. Order is superimposed on order as the natural cycles are broken and interrupted, the face of the earth is populated with teeming billions, and the natural resources of the earth are dissipated in a glut of pseudo-wealth. In this third stage, the natural environment is viewed as an enemy to be conquered while the artificial environment becomes the mark of progress that is synonymous with the word civilization.

Forests gave way to grasslands and grasslands were transformed into deserts. Plentiful wild game became scarce and then extinct. Rivers were dammed and the courses of their natural river beds were altered. Each step of civilization exacted its toll on the natural environment. Societies struggled to overcome the negative effects of each succeeding desecration with additional desecrations —both to compensate for the deleterious results and to further the conquest of the natural environment.

The conquest and desecration of nature were integral aspects of the religious heritage of Western man. The despoilment and mastery also touched upon the deepest emotional and instinctual characteristics of mankind. Mastery over nature was propelled by the fear of crackling lightning flashing in darkened skies, setting fire to forests and grasslands, tearing asunder giant trees. The taming of streams and floods was recompense for the fear and helplessness arising from being victims of rampaging waters. The earthquake and erupting volcano inspired fear in the beholder and brought death and destruction to those in the area of disturbances. Uncertainty and hunger inspired the search for domesticated crops and animals to mitigate the vagaries of nature.

The conquest of nature is also replete with the symbolism of male virility: the huntsman, the mastery of Mother Earth. The "virgin" forests were felled, and the sexual symbols represented by streams, caves, valleys and mountains were brought under male dominance. The "rape of the earth" arises in the innermost memory of evolving man.

Even as man conquered nature, his attitude was ambivalent. He stood in awe before the mighty falls with its roaring cataracts and splashing waters, even as he feared the untamed powers of the water. He communed with the mystery of being in the solitude of the forest and feared the encroaching darkness and the strangeness of the pulsating web of life that spread in profusion in, around, and under the arboreal canopy. He revered the nature that was full of life and growth; he detested the death and decay of rotting organisms with their putrefying images and offending odors. He responded to the freedom of nature; the wildness of its creatures, the meandering of its streams, the lapping of the ocean waves, the wind rustling in the foliage—and sought to tame them all. He loved Mother Earth and the communion with his fellow creatures, but recoiled in horror from the sight and activity of his body as a member of the community of nature. He sought to conquer the thriving life in nature in a desperate attempt to stave off his own death.[1]

Up to the present, all living organisms have been the product of evolutionary processes that have extended over millions of years. Evolutionary development has taken place against the background of the natural environment with its life-maintaining cycles. In a sense, evolution has been an undirected and random experiment in which organisms developed and maintained themselves according to their ability to adapt to the "allness" of their environments. There have been successes and failures, for the criterion has been only that of survival. But survival and the challenge of adaptation have been long-term experiments that have been tested in time spans of millions of years. Genetically, man has been programmed for a life within nature. In his conquest of nature, he is undermining the environment without altering the genetic program. As a consequence, in the short space of several hundred years man has been in the process of destroying those conditions under which

[1] Paul Shepard, *Man in the Landscape: A Historic View of the Esthetics of Nature* (New York: Knopf, 1967), Chs. 3, 7.

he evolved and which have led to the successful survival of the species.

In the third stage of civilization, under the impact of science and technology, man is establishing a wholly artificial environment that seeks to replace the natural one in which he evolved. This process is taking place on a world-wide scale and at an exponential rate. In a mere three hundred years the entire earth has been transformed. In the past twenty-five years the rate of environmental change has accelerated at so dizzying a pace that it now threatens the vestigial natural environment. Even as the rate of change increases, the time span in which the harmful effects of the changes are noted is decreasing. Strontium-90 was appearing in the bones and teeth of children before a decade of atomic bomb tests had elapsed. DDT was appearing in the body tissues of the earth's inhabitants and in mother's milk within the same short period.

Our interference with the natural environment is occurring not on a local scale or in connection with only single aspects of the environment. Our interference is global and all-encompassing. We change the course of rivers and remake the land surface. We change the soils and the biota they sustain. We influence the climate and meddle in the oceans. We deplete all resources at accelerating rates and pollute the air, earth, and waters on a world-wide scale. Indeed, technological civilization may well be defined as the cumulative defacement, destruction, despoliation, and pollution of the earth.

We transform the earth which supports our life to accomplish an unknown end. We hasten toward this end on uncharted paths. We cannot recognize short-term effects of our hasty passage, nor do we comprehend the long-term consequences. We do not know if man and life on earth can be sustained in a totally artificial environment, though we bend all our energies to expand the artificial. We do not know if the natural environment can coexist with the artificial, although we destroy the former.

Technological civilization has called upon science and technology to lead mankind pell-mell into a nebulous future. But twist and turn as we may, we have seen that each solution propounded or effected to solve the major techno-social problems of our time proves to be a quasi-solution that leads to a host of proliferating residue problems.

Science and technology can provide no values to choose and guide the development of more science and technology, nor can scientists define "safe" or "permissible" limits for any of the negative aspects of technology such as radiation or pollution levels. The establishment of any limits affecting humans are not discoverable *a priori* for, according to the axioms of science, these limits can only be derived from observation and measurement. In a sense, technological civilization is an experiment to test the "limits of the earth." But this grand and massive experiment is being conducted *in vivo* on a biological ecosystem. The peoples of the earth are guinea pigs for a vast outpouring of chemicals that enter our soils and foods and bodies; for pollutants that act upon inorganic and organic matter in unknown ways; for biocides that interact with the allness of nature in an unperceived manner; for an artificial environment that may be antithetical to the evolutionary process through which man reached his present state.

Are not the objects of observation perturbed by the act of measurement as were the inhabitants of Uravan; the ducks of Hanford, the marine life of Biscayne Bay; the wildlife off the coasts of Santa Barbara, the Cliffs of Dover, and Chedabucto Bay; the urban dwellers with "pollution" diseases; the irradiated humans of Hiroshima and Nagasaki? Are not the subjects of the experiments being transformed by the physico-chemical and social effects of the technology that act upon them?

In 1959 in West Germany, several instances of a rare congenital malformation, phocomelia, were observed. Babies were born with shortened arms and legs and, in extreme cases, the appendages were reduced to completely functionless nubbins similar to miniature seal flippers. Some babies were born without one, two, three, and sometimes all four limbs. In 1960 the instances of malformation increased, and by 1961 had reached epidemic proportions.

What was it, a worried world inquired? Heredity, radioactive fallout, a mysterious bacillus or virus? It turned out to be that alpha-(N-phthalidomido)glutarimide, a drug popularly known as thalidomide and sold in Europe under at least fourteen trade names, had become a popular tranquilizer and, unwittingly, an unseen and unknown deformer of the human embryo. Discovered in 1954, the drug had lain dormant for four years until its tranquilizing effect was noted; it was widely marketed because it was considered harmless, produced no hangover, and was not dangerous in overdosage. By the time the mystery had been unraveled, nearly four thousand deformed babies had been born in Germany and other European countries.

Thalidomide was a well-publicized case of a drug whose obvious and short-range effects were perceived but whose effect on the whole body was unknown. It is but one of thousands of drugs on the market whose safety and effectiveness have not always been demonstrated and whose effects in the long run and in combinations are generally unknown. Such is the case of the birth control pills which are being used—and tested—by millions of women throughout the world. No one knows their long-range effects, and some of their the short-range side effects have only recently been recognized and are the subject of medical controversy. The banning of cyclamates came years after cyclamates had appeared as an artificial sweetener and became a staple in the diet of millions of people. The presence of thousands of food additives that preserve, color, and alter taste, smell, and consistency of food is a vast experiment wherein short-range negative results are viewed as positive proof of harmlessness. The National Academy of Sciences has recently recommended that the Food and Drug Administration make mandatory the ban on monosodium glutamate in baby foods for, though the health risk to infants appears to be very small, no benefits are gained by usage. A voluntary ban was put into effect after several studies showed brain and eye damage in mice and rats—and after millions of babies had been fed foods containing the additive. The revocation of approval by the Food and Drug Administration for nearly one thousand additives that were not generally regarded as safe raises the question of the harm they may have caused during the period before they were interdicted. The ban on DDT and other persistent pesticides raises the same question, but with greater intensity.

In December 1969 three children in New Mexico were taken

seriously ill and suspected of having encephalitis. Further diagnosis indicated mercury poisoning from eating pork from hogs which had in turn eaten seed grain that had been treated with methyl mercury, a fungicide. With more than 800,000 pounds of mercury compounds used each year for agricultural purposes in the United States, primarily for fungicides, the illness of the children could be the warning of another thalidomide tragedy. Equally serious is the detection of mercury in the waters of thirty-three states in quantities sufficient to cause bans on eating fish taken from these waters in seventeen states. The fish can concentrate mercury in their bodies and cause serious harm to humans who eat them, as has been experienced by the Japanese who have partaken of mercury-tainted fish. Health officials have designated areas of Lakes Erie, St. Clair, and Champlain, the St. Lawrence and Wisconsin rivers, the Tennessee River in Kentucky, the Mobile, Tombigee, and Tensaw rivers in Alabama, Oxbow Lake in Alabama, and Lake Calcasieu in Louisiana as having mercury-tainted fish. High levels of mercury have been found in portions of the Mississippi and Rio Grande rivers and in Lake Onondaga in New York. The mercury pollution in each case has been traced to industrial waste.

Mercury is an insidious poison because it affects the body slowly and is undetected until its effects may be irreversible. In most cases of mercury poisoning, some permanent damage remains. Although industry has been dumping mercury into open waters for thirty years, it wasn't until a young Canadian student, in March 1970, detected mercury in fish caught in Lake St. Clair that an alarmed nation was alerted to the potential hazards.

The cited instances of the unperceived effects of technological society can be extended to countless cases. The spraying of asbestos has been challenged as a cancer-inducing agent; potential dangers arising from the mass feeding of antibiotics to animals have been discovered; the use of polychlorinated biphenyls, an industrial chemical used as a coolant in electric transformers, in paints, adhesives, floor waxes, and pesticides, is being discouraged after concentrations of the chemical were found in the food chain and evidence arose that it could induce birth defects in animals.

The unforeseen consequences detected on unsuspecting people

emphasize the laboratory aspects of the artificial environment that is being established. The creators of the environment are clever in the short run and ignorant in the long run. They do not know if a substance is persistent until it persists; they do not know if a substance is harmful until it harms; they do not know if something they have initiated is irreversible until it cannot be reversed. They do not know the path and degree of accumulation of a substance in the food chain until the substance has accumulated.

Laboratory experiments can reveal some effects, but the effects of thousands of chemicals and food additives and drugs and pollutants acting in myriads of combinations on a spectrum of human populations with their variations in body chemistry, metabolism, allergies, and other sensitivities are beyond laboratory control. The interaction of fertilizers and biocides singly and in combinations on man, the flora, and the fauna are relatively unknown since these effects, too, are largely beyond laboratory control.

There exists, finally, the laboratory of last resort, the laboratory *in vivo*, where the complex interrelationships of the artificial environment are tested directly on human populations. Such was the experience of the pitchblende miners in Schneeberg and Joachimstal who died of cancer from radium poisoning. And the watch-dial painters who wetted their radium-tipped brushes with their tongues. And the asbestos workers who suffered a similar fate. And the thalidomide-deformed babies, and all of the children of the world who have strontium-90 in their bones and DDT in their tissues. And the people who gasp and choke on noxious fumes that spill from busy factories and bustling highways. And the people on whose bodies and in whose progeny the long-range effects of nuclear bomb testing and nuclear energy production will be tested.

A comprehensive and extensive government report on pesticides and their relationship to environmental health[2] presents a sobering and worrisome corroboration of but one aspect of the mass experiments being conducted on society in the face of monumental ignorance leavened only by a smattering of known "facts." Although each solution of pesticide and herbicide control was acclaimed a great success, the laboratory of the earth has shown them to be

[2] *Report of the Secretary's Commission on Pesticides.*

quasi-solutions with the set of generated residue problems remaining unknown, although dimly perceived.

The introduction to the pesticide report is reasonable: "Our need to use pesticides and other pest control chemicals will continue to increase for the foreseeable future. However, recent evidence indicates our need to be concerned about the unintentional effects of pesticides on various life forms within the environment and on human health. . . . We must consider the total problem of pesticide usage not only in the context of what is presently known but also in the context of *the many unknowns still to be determined. Some of these unknowns may never be precisely determined*" (italics added).[3] In recommending "minimal" exposure to pesticides considered a "hazard" to man, the report stresses: "It is of utmost importance that the results of screening tests be scientifically and rationally considered. *The correct interpretation of hazards to human health is sometimes extraordinarily difficult.* . . . However," the report cautions, "it is not in the best interest of the public to permit unduly precipitate or excessively restrictive action *based only on anxiety.*" "Indiscriminate imposition of zero tolerances may well have disastrous consequences upon the supply of food and threaten the welfare of the entire nation. . . . Currently our national resources of funds, manpower, and facilities *will not permit* the concurrent testing of all pesticidal compounds. Priorities for testing must be established. . . . Additional chemicals are being or *should be investigated* and evaluated for potential hazards to human health, *as resources permit.*"[4] Recognizing the dangers, the report suggests caution but indicates that the experiment on the earth should continue, with hazards investigated as "resources permit."

What are the effects of nine hundred pesticides on the 200,000 living species in the United States, most of which species are "considered to be essential to the well-being of man"? *"The scarcity of information concerning the influences of pesticides on natural populations prevents adequate assessment of their total effects,"* is the

---

[3] *Ibid.*, p. 5. In following quotations from this report, italics have been added unless otherwise noted.
[4] *Ibid.*, pp. 10–11.

reply. "Less than 1 per cent of the species in the United States have been studied in this connection, and very few of these have been subjected to adequate observation..... Little data exists [sic] on the distribution, location, and impact of various pest control chemicals in the natural living systems of the world."[5]

World-wide, nearly two million plant and animal species exist. "*No one knows* how many of these estimated two million species are necessary in man's environment for his survival and welfare. . . . Understanding the biology of the life system is most difficult because of the vast number of species and the complexity of their interactions. Current knowledge suggests great caution in evaluating the worth of a particular species."[6] "Until the mode of pesticide action is completely known and the differing responses of species and individuals are understood," it is stated, "it will continue to be impossible to generalize on either the pathophysiology or population effects of pesticides on other target and nontarget species."[7]

What are the effects upon man, an anxious world inquires? "*There are formidable inherent difficulties in fully evaluating the risks to human health* consequent to the use of pesticides," we are told. The difficulty arises from the complex nature of the problems "and the general backwardness in this area of research in *man*,[8] as distinct from work in laboratory animals." Citing a number of variables that are involved, the report continues, "*Little is known about the effect of these variables in practice.* . . . One must realize that the components of the total environment of man interact in various, subtle ways, so that the long-term effects of low-level response to one pesticide are greatly influenced by universal concomitant exposure to other pesticides as well as to chemicals such as those in air, water, food, and drugs."[9] "The consequences of these prolonged exposures on human health *cannot be fully elucidated at present*," the report informs us, although it is known that pesticide residues do accumulate in tissues and body fluids. "No

[5] *Ibid.*, p. 25.
[6] See page 83 for the extinction rate of endangered species.
[7] *Report of the Secretary's Commission on Pesticides*, pp. 188–190.
[8] Italics in report.
[9] *Ibid.*, p. 32.

reliable study has revealed a causal association between the presence of these residues and human disease," we are assured, although such reassurance should be tempered by "the *paucity* of our knowledge in this area."[10]

Much of the discussion of this topic is taken up with "*a frank acknowledgment of the vast areas of ignorance in our understanding of the effects of pesticides on man.* . . . The problems have to be considered from the standpoint of a *reasonable* man, fully cognizant of the present state of the art in the sciences that constitute the basis of safety evaluation and apprised of all the facts at present available on the pesticide problem. Actions based on conclusions that go beyond the available information are only warranted if there are *reasonable* grounds for the belief that the risks of present practices outweigh the benefits."[11]

Reason is appealed to in the face of "vast areas of ignorance" and the "paucity of knowledge," but we are not told on what to base this reason. Indeed, we are assured that much of this reason may be invalid because of the "frank recognition that *many earlier decisions on safety must inevitably be proved wrong* as scientific knowledge grows."[12]

In reviewing studies on animals and other experimental complexities, the report notes that "the question as to whether long-term exposure to pesticides is harmful *remains unanswered.*"[13] It is stated that "only studies in man can provide definitive answers to the questions posed by human exposure to pesticides and other chemicals," and observes that "for a variety of reasons, such tests have not yet been applied in man. . . ."[14] Citing the "indisputable" argument that we should not add to the burden of carcinogens and mutagens in our environment in our present state of ignorance, it is suggested that by some unexplained clairvoyance, "If the additional hazard represented by a compound *is in all probability trivial,* then we have a responsibility to weigh those benefits conferred

[10] *Ibid.,* p. 35.
[11] *Ibid.,* p. 244.
[12] *Ibid.,* p. 245.
[13] *Ibid.,* p. 256.
[14] *Ibid.,* pp. 259–260.

by the use of the compound against the hazard that its presence in the environment represents."[15]

What are the major hazards, some of which may be trivial in relation to specific pesticides? The chief hazards to man are carcinogenicity (capability to produce cancer), mutagenicity (capability to produce genetic damage), and teratogenicity (capability to produce congenital malformations). On carcinogenicity, the report points out "the potential for interaction of pesticides, of pesticides with other chemicals, and suspected factors in cancer causation. The panel recognized the complexity of the potential interactions in carcinogens and *the extreme difficulty of the task of unravelling them.* . . . The sparseness of reports on the topic of interactions among pesticides . . . indicates a need for further attention to this aspect of the problem."[16] "Suspicion as to the carcinogenicity of DDT has been aroused and it should be confirmed or dispelled."[17]

While recognizing the presence of small residues of persistent organochlorine pesticides and their metabolites in human body tissue, the report concludes: "It *appears* that a steady state equilibrium has been established and this is *likely* to be maintained *under current patterns* of pesticide usage. These low levels of tissue residues are *unlikely* to present any hazard to the general population."[18]

As to mutagenic pesticides, the report notes that "although we can point to no pesticide now in wide use that has been demonstrated to be mutagenic, *the overwhelming majority have, however, not been adequately tested.* . . ." Whether damage is occurring "and if so, what is the magnitude of the effect, *is regretably unknown.*"[19] The likelihood that some highly mutagenic chemicals may be developed or already are being used "*is great enough to be a cause for real concern.*" However, "*our ignorance of chemical mutagenesis will not allow the assumption of safety without spe-*

[15] *Ibid.*, p. 262.
[16] *Ibid.*, p. 463.
[17] *Ibid.*, p. 471.
[18] *Ibid.*, p. 530.
[19] *Ibid.*, pp. 567–568.

cific mutagenic tests."[20] And, the report adds, *"it is not possible to predict in detail the kinds of effects that would occur* following an increased mutation rate, nor their distribution in time. Nor *can we be at all accurate* in any quantitative assessment of the total harmful impact of mutation on the population in comparison with other hazards. So, in weighing benefits against risks . . . we have only a *vague idea* of the nature and magnitude of the risk. We must remember, however, that *genetic damage is irreversible by any process that we know of now."*[21]

As for the capability of pesticides and herbicides to cause congenital malformations in humans, the report concludes, "Epidemiologic data on possible effects of pesticides on human reproduction and teratology *are grossly inadequate.* Prospective studies on this subject *are difficult to design and almost nonexistent."*[22]

Vietnam has furnished a large *in vivo* laboratory for what has perhaps been the greatest planned use of herbicides for foliage, crop, and food destruction the world has ever witnessed. Defoliants have been dumped on that unfortunate country by the United States military since 1962, with the bulk of the spraying directed against forest and brush and croplands. In a project ironically called "Ranch Hand," an increasing area of land was sprayed each year, reaching over 1.7 million acres in 1967 with a small reduction in succeeding years. Serious effects upon the forests, especially mangrove and rubber trees, were noted by observers together with deleterious effects upon birds and wildlife.[23] More serious was the use of the herbicide 2,4,5-T of which 3.3 million gallons were sprayed in 1969, when there was evidence that this compound "was clearly teratogenic" as shown by tests on laboratory animals.[24] The *New York Times* reported in early 1970 that villagers had been reporting miscarriages and birth defects which they attributed to defoliants, and the Defense Department, in response to mounting

[20] *Ibid.*
[21] *Ibid.,* p. 571.
[22] *Ibid.,* p. 674.
[23] Fred H. Schirley, "Defoliation in Vietnam," *Science,* CLXIII (February 21, 1969), 779–786; Gordon H. Orians and Egbert W. Pfeiffer, "Ecological Effects of the War in Vietnam," *Science,* CLXVIII (May 1, 1970), 544–554. The latter paper first appeared in the Newsletter of the Society for Social Responsibility in Science.
[24] *Report of the Secretary's Commission on Pesticides,* p. 665.

clamor, restricted spraying of the chemical to areas "remote from population." On April 15, 1970, the United States took steps to halt the use of 2,4,5-T in both the United States and Vietnam, indicating the compound "may produce abnormal development in unborn animals." A scientist from the Food and Drug Administration stated that the dioxin contaminates found in this herbicide and related compounds "were more potent than thalidomide."[25]

Thus does the United States and the rest of the world serve as a laboratory to test the products of technological civilization in peacetime. An appropriate victim is always found upon whom to test special products in wartime. In the words of the pesticide report: "The field of pesticide toxicology exemplifies the absurdity of a situation in which 200 million Americans are undergoing life-long exposure, yet our knowledge of what is happening to them is at best fragmentary and for the most part indirect and inferential."[26]

The peoples of the world cannot be blamed for being quizzical regarding the exploitation of nuclear energy. It was born in the obscene destruction of Hiroshima and Nagasaki, and the mushroom cloud rising over those doomed cities will remain in the collective memory of mankind forever. In the period during which the Hiroshima-size bomb grew into the multi-megaton hydrogen bomb and atmospheric tests were conducted by three nations, both governments and scientists either ignored or underestimated the effects of the tests and the long-range consequences of radioactive contamination of the biosphere. Their ignorance in this area parallels that in the area of biocides—except that the dangers from radioactivity are many times greater than those arising from the use of biocidal chemicals.

The continued testing of atomic bombs, even those used ostensibly for peaceful experiments, and the use and planned construction of nuclear plants for electrical energy production, constitute a vast experiment that encompasses the entire biosphere and all the peoples of the earth. The people have not been asked whether they wish to participate in the experiment, although it will be their

[25] *New York Times,* April 16, 1970.
[26] *Report of the Secretary's Commission on Pesticides,* p. 37.

bodies and their offspring upon whom the results of the experiment will be noted. If DDT has been found to be dangerous and persistent, its manufacture can be halted and the incidence of DDT in body tissues will gradually decrease; it will not enter into new tissues, and its effects will not be passed on to future generations. If radioactivity levels that will be attained as nuclear energy plants increase in number are found to exceed "safe" limits by testing *in vivo* on the populations of the earth, although the nuclear plants can be shut down, their effects will continue for hundreds and thousands of years—directly on the bodies of the living and indirectly on the genes of the unborn.

The nuclear plants now being operated and constructed are a vast experiment in materials, in processes, in controls, in power generation, in reprocessing of fissionable material, in effluent discharge, in waste disposal and storage, in radioactive monitoring, and in determination of the short- and long-range effects upon the biota of the earth. (Some of these problems have been discussed in Chapter 7.) The most serious problem is the long-range effect of radioactivity upon organisms and is manifested in the quest for "safe" or "permissible" limits. The search for any limits is recognition of the transformation of another aspect of the natural into an artificial environment.

Although hundreds of millions of dollars have been expended by the Atomic Energy Commission in monitoring and studying the effects of radioactivity on man, plants, and animals, the areas of ignorance are more numerous than the areas of knowledge, and many of the latter are subject to controversy even among scientists. Yet the effects of man-produced radioactivity on man and other organisms is being tested *in vivo* without full understanding of the possible consequences. Limits are being set on the basis of judgment about technological requirements, not on what is best for continued survival of life on earth.

That nuclear energy production is an experiment is attested by Lee A. DuBridge, former Science Adviser to the President, in testimony before a congressional committee: "As more plants are built and operated, we will gain more experience and, I am certain, added confidence in our ability to design safe plants and perhaps locate them closer to metropolitan areas. . . . The radioactive wastes which are incident to routine operation of nuclear power plants are also a major area of concern. This is a very complicated subject where *there are and continue to be a lack of adequate data to eliminate completely all sources of concerns over the effects of low-level concentrations."* Speaking of exposure to radioactivity, Dr. DuBridge stated, "Sometimes these levels are unknown, particularly when long-term exposure is involved because *it is hard to expose people 25 years to some particularly radioactive material and see if they are harmed at the end of that time.* . . . One has to judge how safe is safe enough in a particular situation. . . . Nuclear physics information and radiological health information should be considered in establishing safety levels. But it is a matter of judgment" (italics added).[27]

Glenn T. Seaborg, chairman of the AEC and a Nobel laureate in chemistry, confirmed these risks when he testified that "there are inherent in this technology certain risks and potential hazards to health and safety, as there are risks in many other activities."[28] The concluding phrase sets the tone for the judgmental criteria for establishing limits—and these criteria often leave science by the wayside. Congresswoman Catherine May queried the commissioners about the *in vivo* experiment. "Have you made any other studies in this area [health statistics], which is a *guinea pig* type of area, where people have been exposed to this particular environment for so many years?" Dr. Seaborg replied, "These studies are not yet complete. They so far show that there is no *measurable* adverse effect."[29]

The subject of permissible or safe limits on exposure to radioactivity is a highly complex one. Limits were established as early as 1928, based upon known effects of small doses to a few people, and were set at 0.1 to 1 roentgen per daily dose. The International Commission on Radiological Protection (ICRP) formally recommended 0.2 roentgen per day as a limit in 1934, and the permissible

---

[27] *Environmental Effects of Producing Electric Power*, p. 13. Italics in quotations from this source have been added by the author unless otherwise indicated.
[28] *Ibid.*, p. 88.
[29] *Ibid.*, p. 96.

radium level inside the human body was set at 0.1 micrograms based primarily on the studies of radium-dial painters. This quantity of radioactivity was estimated to deliver a biologically effective dose of between 15 and 50 rem.* In 1947 the National Committee on Radiation Protection (NCRP), which is the United States counterpart to the ICRP, lowered the permissible external exposure by approximately half, to 0.3 rem per week or about 15 rem per year. In 1953, "maximum permissible concentrations" for one hundred radioactive isotopes were established by the NCRP. By 1956, recognition of the genetic hazard caused the ICRP to recommend reducing the external exposure limit by one-third to 0.1 rem per week, and three years later revised the internal limits to correspond to this figure.

From the above figures, it becomes clear that limits are temporary adjustments to an ongoing experiment on people. The limit was originally established to protect relatively small numbers of persons occupationally exposed to radiation and was not intended for general populations. As such, the limits were originally called "tolerance dose," but this implied a safe threshold level which was not actually true. The term was changed to "maximum permissible dose," then altered again to "maximum permissible exposure." The Federal Radiation Council (FRC), established in 1959 by Executive Order to advise the President on radiation measures, now calls the fluctuating limits the "radiation protection guide" to avoid the implication that there is a single permissible or acceptable level of exposure and also to obscure the potential hazards arising from exceeding the stated limit.

Running continuously through the evolving permissible limits are three factors: (1) experimental data, (2) recognition of our ignorance in many areas, and (3) a politico-technical motivation. It will be seen that the third factor predominates. In announcing the maximum permissible dose in 1954, the NCRP defined it as "the dose of ionizing radiation that, *in the light of present knowledge*, is not *expected* to cause *appreciable* bodily injury to a person at

---

* Rem = roentgen equivalent man. It is the quantity of ionizing radiation that produces, when absorbed by man, an effect equivalent to the absorption of one roentgen of X rays or gamma radiation.

any time during his lifetime" (italics added).³⁰ Factors 1 and 2 weigh heavily in this definition. By 1959 the NCRP had introduced the third factor: "Occupational exposure for the working life of an individual at the maximum permissible values recommended in this report is not *expected* to entail *appreciable* risk to the individual or to present a hazard *more severe* than those commonly accepted in other present day industries" (italics added).³¹ Thus radiation injury could be tolerated as a normal industrial hazard along with black-lung disease, asbestos poisoning, and exposure to other toxic substances.

The Federal Radiation Council extended factor 3 in its 1960 statement: "Radiation Protection Guide, RPG, is the radiation dose which should not be exceeded *without careful consideration of the reasons for doing so.*"³² In its initial memorandum to the President, the FRC stated: "The fundamental problem in establishing radiation protection guides is to allow as much of the beneficial uses of ionizing radiation as possible while assuring that man is not exposed to *undue* hazard. To get a true insight into the scope of the problem and the impact of the decisions involved, a *review of the benefits and the hazards is necessary.*"³³ Recognizing the ambiguity of establishing a limit that might appear to be an acceptable level of radiation, the FRC continued, "Every effort should be made to encourage the maintenance of radiation doses *as far below this guide as practicable*" (italics added).³⁴

The basis for "as low as practicable" was clarified by AEC Commissioner James T. Ramey in 1969 congressional hearings on the subject: "How one achieves levels which are 'as low as practicable' is *always a matter of degree and judgment.*" Later in the hearings,

---

³⁰ "Permissible Dose from External Sources of Ionizing Radiation," National Bureau of Standards, U.S. Handbook No. 59, published for NCRP, September 1954; cited in Karl Z. Morgan, "Permissible Exposure to Ionizing Radiation," *Science*, CXXXIX (February 15, 1963), 566.
³¹ "Maximum Permissible Body Burdens and Maximum Permissible Concentrations of Radionuclides in Air and in Water for Occupational Exposure," National Bureau of Standards, U.S. Handbook No. 69, published for NCRP, June 1959; cited in *ibid.*, p. 567.
³² "Background Material for the Development of Radiation Protection Standards," Federal Radiation Council Report No. 1, May 1960, in *ibid.*, p. 567.
³³ Quoted in *Environmental Effects of Producing Electric Power*, p. 134.
³⁴ See note 32, above.

Dr. Ramey introduced two additional criteria for permissible limits: "You have to take into account the *desirability and the public need* for an activity as well as the risks" (italics added).³⁵ These additional factors were spelled out in greater detail by another commissioner, Theos. J. Thompson: "It is imperative that we examine environmental matters with a sense of overall perspective. *We must start with the basic premise that power plants and associated facilities will have to be built to meet the energy needs of this country.* . . . Accepting the premise of a continually increasing need for more power, we must then look for ways to *minimize* any detrimental impact they will have on the environment . . . recognizing that *advances in technology that contribute to a better life for mankind are necessary for continued human progress*" (italics added).³⁶

A third criterion for establishing limits, and one fraught with the greatest potential for obfuscation of the dangers inherent in nuclear energy production, was enunciated by Raymond T. Moore, associate commissioner of the Environmental Control Administration: "In applying these standards [of the ICRP, NCRP, and FRC] to waste discharges from nuclear powerplants, we support the philosophy that radioactivity levels should be maintained at the lowest practicable amount *consistent with the best available engineering technology*" (italics added).³⁷

Thus the Radiation Protection Guide derives the extent of the hazards to which people will be exposed from political and technological criteria, based upon the desirability and public need for more power plants in the name of progress and the availability of technology to attain whatever limits are temporarily established; these criteria are a matter of human judgment, in accordance with meager knowledge regarding both the short- and long-range effects of ionizing radiation.

The working of the Radiation Protection Guide in the laboratory *in vivo* is well demonstrated by the two broad activities of nuclear bomb testing and electrical energy production. Atmospheric test-

---

³⁵ *Environmental Effects of Producing Electric Power*, pp. 136, 148.
³⁶ *Ibid.*, p. 184.
³⁷ *Ibid.*, p. 426.

ing of hydrogen bombs was undertaken with little awareness of the characteristics of radioactive fallout, or the short- and long-term effects of this fallout. The fallout products, their dispersion and retention in the atmosphere, the rate and methods of fallout, the location of fallout, and their effects on human and animal populations were relatively unknown until data were collected on human guinea pigs throughout the world. The response of scientists in government was a clear indication of the political motivation of the testing, while their scientific response was colored by the politico-technological bases of the Radiation Protection Guide. At first they denied danger from fallout, then belittled its possible effects, and challenged accumulating scientific data which undermined any defense of continued atmospheric testing. Perhaps the most memorable of the defenses of continued testing was by Edward Teller, the prime mover behind the development of the hydrogen bomb, who, in his book *Our Nuclear Future*, wrote: "The reader will see that the worldwide fallout is as dangerous as being an ounce overweight or smoking one cigarette every two months."[38]

Linus Pauling has discussed the falsifications and erroneous analyses of spokesmen for the Atomic Energy Commission and other defenders of the government's political position in *No More War*.[39] Dr. Pauling and Barry Commoner were pioneers in establishing the truth about the great somatic and genetic dangers attendant upon fallout. The accumulation of strontium-90 in the teeth of young children, collected and measured by Commoner's Citizens Committee for Nuclear Information in St. Louis, was irrefutable evidence obtained from thousands of innocent and unsuspecting human guinea pigs. The rising strontium-90 level in cow's milk in the early 1960's led to world-wide concern for children's health. The shorter-lived iodine-131, which was a thyroid carcinogen, also caused growing alarm.

[38] Cited in *Fallout: A Study of Superbombs, Strontium-90, and Survival*, ed. John M. Fowler (New York: Basic Books, 1960), p. 6.
[39] Linus Pauling, *No More War* (New York: Dodd, Mead, 1958), esp. Ch. 6. Details of the fallout controversy are also contained in *Fallout*, and in Jack Schubert and Ralph E. Lapp, *Radiation: What It Is and How It Affects You* (New York: Viking Compass, 1957), esp. Ch. 11.

Reviewing the earlier position of the government, Paul C. Tompkins, executive director of the Federal Radiation Council, spoke at the 1969 congressional hearings of the government's and scientific community's growing concern over the accumulation of radioactive iodine that was fast approaching the "allowable" dosage in 1962 as a result of continued atmospheric testing. Tompkins posed the alternatives before the government as accepting the current level of iodine with its dangers to children, or causing malnutrition among these children by removing milk from the market. In explaining the government's position as favoring the former alternative, Dr. Tompkins added yet another criterion for the establishment of permissible limits of radioactivity. "In real practice," he testified, "we are usually dealing with things that are not a matter of balancing benefit against a risk. We are balancing one risk against another and selecting the lesser of two evils."[40]

The members of the AEC, the FRC, and the government did not pose a third and more reasonable alternative: stop testing. The scientific evidence continued to accumulate, pointing to the necessity of this third alternative. As early as 1958, the United Nations Scientific Committee on the Effects of Atomic Radiation had indicated the concern of mankind with the mounting danger of fallout and, although stoppage was demanded by people everywhere, the Nuclear Test Ban Treaty prohibiting atmospheric tests was not achieved until five years later. That which had been heretofore considered acceptable and within limits was in 1963 termed unacceptable; the rising levels of radioactive contamination of the earth from fallout could no longer be defended—even on politico-technological grounds. The outraged guinea pigs from the laboratory *in vivo* had revolted, demanding human survival as the prime criterion superseding the nonscientific and bogus criteria of desirability, assumed need, lesser of arbitrarily defined evils, and political expediency.

The laboratory experiment continues, however, with the planned expansion of nuclear reactor plants for the production of electricity. As with biocides and fallout, the effects upon man and his environment of discharged radioactive materials from the nu-

---

[40] *Environmental Effects of Producing Electric Power*, p. 409.

clear plants remain largely unknown. Clarence E. Larson, AEC commissioner, testified at the 1969 hearings: "We must acknowledge that measurements of ecological changes due to radioactivity are almost impossible to make, for the simple reason that our measurements are superimposed on the overall changes in our environment due to the ever-increasing presence of man himself."[41] Regarding biological effects of radiation, the commissioner testified that "we have not been able to detect nor do we expect on the basis of biological data in hand, to be able to detect any effect upon man or his environment from current or projected radiological discharges from power reactors. *This is not to say there are no effects. If there are effects, they are not observable.* Our procedures and regulations are based on conservative assumptions which anticipate effects that may or may not occur" (italics added).[42] The assumptions, we have seen, are based largely on nonscientific criteria, and to base procedures on effects that are unknown or unobserved is indeed an act of clairvoyance. The effects may be unobserved because the *in vivo* experiment must run its course through time. Speaking of the somatic effects of radiation, Dr. Larson notes: "With sub-lethal doses of radiation the severity of tissue damage is limited and regeneration of the affected tissues occurs by cell replacement. After a few days or weeks it is difficult to ascertain any biological effects. However, later in the animal's life other types of damage may be expressed, such as shortening of life-span, increased incidence of a number of types of cancer, as well as development of lens opacities of the eye. It is not clear at this time whether any of the late effects of radiation have a threshold dose for their production—a dose below which no biological response can be detected."[43]

But it is assumption of a threshold that is basic to the scientific criterion of a permissible dose, and over which a controversy now rages in the scientific community. John W. Gofman and Arthur R. Tamplin, radiation scientists with the AEC, have contended in a series of papers that there is no threshold, stating that existing per-

[41] *Ibid.*, p. 246.
[42] *Ibid.*, p. 239.
[43] *Ibid.*, p. 247.

missible levels are too high and will result in seventeen thousand additional cancer deaths per year.[44] They cite Brian MacMahon, Professor of Epidemiology at Harvard University, who wrote in 1969: "While a great deal more is known now than was known twenty years ago, it must be admitted that we still do not have most of the data that would be required for an informed judgment on the maximum limits of exposure advisable for individuals or populations."[45]

Gofman and Tamplin believe it to be "irresponsibility of the highest order . . . to set Public Health standards based upon a *hope*, unfounded in evidence, that somehow a poison will turn out to be less toxic than conservative sound estimates would indicate."[46] In urging the lowering of present guidelines by one-tenth, the scientists urge this be done now "since a later revision downward can lead to excruciatingly costly retrofits in a developed industrial application." They do not add that one can not retrofit a human with cancer or leukemia.

Gofman and Tamplin also suggest that "injury to other members of the ecosystem may be of greater long-range relevance to man than the immediate attention to man with extensive neglect of the ecosystem which supports his life." Dr. Tompkins, in his testimony on the ecological effects of radiation, had stated he didn't know how anybody would investigate these effects. "There is no guidance available," he claimed. "That which is important has to do with the multiplicity of various pathways through a rather complex web."[47] Dr. Larson admitted that knowledge on thermal pollution is meager: "A major problem faced in planning for control of the effects of heated water discharged on water quality is the lack of soundly based information on the actual temperature requirements of aquatic organisms of interest. Each biological organism has a temperature optimum at which it reproduces and grows

---

[44] "Low Dose Radiation, Chromosomes and Cancer," *ibid.*, Appendix 12, pp. 640–651; "Federal Radiation Council Guidelines for Radiation Exposure of the Population-at-Large—Protection or Disaster," *ibid.*, pp. 655–683; "Studies of Radium-Exposed Humans," *ibid.*, pp. 695–706.
[45] *Ibid.*, p. 641.
[46] *Ibid.*, p. 705.
[47] *Ibid.*, pp. 411–413.

best. Quantitative effects of above-optima temperatures on important sport and commercial fish and on fish-food organisms are not known in sufficient detail."[48] Raymond Johnson, assistant director of the Bureau of Sport Fisheries and Wildlife, attested to the immensity of the problem: "The difficulty arises from the fact that while man represents a single species with a rather simple life history, aquatic organisms include hundreds of thousands of species with a wide variety of life histories."[49]

The *in vivo* experiment poses its greatest hazard in possible effects on the genetic pool of man. Each dose of radiation, no matter how small, to the reproductive cells of a man or woman increases the chances of mutation in these cells. Since almost all radiation-induced mutations that have effects large enough to be detected are harmful, the possibility of passing on hereditary defects to future generations is increased.

Excessive use of X rays, especially in the genital regions, has been cautioned against as a result of this knowledge. Irradiation of humans, however, from nuclear fallout and from the effluents of nuclear power plants, continues and, in the case of the power plants, the effluent products are planned to increase. Krypton-85, for example, is seen as a serious problem that will require eventual removal of this gas from the effluent. Dr. Larson, in his testimony at the congressional hearings, conceded that "little is known about radiation-induced genetic effects in humans."[50]

The nebulous state of knowledge concerning effects of ionizing radiation that will be experienced by increasing segments of the population is demonstrated in two scientific papers included in the congressional hearings. In one of these papers, Robert Cunningham Fadely, director of research of the Foundation for Environmental Research at Golden, Colorado, analyzes the incidence of cancer and leukemia in counties bordering on the Columbia River in the vicinity of the Hanford Washington Atomic Storage Preserve for the years 1959 through 1965. Fadely points out that the Hanford plant has been producing plutonium for weapons since

[48] *Ibid.*, p. 268.
[49] *Ibid.*, p. 343.
[50] *Ibid.*, p. 249.

1940, and a portion of the radioactive wastes has been stored in pits located within the preserve. He states that a very gradual but chronic seepage and diffusion from these pits have led to radioactive contamination of the Columbia River as it traverses the border of the preserve. Noting that the quantities of radioactive materials in the river had previously been judged safe, Fadely concludes: "A physiographic pattern of malignancy, including leukemia, is now apparent in Oregon. This pattern is in conformity with the assumption that Hanford produced and Columbia River transported radioactive wastes that are significantly carcinogenic in quantities chronically present in affected human food and water supplies."[51]

Not so, retort John C. Bailar III and John L. Young, Jr., of the National Cancer Institute. Reviewing the same data and other data, these authors have written a rebuttal to Fadely's paper. They concluded that in both Washington and Oregon, "mortality rates for all forms of cancer combined have been consistently below the mortality rate for the U.S. white population. Both states have had consistent excess of leukemia mortality, but the excess was present before the Hanford Preserve began operation. . . . No evidence was found that persons living downstream from the Hanford Preserve or along the Pacific Coast of Oregon have had an excess risk of death from cancer in general or from leukemia in particular."[52]

The significance of this controversy and the controversy over a threshold limit is that honest men do not know the answers to the complex problems with which they are dealing. "Prudence and common sense require us to learn a great deal about the quantitative effects of radiation *in humans*," Robley D. Evans cautions.[53] Post-operational surveillance of the effects of effluents from an increasing number of nuclear plants, many of them in close proximity, cannot be accepted as a laboratory experiment, given the unknowns and the tremendous potential danger to the entire ecosystem and to humans in both the short and the long runs. We do not understand effects in isolation and have little idea of even how

---

[51] "Oregon Malignancy Pattern Physiographically Related to Hanford Washington Radioisotope Storage," *ibid.*, pp. 594–608.
[52] "Oregon Malignancy Pattern and Radioisotope Storage," *ibid.*, pp. 609–615.
[53] *Ibid.*, p. 744.

to measure effects in combination or cumulatively. Present arbitrary "limits" cannot be accepted as a cushion for experimentation with survival of the human race at stake.

"In a world we so little understand," Raymond F. Dasmann writes, "a biologist with a conscience can only feel uneasy with the assumptions of our military experts and nuclear technologists."[54] The laboratory *in vivo* that is technological civilization today imperils all mankind with every new step that is taken to render man's environment more artificial. The simplest acts of civilization have interfered with the natural environment in which man has evolved. Civilization has been built at the expense of living communities of organisms and the natural earth and its covering. By destroying the natural environment, man has nurtured the weeds and harmful insects, parasites, and animals that have moved into the niches from which they had previously been excluded. By interfering in all of nature's finely balanced cycles, man destroys the environment that sustains all life without comprehending the threat to his own survival. By destroying ecosystems out of his ignorance and hubris, man threatens to extirpate the natural environment without assurance that he can survive in an artificial one. Who will bear the responsibility if the experiment proves to be negative in the long run and the limits of the earth are exceeded? Will there be survivors to record the negative results?

[54] Raymond F. Dasmann, *op. cit.*, p. 157.

11 *The Decline of Technology*

*T*HE TECHNOLOGICAL process as the motive force of Western civilization, which seeks to extend its sway over all the earth, is proclaimed by its prophets as inevitable. These prophets declare technology's achievements to be steps on the ladder of progress, and view current technological society as rising to higher and higher stages according to a presupposed calculus of ascension. The task of the scientists and engineers is to guide and hasten the technological revolution on a world-wide scale—a revolution invoked in the name of man and aimed at his betterment. The inevitability doctrine and the scientists' and technologists' view of themselves in the role of a vanguard are strikingly similar to the Marxist doctrine of laws "working with iron necessity toward inevitable results" and the new order that will serve mankind. This is perhaps not so surprising, for both science and Marxism derived from the same Judeo-Christian tradition with its messianic fervor to redeem man. Science became a secular faith, Marxism a political one, and both are united in their obeisance to technology.

Some critics of technology have also proclaimed its inevitable triumph. "We are today," Jacques Ellul writes, "at the stage of historical evolution in which everything that is not technique is being eliminated. . . . All else is swept away. . . . Technique reigns alone, a blind force and more clear-sighted than the best human intelligence." "The weight of technique is such that no obstacle can stop it." "The victory of technique has already been secured. It is too late to set limits to it or to put it in doubt." "The autonomy of technique forbids the man of today to choose his destiny."[1]

Yet both the critics and the prophets of technology overlook the dialectical characteristics of the technological process. The build-up of unsolved problems has begun to set into motion forces that may destroy technological civilization, just as every other civilization has been undermined from within itself. "The obstacles which finally bring progress to a halt," Henry George wrote in *Progress and Poverty,* "are raised by the course of progress"; and "what has destroyed all previous civilizations has been produced by the growth of civilization itself." The rape of the earth, the pollution of the environment, the squalor of the urban environment, the uprooting of the old in favor of the new simply because it is new, the degradation and alienation of industrial man, and the increasing militarization of the nations of the world in a state of anarchy are the seeds of destruction now sprouting in technological civilization.

In Chapter 3 technology was shown to be a social process, dynamic and changing, and only partially autonomous in generating its own motive force. Chapter 4 introduced the concept of a quasi-solution and the sets of residue problems that militated against achieving a total or complete solution. The limitations on what man can know and do with science and technology were detailed in Chapter 5, where it was shown that the technological process would become more complex and more costly, with diminishing alternatives and depleted resources to operate with. Some of the consequences of the failure of technology *qua* technology were reviewed in Chapters 6 through 9. The proliferating chains of residue problems combining and interacting in unforeseen and unpredictable ways belie a force that is "more clear-sighted than the best human intelligence." Moreover, the unsolved problems are

[1] Ellul, *op. cit.,* pp. 84, 94, 106, 130, 140.

beginning to accumulate at a faster rate than quasi-solutions can be found for them, and some major techno-social problems are becoming unsolvable. In the preceding chapter the vast experiment *in vivo* to test the limits of the earth using the peoples of the earth as guinea pigs was reviewed, with descriptions of the hazards and dangers of this experiment.

Science and technology are dialectical in nature, characteristics well expressed in the dialectical pattern of the unity of opposites, the transformation of quantity into quality and quality into quantity, and the negation of the negation. Dialectics preclude inevitability but presage radical change and transformation.

The unity of opposites was aptly expressed by Heraclitus who pointed out that the way up and the way down are one and the same way. Biologically, the processes of anabolism and catabolism, the building up and the breaking down of complex organic compounds, go on side by side in the living organism, with life resulting from the interplay of these two opposing yet complementary processes. In nature's life cycle, decay is essential to life; microbes destroy in order to create. In a physical system matter represents a statistical system at rest while its molecules are in constant motion.

In a psychological system a child both loves and hates its parents, because during a long period of helpless infancy he has been both repressed and socialized. The interaction of patterns of compliance and intiative, sociality and self-assertion, cooperation and competition, introversion and extroversion, fear and defiance, give rise not to a dualistic personality but a many-textured one, as behavior patterns are structured and restructured. In every region of mental life, Kurt Lewin observed, "one finds every possible transition between greatest indeterminateness and full determinateness."

The static absolutism that transforms attributes into dichotomies fails to see the gradient existing between the duality, the complementary action of bipolarities leading to a harmony arising from dissonance. Day-night obscures the dawn

> Where the great sun begins his state,
> Robed in flames and amber light . . .

and the dusk when

> The crimson bands of sunset fall,
> Flicker and fade from out the west.

Blackest midnight gives way to moon-drenched light, even as the brilliant sun is covered by drifting cloud. The golden rays of the sun pouring down upon the earth are neither continuous nor golden, for as the rays pass through the falling raindrops the spectrum of the rainbow is reflected in the arch of the heavens. Big-small is not Lilliput or Brobdingnag but the vast expanse between the infinitesimal of atomic particles and the furthest reaches of the cosmos. Good-bad is not heaven-hell but the gamut of human behavior in a cultural pattern.

The interplay between quantity and quality is a continuous and reciprocal process. Since quality represents a difference in quantity, the qualities that science would deny are reintroduced through measurement. Those aspects of the universe that are most readily measured are those that are most understood in a qualitative sense. We have seen how the transformation of quantity into quality has been the factor that evoked a public response to the environmental problem. As was noted, quality is transformed into quantity as the effect of anti-pollution devices on autos, for example, is canceled by an increase in the number of autos.

The third aspect of dialectics, negation of the negation, is illustrated by the metamorphic changes of a larva developing into a butterfly. The formation of the chrysalis destroys the larva in an irreversible process just as it, in turn, is destroyed when it is transformed into the butterfly that has developed within the pupa. The flower offers its beauty to attract the bee so that the act of pollination may be consummated. When the seed is fertilized the plant withers and dies, and in dying brings forth the seed that in the following year will continue the existence of the species. "So our human life dies down to its root," the poet of Walden writes, "and still puts forth its green blade to eternity."

The dialectical process provides a conceptual framework for dealing with change and transformation, with growth and decay. It makes possible the comprehension that even as a thing is, it is changing in an unceasing process. The subatomic particles are in constant vibration and oscillation, as is the cosmos. The human body is constantly undergoing change as bodily components are renewed and exchanged. Civilizations grow and in their growth propagate the factors of decay.

The dialectical process is pervasive and abolishes the categories of formal logic and cause and effect. Causality as a simple "if A then B" proposition is not a scientific fact but a convenience that makes prediction and control possible. But the control is not always wise. One of the difficulties of science, as noted by Hume and Russell, is that induction cannot be logically substantiated. The ecological crises have arisen, in part, because of the concept of causality which abhors a cycle.* The thought that an effect can precede a cause sends shudders down the spines of scientists and logicians. But it is the great cycles of nature that have made life possible and have sustained it in a heretofore unending cyclical pattern.

The ecosystem is a complex interweaving of numerous intricate cycles. The fundamental cycle is one in which plants absorb inorganic material from the soil and carbon dioxide from the air to create organic materials by photosynthesis. The plants are eaten by herbivores who transform the plant substances into living organisms. The organism breathes the oxygen given off by plants and returns the carbon dioxide to the plant. The excretory products of the animal are returned to the soil where they are broken down into organic substances that nourish the plant. Upon death, plant and animal are decomposed into inorganic substances and the fundamental cycle is maintained.

But science has arbitrarily broken the natural cycle to formulate causal laws to satisfy its quest for order. Phillip Frank explains: "The cycles of physical facts that are interpreted as examples of causal laws are small cycles within the whole world process that as a whole is probably no cycle. . . . As a matter of fact, all causal

---

* The alteration of the concepts of "before" and "after" has been discussed in Chapter 5 in relation to cosmology. The postulation of the existence of tachyons, superluminal particles that travel faster than the speed of light, also would destroy the concept of causality. The recognition of cycles in which signals would move backward in time would give rise to effects preceding their causes. The discovery of tachyons would create a new revolution in that the controllability of events would have to be abandoned. See, for example, Roger G. Newton, "Particles That Travel Faster than Light," *Science*, CLXVII (March 20, 1970), 1569–1574.

laws are found by dissecting the world process into such incomplete cycles or, in other words, by finding out what state variables can be, and must be disregarded in order to see in the world process a great many incomplete cycles. . . . The more variables we can disregard and the fewer of them we keep, the more frequently recurrences take place and the closer we come to the causal laws of physics in which, as we have learned, the essential point is the recurrence of states that are defined by a small number of variables."[2] Thus dialectics is sacrificed to order and causality is wrung from cycles. But a causal chain is circular, like an endless belt with no beginning and no ending but only changing phases in a process. A variable is itself a function of other variables in which the past determines the future as the future determines the past. Abstracting an event from a process to make it discrete and subject to empirical study obscures the totality and the richness of the process. In complex systems, as in a biological organism, the circularity of the process makes it more difficult to isolate variables and restrict the initial conditions. The many variables that characterize human behavior are even more difficult to isolate and measure.

Sherrington has termed the concept of causation an anthropism. "To pick out this or that particular contributory condition in a conspiracy of reaction and to label that particular one the 'cause' is arbitrary and artificial."[3] Occam's razor has impoverished man's conception of the universe by the process of arbitrary selection and acceptance of the simplest explanation. Science has disconnected the unified world of nature, where all is connected to all, and has intervened in the natural cycles with a blind causality that threatens to destroy irreversibly the cycles that sustain life. The life system creates and maintains the conditions for its own existence. *A break at any point in the cycle destroys the cycle.* A sustained breakdown, Barry Commoner warns, may prevent a restart of the cycle. Just as the breakdown of civilization may preclude its restart because civilization will have destroyed the necessary conditions for a new beginning, so the restart of the cycle that main-

---

[2] Phillip Frank, *Philosophy of Science* (Englewood Cliffs, N.J.: Prentice-Hall, 1957), p. 285.
[3] Sherrington, *op. cit.*, p. 169.

tains life may be prevented because the conditions that originally existed and gave rise to the cycle will have been destroyed. The depletion of nitrogen-fixing bacteria in the soil has been noted. The destruction of the phytoplankton of the oceans would destroy a primary source of oxygen. In his efforts to create conditions to sustain more life, man endangers all life.

Can an ecological conscience help to set man on a new path that will harmonize technology with the imperatives of nature? The current interest in ecology manifests a belated recognition that the technological experiment has gone awry. Some people are looking for reprieve. The answer, however, cannot be found in ecology, for ecology is a science like other sciences. In its presuppositions it ignores values and qualities that are extraneous to observation and measurement. In a review of a British symposium on an aspect of ecology, for example, pride was expressed that ecology was becoming a science in the image of physics. "The symposium reflects the increasing interest of ecologists in quantitative measurements of environmental components while at the same time they are increasingly bringing the principles of systems analysis to bear on the study of environment as a whole. Obviously, the demand for informational synthesis, along with increasingly better instrumentation, is providing unusual opportunities for analysis of the aquatic, terrestrial, and space environments."[4]

Like the other sciences, ecology is anthropocentric but wishes to serve man better in the conquest of nature by conquering through paternalistic regard for the symbiotic relationships that exist between man and nature. Ecology cannot grant to other species the same rights to the earth as it does to man. The cockroach is much better adapted to the earth than man because its resistance to radiation is many times greater. Some plants are much hardier than man and can survive under circumstances where man will die. Will not ecology seek mastery over the cockroach, the plant, and other living organisms for the benefit of man, although it will seek to do so with scientific knowledge of the relationships that exist between man, plant, and cockroach? Recall that "No one knows

[4] Karl F. Lagler, *Science*, CLVIII (November 24, 1967), 1037.

how many of these two million species are necessary in man's environment for his survival and welfare." To the extent that ecology will introduce value judgments into its schema of knowledge, it will cease to be a science. "The ethical attitude to phenomena," Charles Singer stated, "is inconsistent with the effective advancement of knowledge and has been one of the great enemies of science."[5] To the extent that ecology remains a science, it can only regard man as a mechanism living in a complex ecosystem whose "inexorable laws" must be determined and mathematically formulated. Can the order of the ecosystem be of a different order than the rest of the mechanistic universe in which man plays no part?

The dialectical process, with its incessant folding and unfolding, creating and destroying, negating and affirming, continues to act upon science and technology as social processes mediating between man and nature. In addition to the processes described in preceding chapters, seven other distinct processes can be identified that are transforming science and technology, are operating to nullify their effects, and will, in the long run, lead to their decline. These transforming processes are:

1. The technological process is undermining the premise of science that it is an independent enterprise that is free to discover the truths of the universe.

2. The technological process is undermining and destroying man's creative forces, which are the mainspring of science and technology.

3. The increasing complexity of techno-social problems in a finite and closed system tend to negate a "best" solution.

4. The efforts to control technology through prediction, planning, and assessment will hasten the decay of creative forces and increase human alienation without helping to resolve the crises of science and technology.

5. The organizational structure and institutions of science and technology impose constraints on their further development.

6. Technological civilization becomes counterproductive—

---

[5] Singer, *op. cit.*, p. 108.

wealth becomes anti-wealth, and technology becomes a victim of its own development.

7. The destruction of the creative forces generates a counter-revolution to stave off mastery of the machine over man.

Technology Undermines Science

The varying relations that have existed between science and technology in their respective histories has already been noted. Technology, or technics as I have called the process in its early stages, was first employed by man in the era of pre-civilization. Starting much later, science borrowed much from technology in the period before and following the Industrial Revolution. In the modern period the relationship between the two has been symbiotic. As technological civilization develops, however, this symbiotic relationship may be destroyed, and technology may become the master.

Science in a technological civilization faces many problems. As a discipline it is fragmented into many specialties, each one distinct with its own language, concepts, and techniques. The division of labor theory has been perfected under science to as large an extent as under industrial manufacturing. This fragmentation was responsible, in part, for the failure to perceive the natural world as an integrated and unified closed system.

Since science rejected philosophy as early as the Galilean period, and eschewed a *Weltanschauung* that betokened metaphysical *a priorism*, the fragmentation was inevitable. Those who speak of the "unity of science" are, in the main, logical empiricists who seek to create unity through an all-embracing method of scientific analysis and synthesis. They seek unification of scientific language by defining reality in terms of those operations by which that reality is observed and measured.[6] (This method has been discussed in Chapter 2.) A second approach to unity was taken by Albert Einstein, who sought to integrate the laws of relativity with those of electrodynamics into one grand equation. Einstein was not suc-

[6] *International Encyclopedia of Unified Science*, ed. Otto Neurath, Rudolf Carnap, and Charles W. Morris, 2 vols. (Chicago: University of Chicago Press, 1955).

cessful in this approach. Thus unity has remained an elusive quest, as it must be according to science's own fundamental axioms.

Instead of unification, fragmentation has intensified and science is tending to develop a hierarchical structure, as has been pointed out by Eugene P. Wigner and Alvin M. Weinberg. A division of labor is established between those who create or discover facts and those who evaluate, analyze, and correlate them. The resulting parochialism tends to make criticism and evaluation self-serving.[7] The collection of facts through observation and measurement is, to a large extent, dependent upon the tools that technology can provide. The analysis and evaluation of the data are also coming to be increasingly dependent upon the tools of analysis—among them, specialized analytical devices and computers. The storage, sifting, and retrieval of data are also increasingly dependent upon technology. As a result, science has become heavily dependent upon technology for its own development.

As the ties between science and technology grow closer under the impetus of increasing specialization and growing complexity, the rate of development of each is far from equal. According to Weinberg, "Society's standards of achievement are set pragmatically; what works is excellent."[8] An expanding economy and the technical needs of the state demand that which works. But it is technology that fills pragmatic needs, not science. Technology demonstrates its own workability and makes science work. In this sense, as noted by Ellul, "technique cannot tolerate the gropings and slow tempo of science."[9] The pragmatic utilitarianism of technology tends to demote science, indeed threatens to make it sterile.

The increased emphasis on utility as derived pragmatically from technology is a concomitant of several factors. The pursuit of knowledge for the sake of knowledge or to satisfy curiosity or to fulfill a creative urge is fast becoming a luxury where pragmatic considerations of cost and resources impose serious constraints. The increasing utilization of science and technology by the state, for the purposes of the state, demands quick workable returns on

[7] Alvin M. Weinberg, *Reflections on Big Science* (Cambridge, Mass.: M.I.T. Press, 1967), pp. 47, 51, 119.
[8] *Ibid.*, p. 145.
[9] Ellul, *op. cit.*, p. 313.

investment. The sole drive of industrial civilization is utility in its economic manifestations—that is, products for consumption.

The purposes of the state are mission-oriented rather than discipline-oriented. The state is interested in atomic weapons and atomic energy, not high-energy physics *per se*. The state is interested in epidemiological controls and problems of public health, not the mechanisms of disease or the concepts of molecular biology *per se*. The disciplines are supported only to the extent that they further the general mission. On the surface it would appear that large-scale missions are introducing an inderdisciplinary aspect to the scientific endeavor, and hence are leading to a unification of science. But the appearance is superficial, for in most instances the integration occurs at the level of a product not at that of the disciplines or process.

Finally, mission-oriented programs have called into being what is termed Big Science.[10] The large national laboratories, funded in the hundreds of millions of dollars and employing thousands of workers in an environment of sumptuously equipped laboratories, characterize Big Science. This science is far removed from dispassionate study that seeks truth. It is a science paced by technology and motivated by rapid realization of utility, which is a technological concept rather than a scientific one. Indeed, as Weinberg states, *"Science is simply a means to help achieve nonscientific, politically defined aims"* (italics added).[11] This political orientation has also been carried over into technology. "The piecewise approach to large engineering development," J. J. G. McCue writes, "is being replaced by a systems approach, and the system . . . may be the whole nation. As a result, *engineering problems have become political problems*" (italics added).[12]

Thus the scientific enterprise has added another axiom: science is a means to achieve politically defined aims. The science that was to have renounced philosophy has now accepted politics. The science that once renounced religion and the authority of the state has now reaccepted the latter while continuing to eschew the

[10] Derek J. DeSolla Price, *Little Science, Big Science* (New York: Columbia University Press, 1963).
[11] Weinberg, *op. cit.*, p. 123.
[12] J. J. G. McCue, *IEEE Spectrum*, VII (March 1970), 31.

moral and ethical propositions of the former. Technology, the pragmatic and utilitarian process, is devouring science and, in the long run, hastens its own destruction.

Technology Undermines Creative Forces

Technology has no philosophy; it knows no values; it is free of morality and ethics. It is its own truth, for its truth is that which satisfies a need. And the body of truths is knowledge—but knowledge divorced from thought. Knowledge is used to expand the artificial world in order to conquer and reconquer nature.

Technological rationalism has eliminated sense qualities and has mechanized time, as Friedrich Georg Juenger has observed.[13] It has abstracted from the richness and variety of the universe only that which it can utilize, and has expressed these abstractions in a mathematical language that, along with being impoverished in expression, is foreign to public discourse. Man has abnegated his senses in favor of instruments and his perceptions in favor of formulae. Technology has plundered the earth to produce commodities and has transformed man himself into a commodity. It has harnessed man to the machine and threatens to free him from labor. It has provided leisure for man to consume.

Together with science, technology has deprived the natural order of any meaning but that circumscribed order amenable to manipulation and control. It has transformed the world into a universe of "appalling meaninglessness" (Julian Huxley). Man, a "bit of star dust gone wrong" (Arthur Eddington) has been deprived of mysteries and has had them replaced with problems—but only solvable problems. Man, the commodity, is consumed by technology as engineers concern themselves with nonpeople and artificial intelligence and automata. Man becomes a "meat-type" computer in this obscene transmogrification of values, while human thought becomes "wetware" in conformity with the software of computer programming.

Choice as creation rather than mechanical selection from pre-

[13] Friedrich Georg Juenger, *The Failure of Technology* (Chicago: Henry Regnery, 1956).

determined alternatives is obscured as system analysis seeks to quantify decision-making, rendering it amenable to machines. Data are reified and become real, as computers, telephone networks, and human brains become associated through a process of misplaced analogy. The algorithm, a sequence of programmed steps, becomes the technique for solving problems with computers, for educating children, for conducting wars, for managing cities, for manipulating humans.

The creativity of engineering design is succumbing to the uniformity of materials and kits and packages that are replaceable and can be thrown away. Standard circuits, standard measurements, standard sizes dictate among diminishing choices. Mass production requires standardization; it operates according to its own Occam's razor that seeks to minimize meaningful choice. Craftsmanship, beauty, and aesthetics disappear under the dictates of interchangeability and the requirements of mechanical production. The last vestige of creativity and beauty must be relinquished before the insatiable demands of automation. There can be no deviations, no exceptions, no disorder in the automated factory.

Fallible man, too, must be standardized to coexist in the standardized production process. His aptitudes are measured, his intelligence is tested, and his skills are adapted to the requirements of the machine. Man, the robot, is the factory's aim, not man the creative individual. Tyrannized by subjugation to mechanical time, mechanical machine, and a mechanical world, man produces and consumes as the charts of progress demand, as economics prescribe, and as technology dictates. The "lives of quiet desperation" merge into what Karl Jaspers has termed "Nothingness." Out of Nothingness comes despair, and anxiety, and rootlessness, and disorganization—in short, alienation.

But out of Nothingness cannot come the creativity about which science has boasted and which produces technology. Technology can create nothing by itself, for it is a means available to anybody's ends. Creativity is an end—a striving, a goal, a quest. Technology has no goals or quests. It produces nothing but more technology which further subjugates man and diminishes his creative potential.

Technological rationalism has produced a knowledge that is sterile and as lifeless as the machines it has created. It is not to be wondered that engineers debate whether computers think or whether they can create art and poetry and music. The standards of creativity the engineers use for comparison are those of the machine. Technological rationalism may succeed in making possible prediction of man's behavior—by controlling it. Technology may aid in stamping out the creative disorder that characterizes human behavior by molding man in the image of the machine. Technology may build, as Ellul foresees, the universal concentration camp. But every step technology takes destroys a bit more of that creative capacity of man upon which technology and science ultimately depend.

Slaves cannot create—nor can machines. "When human atoms are knit into an organization in which they are used not in their full right as responsible human beings," Norbert Wiener wrote in *The Human Use of Human Beings,* "but as cogs and levers and rods, it matters little that their raw material is flesh and blood. What is used as an element in a machine, is in fact an element in the machine."[14] With every step technology takes to restrict man, to control him, to transform him into a robot, that much closer does it bring its own ultimate destruction.

There Is No "Best Way"

The conditions under which the technological process interacts with man and nature undergo profound changes when operation is required within the framework of a closed system where all is connected to all and there are no exogenous costs. System analysis, establishment of measures of effectiveness, and determination of cost effectiveness will become less precise, fuzzier, and less amenable to a semblance of scientific inquiry. "The quest of the one

[14] Norbert Wiener, *The Human Use of Human Beings* (Garden City: Doubleday Anchor, 1954), p. 185.

best way in every field . . . in the absolute sense,"[15] as Ellul states the problem, will become less possible with every new technological development. The larger the computers, the more sophisticated the mathematical techniques and accounting methods, the less able will they be to select a "best" way.

We have already seen that technology's sophisticated techniques are at present based upon human value judgments at every stage of analysis, calculation, and decision-making. As the complexity of technology grows and costs mount against a background of diminishing resources, all stages of analysis and decision-making will become increasingly subject to human judgment and evaluation. Social and political values will predominate over technical criteria, because the decisions will involve primarily human choices.

The use of the criterion of optimality reflecting on ultimate efficiency will wane as efficiency loses its validity in a closed system where myriads of competing demands will have equal legitimacy. The demands will continue to increase in number and in intensity as the "limits of the earth" press upon growing populations living under ever more artificial conditions and subject to ever more restrictive controls. There will be no "best" solution to techno-social problems because every decision will please some while displeasing and alienating others. Urban renewal, for example, drives some people from their homes and neighborhoods to make way for new homes or apartments or, perhaps, a highway. The path of a highway destroys farm land or neighborhoods, or possibly a forest. The autophile will be happy, the removed city-dweller or the conservationist unhappy.

As competing demands for the same resource grow, the possibility of dividing the resources to satisfy competing demands decreases. Given a wilderness, a rural park, and urban recreation areas, the needs of multiple users can, in principle, be satisfied. As land is used up and, say, only the wilderness remains, the demands to transform it into a park with highway access and recreational areas will grow. Given a large capacity for steel production and the requisite raw materials, steel can be allocated to schools, hospitals, dwellings, industry, and to war for immediate waste and

[15] Ellul, *op. cit.*, p. 21.

destruction. When steel becomes a scarce resource and an extremely costly commodity, the use of it for one purpose will deny its use for other purposes. There can be no "best way" to allocate the use of a river for irrigation, navigation, recreation, power production, conservation, water supply, industrial use, flood control, fishing, mosquito abatement, and so forth.

Even in pure technology the experts can disagree—a situation that is becoming more common. When this happens, the state is forced to intervene and impose a decision. "It has had to resolve the quarrels of technicians and scientists," Ellul notes, "as formerly it resolved the debates of theologians."[16] This is to be expected, given that science has become a means to achieve political ends. In the case of supersonic transport in the United States, the government overruled and suppressed the report of the experts who opposed its development. The decision to develop the anti-ballistic missile, as with almost all other military systems, was taken on political not technical grounds, although the technical grounds are cited as being decisive. In intervening, the state can even take both sides, as when the Department of Agriculture promotes the growing of tobacco and the Department of Health, Education and Welfare seeks to dissuade the public from smoking because of the carcinogenic properties of tobacco. Or when the Department of Agriculture promotes the use of pesticides and the Department of the Interior seeks to prevent their use. Or when the Bureau of Wildlife and Fisheries seeks to clean up a lake and the Corps of Engineers dumps dredgings in it.

The reliance upon giant computers and system analysis to provide guidance and point to a "best way" is a chimera. It is even more than that; for this reliance poses a growing danger that man will be fooled by the apparent scientific analysis of machines grinding out complex formulations after digesting vast quantities of data. The data will have been pre-selected and the processing instructions that combine and evaluate the variables will be based on human, not technical-scientific, judgments.

It may be more desirable in the long run to note and cope with behavior of fallible man than with the behavior of the infallible

[16] *Ibid.*, p. 311.

machine. Earlier devices, E. E. Morison relates, "looms, engines, generators—resisted at critical points human ignorance and stupidity. Overloaded, abused, they stopped work, broke down, blew up, and there was the end of it. Thus they set clear limits to man's ineptitudes. For the computer the limits, I believe, are not so obvious. Used in ignorance or stupidity, asked a foolish question, it does not collapse; it goes on to answer a fool according to his folly. And the questioner being a fool will go on to act on the reply."[17]

The techniques that are used to develop giant military systems and space ships are not readily transferable to the problems of the city, the ghetto, poverty, the unindustrialized societies. None of these techno-social problems or others that involve people are amenable to the fixed parameters, the fixed constraints, and the fixed objectives that seemingly characterize these other large-scale mechanical systems. The military can utilize the monster weapons systems it produces and the myriads of other products and techniques it provides its users, because these users bear the closest resemblance to robots that man has yet produced. They are offered no choices; they are presented with decisions. Weapons systems and robots to use them epitomize the technological society.

Suboptimization, in which parts of a complex system are "optimized" according to a specified criterion, cannot produce an "optimal" solution in a social framework. Theoretically, even in rigidly determined systems, suboptimization cannot guarantee optimality over the whole system. The technological society is living proof of the failure of suboptimization. Each technological process is supposedly suboptimized: raw material extraction is optimized; industrial production seeks maximum efficiency; the economic system supposedly allocates resources in an optimum manner, and the market economy supposedly distributes commodities in an optimum manner. Autos produced with maximum efficiency operate on highways designed to maximize mobility. Yet economic ills abound and transportation crawls to a halt.

Today technological society faces manifold crises as optimality has brought mankind to the verge of catastrophe. The computers

[17] E. E. Morison, *Men, Machines and Modern Times* (Cambridge, Mass.: M.I.T. Press, 1966), p. 91.

and system analysis can "answer a fool according to his folly." They cannot aid man in making choices, for they cannot deal with the qualitative and creative aspects of human and social life which must be called upon to re-create the present and fashion the future. They cannot select the one "best way," because this way is an artifact and only mirrors the wish of those who would control man rather than establish the conditions to make him free.

Technology Control Hastens the Destruction of Technology

Technology can be controlled, its proponents assert, through planning, and its effects can be channeled positively through assessment and evaluation. Both of these premises are based upon the art of futurism, a putative technique for predicting and controlling the future. Futurism is an attempt to visualize the real world in advance of its time. It is an attempt to select from this vision those aspects most desirable and to plan to achieve them through available techniques and by devising other techniques that require development. It is an attempt not only to foresee but to control history. Yet the future, according to R. G. Collingwood, "is an object not of knowledge but of hopes and fears; and hopes and fears are not history."[18] The hopes expressed in the form of plans are either self-fulfilling, as energy and resources are directed toward their attainment, or they are unfulfilled as the best laid plans "gang aft agley." The latter course is much more probable because the long-range effects of technology cannot be predetermined. The proliferating chains of residue problems arising from quasi-solutions interact and interrelate in a manner that is intrinsically unknowable.

It was the unknowability of these long-range interactions and interrelationships that has contributed to the richness and diversity of technology *qua* technology. To the extent that the interactions and relationships are foreseen or controlled, the richness and diversity will disappear, and another creative source of technology will dry up.

The future, moreover, is aborning in the present after having

[18] R. G. Collingwood, *The Idea of History* (New York and London: Oxford University Press, 1956), p. 120.

been conceived in the past. A plan made in the present to forecast the future is thus inevitably fettered by the past. We are today hostages of the past bequeathed us by our predecessors on earth, just as our progeny will be hostages of our present and the future we would like to plan. Since these processes have their origins in the past, prediction of the future as an extrapolation of the present is fraught with great difficulty. "The future is already in the past," John Wilkinson observes in discussing futuribles, "in almost every event that acts to change the social order when effective constraints have not been introduced simultaneously and consciously with the innovation."[19]

The introduction of constraints, as in the technique of technology assessment, raises two serious problems. To constrain is to prevent something from happening or to channel events in such a manner that something will happen. The constraint must be based on a prediction of possible effects and on the feasibility of controlling these effects. The second problem arises in regard to the effect of the constraint itself on the development of the technology being assessed and on the technology that will bring about the constraint. Ultimately both problems can be reduced to the problem of planning science and technology. But planning science is a contradiction in terms: the freedom of inquiry and the autonomy of the scientific enterprise are both reduced to the extent that science is planned.

Technology assessment is "sociotechnical research that discloses the benefits and risks to society emanating from alternative courses in the development of scientific and technological opportunities."[20] A Technology Assessment Board, as has been proposed in Congress, would promote ways and means to transfer technological research into practical use, would identify undesirable by-products and side effects in advance of their crystallization, and would in-

[19] John Wilkinson, "Futuribles: Innovation vs. Stability," *Center Diary*, XVII (March–April 1967), 16–24.

[20] U.S. Congress, House, Committee on Science and Astronautics, *A Study of Technology Assessment*, Report of the Committee on Public Engineering Policy, National Academy of Engineering (Washington, D.C.: Government Printing Office, 1969), p. 1.

form the public so that appropriate steps might be taken to eliminate or minimize these undesirable effects. Technology assessment would be expected to: (1) clarify the nature of existing social problems as they are influenced by technology; (2) provide insights into future problems for the establishment of long-term priorities and allocation of natural resources; (3) stimulate sectors of society to develop technology that is most socially desirable; and (4) educate the government and public concerning short- and long-term effects of alternative solutions to current problems.

The National Academy of Engineering makes a distinction between assessment of problem-initiated and technology-initiated analyses. Assuming causal chains that converge toward a few end points, the academy suggests that system analysis can handle such problem-initiated analysis as is involved in the reduction of subsonic aircraft noise, air pollution, low-cost urban housing, etc. Technology-related analysis is concerned with prediction of the possible consequences of a given technology. Here analysts concede that as the causal chains diverge, predictability decreases with the degree of divergence. The number of foreseen consequences becomes multiplied by the number of policy alternatives considered for each. This is essentially a restatement of the thesis of Chapter 4, except that the proliferating chains of residue problems create many unforeseen consequences in addition to foreseen ones. The significance and number of the unforeseen problems are such that the efficacy of technology assessment is brought into question.

The dialectical character of technological change, rather than direct causality, also mitigates against realistic assessment. The academy perceives the element of hope and fear in forecasting the future in its summary: "The cause-effect chain does not lead easily to well-defined conclusions when applied to technology-initiated studies unless the study is modified by focusing the technology toward one or at most a few potential areas of social concern or of social opportunity that might be significantly affected by the subject technology. If the selection of these areas is perceptive, the most significant future impacts (even second- and higher-order impacts) of the technology will be identified. The uncertainty of this approach is that in making the selection of problems to be addressed, important social and political impacts could be overlooked."[21]

The above is, in effect, a recipe for selecting the obvious problems for which solutions can be conceived, and putting the others aside—which is science's universal approach. We have already noted that the methods for analyzing even these selected problems are based on other assumptions derived from human judgment. A congressional committee has recognized that aside from economic and easily quantifiable social effects, "other results are not so easily calculated into risk-benefit equations."[22]

The difficulty of establishing criteria for controlling technological development is also recognized by a study of the National Academy of Sciences: "The fact is that, with respect to major technological applications, we lack criteria to guide the choice between efficient resource allocation . . . and other objectives."[23] Lacking a calculus of current costs and benefits, the Academy of Sciences casts about instead to determine on whose shoulders to place the uncertainties. "In any situation of imperfect knowledge," the report states, "when the consequences of a contemplated action can only be surmised and when its costs and benefits cannot confidently be reduced to a net quantity, it becomes critical to decide where the burden of such uncertainties should fall."[24]

Harold P. Green, Professor of Law and director of the Law, Science, and Technology program at George Washington University, has aptly summarized the process of technology assessment. The problem of risks tends to be minimized by reasoning that "1) We do not have enough scientific knowledge to tell us whether or not the risks are really significant, but our present judgment is that the risks

[21] *Ibid.,* p. 17.

[22] U.S. Congress, House, Committee on Science and Astronautics, Subcommittee on Science, Research, and Development, *Third Progress Report, Policy Issues in Science and Technology, Review and Forecast,* 90th Congress, 2d Session (Washington, D.C.: Government Printing Office, 1968); quoted in *A Study of Technology Assessment,* p. 18.

[23] U.S. Congress, House, Committee on Science and Astronautics, *Technology: Processes of Assessment and Choice,* Report of the National Academy of Sciences (Washington, D.C.: Government Printing Office, 1969), p. 31.

[24] *Ibid.,* p. 33.

are insignificantly small; 2) as the project goes forward, further research will be undertaken to verify our judgment that the risks are insignificantly small; 3) whatever risks do exist can be reduced to tolerable dimensions through technological devices; 4) if the risks indeed are found to be, and remain, significant, the program will of course be abandoned or drastically reduced or controlled to protect the public interest."[25] This approach we readily recognize as the laboratory *in vivo*, with the uncertainties tested on the peoples of the earth.

The Academy of Sciences recognizes the inherent difficulty in foreseeing "the convergence of several technological developments or of one such development with other trends"; in "imagining the effects of scale"; in the "perception of social and environmental consequences"; in "imagining the supporting systems new technologies would demand"; and in "reliably foreseeing advances in technology itself."[26] But in spite of the problems and caveats raised by the academies' reports, technological civilization is impelled to do something to constrain and control the rush of technology to dominate all human endeavor. So technology assessment, planning, and prediction are considered among the techniques of control.

The most desirable method of prediction is to design the future so that control is built into development. Uncertainty is replaced with certainty, randomness by order, diversity by conformity. This is the master plan of technique foreseen by Ellul. But this is also the surest path to destroying science and technology. For a controlled future in the interests of constraining technology must ultimately seek to control people. Controlled people, or robots, cannot, however, create a science or maintain a technology.

Organizations and Institutions Impose Constraints

Although a lone scientist can conceivably still make a scientific discovery through his own resources and a lone technician can originate a new enterprise in his basement, the overwhelming trend in both science and technology is toward the organization and in-

[25] *Science*, CLXVI (November 14, 1969), 850–852.
[26] *Technology: Processes of Assessment and Choice*, pp. 44–47.

stitutionalization of both science and technology. This trend has developed out of the growing complexity and increasing cost of scientific and technological enterprise. Medical research and medical technologies require laboratory facilities, complex equipment, access to hospitals. Physical research requires huge atom-smashers, giant telescopes, costly furnaces, microscopes, and analytical and computational devices. The conquest of the atom and the moon required the formation of large-scale enterprises to manage, organize, and develop the intricate complex of men and machines involved in these programs.

Organizations have their own dialectic. They have a distinctive birth and death process. Once established, their primary objective is to maintain themselves. Bureaucracy, self-serving actions, and growth come to substitute for their original *raison d'être*. To the extent that an organization accomplishes its mission, it changes the conditions that gave rise to the organization. Hence organizations should be ephemeral and transient, continually giving rise to new organizations that take on new tasks. But once a system becomes organized, it tends to inhibit and restrict efforts to reorganize. With the passage of time, organization becomes a block to further organization. It also becomes a deterrent to cope with the new situation it has helped to bring into existence.

The organization of science to carry out "politically defined aims" is a contradiction. Organization requires authority, and authority requires hierarchy, and hierarchy demands obedience. The administrative authority in scientific organizations cannot be divorced from scientific authority, for the power to make decisions invests the decision-maker with intellectual authority as well. Freedom to seek the truth must by necessity be circumscribed by the demand to perform an assigned task. In Big Science the assigned task may well be the equivalent of the fractionated piece-labor that is the lot of the worker on an industrial assembly line.

The Manhattan Project has become the prototype of the scientific organization formed to carry out political ends: compartmentalization of work, divorce of scientist from ends, renunciation of personal responsibility, obedience to authority—and secrecy. The close association between science and war and the support of science by government for national purposes has become a major component of Big Science in all technological societies. Fritz Haber, the German Nobel laureate who synthesized ammonia and created the chlorine gas used in World War I, stated this point bluntly when he affirmed that a scientist belongs to the world in peace but to his country in time of war. The secrecy that surrounds a great deal of Big Science stultifies science even as science stultifies man. The mistakes made in secrecy persist because there is no corrective policy. Natural laws, however, cannot be secret.

Increasingly, science is based not on human needs or curiosity but on political aims. Scientific institutions begin to assume roles that are intrinsically inimical to science. Science becomes more and more "extrinsically motivated," according to Weinberg, as it seeks to explain questions that were originally part of some other problem, "usually, though by no means always, some practical matter." Even these practical matters are usually selected not on the basis of importance but on solvability. "Unfortunately," Weinberg adds, "the important questions are often the most intractable ones," and "the strategy of pure science is always to deal with soluble problems which by their nature, tend to be narrow in impact."[27] The postponement of delving into Wigner's "nth layer of concepts" is a constant danger to a Big Science engrossed in practical matters that are solvable and are assigned to it as political missions. The acceptance of secrecy and the abandonment of democracy to the decisions of experts undermines the integrity of science.

As science and technology expand, the number of organizations formed to carry out assigned tasks grows apace. When the countertechnologies begin to proliferate, the number will expand still further. Is there a right size for the number of scientific organizations or the individual organization? Shall the proliferating federal agencies, for example, be centralized under a science "czar" or shall they be decentralized? Shall agencies be expanded to encompass more assignments or broken up into autonomous sub-agencies? Can truth and freedom be achieved under hierarchy, whether small or large?

[27] Weinberg, *op. cit.*, pp. 148–150.

Size, however, appears more significant in relation to the technological process. Is there a right size for social units in which technology functions? Is there a right size for units of technology? Although technology has made possible the development of large urban conglomerations, are not these conglomerations now facing breakdown? The impending breakdown is twofold, encompassing both the social unit and the technological unit. The physical decay and deterioration of cities is a result of many factors, including the inability of technology to provide essential services. The social disintegration arises from the anomie welling from the sense of Nothingness, the feeling of people that they are unwanted and unneeded, the boredom that comes from the satiety of the few, the despair that comes from deprivation of the many.

As populations and economies continue to expand, social and political organizations must expand together with the technologies they support. Conceivably, a size can be reached when neither centralization nor decentralization nor combinations of both can be developed that will effectively fill the needs. Numbers create their own problems whose logic of solution cannot be decentralized. Traffic control and highways must be increasingly centralized and coordinated because of the characteristics of vehicles. However, a point in size may be reached when, as we have seen, autos and highways can only produce immobility. Sewage and waste treatment in a closed system cannot be decentralized but must become increasingly centralized. There is a point, however, beyond which a sewage system cannot function.

When size, under the impact of exponential growth, reaches a certain point, there can be no balance between centralization and decentralization. There may be a critical queue size, John Riordan notes, beyond which order is impossible: a waiting line becomes a mob. A telephone system begins to malfunction when it is excessively overloaded, as the inhabitants of New York City are learning. The malfunction arises not only from failure to keep up with demand but from technological circumstances. The system becomes so complex that it becomes incomprehensible, in part, to its engineers. Unsuspected relationships arise, synergistic effects occur, backpaths and race conditions are established. Reliability of tens of thousands of components in tandem is reduced by the multiplicative law of associated components. The great Northeast blackout of November 1965, during which one-sixth of the United States with a population of more than thirty million people was deprived of all electric power, is a harbinger of the results of technological solutions that ignore asymptotes. The aborted mission of Apollo 13 is a reminder that mechanism is as fallible as man.

The increasing role that technology is taking in the development of all facets of modern civilization will make increasing demands upon the social organization and the individual. Component parts of the industrial system are being integrated into more comprehensive forms of organization as centralized government is forced to coordinate, regulate, and integrate individual and social action. The social system is being forced into ever more rigid patterns of static organization as control of the environment becomes more difficult and complex, as resources are further depleted, and as the results of past waste and excesses are felt. "With the consumption of each additional barrel of oil and ton of coal, with the addition of each new mouth to be fed, with the loss of each additional inch of topsoil," Harrison Brown has warned, "the situation becomes more inflexible and difficult to resolve. Man is rapidly creating a situation from which he will have increasing difficulty extricating himself."[28]

Failure to fulfill the requirements of technological civilization in maintaining the artificial environment or bringing about the degree of social organization required to achieve stability in the national and international spheres will result in dislocations and upheavals that will tend to disintegrate civilization. An inflexible pattern of social control may lead to a society of ants; the failure to develop social sanity may lead to a society of savages. The first may well culminate in the second and, in either case, man may reach the point of no return, for the chance circumstances that made possible the development of technological civilization will have been de-

[28] Harrison Brown, *The Challenge of Man's Future* (New York: Viking, 1956), p. 265.

stroyed by that civilization. So ravaged and wasted will the earth be that only a highly organized technical society could again create an advanced civilization. A modern savage whose forebears thousands of years ago discovered how to work with metallic ores and make primitive tools will be unable to re-create the vast complex of engineering techniques that would make possible a new start. Without medical techniques the population would be subject to epidemics and pestilence, and thousands would perish in the absence of the many life-saving niches developed by modern medicine.

To prevent breakdown while fomenting continued exponential growth, the path of all technological societies must lie in the direction of rigid controls. Big economy requires big government which requires big science and big technology. Tendencies toward disorder and anarchy will be stomped out as being inimical to "progress." "Law and order" becomes the shibboleth of technological society in decline. Decision-making will become the province of experts who will substitute for the democratic political process. Experts who live in the unreal world of science and technology will become the arbiters of the fate of the real world. Decision-making based on facts with a criterion of efficiency becomes the task of a new priesthood. Until printing was invented, Lewis Mumford has observed, the written word was primarily a class monopoly. The priests and the court were privy to the secrets of nature and the wisdom of man. "Today the language of higher mathematics, plus computerism, has restored both the secrecy and the monopoly, with a consequent resumption of totalitarian control."[29] The world of *1984* draws nigh both in time and concept. "What was once Orwell's science fiction," Allan F. Westin relates, "is now current engineering."[30] Even in technology assessment, whereby attempts may be made to ameliorate the harmful effects of technology in the laboratory *in vivo*, it is believed neither "practical [nor] desirable to open the whole process of technological assessment and decision-making to public view." The academicians believe it important to "assure that the evidence and arguments on which major decisions are

[29] Mumford, *op. cit.*, p. 199.
[30] *Business Automation*, XVII (January 1970), 62.

based will be open to public scrutiny and will be subject to timely review in appropriate public hearings."[31]

The totalitarian society is a rigid and efficient society. The military flourishes in such a society. Science grows—in size not depth, until its wellsprings dry up. Technology can flourish, for it is required to establish the instruments of social control, until it too languishes under the loss of the creative impulse.

Technological Civilization Becomes Counterproductive

The technological society has planted the seeds of its own destruction, and the shoots are now beginning to appear. Putting the metaphor closer to the violence of the age, Elmer W. Engstrom has written in the *American Scientist*: "The introduction of a new technology without regard to all of the possible effects can amount to setting a time bomb that will explode anywhere from a month to a generation in the future."[32] The bombs have now begun to explode, and the repercussions are mounting. The proliferating residue problems have begun to combine and establish interrelationships that could not be anticipated or foreseen. These residue problems have begun to interfere with the solutions to some problems and to reinforce the intensity of others. Many of the problems, including the major problems that beset a technological society, are becoming unsolvable.

If man now shows a belated awareness of the unfolding drama of survival it is because the progress of technology itself has been impaired. Autos can no longer provide the mobility for which they were intended. The air and waters have been polluted to the point where human life has become endangered. The arms race makes the search for peace more elusive. Depleted resources have begun to manifest themselves in the search for substitutes, in the mining of low-grades ores, in efforts of reclamation. The larger cities are becoming ungovernable; the sprawling urban conglomerations are rotting at the core, and people seeking escape destroy the land and

---

[31] *Technology: Processes of Assessment and Choice,* p. 67.
[32] Elmer W. Engstrom, "Science, Technology, and Statesmanship," *American Scientist,* LV (March 1967), 72–78.

its resources to establish new centers of habitation—which promptly come to resemble the old.

To maintain technology, technology itself must be increasingly devoted to the production of counter-technologies which are attempts to mitigate the more serious and harmful effects that technology brought about in the first place. Anti-pollution devices on autos, factories, and power plants are becoming mandatory. Noise silencers will become mandatory on jet planes, pneumatic tools, and heavy equipment. Scrap and waste will begin to be collected, sorted, cleaned, and reclaimed. Biodegradable, nonphosphate, nonnitrate detergents, "harmless" food additives and drugs and pesticides will be sought. The green revolution will have to undergo profound changes to mitigate the effects of excessive nitrogen and phosphates on the soil. Sewage will come to be regarded as a precious commodity, and recycling and reclamation will have to be introduced in its treatment. The mineral and chemical wastes poured into the waters will have to be screened out and removed for reclamation. The slag heaps and mine tailings will have to be reclaimed and redeposited without scarring and marring the landscape.

Since every technological process is a consumer of resources and energy, and every process is a polluter and gives rise to waste, every process will be caught up in the necessity to provide counter-technologies for the purpose of maintaining itself. Technology will be compelled to produce more technology not to progress but in order not to retrogress. Like Alice, technology will have to run faster merely to stand in the same place. Counter-technology will begin to predominate over technology as the time bombs set during three hundred years continue to detonate at a faster pace. Moreover, little will be accomplished if the technologies are despoiling and polluting even as the counter-technologies operate to reverse and redress the effects. For counter-technologies to be effective, some technologies will have to be modified, some will have to be discarded. The entire technological process will be forced to undergo scrutiny and evaluation.

If counter-technologies are to achieve anything, they cannot be treated as isolated processes. They must operate in the closed system where all is connected to all. Every counter-technology must become a part of a large and complex cycle, a cycle that embraces the entire world. The rivers of the continents flow into one large body of water, the earth's oceans. The layer of air that surrounds the earth sustains all life; the thin layer of soil on the earth's surface nourishes all life.

The world-wide counter-technology will be a super-system. It will require a giant organization to monitor world-wide technology, collect and evaluate data, and devise and operate appropriate counter-technologies. The system will be complex and intricate. It will require large forces for proper maintenance. It will have to rely on giant computers and sophisticated analytical and decision techniques. It will have to include a world-wide political system to police and enforce the counter-technology.

One begins to sense that we have been here before. Counter-technology is at base still a technology, and as such is subject to the dialectical laws of change as is its counterpart. Each solution to a counter-technology problem will prove to be a quasi-solution. Each will generate a set of residue problems arising from incompleteness, augmentation, and secondary effects. The residue problems will proliferate in endless chains with unforeseen results until they too become unsolvable. The organization required to maintain world-wide control of the environment will exhibit the anti-creative tendencies of all organizations; the bigness will undermine freedom and install authority; the missions and goals will be politically defined; the computers will be unable to determine a "best way." In short, the measures taken to insure that technology will remain in the same place will hasten its destruction.

The attempt to externalize costs in a closed system has piled up an enormous debt that society is now compelled to repay. The coal and oil and minerals that have been consumed are, to a very large extent, irretrievably gone. Some of the excesses committed against the earth and the environment may be irreversible. Even now it can be asked whether technology has not already brought the world beyond the point of no return.

To stave off a possible future that may be the result of three hundred years of unexcelled greed and blindness, the world faces the accumulated debt incurred during this period. The wealth that has also been accumulated must now be dissipated to pay for the anti-wealth that does not appear in the balance sheets of Gross National Product. The question as to who will pay the bill is largely academic. Producers state that the cost of counter-technologies will ultimately be paid by the consumer. Some costs will be borne by industry. Governments will be called upon to provide an ever-growing financial support for the counter-technologies. To the extent that the debt is payable, it will be repaid ultimately by all the peoples of the earth.

The cost of cleaning up the waters of the United States is estimated to be some amount exceeding $100 billion. The cost of saving Lake Michigan is estimated at more than $10 billion. These costs do not include all the counter-technologies that will be required to prevent pollution while other counter-technologies attempt to undo the damage that has already been done.

Already straws are in the wind. A New York utility has stated that restriction on sulfur content of fuel oil has added $15 million to its annual fuel bill, with an equal amount to be added when tighter restrictions are imposed. A special $120 million refinery is nearing completion in Judibana, Venezuela, to desulfurize the oil it ships to the New York area to comply with its restrictions. Representatives of the coal-mining industry are protesting federal regulations aimed at protecting the health and safety of mine workers. These regulations, it is contended, are paralyzing production and forcing marginal mines out of business. Canada is beginning to take steps to preserve the tundra and the permafrost against the proposed rape of its newly found mineral treasures in the Arctic region—steps that will slow or impede exploitation and eventually raise costs.

The costs to maintain an ever more artificial environment under the highly restricted conditions that will be imposed on technology, together with the financial burden of developing, operating, and maintaining counter-technologies, will soon begin to dwarf the huge sums spent on wars and preparation for wars. But wealth is derived from labor and resources. The resources are nearing depletion, and counter-technologies will only hasten their disappearance. Labor expended to disperse the accumulation of anti-wealth will not produce wealth. How, then, will the cost be paid?

The counter-technological society will bring into play more decisively the limitations and constraints discussed in Chapter 5. Impotence principles of conservation of energy and thermodynamics will impose roadblocks; the perimeters of the possible will establish constraints; the limitations on man's knowledge and understanding will have grown. The further development of technology and counter-technology will be more difficult and more complex. Greater control will be required of larger aggregates against a background of problems becoming increasingly unsolvable. Like the sorcerer's apprentice, research and development will be inundated by more and more data and information.

Research will be the key to the counter-technological society as it was for the technological society. Anti-wealth will have to be considered, and those areas formerly ignored will have to be explored. Unfortunately, the scientific community will have to start nearly at the beginning in this new venture. William D. McElroy, director of the National Science Foundation, reviewed the situation as the 1970's began and found that: "Thirty years ago much of the science was available and was used to devise urgently needed technologies; today much of the requisite knowledge simply does not exist. We know far too little about the interactions that occur within any ecological system. We do not really understand the dynamics of our environment or the effects of technology upon it. We know little about the more subtle effects of pollution. We cannot predict with confidence the behavior of individuals nor that of social groups and institutions. We are not in a position to assess adequately the relative costs and benefits to society of any technology or any course of action."[33]

Little will be gained by belaboring the point that this statement is virtually a confession of intellectual bankruptcy, or by asking the identity of the "urgently needed technologies" that were pursued so arduously. Yet this theme occurs in almost everything that touches

[33] William D. McElroy, "A Crisis of Crises," *Science,* CLXVII (January 2, 1970), 9.

upon our present crises. One is overwhelmed by the tasks that await us on the frontiers of research—and the depths of our ignorance.

The research required for water quality control, for example, includes:

—determination of limits and desirable levels of dissolved oxygen and temperature necessary to protect aquatic life.

—definition of acceptable bacteriological limits in waters used for recreation.

—determination of chronic effects which minute concentrations of pollutants might have on life systems.

—determination of the causes and cure of eutrophication.

—study of the relation of water treatment to pollution and eutrophication.

—study of the economic effects of water quality.

—study of the problem of storm waters, and the establishment of dual sewage systems.

—study of the distribution of toxicants in the food chain of wildlife.

—study of the mechanism of absorption and deposition of toxicants in body tissues.

—etc.[34]

The research required to investigate the effects of nine hundred pesticides and herbicides on the two million living species of animals in the world, as recommended by the Secretary's Commission on Pesticides, is staggering. The report on solid waste management similarly identified a sufficient number of research projects to occupy all the scientists and engineers in the country for many years. The report recommended a national program that would "optimize" solid waste management systems to provide "efficient, economic pathways" to deal with waste from the point of origin to the point of ultimate disposal in a total system. More than 140 distinct and comprehensive studies were recommended.[35] Some of the studies would have to be multiplied by the number of industries

---

[34] *Research Needs*, companion report to *Water Supply Criteria*, prepared by the National Technical Advisory Committee of the Department of the Interior; summarized in *Environmental Science and Technology*, II (November 1968), 998–999.

[35] *Solid Waste Management*, pp. 87–102.

producing wastes and by all the categories of waste products, to make a total of many thousands of projects. The study on oil spillage recommended more than 115 distinct research studies, many of which encompassed numerous subresearch projects.[36]

Each research study will become a technological process embodying the six steps of the process described in Chapter 3. Each step of each process will pose its own set of subproblems. Each problem and subproblem will become a quasi-solution once again generating new sets of residue problems.

Thus every step in the further development of the technological society, and every step in the building of counter-technology, will serve to destroy those vital and creative forces that alone can sustain science and technology. With every step that is sought to carry technological society forward, its ultimate decline will be assured. As decline sets in, organization and control will be increased, a shield of "law and order" will mask the increasing repression that will come about as societies confront the limits of the earth and face the perimeters of the possible. The universal concentration camp may be the ultimate fate of technological civilization, instituted as the final desperate act as societies strive to stave off collapse.

This decline will be accompanied by a number of paradoxical circumstances arising from the dialectical process. Science had sought to eliminate sense qualities and observe the reality of nature through measurement and experiment, but the reality has receded and become an abstraction of abstractions which rest more and more on the faith to believe them. Science had sought to remove the idea of sin from its activities, and sin has returned by way of the consequences of those activities. Science began as a passion to find truth and promote freedom and has become a tool for carrying out politically assigned tasks in an atmosphere of secrecy and narrow parochialism. Science, which sought to establish reason as authority, finds that the reason is at base human judgments and values that seek to justify human behavior. Science attempted to free man from the laws of Providence and instead shackled him to the "inexorable" laws of physics. Both science and technology have sought

[36] *Oil Spillage Literature.*

to free man from the control of nature, but have submitted instead to the control of society. Finally, after bringing mankind to the verge of catastrophe by exploiting and dominating nature, science and technology have brought themselves to the state of decline as they become increasingly dominated by a nature that threatens to bring the scientific-technological experiment to a halt.

The seventh and last transforming process that was cited as contributing to the decline of the technological society—the resistance of and counteraction by man—is perhaps the most powerful of the forces against technology. Discussion of this force has been incorporated in the next, and concluding, chapter because resistance and counteraction are not conceived as negative forces. They are, instead, the sole hope that the decline of technology will bring about a positive transformation, and that the decline will lead to the *inefficient society* where man can once more become a part of nature and where man, not the machine, will create the outlines of a new society.

12   *Toward the Inefficient Society*

THE ANALYSIS throughout this book has led to the conclusion that science and technology cannot help to solve the problems that confront the world because they are major contributors to the problems. The analysis has demonstrated that science and technology have helped to foment the major crises that confront mankind and that their further development will increase the unsolvability of the problems they helped to create. Although science and technology have become major, if not *the* major, motive forces of modern civilization, they tend to negate themselves and to doom the civilization they drive but are unable to save.

The freedom that technological society promised is negated as mastery over man accompanies the drive toward mastery over a recalcitrant nature. Obedience is demanded by experts, with the state becoming adept in the dynamics of control. Under the cloak of technological rationalism man, who is becoming a robot, willingly accepts his role as servant to the machine in return for a cornucopia of pecuniary consumption. He accepts the control of the state in payment for this bounty. Technology, as Herbert Marcuse explains, "provides the great rationalization of the unfreedom of man and demonstrates the 'technical' impossibility of being autonomous. . . . Technological rationalism thus protects rather than cancels the legitimacy of domination. . . ."[1] Man is enslaved and revels in his slavery although he calls it other names. Man becomes what the market wishes him to become—a consuming animal, one that never really experiences what is his but experiences himself as the person he is expected to be, as Erich Fromm has explained. At the end of the technological experiment lies Ellul's universal concentration camp.

We have been deprived of the solace of believing that no matter how black today the morrow will be brighter. Tomorrow, in fact, may be worse to a degree that is difficult to envision. We can no longer sustain the myth of "progress," although history should have revealed the wish content of this myth long ago. The dialectical transformation of society, paced by man's genius and irrationality, has turned the hope and vision of a better society into the cynical foreboding of *Brave New World* and *1984*. The test-tube creations already have their counterparts in the dehumanized automata of the cybernetic age. Social organization has become a vast conspiracy. Guernica, Lidice, Dachau, Hiroshima, BenSuc, and MyLai have become symbols that sear man's heart and burn his conscience —where conscience is still able to evaluate human dignity.

Pessimism, where warranted, is a necessary antidote to mistaken euphoria as a first step to realism. In the foreword to his book, Jacques Ellul refrained from providing solutions. He asserted that "in the present social situation there is not even the beginning of a solution, no breach in the system of technical necessity."[2] I have tried to show that the breach is being established by technology, which is negating itself. The technological society and the countertechnological society both unleash forces that are self-destructive. Whereas Ellul sees the destruction of man in the triumph of technology, I see it in the *decline* of technology.

But in this decline can we not see a tiny glimpse of hope? Slaves have lost their freedom and cannot create. They are harnessed, albeit unwillingly, to the machines they serve, but as long as life

---

[1] Herbert Marcuse, *One-Dimensional Man* (Boston: Beacon Press, 1968), pp. 158–159.
[2] Ellul, *op. cit.*, p. xxxi.

remains in them, they can revolt. Not all are willing. The youth of this generation refuse to accept the possibility of no future. They have looked askance at the slavery of mankind and his worship of what they deem false gods. They question the secular faiths that have enthroned a reason of lies and deceit, greed and selfishness, arrogance and ignorance. Every time a baby is born, R. D. Laing reminds us, there is a possibility of reprieve. "Each child is a new being, a potential prophet, a new spiritual prince, a new spark of light precipitated into the outer darkness."[3] Here is at once the power that can hasten the decline of the technological society and can transform it into a society based upon man in nature, with human values and with faith in a sensate, erotic, and creative individual.

I am pessimistic because, though I can foresee a different future, I am not sure that a way can be found to arrive at that future. Nor can I help to wonder whether in that future man will despoil his environment and endanger his future once again, not because he had created a new technology, but because he is man. Science and technology cannot be viewed as aberrations but represent vital drives of man—for mastery over nature to overcome fear and want, for mastery over man to display domination and power. Capital accumulation under all civilizations has been cruel. Earth and animal and man have been ruthlessly exploited to build an economic base consistent with technical means. The excesses of capitalism have been mirrored by the excesses of communism. In every age the masses of people have been mobilized in voluntary and involuntary slavery on behalf of techniques.

We have now arrived at a time where to continue the path of technological society will lead either to the universal concentration camp or to biological extermination. We can seek to extricate ourselves from impending disaster by denying the crisis and calling for more and better techniques—a policy that will exacerbate, not ameliorate, the situation. Or we can recognize the basic error in the technical vision and seek new directions to an unknown future. It is simple to state that the growing population and their growing needs

---

[3] R. D. Laing, *The Politics of Experience* (New York: Pantheon, 1967), p. 14.

can only be accommodated by more techniques. But this continues a vicious cycle, a treadmill that runs faster toward dissolution. The technological society is being transformed by the dialectical processes it has set into being. There is no inevitable outcome but many different possibilities.

The present crisis is not one of techniques but of philosophy. The age of reason and its flowering in science and technology have been shown to be unreasonable. Unreasonable not because science and technology were misused by evil man but because science can achieve only a partial vision of the universe and, in achieving it, has dehumanized man. The exponential path to an undefined and undiscovered progress is a lemming-like compulsion toward suicide. Technology has been a means to tame the land, the plants, the animals, nature, and finally man. The earth has been impoverished in the process, and man has become alienated and enslaved. To turn aside from a philosophy based on the assmuptions of technological civilization, therefore, is not an act of retrogression, a utopian discursion, an escape from reality. It is an act of necessity. By recognizing this necessity man can reach for freedom along new paths. Man can transcend alienation and rediscover himself in the philosophy of post-technological man.

This philosophy is based on eight principles which are value-oriented and derived from moral and ethical considerations of man. They are points of view that can be interpreted and reformulated in many diverse ways along many paths of wisdom—for wisdom is not an answer but an open pathway that leads to enlargement of the human spirit.

Man Is Within Nature, a Part of the Mystery of Life

Man is not the apogee of life nor the special object of evolution. He is the cosojourner on earth with millions of living species. The earth was not made for his domination, and he dominates it at his peril. Though he seeks to dominate nature through force or understanding, the results are the same. Nature is a living, not a domination—a complex web of organic relationships bound together in endless and ageless cycles. The rich tapestry of organic life must ever elude man's analysis, for once the mystery of the tapestry is decoded, man will perish.

Nature is a mystery to be worshiped, not a problem to be solved. A natural scene embodies a mysterious infinity, according to John Ruskin, an infinity of variety not numbers. Nature is a setting to be a part of, not an environment to be mastered. The bounty of the earth is a treasure to be guarded, not a commodity to be dissipated. Land is to be loved and respected, according to Aldo Leopold—a view which to Leopold was an extension of ethics. The lightning-shattered oak that he sawed for cordwood was a rich repository of living history. The fields that he tramped through also were a testament to bygone people and bygone animals, many extinct. Thoreau, too, farmed the dust of his ancestors, although he doubted that chemists could detect their traces. He went forth "to redeem the meadows they have become. I compel them to take refuge in turnips." Nature inspires reverence of the past and imparts immortality to the departed. Nature is not an accessory to which man can become accustomed, or get along without, according to one's taste. It is a vital part of man's being. It nourishes his life, and its absence imposes sterility. Life outside of nature is a form of cruelty, for it transforms vibrant, living people into artificial organisms whose prototype is the feedlot of industrialized farms. This is the direction toward which man is moving. In his billions he will vegetate in assigned feedlots, consume assigned rations (perhaps in the form of pills), and perform assigned tasks. The feedlot creature has no need of pasture and woods, sun and rain. Chemical foods and fibers, artificial light and artificial climate can serve his needs—but only for a time. The ultimate source of the foods and fibers, the water and climate, is the real world of nature of which more and more of civilization is deprived.

Man requires nature not as a resource to be exploited, even if wisely, but as a source of inspiration, humility, and human renewal. Nature is not a psychological renewer to be recommended for weekends or vacations. It is a necessity without which human life cannot be lived, though organic life may continue. Man in nature can transcend nature through confrontation with the cosmos and thereby overcome the alienation that saps his being.

Man must preserve and protect nature as he values his own life. The cycles are to be guarded, the stores to be replenished. Each plant cut from its stem must have meaning, each tree felled must be justified in the laboratory of nature, not the counting houses of man. Each grain of sand moved or stone quarried must be viewed as an act that disturbs a process that has proceeded for millions of years. Nature is not an enemy when its ways are understood, and when man assumes his humble role within its grandiose framework. Community within nature can be the prelude to community with man. Man is truly an infant among the earth's living organisms, though his development was among the longest. It is in the spirit of humility that man seeks to comprehend the agelessness of the earth and its fellow creatures. The rhythm of spring returning to a hillside at Walden is inspiration for man as it was for Thoreau, America's bachelor of nature: "The earth is not a mere fragment of dead history, stratum upon stratum like the leaves of a book, to be studied by geologists and antiquaries chiefly, but living poetry like the leaves of a tree, which precedes flowers and fruit—not a fossil earth, but a living earth; compared with whose great central life all animal and vegetable life is merely parasitic."[4]

There Are Infinite Paths to the Apprehension of Reality

The infinite variety of the world has been diminished by the selectivity of science, the narrowness of its fields of study, and the circumspection of these fields by a Pythagorean mentality. Occam's razor has destroyed the multiplicity of explanations for phenomena, all of which may be equally valid and real. The telescope and microscope have aided man's external vision but have blinded him to that which stands before him—the natural world. The absurd, meaningless world of technological man sees information in mathematical formulae and fails to see that more information is contained in a seed that grows into a plant, bears fruit, and withers.

Man is saint and devil, rich and poor, intelligent and ignorant,

[4] Henry David Thoreau, *Walden* (New York: New American Library, 1958), p. 205.

loving and hateful, warlike and peaceful, savage and tender, noble and despicable, exalted and depressed—an infinitude of beings of which only glimpses can be noted by sociologists and psychologists. The artist with his inner vision can more readily probe and describe the real world of man, for he reveals himself in the purview of his life and behavior. The Bible, *Hamlet* and *Macbeth*, *The Prince* and *Faust* reveal greater understanding of the conditions and predicament of man than do the findings of modern-day scientific analysts. *War and Peace* and *MyLai 4* are more apt representations of the inanities of war and the struggle for peace than the military strategists and political statesmen can depict. Thoreau, Audubon, Muir, and Aldo Leopold have rendered a more exciting, humane, and inspiring view of the natural world than have the natural scientists. Compare, for example, Leopold's description of the cutleaf Silphium with Asa Gray's technical description à la Linnaeus. One is redolent with the history of the Wisconsin prairies, beginning more than a century ago, and describes the tenacity of the plant to stave off encroaching civilization. The other is a coldly analytical botanical description. The former description would make one hesitate, even momentarily, before destroying any plant. The latter's lifeless and impersonal description is an invitation to the mower from the highway department to cut off the living remnant of a bygone prairie.

In painting, too, the inner vision of the artist reveals the many-textured human, not the statistical model of the scientific mind. The paintings and sculptures of Michelangelo have remained as expressions of a vision of man through centuries. The latest analytical table of the scientist of man will be outdated with the accumulation of the next bit of data. Sociologists study the slums and ghettos and human refuse heaps of the technological megalopoli as empirical facts. Authors and painters and actors depict the human degradation and depravity of these abominations. Concern and protest are writ large on their canvases and pages and stages.

Unfortunately, the technological mind has made its inroad into art as well as other facets of life. The artist has become entranced with technique, with materials. The inner vision of truth-seeking has been replaced by cleverness at manipulating media. The architectural monstrosities that blight urban centers are the creation of technicians, not artists.

The artist reveals the heart and soul of man. The scientist observes and measures externalities. The artist presses all his senses into the quest for truth; the scientist presses lifeless instruments into the search while denying his senses. The technician manipulates matter and energy for immediate utility.

The bias against the human truth-seeker is not a product of this age. Throughout human history the humanist and artist have waged an unceasing struggle against the experts of power, knowledge, and manipulation. David Hume aired the ascending scientific viewpoint with the remark that "Poets are liars by profession who always endeavor to give an air of truth to their fictions." Abbé Nicolas Trublet, imbued with the spirit of the Enlightenment, echoed Hume's words a few years later in even stronger language. "As reason is perfected, judgment will more and more be preferred to imagination, and, consequently, poets will be less and less appreciated."

The retort to the rationalists who were not aware of their irrationality was well rendered by Madame Sophroniska in André Gide's *The Counterfeiters:* "I really think there's more truth to be found in the poets; everything which is created by intelligence alone is false." Scientific intelligence cannot provide a complete picture of the world. It is too fragmented, too specialized, too selective, too falsely objective. The resurrection of the natural world as a home for man to be sensed by all the modalities and buttressed by imagination and love and hate and fear and dread and understanding is necessary if we are to remain human. "It must be poor life that achieves freedom from fear," Leopold wrote in assessing the growing impoverishment of the American continent.

Can man, who no longer fears the lightning or the roaring winds or the energy in the sun and earth, fear himself? And if he fears not himself, can he master the suicidal urge that impels him to his own destruction? If power and mining the earth are his profane religion, the bulldozer his cross, will Mother Earth not succumb to the technological barbarism that is parasitic on all life, that sucks the earth's sustenance and returns nought to it?

The spiritual crudity of man's ascent to the moon was manifested in the military aspects of the venture. It is sufficient commentary on the society whose vital resources were expended that neither scientist nor artist participated in this breaking loose from the earth's habitat. The response of the people of the earth has been laconic; the second visit to the moon elicited boredom. It was a technological feat superbly performed by robot engineers. The computers plotted accurate trajectories, the rockets fired as required, the mission was accomplished with military precision, the astronauts were brave and heroic pioneers. But it was an event in which the technological spirit triumphed over the human spirit.

The human spirit is not numbers, nor atoms, nor enzymes, nor brain cells. It is life itself, and the attempt of science to reduce life to a mechanical system, even though life be extended and perhaps created in the test tube, must end as Aldous Huxley foresaw in his Greek-designated automata. Human life transcends science and towers over technology. The forces that strive to encompass man in an artificial cocoon supplied with sensitive sensors stifle and suppress man. His confrontation with both the sacred and the profane must be with all the modalities and sensitivities, imagination and dreams, of which man is capable.

The Inefficient Society Is a Humane Society

The myth of the efficient society has led man to a blind pass. In pursuit of technology he has enslaved himself to a heartless master. The question is not to redirect technology but to inquire whether efficiency and the technology directed toward its achievement are not false goals foreign to nature and man. The question is whether efficiency is not an artifact embedded in a mythological framework that has masked the real world and subverted the goals of mankind?

The problem of efficiency may be posed in apocalyptic fashion by considering two suicides. In the first, the victim slashed his wrists, whereupon he slowly bled to death. In the second, the victim put a bullet neatly through his brain and died instantaneously. Which was the more efficient suicide, it might be asked. Immediately one senses that there is something amiss in the query. It is a valid

English sentence; each word has meaning, and yet one perceives a quality that renders it unrealistic. This quality becomes more evident if we consider the Nazis' answer to the "final solution" for the "inferior" races during World War II. Scientists and engineers labored to create a gigantic extermination machine that culminated in the gas chambers and crematoria. If the question is asked, "Were not Dachau and Auschwitz *efficient* extermination centers?" the grotesqueness and obscenity of the question becomes apparent, as does the quality we are seeking.

In science, efficiency considers the ratio of outputs to inputs. Efficiency is a measure of the loss of energy in an energy transformation. In a broader sense, it is the ratio of an achieved goal compared to an ideal or an objective maximum measure. When humans are introduced into the equation of efficiency, however, it is not sufficient to consider only input and output in terms of energy, or to compare material achievements with objective goals. It is necessary to consider human life with a past, a present, and an infinite variety of potential futures.

Efficiency is a measure for evaluating means but omits all reference to ends. But in human terms we must always consider the full question, "Efficient for what?" When the problem is looked at in this light, the incongruity of evaluating the efficiency of techniques of suicide or mass murder is evident. Efficiency in destroying human life is a fundamental negation, and the failure to consider the ends to which means are directed transforms the term itself into a contradiction. That engine is best that recovers the maximum number of calories from burning fuel, but an engine that burns human beings is an atrocity. That machine is best that produces a product with minimum expenditure of energy at lowest cost. But if the machine dehumanizes man or disemploys him, it too is an abomination.

Western civilization has placed great emphasis on efficiency, and through what superficially appears to be objective analysis has developed and nurtured economic man in a commodity-centered society. Paradoxically, the greater the objective efficiency, the more the problems attendant upon a technological society have multiplied. The climax of such a society may well be reached when automatically programmed factories capable of producing an economic manna are forced to close because no one works and cannot purchase the output. To protect the facilities from those who want, the destructive genius of the technician and scientist is increasingly employed to prepare and use instruments of mass annihilation.

Efficiency is the goal of all technique, the implicit and explicit calculus of cost-benefit evaluations. It is the juggernaut that threatens man's destruction along with the world he rushes to destroy. Man is an ephemeral creature who pays homage to posterity, and in his myopic quest for efficiency reduces the possibility of there being a posterity. Extend the calculus in evolutionary time and efficiency is dialectically transformed into a force that hinders survival and destroys the conditions of life. Focus on the short-range and the calculus is found to stultify and degrade life.

The unending process of evolution has been without direction, meandering along countless paths in the course of geological time. There are marvels and monsters, pygmies and giants, long-lived and short-lived species. Some species have survived unchanged over millions of years, others are relatively new. Evolutionary history is rich with experimentation, dead-ends, false starts. It has led to man and will continue—if we do not bring it to a halt. Evolution has never been efficient. Its course has been marked by waste and redundancy and proliferation of species. An efficient evolution would long since have rendered the planet lifeless. An efficient evolution anticipated by technological man leads to Capek's robots and Huxley's epsilons.

Efficiency is an engineering measure, not an evolutionary measure. Efficiency is a technique, not an end. Efficiency is a calculus applicable to machines, not to people or their societies. Efficiency calls for the cheapest way, not the humane way. It demands the greatest production, though it begs the question about whether production is the primary goal. It necessitates standardization to maximize utilization of more inputs and effect interchangeability of component parts. It requires homogeneity to ensure predictability and make control possible. Each criterion is based, even in an engineering sense, on human values in the guise of mathematical objectivity. The true criteria of the efficient society and efficient humans are greed and lust and power. These are the criteria that led to the efficient crematoria, the efficient hydrogen bombs, and the efficient production system.

The rationalization of business and government is a shared goal of entrepreneur and politician. The cost-benefit ratio is the decisive element of choice, although the exogenous costs may be unknown, the endogenous costs are too often miscalculated, and the benefits are restricted and short-range. Dialectically, even within the context of the technician's narrow vision, the efficient becomes inefficient, the unexpected is victorious over the predicted, the random element upsets the ordered pattern. Underlying the rational characteristics of the capitalist economy is a foundation of irrationalism that borders on anarchy as the invisible hand seeks to mediate between the greed and avarice of competing overlords of production and finance. Underproduction, overcapacity, recession, an economy of waste and planned obsolescence, bankruptcies, speculation, a wasteful multi-billion-dollar advertising industry, manufacture of a surfeit of unneeded commodities, payment of billions to farmers for not growing crops—these are among the blatant irrationalities that belie the efficient society. This pattern will become more prevalent as applied quasi-solutions fail to achieve their intended objectives. As residue problems accumulate, the efficient society will become more destructive and even less humane.

The answer to this paradox is not to develop a more powerful calculus to ensure efficiency but to relegate the concept of efficiency to history as a false quest. There is no calculus that can demonstrate that the businessman's obsessive infatuation with production is more efficient—in a living sense—than is the quiet contemplation of the world by a contemporary Siddhartha. What are the inputs and outputs of a worker bound by time-and-motion studies in a clanking, noisome factory in terms of human creativity? What are the inputs and outputs of a camper who canoes across a wilderness lake in a solitude of clean waters and virgin timbers? What are the inputs and outputs of a strip miner on the hills of Kentucky? Or the outputs and inputs of 93 million autos in the United States? What are the inputs and outputs of the breeding and summer residences of the vanishing group of whooping cranes, and of the cranes themselves?

Production and pecuniary gain will produce one calculus, appreciation of nature and enrichment of life another. The former tends to strangulation of life; the latter, based on human needs, enriches that life through inefficiency that promotes disorder, heterogeneity, spontaneity, and unpredictability.

The efficient society is a threat to survival. "Man has been highly successful as a biological species because he is adaptable . . . ," René Dubos relates. "History shows . . . that societies which were once efficient because highly specialized rapidly collapsed when conditions changed. A highly specialized society is rarely adaptable. Cultural homogenization and social regimentation resulting from the creeping monotony of technological culture, standardized patterns of education, and mass communication will make it progressively more difficult to exploit fully the biological richness of our species and may constitute a threat to the survival of civilization. We must shun uniformity of surroundings as much as conformity in behavior and strive to create diversified environments. This may result in the loss of some efficiency, but the more important goal is to provide the many kinds of soil that permit the germination of the seeds now dormant in man's nature."[5]

The inefficient society is the goal of the post-technological society. Efficiency is for robots and those who would master the earth though it hasten the end of their petty lives. Efficiency jeopardizes human survival and all life on earth. The humane society, not efficiency, is a fitting goal for man.

The Achievement of Asymptotes Is the First Step Toward the Inefficient Society

Growth, like efficiency, is not a product of reason but irrationality. The continuing rise of the exponential curves that mirror growth in a technological society are based not on human need but upon the metaphysics derived from science and technology. That the growth must end has been the thesis of the chapters on technological real-

---

[5] René Dubos, "Man and His Environment: Adaptations and Interactions," in *The Fitness of Man's Environment* (Washington, D.C.: Smithsonian Institution Press, 1968), p. 240.

ism. That growth may be brought under control before it leads to universal destruction is a hope. The achievement of asymptotes, limiting plateaus of growth, is the first step in attaining this hope.

According to the Book of Genesis, man lived in the Garden of Eden and was compelled to labor following his expulsion after eating of the tree of knowledge. But work, we have noted, was to primitive man a part of his life imbued with the richness of the myths and symbolism he created. The element of play and creative activity in work has also been noted. Mumford remarks that the trait of industriousness first appeared in Neolithic culture when man learned to dedicate himself to a single task that sometimes extended over a long period of time. The development of work in history has been traced by Adriano Tilgher in *Homo Faber: Work Through the Ages,* in which he describes the development and transformation of the concept from the time of the Greeks to the onset of the Great Depression in 1929. The Greeks considered work to be demeaning, an activity fit for slaves not free men. The early Hebrews also considered labor as drudgery, but saw its necessity in the duty to expiate the sin committed in Paradise; they also saw labor as the process of restoring the primal harmony. The early Christians, too, regarded work as punishment for original sin. The goals of labor were restrictive, and strictures were placed on dishonest dealings in the marketplace and on usury.

Max Weber saw in Protestantism the motive force that canonized labor and the pursuit of wealth into virtues, whereas they had previously been denounced as vices. The rise of economic man, who labored as an act of religious faith, was the product of capitalism —an economic system that established the compounding of profit as its end goal. Compound profit requires unlimited growth and, hence, economic growth became the basis of the new economy. The coincidence of the Reformation and the rise of capitalism tends to obscure the Industrial Revolution, another causal factor that greatly contributed to the changing economic order. It was the Industrial Revolution that led to the triumph of capitalism in the industrializing nations and that inspired necessary adjustments in then prevalent religious and social mores.

Technology begets more technology, becomes ever more complex, more costly. The search for new techniques, new materials, and new processes as technology strives to complete a quasi-solution and then augments the solution at a higher level, fuels the treadmill of economic development, expansion, and growth. It is no accident that communism has accepted the technological thesis *in toto* and has elevated technological efficiency and economic growth to the pantheon of communistic idols. Both orthodox Christianity and orthodox communism were bent to worship the same idols of labor, efficiency, and growth, even though the revolution that Lenin inspired occurred exactly four hundred years after Luther nailed his ninety-five theses to the church door at Wittenberg. In fifty years, Russia has bridged the lag of centuries and through a process of capital accumulation which was as brutal, but probably not more so, than the capitalist countries, has become a technological society. Its economic and ecological problems are similar to those experienced by the older technological societies, although its moment of truth may be slightly delayed because of its late start in the massive destruction of its environment.

There are two primary requisites for economic growth: there must be something that can grow, and the conditions for nurturing growth must be favorable. It has been shown that neither of these requisites is valid today and they will become less so as time passes. Both local and global resources are becoming depleted; the proliferating chains of residue problems are generating pollution and waste products that threaten to inundate the earth. The earth has already become uninhabitable for thousands of living species, and the roll of extinction grows each year. The dialectical effects of technology have set into motion counter-technology and have initiated the degeneration of the creative forces that drive the technological process. Finally, economic growth based upon the continued rise of all the exponentials that have been discussed is leading to a continuous downgrading of the quality of human life. In the period of counter-technology, increased efficiency is another quasi-solution hastening decline.

There is an increasing recognition that the treadmill concept of *Homo economicus* is due for revision. "We have tended to become convinced that the health and progress of our society and of the world society for that matter," Richard A. Falk stated at a congressional hearing on growth, "are directly and uncritically correlated with the expansion of the gross national product [and] we lose sight of the fact that such an index of 'development' may, after a certain level, serve more as a sign of societal deterioration than improvement."[6] Professor Falk testified that a doubling of the GNP would produce a less livable society. Stewart L. Udall, former Secretary of the Interior, has voiced his belief that limits will have to be put on this country's growth in the next decade to prevent destruction of the environment. "We have to get off this train of galloping growth," he said.[7] J. Alan Wagar has echoed Udall's admonition: "In its time the treadmill pattern of growth was progress enough and served us well. But as the relationships change between human numbers and the total environment, we must abandon unregulated growth before it strangles us."[8] Philip H. Abelson, editor of *Science*, also recognizes the necessity of achieving asymptotes: "Society has been, and still is, on a great growth kick. If we are interested in a long-term future for man, we will regard rapid growth with suspicion. We will look for, and point out, the unexpected and unpleasant consequences of exuberance long-continued, and seek to moderate it before irreparable damage has been done."[9] Udall reminded the economists that the ecologists are the final accountants regarding life on this planet. This transfer of accountability for all life from the hands of one set of experts to another is unacceptable, however. Accountability belongs to the people, for they will pay the ultimate bill in survival or annihilation, and it is with the quality of their lives that the experts play.

We no longer have a choice in the matter. The consequences of technological meddling in nature require immediate redress. Man must hurriedly achieve asymptotes in population, in production, and in consumption—or face extinction. The asymptotes must be achieved in a number of steps, all of which will be extremely diffi-

[6] *Effects of Population Growth on Natural Resources and the Environment*, p. 10.
[7] *New York Times*, January 16, 1970.
[8] J. Alan Wagar, "Growth Versus the Quality of Life," *Science*, CLXVIII (June 5, 1970), 1179–1184.
[9] *Science*, CLXII (October 11, 1968), 221.

cult but not more difficult than attempting to solve the problems that will attend continued growth. In the attempt to achieve asymptotes there is hope; any other way is without hope.

The first and most pressing need is to stabilize population. Zero population growth is not a necessary slogan only for the rest of the world but is applicable to the United States as well. A doubling of the world population and a 50 per cent increase in the United States population by the year 2000 may well push the growth curves beyond a point where they can be controlled without major catastrophe.

The utilization of resources must be cut back drastically. Waste of resources must be considered a criminal attack against the future of mankind. Planned obsolescence joins waste as a criminal practice. These are luxuries that could never be afforded, and now can no longer be tolerated. Remaining resources must be conserved and hoarded for essential use only.

There will be a price to pay for getting off the escalator of progress. The standard of living must be reduced to accord with the perilous state of the world's resources and its polluted environment. Frills and extravagances and luxuries must be eliminated. Society must be restructured to accommodate to new requirements for frugality. Transportation will have to become public- rather than private-oriented. The private auto must become an anachronism, mass transportation the travel mode of the future. Further, communities must be structured in such a way that private transportation, at least as we know it today, is not required. The availability of resources and true needs must dictate production requirements, not the spurious needs created by advertising.

It may well be that we shall be compelled to return, in part, to a style of life and production in which we can become more self-sufficient than the dependent urban dweller of today. As an ecological unit in a finite and depleted world, a maintenance economy is the goal. This change in economic behavior and style of living need not be a retrogression, however. It can restore a sense of balance to a robotized, consuming animal and free him once again to be human. It can elevate the quality of life which is deteriorating through incessant growth and expansion. To the extent this redirection is achieved willingly and with understanding, destruction and tragedy will be diminished.

In Labor Man Creates Himself

As the narrow and shortsighted economic policies of unceasing growth and conspicuous waste and resource depletion are transformed into a maintenance economy, attitudes toward labor must also be changed. Labor must become productive in filling human needs directly in the act of labor—not indirectly in the by-products of consumerism.

The goal of an automated society in which human labor will be unwanted and irrelevant does not meet the human needs of society. The headlong rush to render man obsolete has been spurred by a technology that strives to become autonomous to the extent man relinquishes control over it. The statistics of unemployment and employment are not measures of a healthy society, for while the former is an indicator of the extent to which society denies its members fruition through productive labor, the latter masks, in the main, the slavery of tens of millions who perform tasks they abhor. Millions who are nominally employed are little more than robots, their human, creative potential is submerged by work in which they have no interest or concern.

Two of man's age-old dreams, surcease from toil and an economy of abundance, are subject to the dialectical process that transforms the dreams into nightmares in which humanity is disintegrated and institutions are corroded. The manna of abundance entangles man in a treadmill of production and consumption that corrupts his spirit.

In a maintenance economy labor must be restored to its role as one of life's creative, playful, artful enterprises. A sense of craftsmanship and pride in the product of one's labor must supplant the wage-slavery of technological civilization. Human labor cannot be regarded as a commodity to be bought and sold on the labor exchange. "The rule and root of all labor," according to Ruskin, is "that every atom of substances, of whatever kind, used or consumed, is so much human life spent; which, if it issue in the saving present life, or gaining more, is well spent, but if not is either so much life prevented, or so much slain."[10] Thomas Carlyle believed much the same: "Even in the meanest sort of Labor, the whole soul of man is composed into a kind of real harmony the instant he sets himself to work." The self-renewal of the man through labor was one of the prime motivations of the pioneer movement in rebuilding Israel. The pioneers sang of the return to the soil, *"livnot u'lehibanot ba,"* to build and to be rebuilt in the reclamation of the land.

Ruskin pointed out that the meaning of the word *valorem*, from which value is derived, is to be well or strong. "To be 'valuable,' therefore, is to 'avail toward life.'" That which leads away from life "is unavailable or malignant." Wage-slavery in which man is harnessed to the needs of the machine contributes to what Ruskin termed "illth." It is upon the illth of labor that technological societies build their bogus wealth that impoverishes their people and their land.

The Machine Must Be Made Subservient to Man

The evaluation of the production process ignores the human condition, for efficiency is a fraudulent measure of the health of a society. In this sense economic society has failed. To serve man the machine must be operated to meet neither the criteria of engineering nor economic efficiency. The machine must be utilized to aid man in satisfying his basic needs, among which is realizing his creative potential. Production for private or public profit is equally illth-producing. In capitalist society, an automated factory is strike-proof; under communism, it is independent of the labor attitudes of the workers. No piece wages or shop stewards or Stakhanovite exemplars are required to pace production under automation.

To the extent man is utilized in the productive process he is treated as part of the man-machine system. His needs as a creative laborer are subordinated to the requirements of the machine. The pace of production is set by mechanical time: the period of a con-

---

[10] John Ruskin, *Unto This Last* (Lincoln: University of Nebraska Press, 1967), p. 96.

veyor belt's passage, the meshing of gears and cams, the limit point on a dial. Technology and the machine have no objection to man, Juenger writes, "so long as he surrenders unconditionally to the technical organizaton." The responsibilities and challenges to the individual are reduced by rationalization and the drive for efficiency. Decisions are made directly by the machine and indirectly by unseen experts.

If man is a mechanism, he cannot act in an indeterminate way. If he is granted a will and choice, then he cannot be treated as a mechanism. To transcend the machine he must cease to be an object or a commodity in the productive process. But it was the transformation of man into a commodity that was the major contribution of capitalist economics to the science of political economy. Human association was not direct but indirect through the industrial society. Individuals were separated, except where separation was broken by the demands of the machine. Human relationships were masked by the intervention of the productive process.

Karl Marx termed this development the "fetishism of commodities." The value relation between the products of labor which define commodities seems to have no connection with the social relations arising from them, and appears only as the relation of things. The value of a commodity is regarded as its objective characteristic instead of the embodiment of human effort. Dead capital thus becomes the master of living labor.

The division of labor subdivides a man and often even denies him the status of a cog in a machine. It homogenizes the personality, seeking to suppress any differentiation. Productive effort was divorced from enjoyed satisfactions. The things made were not man's, either as a product or as craftsmanship. While the total productive process was essentially social and collective, the division of labor fragmented social labor and made it appear as an objective factor. This fragmentation aided the concept of individualistic and subjective consumption.

The division of labor and the fetishism of commodities were seen by the young Marx as the basis for alienation. "The more commodities a worker produces, the cheaper a commodity he becomes," he wrote. "The devaluation of the world of man proceeds in direct proportion to the exploitation of the values of the world of things. Labour not only produces commodities, but it turns itself and the workers into commodities."[11] Marx did not need to see the effects of automation on the laborer; the mines and mills of nineteenth-century England told the same story.

It would be simple to point an accusing finger at the capitalist economy as the creator and manipulator of the inhuman side of technology. Yet the Marxists would have us believe that technology is ultimately a benefactor of mankind. Freed from capitalist exploitation, they say, technology can lead to a better life under socialism. This too, however, is part of the mythology of technology. Both capitalism and communism are identical in their pursuit of economic efficiency, and their competition lies primarily in the race to superiority in commodity production. The fetishism of commodities is as characteristic of communism as it is of capitalism. By accepting the forces of production as autonomous forces, and seeking to alter only the relations of production, Marxism has nurtured commodity fetishism and subordinated socialist man to the technological machine. The division of labor, the quest for efficiency, and consumerism have become staples of both the Russian and the American economies.

The increasing collectivization of society and the growing bureaucratization of the state apparatus have not arisen from ideological concepts, but from the pragmatic needs of industrial society to manage an increasingly complex technology. Technology and factory production are the supreme collectivizers. Although the United States preceded Soviet Russia in its total allegiance to the dictates of the technological society, the two nations' mutual acceptance of progress through technological collectivization attests to the apolitical and nonideological nature of technology. There is universal agreement on its purported benefits and its inevitability. The struggle is over which nation can exploit it better.

A divorce between ownership and management has occurred under capitalism, whereas in Russia a professional and bureaucratic elite has *de facto* control of the productive system. The state

---

[11] Karl Marx, *Economic and Philosophic Manuscripts;* quoted in *Socialist Humanism,* ed. Erich Fromm (New York: Doubleday Anchor, 1966), p. 121.

as employer and owner in the welfare society with its nationalized industry has shown the same obeisance to what Marcuse has called the "logos of technics." The ecological and sociological effects of growth and technology are evident wherever technology has become the driving force. Western Europe and Japan have joined the United States with smog-shrouded cities, polluted waterways, poisons in food and air and water, traffic jams, and the psychological effects of crowding, noise, and machine-paced living. The Soviet Union, under the impetus of the same imperative, is beginning to suffer the same sets of problems, for a quasi-solution to a techno-social problem has the same impact whatever the relations of production.

The 242 million inhabitants of the Soviet Union are discovering that dialectics apply to the conquest of nature as well as to the realm of political economy. In early 1970, for example, a comprehensive Soviet legislative program was announced to conserve water and curb increasing pollution of the hundreds of thousands of Russian lakes and rivers. The Moscow and Volga rivers are polluted, as are Lake Baikal and the Caspian Sea. The Aral and Azov seas have experienced declining water levels resulting from the diversion of rivers that flow into them to provide water for irrigation and hydroelectric plants. Soviet conservationists have begun to note the greedy appetite of mines and factories for fertile land and the pollution the factories produce. Chemical pesticides devastate the geese and cranes and partridges of Russia, as they have done in the United States. As if echoing Rachel Carson, Vladimir Peskov notes that "Our woods, gardens, and fields are becoming quieter and quieter."[12] DDT was banned in May 1970, and other pesticides that were producing a Russian "Silent Spring" were restricted.

In the Soviet journal *Agricultural Economy*, Boris N. Bogdanov wrote that "the damage being done to the economy through improper use of natural resources and environmental pollution is immense." Calling for a "nature-protecting" law for the entire Soviet Union, Bogdanov stated the law should formulate "general principles governing the treatment of nature, laying down rules and regulations for the preservation and exploitation of the natural

[12] *New York Times*, May 14, 1970.

riches. . . ."[13] Smoke and dust, mountains of waste and industrial refuse, problems with detergents, fish kills, oil slicks, and the beginning of a counter-technology have manifested themselves in the world's second largest technological society.

It will be an irony of historic dimension if communistic and capitalistic nations destroy not only each other but a large portion of mankind as well over contending philosophies of ordering the relations of production when in fact the forces of production are the basic cause of the problems each nation faces. The common denominator of human subservience and exploitation in both nations stems from an identical quest for efficiency. Alteration of the relations of production cannot alone humanize the technological industrial system, although it is a precondition. The forces of production, the industrial production system, are the great enslavers of mankind. The search for emancipation and freedom through technology is a contradiction under any economic system.

A restructuring of society in terms of man's needs, rather than those of the machine, and the recognition of labor as a creative activity essential to man's integrity and well-being, are required. Control of the machine in the interest of a humane society is now a critical necessity. The machine cannot be dislodged from its semiautonomous position of power over man through scientific research, planning, or principles of ownership. What is needed is to reorient the nature and function of the production process. Any such reorientation in consonance with a maintenance economy and the achievement of asymptotes on growth will require a rollback of the machine. This need not imply the Luddite tactic of machine-smashing, but machine control and restriction. The productive process must become human-oriented, not machine-oriented. To perform meaningful labor will necessitate factories that are paced to the physical, social, and psychological needs of humans. There will have to be a diminution of production in accord with the diminution of resources. There will have to be a growth in the quality of labor and the freeing of man from bondage to an insensate machine. Not production but human productivity will become the goal. Commodities, not people, will be the by-products.

[13] *New York Times*, April 9, 1970.

Smaller creatories owned and operated by artisans, designers, and craftsmen will replace the monstrous and humanly destructive factories. Group participation and decision-making, not division of labor and unseen management, will set goals and establish work conditions. Combined creatories, recreational centers, schools, and lifelong educational centers will replace the fragmented social roles of the worker. Fragmented life will become integrated in meaningful activities performed during the laboring day and not relegated to leisure hours. Work and leisure will merge, with human potentials freed to serve man.

The dialectical materialists have been strangely silent regarding the antithesis to communism and analysis of a succeeding stage. May not the radical change in the forces of production be the next stage in the development of man? May not the freeing of man from the tyranny of the technological society be an appropriate step for post-technological man? May not the illths of capitalism and communism be superseded by a true wealth of humans living and creating in harmony with their natural environment?

The Post-technological Society Must Be Free and Open

A madness is creeping over the face of the earth. A bit of madness has always raged within man, but today this madness seizes whole communities and nations. The blood-drenched fields of Flanders and trenches of Verdun; Mussolini, Hitler, Franco, and Stalin, Dachau and Auschwitz, Hiroshima, the six years of World War II, the armed truce of the past twenty-five years with its hundreds of minor conflagrations attest to this madness. The twenty-five year war in Indochina, the twenty-two-year-old war of Palestine-Israel, a divided Korea, a divided Germay, continents in revolt, continents groaning under exploitation—all attest to this madness.

Billions become ever more fearful as they see the earth shrink before their eyes, the resources depleted, the land despoiled, the air and water fouled. Nations strive to maintain what they have, seeking to ensure a firm hold on remaining supplies. Strangers jostle each other in over-crowded cities, fearful of their possessions and their lives. Youth fights the established society, the colored and underprivileged fight for the right to live a decent life. Cities are burned and looted, policemen are shot, innocent citizens are slain in the crossfire of hunted and hunter. "Law and order," the legacy of a world that has made armed might the arbiter of decisions, is descending upon the world.

This violence is an act of rebellion and defiance and despair. To some it is the only means they have to assert their freedom; to others it is a means of holding on to that which they possess. To some it is a declaration of emancipation; to others it is an attempt to hold back history.

The decline of technology can help lift the pall of fear that enshrouds the earth if the decline is achieved as a positive step in humanizing a world that has strayed from nature and human values. Community, love, respect for all individuals—these are the ingredients that can help to harness the machine to man's needs. A free society of cooperating individuals is a prerequisite for a sane post-technological society. It will be an open society with freedom of choice from a multitude of alternatives. There will be no experts in life—only people who live.

Revolt and dissent are necessary stabilizing forces in the steps leading toward the future. Institutions must be brought into harmony with conditions that exist in the present, even though they arose under different conditions in the past, or repression must occur. But repression is the path toward the concentration camp not stability. Dialectically, the disorder stemming from freedom and diversity can beget order, while planned order can beget disorder. A leaven of disorder is the pathway to a free and humane society.

The Paths to the Future Are Varied and Infinite

There can be no blueprint to the future. There is no "best way," for there must always exist many ways to the many possible futures. A stable community in nature is one that is diverse and complex; an unstable community is homogeneous and ordered. To assure a future for man the earth must remain a stable community by remaining diverse and heterogeneous.

The imperative of technology to level all before it to a monotonous sameness constitutes the greatest menace of technological society. Already the United States shows this homogenized appearance. Over a highway network criss-crossing three million square miles one can travel without seeing the country, guided by ubiquitous and alike signs over thousands of miles of concrete pavements. One can alight at any airport and not know in what city one is, so much are the airports alike. The cities, too, in their architecture, their chain stores, their mass-produced products, the sameness of the automobiles, have an obsessive homogeneity.

Sameness permeates the land like a blight. Wherever technological society has been established the blight follows. In Western Europe and the industrialized areas of Asia, in island possessions and neo-colonial empires, the homogenizing blight appears. Under the guise of tourism, it inundates places heretofore relatively untouched by technological civilization.

The blight of homogenization is part of the creeping madness engulfing the world. It precedes the smog, pollution, and despoliation that are the hallmarks of what is becoming the universal civilization. In this way lies tragedy, however, the ultimate tragedy. Every nation that succumbs to the universal civilization reduces the potential for man to survive an unknown future.

The concern of the Third World to join the universal civilization is a real one. People in the Third World Nations believe the priority of economic progress must take precedence over all concern about the environment or sameness. "We are not concerned with pollution but with existence," an Indian leader declared at a conference of the International Organization of Consumer Unions in Vienna. "The wealthy nations worry about car fumes. We worry about starvation." A Malayan said: "Some of us would rather see smoke coming out of a factory and men employed than no factory at all."[14] These views were mirrored by youthful delegates to a United Nations symposium on urbanization and the environment in Michigan. A delegate from Seoul, Korea, remarked, "Smog is the mark of progress." "The legacy we can give the future of great masses in poverty is worse than the legacy of an untouched environment," said the director of a natural resources development center in Vene-

[14] *New York Times*, July 3, 1970.

zuela. Stating that industrial pollution is not the problem of the Third World nations, a United Nations official expressed a desire "for more of it." Sadly, the fetish of progress was worshiped by a young Ceylonese who, speaking on behalf of the two-thirds of mankind living in the Third World nations, said: "They have little interest in the purity of the air they breathe, the freshness of the waters of their lakes and rivers, the natural beauty of their mountains."[15]

While recognizing the concern of Third World leaders with the welfare of their peoples, we must caution them that the universal technological civilization may well be a trap rather than a panacea. The laboratory *in vivo* is a many-edged sword. Quasi-solutions and technological fixes are boomeranging all over the world. The Egyptians have sacrificed the Nile system, with a five-thousand-year history of successful service, for the Aswan Dam whose benefits appear to be dubious indeed. Other major irrigation and hydroelectric projects are producing unforeseen negative effects.

The American technological society is not a model for saving the world. It may well destroy America and help blight the rest of the world. It has solved the problem of production but at horrible cost —and even this only in the short term. Thus, the rush of the newly emancipated nations of the world to jump on the bandwagon of the universal civilization is doubly shortsighted. Although offering benefits to the peoples of the Third World in the short term, technology's proliferating chains of residue problems cannot but redound negatively in the near future—a generation or two—with the counter-technological society to follow. More serious, the cooperation of the Third World in joining the universal civilization reduces the stability of the world community while tending to further homogenize it.

The value systems, social institutions, even the tribal loyalties, mores, and customs of the peoples of the Third World can be significant forces for human survival. They should not be thoughtlessly overthrown in favor of forced urbanization which brings with it the growth of slums and the creation of an army of rootless and alienated people. The preservation of the autonomy and distinctive

[15] *New York Times,* June 19, 1970.

characteristics of pre-technological societies is of prime importance to human survival. "It could well prove the most important contribution of the second half of the twentieth century, a heritage of human variety for the world of tomorrow," Dasmann pleads. The imposition of modern techniques on these peoples, Dasmann further warns, would "crush out old ways of living. In saving the lands and providing 'freedom from want' for their peoples, we must imprison the lands with fences and rigid boundaries, and enslave the people with new rules and restrictions. And we are not certain that we know enough about land or people to justify such change."[16]

Max C. Otto offers even more cogent advice. Unless scientific progress and the interests of humanity can be reconciled, he claims, we can "look forward to the eventual downfall of the scientifically advanced races. Only those which are backward with regard to science can hope to survive—providing they can keep out of the way of their scientific neighbors."[17]

The affluence of the advanced technological societies, it should be recalled, was—and, to an extent, still is—based on world-wide plunder and exploitation. The Third World nations cannot hope to emulate this plunder, nor need they do so. Theirs are wealthy countries, and their wealth includes the "purity of the air they breathe, the freshness of the waters of their lakes and rivers." This, moreover, is wealth of a kind without which people cannot live—a truth only now being recognized by the affluent technological societies. The natural wealth and resources of the Third World should be retained and preserved for its peoples, not bartered away to swell further the opulence of the advanced nations.

Milk and meat are more important to the welfare of the Third World than nuclear reactors, superhighways, and the other gaudy paraphernalia of technology. A new plow, a simple solar heater, small cottage industries, intensive agriculture within a natural environment—these can help lay the foundations for a sane economy based upon human needs and aspirations. Ernst D. Bergmann, a leader in Israeli science, found it curious that "in many of the new

[16] Dasmann, *op. cit.*, pp. 233, 41.
[17] Max C. Otto, *Science and the Moral Life* (New York: New American Library, 1952), p. 71.

nations there is the attitude that agriculture is a lower form of civilization, that to be truly modern you must be an industrial state." The most reasonable course for a new nation, he advised, "is to begin with agriculture and develop it to the point where (1) the country is self-sufficient in food and (2) agriculture becomes the basis for new industries that would use agricultural products as the source of raw materials."[18] For example, Cuba, after its revolution, at first sought to repudiate its sugar industry in favor of forced industrialization. In subsequent years the sugar crop again became the principal mainstay of the economy and a basis for improvement of the life of the Cuban people.

The natural environment is a luxury only to those who possess it. Lose it and one's own existence is imperiled. This is the lesson to be learned from the technological societies. Affluence is ephemeral, and often reduces the quality of life. Wealth is people and land and pure air and pure water. Here is a challenge to the Third World. Go your separate ways. Create new ways to harness the largesse of your countries and the creative labor of your peoples. Do not become lost in the monoculture of a universal civilization which will destroy you together with those you imitate. Your heritages and your histories are unique pointers to a different tomorrow.

Here, too, is a task for the United Nations. Not technical assistance to ensure the monoculture, not a decade of development to hasten the universal technological civilization, but an effort to preserve the diversity of the peoples of the world is the path of true statesmanship. While Third World nations preserve and carefully utilize their resources, protecting their natural environments within the context of their unique histories, the advanced technological nations can seek to reclaim that which has been wasted and despoiled during three hundred years of worship of a false progress.

[18] Ernst D. Bergmann, "Technical Strength for a New Nation," *Science and Technology,* December 1967, pp. 62–69.

Suggestions for Further Reading

THE SOURCES cited in the foregoing chapter notes represent only a small portion of the large collection of literature on the topics and ideas discussed in *Overskill*. This section introduces the reader to the most important of these works, omitting journal articles and reports already referred to in the Notes. The books are classified by themes and concepts, though pertinent chapters are indicated throughout.

In the Beginning: Early Man (CHAPTER 2)

Gordon Childe. *What Happened in History.* Harmondsworth: Pelican Books, 1957. An account of man from prehistoric times to the fall of the Roman Empire.

Henri and H. A. Frankfort, John A. Wilson, and Thorkild Jacobsen. *Before Philosophy: The Intellectual Adventure of Ancient Man.* Harmondsworth: Pelican Books, 1954. "A study of the primitive myths, beliefs, and speculations of Egypt and Mesopotamia, out of which grew the religions and philosophies of the later world." Penetrating analysis of mythopoeic thought, its logic and emotional characteristics. Also considers the role of the Hebrews in replacing myth with religion and the genesis of critical thought among the Greeks.

Jacquetta Hawkes and Leonard Woolley. *Prehistory* (Part 1) and *The Beginning of Civilization* (Part 2). New York: Harper and Row, 1963.

The first volume of UNESCO's History of the Scientific and Cultural Development of Mankind reviews man's prehistoric past and the beginning of civilization, covering the period up to the thirteenth century.

Three books that deal with man in the pre-technological cultures of primitive societies are:

Bronislaw Malinowski. *Magic, Science and Religion and Other Essays*. Garden City: Doubleday Anchor, 1954.

Paul Radin. *The World of Primitive Man*. New York: Grove Press, 1960.

Robert Redfield. *The Primitive World and Its Transformations*. Ithaca: Cornell University Press, 1953.

Two works that contrast the Western and Eastern worlds are:

Joseph Campbell. *The Masks of God* (Vol. 1, *Primitive Mythology*; Vol. 2, *Oriental Mythology*). New York: Viking, 1959, 1962. Describes and analyzes the bifurcation of Eastern and Western mythologies that began at Sumer in the third millennium B.C.

R. G. H. Siu. *The Tao of Science: An Essay on Western Knowledge and Eastern Wisdom*. Cambridge, Mass.: M.I.T. Press, 1958.

History of Science and Technology (CHAPTER 2)

The history of science and technology has become a "scientific" discipline and has spawned a plethora of books and journals. Three helpful books are:

J. D. Bernal. *Science in History*. London: Watts, 1954.

Herbert Butterfield. *The Origins of Modern Science, 1300–1800*. New York: Macmillan, 1958.

Charles Singer. *A Short History of Scientific Ideas to 1900*. New York and London: Oxford University Press, 1959.

A more detailed and extensive history is found in:

George Sarton. *Introduction to the History of Science*. 3 vols. Baltimore: Williams and Wilkins, 1927–1948.

Greek science is well covered in:

Benjamin Farrington. *Greek Science*. Harmondsworth: Pelican Books, 1953.

George Sarton. *A History of Science: Ancient Science Through the Golden Age of Greece*. New York: Wiley, 1964.

An overview of the development of science in the Orient is presented in:

Joseph Needham. *Science and Civilization in China*. London: Cambridge University Press, 1954.

For the roots of the "conquest of nature" by science, see:
Max Weber. *The Protestant Ethic and the Spirit of Capitalism*. Translated by Talcott Parsons. New York: Scribner's, 1958.

Lynn White, Jr. "The Historical Roots of Our Ecologic Crisis." *Science*, CLV (March 10, 1967), 1203–1207. Also reprinted, together with other historical essays, in the author's *Machina Ex Deo: Essays in the Dynamism of Western Culture* (Cambridge, Mass.: M.I.T. Press, 1969).

General histories of technology can be found in:
T. K. Derry and Trevor I. Williams. *A Short History of Technology: From the Earliest Times to A.D. 1900*. New York and London: Oxford University Press, 1961. A summary and condensation of the massive work by Singer and others, cited below.

Friedrich Klemm. *A History of Western Technology*. Cambridge, Mass.: M.I.T. Press, 1964. An interesting book containing excerpts from many sources with a running commentary.

Melvin Kranzberg and Carroll W. Purcell, Jr., eds. *Technology in Western Civilization* (Vol. 1, *The Emergence of Modern Industrial Society*; Vol. 2, *Technology in the Twentieth Century*). New York and London: Oxford University Press, 1967.

Charles Singer, E. J. Holmyard, A. R. Hall, and Trevor I. Williams. *A History of Technology*. 5 vols. Oxford: Clarendon Press, 1954–1958. A comprehensive survey of technology.

Abraham Wolf. *A History of Science, Technology and Philosophy in the Sixteenth and Seventeenth Centuries*. 2d ed. London: Allen and Unwin, 1950.

―――― *A History of Science, Technology and Philosophy in the Eighteenth Century*. 2d ed. New York: Macmillan, 1939.

Both Klemm, pp. 389–392, and Bernal, pp. 934–948, have extensive bibliographies, as does Kranzberg, Vol. 2, pp. 709–739, and Derry, pp. 750–758.

History, Philosophy, and the Enlightenment (CHAPTERS 1, 2)

Francis Bacon. *The Works of Francis Bacon*. Edited by J. Spedding, R. L. Ellis, and D. D. Heath. 14 vols. London: 1857–1874.

Carl L. Becker. *The Heavenly City of the Eighteenth-Century Philosophers*. New Haven: Yale University Press, 1961.

Edwin Arthur Burtt. *The Metaphysical Foundations of Modern Physical Science*. Garden City: Doubleday Anchor, 1954.

Ernst Cassirer. *The Philosophy of the Enlightenment.* Princeton: Princeton University Press, 1951.

R. G. Collingwood. *The Idea of History.* New York and London: Oxford University Press, 1956.

René Descartes. *Discourse on Method and Meditations.* Indianapolis: Bobbs-Merrill, 1960.

Galileo Galilei. *Dialogue Concerning the Two Chief World Systems.* Translated by Stillman Drake. Berkeley: University of California Press, 1962.

Peter Gay. *The Enlightenment: An Interpretation* (Vol. 1, *The Rise of Modern Paganism;* Vol. 2, *The Science of Freedom*). New York: Knopf, 1966 and 1969.

David Hume. *An Inquiry Concerning Human Understanding.* New York: Liberal Arts Press, 1955.

J. J. Rousseau. *The Social Contract and the Discourses.* Translated by G. D. H. Cole. New York: Dutton, 1950.

Oswald Spengler. *The Decline of the West.* 2 vols. New York: Knopf, 1961.

Arnold Toynbee. *A Study of History.* 10 vols. New York and London: Oxford University Press, 1935–1959.

———. *A Study of History.* Abridged by D. C. Somervell. 2 vols. New York and London: Oxford University Press, 1958.

Progress (CHAPTERS 2, 11, 12)

The concept of progress as a driving force of Western civilization is traced in a number of sources:

J. B. Bury. *The Idea of Progress.* New York: Dover, 1955.

Ludwig Edelstein. *The Idea of Progress in Classical Antiquity.* Baltimore: Johns Hopkins Press, 1967.

George H. Hildebrand. *The Idea of Progress: A Collection of Readings.* Berkeley: University of California Press, 1949.

S. Pollard. *Idea of Progress: History and Society.* New York: Basic Books, 1964.

W. Warren Wagar, ed. *Idea of Progress Since the Renaissance.* New York: Wiley, 1969.

Commentary on history and progress as critique and plan can be found in:

Raymond Aron. *Progress and Disillusion: The Dialectics of Modern Society.* New York: Praeger, 1968.

Carl L. Becker. *Progress and Power.* Stanford: Stanford University Press, 1935.

M. de Condorcet, *Sketch for a Historical Picture of the Progress of the Human Mind*. New York: Noonday Press, 1955.

Georges Sorel. *The Illusions of Progress*. Translated by John and Charlotte Stanley. Berkeley: University of California Press, 1969.

The Scientific View

THE PHYSICAL WORLD (CHAPTERS 2, 3, 5)

Max Born. *The Restless Universe*. New York: Harper, 1936.

Albert Einstein and Leopold Infeld. *The Evolution of Physics*. New York: Simon and Schuster, 1968.

Martin Gardner. *Relativity for the Million*. New York: Macmillan, 1962.

Henry Margenau. *The Nature of Physical Reality*. New York: McGraw-Hill, 1950.

Hans Reichenbach. *The Philosophy of Space and Time*. New York: Dover, 1958.

EVOLUTION AND LIFE (CHAPTER 5)

Walter B. Cannon. *The Wisdom of the Body*. Rev. and enlarged ed. New York: Norton, 1963.

Loren C. Eiseley. *The Immense Journey*. New York: Random House, 1957.

Garrett Hardin. *Nature and Man's Fate*. New York: Rinehart, 1959.

A. I. Oparin. *Origin of Life*. 2d ed. New York: Dover, 1953.

George Gaylord Simpson. *The Meaning of Evolution*. New Haven: Yale University Press, 1949.

Ernst Schrödinger, *"What Is Life?" and Other Scientific Essays*. Garden City: Doubleday Anchor, 1944.

LOGICAL POSITIVISM (CHAPTERS 2, 5, 11)

Alfred Jules Ayer. *Language, Truth and Logic*. New York: Dover, n.d.

———, ed. *Logical Positivism*. New York: Free Press, 1959.

Percy W. Bridgman. *The Logic of Modern Physics*. New York: Macmillan, 1927.

Leszek Kolakowski. *The Alienation of Reason: A History of Positivist Thought*. Translated by Norbert Guterman. Garden City: Doubleday Anchor, 1969.

Otto Neurath, Rudolf Carnap, and Charles W. Morris. *International Encyclopedia of Unified Science*. 2 vols. Chicago: University of Chicago Press, 1955.

SUBJECTIVITY IN SCIENCE (CHAPTERS 2, 3, 11)

F. S. Kuhn. *The Structure of Scientific Revolutions*. Chicago: University of Chicago Press, 1962.

Michael Polanyi. *Personal Knowledge: Towards a Post-Critical Philosophy*. Chicago: University of Chicago Press, 1958.

Israel Scheffler. *Science and Subjectivity*. Indianapolis: Bobbs-Merrill, 1967.

SCIENCE AND VALUES (CHAPTERS 2, 3, 11, 12)

J. Bronowski. *Science and Human Values*. New York: Harper Torchbooks, 1959.

Theodosius Dobzhansky. *The Biological Basis of Human Freedom*. New York: Columbia University Press, 1960.

Everett W. Hall. *Modern Science and Human Values: A Study in the History of Ideas*. New York: Delta, 1966.

A. V. Hill. *The Ethical Dilemma of Science*. London: Scientific Book Guild, 1962.

Max C. Otto, *Science and the Moral Life*. New York: New American Library, 1952.

Moritz Schlick. *Problems of Ethics*. Translated by David Rynin. New York: Dover, 1962.

C. H. Waddington. *The Ethical Animal*. New York: Atheneum, 1961.

Cybernetics (CHAPTER 3)

W. Ross Ashby. *An Introduction to Cybernetics*. New York: Wiley, 1958.

Ludwig von Bertalanffy. *Robots, Men and Minds*. New York: Braziller, 1967.

Norbert Wiener. *Cybernetics*. Cambridge, Mass.: M.I.T. Press, 1948.

———. *The Human Use of Human Beings*. Garden City: Doubleday Anchor, 1954.

General Readings in Science (CHAPTERS 2, 3, 5, 11, 12)

Eric Temple Bell. *Magic of Numbers*. New York: McGraw-Hill, 1946.

Richard B. Braithwaite. *Scientific Explanation: A Study of the Function of Theory, Probability and Law in Science*. London: Cambridge University Press, 1955.

Robert E. Egner and Lester E. Dennon, eds. *The Basic Writings of Bertrand Russell*. New York: Simon and Schuster, 1961.

Herbert Feigl and May Brodbeck, eds. *Readings in the Philosophy of Science*. New York: Appleton-Century-Crofts, 1953.

Lewis S. Feuer. *The Scientific Intellectual: The Psychological and Sociological Origins of Modern Science*. New York: Basic Books, 1963.

Phillip Frank. *Philosophy of Science*. Englewood Cliffs, N.J.: Prentice-Hall, 1957.

Charles Coulton Gillispie. *The Edge of Objectivity: An Essay in the History of Scientific Ideas*. Princeton: Princeton University Press, 1960.

Julian Huxley. *Man in the Modern World*. New York: New American Library, 1955.

Alfred Korzybski. *Science and Sanity: An Introduction to Non-Aristotelian Systems and General Semantics*. 3d ed. Lakeville, Conn.: International Non-Aristotelian Publishing Company, 1950.

Marshall McLuhan. *Understanding Media: The Extension of Man*. New York: New American Library, 1966.

James R. Newman, ed. *The World of Mathematics*. 4 vols. New York: Simon and Schuster, 1956.

Derek J. De Solla Price. *Little Science, Big Science*. New York: Columbia University Press, 1963.

Charles Sherrington. *Man on His Nature*. 2d ed. Garden City: Doubleday Anchor, 1951.

Herbert A. Simon. *Models of Man, Social and Rational*. New York: Wiley, 1957.

Alvin M. Weinberg. *Reflections on Big Science*. Cambridge, Mass.: M.I.T. Press, 1967.

Alfred North Whitehead. *Science and the Modern World*. New York: New American Library, 1953.

Benjamin Lee Whorf. *Language, Thought and Reality: Selected Writings of Benjamin Lee Whorf*. Edited by John B. Carroll. Cambridge, Mass.: M.I.T. Press, 1966.

The expansion of information sources in Russia is summarized in:

Leonard N. Beck. "Soviet Discussion on the Exponential Growth of Scientific Publications." Pages 5–17 in *Proceedings of the American Society for Information Science*, Vol. 7. Edited by Jeanne B. North. Thirty-third Annual Meeting, October 11–15, 1970. Washington, D.C.: American Society for Information Science, 1970.

Land, Food, and Population (CHAPTER 6)
POPULATION

Renée Dumont and Bernard Rosier. *The Hungry Future*. New York: Praeger, 1969.

Paul Ehrlich. *The Population Bomb*. New York: Ballantine, 1968.

Paul R. and Anne H. Ehrlich. *Population, Resources, Environment.* San Francisco: Freeman, 1970.

Philip Morris Hauser, ed. *The Population Dilemma.* Englewood Cliffs, N.J.: Prentice-Hall, 1963.

William and Paul Paddock. *Famin —1975! America's Decision: Who Will Survive.* Boston: Little, Brown, 1967.

RESOURCES

Fairfield Osborn, *The Limits of the Earth.* Boston: Little, Brown, 1953. A pioneering book.

Energy (CHAPTER 7)

Sam H. Schurr, et al. *Energy in the American Economy, 1850–1975: Its History and Prospects.* Baltimore: Johns Hopkins Press, 1960.

ATOMIC ENERGY

Richard Curtis and Elizabeth Hogan. *Perils of the Peaceful Atom.* New York: Ballantine, 1969.

Sheldon Novick. *The Careless Atom.* New York: Delta, 1970.

Jack Schubert and Ralph E. Lapp. *Radiation: What It Is and How It Affects You.* New York: Viking, 1957.

THERMAL POLLUTION

Peter A. Krenki and Frank L. Parker, eds. *Biological Aspects of Thermal Pollution.* Nashville: Vanderbilt University Press, 1969.

Oil Pollution (CHAPTER 8)

Stanley E. Degler, ed. *Oil Pollution Problems and Policies.* Washington: BNA Books, 1969.

Richard Petrow. *In the Wake of Torrey Canyon.* New York: McKay, 1968.

J. E. Smith, ed. *Torrey Canyon Pollution and Marine Life.* London: Cambridge University Press, 1968.

Pollution, Waste, and Ecology (CHAPTERS 6, 7, 8)

N. C. Brady. *Agriculture and the Quality of Our Environment.* Washington, D.C.: American Association for the Advancement of Science, 1967.

Rachel Carson. *Silent Spring.* Boston: Houghton Mifflin, 1962.

Pierre Dansereau, ed. *Challenge for Survival: Land, Air, and Water for Man in Megalopolis.* New York: Columbia University Press, 1970.

F. F. Darling and J. F. Milton, eds. *Future Environments of America*. New York: Natural History Press, 1969.

Raymond F. Dasmann. *A Different Kind of Country*. New York: Collier-Macmillan, 1968.

———. *The Last Horizon*. New York: Macmillan, 1963.

Environmental Quality. First Annual Report of the Council on Environmental Quality. Washington, D.C.: Government Printing Office, 1970.

John C. Esposito. *Vanishing Air*. New York: Grossman, 1970.

The Fitness of Man's Environment. Washington, D.C.: Smithsonian Institution Press, 1968.

S. F. Freeman. *Resources and Man*. Washington, D.C.: National Academy of Sciences, 1969.

Frank Graham, Jr. *Since Silent Spring*. Boston: Houghton Mifflin, 1970.

Harold W. Helfrich, Jr., ed. *The Environmental Crisis*. New Haven: Yale University Press, 1970.

Ron M. Linton. *Terracide*. Boston: Little, Brown, 1970.

Gene Marine. *America the Raped: The Engineering Mentality and the Devastation of a Continent*. New York: Simon and Schuster, 1969.

Wesley Marx. *The Frail Ocean*. New York: Ballantine, 1967.

Donald Michael. *The Unprepared Society: Planning for a Precarious Future*. New York: Basic Books, 1968.

A. Q. Mowbray. *Road to Ruin*. Philadelphia: Lippincott, 1969.

Max Nicolson. *The Environmental Revolution: A Guide for the New Masters of the Earth*. New York: McGraw-Hill, 1970.

E. C. Pielow. *An Introduction to Mathematical Ecology*. New York: Wiley-Interscience, 1969.

G. G. Polikarpov. *Radioecology of Aquatic Organisms: The Accumulation and Biological Effect of Radioactive Substances*. Translated from the Russian. Amsterdam: North-Holland, 1966.

Roger Revelle and Hans H. Landsberg, eds. *America's Changing Environment*. Boston: Houghton Mifflin, 1970.

James Ridgeway. *The Politics of Ecology*. New York: Dutton, 1970.

Robert Rienow and Leona Train. *Moment in the Sun*. New York: Ballantine, 1967.

Henry Savage, Jr. *Lost Heritage*. New York: Morrow, 1970.

Paul Shepard and Daniel McKinley, eds. *The Subversive Science: Essays Toward an Ecology of Man*. Boston: Houghton Mifflin, 1969.

George R. Stewart. *Not So Rich As You Think*. Boston: Houghton Mifflin, 1967.

John and Mildred Teale. *Life and Death of the Salt Marsh*. Boston: Atlantic–Little, Brown, 1969.

James S. Turner. *The Chemical Feast*. New York: Grossman, 1970.
Philip L. Wagner. *The Human Use of the Earth*. New York: Free Press, 1960.
Kenneth E. F. Watt. *Ecology and Resource Management*. New York: McGraw-Hill, 1968.
William H. Whyte. *The Last Landscape*. Garden City: Doubleday, 1968.

A review of the problems of pollution in Russia taken from Russian sources is found in:
Marshall I. Goldman. "The Convergence of Environmental Disruption." *Science*, CLXX (October 2, 1970) 37–42.

War (CHAPTER 9)

Saul Aronow, Frank R. Ervin, and Victor W. Sidel, eds. *The Fallen Sky: Medical Consequences of Thermonuclear War*. New York: Hill and Wang, 1964.
Nigel Calder, ed. *Unless Peace Comes*. New York: Viking, 1968.
Henry M. Christman, ed. *Peace and Arms*. New York: Sheed and Ward, 1964.
Fred J. Cook, *The Warfare State*. New York: Macmillan, 1962.
Norman Cousins. *In Place of Folly*. New York: Harper, 1961.
John M. Fowler, ed. *Fallout: A Study of Superbombs, Strontium-90, and Survival*. New York: Basic Books, 1960.
Irving Louis Horwitz. *The War Game: Studies of the New Civilian Militarists*. New York: Ballantine, 1963.
Ralph Lapp. *Kill and Overkill: The Strategy of Annihilation*. New York: Basic Books, 1962.
Seymour Melman. *The Peace Race*. New York: Ballantine, 1961.
Linus Pauling. *No More War*. New York: Dodd, Mead, 1958.
Jack Raymond. *Power at the Pentagon*. New York: Harper and Row, 1964.
Lewis F. Richardson. *Arms and Insecurity: A Mathematical Study of the Causes and Origins of War*. Edited by N. Rashevsky and E. Trucco. Pittsburgh: Boxwood Press; Chicago: Quadrangle Books, 1960.
———. *Statistics of Deadly Quarrels*. Edited by Quincy Wright and C. C. Lienau. Pittsburgh: Boxwood Press; Chicago: Quadrangle Books, 1960.
Bertrand Russell. *Common Sense and Nuclear Warfare*. New York: Simon and Schuster, 1959.
Quincy Wright. *A Study of War*. 2 vols. Chicago: University of Chicago Press, 1942.

Technology Assessment and Futurism (CHAPTERS 1, 11)

As science and technology move toward the goals of prediction and control, the development of evaluative measures for technology is becoming a technique in its own right and a rationale for dealing with the future is being sought. Although some of the books dealing with this subject profess to consider values and goals, these are always embedded within the *a priori* assumptions of the scientific enterprise, and the frame of reference is invariably the *status quo* of social, political, and economic institutions.

The following books deal with the evaluation and impact of technology:

Robert U. Ayres. *Technological Forecasting and Long-Range Planning.* New York: McGraw-Hill, 1969.

Kurt Baier and Nicholas Rescher, eds. *Values and the Future: The Impact of Technological Change on American Values.* New York: Free Press, 1969.

Raymond A. Bauer. *Second-Order Consequences: A Methodological Essay on the Impact of Technology.* Cambridge, Mass.: M.I.T. Press, 1969.

Emmanuel Mesthene. *Technological Change: Its Impact on Man and Society.* New York: New American Library, 1970.

Donald A. Schon. *Technology and Change: The New Heraclitus.* New York: Delacorte Press, 1967.

Two major studies of technology assessment are:

U.S. Congress, House, Committee on Science and Astronautics. *A Study of Technology Assessment.* Report of the Committee on Public Engineering Policy, National Academy of Engineering. Washington, D.C.: Government Printing Office, 1969.

———. *Technology: Processes of Assessment and Choice.* Report of the National Academy of Sciences. Washington, D.C.: Government Printing Office, July 1969.

Futurism, the art of bringing the future out of the realm of the seers and prophets into the keeping of science, is discussed in:

Arthur C. Clarke. *Profiles of the Future.* New York: Harper and Row, 1962.

Dennis Gabor. *Inventing the Future.* New York: Knopf, 1964.

John McHale. *The Future of the Future.* New York: Braziller, 1969.

The futures are predicted in a lengthening list of books among which the following are cited:

Burnham P. Beckwith. *The Next 500 Years*. New York: Exposition Press, 1968.

Harrison Brown. *The Challenge of Man's Future*. New York: Viking, 1956.

Harrison Brown, James Bonner, and John Weir. *The Next Hundred Years*. New York: Viking, 1963.

W. R. Ewald, ed. *Environment and Change: The Next 50 Years*. Bloomington: Indiana University Press, 1968.

Gerald Feinberg. *The Prometheus Project: Mankind's Search for Long-Range Goals*. Garden City: Doubleday Anchor, 1969. A frightening book that epitomizes the hubris of the scientific mind ("a man is anything we call one").

Foreign Policy Association. *Toward the Year 2018*. New York: Cowles Education Corporation, 1968.

Herman Kahn and Anthony J. Weiner. *The Year 2000: A Framework for Speculation on the Next Thirty-three Years*. New York: Macmillan, 1967.

Gordon Rattray Taylor. *The Biological Time Bomb*. New York: World, 1968. Here, as in the Feinberg book, the Huxleyan implications of the biologists are seen to parallel the dangers brought about by the physical scientists.

Stewart Udall. *1976: Agenda for Tomorrow*. New York: Harcourt, Brace and World, 1956.

C. S. Wallis, ed. *Toward Century 21: Technology, Science and Human Values*. New York: Basic Books, 1970.

Interpretation, Assessment, and Critique of Science and Technology (CHAPTERS 2, 3, 5, 11, 12)

Lorin Baretz. *The Servants of Power: A History of the Use of Social Science in American Industry*. New York: Wiley, 1965.

Jacques Barzun. *Science—The Glorious Entertainment*. New York: Harper and Row, 1964.

Isaiah Berlin. *Historical Inevitability*. New York and London: Oxford University Press, 1954.

J. D. Bernal. *The Social Function of Science*. New York: Macmillan, 1939.

Norman Birnbaum. *The Crisis of Industrial Society*. New York and London: Oxford University Press, 1969.

Robert Boguslaw. *The New Utopians: A Study of System Design and Social Change*. Englewood Cliffs, N.J.: Prentice-Hall, 1965.

Norman O. Brown. *Life Against Death*. New York: Vintage, 1959.

John G. Burke, ed. *The New Technology and Human Values*. Belmont, Calif.: Wadsworth, 1967.

Barry Commoner. *Science and Survival*. New York: Viking, 1966.

S. Demczynski. *Automation and the Future of Man*. London: Allen and Unwin, 1964.

Jack D. Douglas, ed. *Freedom and Tyranny in a Technological Society*. New York: Random House, 1970.

René Dubos. *Dreams of Reason: Science and Utopias*. New York: Columbia University Press, 1961.

———. *Reason Awake*. New York: Columbia University Press, 1970.

Jacques Ellul. *The Technological Society*. Translated by John Wilkinson. New York: Knopf, 1965.

Victor C. Ferkiss. *Technological Man: The Myth and the Reality*. New York: Braziller, 1969.

Erich Fromm. *Revolution of Hope: Toward a Humanized Technology*. New York: Harper and Row, 1968.

———. *The Sane Society*. New York: Rinehart, 1955.

———, ed. *Socialist Humanism*. New York: Doubleday Anchor, 1966.

Daniel S. Greenberg. *The Politics of Pure Science*. New York: New American Library, 1968.

Edward C. Higbee. *A Question of Priorities: New Strategies for Our Urbanized World*. New York: Morrow, 1970.

Johan Huizinga. *Homo Ludens: A Study of the Play Element in Culture*. Boston: Beacon Press, 1955.

Pierre Jalée. *The Third World in World Economy*. New York: Monthly Review Press, 1969.

Karl Jaspers. *The Future of Mankind*. Chicago: University of Chicago Press, 1961.

Friedrich Georg Juenger. *The Failure of Technology*. Chicago: Regnery, 1956.

Petr Kropotkin. *Mutual Aid: A Factor in Evolution*. Boston: Extending Horizon Books, 1955.

R. D. Laing. *The Politics of Experience*. New York: Pantheon, 1967.

Arthur O. Lewis, Jr. *Of Men and Machines*. New York: Dutton, 1963.

Herbert Marcuse. *Eros and Civilization: A Philosophical Inquiry into Freud*. New York: Vintage, 1962.

———. *One-Dimensional Man: Studies in the Ideology of Advanced Industrial Society*. Boston: Beacon Press, 1968.

Floyd W. Matson. *The Broken Image: Man, Science and Society*. New York: Braziller, 1964.

Margaret Mead. *Cultural Patterns and Technical Change*. New York: New American Library, 1955.

P. B. Medawar. *The Art of the Soluble.* New York: Barnes and Noble, 1967.

Jean Meynaud. *Technocracy.* New York: Free Press, 1970.

E. E. Morison. *Men, Machines, and Modern Times.* Cambridge, Mass.: M.I.T. Press, 1966.

Herbert J. Muller. *The Children of Frankenstein: A Primer on Modern Technology and Human Values.* Bloomington: Indiana University Press, 1970.

Lewis Mumford. *The City in History.* New York: Harcourt, Brace and World, 1961.

―――. *The Myth of the Machine: Technics and Human Development.* New York: Harcourt, Brace and World, 1966.

―――. *Technics and Civilization.* New York: Harcourt Brace, 1934.

John U. Nef. *Cultural Foundations of Industrial Civilization.* New York: Harper and Row, 1960.

Richard F. Nelson, Merton J. Peck, and Edward A. Kalachek. *Technology, Economic Growth and Public Policy.* Washington, D.C.: Brookings Institution, 1967.

Harold L. Nieburg. *In the Name of Science.* Chicago: Quadrangle Books, 1966.

F. S. C. Northrop. *The Logic of the Sciences and Humanities.* New York: Meridian Book, 1960.

Fritz Pappenheim. *The Alienation of Modern Man.* New York: Monthly Review Press, 1959.

Don K. Price. *The Scientific Estate.* New York and London: Oxford University Press, 1965.

Catherine Roberts. *The Scientific Conscience.* New York: Braziller, 1967.

Theodore Roszak. *The Making of a Counter-Culture: Reflections on the Technocratic Society and Its Youthful Opposition.* Garden City: Doubleday Anchor, 1969.

John Ruskin. *Unto This Last.* Lincoln: University of Nebraska Press, 1967.

Bertrand Russell. *The Impact of Science on Society.* New York: Simon and Schuster, 1953.

John R. Seeley. *The Americanization of the Unconscious.* New York: International Science Press, 1967.

Richard Sennett. *The Uses of Disorder: Personal Identity and City Life.* New York: Knopf, 1970.

Thorstein Veblen. *The Instincts of Workmanship and the State of the Industrial Arts.* New York: Viking, 1943.

Aaron W. Warner, Dean Morse, and Alfred E. Eichner, eds. *The Im-*

pact of Science on Technology. New York: Columbia University Press, 1965.

Norbert Wiener. *God and Golem.* Cambridge, Mass.: M.I.T. Press, 1964.

Nature and Man (CHAPTERS 11, 12)

Marston Bates. *Man in Nature.* Englewood Cliffs, N.J.: Prentice-Hall, 1964.

John Burroughs. *The Writings of John Burroughs.* 15 vols. Boston and New York: Houghton Mifflin, 1904–1908.

Rachel Carson. *The Sense of Wonder.* New York: Harper and Row, 1965.

Wilson O. Clough. *The Necessary Earth: Nature and Solitude in American Literature.* Austin: University of Texas Press, 1964.

R. G. Collingwood. *The Idea of Nature.* New York and London: Oxford University Press, 1960.

Loren Eiseley. *The Firmament of Time.* New York: Atheneum, 1960.

Arthur A. Ekirch, Jr. *Man and Nature in America.* New York: Columbia University Press, 1963.

Frederick Elder. *Crisis in Eden: A Religious Study of Man and Environment.* Nashville: Abingdon Press, 1970.

Paul L. Errington. *Of Predation and Life.* Ames, Iowa: Iowa State University Press, 1967.

Norman Foerster. *Nature in American Literature.* New York: Russell and Russell, 1923.

Y. H. Krikorian, *Naturalism and the Human Spirit.* New York: Columbia University Press, 1944.

Aldo Leopold. *A Sand Country Almanac.* New York and London: Oxford University Press, 1949.

Roderick Nash. *Wilderness and the American Mind.* New Haven: Yale University Press, 1967.

Marjorie Hope Nicolson. *Mountain Gloom and Mountain Glory: The Development of the Aesthetics of the Infinite.* Ithaca: Cornell University Press, 1959.

Samuel R. Ogden, ed. *America the Vanishing: Rural Life and the Price of Progress.* Brattleboro, Vt.: Stephen Greene Press, 1969.

Paul Shepard. *Man in the Landscape: A Historic View of the Esthetics of Nature.* New York: Knopf, 1967.

Wylie Sypher. *Literature and Technology: The Alien Vision.* New York: Random House, 1968.

Henry David Thoreau. *Walden.* New York: New American Library, 1958.

Index

Abelson, Philip H., 296
Aberfan mud slide, 197
Adaptation: in evolution, 222; to change, 93–94
Africa: colonization in, 127; settlement and clearing, 134
Agriculture, 55, 74, 125, 196; as chemical mining, 138, 142, 145; basis for economy of Third World, 308–309; creation of new plant diseases, 145; effects of Aswan Dam, 136–137; energy use in, 149; exploitation in Third World, 129; in Paleolithic and Neolithic societies, 42–43; in primitive societies, 19; use of fertilizers in, 76
Alaska, oil exploration in, 190–191
Alewives, 112–114, 122, 143
Alexander the Great, 78
Algorithm, 258
Alienation, 8, 247, 253, 258, 284, 285, 300, 307
Amazon rainforest, 134
American Association for the Advancement of Science, 9
American Chemical Society, 105

American Gas Association, 157–158
American Institute of Physics, 105
Amosov, Nicholai M., 59
Ampère, Marie André, 57
Anaxarchas, 78
Anaximander, 20
Angola, oil discovery near, 186
Anti-wealth, 276–277
Arabia, misuse of land in, 134
Arabian Gulf, oil discovery in, 186
Arabs, language, 42
Archimedes, 34, 44, 88
Aristotle, 20, 21, 24
Artificial environment, 119, 220–221, 222–223, 227, 234, 245, 257, 260, 271, 276
Artificial intelligence, 59, 61, 257
Ashby, W. Ross, 57
Asia, 204; blight of universal civilization, 306; clearing of land in, 134; colonization of, 127
Aswan Dam, 136–137, 307
Asymptotes, 118, 119, 121, 185, 271, 294–298, 303; definition of, 115
Atomic energy, 69, 74, 80–81, 94, 148, 165–179, 195, 227, 233, 238,

326 *Index*

Atomic energy *(Con't.)*
240, 256; exposure limits, 233–245; fusion process, 64; generation in 1980, 150; generation in 1990–2000, 155, 156
Atomic Energy Commission (AEC), 166, 173, 176, 239, 240, 241; bomb tests, 240; breeder reactor funding, 167–168; promoter of bombs and energy, 165–166; research on waste storage, 169–170; testimony on limits, 234–235, 237–238, 241–242; testimony on risks, 235, 241
Atomic weapons. *See* Nuclear weapons.
Audubon, John James, 287
Auschwitz, 290, 304
Auto, 8, 55, 69–70, 73, 74, 75, 83, 94, 120, 180, 195, 262, 270, 273, 292, 305; accidents, 184–185, 193; anachronism, 297; antipollution devices in, 192–193, 249, 274; energy requirements of, 149, 153; gasoline consumption of, 157, 185; pollution from, 161, 191–193; production of, 181; scrapped, 181, 183; U.S. registration of, 181–182, 184; world registration of, 184
Automation. See Cybernetics.

Babbage, Charles, 5
Bacon, Francis, 10, 23–24, 30, 46
Bailar, John C., III, 244
Becker, Carl L., 32
Behavioral science. *See* Social science.
Belgium, pollution in, 140, 160
Bell, Eric Temple, 108
BenSuc, 282
Bergmann, Ernst D., 308
Bering Sea, oil find in, 186
Bernal, J. D., 43
Bertalanffy, Ludwig von, 36
Bhagavad-Gita, 39
Bible, 18, 20, 22, 23, 24, 120, 122, 287, 294

Big Science, 106, 256, 268, 269, 272
Biocides, 136, 195, 224, 227, 233, 240
Biology: anabolism and catabolism, 248; body as quasi-machine, 36; disturbing effects of measurement in, 98; impotence principles in, 89; of life system, 229; promise of immortality, 7; transformation to mathematics, 98–99
Bogdanov, Boris N., 135, 302–303
Bohr, Niels, 98
Born, Max, 102
Bouillenne, Raymond, 82, 134
Brain, 102–103, 104–105, 106, 289; analogy with computer, 104, 258; growth to maturity, 117–118
Bremerman, H., 99
Broecker, Wallace S., 162
Brooks, Harvey, 11–12
Brown, Harrison, 126, 128, 157, 158, 166, 167, 271
Brown, Norman O., 39
Bureaucracy, 268; independence from ideology, 301
Bury, J. B., 29–30, 33

Canada: ban on poisoned fish, 199; diversion of rivers, 138; oil prospecting in, 190; preservation of Arctic, 276
Capek, Karel, 53–54, 291
Capital accumulation: cruelty of, 283; in Soviet Russia, 295; in Western nations, 126–127
Capitalism, 46, 126; anarchy in, 292; and automated factory, 299; and "forces of production," 135–136, 304; and Galilean physics, 46; excesses of, 283; and Reformation, 294; share of world GNP, 129; similarity to communism in technological goals, 301–302
Carbon dioxide: carbon cycle and climatic changes, 161–164, 165; in photosynthesis, 82, 161, 250; in residue chain, 70

Carbon monoxide: and antipollution devices, 192; effects in blood, 191; in auto exhaust, 55, 69, 191
Carcinogenicity: and radiation exposure, 241, 243; of asbestos, 226, 227, 237; of iodine-131, 239; of oil, 188; of pesticides, 230–231; of radium, 227; of tobacco, 261
Carlyle, Thomas, 299
Carson, Rachel, 7, 142, 302
Caspian Sea, pollution of, 302
Causality, 250–251, 265
Centralization of organizations, 269–271
Chemical Abstracts Service, 105–106
Chemical-biological weapons, 212; availability in Vietnam, 213; dumping in oceans, 171
Chess, 86
Chicago: and alewives, 112; fountain at Illinois Institute of Technology, 3; optimal expressway pattern in, 184; proximity to Dresden atomic plant, 173
China, 207; atomic weapons tests, 175; deforestation in, 133, 134; industrial potential of, 128; technology in early civilizations, 23
Christianity: and after world, 16; and capitalism, 295; and neoplatonism, 22; attitude toward labor, 294; attitude toward progress, 29
Civilization, 8, 10, 17, 43, 81; ancient, 133, 254; as artificial environment, 220–221, 223; breakdown of, 5, 30, 247, 250, 251, 271; industrial, 28, 256; technological, 6, 75, 93, 223, 233, 245, 254, 267, 271, 273, 279, 282, 284, 298, 305; universal, 306, 307–309
Climate: effects of deforestation, 133; effects of environmental changes on, 55, 70, 158, 161–164, 178, 223, 285; modification in war, 217

Cloud, Preston E., 130
Coal, 46, 166, 168, 271; burning of, as quasi-solution, 64–65; complexity of residue problems, 72–73; depletion of, 82, 275; equivalent energy in granite, 167; formation of, 81–82; nonrenewable resource, 83; pollution from, 158–161, 168; production of energy from, 148–150, 155, 156; regulations for mining of, 276; relation to other energy sources, 74; removal of sulfur from, 160; reserves of, 158; wastes from, 197
Cold War, 79, 127
Collectivization of society, 301
Collingwood, R. G., 263
Columbia River: health effects from radioactivity in, 243–244; tritium in, 173
Commoner, Barry, 138, 165, 239, 251
Communication, 85, 90, 293; and limitations of intellect, 103–104; and McLuhan's thesis, 58
Communism: acceptance of technology, 295; and "forces of production," 135–136, 304; and automated factory, 299; antithesis of, 304; excesses of, 283; similarity to capitalism in technological goals, 301–302
Competitive exclusion principle, 89
Computer, 52, 86, 91, 94, 106, 162–163, 255, 257, 272, 289; analogy with brain, 104, 258; and automation, 5, 57–58; and "best way," 61, 260, 261, 262, 263, 275; and creativity, 259; and humans, 52; entropy in, 99; fallibility of, 261–262; in automatic indexing, 59; in automatic translation, 58; in counter-technology, 275; in decision-making, 258; in measuring Vietnam pacification, 214; in war games, 208; increasing complexity of, 73; size limitations in system analysis, 60–61; speed of, 63, 91

Comte, Auguste, 30, 36
Condorcet, de M., 27
Copernicus, Nicholas, 24, 80
Corps of Engineers, U.S. Army, 137, 261
Cosmology, 87; Egyptian, 18–20; Mesopotamian, 18–20; scientific, 80, 100–101
Council on Environmental Quality, 171
Counter-technology, 215, 274–277, 279, 295; in Soviet Russia, 303; in Third World, 307; proliferation of organizations for, 269; self-destruction, 282
Creation of world, 119–120
Creatory, 304
Cuba: Bay of Pigs, 79; agriculture in economy, 309
Cybernetics, 5, 10, 52, 59, 107, 282; and automation, 73, 94, 258, 299, 301; and communication, 58; and social system, 58; automata and artificial intelligence, 59; claims for, 56–57; man in system, 52
Cyclamates, 225
Cyrenaica, deforestation in, 134

Dachau, 7, 282, 290, 304
D'Alembert, Jean le Rond, 28
Darwin, Charles, 30, 31, 101–102
Dasmann, Raymond F., 133, 245, 308
DDT, 143–144, 190, 223, 225, 227, 231, 234, 302
Decentralization of organizations, 269–271
Dedijer, Vladimir, 214
Deforestation, 133–134, 135
Democritus, 20
De Re Metallica, 44
Descartes, René, 10, 26–27, 28, 30
Detergents, 50, 159, 195, 274, 303; and eutrophication, 139–141; consumption in U.S., 139; removal of phosphorus from, 141–142; use of NTA in, 141
Deuterium, 81, 126, 168

Dialectics, 12, 48, 61, 76, 78, 122, 146, 165, 265, 279, 295; and causality, 250–252; and decline of civilizations, 247, 250; and "forces of production," 304; and order, 305; in calculus of efficiency, 291; in capitalist economy, 292; in counter-technology, 275; in health, 67–68, 93, 122; in organizations, 268–273; laws of, 248–249; leading to decline of science and technology, 253–280; negation of quasi-solutions, 74, 138, 144, 145; of nature in Soviet Russia, 302–303; transformation into opposites, 114, 298; transformation of quantity into quality, 63, 69, 74, 191, 193; transformation of society, 282, 284
Diderot, Denis, 16, 28
Division of labor, 181, 254, 255, 268, 300, 301, 304
Donora, pollution sickness in, 160
Doomsday, 119, 124, 206
Drugs, 224–225, 227, 229, 274
Dubos, René, 94, 293
DuBridge, Lee A., 234–235

Eastern civilizations, 23, 109
Ecology, 121, 186, 277, 295, 302; as science, 252–253; cycles in ecosystem, 250; definition of, 84; effects of ash accumulation on, 159; effects of biocides on, 146; effects of dams and reservoirs on, 138; effects of oil spills on, 190; effects of radioactive wastes on, 169, 241; in Arctic regions, 190
Economics, 31, 46, 75, 114, 138, 278, 283, 299, 303; and assault on nature, 12; and cybernetics, 58; and subjugation of man, 258; as scientific discipline, 39; constraints in technological process, 49; costs of counter-technology, 276; debts from despoliation, 275–276; external costs, 56; failure of suboptimization, 262; in efficiency cal-

culus, 291, 292; in technology assessment, 266; of consumption, 256; of waste, 183, 194–195; shared concepts under capitalism and communism, 295, 301–302; transformation of man into commodity, 300
Eddington, Arthur, 257
Efficiency, 4, 10, 12, 38, 39, 47, 138, 289–293, 295, 300; and optimality, 260, 262; as capitalist and communist goal, 301, 303; as communist goal, 295, 301; as measure of society, 299; false quest for, 272, 289, 292; in economic system, 291–292; in physical processes, 63, 89; in resource allocation, 266; in scientific management, 51–52; in waste reduction, 194, 278; opposed to adaptability, 291, 293
Egypt: Aswan Dam, 136–137, 307; cosmology, 18; transformation into desert, 134; view of nature, 19–20
Einstein, Albert, 39, 96, 102, 254
Eisenhower, Dwight D., 203, 204, 206
Electricity, 45, 158, 177, 271; generation in nuclear plants, 165, 166, 172, 178, 233, 238, 240; growth in generation, 154–157; pollution from generating plants, 158–161
Ellul, Jacques, 47–48, 58, 71, 246–247, 255, 259, 260, 261, 267, 282
Elsasser, Walter M., 98–99
Energy, 45, 63, 69, 74, 76, 82, 83, 84, 88–89, 98, 99, 141, 148, 160, 165, 195, 198, 238, 263, 274; production in U.S., 148–150, 154–157, 166
England, 147, 178, 301; atomic accidents in, 172; atomic bomb tests, 175; expropriation of church estates in, 127; in Industrial Revolution, 126–127, 150; oil spills off coast of, 187
Engstrom, Elmer W., 273

Enlightenment, 5, 11, 27, 28, 30, 32, 288
Entropy, 84, 85, 89, 99
Epistemology, 80, 107–110
Erosion, 135, 146, 187
Ethics, 13, 16, 17, 26, 33, 34, 39, 47, 56, 107, 253, 257, 284, 285
Euphrates River, 18
Eutrophication, 138–139, 140, 141, 278
Evans, David M., 170
Evans, Robley D., 244
Evolution, 10, 30, 31, 84, 92, 94, 144, 222–223, 224, 284, 291
Exponential growth, 74, 121, 148, 155, 200, 223, 270, 272, 284, 293, 295; defined, 115; of auto accidents, 184; of autos, 181–182, 184, 185, 193; of coal production, 150; of electrical energy, 155–156, 177, 178; of pesticides, 142; of population, 120, 122–124, 146; of world oil production, 153
Extinct species, 83, 202, 295

Fadely, Robert Cunningham, 243–244
Falk, Richard A., 295–296
Fallout, radioactive, 175, 223, 233, 238–240, 243
Fallout shelters, 212
Farrington, Benjamin, 21
Federal Power Commission, 161, 166
Federal Radiation Council (FRC), 236, 237, 238, 240
Federal Water Pollution Control Agency, 178
Fermat, Pierre de, 86
Fertilizer, 6, 55, 73, 76, 125, 136, 138, 142, 145, 146, 165, 195, 227
Fetishism of commodities, 300, 301
Finitude, 73–74, 80–84, 86, 121
Flawn, Peter, 157–158
Foerster, Heinz von, 119, 120, 124
Food, 8, 9, 74, 81, 85, 86, 94, 122, 132, 133, 138, 142, 146, 177, 195, 224, 227, 229, 232, 309; additives, 225, 227, 274, 278, 285,

Food *(Con't.)*
302; and Malthusian theory, 115; and population limits, 124–126; as renewable resource, 82; dialectics of production, 138, 165; effects of dams on, 137; effects of DDT on seafood, 143; effects of oil spills on, 186, 188; food chains in Arctic, 190; pesticide tolerances for, 228; production and processing of, 75–76; waste products of, 196, 199

Food and Drug Administration, 225, 233
Forces of production, 135–136, 301, 303, 304
Ford, Henry, 180
Forests, 7, 82, 133, 146, 199, 220, 221, 222, 232, 260; in Amazon, 134; in Africa, 134; in China, 134; in Soviet Russia, 135; in United States, 134
Foster, John S., Jr., 215–216
France: atomic bomb tests, 175; importation of minerals, 130; pollution in, 140
Franco, Francisco, 304
Frank, Phillip, 250
Frankfort, Henri, 19
Freud, Sigmund, 8
Fromm, Erich, 282
Fulton, Robert, 45
Futurism, 11, 263

Galilei, Galileo, 6, 10, 21, 22, 28, 35; and "fatal estrangement," 39; effects of concepts on man, 46, 53; methodology of new science, 24–25; on growth limits, 92; rejection of philosophy, 254; scientist and technician, 44
Gardner, Trevor, 215
Gas, natural, 74, 148, 160–161, 186; as source of energy, 149, 155, 156; depletion of, 83, 157–158; production in U.S., 150
Gay, Peter, 27
Gell-Mann, Murray, 101

Geometrodynamics, 100–101, 102, 109
George, Henry, 247
Germany, 127; division of, 304; pollution in, 140; thalidomide use in, 224–225
Gide, André, 288
Gillispie, Charles Coulston, 26, 39
Goedel, Kurt, 86
Gofman, John W., 241–242
Golbach's conjecture, 86
Gray, Asa, 287
Greece: attitude toward knowledge, 24; attitude toward labor, 44, 294; beginnings of science, 20–21, 87; deforestation in, 134; dethronement of, 29; fate in classics, 88; machinery in war, 35; separation of philosophy and crafts in, 21, 44; view of change in, 29
Green, Harold P., 266–267
Greenland, nuclear accident near, 172
Gross domestic product (gross national product), 56, 119, 120; and anti-wealth, 276; and quality of life, 296; compared to arms expenditures, 205, 206; of world, 129
Guernica, 282
Guerrilla war, 204, 214
Gulf of Mexico, 121, 187, 188

Haber, Fritz, 269
Haldane, J. B. S., 92
Haldane-Muller principle, 89
Hardin, Garrett, 88, 89
Harmon, Leon, 104
Hauser, Philip M., 128
Hawthorne experiment, 54
Health: and population growth, 67–69; deterioration in, 93; dialectical effects of, 67–68, 93, 122; effects of pollution on, 158, 159–160, 191, 192; interest of state in, 256; radiation hazards to, 235; relation of pesticides to, 228, 229–230

Hebrews: attitude toward labor, 294; attitude toward nature, 20, 22
Heisenberg, Werner, 98
Helmholtz, Hermann von, 88
Heraclitus, 20, 248
Herbicides, 125, 138, 142, 145, 146, 227, 232–233, 278
Highways, 55, 74, 83, 120, 184, 193, 260, 262, 308; and urban land consumption, 132, 133; effect on drainage, 146; growth of, 186; homogeneity of federal network, 306; need for centralized control of, 270
Hiroshima, 7, 17, 38, 165, 224, 233, 282, 304
Hitler, Adolf, 304
Hofstadter, Richard, 31
Hooke, Robert, 34
Hopi, influence of language on world view, 109
Hubble, Edwin Powell, 101
Hudson River, radioactive contamination of, 175
Huizinga, Johan, 43
Hume, David: argument against induction, 109–110, 250; attack against poets, 288; on primacy of mathematics, 28
Huxley, Aldous, 282, 289, 291
Huxley, Julian, 92, 257
Hydrocarbons, 69, 191, 192

Illinois Institute of Technology, 3
Illth, 299, 304
Iltis, Hugh H., 94
Imperialism: and "have not" nations, 132; and oil, 186
Impotence principle, 80, 86, 115, 193, 200; and entropy in measurement, 99; definition and examples, 87–90; in counter-technological society, 277
India: ancients' reverence of earth in, 19; early technology, 23; industrialization in, 306; negative effects of irrigation systems in, 137

Induction: and biological explanation, 98–99; and Newton's rejection of hypotheses, 90; and orderliness of universe, 16; as form of conditioned reflex, 35; Hume's argument against, 109–110; problem of logical substantiation, 250; Russell's challenge, 110
Industrial Revolution, 33, 46, 254; and rise of capitalism, 294; assault on man, 38–39; effects of technics on science during, 44–45; effects on England, 126–127; effects on other nations, 127; reliance upon coal and iron in England, 150; stage in artificiality of environment, 220–221; violence of, 10
Inefficiency: and thermodynamics, 89; in inefficient society, 280, 293; of nature, 10, 291
Inexhaustibility theorem, 104
Infinitude, 80–81
Information, 84, 103, 104, 277; capacity of brain, 106; in measurement, 99; retrieval of, 255
Ingersoll, Alfred C., 59
Innovation, 73
Institute of Electrical and Electronics Engineers, 105
International Commission on Radiological Protection (ICRP), 173, 235–236, 238
Invention, 73
Invisible hand, 13, 292
Irrigation, 74, 146; negative effects of, 136–137, 145, 302, 307
Israel: agricultural experience in, 308–309; labor ethic of, 299; war in, 304

Jalée, Pierre, 129–130
Japan, 127; mercury contamination of fish in, 226; pollution in, 140, 302
Jaspers, Karl, 258
Jet planes, 75, 91, 163, 261, 274
Johnson, Lyndon B., 204
Johnson, Raymond, 243

Joule's law, 88
Judeo-Christian ethic, 16, 46, 246
Juenger, Friedrich Georg, 257, 300

Kant, Immanuel, 96
Kelvin, Sir William Thomson, 28
Kennedy, John F., 203
Kepler, Johannes, 17, 24, 28
Khrushchev, Nikita, 303
Klemm, Friedrich, 35
Korea: demand for industrialization, 306; division of country, 304; operations research in war, 213
Kranzberg, Melvin, 45
Krypton, 173, 174, 243
Kuriba dam, 137

Labor, 257, 276–277, 294; as creative activity, 303; changing concepts of, 43–44, 294; in Third World, 309; self-realization of man through, 298–299
Laing, R. D., 283
Lake Baikal, pollution of, 140, 302
Lake Erie: destruction of, 113, 139, 140, 188; mercury-tainted fish in, 226
Lake Michigan: alewives kill-off in, 112–114; effects of pesticides in, 143; estimated cost of saving, 276; eutrophication in, 140
Lake Superior, dumping of iron ore tailings in, 197, 200
Lamarck, Jean Baptiste, 89
Land, 74, 81, 84, 86, 124, 136, 137, 146, 181, 183, 184, 193, 195, 197, 200, 217, 223, 260, 273, 284, 285, 304, 308, 309; converted to deserts, 134; decrease of farmland, 133, 135, 137; in ancient Greece, 134; in Soviet Union, 135, 137, 302; loss through erosion, 135; loss through mining, 135; population density in U.S., 132–133; reclamation of, 138
Landauer, Rolf, 99–100
Language, 41–42, 103; as communication without meaning, 58; development in prehistoric society, 42–43; origins by Paleolithic man, 42; Whorfian hypothesis of, 108–109
Larson, Clarence E., 241, 242, 243
Law and order: legacy of war, 305; mask for repression, 279; shibboleth of declining society, 272
Lead, 128; pollution from, 69, 144, 191–192
Leibniz, Gottfried Wilhelm, Baron von, 5
Lenin, Vladimir Ilyich, 295
Leopold, Aldo, 285, 287, 288
Lewin, Kurt, 248
Libraries, contents of world's, 105
Lidice, 282
Life, 8, 29, 37, 84, 85, 92, 94, 109, 119, 122, 134, 165, 166, 171, 177, 188, 189, 217, 219–220, 222, 223, 228, 234, 242, 245, 248, 251–252, 263, 273, 275, 289, 290, 291, 293, 295, 296, 297, 299, 304, 305; origin of, 101–102, 106
Lightfoot, John, 119–120
Limits, 165; in biology, 82, 92–93, 122; in discovery, 78; in human growth, 115–119; in nature, 119; of absolute zero, 91; of communication, 90; of earth, 9, 82–84, 224, 245, 248, 260, 279; of energy, 83; of exposure to pesticides, 229–231; of human survival, 84–86; of intelligence, 107; of man's ineptitude, 262; of mathematics, 98–100; of pollution, 224; of the possible, 90; of production, 83; of radiation exposure, 224, 234–238, 240, 241–243, 245; of sound, 91; of technology, 247; on economic growth, 295–296; on knowledge, 100–107, 277; on measurements, 95–100; threshold for pollution, 159; to sense apprehension, 25; types of, 80
Logical positivism (empiricism), 35–36, 109, 254
London, air pollution in, 160

Index

Longevity, 92–93, 160
Lotka, A. S., 103
Luther, Martin, 295
Lysenko, Trofim Denisovich, 89

MacArthur, Douglas, 203
Machiavelli, Niccolò, 32
MacMahon, Brian, 242
Maintenance economy, 297, 298, 303
Malaria, 143
Malthus, Thomas Robert, 115, 125
Man: adaptive capabilities of, 93–95; as commodity, 257, 300–301; as computer, 257; as consuming animal, 282; as mechanism, 300; as a subsystem, 52–55; post-technological, 284, 304; standardized, 258
Mandelbaum, Leonard, 79
Manhattan Project, 39, 268
Marcuse, Herbert, 281, 302
Mariner's compass, 23, 44
Marx, Karl, 126, 300–301
Marxism, 246, 301
Massachusetts Institute of Technology, 163
Mathematics, 108, 253, 257, 286, 291; abets totalitarianism, 272; and "best way," 60, 61, 260; attitude toward by: Copernicus, 24; Galileo, 25; Hume, 28; Kelvin, 28; Kepler, 24; Newton, 27; Plutarch, 34–35; Voltaire, 28; axiom of science, 15; human origins of, 108; in behavioral science, 37; in biology, 98–99; in industrial engineering, 55; laws of universe, 95; of brain, 104–105; of growth, 114–115; of war, 208–212; physical limitations of measurement, 98–100; unsolvable problems in, 86–87
Maxwell, James Clerk, 95
May, Catherine, 235
McCue, J. J. G., 256
McElroy, William D., 277
McLuhan, Marshall, 58
Mead, Margaret, 93
Mercury, 82, 144, 199, 225–226

Michelangelo Buonarroti, 287
Michelson-Morley experiment, 95–96
Millennium, 6, 9, 16, 27, 29, 246
Minerals, 46, 82; depletion of, 74, 82–83, 166, 275; imports in U.S., 130; in new technologies, 64; reserves of, 130
Mining, 64, 72–73, 76, 133, 135, 194, 196, 197–198, 273, 288; in Soviet Russia, 135
Minkowski, Hermann, 96
Mississippi River, 121, 198, 226
Monosodium glutamate, 225
Moon landing, 5, 7, 10, 13, 17, 38, 79, 80, 81, 289
Moore, Raymond T., 238
Morison, E. E., 262
Morocco, deforestation in, 134
Muir, John, 287
Mumford, Lewis, 33, 43–44, 45, 48, 58, 272, 294
Mussolini, Benito, 304
Mutation, 89, 230–232, 243
MyLai, 282, 287

Nagasaki, 38, 165, 224, 233
Napalm, 190, 213
Napier, John, 5
National Academy of Engineering, 265
National Academy of Sciences, 58, 125, 157, 170, 225, 266, 267
National Air Pollution Control Administration, 162
National Committee on Radiation Protection (NCRP), 236, 237, 238
National Library of Medicine, 105
National Science Foundation, 277
Nature, 21–22, 23, 27, 28, 31, 34, 37, 44, 103, 114, 121, 143, 164, 190, 219, 272, 279, 280, 296, 302, 305; as closed system, 55, 254, 259, 260, 275, 281; as manifestation of God, 22; conquest of, 12, 13, 27, 30, 38, 46–47, 81, 87, 107, 122, 148, 166, 186, 218, 221–222,

Nature *(Con't.)*
  252, 257, 283, 302; control by magic, 19; cycles in, 119, 219–220, 222, 245, 248, 250, 251, 284; laws of, 30, 87, 90, 100, 108; man in, 18–20, 42–43, 107, 283, 284–286
Nazi (German National Socialist Party), 38, 171, 290
Nef, John U., 26
Neolithic period, 42–44, 220, 294
Neoplatonism, 22
Newton, Isaac, 24, 28, 102, 109; and absolute time, 95; and capitalism, 46; attitude toward God, 87–88; contributions to science, 27; determinism of, 98; induction, 90
New York: abandoned autos in, 183; breakdown of telephone system, 270; pollution control in, 276; rise in pulmonary emphysema in, 160
Nigeria: deforestation in, 134; oil finds near, 186
Nile River, 134; and Aswan Dam, 136–137, 307; in Egyptian cosmology, 18
Nitrates, counter-effects on soil, 138
Nitrogen oxides, 69, 159, 160, 191–192
Northeast power blackout, 271
Norway, oil finds near, 186
Nuclear accidents, 172–173, 217
Nuclear energy. *See* Atomic energy.
Nuclear test ban treaty, 175, 240
Nuclear weapons, 4, 11, 38, 64, 165, 203, 208, 211, 227, 233, 256, 292; accidents with, 172, 217; and radioactive fallout, 175, 223, 233, 238–240, 243; arsenals of, 205; effects in war, 211–212; first test explosion, 39; test ban treaty, 175, 240; tritium from tests, 174, 175

Occam's razor, 251, 258, 286
Oil, 74, 81, 83, 148, 149–150, 156, 159, 185, 271, 275, 276; depletion of, 157, 193; exploration for, 186, 190–191; production in U.S., 150, 152, 157; production in world, 150, 153–154; spills, 165, 186–190, 217, 224, 279, 303; transportation of, 186–187
Operations research, 59, 208, 212–213
Oppenheimer, J. Robert, 17, 39–40
Orwell, George, 272, 282
Otto, Max C., 308
Oxygen, 94, 161, 250, 252, 278; in eutrophication, 139; in tritiated body water, 173; nonrenewable resource, 164; supply in atmosphere, 161–162

Paleolithic period, 42–43
Pascal, Blaise, 5
Pauling, Linus, 239
Peace, 8, 202, 205–206, 215, 216, 269, 273, 287
Perimeters of the possible, 80, 90–95, 115, 277, 279
Pesticides, 125, 138, 142–145, 146, 159, 165, 201, 225, 226, 227–232, 261, 274, 278, 302
Peterson, Eugene K., 163–164
Petroleum. *See* Oil.
Philosophes, 22, 27, 32
Phosphates, 274; in eutrophication, 139–141; in fertilizers, 55
Photosynthesis, 82, 161, 250
Plato, 21, 24, 96, 108, 134
Pliny, 22, 35
Plutarch, 34–35, 78
Plutonium, 167, 174, 175, 217, 243
Poison gas, 212, 213, 269, 290; dumping in ocean, 171; killing of sheep, 171
Pollution, 7, 8, 56, 69, 94, 135, 136, 140, 141, 143, 144, 146, 147, 162, 164, 165, 171, 195, 196, 197, 198, 199, 200, 216, 218, 223, 224, 226, 247, 249, 265, 273, 276, 277, 278, 295, 302, 306, 307; disease of, 160, 192, 224; from autos, 55, 69–70, 73, 75, 180, 191–193; from fossil fuels, 158–161, 168; from jet planes, 163; of oceans, 171, 186,

187–188, 200; radioactive, 166, 168, 173–176, 178; thermal, 69, 168, 176–178, 217, 242–243
Population, 8, 43, 69, 74, 76, 84, 119, 128, 129, 132, 136, 138, 142, 146, 148, 155, 160, 164, 181, 216, 221, 260, 270, 283; dialectics of, 93; infinite growth of, 119, 120; Malthusian theory, 115; problems of, 122–126; zero growth of, 124, 297
Positivism. *See* Logical positivism.
Post-technological society, 114, 304–305
Pragmatism, 47, 87, 255
Price, Don K., 9
Progress, 4, 6, 9, 13, 14, 28–33, 81, 146, 238, 246, 247, 258, 273, 284, 296, 297, 308, 309; and artificial environment, 221; and capital accumulation, 127; and efficiency, 10; and exponential growth, 115; in Third World, 306, 307; myth of, 282; religion of, 6–7
Propositions, formally undecidable, 86
Psychology, 36–37, 54, 248
Ptolemy, Claudius, 87
Pyrotechnica, 44
Pythagorean philosophy, 4, 16, 24, 286

Quantum theory, 37, 97–99, 100
Quark, 101, 102, 109
Quasi-solution, 113, 114, 121, 136, 141, 142, 148, 157, 158, 160, 166, 168, 170, 171, 177, 178, 180, 181, 184, 187, 188, 191, 192, 194, 195, 198, 200, 224, 228, 247, 263, 275, 279, 292, 295, 302, 307; augmentation in, 63–64, 70; defined, 63–64; difficulty of solution, 71–76; effects of, 67; incompleteness in, 63, 70; no solution of, 76; proliferation of, 67, 70, 71, 74, 114, 121; residue generations in, 65–68; residue of unsolved problems in, 64–65; secondary effects of, 64–65, 70–71, 72; unforeseen results of, 67, 114, 126, 137, 138, 226, 247, 265, 273, 275, 307

Radford, Edward P., 174
Radiation protection guide (RPG), 236–238, 239
Ramey, James T., 237–238
Rapoport, Anatol, 215
Rathjens, George W., 216
Reactors, nuclear, 167, 176, 177, 308; accidents with, 168, 172–173; breeder, 167–168, 176
Relations of production, 10, 135, 301, 302, 303
Relativity, 37, 87, 96, 100, 254
Renaissance, 30, 35
Research, 46, 120, 204, 277, 303; on biocides, 278; on oil spillage, 279; on wastes, 278; on water quality, 278
Resources, 17, 75, 76, 86, 122, 124, 126, 127, 147, 183, 184, 195, 202, 204, 215, 217, 255, 260, 262, 263, 265, 266, 274, 289, 302, 309; availability of, 128–129, 130–132; consumption of in U.S., 128–130; cutback in use of, 297; depletion of, 7, 13, 51, 56, 69, 74, 82–83, 115, 120, 130, 146, 148, 157, 165, 166, 168, 185, 193, 198, 216, 221, 223, 247, 271, 273, 275, 276, 295, 298, 303, 304; finitude of, 81; nonrenewable, 82; renewable, 82–83
Reuss, Henry S., 129
Rhine River, pollution of, 140
Rhodesia, 137
Richardson, Lewis F., 206
Riordan, John, 270
Robot, 36, 54, 217, 258, 259, 262, 267, 281, 289, 291, 293, 297, 298
Rome, 29, 34, 35, 44, 134, 200
Roosevelt, Theodore, 51
Rousseau, Jean Jacques, 30
Royal Society, 34, 46, 147
Ruskin, John, 285, 298–299
Russell, Bertrand, 15, 75, 110, 202, 250

336 *Index*

Russia, 59, 89, 137, 138, 175, 203, 205, 207; arms expenditures in, 205; as technological society, 295, 301; casualties in nuclear war, 212; loss of land in, 135; pollution in, 140, 302–303

Sahara Desert, resultant of deforestation, 134
Saint Augustine, 23, 32
St. Lawrence Seaway, 113–114
Santa Barbara oil spill, 187, 188, 190, 224
Santayana, George, 15
Satellites, 90–91, 208, 217
Schistosomiasis, 137
Schrödinger, E., 39, 84
Schurr, Sam H., 157
Science: and law, 11; and secrecy, 268–269, 279; and values, 16, 33, 34, 36, 38, 39, 53, 103, 107, 224, 253, 279; and war, 34–35, 45, 202, 204, 208, 268; as commodity, 34; as myth, 4; as religion, 6–7, 15, 17, 39, 87, 246; axioms of, 15–16, 32, 33, 35, 38, 51, 57, 60, 87, 107, 224, 255, 256; "fatal estrangement" of, 26, 39; history of, 14–15, 20–22, 78; ignoring the difficult, 37, 62, 108, 257, 265–266, 269; institutionalization of, 267–268; laws of, 17, 25, 88–89, 95, 102, 110, 250–251, 279; limits of, 78, 80, 98, 103–105; medieval, 23; metaphysics of, 16, 17, 22, 35, 47, 98, 109; planning of, 264; rejection of qualitative values, 16, 24, 25, 26, 33, 37, 252, 253, 279; relation to technics, 44–45, 254; relation to technology, 4, 6, 45, 49, 255
Scientific literature, 105–106
Scientific management, 51, 54
Seaborg, Glenn T., 235
Seeley, John R., 60, 104
Seidenberg, Roderick, 90
Seine River, pollution of, 140
Seismograph, 23

Seneca, 44
Sewage, 56, 139, 140, 141, 142, 159, 194, 200–201, 270, 274, 278
Shadwell, Thomas, 147
Shakespeare, William, 103, 287
Sheaffer, John R., 141
Sherrington, Charles, 102, 251
Siddhartha, 292
Simpson, George Gaylord, 106, 110, 145
Singer, Charles, 21, 253
Smith, Adam, 46
Social institutions: breakdown of, 8; dialectics of, 268–273; inertia of, 76
Social science: Comte's three laws of, 30–31; contributions to Vietnam war, 213; implicit value judgments, 60; in the image of physics, 36–38
Sociology, 30, 36–37, 55, 60, 287
Somervell, D. C., 8
Soviet Union. *See* Russia.
Spain, nuclear bomb accident in, 172
Spencer, Herbert, 31, 32
Spengler, Oswald, 5, 15
Sprat, Thomas, 46
Sputnik, 79
Stalin, Joseph, 304
Stephenson, George, 45
Stoicism, 22
Strontium-90, 173, 223, 227, 239
Sulfur dioxide, 159, 160–161
Sweden, mercury contamination of lakes in, 199
Switzerland, pollution in, 140
System analysis, 56, 59–60, 61, 106, 252, 256, 258, 259–263, 265

Taconite process, 74
Tamplin, Arthur R., 241–242
Taylor, Frederick Winslow, 51
Technics, 42, 44, 254; in Neolithic period, 42–44, 220; in Paleolithic period, 42; merge into technology, 44–45
Technological impulse, 79

Technological process, 56, 60, 71, 78, 194, 270; "best way," 49, 60, 260–263, 305; breakthrough in, 70, 72; defined, 47; perpetual motion in, 50–51, 65, 72; steps of, 48–50
Technological rationalism, 257, 259, 281
Technological realism, 9, 121, 293–294
Technology: and values, 47, 224, 257, 301; assessment, 11, 253, 263–267, 272; autonomous force, 10, 47, 48, 49, 50, 61, 114, 247, 298, 303; decline of, 79, 114, 253–280, 282, 305; defined, 41; dialectics of, 12, 63, 69, 114, 122, 193, 247, 248, 253–280; homogenization resulting from, 305–306; obsolescence in, 51, 92, 195, 216, 292, 297; relations with science, 6, 45, 254, 255; social process, 49; utilitarian aspect, 47, 255–256, 288
Technology Assessment Board, 264–265
Technosocial problem, 71
Teller, Edward, 239
Teratogenicity, 231, 232–233
Thales, 20
Thalidomide, 224–225, 227, 233
Theoretical performance limitations, 91
Thermodynamics, second law of, 88–89, 99
Third World, 127, 213; imports of oil from, 186; medical science and population growth in, 69, 124; military expenditures in, 204–205; problems of universal civilization, 306–309; raiding of, by industrial nations, 129–132; technical aspirations of, 6; threat of famine in, 146
Thompson, Theos. J., 238
Thoreau, Henry David, 249, 258, 285, 286, 287
Thorium, 167

Tibet, effects of deforestation in, 134
Tigris River, 18
Tilgher, Adriano, 294
Time, 7, 12, 95–96, 108, 109, 257, 258, 299
Tokyo: burial sites in, 136; pollution of rivers in, 140; mercury poisoning in, 199
Tompkins, Paul C., 240, 242
Torrey Canyon, 165, 187, 188
Toynbee, Arnold, 5, 8
Tredgold, Thomas, 46
Tritium, 173–174, 175
Trobriand Islanders, 42
Trublet, Abbé Nicolas, 288
Twain, Mark, 120–121

Udall, Stewart L., 296
Uncertainty principle, 98, 99
United Nations, 203, 240, 306, 307, 309
United States Constitution, 11
University of Illinois, 119
Unsolvable problems, 36, 62, 80, 86–87, 202, 216, 248, 273, 275
Ural River, pollution of, 140
Uranium, 167, 198; mine tailings of, 176; reserves of ores, 166
Utopia, 6, 10, 16, 284

Van der Rohe, Mies, 3
Venezuela: attitude toward poverty and pollution, 307; plant for desulfurizing oil, 276
Vietnam, 204, 205, 210, 213, 215, 232–233, 304
Vinci, Leonardo da, 35
Vitruvius, 35
Voltaire, 27, 28

Wagar, J. Alan, 296
Wald, George, 102
War, 8, 34–35, 45–46, 47, 58, 79, 172, 186, 202–205, 258, 260, 268, 276, 287, 304; armed forces for, 205; casualty estimates from, 211–212; expenditures for, 204–205, 216; gaming, 208–210, 213,

War *(Con't.)*
218; of the future, 217; operations research in, 208, 212–213; physics of, 209–210, 214–215
Warheit, I. A., 59
Waste, 55–56, 83, 89, 120, 133, 135, 139, 142, 160, 183, 185, 191, 197, 218, 270, 271, 274, 291, 295, 298, 303; as crime, 297; disposal of, 197, 200, 274, 278; in consumer goods, 199–200; in food production, 196, 199; in mining, 194; radioactive, 169–171, 235, 238, 241, 244; research on, 278–279; types of, 194–195; volume of, 195–196, 198–200
Watt, James, 45
Weber, Max, 294
Weinberg, Alvin M., 255, 256, 269
Weismann, August, 89

Westin, Allan F., 272
Wheeler, John A., 100–101, 102
Whitehead, Alfred North, 15, 16, 33, 37, 88, 107
Whittaker, Edmund T., 88
Whorf, Benjamin Lee, 108–109
Wiener, Norbert, 57, 58, 107, 259
Wiesner, Joseph E., 215
Wigner, Eugene P., 103–104, 106, 255, 269
Wilkinson, John, 264
Wingfield-Stratford, W., 126
Wittgenstein, Ludwig, 109
Work. *See* Labor.

York, Herbert F., 215, 216
Young, John L., Jr., 244

Zambia, 137

A Note on the Author

EUGENE S. SCHWARTZ was until recently Senior Scientist specializing in computer and information sciences at the Illinois Institute of Technology Research Institute in Chicago. In addition to many scientific papers and articles, he has published essays and stories in such magazines as *The Nation* and *The New Leader*.

ROME, OSTIA, POMPEII

Rome, Ostia, Pompeii

Movement and Space

Edited by
RAY LAURENCE AND DAVID J. NEWSOME

OXFORD
UNIVERSITY PRESS

UNIVERSITY OF WINCHESTER
LIBRARY

OXFORD
UNIVERSITY PRESS

Great Clarendon Street, Oxford OX2 6DP

Oxford University Press is a department of the University of Oxford.
It furthers the University's objective of excellence in research, scholarship,
and education by publishing worldwide in

Oxford New York

Auckland Cape Town Dar es Salaam Hong Kong Karachi
Kuala Lumpur Madrid Melbourne Mexico City Nairobi
New Delhi Shanghai Taipei Toronto

With offices in

Argentina Austria Brazil Chile Czech Republic France Greece
Guatemala Hungary Italy Japan Poland Portugal Singapore
South Korea Switzerland Thailand Turkey Ukraine Vietnam

Oxford is a registered trade mark of Oxford University Press
in the UK and in certain other countries

Published in the United States
by Oxford University Press Inc., New York

© Oxford University Press 2011

The moral rights of the author have been asserted
Database right Oxford University Press (maker)

First published 2011

All rights reserved. No part of this publication may be reproduced,
stored in a retrieval system, or transmitted, in any form or by any means,
without the prior permission in writing of Oxford University Press,
or as expressly permitted by law, or under terms agreed with the appropriate
reprographics rights organization. Enquiries concerning reproduction
outside the scope of the above should be sent to the Rights Department,
Oxford University Press, at the address above

You must not circulate this book in any other binding or cover
and you must impose the same condition on any acquirer

British Library Cataloguing in Publication Data
Data available

Library of Congress Cataloging in Publication Data
Data available

Typeset by SPI Publisher Services, Pondicherry, India
Printed in Great Britain
on acid-free paper by
MPG Books Group, Bodmin and King's Lynn

ISBN 978-0-19-958312-6

3 5 7 9 10 8 6 4

Dedication

As this volume was being readied for delivery to the Press, we were saddened to learn of the death of Professor William L. MacDonald. His *The Architecture of the Roman Empire II: An Urban Appraisal* (1986) was an important stepping stone in introducing Roman studies to some of the now familiar themes from urban geography. Most notable of these in the context of the present volume was an emphasis on the street as a venue for social interaction and the development of the concept of the 'urban armature', by which notions of the urban image first moved into our vocabulary for studying the Roman city. His influence, whether implicit or explicit, pervades the pages that follow.

Preface

The subject of this volume is movement as a variable in the Roman city. Its broad objective is to make movement meaningful: meaningful for understanding the economic, cultural, political, religious, and infrastructural behaviours and practices that produced different types and rhythms of traffic and interaction in the Roman city. The chapters assembled here are interested in how patterns of movement both responded to and generated particular configurations of urban and societal development, and resulted in particular representations of the city. Movement did not simply occupy space, it shaped and redefined it.

The study of the Roman city, during the late 1980s through to the first decade of the twenty-first century, experienced a shift in focus away from the architecture *of* the city towards an understanding of activities *within* the city. This has been characterized as a parallel development to what is often termed 'The Spatial Turn' in cultural studies. In recent years, a new focus has emerged that places the subject of movement at the very centre of research on the nature of ancient urbanism; we cannot understand the relationship between space and society—the two sides of the core dialectic of the 'Spatial Turn'—if we do not first understand the way society shaped and reshaped its movements to and through those spaces. The chapters in this volume demonstrate how movement contributes to our understanding of the way different elements of society interacted in space. In so doing, we gain new insights into a uniquely broad range of historical issues: the commoditization of movement in patronage relationships; the appropriation of 'architectural space' by 'movement space'; an understanding of how urban spatial developments were generated by patterns of movement, rather than vice versa; the importance of movement and traffic in influencing representations of ancient urbanism and the Roman citizen. This volume studies movement as it is found both at the city gate, in the forum, in the portico, on the street, and as it is represented in texts. Throughout, we are concerned with the residues of movement—the impressions left by the movement of people and vehicles, both as physical indentations in the archaeological record and as impressions upon the Roman urban consciousness.

Rome, Ostia, Pompeii: Movement and Space is intended as a contribution to several areas engaged with the Roman city. Its interdisciplinary approach will be of interest to classicists, ancient historians, and archaeologists as well as wider audiences in cultural studies, urban geography and sociology, spatial analyses, and town planning. The volume contributes to the study of the ancient city at large by combining chapters that are theoretical and methodological with empirical case studies. As such, it provides new interpretations of existing data whilst stimulating future advances in the field. Although our case studies are concerned primarily with the three cities named in the title, the implications of these studies will appeal to those working on other sites and other periods.

This volume draws together leading research on movement and traffic in the Roman city to define a new paradigm in the study of ancient urbanism: one in which movement is meaningful.

Acknowledgements

The editors would like to thank Hilary O'Shea and the readers at Oxford University Press for their enthusiastic response to our initial proposal and their work since. For their work on production we thank Kathleen Fearn and the team of editors, readers, and typesetters. We thank Céline Murphy (University of Kent) for compiling the index. Time, money, and space was provided for both editors in the Institute of Archaeology and Antiquity, University of Birmingham, where the bulk of this work was developed. We would like to thank all of the contributors to this work for keeping to time and helping to make the whole process remarkably uncomplicated, as well as thank those whose work is not represented here but who have informed the debate. Timothy O'Sullivan was kind enough to share the introduction to his, then forthcoming *Walking in Roman Culture* (2011)—the scope of which precluded him from contributing here.

Ray Laurence would like to thank Andrew Wallace-Hadrill, with whom he has discussed movement and space for a number of years; his former colleagues at the University of Birmingham, in particular Simon Esmonde-Cleary and Gareth Sears; and would like to acknowledge lengthy discussions on this topic over the last three years with David Newsome and Francesco Trifilò; these have been rewarding and have revealed new ways to understand space in antiquity.

David J. Newsome would like to thank Ray Laurence who, as his PhD supervisor, might have warned against undertaking this project and who, when the thesis started to drop behind schedule, was willing to take on some additional administrative burdens; Simon Esmonde-Cleary; Gareth Sears; Diana Spencer; Andrew Wallace-Hadrill; Francesco Trifilò, in whose company I have spent many thought-provoking hours, but not as many as I should like, discussing the fundamentals; Hanna Stöger; Neil Allies; Amy and my family.

While this volume was in production, the editors were fortunate enough to take part in the workshop *The Moving City: Processions, Passages and Promenades in Ancient Rome* at the Swedish and Norwegian Institutes in Rome. We would like to thank the organizers, Jonas Bjørnebye, Simon Malmberg, and Ida Östenberg. We expect the publication of this workshop to make further important contributions to the subject. Particular thanks to Simon Malmberg for his memorable guided walk of the course of the ancient Clivus Suburanus.

Contents

List of Figures   xiii
List of Tables   xvii
Notes on Contributors   xix

Introduction: Making Movement Meaningful   1
David J. Newsome

Part I: Articulating Movement and Space

1. Movement and the Linguistic Turn: Reading Varro's *De Lingua Latina*   57
Diana Spencer

2. Literature and the Spatial Turn: Movement and Space in Martial's *Epigrams*   81
Ray Laurence

3. Measuring Spatial Visibility, Adjacency, Permeability, and Degrees of Street Life in Pompeii   100
Akkelies van Nes

4. Towards a Multisensory Experience of Movement in the City of Rome   118
Eleanor Betts

Part II: Movement in the Roman City: Infrastructure and Organization

5. The Power of Nuisances on the Roman Street   135
Jeremy Hartnett

6. *Pes Dexter*: Superstition and the State in the Shaping of Shopfronts and Street Activity in the Roman World   160
Steven J. R. Ellis

7. Cart Traffic Flow in Pompeii and Rome   174
Alan Kaiser

8. Where to Park? Carts, Stables, and the Economics of Transport in Pompeii   194
Eric E. Poehler

9. The Spatial Organization of the Movement Economy: The Analysis of Ostia's *Scholae*   215
Hanna Stöger

Part III: Movement and the Metropolis

10. The Street Life of Ancient Rome 245
    Claire Holleran

11. The City in Motion: Walking for Transport and Leisure in the City of Rome 262
    Elizabeth Macaulay-Lewis

12. Movement and Fora in Rome (the Late Republic to the First Century CE) 290
    David J. Newsome

13. Movement, Gaming, and the Use of Space in the Forum 312
    Francesco Trifilò

14. Construction Traffic in Imperial Rome: Building the Arch of Septimius Severus 332
    Diane Favro

15. Movement and Urban Development at Two City Gates in Rome: The Porta Esquilina and Porta Tiburtina 361
    Simon Malmberg and Hans Bjur

    Endpiece: From Movement to Mobility: Future Directions 386
    Ray Laurence

Bibliography 402
Index 441

List of Figures

3.1. How global integration is calculated — 103
3.2. Global integration analyses of Pompeii with the location pattern of shops — 106
3.3. Topological distances with a radius of 80 m with the location pattern of fountains — 108
3.4. Diagrammatic principles on the topological depth between private and public space, and the difference between constituted and unconstituted streets — 110
3.5. Constituted and unconstituted streets in Pompeii — 111
3.6. Diagrammatic principles on the relationship between intervisibility and density of entrances — 112
3.7. Example of a side street with low intervisibility and density of entrance, and a main street with high intervisibility and density of entrances in Pompeii — 113
3.8. Entrance density, and the degree of intervisibility between entrances — 114
3.9. Agent-based modelling of Pompeii's public spaces — 116
4.1. Soundscapes in the Forum Romanum — 127
5.1. The bench outside a Pompeian tavern (VI.1.2) — 138
5.2. The narrow sidewalks and roadbed of the Via degli Augustali, Pompeii — 139
5.3. A procession in honour of Cybele graced the doorway of a Pompeian shop (IX.7.1) — 140
5.4. Wheeled traffic was restricted in and around Pompeii's forum (shaded); ramped sidewalks fronted several properties throughout the city (dotted) — 142
5.5. Herculaneum's Decumanus Maximus — 149
5.6. Fountains at Pompeii straddling the edge of the sidewalk — 150
5.7. The eastern side of the Stabian Baths at Pompeii — 152
5.8. The ramp fronting the Caserma dei Gladiatori at Pompeii (V.5.3) — 154
5.9. The sidewalk in front of the Casa di L. Caecilius Phoebus, Pompeii (VIII.2.36–7) — 156
6.1. A fairly 'typical' street-front along the eastern side of the Via Stabiana at Pompeii (I.3.2) — 161
6.2. Threshold with shuttered groove and 'night door' into the shop at VIII.7.4, seen from within the shop — 165

List of Figures

6.3.	The plaster cast of the shopfront doorway at IX.7.10, Pompeii	165
6.4.	Shopkeeper on a relief at Ostia	168
7.1.	Information of interest to Pompeian cart drivers	178
8.1.	Paved Ramp at I.1.4, Pompeii	198
8.2.	Rut Ramp at II.5.4, Pompeii	198
8.3.	Side Ramp at VIII.6.10, Pompeii	199
8.4.	Distribution map of transport properties, Pompeii	202
8.5.	Plan of the Casa del Menandro, Pompeii	207
8.6.	Plan of the inn at I.1.3–5, Pompeii	211
8.7.	Rounding on brick pier in I.1.4, Pompeii	212
9.1.	Ostia in the second century CE (excavated areas only), location of guild buildings along the main roads	220
9.2.	The Forum in the Antonine period	222
9.3.	Terme dei Cisiarii (II.ii.3): detail from *frigidarium* C	223
9.4.	Ground plans of guild buildings display diversity of layout	226
9.5.	The spatial geometry of human activity	230
9.6.	Casa dei Triclini, I.xii.1; access analysis (j-graph)	231
9.7.	Street network, excavated areas only; axial analysis, integration (HH, n-streets, 150)	236
9.8.	Street network, extended area; axial analysis, integration (HH, n-streets, 476)	236
9.9.	Isovist areas visible from the entrances to the Casa dei Triclini (Isovist 1, 2, and 3)	239
9.10.	The Casa dei Triclini and its external circulation space	240
11.1.	A wall painting from the so-called House of the Baker (VII.3.30), Pompeii	271
11.2.	Reconstruction plan of the Templum Pacis	280
11.3	Plan of the Templum Pacis restored on the basis of recent excavations	281
11.4.	Possible traffic patterns in the Templum Pacis	283
11.5.	Plan of the Templum Divi Claudi restored on the basis of the *FUR*	285
11.6.	Possible traffic patterns in the Templum Divi Claudi	286
11.7.	Possible traffic patterns in the Divorum	287
13.1.	Distribution of game boards within the forum of Timgad	314
13.2.	Map of the carved game boards recorded in the space of the Forum Romanum	316
13.3.	Carved *duodecim scripta* game board	320
13.4.	Three variations of the *merels* game board	321
13.5.	Example of a *mancala* game board	322

List of Figures

13.6. Carved *ludus latrunculorum* game board	322
13.7. Game of *duodecim scripta* dedicated by Photius on a reused statue base	327
14.1. Digital reconstruction of the Forum Romanum, third century CE	333
14.2. Arch of Septimius Severus in the Forum Romanum	336
14.3. Plan of Forum Romanum in the Severan age	336
14.4. Map of Severan Rome	338
14.5. Recreation of capstan and ropes used to lower a block on a steep urban street in Rome	341
14.6. Fragments of the *Forma Urbis Romae* marble plan showing the Clivus Suburanus and Clivus Pullius	345
14.7. Map of central Rome in the Severan period showing possible routes for the movement of construction traffic from the Tiber	347
14.8. Digital reconstruction of a boom arm crane during phase 2 of constructing the Arch of Septimius Severus	350
14.9. Late antique depiction of construction activities, from transport to the mixing of mortar and carving of stones on a mosaic from Ste-Marie-du-Zit, Sufetula	351
14.10. Construction workers operate the pulleys of a crane while other labourers move building materials and cut blocks; first-century fresco from the Villa of San Marco at Stabia	352
14.11. The four main building phases of the Arch of Septimius Severus	353
14.12. Five men operate a treadwheel crane while four others deal with the pulleys and guy ropes on a relief from Tomb of the Haterii	354
14.13. Estimated area around the Arch of Septimius Severus encumbered by guidelines during phase 3	355
14.14. Digital reconstruction of a lifting tower for the Arch of Septimius Severus	356
15.1. The south-eastern city edge of Rome with locations mentioned in the text	362
15.2. The Via del Corso today: an example of vehicle traffic on a 'pedestrian street' within the Zone of Limited Traffic	371
15.3. The Porta Esquilina in 1756	373
15.4. The monumental fountain outside the Porta Esquilina in 1606	377
15.5. Plan of the Porta Tiburtina	378
15.6. The Porta Tiburtina, inner side	379
15.7. The Porta Tiburtina in 1747, outer side	380
E1. Movement around the Fourth Plinth in London's Trafalgar Square	396

List of Tables

5.1.	Primary uses and area of properties fronted by ramped sidewalks	157
7.1.	Sample of rut widths from within Pompeii	183
8.1.	Distribution of ramps across property types	199
8.2.	Spatial distribution of transport properties across property types	200
8.3.	List of all transport properties	203
9.1.	Sample for spatial analysis: five selected guild buildings, Scavi di Ostia	224
9.2.	Access values for spaces with high interaction potential	232
9.3.	Overall size of ground floor space, including *tabernae*	233
9.4.	Global integration values for all streets [n-150] within the excavated areas	237
9.5.	Global integration values for all streets [n-467] including the unexcavated areas	238
11.1.	Summary table of the widths of the identified and unidentified streets depicted on the *FUR*	268
11.2.	Dimensions of Rome's monumental porticoes and portico-temples	276
11.3.	Table of widths of porticoes and walks in Rome's monumental porticoes and portico-temples	277

Notes on Contributors

Eleanor Betts is an Associate Lecturer in Archaeology and Classical Studies at the Open University.

Steven J. R. Ellis is Associate Professor in the Department of Classics at the University of Cincinnati.

Diane Favro is Professor of Architecture at University of California Los Angeles and former President of the Society of Architectural Historians.

Jeremy Hartnett is Assistant Professor of Classics at Wabash College, and Associate Professor at the Intercollegiate Centre for Classical Studies in Rome.

Claire Holleran is a Leverhulme Early Career Research Fellow at the University of Liverpool.

Alan Kaiser is Assistant Professor in Archaeology at the University of Evansville.

Ray Laurence is Professor and Head of the Section of Classical and Archaeological Studies at the University of Kent.

Elizabeth Macaulay-Lewis is a Graduate Teaching Co-Ordinator for the School of Archaeology at Oxford University.

Simon Malmberg and **Hans Bjur** are research associates at the Istituto Svedese di Studi Classici a Roma.

David J. Newsome completed his PhD at the Institute of Archaeology and Antiquity at the University of Birmingham.

Eric E. Poehler is Associate Professor in the Classics Department at the University of Massachusetts, Amherst.

Diana Spencer is Senior Lecturer in Classics at the Institute of Archaeology and Antiquity at the University of Birmingham.

Hanna Stöger is a PhD Research Student in the Faculty of Archaeology at the Universiteit Leiden.

Francesco Trifilò is a Leverhulme Trust Research Fellow in Classical and Archaeological Studies at the University of Kent.

Akkelies van Nes is Assistant Professor in Urbanism (Spatial Planning and Strategy) at the Technische Universiteit Delft.

Introduction

Making Movement Meaningful

David J. Newsome

Now I'll examine the footprints here... He went this way... here's the mark of a shoe in the dust... I'll follow it up this way. Now here's where he stopped with someone else... Here's the scene of some sort of fracas... No, he didn't go on this way... he stood here... from here he went over there... A consultation was held here... There are two people concerned, that's clear as day... Aha. Just one person's tracks... He went this way... I'll investigate... From here he went over here... from here he went—nowhere! It's no use. What's lost is lost.[1]

Studying movement in the Roman city might remind us of the scene in Plautus' *Cistellaria* in which the maid Halisca searches in the street for the casket she had dropped some time earlier. She studies the dust, following footprints, in order that she might locate the movement and direction of whoever has picked up the casket. She recognizes variations between footprints made in motion and footprints made standing still, between the convivial sharing of space and a physical altercation, and the point at which the movement of two people together diverged into two separate paths. All of these details of movement are, she hopes, retrievable from their impressions in the ground. Underlying her faith in this detection, which she compares to augury, is the presupposition that the impressions of movement in the ground can be related to individuals, to purpose, and to interaction in time and space. They are signs which might be read.

---

[1] Plaut. *Cist.* 682, 697–703, 'nunc vestigia hic si qua sunt noscitabo [...] Sed is hac iit, hac socci video vestigium in pulvere, persequar hac. in hoc iam loco cum altero constitit. hic meis turba oculis modo se obiecit: neque prorsum iit hac: hic stetit, hinc illo exiit. hic concilium fuit. ad duos attinet, liquidumst. attat, singulum video vestigium. sed is hac abiit. contemplabor. hinc huc iit, hinc nusquam abiit. actam rem ago. quod periit, periit.'

Halisca's monologue is salutary; she perhaps should have known better. The Latin for footprint or track—*vestigium*—has the figurative meaning of pertaining to a moment, or an instant. It denotes something that is transitory: footprints indicate activity, but only when that activity has moved elsewhere. Following footprints in the sand is to be always, at the very least, one step behind. Halisca is following fleeting moments, which through accumulation become entangled and lost in a congregation of impressions in the dirt. Locating the specific act of a specific individual, at a specific time or moving to a specific place, is impossible. The numbers of impressions confuse and following the most promising tracks leads nowhere. Rather pessimistically, she concludes that what is lost, is lost ('quod periit, periit').

The chapters in this volume aim to assert movement as a key variable in the study of the ancient city. The contributors are united by the common thread of examining the impressions of movement in the historical record (archaeological and textual) and using those impressions—in both physical and figurative senses—to understand the society that produced them. This volume has the broad objective of making movement meaningful: meaningful for understanding the economic, cultural, political, religious, and infrastructural behaviours and practices that produced different types and rhythms of movement and interaction in the Roman city, and how those patterns of movement both responded to and generated particular configurations of urban and societal development. The chapters assembled here examine movement and traffic flow not just as the motion of pedestrians or vehicles within an infrastructural network but also as a social network in which strategies of inclusion/exclusion, proximity/distance, commodity, and hierarchy can be observed.

The springboard for this volume was provided by several conference sessions which confirmed a wide engagement with the volume's broad theme, from scholars working on a range of questions and with a range of evidence types.[2] Yet despite this, and a flurry of doctoral theses and research projects which engaged with movement, no single volume has pulled together these various approaches to demonstrate how, together, they consolidate to form a new feature of the study of Roman urbanism: from space to movement in space.

The editors have aimed to bring together both textual and archaeological studies of movement relating to the city in Roman Italy, in particular at the

---

[2] Sessions organized by those involved in the present volume include: Newsome: 'Movement in the Ancient City: New Approaches to Urban Form and Theory', Theoretical Archaeology Group Conference, York (Dec. 2007); Laurence: 'Spatial Organisation and the Roman City', Classical Association Conference, Birmingham (Mar. 2007); 'Interaction in the Roman City: Understanding Movement and Space', European Association of Archaeologists Annual Conference, Malta (Sept. 2008). Hartnett and Stöger: 'The "Spatial Turn" and Beyond: Roman Cities and the Archaeology of Daily Life', 18th Theoretical Roman Archaeology Conference, Amsterdam (Mar. 2009).

cities of Rome, Ostia, and Pompeii. Rome—although its archaeology is complex—is a city populated by ancient texts that shed light not only on its topography but also on the reception of that topography and patterns of the representation of urban space. Pompeii provides the richest archaeological data we can exploit, while Ostia has been subject to increasing analyses over the previous decade. These three cities therefore represent a particular focus on Italy while also reflecting the best of recent scholarship, itself prompted by the evidence available.

This volume brings together classicists, archaeologists, and architectural historians. This is not only because of a fundamental belief in benefits of interdisciplinary inquiry, nor simply because this reflects the recent, current, and ongoing research on the Roman city. It is also because movement and traffic are subjects that must be studied for both their physical presence in archaeology and their imprint on the representation of urban space in texts. Movement can be studied from the archaeological signatures of wheel ruts impressed in paving stones, as well as from the textual descriptions of wagons clattering along basalt streets. One helps provide further context for the other. In theoretical terms, we can study movement and traffic from both the perspective of spatial practice (the physical behaviour of vehicles and pedestrians moving about the city streets) and the spaces of representation (the ways in which movement and traffic were placed within descriptions of and responses to the city). Moreover, by considering the mediation of movement and traffic through, for instance, the consciously conceived and implemented hierarchies of accessibility or traffic regulations, we can see how movement and traffic fit more broadly within the representations of space—the way in which the city was organized within (or, perhaps, chiefly to produce) a particular paradigm of an urban, imperial society. Few facets of urbanism are so directly amenable to comparative study of this kind, which allow us to speak so loudly about the methodological, the evidential, and the theoretical. It is this kind of comparative analysis, the sum of the parts of this volume, which allows us to make movement meaningful.

The purpose of the remainder of this introduction is to address some of the core themes that are developed throughout the chapters, and to discuss the evidence at our disposal for making movement meaningful. It first sets this work in context by discussing recent studies of the Roman city that have engaged with streets, traffic, and movement, before then considering the evidence at our disposal—both archaeological and textual. Following this, this introduction details why movement should be seen as important within the context of Roman perceptions and definitions of space, and details the use of the term *locus celeberrimus* ('the busiest place') within a relative hierarchy of places within the city—a hierarchy based on movement. Then, by way of an example at the Porta Capena, the principle of the 'movement economy', which underpins many of the chapters in this volume, is discussed. A summary of each chapter concludes this introduction, with attempts made to set the individual contributions in their wider context of contributing to the task of making movement meaningful.

RECENT STUDIES OF MOVEMENT AND THE ROMAN CITY

Urban space has been firmly on the agenda in Roman studies since the late 1980s, after which we can characterize a paradigm shift in line with developments in broader cultural studies. We can recognize this shift from at least MacDonald's study of Roman architecture in 1986.[3] Although not a detailed engagement with spatial theory, the work was inspired by emergent trends in urban geography, particularly the focus on streets and open spaces as venues for social interaction.[4] At the heart of MacDonald's survey of urban form was the concept of the urban armature—'a clearly delineated, path-like core of thoroughfares and plazas'. This type of space provided 'uninterrupted passage throughout the town' and 'gave ready access to its principal public buildings'.[5] This emphasizes connective architectures rather than architectural typologies per se and ostensibly asserts movement as an important variable in urbanism. However, because MacDonald's aim was to identify the elements that made Roman urban form specifically Roman, his view was essentially normative—to perceive common infrastructure and infer common function. There is no notion that movement and space were social variables. Although a useful first step, MacDonald's urban armature is not a theoretical tool for interpreting the role of movement in the city.

Perhaps unsurprisingly, Pompeii has received the most sustained attention to its street network and the activities associated with movement and traffic. Many of the authors of those recent works are represented here. One can observe a shift away from interpretations of urban space based on models of economic or social zoning towards interpretations based on interaction around the organizing framework of the city's street network.[6] The study of

---

[3] MacDonald (1986). However, Raper (1977, 1979) provided important milestones in the development of an analytical discipline of Roman urban space.

[4] Notable in this regard are MacDonald's references to Anderson (1986 [1st pub. 1978]); Cullen (1971); Rudofksy (1969); Whyte (1980, 1981). Also evident are steps to incorporate 'the urban image' into Roman architecture, based on Lynch (1960) and Gould and White (1974). This would later influence Kostof (1991, 1992), itself the inspiration for Favro's study of Augustan Rome (1996).

[5] MacDonald (1986: 5–31).

[6] For earlier approaches to Pompeian spatial organization, see e.g. Raper (1977, 1979) on socio-economic zoning; Eschebach (1970). Laurence (1994a) pre-empted a step change.

issues associated with the street network was prompted by a greater understanding of the variable nature of the network itself. Rather than seeing the network as a homogenous and undifferentiated web of streets, the empirical observations of the wheel ruts left by vehicle traffic have contributed to a greater understanding of the different levels of use at different parts of the city (discussed in more detail below).[7] Recently, Poehler has attempted to reconstruct the change of direction at street junctions, based on the angle of wear on the kerbstones, and has posited a complex, if at times rather inflexible, system of one- and two-way streets.[8] Focusing on the pedestrian rather than the vehicular, Ellis has interpreted the location of *tabernae* according to the expected patterns of the footfall of pedestrian traffic.[9] Some attempts to measure the discrepant levels of interaction by parameters other than wheel ruts have been attempted, based on space syntax analyses of the built environment. Such methods establish which streets, all things being equal, are the most integrated in the network, and assume a positive correlation between integration and levels of activity.[10] Of course, the city is not so straightforward that all things are equal, and those studies that regard the result of a space syntax analysis as an end in its own right have arguably been of less value for the development of Pompeian studies.[11] As such, the value of space syntax analysis is in supporting other, independent data on the ground.[12] Other recent studies have plotted the location of public amenities in urban space, which are contextualized according to reconstructed patterns of movement and traffic.[13]

In addition to such studies, there has been a recent emergence of studies discussing movement as a social practice, based on other forms of cultural evidence and conceptual interpretations of the pedestrian in space.[14] In these terms, access is important for understanding social relationships within

---

[7] The pioneering work was that of the Japanese in the early 1990s, who mapped the depth of wheel ruts across the site, see Tsujimura (1991).

[8] Poehler (2006).

[9] Ellis (2004).

[10] Laurence (1994a); Newsome (2009a).

[11] e.g. Fridell-Anter and Weilguni (2003). A further caveat is that space syntax does not reveal patterns of use, but statistically probable patterns based on ahistorical concepts of spatial practice; see Newsome (2009a: 124 for a critique). See Grahame (2000: 24–36) for an overview of the main theoretical suppositions. The space syntax methodological handbook remains Hillier and Hanson (1984). For space syntax in other Roman urban contexts, see Kaiser (2000) on Empúries and Stöger (2009) on Ostia.

[12] See Newsome (2009a). Laurence (1994a, 1995) correlated space syntax results with other empirical data, such as the frequencies of graffiti or doorways. Grahame (2000: 40) criticized Laurence for relegating space syntax to an 'ancillary method', but this is arguably as it should be.

[13] See e.g. Hartnett (2008) on the location of benches.

[14] Corbeill (2002, later republished as [2003: 107–39]) highlighted cultural attitudes to walking and emphasized the variable political aesthetic embodied in movement; O'Sullivan (2006, 2007) has focused on the philosophical *theoria* and contemplative elements underlying *ambulatio*. See below, on the discussion of the chapter by Macaulay-Lewis.

Roman space. As such, movement and access are politically and culturally significant: divergent privileges of access are one of the most conspicuous and readily understandable differentiators between members of the same society. As we will see with examples from the Roman period, architecture enables the embodiment of cultural concepts and hierarchies of accessibility into the physical space of the city. As movement takes place within specific architectural milieux, which are conceived and designed before they are used, we can link movement to idealism: the control of movement and the restriction of accessibility reveal not only spatial practice but also the permissible forms of practice.[15] In such work, movement has become 'a sociocultural metaphor' and it is studied from a variety of complementary research angles.

In terms of theoretical positions which have influenced Roman scholarship, Henri Lefebvre identified the street as central to urban and social life. It is here that his work on rhythmanalysis is intriguing, though it has received less attention than his work on the social production of space. Rhythm is found 'in urban life and movement through space'.[16] Lefebvre's concept of centrality, constructed in the rhythms of urban life and movement through space, leads to Hillier's theories of urban morphology and movement. Movement is at the core of Hillier's approach to centrality as a process. We can usefully divide his theories into two overlapping parts: natural movement and movement economy. The theory of natural movement holds that because movement in urban contexts can, within reason, be from anywhere to anywhere else, short cuts are created not because of 'attractors' but because of the layout of the street network itself.[17] Accordingly, movement is based on distance minimization. Hillier summarizes his approach to natural movement as: 'Good space is used space. Most urban space use is movement. Most movement is through movement'. Hillier's theory suggests that the prime motive for through movement is distance minimization. This links to accessibility. Indeed, where we talk of a centre being the most integrated space, we can also call it the space with the most potential for through movement (a theme developed in Newsome's chapter).

In short, many recent studies have moved from investigations of urban space to investigations of the street. In turn, focus on the street as infrastructure has led to studies of movement and interaction within shared space, whether this be the congestion of vehicle traffic or the creation of place-

---

[15] Frederick (2003: 222) notes how elite involvement in religious and political institutions was based on the ability to interact with others in prescribed spaces. As such, it was dependent on access. Controlling access is therefore one of the principal ways to demonstrate authority. Valerius Publicola's house on the Velia was noteworthy not only because it commanded a view over movement that passed by it, but because that movement could not access the house itself, for it was surrounded by steep slopes (Plut. *Publ.* 10.2). This emphasis on space as a metaphor for constraining power has much in common with Foucault (1975).

[16] Lefebvre (2004: p. viii).

[17] Hillier et al. (1993); Hillier (1996a: 120).

making through the 'performative act' of pedestrian walking.[18] The immediate research context of this volume is therefore one in which movement is on the agenda as an issue in its own right, rather than an inferred but not investigated corollary of urban form.[19]

EVIDENCE OF MOVEMENT: ARCHAEOLOGY AND TEXTS

Movement and space are subjects for which there is a broad range of historical evidence, from the physical and archaeological to the representational and textual. It is therefore a subject that allows one to investigate disparate but complementary data in pursuit of a common theme. In terms of the research questions asked, we can frame them around practice and representation. Archaeology and text work together to provide a spatial analysis of movement—its design, its practicality, and its influence on perceptions of the city. This is not necessarily to say that spatial practice is dictated by spatial form, but to emphasize that representations of space are constituted by spatial processes. This follows the work of Henri Lefebvre, who was influential in developing a conceptual triad of space, the framework of which helps to compartmentalize some of the issues in this volume.[20] We can think in terms of: i) spatial practice—the movement of traffic through the reality of physical urban space; ii) representations of space—maps and plans, including architectural designs, showing how space was conceived; iii) spaces of representation—writings about space, showing how space was perceived.[21] Spatial infrastructure (the physical objects: streets, sidewalks, steps, doors) allows us to infer spatial practice, if at least by excluding those types of user that were prevented from accessing certain spaces. Architectural forms allow us to see

---

[18] On traffic congestion, see van Tilburg (2007), reviewed by Newsome (2008). On the 'performativity of place' see C. F. Weiss's (2010) discussion of movement through Ephesus.

[19] Recent PhD theses on movement and space include Macaulay Lewis (2008); Trifilò (2009); Newsome (2010); Stöger (2010). Current and ongoing research projects focused on streets and urban movement include Rome: the 'Via Tiburtina Project' (Istituto Svedese di Studi Classici a Roma); Pompeii: 'The Via Consolare Project' (San Francisco State University); Ostia: 'Investigating the Mediterranean City in Late Antiquity (AD 300–650)' (Kent and Berlin universities).

[20] Lefebvre (1991). Given the limits on this introduction, it is not desirable to review the formation of this theory or its criticisms in subsequent social theory here. For responses to Lefebvre, see Shields (1999) and Elden (2004). In the context of classical scholarship, see Hitchcock (2008: 164–8). Newsome (2009b: 25–9), with bibliography, critiques Lefebvre's concepts of centrality in ancient Rome.

[21] The English translation of Lefebvre's spatial triad leaves something to be desired in terms of the distinction between 'representations of space' and 'spaces of representation'. In short, the former is space as it is conceived, the latter is how it is perceived. This distinction is more immediately comprehensible in Lefebvre's French terms: *l'espace conçu* and *l'espace vecu*.

representations of space—they demonstrate how patterns of space were conceived and built. To this we can also add such evidence as the marble plans of Rome, both the Severan *Forma Urbis Romae* and its antecedents, which show how the space of the city was conceived for the purposes of representation.[22] Finally, our knowledge of how space was perceived can be found in literary and epigraphic texts and in visual depictions of urban space.

Our evidence is varied and cumulative, and seeks to examine the relationship between urban space as a physical state and as a conceptual and perceptual construct. Traffic flow is the process of physical agents using a network, or infrastructure, of movement. In so being, traffic is a social as well as spatial network. This network defines relations between people and things. It sets the physical boundaries and the permitted frameworks of interaction, and it manifests a concept of space—an urban disposition—that reveals not only how people structure physical space but their place within it.

As this volume is concerned with urban movement, 'traffic' constitutes a key element. It is necessary to clarify what is meant by this seemingly uncomplicated term. As I noted in response to one of the recent works on traffic in the Roman city, we do not have a detailed definition of traffic from our ancient sources.[23] Instead of a single concept, we find an array of verbs relating to movement (e.g. *ambulare*; *currere*; *transire*; *vehere*). Where road users are discussed, more commonly we find either the plural form of different types of vehicles (though curiously, we rarely have a single passage that mentions more than one type of vehicle in the *same* street) or, for large numbers of mobile pedestrians, we find their compression into a single, homogenous group—the *turba*.[24] The sheer number of road users is a familiar trope in the dystopian visions of ancient Rome, but volume of users is not necessarily the same as our use of the term traffic, which is not only descriptive but perceptual. In part, this reflects the difficulty of applying a modernist term (and its associated urban sensibilities) to the ancient city in lieu of an identifiable ancient equivalent. We can recognize infrastructures and infer practice accordingly but the blanket transference of terms and themes is problematic.[25] The issue is not so much that there were not equivalents in the ancient city but that, by looking for those equivalents within frameworks of urbanism that are entirely modern, we risk losing sight of the dominant paradigms of urban movement and the management of infrastructure in antiquity.

---

[22] See Wallace-Hadrill (2008: 301–12) for a recent overview. Rodríguez-Almeida (2002) provides the most detailed discussion of the Severan and pre-Severan marble plans of the city. For possible functions, see Taub (1993) and Trimble (2007).

[23] Newsome (2008: 444) on van Tilburg (2007).

[24] Juv. 3.239; Sen. *Cl.* 1.6.1. See Sofroniew (2006) for an overview of *turba*, largely playing on Juvenalian negative associations.

[25] Newsome (2008: 444) broadly defines traffic as the aggregation of pedestrians and vehicles in a particular locale at a particular time.

Theoretically, the assertion of the pedestrian in narratives about urban space might be seen as a return to 'the human scale' of the city.[26] This volume is not a study of the human scale of the Roman city, but our interest in movement, space, and small-scale change resonates with some of the issues in human scale planning. In terms of urban planning, this means that the urban space is structured in such a way that it is accessible to the human pedestrian, in contrast to the 'automotive scale' which now governs the organization of urban space. Recent work has also begun to address the city of Rome as a patchwork of smaller scale places and local identities—the 'cellular structure' of the city.[27] We might consider the human scale to be an antidote to the problem of applying modern concepts from urban geography to the city in antiquity. The human scale is not a byword for the reassertion of the pedestrian in our historical narrative (and indeed some of the most important assertions in this volume are based on the interpretation of vehicle movement), but it helps us to consider the importance of concrete spatial relations—and changes to those relations—at what might seem to be a trivially small scale.[28]

We can now consider some of the evidence which is most commonly used to understand movement and traffic in the Roman city, and consider too its associated caveats. This section does not intend to discuss all the evidence at our disposal, for this is the concern of the individual chapters themselves. Rather, this brief overview highlights some of the most common assumptions and queries, primarily focusing on the most familiar evidence that has been used in previous studies: for archaeology, wheel ruts as evidence for vehicle movement in Pompeii; for texts, the legal restrictions on vehicle movement.

Archaeology: wheel ruts

Like Halisca, our study of movement is a study based on proxy evidence. We see action only indirectly—through the observation of data in which we recognize the residue of behaviours that cause that particular form of evidence.

---

[26] On the human scale, see Sale (1980: 38). In modern cities this has been fiercely criticized for failing to acknowledge that scales change across time and space (Lefebvre 1996: 149). This concept has not been applied in detail to the Roman city, but see Gros (2005: 209) and below in this introduction, on the chapter by Laurence and the human scale of Martial and his movements through the city.

[27] Wallace-Hadrill (2003; 2008: 264–9); Lott (2004). See earlier thoughts in Laurence (1991). This has been fostered by an increased awareness that one of the more frequent urban toponyms, the *vicus*, can refer not only to single streets but to the local neighbourhood unit. See the influential work of Tortorici (1991) and Tarpin (2002: 92).

[28] In today's world where the human scale is akin to global scale thanks to ever wider and ever faster transportation, and in which the Internet has compressed space and time and revolutionized social interaction, it can be all too easy to forget that the scale of the ancient city revolved around concrete, spatial settings and movements that may to us seem trivial or insignificant.

The example from the Roman street which has attracted most attention in recent years is the wheel rut: the impressions left in the paving stones of the street by the pressure made by passing vehicles, most fervently studied at Pompeii.[29] Ostensibly this evidence presents few dilemmas; the movement of vehicles over the paving stones produces the wear which, in previous studies at least, is considered consistent and predictable for the purposes of diagnosing vehicle movement. Mapping the relative depth of wheel ruts across the city, such as the influential work done by Sumiyo Tsujimura at Pompeii, maps the relative intensity of the use of streets by vehicle traffic, according to the logic that deep ruts indicate a high volume of vehicle traffic while, conversely, shallow ruts indicate a lower degree of vehicle traffic.[30]

Tsujimura's work took Pompeian wheel ruts from the deductive to the quantitative, and was an important step in transforming what had simply been evocative—think of Mark Twain's memorable critique of the wheel ruts in which he caught his foot and harangued against the street commissioner much like Cicero harangued against Verres for his neglect of the Vicus Tuscus—to being potentially informative for understanding a whole range of correlated, spatial data.[31] An understanding of the relative distribution of wheel ruts theoretically enables us to say which streets were busiest and which, conversely, were free of vehicles. This evidence, itself based on a proxy measure, then becomes the criterion for other proxy evidence—the route of processions, for example.[32]

Eric Poehler, who contributes a chapter to this volume, has revived interest in Pompeian wheel ruts with his detailed recording and interpretation of directional change at junctions.[33] Building on Tsujimura's work, Poehler's evidence allows one to discern the angles of movement from one street to the next and, when read together with criteria such as street width or Tsujimura's depth maps, allows one to discern whether a street was bidirectional or was one-way and, if so, in which direction. This produces a refined version of earlier Pompeii wheel rut maps, and one in which directionality reconstitutes

---

[29] But see also studies in Rome, e.g. Newsome (2010: 140–1, fig. 30) on the paving, wheel ruts, and the separation of vehicles and pedestrians in the Forum Romanum; Ungaro (2005: 212) on the lack of wheel ruts on the Via Biberatica which, alongside the use of steps, marks this out as a distinctly pedestrian space.

[30] Tsujimura (1991).

[31] For Twain on Pompeii, see *Innocents Abroad* (1875). Cic. *In Ver.* 2.1.154, 'quam tu viam tensarum atque pompae cius modi exegisti ut tu ipse illa ire non aurea' ('for that road, the road of sacred cars and of such solemn processions, you have had repaired in such a way that you yourself do not dare go by it').

[32] Most well known is Wallace-Hadrill's (1995) assertion that the western stretch of the Via dell'Abbondanza was used for processions from the Foro Triangolare to the forum, because it lacked wheel ruts despite being, to all other estimations, a 'busy' street.

[33] Poehler (2006). Van Tilburg (2007: 137–43) and Laurence (2008: 89–92) also discuss wheel rut data in Pompeii but both were written before Poehler's findings appeared in print.

movement through space rather than simply plots relative intensity. Moreover, Poehler's work enables further consideration of the dynamics of routes, of the system of governance and knowledge transfer which informed and directed—literally—the movement of vehicles on the city's streets.

Subsequently, Weiss has mapped the location of the holes cut into Pompeian sidewalks in Regions VI and VII, showing that they cluster on busy streets—busy in the post-Tsujimura/Poehler sense of having deep ruts on two lanes of traffic moving in both directions, such as the Via del Foro.[34] Weiss's study again replaces intuition, which guided earlier studies, with the measures of empiricism and further demonstrates the potential of correlated evidence in Pompeii. However, as Hartnett's chapter in this volume ably reminds us, the more we populate the street with the variety of 'nuisances' that occurred there, the less clear it is how they functioned as traffic arteries. In her study, Weiss agrees with the earlier theory that the holes cut into Pompeian sidewalks were for the tethering of animals, but the correlation with the busyness of streets implies that these features occur in spaces of movement, rather than stasis. Like at the example of the Casa del Marinaio (VII.15.1–2), which I have described in detail elsewhere, having an animal stop in the street is inconvenient for the logistics of traffic flow and often demands a reconfiguration of infrastructure to accommodate this need or to remove it.[35]

Weiss's suggestion that a cluster of sidewalk holes on the Via del Foro—the street with the most holes and the greatest frequency—is evidence of a sort-of 'parking lot' fits well with what we might say about the restriction of animals and vehicles from the forum itself. Indeed, the first of the frescoes from the series in the *praedia* of Iulia Felix (II.4.3) may depict such a scene.[36] This shows a large wagon—a *plaustrum*—pulled by two mules and, facing the opposite direction, an ass carrying a load. Columns are visible in the background, and a figure in the centre of the image appears to be wearing a tunic or toga, both of which imply a civic space of some kind. This arguably represents the arrival of products to the space of the forum, although this is a space *outside* the forum, at its borders and where, in the reality of Pompeii, we might find the 'parking lot' on the Via del Foro.[37] What does this mean for the reality of the Via del Foro, a street often considered in terms of ostentatious movement towards the glittering civic centre, channelled through arches and under insignia? On the ground the reality seems rather more mundane, with tethered animals (and the vehicles they were pulling) cluttering the street on both sides,

---

[34] C. L. Weiss (2010). There are 268 sidewalk holes in her study, 107 of which are intact, 153 of which are broken, and 8 of which are unfinished. These figures clearly indicate a city in process.

[35] Newsome (2009a).

[36] Newsome (2010: 278–83). See also Nappo (1989: 79–80, fig. 1).

[37] A subsequent image in the series shows a mule being held by its reins by an attendant, not tethered to the sidewalk. See Nappo (1989: 80, fig. 4).

not only inconveniencing movement but also implying the presence of other nuisances: noise, perhaps animal food, and almost certainly animal urine and excrement.

Previous studies have examined streets and impressions for evidence of repetitive and consistent actions by vehicular and pedestrian traffic. This, then, is an interpretation of the use of the city based on the evidence of traffic infrastructure acting as proxy for traffic behaviour. Descriptions of such evidence as 'diagnostic' are perhaps semantically overreaching; such evidence cannot diagnose types of movement per se but can be read as a symptom of them. Poehler's rather more restricted and pragmatic use of 'diagnostic' wear at junctions, where the angle of wear can diagnose the angle of turns, cannot disguise the fact that we cannot diagnose the types of movement or the journeys that led to such turns being made in the first place. As such, this body of evidence is a symptom of urban movement rather than a diagnosis of its particular properties, its motivations, or the conditions which provoked it. The study of traffic management within the Roman city has a problem in the apparent lack of street signs which might convey instructive information to those moving through the city. As such, we lack detailed understanding of how such movement—as a culturally specific behaviour—was learned and practised. A possible street sign relating to the use of a street comes from the aptly (modern) named Semita dei Cippi in Ostia. There, on the east side of the street and within a few metres from one another, are two inscribed *cippi* (small stone pillars usually carrying information) a little over half a metre in height, bearing 'HAEC | SEMITA HOR | P R I | EST'—which Laurence suggests might indicate that the road was private (*privatus*).[38] Such evidence is rare and, in any case, it is not clear how such privatization would be backed or enforced by law. In the absence of such evidence for the management of traffic in Pompeii, Poehler's reconstructed traffic system is also a reconstruction of shared behaviour—a particular *habitus* of movement—in which wheel ruts inform the decision making of road users (they are indicative of existing actions of movement) and generate repeated behaviours. This positions wheel ruts as the driving force behind, rather than the result of, decision making.

On decision making, in his chapter Alan Kaiser notes how the Vicolo del Mercurio was 'preferred' to the Via della Fortuna. This assumed preference is based, as above, on the depth and frequency of wheel ruts on the former street, which was narrower than the latter and was only wide enough for one lane of traffic. Poehler's analyses have indicated that the Vicolo del Mercurio was, in 79 CE, a westbound street but one that had undergone a reversal from eastbound sometime in the first century. Given that the Via della Fortuna seems more capable of sustaining a higher volume of traffic (it was wider and

---

[38] Laurence (1998a: 444), in response to Bakker (1994: 197–8), who did not define PRI within his interpretation.

accommodated movement in both directions), Kaiser considers this an 'odd choice', thereby implying that such evidence is the by-product of ad hoc decision making (Poehler's arguments, on the other hand, emphasize the city-wide administration of the traffic system). However, this choice may be perfectly sensible. The Via del Mercurio may have been preferred precisely because it was not a wide, busy, two-lane street; it may have been regarded as an easier route along which to move, with minimal interference from activities either side of, or in, the vehicle carriageway. The busier, parallel street one block south would have had much more potential for the kinds of nuisances identified by Hartnett or, speaking of the city of Rome, Holleran in this volume—deliveries to the baths, or animals tethered to one of the many holes along the sidewalk.[39] We should not ignore the principle of the 'double bind'—narrow streets get choked with traffic, whereupon new, wider streets are constructed to relieve this congestion but which, upon opening, are used by all precisely because they are an improvement on existing infrastructure, with the result that it ceases to function as intended almost immediately. Paradoxically, in such a scenario, the narrower streets are those which become the most suitable for uninterrupted movement. Furthermore, in a city with vehicles of different speeds and manoeuvring capabilities, the heavier and more cumbersome traffic may have avoided, or been diverted from, the busier traffic arteries. By virtue of their increased weight and increased pressure on the basalt, these vehicles may have left wheel ruts that are disproportionately deep in relation to their infrequency.

The more we measure and quantify, the less we can generalize about traffic in Pompeii. The evidence is not a simple repository of patterns of movement in 79 CE. Post-excavation factors will have reshaped the evidence. Most straightforwardly, the process of excavation itself would have led to the use of the Pompeian streets for the removal of debris and objects. The deep ruts along Pompeii's main east–west streets, the Via di Nola and Via dell'Abbondanza, reveal as much archaeological as historical use; the patterns of movement to which those ruts are ascribed relate as much to the position of the earlier Ferrovia Circumvesuviana station outside the Porta di Nola, before it was relocated to outside the Porta Marina on the other side of the city, and the routes used in the earliest itineraries around the *scavi*.[40] More specifically, paving and stepping stones might have been removed altogether, as was the case when Pope Pius IX visited in October 1849. Rather than walk, the pope used a cart, at least in places; in order that the cart could move through the

---

[39] Interestingly, there are no holes along the marble pavement on Via della Fortuna contiguous with the Temple of Fortuna Augusta. There are, however, a cluster of holes on the pavement directly opposite.

[40] Laurence (2005: 92–3) on these routes in relation to the Circumvesuviana station.

streets, ancient stepping-stones were removed and, importantly, not all of them were replaced.[41]

These instances demonstrate that the 'Pompeii premise' applies no more to the streets and movement infrastructure of the city than it does to houses and the distribution of artefacts. Wheel rut data is a powerful indicator of use but it is problematic for chronological studies and, moreover, for understanding the relative use of contemporaneous streets. Wheel rut depth is not a proxy for volume of how much traffic used that street relative to others, since a range of other factors might influence the depth of a rut: the weight and loading of vehicles; the drainage of water; uncertainty over chronologies of repaving; and the modern problems of the use of that street for the removal of material from the excavations, or the alteration of streets for reasons like the visit of Pius IX. On methodological caveats, though we know that we are missing some stepping stones (identifiable through gaps in the paving) without corroborating evidence we cannot easily ascribe causation to any of those absences. Furthermore, while some things may be identifiable when removed from the street, others may not be; we do not know all that we are missing.

Texts: traffic regulation

A number of papers in this volume discuss the regulation of traffic in the Roman city.[42] Our knowledge of traffic regulation in the Roman city derives from a combination of legal texts (*leges*) and disparate references peppered throughout the surviving textual corpus to edicts or decrees issued by particular emperors. The most pertinent surviving legal code that is of interest for understanding movement and the ancient city is the *Tabula Heracleensis*, discovered in Heraclea, in southern Italy. Its wording relates directly to the city of Rome and is accepted as preserving the text of the *Lex Iulia Municipalis*, enacted by Caesar in *c*.45 BCE.[43] Of particular interest to those interested in

---

[41] See Harris (2007: 178–9), citing the *Diario della venuta e del soggiorno in Napoli di Sua Beatitudine Pio IX*. See also Conticello (1987), with discussions on the scholarly community forum, *Blogging Pompeii* (http://bloggingpompeii.blogspot.com/2009/02/pius-ixs-visit-to-pompeii-in-1849.html), with maps by Eric Poehler linked at (http://www.pompeiana.org/misc/Blogging_Pompeii/Removed_Features_SS.htm). Thanks to Jo Berry for discussing this issue.

[42] For a recent overview of traffic regulation, with appropriate references, see van Tilburg (2007: on Rome, 127–31; on the empire more generally, 132–6).

[43] On the context of the *Tabula Heracleensis* as well as text and translation, see Crawford (1996). References to 'the municipal law' (most likely the *Lex Iulia Municipalis*) are made by numerous sources in later legal sources (*Dig.* 50.9.3; *Cod.* 7.9.1). Other legal codes which are important for understanding the regulation of changes to public space, including streets, include the *Lex Colonia Genetivae* and the *Lex Tarentina*. These do not, however, discuss the use of streets by particular types of movement.

movement and traffic are the following sections which detail the restrictions—and related exemptions—on vehicle traffic within the city of Rome:

> After January 1 next no one shall drive a wagon along the streets of Rome or along those streets in the suburbs where there is continuous housing after sunrise or before the tenth hour of the day, except whatever will be proper for the transportation and the importation of material for building temples of the immortal gods, or for public works, or for removing from the city rubbish from those buildings for whose demolition public contracts have been let. For these purposes permission shall be granted by this law to specified persons to drive wagons for the reasons stated.
>
> Whenever it is proper for the vestal virgins, the king of the sacrifices, or the flamens to ride in the city for the purpose of official sacrifices of the Roman people; whatever wagons are proper for a triumphal procession when any one triumphs; whatever wagons are proper for public games within Rome or within one mile of Rome or for the procession held at the time of the games in the Circus Maximus, it is not the intent of this law to prevent the use of such wagons during the day within the city for these occasions and at these times.
>
> It is not the intent of this law to prevent ox wagons or donkey wagons that have been driven into the city by night from going out empty or from carrying out dung from within the city of Rome or within one mile of the city after sunrise until the tenth hour of the day.[44]

Kaiser notes that these restrictions on 'vehicle traffic' related only to a particular form of vehicle—the *plaustrum*—a large wagon familiar from literary texts as a slow, noisy and dangerous form of transport.[45] Kaiser contends that the specificity of the *lex* must be accounted for, and since *plaustrum* is never used to refer to 'vehicles' more broadly, when we speak of restrictions on

---

[44] *Lex Iulia Municipalis* lines 54–69: 'vacat | quae uiae in u(rbem) R(omam) sunt erunt intra ea loca, ubi continenti hab<i>tab<i>tur, ne quis in ieis uieis post k(alendas) Ianuar(ias)/ primas plostrum inte<r>diu post solem ortum neue ante horam decimam diei ducito agito, nisi quod aedium | sacrarum deorum inmortalium caussa aedificandaru<m> operisue faciumdei causa adu<e>hei porta- |ri oportebit, aut quod ex urbe exue ieis loceis earum rerum, quae publice demolienda<e> loca<tae> erunt, publice exportarei oportebit, et quarum rerum caussa plostra h(ac) l(ege) certeis hominibus certeis de causeis agere | ducere licebit. vacat | quibus diebus uirgines Vestales re<gem> sacrorum flamines plostreis in urbe sacrorum publicorum p(opuli) R(omani) caussa | uehi oportebit, quaeque plostra triumphi caussa, quo die quisque triumpha<b>it, ducei oportebit, quaeque | plostra ludorum <caussa>, quei <urbei> Romae <p(ropius)ue> urbei Romae <p(assus) m(ille)> publice feient, inue pompam ludeis circiensibus ducei agei opus | erit, quo{ue} minus earum rerum caussa eisque diebus plostra interdiu in urbe ducantur agantur, e(ius) h(ac) l(ege) n(ihilum) r(ogatur). | quae plostra noctu in urbem inducta erunt, quo minus ea plostra inania aut stercoris exportandei caussa | post solem ortum h(oram) (decimam) diei bubus iumenteisue iuncta in u(rbe) R(oma) et ab u(rbe) R(oma) p(assus) mille esse liceat, e(ius) h(ac) l(ege) n(ihilum) r(ogatur).'

[45] On their slow movement, see Verg. G. 1.138; on their creaking noise, Verg. G. 3.536; on the danger of them overturning, Juv. 3.241–3 and Favro, in this volume. The tendency for *plaustra* to topple over led to the idiom 'plaustrum perculi'—literally, 'I have overturned my *plaustrum*', to mean 'I have had a misfortune' (Plaut. *Epid.* 4.2.22).

vehicles in the city of Rome we should instead think of restrictions only on a particular vehicle type. The *plaustra* related to the public building construction industry were exempt from the regulation, and would have been free to move through the city at all times, as would those wagons removing waste from the city (human and animal detritus as well as building debris), or involved in sacral or triumphal processions, or the provision of games. We should not underestimate the scale of movement this would have permitted. As demonstrated in Favro's chapter, the movement of goods for the construction of public buildings was a monumental undertaking in its own right. It would have taken many wagons to deliver the necessary materials from outside the city of Rome, working for a considerable period of time (see below).

The restrictions on *plaustra* in the city of Rome were an action ratified by the law. They are one clause in a bigger scheme of municipal administration, but those stipulations do not themselves discuss or reveal intent.[46] What was the intention of restricting *plaustra* from the city after sunrise until mid-to-late afternoon? The law does not tell us why traffic was to be managed in this way. It may be that the law was a reaction to the sharp rise in vehicle traffic in the mid-first century BCE (a rise that requires satisfactory explanation). Accordingly, these are traffic management laws, based on regulating the use of space by a particular type of user, with the intention (by inference of depleted numbers of vehicles on the streets) being to reduce congestion in the urban area. However, the volume of traffic that would have continued to move into, through, and out of the city, at all times of day according to the permitted exceptions, would have been significant; certainly enough vehicles would have continued to use the city to create congestion. Therefore, it may be more prudent to think of the intentions of this traffic management not in terms of restricting some traffic but of facilitating others.

In this sense, the restriction on some vehicles and the concomitant reduction in congestion are both actions, where the latter has previously been seen as the intention. Instead, the intention of reduced congestion was to facilitate the movement of exempt vehicles. The action enables the intention: by banning certain vehicles it makes movement easier for certain others. Again we are reminded that in examining movement (whether it be infrastructure or legal or textual descriptions) we are working with proxy evidence, leaving traces in the archaeological or historical record as consequences of action. The motivations for those actions remain elusive and in need of explanation. In understanding the intention of managing the movement of vehicles in the city, the exceptions to the restrictions are perhaps more significant than the restrictions themselves: the legislation includes exemptions for circumstances

---

[46] The *lex* does note that to which it is not relevant, through the formula *eius hac lege nihilum rogatur*—'regarding that, nothing is proposed by this law'.

judged to be whatever and whenever proper (*oportet*). The law itself does not prescribe those conditions, and the extent to which something must have been deemed 'proper' may have been more subjective than legally objective.

This brings us to another issue which cannot be ignored: the semantic variation in the sources which we tend to conflate as relating to the 'regulation' of 'traffic'. The disparity of the evidence must be accounted for: anecdotal reference to regulating behaviours of movement (which is not to say, 'traffic restriction') vary across time and space, as well as between the context of their discussion and whether or not they are mentioned as the telling exceptions. In other words, we may only know of restrictions on traffic when we hear of a response to some contravention. What is more significant here—that the law was subsequently renewed or enforced, or that the contraventions occurred at all? The banning of vehicles from the city of Rome, the fulcrum of traffic regulation from the *Lex Iulia Municipalis*, was then seemingly reapplied by Hadrian, who also forbade the riding on horseback in cities, a decree ostensibly first heard of in an edict of Claudius from the previous century, who forbade movement through the towns of Italy unless on foot, in a chair, or in a litter.[47] Marcus Aurelius is then said to have revived the (Hadrianic) restriction, so that riding or driving *in civitatibus* was not permitted.[48] This, like the repeating of legislation governing the *discrimina ordinum* in the Roman theatre, should remind us that while laws were passed they may not have been followed, and renewals had to be issued in successive generations.[49] Marcus Aurelius' edict is noteworthy because it would have come just a few decades after Hadrian, implying that the social adherence to regulations did not survive even a generation before it needed restating by the new emperor. It should be noted that restrictions like those advocated by Marcus Aurelius do not necessarily imply a formal restating of an original edict or the creation of a new one, although the very fact that they are mentioned for some emperors and not for others might imply a certain degree of official action. In the *Historia Augusta*, the almost identical regulation is presented with different verbs for different emperors: *prohibire* for Hadrian and *veto* for Marcus Aurelius. Suetonius' description of Claudius, in contrast, implies a formal edict (*monuit edicto*—from *moneo, to instruct / advise*).

---

[47] SHA *Hadr.* 22.6 ('vehicula cum ingentibus sarcinis urbem ingredi prohibuit. Sederi equos in civitatibus non sivit'). Suet. *Claud.* 25.2 ('viatores ne per Italiae oppida nisi aut pedibus aut sella aut lectica transirent, monuit edicto').

[48] SHA *Marc.* 23.8 ('idem Marcus sederi in civitatibus vetuit in equis sive vehiculis').

[49] The standard text on the *discrimina ordinum* remains Rawson (1987). The social segregation of seating, which stipulated that the first fourteen rows of the theatre were reserved for the *equites*, was first enshrined in the *Lex Roscia Othonis* of 67 BCE (see Cic. *Mur.* 19; Hor. *Epod.* 4.15–16; Livy *Epit.* 99; Vell. Pat. 2.32.3), which was revived by Domitian (see Suet. *Dom.* 8.3; Juv. 3.153–9; Mart. 5.8.3). The *lex* itself may have been a codification of an ancient privilege (Livy 1.35).

However, such edicts on traffic regulation may not be as straightforward as the reinforcement of lapsed policy. We are again hostage to the specificity (or lack thereof) of the evidence at our disposal. The original meaning of textual references may have been more specific or, conversely, more general than translations tend to admit. The example of Hadrian's reprise of traffic regulation, discussed in the *Historia Augusta*, is a useful one. According to this text, Hadrian banned *vehicula*—a rather more generic term than the specific *plaustrum*, discussed above from the *Lex Iulia Municipalis*—from the city. This is further qualified as vehicles that were carrying a particularly large (*ingentibus*) load. Again, there would be a certain amount of flexibility in this edict, to judge at what point a load became sufficiently substantial that it should not be permitted in the city. However, this is saying more than the text allows. This edict might indicate that the banning of the traditional vehicles for heavy loads—*plaustra*—led people to pile more onto those other *vehicula*, of all types, that were not restricted under the earlier law. As such, Hadrian would not be renewing the law but extending it, in response to those skirting its spirit, if not its letter. A more thorough examination of the lexis of the *Historia Augusta* reveals that semantically wide-ranging *vehiculum* is the common designation for wheeled traffic and is used to describe vehicles of all kinds in both rural and urban contexts.

Occasionally, references appear that clarify or supplement the kind of vehicle to which *vehiculum* might refer, but these are problematic because we do not know if we are reading deliberate specificity, synonymy, or simple misunderstanding. A look at the references which include *vehicula* with some other designation reveals our problems with the consistency of this evidence. Elagabalus apparently travelled with sixty *vehicula* or, when emperor and inspired by Nero travelling with a retinue of 500, would travel with no fewer than 600 (*sescenta vehicula*). But we are told Nero travelled with 500 *carrucae*, a more specific designation than the generic *vehicula*. This is the only reference to *carrucae* in this particular biography and was no doubt derived—the anecdote as well as the vehicle type—from Suetonius' own comments on Nero's travelling habits.[50] In another instance, Maximinius the Younger is said to have climbed into Caracalla's *vehiculo* and had to be routed out of it by *mulionibus carrucariis*, implying that the generic *vehiculo* in question was a specific type, like that of Nero and by inference Elagabalus: an imperial *carruca*.[51] Elsewhere, Maximus' father was referred to as a *raedarius vehicularius fabricator*—one who makes wagons, but more specifically *raeda*.[52]

---

[50] SHA *Heliogab.* 31.4–6 varies from Suet. *Nero* 30.3, which says Nero travelled with a thousand, but both sources refer to *carruca*.

[51] SHA *Max. Duo.* 30.6 ('subito per publicum veniente vehiculo Antonini Caracalli [. . .] et vix aegreque a mulionibus carrucariis deturbatus est').

[52] SHA *Max. Balb.* 5.1 ('ut alii raedarius vehicularius fabricator').

Similarly, Zenobia is said to have ridden in a *carpentum* (*vehiculo carpentario*) rather than a *pilentum*.[53] Aurelian's decree that commoners could have their *carrucae* decorated with silver contrasted to their earlier *vehicula* in bronze or ivory.[54] In all of these instances, we see the combination of *vehicula* with a more specific type: *carruca, raeda, carpentum*. What is this telling us? Simply that the authors of the *Historia Augusta* preferred to avoid repetition, by using both the generic and the specific designations for wheeled traffic in instances which called for more than one reference? Or that these terms are more interchangeable than we might think? Martial, in the first century CE, seems to present *carrucae* and *raedae* as synonymous when talking of Bassus' heavily loaded vehicle at the Porta Capena.[55]

There are three 'legislative' references to *vehicula* throughout the *Historia Augusta*, which might influence our understanding of intra-urban movement: Hadrian forbidding the entry into the city of Rome of heavily laden vehicles, discussed above; Marcus Aurelius forbidding riding and driving within the limits of the city; and Severus Alexander permitting senators to use vehicles in the city, a dispensation that implies exclusivity.[56] In addition, Elagabalus gave orders (*iussit*) for the Senate to leave the city—which they did promptly even if they had no *vehicula*. This should not be considered legislative because it did not pertain to the regulation of traffic more broadly.[57] We can also add an anecdote from the biography of Septimius Severus, wherein the unwanted attention of an old comrade in Lepcis Magna—who embraced the then proconsul in the public space of the street—prompted the herald to pronounce that this was not permitted. Thereafter, *legates* rode in *vehicula* where they had previously gone on foot.[58] This need not have been a behaviour ratified in law, but may simply have been a collective response to separate themselves from the rest of the streetscape. This would create, or reinforce, a hierarchy of movement—between the pedestrian and those travelling on wheels.

We must be cautious, therefore, when we read translations of *vehiculum* as (both) 'carts and waggons'.[59] Similarly, we must pay attention to the specific

---

[53] SHA *Zen.* 30.17 ('usa vehiculo carpentario, raro pilento'). *Carpenta* and *pilenta* are also grouped in Livy 5.25.9, who says that matrons were granted the honour of riding the former to sacred festivals and the latter on holy days and work days ('honoremque ob eam munificentiam ferunt matronis habitum ut pilento ad sacra ludosque, carpentis festo profestoque uterentur').

[54] SHA *Aur.* 46.3 ('ut argentatas privati carruchas haberent, cum antea aerata et eburata vehicula fuissent').

[55] Mart. 3.47 ('plena Bassus ibat in reda [...] nec feriatus ibat ante carrucam').

[56] SHA *Hadr.* 22.6; *Marc. Aur.* 23.8; *Aur.* 5.4. That Hadrian's edict related only to those entering the city reminds us of the specificity of the honour that allowed matrons to ride in *pilenta*—only to (*ad*) the sacred festivals. This is a specific spatial act, rather than a more general condition of urban movement.

[57] SHA *Heliogab.* 16.1.

[58] SHA *Sev.* 2.7.

[59] Magie's (1921) Loeb translation of SHA *Marc. Aur.* 13.3 ('tanta autem pestilentia fuit, ut vehiculis cadavera sint exportata').

spatial context of the restrictions. Hadrian's edict was not a restriction of heavily loaded vehicles in the city of Rome, per se, but those entering the city (*ingredi*). Again, we might infer such *vehicula* already within the city would not have been restricted. This says more about the transportation of goods between city and hinterland than it says about movement within the city itself. This is a significant difference in understanding the quantities and spatial distribution of vehicle traffic in Rome in the early-to-mid second century CE. These are methodological issues. When it comes to understanding the particularities of movement in the Roman city, we cannot be blasé with how we interchange the generic and the specific in textual evidence. The specificity of ancient terminology must not be abandoned in favour of more convenient modern translations that imply generality; conversely, the use of general terminology in ancient texts should not lead us to conclude that the actual laws thereby described were any less specific than others.

MOVEMENT AND PLACE: THE *LOCUS CELEBERRIMUS*

We can now discuss the evidence that movement was an important organizing principle in antiquity and that urban space was perceived and represented based on relative patterns of movement through the discrete locales of the city. As Spencer discusses in her chapter, the notion that movement defined place is expressed by Varro: 'where anything comes to a standstill is a place'.[60] Similarly, in attempting to tie motion, place, and body together in their proper relationship, he states: 'nor is there motion where there is not place and body, because the latter is that which is moved, the former is where to'.[61] This concept of place is intrinsically linked to movement, in so far as place is defined as wherever movement is halted. In this sense, place is a destination, defined by the absence of movement but presupposing movement *to*. This parallels the thoughts of the humanist geographer Yi Fu Tuan, who argued that 'if we think of space as that which allows movement, then place is pause; each pause in movement makes it possible for location to be transformed into place'.[62] This definition establishes space and place as two separate categories, defined by movement, albeit co-dependent. However, does Tuan—or Varro for that matter—thereby deny the street its role as a 'place' and render it instead merely the passage between places? This establishes a peculiarly modernist sense of space that is out of step with Roman attitudes in which

---

[60] Varro *Ling.* 5.15, 'ubi quodque consistit, locus'.
[61] Varro *Ling.* 5.12, 'neque motis, ubi non locus et corpus, quod alterum est quod movetur alterum ubi'.
[62] Tuan (1977: 6). This is discussed further in C. F. Weiss (2010: 66–7).

the street itself could be a *locus* (see below). A similar bias may explain the definition of urban centrality as tied to specific, architecturally defined places of activity (activity within spaces where movement stopped)—the forum, for example. Simply, if *loci* are destinations, then streets are not *loci*; if they are not *loci*, they cannot be central places.[63] Yet this would be inaccurate and would deny movement its role in place making, other than in enabling the movement from one place to the next.

In a preceding passage, Varro alternatively defines place as 'where there is motion is a place'.[64] Place is thus alternatively defined as both where things move and where things have moved to. These two definitions, seemingly at odds with one another, cover the whole spectrum of movement and emphasize the importance of understanding both movement *through* and movement *to* a particular space. This emphasis on movement is also evident in works where *locus* is considered but in which the definition of the term is not the primary aim of the author. We can see this in two similar treaties on the art of rhetoric from the first century BCE and CE. Cicero's *De Inventione* contains the following discussion on place:

> In considering the place where the act was performed, account is taken of what opportunity the place seems to have afforded for its performance. Opportunity, moreover, is a question of the size of the place, its distance from other places, whether remote or near, whether it is a solitary spot or more crowded, and finally it is a question of the nature of the place, of the actual site, of the vicinity, and of the whole district.[65]

Place is thus considered according not only to its absolute but also to its relational properties. Such properties are not purely physical but are also socially constructed. It is in this context that the consideration of whether the space is solitary or crowded is important. Movement is thus implied in terms of both the physical accessibility of a place and the accumulation of people there. Quintilian follows *De Inventione* closely when discussing the appropriateness of particular actions for particular places: 'Time and location also need special consideration [...] is it a public or private place, crowded or secluded'.[66] These brief considerations demonstrate that *celebritas* was

---

[63] On central places based on movement, see below for Roman concepts and Tuan (1977: 182) on how personal centres are reinforced through 'a complex path of movement that is followed day after day'. Newsome (2010) engages with the links between movement and centrality in more detail than is possible here.

[64] Varro *Ling.* 5.11, 'ubi agitator, locus'.

[65] Cic. *De Inv.* 1.38, 'Locus consideratur, in quo res gesta sit, ex opportunitate, quam videatur habuisse ad negotium administrandum. Ea autem opportunitas quaeritur ex magnitudine, intervallo, longinquitate, propinquitate, solitudine, celebritate, natura ipsius loci et vicinitatis et totius regionis.' *De Inv.* 1.8 directly contrasts a *locus celeberrimus* with a *locus desertus*.

[66] Quin. *Inst.* 1.1.47, 'Tempus quoque ac locus egent observatione propria [...] et loco publico privatone, celebri an secreto.'

important in forming a relative opinion of the nature of place. This is a perception of space. The importance of this can be understood further by considering the use of the associated term: *locus celeberrimus*.

Given that movement was important for the definition of space, and that levels of movement are not constant but vary from place to place, we might expect some definition of space that is based upon *relative* patterns of movement. For this, we can examine the use of the term *locus celeberrimus*, most conveniently considered a superlative spatial adjective, meaning the 'busiest' or 'most frequented' place. Thomas defines *loci celeberrimi* as 'places of great concentration [that imply] the high density and volume of the human traffic that congregated there, and so the greater possibility of renown'.[67] This translation is not entirely straightforward. As an adjective, *celeber* can mean busy, frequented, or much used, as this translation follows. But, as Thomas's definition suggests, it can also mean famed, renowned, or celebrated—the origin of our 'celebrity'. The use of *celeberrimus* to mean 'most renowned' seems to be a natural result of patterns of movement. It relates to the exposure of a given item (statue, building, site) in urban space. However, in such instances, another adjective often seeks to clarify the distinction, wherein one relates to movement and the other relates to renown. Cicero, for example, says that his house was in the busiest (*celeberrimus*) and finest (*maximus*) part of the city.[68] The former does not presuppose the latter, nor vice versa.[69] This can be seen in the example of the Vicus Tuscus, running from Forum Romanum to the Circus Maximus. The busy nature of this thoroughfare was emphasized in antiquity.[70] But Plautus had earlier characterized this area as less than reputable, and this character persisted into the first century CE.[71] This suggests that the description of the area as a *locus celeberrimus* was based on the intensity of social activity there, regardless of the base nature of that activity. It is a spatial, not a social, superlative. Still, some explanation remains as to why there are two terms for referring to crowded space: *celeberrimus* and

---

[67] Thomas (2007a: 117). This interpretation of the term is also found in Stewart (2003), Gros (2005), and Trifilò (2008).

[68] Cic. *Dom.* 146, 'urbis enim celeberrimae et maximae partes'. See also Cic. *In Verr.* 1.129, on the *aedes Castoris* a 'celeberrimo clarrisimoque monumento'.

[69] In any case, a different adjective altogether might be preferred, as Amm. Marc. 16.10.13 on the Rostra as a *locus perspectissimus*.

[70] Cic. *In Verr.* 2.1.154; Prop. 4.2.5–6; 49–50.

[71] Plaut. *Curc.* 480-4 mentions the Vicus Tuscus explicitly, but also the immediate local area of behind the Aedes Castoris and the *tabernae veteres* (the site of the later Basilica Iulia), as being the location of undesirables—loan-sharks, usurers, and male prostitutes: 'sub veteribus, ibi sunt qui dant quique accipiunt faenore | pone aedem Castoris, ibi sunt subito quibus credas male | In Tusco vico, ibi sunt homines qui ipsi sese venditant | vel qui ipsi vorsant vel qui aliis ubi vorsentur praebeant.' Horace *Sat.* 2.3.228 talks of the impious crowd of the Vicus Tuscus—'Tusci turba impia vici'—and the *salax taberna* of Catullus 37 was in the vicinity. See Papi (2002) for an overview of this area.

frequentissimus ('most frequented'). The link between 'busyness' and 'renown' might be underscored by the alternative use of *frequens*, and the associated *locus frequentissimus*. However, we might argue that while *frequentissimus* is a description of the space, *celeberrimus* is the description of the perception of that space. It is the difference between noting that a site 'was crowded', and noting that a site is 'a crowded space'. This is further demonstrated by the fact that *frequentissimus* stems from a verb—*frequentare*—while *celeberrimus* stems from an adjective—*celeber*. The former relates to what is or was done in space; the latter relates to how that space is perceived.[72] That *celeberrimus* is a superlative is important in that it relates to relative, not absolute, spatial practice, because a *locus celeberrimus* is *busier* than others. This helps us to form a perceptual hierarchy of place within the otherwise undifferentiated space of the Roman city, based above all on movement and traffic.

Stewart's study of Roman statuary included a brief note on the significance of *locus celeberrimus*.[73] This was included in an examination of how statuary related to the lived space of the urban population, and to urban movement. Stewart began by noting an example from Cirta, Numidia, where a decree was passed ordering the removal and rearranging of statues that had clogged the road through the forum.[74] This is an important example of the recognition and amendment of a specific problem in the use of urban space, and demonstrates the need to maintain movement over other considerations. The problem had evidently been caused by the zeal with which honorary statues jostled for position around the passage through the forum. This is because this location would have been a *locus celeberrimus*: the volume of traffic attracted the visual display of honorary statuary, and the persistence of that praxis meant that steps had to be taken to maintain a functioning traffic artery.[75] There are similar examples from Rome. In 158 BCE, M. Aemilius Lepidius was said to have removed statues from the Aedes Iuppiter Capitolinus, because

---

[72] Ov. *Ars Am.* 1.147 refers to *frequens* while *F*. 4.391 refers to *celeber*. The former describes a crowded procession: 'at cum pompa frequens caelestibus ibit eburnis'. The latter describes the nature of the circus once it is crowded with a procession: 'Circus erit pompa celeber numeroque deorum'. *Celeber* does not simply describe the activity but describes the perception of space based on that activity.

[73] Stewart (2003: 136–40).

[74] Stewart (2003: 135). *CIL* VIII.7046: 'VIAM COM[MEANTI] | BUS INCOMM[ODAM PAR] | TIM ADSTRUCT[IS CREPI] | DINIBUS AEQUA[TISSQUE] | STATUIS QUAE IT[ER TOTIUS] | FORI ANGUST [ABANT] | EX AUCTO[RITATE] | D. FONTEI FR[ONTONIANI]'. Stewart notes that '[par]tim adstruct [is]' might also be reconstructed as 'raptim adstructis', 'with rapid construction'. This might appear more suitable in the dedication and would demonstrate the expediency with which this spatial change was demanded or expected.

[75] Placing honorific statuary at junctions relates to Lynch's concept of heightened attention at places of intersection and decision making in the urban landscape (1960: 72–3). Because one is required to make decisions at junctions, one pays more attention to that space, giving it 'special prominence'.

they were obstructing the surrounding colonnade.[76] Later, Suetonius tells of Caligula clearing statues from the Campus Martius. These in turn had been brought there by Augustus from the Area Capitolina, ostensibly because that space had become too congested.[77]

Following Stewart, Trifilò has further investigated the interaction between social display and spatial practice, again framed around the relationship between honorary statuary and *loci celeberrimi*.[78] Trifilò demonstrates how in imperial fora, *loci celeberrimi* would be reserved for imperial representation, forcing other groups to cluster in other spaces. This notion is supported by evidence where non-imperial citizens have been honoured with statuary, notable in the inscription because they have been given a location in the *locus celeberrimus* of the forum.[79] Again, this emphasizes the link between spatial use and status. This differs from the Greek concept of the city and the location of monuments therein. Thomas notes how the Greek equivalent of *loci celeberrimi—topoi episēmotatoi*—derived from the perceived 'monumentality' of a space.[80] This is rather closer to Lefebvre's *l'espace conçu* than the Latin, derived from the perception of the concentration of human traffic and associated with praxis: *l'espace perçu*.

The link between status and a *locus celeberrimus* can most immediately be seen in Suetonius' biography of Augustus. This opens with: 'There are many indications that the Octavian family was in days of old a distinguished one at Velitrae; for not only was a *vicus* in the busiest part of the town [*celeberrima parte oppidi*] long ago called Octavian, but an altar was added there, consecrated by an Octavius'.[81] It is telling that Suetonius chooses to open his account of the emperor by demonstrating the influence of his family on the *locus celeberrimus* of their town. It is for the status inferred by influence in such a location that Cicero wished to buy into a *locus celeberrimus*, as his detailed searches for *horti* near Rome clearly demonstrate.[82] These *horti* had to

---

[76] Livy 40.51.3. In 179 BCE, L. Piso, via Pliny (*HN* 34.30), recorded the clearance of the *statuas circa forum*, except those set up by the will of the people or the Senate, by P. Cornelius Scipio and M. Popilius in 158 BCE. However, this was to curb the ambition and self-aggrandizement that led to such unsanctioned statuary, not to facilitate movement.

[77] Suet. *Cal.* 34: 'statuas virorum inlustrium ab Augusto ex Capitolina area propter angustias in campus Martius'.

[78] Trifilò (2008).

[79] e.g., the *Decretum Tergestinum* (*CIL* V.532) from the reign of Antoninus Pius. Trifilò convincingly argues that the right to have a statue erected 'in celeberrima fori nostri parte' is a particular honour, being out of sorts with standard practice.

[80] Thomas (2007a: 117).

[81] Suet. *Aug.* 1, 'Gentem Octaviam Velitris praecipuam olim fuisse, multa declarant. Nam et vicus celeberrima parte oppidi iam pridem Octavius vocabatur et ostendebatur ara Octavio consecrate'. On the significance of honorific names for *vici*, see Fallou and Guilhembet (2008).

[82] Numerous correspondences with Atticus in 45 BCE detail his perceptions and aspirations: Cic. *Att.* 12.19.1, 'nihil enim video quod tam celebre esse posit'; 12.23.3, 'Ostiensi Cotta celeberrimo loco sed pusillum loci'; 12.37.2, 'sed nescio quo pacto celebritatem requiro';

be in public view in order that they attracted the most renown. But this public view, like that sought by Rome's suburban *villae*, was based on transitory traffic, which brings us back to the earlier discussion on the overlap of spatial use and social prominence.[83] We need not then think it incongruous that *horti* in the peripheral Transtiberim might be considered *loci celeberrimi*, even if we consider *loci celeberrimi* to be analogous with centrality. Centrality is defined in terms of praxis, not geography. Nor in the example of Cicero's *horti* was *celebritas* generated by the inherent fame of the land he wished to purchase. As has been noted, 'Cicero's list of *horti* is no roll-call of the high society of the day'.[84] His search for a *locus celeberrimus* was not the search for land that was already renowned, but land that had potential for renown because of its physical disposition in or around the city and its road network.[85] This is aptly demonstrated by the gardens of Cotta, which Cicero described as sordid (*sordida*) and small (*pusilla*) but which were nevertheless a *locus celeberrimus* because of their prominent location on the Via Ostiensis.[86]

Although both Stewart and Trifilò identify a common concept in the decisions behind the location of honorary statuary, epigraphic evidence attesting to the concept in Rome itself is rare. There are no surviving inscriptions from the Republic or early imperial period that employ the term *locus celeberrimus*.[87] Instead, we are reliant on literary sources to understand how this term was used. This is itself useful for understanding how the term was employed to represent space. The relative lack of epigraphy is not necessarily a lacuna in our evidence but reflects culturally divergent ways of classifying urban spaces. In this sense, the absence of *locus celeberrimus* in epigraphic sources from Rome reflects an absence of *locus celeberrimus* in official conceptions of space, as manifest in formal dedications. Instead, we find it used in what we might call unofficial, written representations of space. To frame this in Lefebvre's terms, the term *locus celeberrimus* is a space of representation, based on the response to specific spatial practices. It is not a concept of space, reified in urban planning or the formal recognition of the nature of a given space. The use of *locus celeberrimus* is responsive.

13.29.2, 'sed celebritatem nullam tum habebat, nunc audio maximam'. This last reference includes the observation that while a site had not formerly been a *locus celeberrimus*, it now was, reminding us that the use of space is a process.

[83] On suburban *villae* positioned in *loci celeberrimi*, see Griesbach (2005: 115).
[84] Wallace-Hadrill (1998a: 5).
[85] The locations of the *horti* were: on the Esquiline: Lamiani; Transtiberim: Cassiani, Clodiae, Cusinii, Damasippi, Drusi, Scapulani, Siliani, Trebonii; Via Ostiensis: Cottae; Via Appia: Crassipedis.
[86] Cic. *Att.* 12.23.3; 12.27.1.
[87] The five inscriptions that may, with varying degrees of confidence, refer to *loci celeberrimi* are from the late 4th or 5th centuries CE.

This has recently been discussed by Gros, in relation to the *locus celeberrimus* of the Septizodium, built at the start of the third century CE by Septimius Severus, at the south-east of the Palatine.[88] Movement has a generative function in creating certain *types* of space. In this sense, Ammianus Marcellinus' concept of the *locus celeberrimus* of the Septizodium is based on the same principles as Hillier's theory of the movement economy. This example also serves to highlight the overlap between practice and representation. The Septizodium was not designed as a 'centre'. Indeed, if we follow the line of the *Historia Augusta*, it was conceived as an entrance for those approaching the city from Africa, by the Porta Capena.[89] But, whatever its conception, it was later perceived and represented as a centre because of the spatial practice that surrounded it. This is the social production of space exemplified.

STREETS, GATES, AND PEOPLE: THE MOVEMENT ECONOMY OF THE PORTA CAPENA

Many of the themes in this volume can be observed around one example in the city of Rome: the Porta Capena, which carried the Via Appia into the city from the south.[90] We have numerous references that demonstrate the importance of movement in the development of this space—both inside and outside the walls. It was at the Porta Capena that Juvenal's Umbricius paused while his *raeda* was loaded with his possessions.[91] This likely borrows a theme from Martial, in which Bassus takes his well-loaded *raeda* (or, later in the text, *carruca*) out through the Porta Capena to his country properties.[92] Umbricius waited beneath the old arch (*veteres arcus*), which may refer to an earlier single-span arch that was incorporated into a later, larger structure. This may

---

[88] Amm. Marc. 15.7.3 ('Septemzodium convenisset celebrum locum'). See Gros (2005: 212) for discussion.

[89] SHA *Sept. Sev.* 24.3.

[90] On the Via Appia starting at the Porta Capena, see Front. *De. Aq.* 1.5 ('Viam Appiam a Porta Capena usque ad urbem Capuam').

[91] Juv 3.10–11 ('sed dum tota domus raeda componitur una, substitit ad veteres arcus madidamque Capenam'). The dripping can be understood because it carried the Rivus Herculaneus, a branch of the Aqua Marcia (Front. *de. Aq.* 1.19, 'Marcia autem partem sui post hortos Pallantianos in rivum qui vocatur Herculaneus deicit [...] finitur supra Portam Capenam'). Dripping could cause its own hazards that might dissuade movement—Martial tells of a boy pierced by a shaft of ice that had fallen from an arch by the Porticus Vipsania, probably that which carried the Aqua Virgo across from the Quirinale to the Campus Martius (Mart. 4.18).

[92] Mart. 3.47, who also notes the dripping from the Porta Capena ('Capena grandi porta qua pluit gutta [...] plena Bassus ibat in reda [...] nec feriatus ibat ante carrucam'). On Juvenal's borrowing from Martial, see Laurence, in this volume.

be understood as a response to increases in movement: infrastructure responding to traffic patterns.[93]

For Umbricius and Bassus, the gate was an exit from the city. But carrying a road in two directions, the Porta Capena was also an entrance. This had important political and military connotations. Q. Fulvius Flaccus' army is said to have entered Rome in 211 BCE through the Porta Capena and then moved quickly (*contendit*) to the Esquiline. The speed of movement would have required a direct system of roads within the city, skirting the south-east of the Palatine and moving towards the area of the Compitum Acilii.[94] Indeed, excavations over recent years have revealed an important junction of streets at the north-east corner of the Palatine, near the later Meta Sudans: one of these roads led to and from the area of the Porta Capena.[95]

Back at the gate, as an entrance, an altar to Fortuna Redux ('brought-back') was erected by the Senate, in honour of Augustus' return from Syria. The Senate would meet the returning emperor outside the gate at the Senaculum, and the emperor changed from military to civic dress (or vice versa, if departing the city) at the nearby Mutatorium Caesaris. In addition, the *transvectio equitum* would pass through the gate each year on its way to the Capitoline.[96] Cicero returned from exile through the Porta Capena, proudly stating that he was met there by a cheering crowd.[97] Similar crowds lined his route to the Forum Romanum and the Capitoline, but were at their most numerous and most vociferous at the Porta Capena itself, to where people had moved upon hearing of his approach through the hinterland south of the city. The notion of the Porta Capena as the point of arrival to the city of Rome is said to have informed Septimius Severus' decision to locate the Septizodium where he did, on the south-east corner of the Palatine, so that it might be the first thing seen by those entering the city from Africa, having moved through the gate.[98] Those moving to the city on the Via Appia would be using one of the first arterial roads to be paved in durable materials suitable for heavy wheeled transportation: the final mile of the road, outside the Porta Capena, was paved in *silex* fifteen years before the censors afforded the same treatment

---

[93] On the development of the Porta Capena, a 4th-century CE source credits it to Domitian (*Chron.* 146).

[94] Livy 26.10 ('in hoc tumultu Fulvius Flaccus porta Capena cum exercitu Romam ingressus, media urbe per Carinas Esquilias contendit').

[95] This junction appears to have been the boundary between five of the Augustan urban regions. On this important junction, which was continually developed from the 7th century BCE to 64 CE, see Panella (1998); Zeggio (2005); King (2010); Newsome (2010: 115–20).

[96] On the *transvectio*, which originated from the Aedes Martis outside the gate, see Dion. Hal. *Ant. Rom.* 6.13.4.

[97] Cic. *Att.* 4.1.5 describes his return to Rome from Brundisium.

[98] SHA *Sev.* 24.3 ('cum Septizonium fecerit nihil aliud cogitavit quam ut ex Africa venientibus suum opus occurreret'). *CIL* VI.1032.

to the Clivus Capitolinus.[99] That we have more references to the area around the Porta Capena, from the first century BCE until the fourth century CE, should be understood as a measure of the lasting importance of this urban space, based on the intensity of movement there. This was the movement of both pedestrian and vehicular traffic. These two types of road user may have been separated, or at least encouraged to separate, through the provision of a pedestrian sidewalk. Fragments 1a–e of the *Forma Urbis Romae* depict continuous lines between the road and the *tabernae* that flank it, perhaps indicative of a sidewalk. In addition, describing the paving of the Via Appia, Livy notes how it was first provided with a *semita* in *saxo quadrato* (tufa)—and the *via* itself was later paved in *silex* (basalt). The difference between *semita* and *via*, presented by the same author to describe the same road, may be understood as referring to distinct parts of that road: the sidewalk and the carriageway, which appear also to be differentiated on the *Forma Urbis Romae*.[100]

The area around the Porta Capena was an important point in the landscape of movement at the 'periphery' of the city. The importance of the Porta Capena in the infrastructural organization of the city of Rome might be inferred from the Regionary Catalogues, which indicate that Regio I started at and was named after the gate. Inside the walls, and before the Septizodium, was a large open space where numerous routes converged—a node rather than a piazza—at which there would have been an intensity of movement that exceeded that on the linear streets within the city. In the Middle Ages, the area around the Septizodium was referred to by the toponym 'Septem Viae', and the *Forma Urbis Romae* allows us an impression of the space immediately inside the gate, on which we can trace the confluence of several roads.[101] As noted, Ammianus Marcellinus characterized the area of the Septizodium as a *locus celeberrimus*. In addition to the convergence of roads at this point, we can also note the presence of nearby *tabernae* and a *macellum*, as well as fountains, baths, and temples—the stairs of which were crowded upon Cicero's return. These are things that respond to the movement of people and, in turn, attract people together in space, so that the movement around the area of the Porta Capena is not just movement *through*—in the manner which we

---

[99] The first mile from the Porta Capena was given a sidewalk (*semita*) in *saxo quadrato* in 296 BCE (Livy 10.23.12), with the road to Bovillae paved in *silex* three years later (Livy 10.47.4). On the paving in 189 BCE, see Livy 38.28.3. On the paving of the Clivus Capitolinus see Livy 41.27.7. We might suggest that the introduction of paving reflects the importance of this area to the community as a whole, but see Muccigrosso (2006: esp. 189–91) on the political rivalries and personal agendas which helped to determine the treatment of this stretch of the Via Appia in the middle Republic. I am grateful to OUP's anonymous reader for providing this reference.

[100] Inside the gate and towards the Septizodium, there was an arcade on the south side of the street which would have further separated pedestrians from the many vehicles using the carriageway (visible on *Forma Urbis Romae* fragments 7a–d).

[101] See also Thomas (2007b: fig. 1), and Newsome (2010: 57–9 and fig. 2).

typically consider city gates as places of passage from one area to another—but movement *to*.

In this area we can imagine an intensity of movement and interaction. Juvenal suggests that beggars (*mendicanti*) occupied the area around the gate and nearby groves.[102] This would have been a choice location not only because many different people moved through there, but also because vehicles would have stopped to pay customs duties. This is a space of economic transaction, alongside which begging would try to position itself. Indeed, the frescoes from the *praedia* of Iulia Felix in Pompeii, which depict economic transactions and the congregation of people moving through the space of the forum, include beggars at the edge of the scene. Though Juvenal is hyperbolic, the fact remains that begging in a space of movement would be advantageous. As Holleran notes in her chapter on the activities that populated the Roman street: 'By its very nature, begging must take place in the public arena and the most popular spots were unsurprisingly some of the busiest; places where pedestrian and vehicular traffic was forced to slow down and even stop'.

Although both Juvenal and Martial present the Porta Capena as a point of departure, the linear movement under and through the gate was not a linear movement from city to countryside. As with the Porta Esquilina, we should not see the gate as separating the city from its hinterland but instead joining distinct parts of the urban economy together through movement. In the area outside the Porta Capena were several *areae* (open spaces at which particular activities or groups were congregated): the Area Carruces, the place at which travellers would leave their carriages before entering the city, as well as the associated *scholae* of the *collegium carrucariorum*; the Area Radicaria, depicted on the *Forma Urbis Romae* and understood as the customs area associated with the gate; and the Area Pannaria, known only from the Regionary Catalogues in Regio I. Although *areae* are defined as empty spaces (*locus vacuus*), this should be understood as empty only of buildings: they would have been spaces of significant movement of goods, people, and traffic.[103]

The relationship between movement and urban development is described by the 'movement economy'—the reciprocal effects of movement and space on one another. In short, the configuration of urban space generates movement, which leads to movement-seeking types of land use (the construction of permanent *tabernae*, for instance) developing at spaces that are rich in movement and traffic, which produce multiplier effects on movement and attract more use.[104] This is a dynamic process, in which movement has a

---

[102] Juv. 3.15–16.
[103] For definitions of *areae*: Fest. Paul. Fest. 11 ('area proprie dicitur locus vacuus') and Varro *Ling.* 5.38 ('in urbe loca pura areae').
[104] For a fuller discussion of the theory of 'movement economies' see Hillier (1996a: esp. 113–14).

generative function. Movement and traffic help to make particular kinds of spaces—they are not simply the things which pass through those spaces. Movement has, and is, agency.

The economic exploitation of traffic at city gates is enabled because traffic becomes compressed and slows or even stops. The reason we find commercial properties such as inns with stables near the city gates at Pompeii is the same reason we find beggars at the Porta Capena: both are spatial behaviours located in response to the economic opportunities afforded by traffic. The principle of the movement economy also shows urban space developing in response to the needs of society. We should not talk of the monumentalization of city gates as though they are simply ornamental portals within the circuit of walls. A historical understanding of movement and traffic highlights the extent to which the gates themselves and the area around the gates develop in response to changing patterns of movement—whether pedestrian, vehicular, or both—and the multiplier effects of the movement economy at places of significant traffic intensity. Again, movement makes space.

BETWEEN ROME AND THE RD909

Before we examine the specific interests of the chapters in this volume and the theoretical positions behind them, we can consider a recent situation from a modern metropolis which, although removed in space and time from the main focus of this volume, nevertheless raises congruent issues relating to movement, traffic, and their role in everyday society. These issues prefigure those which are discussed at length in the chapters that follow, for instance: problems over the regulation of traffic and everyday practice; the creation of nuisances; the movement economy; temporal rhythms of traffic; the legal definition of different spaces of movement; the impact of new infrastructure on the 'natural movement' of the city; the delegation of municipal authority; or the articulation of political or personal relationships through movement and strategies of inclusion or exclusion.

In September 2009, a half-mile section of the 27-mile route between Paris and Chantilly—the Route Départementale (RD) 909—became the focus of political feuding between the two mayors of neighbouring communes, c.4 miles north-west of the capital—Levallois-Perret and Clichy-la-Garenne—and between conservative and socialist mayoral systems, respectively. The RD909 is a major route into and out of the centre of Paris, carrying over 20,000 commuter vehicles each day. The mayor of Levallois-Perret, in an attempt to reduce the amount of traffic moving through his town (traffic which did not stop and so brought no reciprocal economic benefit), made a stretch of the road under his jurisdiction (the Rue Victor Hugo) one-way (northbound). Only traffic moving out of Paris could use this section of the road.

This change had the effect of redirecting traffic through, and increasing congestion in, the neighbouring Clichy-la-Garenne, whose mayor responded by making the section of the RD909 under his jurisdiction—the Rue d'Asinières—one-way in the opposite direction (southbound). In addition, the road on which most of the initially redirected traffic would have had to travel eastwards through Clichy-la-Garenne—the Rue Pierre-Bérégovoy—was redirected to be westbound only. As a result, three one-way roads converged at the same intersection. Inevitably, this caused traffic chaos.[105] The management of continuous space failed under the pressures of local conflict. The municipal and national police were needed to direct traffic away from the area, their deployment emblematic of the failure of the localized administration of a traffic system that, while passing through communes, is a system inextricably related to the city of Paris 4 miles away. Though it may seem like a farcical indictment of French municipal administration, the scenario raises important issues which are pertinent to the themes discussed throughout this volume. We can consider some of these here.

The interpretation of the regulation of traffic within the city must account for the fact that while the space of the city is continuous, the supervision of that space could be the subject of discrete, localized authorities. What happens at the boundaries of administrative responsibility remains one of the key unknowns in our understanding of the Roman city, although we might make some informed suggestions based on the relationship between urban administration and the infrastructures of movement. Like twenty-first-century Paris, the administration of the city of Rome in antiquity was divided into smaller metropolitan units, be they the fourteen regions (*regiones quattuordecim*) into which the city was officially divided in 7 BCE, or, a further administrative level down, the 265 *compitum larum* described by Pliny the Elder.[106] In the Flavian city of the late first century CE, these *compita* likely represented the equivalent number of *vici*, although Ovid earlier suggested there were 500.[107] Whether accurate, estimate, or hyperbole, the administration of the city of Rome was tiered and, like Paris, traffic would have cut through jurisdictional boundaries to traverse the whole city and peripheral built up areas. The situation with the RD909 was enabled because the delegation of local responsibilities also provides considerable autonomy for individual communes. Accordingly, the mayor of Levallois-Perret could take the unilateral decision

---

[105] Reported in *Le Parisien* (31 Aug. 2009). The episode was given the title 'La guerre des sens interdits' (loosely, 'the war of one-way streets').

[106] On the 14 regions: Suet. *Aug.* 30.1 ('spatium urbis divisit in regiones vicosque divisit instituitque'). On the *compitum larum*: Plin. *HN* 3.66–7.

[107] Ov. *F.* 5.145 ('Urbs habet, et vici numina terna colunt'). The figure of 500 *vici* is derived from there being two *lares* at each *compital* shrine, alongside a figure of Augustus himself.

to regulate the direction of the RD909 within his own area of jurisdiction, without being required to consult the neighbouring authorities just a stones-throw away in Clichy-la-Garenne, and vice versa. Local delegation conspires to favour local rather than regional use of space, and to minimize movement *through* areas, unconcerned with the practicalities of detours or their effect on the spaces that form the alternative route.

Rome was a metropolis with a system of governance that was divided into local administrative units. This map of jurisdiction was spatial, based around the infrastructures of movement into, out of, and within the capital city. The Augustan reorganization of the administrative units of the city into fourteen regions in 7 BCE was based on the existing infrastructures of movement, so that major thoroughfares formed the boundaries between different administrative areas of the city. Examples of this include the Via Lata (the modern Via del Corso), which divided Regions VII (Via Lata) and IX (Circus Flaminius); the Argiletum, which divided Regions IV (Templum Pacis) and VIII (Forum Romanum vel Magnum); or the precursor of the modern Via dei Cerchi beneath the Palatine, which divided Regions X (Palatium) and XI (Circus Maximus). There is some suggestion that this latter boundary was coloured red on the *Forma Urbis Romae*, which may indicate that regional boundaries were formally marked on the Severan plan, and on its likely Augustan predecessor.[108] Transposed two thousand years ago, the mayors of Levallois-Perret and Clichy-la-Garenne might be seen as squabbling *vicomagistri*, each changing the regulation of traffic where arterial roads passed through their own *vicus*, perhaps necessitating the intervention of the *vigiles*, the *praefectus urbi*, or, at the last, the emperor himself. Interesting in this regard is that the major arteries of the Via Lata or the Argiletum formed the boundaries between regions and therefore existed neither exclusively in one jurisdiction or the other. This may have mitigated the problems encountered with the RD909: the removal of absolute authority over arterial roads, with responsibility instead invested in the *cura viarum*.[109] In considering how movement was regulated, we need to pay closer attention to the fact that while space is continuous the jurisdiction over space is discontinuous and fragmented. We should also consider, too, how different tiers of administrative responsibility came into contact, and under what circumstances. In first-century CE Pompeii, for example, local jurisdiction evidently failed to maintain order over the use of public space and necessitated the intervention of imperial authority from the city of Rome itself, when public land was illegally occupied and had to be restored to the public by the tribune Titus Suedius Clemens. An inscription commemorating this intervention—which should be considered didactic as well as informative—was placed at points of significant volumes of movement,

---

[108] Ciancio Rosetto (2006: 127, 138).
[109] On the *cura viarum* under Augustus, see Cass. Dio 54.8.4.

at at least three of the city gates, the Porta Ercolano, Porta Nocera, and Porta di Vesuvio.[110] The delegation of Titus Suedius Clemens under the authority of the emperor Vespasian can be read in much the same way as the delegation of the national police to redirect traffic on the RD909: his deployment is emblematic of the failure of local administration.

The changing local traffic regulations on the RD909 near Paris raise another interesting aspect of movement that is addressed in this volume—nuisance—and remind us that while the city is 'a living thing' it does not live independent of human agency.[111] Movement through the city can grind to a halt when individuals lose sight of the bigger picture and focus on their local context. Movement in space is made of individuals, traffic is made of individuals in congregation, and nuisances to movement can be made by individuals acting either without consultation or under the duly delegated powers at their disposal. The case of the RD909 outside Paris raises questions about who is to blame for nuisances: the mayor of Levallois-Perret or of Clichy-la-Garenne? According to the resolution of the regional *préfet*, acting on behalf of the central government, it was the latter; the mayor of Clichy-la-Garenne was ordered to re-establish two-way traffic on his section of the road. The ruling judged that the change of direction in Levallois-Perret was coherent, while the subsequent, retaliatory change in Clichy-la-Garenne caused 'serious disorder' ('désordres graves'). That action, the prefect decreed, was simply to create a nuisance. However, the mayor of Clichy-la-Garenne argued that the first action had been a nuisance because it was instigated without prior consultation. Despite being ordered by the regional prefect to revert to a two-way traffic system, the mayor of Clichy-la-Garenne recognized his ability to cause a nuisance under the terms of the law. He threatened that he could push the absurd even further and establish an altogether new one-way street on the as-yet unchanged Quai de Clichy. That road had also been downgraded in the decree which allowed the alteration of traffic on the other local streets. This would be a nuisance that was entirely permissible within the letter of the law, even if against the spirit of it. In this instance, simply, we find two local elites, with different political persuasions, using the heavily commuted traffic artery to make a broader point: you block my street, and I'll block yours. As one commentary at the time asked, where is the motorist in all of this?

Both of the communes involved lie outside the city of Paris but are within its municipal area, in much the same way that the legal codes for the city of Rome extended beyond the city walls to include 'the streets of Rome or along those streets in the suburbs where there is continuous housing' ('quae viae in u(rbem) R(omam) sunt erunt intra ea loca, ubi continenti hab<i>tab<i>tur'). However, in order for the initial change to be made, the legal definition of the

---

[110] *CIL* X.1018, 'loca publica a privatis posseisa [. . .] rei publicae pompeianorum restituit'.
[111] On the city as a living thing: Plut. *Mor.* 559a.

street had to be altered. The RD909 was classified as a county road and officially recognized as 'heavily travelled' ('à circulation intense'). In order to impose a one-way system through Levallois-Perret, it was necessary to declassify the route as a 'town road'—from a *voie départementale* to a *voie communale*. This has the effect of legally fragmenting linear space into its component parts within and without the commune of Levallois-Perret. Within the commune's area of jurisdiction, the RD909 was named the Rue Victor Hugo and had a different legal definition as well as a different odonym. Such changes may have had significance in the city of Rome in antiquity, where name changes for arterial roads were common once those roads entered into the city. Examples of this include the Via Flaminia, which became the Via Lata after the Milvian Bridge, or the Via Tiburtina, which became the Clivus Suburanus once it was inside the circuit of the city walls and passing through the Esquiline. In this sense, the articulation of space shifts from national to local resolutions: from the name of major inter-urban routes—named from their destination—to names of relative local measures (the Via Lata was 'wide' only in relation to other routes within the city) or local districts (the Subura). National space becomes local space; national concerns become local concerns.

Further issues for which we can recognize ancient equivalents throughout this volume include the temporality of movement. Importantly, the reversal of the RD909 in Levallois-Perret has specific temporal effects, as it only relieves the area from the morning rush hour of traffic moving towards the capital. The one-way system would still allow for vehicles moving *from* Paris to pass through Levallois-Perret in the evening. The decision to legislate the street in one direction rather than the other is therefore one which changes the rhythms of movement through the local area and would have created an imbalance for the residents of the commune: in the morning their street would have been clear but in the evening they would have still had to compete with other road users passing through. The temporality of movement and traffic has often been discussed for the city of Rome, most frequently in relation to the 'tenth hour' (*horam decimam*) at which time the *Lex Iulia Municipalis* allowed *plaustra* to enter the city (discussed above). The tenth hour has often been conflated with the less precise period of 'daylight', but in reality the tenth hour corresponded to around 17.00 in summer and 15.00 in winter. This means that at different times of the year this restriction had different effects. In the summer, we might expect another four or five hours of Italian evening sunlight—light enough to move around the unlit city streets—while in winter there would have been perhaps between one and at most three hours of light before sunset. This serves to remind us that the city, and movement through it, is composed of different rhythms, be they rhythms throughout the day or from season to season.

A final thought concerns the impact of infrastructural change on the surrounding areas, and the ways in which changes to patterns of movement influence urban development. As noted, the RD909 was a 27-mile route between Paris and Chantilly. These communes therefore lie within a half-mile section of what we might term the 'natural movement' between Paris from any of the communes or cities to the north of Levallois-Perret and Clichy-la-Garenne. In other words, movement from a different origin and to a different destination will pass through these spaces because they form the most direct and convenient route. The volume of traffic in either commune is not related to the 'attractors' of that commune itself, but is simply a result of their position within a wider network of movement. However, because the route is a regional commuter route on which vehicles move at speed, that traffic does not bring with it the associated benefits of the large-scale movement of people and transport. In other words, while directly within the 'natural movement' between Paris and the north, these areas receive no benefit from the 'movement economy', whereby movement and economic enterprise reinforce one another through multiplier effects: movement brings people (and money), local space exploits that movement by developing movement-seeking land uses (shops, petrol stations), these in turn attract more movement, and so more movement-seeking land use patterns are consolidated (see below). This positions movement and traffic as a driver of urban development, rather than simply something that passes through existing space. However, the principle of the movement economy includes the apparent paradox that it requires movement to stop. As noted, the traffic moving through Levallois-Perret did not and so it brought no benefit to the local economy and had no role in driving urban development. Such themes occur in many of the chapters throughout this volume, most notably in the context of urban development and the movement economy at city gates or at street junctions, and in alterations to the natural movement of the city through large scale topological changes which cause redirections of traffic. Underlying both of these issues is the hitherto neglected criterion of whether movement is *through* movement or *to* movement. This is a significant distinction, and one that is developed throughout the chapters that follow.

OUTLINE OF THE PRESENT VOLUME

Diana Spencer's chapter on Varro's *De Lingua Latina* opens the assortment of contributions, with a particular interest in the relationships between memory, citizen identity, and the terminology of movement within the context of changes to the *res publica* and the cultural traditions of the elite. Varro's second triad of books explores the semiotic relationship between time, space, and movement. This is clear early on, when Varro defines all things according to whether they are in stasis or movement: 'time never exists without motion:

for a break in motion is time too. Nor does motion exist where there is no place or body, because the latter is what is moved and the former is where to. Nor where this movement is, can there be no action. Therefore the four-horse team of the elements are: place and body, time and action'.

As Spencer describes, Varro positions movement as a key epistemological tool in which the right kind of movement was 'a *sine qua non* for well-informed and appropriate participation in the discourse of public life'. Drawing on some of the familiar theorists of spatial relations and walking in particular—de Certeau, Lefebvre, Bachelard—Spencer highlights the ways in which movement was purposely used to establish cultural differences within Varro's construction of Roman citizen identity. This could be achieved through the definition of individual and societal behaviours from the wide-ranging vocabulary pertaining to human movement.

Ray Laurence's chapter examines the representation of movement and the city of Rome in a different sort of text: the *Epigrams* of Martial, written in the late first century CE. Laurence demonstrates how temporal context is imperative for understanding the city which Martial writes. Martial's representation of place is also a representation of time, whether that be the timing of urban redevelopment (in the aftermath of the fires of 64 and 80 CE), the timing of emperors (i.e. the period of Domitian giving way to Nerva), or the timing of diurnal rhythms (the morning *salutatio*, or the afternoon trips to the *thermae*). We might also link it to the timing of Martial's own life cycle. Too often it is forgotten that his complaints about moving up the Clivus Suburanus are the complaints of a man moving through middle age. We can think of Horace's descriptions of the once energetic litigant Philippus, a man whose old age and related physical decline was evidenced by his protestation that the Carinae was too far ('nimium distare') from the Forum Romanum—two spaces that were in reality close to one another (see Newsome, in this volume).[112] If Martial was not quite a *silicernium* nor was he a speedy whippersnapper who careered along the Via Flaminia.[113] Martial's *Epigrams* open up new ways on thinking about not just the representation of space, but also the representation of time across that space. Martial writes a city which is based on the rhythms of the actions of people. All of this opens the possibility of a detailed rhythmanalysis of Martial's *Epigrams*, examining the intersection of place, time, and the exertion of energy by pedestrian or vehicular movements.[114]

A recurring theme in the *Epigrams* is Martial's relationship with his patrons, one that can be summarized as a relationship in which it is demanded

---

[112] Hor. *Epist.* 1.7.48 ff. Newsome (forthcoming) on movement, age, and physical exertion.

[113] On the *silicernium* ('one who looks at the stones [*silices cernit*]'—a reference to the stooped posture of the elderly walker), see Ter. *Ad.* 587. On the speed of youths on the Via Flaminia, see Juv. 1.60–1.

[114] See Lefebvre (2004). The temporal structure of the Roman day is set out in Mart. 4.8.

that he move through the city to them. This establishes a dependency and a control over both Martial's space and his time, both of which are stretched in the context of the sprawling (not to mention undulating) metropolis. As Laurence notes, Martial had to walk 2 miles to the house of Decianus, which became 4 miles when he was denied entry or found the house empty and had to make his way back home (this itself reveals that Martial's movement to Decianus is not part of his wider actions in the city. If he does not see Decianus, he returns to his point of origin; this is not an act of movement which coincides with others).[115] This has the effect of altering the 'human scale' of Martial's city; distance is not only spatial but is temporal, modelled through individual movement in the urban landscape.[116]

On the theme of dependency and the commoditization of movement, we can recall Carlo Levi, writing of a Neapolitan gentleman who each day during his stroll would meet a beggar and give him a few coins:

> When the old gentleman happened to move house, he explained to the beggar that he would be changing the route of his daily walk and, begging his pardon for the inconvenience, suggested that he come over to the other part of town where he would be walking. The beggar showed up on the first day and the second, but on the third day he told him, with a certain note of reproach: 'Your Excellency, you live too far away. Find yourself another beggar'.[117]

The title given to this anecdote—'Il potere dei poveri'—reflects the links between movement, status, and patronage. This relationship works two ways. On the one hand, the beggar is dependent upon the old gentleman for the regular distribution of coins; on the other, the old gentleman is dependent upon the beggar for the image of having one dependent upon him. It is this mutual interdependence that causes that old gentleman to apologize for the inconvenience of his move and to suggest that the beggar seek him out in the new part of town where it would otherwise take time to establish new dependencies. These are dependencies in which movement and status closely intertwine. In Martial's Rome, too, clients need to be seen to have patrons moving to them. This is a fundamental part of the rhythms of movement in the city. As Laurence states: 'Movement, in itself, becomes a commodity measured in miles, and associated with a specific time: the first two hours of the day'. The commoditization of movement as an indicator of status can be seen in the context of funerary processions, in which standard economies of effort—taking the shortest and easiest route—were replaced by deliberately

---

[115] Mart 2.5. Martial mentions distance four times in this relatively short epigram.
[116] On the scale of Rome, see Mart. 4.64. On the dissolution of centrality by routes which link new buildings in the city through movement, see Gros (2005: 209).
[117] Levi (2004: 92–3).

circuitous routes in order to gather more of a retinue.[118] This should remind us that the principles behind the social act of movement are different to those behind the physical act of motion.

A further effect of the movement of Martial in the context of the *salutatio* was that it removed him—and others like him—from their immediate local neighbourhood on a routine basis. This stretches Martial's social network across the city of Rome (if only to the point of destination) but has the effect that he lacks the kind of localized network one might expect to find within his own *vicus*. This is exemplified by an epigram about his neighbour (*vicinus*), Novius, who Martial says lives so close that their hands could meet from each other's windows.[119] And yet, Martial never sees him. Despite being the closest of all Martial's contacts, the rhythms of social movement demand that Novius and Martial leave their *vicus* and traverse the city, with the result that expected relationships within their local network are neglected. Still, it is precisely because Martial's social networks are dispersed throughout the city that we hear so much from him about movement *through* space. In this regard, we might think of the urban sociologist Walter Benjamin's remarks about how having personal networks necessarily make one familiar with otherwise anonymous physical spaces: 'a network of streets that I had avoided for years was disentangled at a single stroke when one day a person dear to me moved there'.[120] The metropolis of Rome is represented in Martial by the movement that connects these discrete worlds in time and space, even if the journeys themselves are less than ideal.

Moving away from texts towards formal spatial analysis, Akkelies van Nes's chapter introduces some of the key concepts of space syntax—one of the methodologies and theoretical frameworks which has become popular over the last two decades as a way to describe, quantify, and interpret Roman spatial organization. Van Nes's chapter is not intended to be an exhaustive space syntax analysis of Pompeii itself, but to explore the potential of methods both in isolation and when correlated with other data. Such analyses prefigure the detailed investigation of Ostian *scholae* by Hanna Stöger, later in this volume. Van Nes's detailed discussion of method forces us to ask challenging questions of our approach. These are methodologies which have been developed through the study of the modern city and implicit in all such methods is the assumption that the results can be correlated with known data for spatial practice, whether that be crime rates, pedestrian footfall, land rent values, or so on. In other words, the results of space syntax analyses require confirmation from

---

[118] Favro and Johanson (2010: 16–18) describe such practices between the Sacra Via and the Forum Romanum.

[119] Mart. 1.86.

[120] Benjamin (1985: 69). As one commentator has remarked, 'a strange geography becomes smaller and tamer as part of personal history' (Tonkiss 2005: 123).

real data; without this, they are simply statistical abstractions of *probable* and *relative* spatial use.

Of course, those of us studying the Roman city lack the kind of evidence which contemporary space syntax scholars routinely employ to make sense of their abstracted analyses. This presents problems. However, van Nes's chapter emphasizes the universal properties of space and spatial relationships. It is then up to archaeologists and historians to fill in the correlations with what we know from our other evidence. We therefore require our evidence to be localized and highly detailed. Moreover, we need to have chronological data wherever possible, to measure the effects of topological change from one period to the next. I have demonstrated something similar for the space around the Casa del Marinaio at Pompeii (VII.15.1–2).[121] The streets that bordered this house at the edge of the insula were noted for their 'oddities' in the most recent monograph on the property, although these oddities could not be explained from a synchronic reading.[122] In order to explain them it was necessary to untangle the modifications and additions to the street network both immediately outside the property and, importantly, at the junctions from which traffic would be directed. In short, one can observe a gradual removal of routes to the west of the forum in the first century BCE and CE, with the effect that traffic was channelled past the Casa del Marinaio. Space syntax analyses allow us to model this change from over successive periods. The Casa del Marinaio previously overlooked relatively segregated back streets, on which little traffic would pass. The use of the streets for stopping and unloading of goods was no longer viable when traffic patterns increased, and so the property was forced to adapt. This not only explains the 'oddities' in this area, but also demonstrates the need for more chronologically based interpretation of Pompeian archaeology, particularly around the heavily developed area of the forum which changed dramatically. These examples highlight the need to account for urban change, often lacking in discussions of Pompeii's urban space, and lacking even more so in traditional applications of spatial analyses.

Space syntax has contributed to the study of the Roman city but, by abstracting streets to lines and points of connection within a topological plan, it necessarily divorces them from the realities of urban life. The street is not just what space syntax terminology would call an 'axial line', and we must account for the great diversity that existed within this infrastructure. Eleanor Betts's paper, like Claire Holleran's later in the volume, focuses on the diversity of the experience of the city street. However, Betts is particularly interested in what sensory experiences the variety of activities might have provoked, and examines the links between movement and the non-visual

---

[121] Newsome (2009a).   [122] Franklin (1990: 38).

senses, which have traditionally been neglected in studies of urban movement and decision making. The sensory experience of movement is clear in a number of sources, whether that be Martial trudging through the mud on the Clivus Suburanus or, rather more indirectly, Vespasian having mud heaped into his toga for having failed properly to maintain the city streets while acting as the aedile (and so, by inference, he should be covered in mud like the pedestrians).[123] Seneca's vivid account of the sounds and smells that disturb him are conveyed from a position of stasis in his study, but they can be read as a rhythmanalysis of movement through the space below and around.[124] In this way, Seneca's observations prefigure those of Lefebvre's 'Seen from the Window', a contemplation on the movements around the busy Parisian junction beneath his apartment, which vividly describes movement not only in space but also in time throughout the day and into the night (the only difference being that Lefebvre intends to observe this interaction and bustle, while Seneca cannot avoid it).[125] More literal movement comes to Seneca at the end of his description, in which he concedes he must move to escape the distractions. Sound and smell permeate in a way that vision does not, and this renders Seneca's lodgings accessible to the effects of movement, if not the physical act of motion. This, Betts reminds us, serves to blur the distinctions between inside and outside, public and private, or even town and country; the non-visual senses tend to dissolve architectural thresholds. Betts's paper positions movement as a sensory experience which is guided not only by one's feet but also by one's responses to a range of—perhaps unseen—stimuli. This is expressed clearly in Suetonius' biography of Claudius:

> He was eager for food and drink at all times and in all places. Once when he was holding court in the forum of Augustus and had caught the savour of a meal which was preparing for the Salii in the temple of Mars hard by, he left the tribunal, went up where the priests were, and took his place at their table.[126]

This anecdote collates movement from one space to another with the awareness of smell. The emperor, quite literally, follows his nose. Although we might say that the Temple of Mars Ultor and the *area* of the Forum Augustum are parts of the same complex, and although Suetonius notes that the temple is close to (*proxima*) where the emperor was, this anecdote nevertheless emphasizes the movement between one and the other. Claudius leaves (*deserto*) one space and moves to another (*ascendit*). Significant here is the way that the smell (*nidor*—the smell of burning animal fat) penetrated between these

---

[123] Suet. *Vesp.* 5.3.
[124] Sen. *Ep.* 56.
[125] Lefebvre (2004: 27–38), beginning with 'Noise. Noises. Murmurs'.
[126] Suet. *Claud.* 33.1, 'Cibi vinique quocumque et tempore et loco appetentissimus, cognoscens quondam in Augusti foro ictusque nidore prandii, quod in proxima Martis aede Saliis apparabatur, deserto tribunali ascendit ad sacerdotes unaque decubuit'.

locales. Betts's chapter reminds us that smells and sounds, unlike vision, can drift between different spaces and influence movement accordingly. The city of Rome is replete with areas for which we can imagine a powerful sensory experience, such as the Velabrum with its many markets of cheeses, oils, and wines, or the nearby Vicus Unguentarius ('street of the perfume sellers').

Noise, like smell, gives definition to particular places and may have influenced the way one moved to and through—or around—them. Movement is inherently a noisy affair, involving as it does the concentration of people in space or the clatter of vehicle wheels on basalt paving stones. Indeed, we may suggest that the clamour of the Argiletum, which at that time remained one of the central thoroughfares to the Forum Romanum, was what led Pliny to suggest that the three monuments of greatest distinction in the city of Rome were the Forum Augustum, the Basilica Aemilia, and the Templum Pacis: three sites which had the common theme of being removed from the busy street.[127] However, their architectural separation—and the related separation of movement from these locales—was little barrier to noise. Pliny, who emphasizes the need for quiet when contemplating art, elsewhere laments how the Templum Pacis was so noisy that one could not properly appreciate the works of art that were displayed there.[128] This was probably because of the intrusive clamour from the Argiletum. The creation of the new forum in the 90s CE will have removed this intrusion and mitigated the effects of intrusive noise within the surrounding monuments. While this had the effect of making movement through the middle of the city more complicated, it had the related effect of making movement within the Templum Pacis a more peaceful activity than before (see Macaulay-Lewis, in this volume, on walking within Flavian porticoes).

The city street is shared space in which are brought together people who do not know one another and who may have no ties other than that they inhabit the same city, move through the same spaces, and belong to the same 'society', however loosely defined. It is therefore often regarded as a space of accord, both as a metaphor and as the place where such accord is practised daily.[129] What does it say, then, when the city street is used in order to disrupt the movement of others? This is a key theme in Jeremy Hartnett's examination of nuisances on the Roman street, which focuses on the obstacles to movement through the streets of Pompeii and Herculaneum as the results of conscious actions from people. These might lead to or reflect confrontations and tie issues of movement to issues of law and inscribed social privilege.

---

[127] Plin. *HN* 36.102.
[128] Plin. *HN* 36.27.
[129] Ballet, Dieudonné-Glad, and Saliou (2009: 10), 'La rue, par sa forme même, n'est pas seulement une métaphore de cet accord, mais un des lieux où il s'exprime et se fait au jour le jour'.

Hartnett details the variety of obstructions that threatened to impede traffic moving through the street: from hitched mules and street-side sales counters to religious processions and wagon-stopping *cippi*. By focusing on the creation and tolerance of impediments, Hartnett is able to suggest that economic optimization was not as much of a structuring factor for urban life and spatial organization as scholars have suggested. Moreover, blockages to traffic around civic and ceremonial areas forced wheeled traffic to take circuitous routes through the city, thus both casting into high relief the symbolic power of traffic (or lack of it) and also suggesting its importance relative to efficient movement. This chapter also reveals how only certain obstructions to traffic were legal, and therefore allows us to see the difference between the codified letter of the law and the everyday bending of its spirit, revealing a disjuncture between rules and practice. Hartnett's chapter reminds us that the street was a venue for social interaction and therefore the performance of status. He successfully marries legal texts and archaeological material to demonstrate how nuisances reveal the symbolic control exercised over movement through the city. One of the most pertinent examples of this is the doors on the house of P. Valerius Publicola. These were allowed to open outwards, into the space of the pedestrian sidewalk and therefore acted as a nuisance for all passing by.[130] This was an inversion of standard practice and, as well as disrupting efficient movement, would have brought street users into direct contact with the different way in which Publicola's doors were hinged, and therefore with the superior status reflected in that nuisance. Hartnett's chapter uses similar examples from the sidewalks of Campania to reveal how the little things that get in the way can be used to understand a much larger image of social relations within the Roman city.

Hartnett's study of nuisances implies that the maximization of the economic opportunities afforded by passing traffic was not the main guiding principle on the Roman street. If it were, then nuisances would be counterintuitive. Following this, albeit from a different point of view altogether, is Steven Ellis's study of the shape of street fronts in the Roman city. Ellis demonstrates how in the second half of the first century CE, increasingly shops were constructed so that earlier arrangements—which reacted to ambulatory traffic and retail competition—yielded to new layouts which positioned entrances almost exclusively on the right-hand side. This, Ellis argues, is part of a wider 'cult of the right' and a response to the *Lex Neronis de modo aedificiorum urbi*s. These changes reveal that the designed manipulation of movement was for purposes quite different from economic exploitation of passing traffic. This chapter serves to remind us that economic rationality is not the only organizing principle in the Roman city, and explores alternatives which would have

---

[130] Plut. *Publ.* 20; Plin. *HN* 36.112.

produced a new-look Roman streetscape, a homogeneity of economic infrastructure, and, through the standardized articulation of entrances, a new way of moving between the street and the commercial properties which lined it.

Alan Kaiser's chapter is primarily concerned with challenging assumptions about the use of vehicles within the Roman city. In a similar vein to the legal charters discussed earlier in this introduction, Kaiser examines the variation that existed between carts using the streets of Pompeii and within texts pertaining to their regulation. By demonstrating that certain regulations and certain restrictions to the movement of vehicles within Pompeii would have related only to a particular type of vehicle, Kaiser makes the broader point that attitudes towards certain types of traffic were different to others. This raises interesting questions about the informal, perceived status of *plaustra* and the manner in which traffic regulation was administered in the ancient city. According to Kaiser, it would have been the prerogative of local neighbourhoods to regulate the movement of vehicles through their streets, acting without consultation with those responsible for, or without an interest in the effects their regulations had on, wider city traffic. Such a reading conflicts with those which consider the traffic system of the Roman city, and of Pompeii in particular, to have been overseen by local magistrates and instead conjures an image of a city in which the routes open to vehicles was determined on a rather ad hoc basis. On the one hand, this removes the need to find how the traffic system of Pompeii was administered; on the other, it rather complicates things by indicating that the movement of goods could be at the discretion of individuals.

Underlying much of Kaiser's argument is the evidence of wheel ruts, already discussed above. Wheel ruts as proxy evidence allows for the following extrapolation: the distance between wheel ruts in the street enables one to measure the gauge of the vehicle and so the vehicle type; this allows one to estimate which type of animal could pull such a vehicle and how much of a burden that animal could move; this, with other wheel rut evidence, allows one to build up an image of the types of transportation, the routes taken, and the delivery of specific kinds and volumes of materials to specific properties. In short, although never without other evidence, this extension of evidence leads to estimates of the delivery of materials based on the axle width on the street. In addition, Weiss's study of sidewalk holes, discussed above, indicates that 51 per cent of such tethering points in the north-west of the city were related to commercial properties. Such evidence from the infrastructures of movement allows us to discuss the dynamics of the transportation economy of the ancient city, and it is this theme which is developed in detail in the chapter by Eric E. Poehler, who distinguishes between Household and Commercial modes of transportation. Poehler charts that investment in transport related services and infrastructure (ramps and stables, which leads to the identification of 'transport properties'). In so doing, his chapter establishes transportation (and therefore movement) as a driving force in urban development.

The most pertinent example of the movement economy at work in the Roman city is the development of space around city gates. They are particularly significant because all traffic entering and exiting the city must pass through them, and is therefore directed on the streets that lead to them, and through the spaces that surround them. Poehler's detailed analysis of Pompeii demonstrates that almost half of all transport properties lie within the 20 per cent of the total urban area of the city, within a 100-m arc from the city gates, and mostly opening onto the main, two-way streets that lead to a gate. At four of the seven city gates at Pompeii, the first doorway is a transport property, accessed by a ramp, and over 90 per cent of all inns with a ramp are in the vicinity of a city gate. As Poehler neatly summarizes: 'the concentration of so many ramps and stables near the gates elides with the expected high volume of traffic here [...] In fact, this traffic establishes the gates as important peripheral generators of urban space'. This prefigures the survey of urban development in the city of Rome, by Malmberg and Bjur (see below).

The theme of the movement economy is discussed in more detail for the city of Ostia in Hanna Stöger's analysis of the distribution of guild buildings (*scholae*) throughout the city. Her chapter seeks to identify the difference between direct or 'purposive' movement—the movement *to* somewhere—and the exploitation of *through* movement within the context of the location of *scholae* and their associated commercial properties. This helps to provide further context for the position of such buildings within the city, and helps us to understand them within the wider urban network. In a sense, Stöger's chapter allows us to replace models of economic zoning in the Roman city with models of movement zoning: properties are located where they are because they intend to exploit high volumes of passing movement. The multiplier effects of the movement economy lead to a high density of land use for 'movement-seeking' activities. It is this which causes similar activities to cluster in similar types of space. Zoning, then, is not based on the concentration of activities per se, but on the concentration of activities as a result of the movement economy.

This links to the theme of natural movement because those streets on which *tabernae* and *scholae* tend to be found are those which were the major thoroughfares of the city, and would be used by higher volumes of traffic, en route to some other destination. Through a detailed space syntax analysis, Stöger is able to show how the *scholae* of Ostia fit within the city's network of movement. This provides powerful evidence for understanding the principles by which the ancient city was organized. Again, as in many of the chapters throughout this volume, such work enables us to move from discussing the location of certain activities as 'attractors' in the urban landscape to considering how their location was based on existing patterns of unrelated movement.

Again, movement *through* is seen to be more influential on urban development than movement *to*. Further studies may indicate the effect of this logic of space on the rental markets within the Roman city, and lead to further clarity over the extent to which the divisions of insulae were conditioned by the through-movement potential of the streets that passed around them.

The third part of this volume shifts focus away from Pompeii and Ostia to the city of Rome, and explores movement within the context of the metropolis: an environment in which social networks are stretched over significantly greater distance and in which movement was tempered not only by geometric distance and individual constraints on time but also by the orography of the city of hills. This is an urban environment in which economies of effort were tested by social dependencies—the need to move through the city for patronage, as discussed in Laurence's earlier chapter—political gradations of access and permeability, regulation and disruption caused by monumental building.

The sheer variety of these activities is discussed by Claire Holleran, whose paper paints a vivid picture of the city in which streets are far from simply the arteries of transportation familiar from modernism. In particular, Holleran's chapter reveals the role of the street within the urban economy of ancient Rome. This discussion emphasizes the centrality of the street for many of those who lived in the city and depended on the street as a venue for economic opportunity. This, in turn, depends upon the users of those streets, and so on movement. Holleran reminds us that the streets of the city were far more than a physical network of routes; they were the support network for the lives of the city's inhabitants.

Linked to Hartnett's study of nuisance, Holleran includes the well-known Domitianic decree that shopkeepers were to be moved back into the confines of their *tabernae*, having habitually spilled over into the space of the street.[131] The evidence for this order comes to us from Martial and it is interesting—if problematic—because it is presented in a vocabulary of movement infrastructure: 'You [Domitian] ordered that the districts grow wide, and what before had been a path become a street'. Martial overlaps three levels of spatial resolution here—*vicus*, *semita*, and *via*. *Vicus*, though imprecise, helps contextualize where this activity had been taking place and clearly distinguishes such economic expansion from the architecturally defined space of, for example, the *macellum*. The change from *semita* to *via* is indicative not only of increased width but also of the accessibility of the *via* for vehicle traffic; by inference the congestion of shops had become so great that they no longer functioned as *viae* but were more like *semitae*, for the pedestrian. This may help us tie the vocabulary of movement to use as well as the physical properties of the infrastructure itself. Interesting too is the way in which Martial describes

---

[131] Mart. 7.61, 'Iussisti tenuis, Germanice, crescere vicos, et modo quae fuerat semita, facta via est'.

the appropriation of urban space by physical activity; using the verb *abstulo* ('to take away', 'to remove') to describe the effect of *tabernae* using space outside their thresholds. It is clear that the motivation for the expansion of such trade would have been to capitalize on passing traffic. This positions 'removal' as, in reality, the outcome of enterprise maximizing the economic potential of movement: these are *tabernae* which are reshaping, bit by bit, their position within the movement economy. Movement is at the heart of this issue, whether that be the *tabernae* responding to passing traffic, the perceptual change from *semita* to *via*, or the fact that cluttered sidewalks forced the praetor to walk in the mud. The shops which lined the Roman street were thus active participants in the complex negotiation of movement through the city. The *Forma Urbis Romae* presents a sanitized version of space in which the boundaries between insulae and *viae* are well defined; the reality was quite different.

Holleran's chapter on the city of Rome reveals that moving in the Roman city—even for the praetor—was a difficult affair. This may have been one reason why the city of Rome saw the development of pedestrian spaces which were public but which were removed from the street itself. These were spaces for walking as an activity rather than as a means of moving between activities, and their detachment from urban infrastructures of movement brings to mind John Gay's discussion of similar features in eighteenth century London: '*but sometimes let me leave the noisie roads, and silent wander in the close abodes; where wheels ne'er shake the ground; there pensive stray, in studious thought, the long uncrowded way*'.[132] Elizabeth Macaulay-Lewis's chapter focuses on walking within such spaces in Rome—the monumental portico complexes of the late Republic and the late first century CE.

Her reading combines literary texts with recent excavation data and one of the most well-known representations of Rome's urban landscape, the Severan *Forma Urbis Romae*, to produce an 'archaeology of walking'.[133] Such porticoes were spaces of *otium*, in which we can discern the behaviour of leisured walking as distinct from walking in order to move purposefully between two or more locations. Within the infrastructures of movement, such complexes existed in a peculiar limbo between public and private, but walking there would have been unambiguously a sign of one's ability to engage in *otium*. As has been noted, 'the Roman *ambulatio* flaunted the economic independence of the walker, who did not need to use his body to earn a wage, and could instead engage in the repetitive act of walking nowhere'.[134] There are few better

---

[132] John Gay, *Trivia: Or, The Art of Walking the Streets of London* (1716), 2.271–4.
[133] See also her earlier doctoral thesis, Macaulay-Lewis (2008: esp. 98–104).
[134] O'Sullivan (2011: 7). We can think, too, of Plautus' (*Curc.* 470–81) description of the 'citizens of repute and wealth' who stroll (*ambulant*) in the Forum Romanum. These are conspicuously defined only by walking, in a scene otherwise populated with those engaged in

examples of this 'walking nowhere' than Macaulay-Lewis's portico complexes, in which the circuit of walks wound around planted or water features but were entirely confined within a space that had no sense of identifiable destination or purpose, other than the purpose of walking itself.

Despite that lack of a satisfactory Latin (or Anglophone) equivalent, recent studies have been keen to interpret Roman urban walking through the analogy of the Parisian *flâneur*—the name given to those whose habit was to stroll (*flâner*) the streets in directionless contemplation, acting both as observers of and participants in the representation of city space.[135] Larmour and Spencer are right to note the similarities but, while the aimless wanderer may have been present in Rome, the lack of a Latin equivalent noun may indicate that such walkers were not part of the representation of space, certainly not compared to the representation of vehicles and congestion, discussed throughout this work. This may surprise us, because the contemplative wanderer, as Macaulay-Lewis's chapter and other studies have shown, were very much part of Roman society. We may explain this because, unlike Paris, ancient Rome had specific spaces for its *flânerie*, identified by Macaulay-Lewis in an urban context as the portico and *templum* complexes of, in particular, Flavian Rome.[136] This moves them from the street to architecturally defined and enclosed (if permeable) spaces of walking. Indeed, Benjamin—the first to investigate *flânerie* in detail—speculated that Paris created the *flâneur* rather than Rome because the latter city was too full of 'temples, enclosed squares, national shrines'—things which prevent the cobblestones, shops signs, and steps from entering into one's mental construct of the city.[137] Anyone who has read Carlo Levi's evocative *flânerie* through the city of *la dolce vita* may disagree with that assessment, but this may certainly help interpret why the ancient city apparently lacked the aimless pedestrians of Parisian boulevards. Simply, such aimless walkers walked not in the streets but in landscapes of *otium* that were enclosed and isolated from the wider movements of the congested, Flavian city. Another interpretation is that because Roman walking was so evidently a conscious social activity, the class filters of representing status means that we rarely hear from those who walked alone.[138]

---

activity—even activity that presupposes movement (indeed, Plautus' description of the forum begins with the imperative to 'go to' (*ito*) the Comitum).

[135] See, most explicitly, Larmour and Spencer (2007a: 17–18). O'Sullivan (2011: 5) reminds us that we rarely encounter the solitary *flâneur* in Latin literature. Even Horace's apparent *flânerie* on the Sacra Via (Hor. *Sat.* 1.9), which he establishes in its opening lines as solitary contemplation, was actually accompanied by a slave, who remains unmentioned until Horace's personal space is invaded (*Sat.* 1.9.9–10).

[136] See also the important work by O'Sullivan (2006) on walking for philosophical contemplation within the porticoes of *villae*.

[137] Benjamin (1999: 417).

[138] O'Sullivan (2011: 5).

It is, of course, interesting to consider the context in which this walking took place. Access to these spaces was restricted, as Macaulay-Lewis describes, and once inside was free of the kind of activities described by Holleran or Hartnett which made movement on the city streets less than straightforward. But, as Larmour and Spencer note, the similarities with Roman walking and *flânerie* centre on the physical phenomena of the city as stimulus for connotative-associative mental walks, and of blending spaces of the present with the spaces of the past. The concentration of leisured walking within controlled architectural environments might be seen to prohibit the connotative freedom of the walker and position them instead within space that is entirely conceived, rather than represented.[139] Certainly the *flâneur*'s freedom to transcend spatial and temporal boundaries is restricted within the portico, and it is this which made Benjamin consider the city unsuitable compared to Paris. Moreover, even within the context of walking for 'leisure' it is clear that this existed only within prescribed spatial conditions, as Macaulay-Lewis notes of the Templum Pacis and Templum Divi Claudi: 'one could not simply wander as one pleased, but had to walk, at least partially, in accordance with the design of the space'. Still, this may not be too far from Benjamin's analysis of the empowerment of strolling in Paris: 'strolling could hardly have assumed the importance it did without the arcades'.[140] There, *flânerie* remained linked to infrastructures of movement, with few wide pavements before Haussmann's urban renovation and with the threat of vehicles in narrow streets. *Flânerie* existed within specific habitats; Rome's porticoes may be the Flavian equivalent of Paris's *passages couverts*.

Macaulay-Lewis's observations prompt us, in future studies, to consider the manipulation of walking within constructed environments. This is neither walking for transport nor is it Baudelairean *flânerie*. Leisured walking within portico complexes is representative of an apparent freedom but one that is restricted to conceived, designed spaces, replete with programmatic schema which could serve political as well as philosophical purposes.[141] Politically speaking this renders the independent mind of the *flâneur* ineffective within a space that is not appropriated by them but is designed for their use. Replete with politically or militaristically motivated material—manifest in didactic art, architecture, statuary, or plantings—these spaces enabled less freedom than

---

[139] Diana Spencer (2011) develops the notion that porticoes function as a condensed version of the diversely associative space of the city street, including imagery that one might expect to find in a streetscape. I thank her for discussing this issue.

[140] Benjamin (1997: 36).

[141] Excavations in the Templum Pacis brought to light statuary including an ivory of Septimius Severus in philosophical pose and a small portrait of the Greek stoic Chrysippus, famous for his axioms including—most likely coincidentally—the detailed discussion of definite and indefinite propositions based on walking (see Diog. *Zen.* 48).

the connotative-associative goldmine of the city street.[142] Movement is meaningful here on more than one level: as spaces of *ambulatio* and *otium*, but spaces in which such freedoms existed in built space which through controlled access and the design of routes within them, made the appropriation of space more difficult than, for example, in the *flâneur*'s favoured *loci*—the city streets. Macaulay-Lewis offers the suggestion that there may have been 'disobedient' walkers who did not follow the *habitus* of movement within porticoes; such walkers warrant further attention.

Macaulay-Lewis's observations on the permeability of porticoes in Rome lead her to conclude that, in terms of their role in urban movement, they were primarily destinations rather than routes. Whether a space was primarily moved to or moved through is also the focus of David J. Newsome's survey of the changing integration of fora over the first centuries BCE and CE. Newsome's interest is in the changing 'through-movement potential' of these spaces over time, and in the progressive redefinition of the relationship between the forum and the city as manifest through the permeability of civic space. His chapter first positions the Forum Romanum within the context of the 'natural movement' of the city—its position within the spatial configuration of routes which led it to be used as a short cut by movement which did not have the forum as its origin or destination. In the broader understanding of the importance of the forum within the Roman city, this positions the forum as the principal urban node but one which, nevertheless, is based on the same principles as others, being sustained by patterns of movement: the most frequented space in the city is that at which most people cluster because the road system leads there. This emphasis on 'natural movement', rather than the Forum Romanum as what spatial analysts would call an 'attractor', prioritizes spatial practice in our interpretation of centrality within the city: it is a topographical construct, based on the movement of people. The through-movement potential of the Forum Romanum contrasts with the segregated spaces of the imperial fora which were constructed alongside it over the first centuries BCE and CE. These did not carry thoroughfares through them and were less accessible than their Republican predecessor. This phenomenon is discussed in most detail regarding the city of Rome, although similar changes to the integration—the accessibility to different types of movement—of existing fora in other cities of Roman Italy can also be observed. Not least of these was Pompeii, which over the course of the hundred years preceding the eruption was transformed from an urban node into and through which numerous routes of cross-city traffic passed, to a clearly delineated piazza which was inaccessible to vehicles and which had several connections with neighbouring streets severed. Something similar can be observed too at Ostia,

---

[142] On the display of politically motivated material within these portico complexes, see Macaulay-Lewis (2009).

and at Volsinii (Bolsena). These changes cluster in time and point to a cultural redefinition of central space: a cultural redefinition based on movement and accessibility.

Considering this politically, the permeability of the Forum Romanum was one of its defining characteristics in antiquity and could be considered a proxy for the accessibility of the ruling elite to the citizen body—the right of Republican, citizen participation is symbolized by the penetrability of the spaces of decision making. This is the process of the redefinition of Republican to imperial public space, and positions movement as an important variable in terms of changing interactions within the Roman state. While this tells its own story of the kinds of traffic that could enter these spaces, significant too is their wider impact on movement across the city of Rome. As Newsome discusses, the Forum Romanum could be used as a short cut by numerous routes, while the imperial fora demanded that traffic was forced into lengthy and often complicated detours. This has the effect of removing a vast area from patterns of urban movement. Newsome's chapter reveals how the 'centrality' of the forum changed over time; this change parallels broader changes to the politics of access in the first centuries BCE and CE.

Following this examination of the changing topology of fora, Francesco Trifilò looks in detail at one particular activity which occurred there: gaming using 'boards' which were inscribed into the paving of the piazza or surrounding buildings. His chapter underlines the significance of movement for wider understandings of aspects of Roman culture. Hitherto considered in the broader context of 'gambling', Trifilò establishes a difference between games of chance and games of skill and locates such gaming within specific types of spaces: spaces of movement, encounter, and visibility. In short, the tendency for carved game boards to be located in busy spaces, at the junctions of movement and at the points of access to the central civic space, can be read as a proclivity for high-visibility. Like Vortumnus, who overlooked the intersection of the Vicus Tuscus south of the Basilica Iulia, the gamers positioned there would have delighted in the crowd that passed by.[143] This is quite contrary to the traditional image of gaming/gambling within the *popinae* and *cauponae* of the back-streets of Pompeii. The location of such game boards around access routes establishes a distinction between the spaces of *negotium* and *otium*. Like Macaulay-Lewis's leisured walkers in Flavian porticoes, using a game board beside the heavily passed and highly visible spaces of fora would have signified having time on one's hands. In this chapter, then, we see how routes of movement and access informed the location of 'static', inscribed activity. Interesting here, also, is the fact that carved game boards cluster at the south-eastern end of the basilica, next to the Vicus Tuscus, rather

---

[143] Prop. 4.2 on the location of the statue of Vortumnus and its sight of the crowd (*turba*) who passed by (*transeo*) on the Vicus Tuscus.

than alongside the Vicus Iugarius. This, again by proxy, may help us understand the relative use of these parallel streets in the absence of footfall itself.

The clustering of game boards within the portico of the Basilica Iulia may also allow us to say something of the rhythms of such activity. The portico is visible, several steps higher, to those passing along the Vicus Tuscus below. Yet the preference to use the internal pavement of the portico rather than the perhaps more ergonomically advantageous steps of the basilica itself (remembering that the game boards are on the floor, and so require stooping or bending which the steps—being staggered—would have helped to ameliorate), may be related to another feature which dictated practice in open civic space: the sun. Played throughout the day, game boards in direct sunlight would have been more visible to passing movement but would have been rather less comfortable spaces for idling *otium*.[144] Their location, then, can be considered a balancing act between passing movement, visibility, and comfort. In any case, all such issues combine to reveal how the users of the space of the forum appropriated it to suit their own conditions. Trifilò's chapter reveals the cumulative act of inscribing game boards as a product within the ongoing process of the social appropriation of architectural space; this is space remade in response to, and in exploitation of, the visibility that is generated by urban movement.

The Vicus Tuscus features prominently in Diane Favro's paper on the effects of monumental construction on movement through the city, as it formed one of the principal approaches to the Forum Romanum for the construction of the Arch of Septimius Severus at the start of the third century CE. Using this particular build as an example, Favro examines the various logistical challenges posed by building within the middle of the city at this time, with wider implications for the transportation of materials along city streets. Her chapter reminds us that monuments arise only following considerable effort and disruption to 'everyday' movement. Moreover, the frequency of monumental building in Rome means that certain periods, such as the Augustan or Flavian, would have been routinely congested from additional, heavily laden vehicles transporting construction materials to and from the site. Whether this arrived by river or through the city gates, the route taken through the city itself would have been difficult and hazardous.

---

[144] On the link between streets, movement, and the comfort of shade, see Tac. *Ann.* 15.43 on Nero's widening of the streets following the fire of 64 CE: 'erant tamen qui crederent veterem illam formam salubritati magis conduxisse, quoniam angustiae itinerum et altitudo tectorum non perinde solis vapore perrumperentur: at nunc patulam latitudinem et nulla umbra defensam graviore aestu ardescere' ('Some, however, thought that its old arrangement had been more conducive to health, inasmuch as the narrow streets with the elevation of the roofs were not equally penetrated by the sun's heat, while now the open space, unsheltered by any shade, was scorched by a fiercer glow').

Is it any wonder, then, that the written Rome of Martial—in the context of a Flavian building programme which paralleled the Augustan—should be one of traffic, congestion, and inconvenience: 'it is scarcely possible to get clear of the long trains of mules, and the blocks of marble which you see dragged along by a multitude of ropes', as well as scaffolding which blocked roadways.[145] These are representations of Rome which colour our overall image of urban realities, but they are borne out of particular urban redevelopments.[146] Also, the permission and ability to disrupt such fundamentals as the way others can move in the public space of the street positions construction traffic as not only an inconvenience but also a reminder of social hierarchies. As Favro notes, when M. Aemilius Scaurus moved large columns to his *private* house, he was made to pay a security deposit to the overseer of the sewers; no such bond was necessary when he later moved them again to a *public* theatre.[147] We cannot say how individuals' responses to new monuments may have been tempered by the disruption caused to them by construction traffic, but we can see in Martial how such disruption imprinted itself on the urban consciousness of Flavian Rome—to be recycled, most famously, in Juvenal's *plaustrum* carrying its threatening load of Ligurian marble over the heads of other road users.[148]

As Simon Malmberg and Hans Bjur discuss in their chapter, much of the building material to the city of Rome would have passed along the Via Tiburtina, the most expedient route to the city for the travertine extracted from the quarries of Tivoli. This would have brought it through the Porta Esquilina and Porta Tiburtina, from where major urban thoroughfares led towards the Forum Romanum through the Subura.[149] The development of the gates—as well as the spaces either side—was driven by high volumes of movement at this bottleneck in the urban landscape. As spaces of intense movement, gates became spaces of delay. It is this delay which proves crucial in the development of urban nodes—replete with inns, coachmen, storehouses, prostitutes. The similarity of nodes at the Porta Tiburtina and the Porta Capena, discussed above, or even as observed in Poehler's chapter on the much smaller city of Pompeii, reveals the persistence of movement as a generative force in the development of the urban periphery.

These 'edges' could be said to function as 'centres', and they demonstrate the importance of the movement economy in creating a polycentric city in

---

[145] Mart. 5.22, 'vixque datur longas mulorum rumpere mandras quaeque trahi multo marmora fune vides'. On scaffolding in *media via*, see Mart. *Spect.* 2.2, 'et crescunt media pegmata celsa via' ('and the scaffolding rises high in the middle of the road').

[146] Similar responses in the Augustan period include Tib. 2.3.43–4 on the 1,000 oxen moving a foreign marble column through the city. Here the spectacle is both the column, the effort needed to move it, and the disruption this movement causes.

[147] Plin. *HN* 36.2.

[148] Juv. 3.255–9.

[149] Malmberg (2009).

which economic and social prominence was increasingly diverted from the earlier fora.¹⁵⁰ In this regard, where the Forum Romanum developed as it did because of its position within the 'natural movement' of the city (discussed in Newsome's chapter), its progressive removal from patterns of urban movement meant that it largely changed from a place of movement *through* to a place of movement *to*: a destination rather than a route. City gates can be seen to have changed in similar ways, with increasing 'attractors' clustered around them. But at all times, unlike the central fora, city gates continued to function as places for movement *through*. At these spaces, the multiplier effects of the movement economy were not diminished through topological change but, rather, were intensified because of it. This contributed to a Rome that was increasingly fragmented.

By stretching its chronological horizon over six centuries, this chapter reveals the movement economy to be, in the authors' words, 'a universal dynamic'. The development of this area is not so much driven by specific local and temporally distinct actions but by the consistency of movement and movement-seeking activities. The development of a new urban node around the basilica of San Lorenzo fuori le Mura is the direct legacy of the processes of Roman movement discussed through this chapter and this volume more broadly. In time, this basilica and six others became the *loci* for movement of a more targeted sense, as routes on the intra-urban pilgrimage. In this sense, while the *sette chiese* came to replace the *septem colles* as the dominant organizing nodes of Rome's landscape, movement remained the key catalyst for urban development.

By happenstance, it is twenty-five years since the publication of William MacDonald's *The Architecture of the Urban Empire II: An Urban Appraisal*. As noted earlier in this introduction, MacDonald's work was influential in attuning scholars of the Roman period to the importance of the street as an organizing unit in the ancient city. The papers in this volume advance such discussions by demonstrating how movement in space was an important social as well as spatial variable. While there was no conscious decision to publish such a volume on the silver anniversary of MacDonald's work, it nonetheless demonstrates the changes in the discipline since the 'Spatial Turn' of the late 1980s. The papers in this volume demonstrate that interest in space has given way to, or at least been caught up by, interest in movement. This represents a natural but significant divergence from earlier studies.

---

¹⁵⁰ More approximating the developmental processes which Hillier describes in the theory of the movement economy: 'emergent patterns of space shape movement, and through this shape land use patterns, leading through feedback and multiplier effects, to the generic form of the city as a foreground network of linked centres' (Hillier 1996: p. vii).

It is perhaps appropriate that the contributors to this volume, while themselves representing a broad cross-section of research traditions, career paths, and ages, all developed their particular research foci within a broader post-'Spatial Turn' environment in which archaeologists, ancient historians, and classicists increasingly talked not only to one another but also to cognate disciplines in human geography and the wider social sciences. The end result is, therefore, a volume which casts new light on the period under discussion and which reflects the trajectory of our discipline. Speaking on behalf of the editors, we hope this encourages not only greater dialogue between the traditional factions of historical studies but also wider engagement with scholars working on similar themes in different periods. This can be seen, for example, in a conference in late 2010—Blocked Arteries: Circulation and Congestion in History—which brought together scholars from a range of geographically and temporally distinct areas, to encourage dialogue about a subject which transcends historical boundaries: the importance of movement. Three of the authors in this volume (Laurence, Newsome, and Trifilò) discussed new research in Roman studies at this conference, bringing such work into contact with scholars of the city in the Renaissance, the Industrial Revolution, or the twenty-first century, and examining movement and traffic not only as the physical circulation of humans and vehicles in the urban environment but also as a representation of contemporary societies. The editors of the present volume sincerely hope, and firmly expect, that this collection will speak not only to scholars and students of the city in Roman Italy but also to colleagues in related disciplines.

Twenty-five years on from MacDonald, it would be wrong to predict where we might be twenty-five years on from the present collection. Still, we can assert the future directions which we hope this volume might inspire; if this collection achieves a paradigm shift in our understanding of ancient urban space and society, it is important that we signal to where that shift might now take us. Appropriately, these thoughts sit at the end of the collection, in a dedicated endpiece by Ray Laurence which signals the implications of this volume for future research, in dialogue with other sub-disciplines in ancient political and social history. For now, we can return to Halisca, feverishly searching for footprints in the dust. While she concluded that 'what is lost is lost', the chapters in this volume demonstrate how movement was, and remains, meaningful.

Part I

Articulating Movement and Space

1

Movement and the Linguistic Turn

Reading Varro's *De Lingua Latina**

Diana Spencer

The late Republican polymath Varro set his second triad of 'books' on the Latin language (*De Lingua Latina*) the task of exploring the semiotic relationship between time, space, and movement in the development of Latin.[1] These books explore how *langue* (the conventions and rules of language) relates to *parole* (the speech act) over time. This chapter identifies and analyses key passages from this section of Varro's study in order to show how the discourse of movement, and in particular urban movement, was an important feature in elite communication and helped to shape a particular view of what society and citizenship meant, in Varro's circle at least.[2] Varro's *De Lingua Latina* offers a unique opportunity. It shows how one influential individual, acutely aware of the tangible changes taking place in the city and the *res publica* on a day-to-day level and also alert to the radical breaches occurring in the cultural traditions that defined elite identity, perceived a connection between linguistics, language in use, and the acts that constituted citizenship.[3] For this reason, this chapter also investigates how the terminology of movement intersects with the terminology of memory. Ultimately, this chapter proposes that Varro makes language, identity, and action into a nexus whereby thinking about citizenship took movement as a key epistemological tool, and made the right kind of movement a *sine qua non* for well-informed and appropriate participation in the discourse of public life.

---

* The editors' welcome invitation to participate in this volume has been followed up by their enthusiasm in providing a whole array of pertinent and helpful comments and suggestions: I am extremely grateful.

[1] Varro *Ling.* 5–7; see *Ling.* 5.1–3.

[2] Dugan (2005) emphasizes the role of self-invention through rhetorical performance—and in particular, textualization of the ideal speaking self—in fashioning an impressive persona.

[3] Rawson (1985) remains an excellent introduction to the era.

MOBILE VOCABULARY

Book 5, the first book of the extant text, explicitly marks a new beginning: Varro is at the halfway point in the first hexad (2–7), has changed dedicatee, and shifted his thematic focus.[4] In Books 5–7 Varro will tackle how terminology was imposed upon the world, and he illustrates how traces of previous generations of endeavour and organic change persist by means of a metaphor matching language to a manmade landscape.[5]

> Sed qua **cognatio** eius erit uerbi quae radices egerit extra fines suas, **persequemur**. S*ae*pe enim ad limitem arboris radices sub uicini prodierunt segetem. Quare non, cum de locis dicam, si **ab agro ad agr*a*rium hominem**, **ad agricolam peruenero**, **aberraro**.[6]
>
> [concerning the primal classes of words] But wherever the **family** of the word we're interested in should be, even if it has forced its roots out beyond its natural territory, **we'll still follow** it. For often the roots of a tree by the property line will have advanced out under a neighbour's cornfield. For this reason, if, when I speak of places, **I move from field to** an **agrarian man**, and **arrive at** a **farmer**, **I** still **won't have gone astray**.

The act of speech, in particular for Varro's rhetorically trained audience, who in effect 'spoke' for a living, equals a purposeful movement towards a destination. This passage presents the terminology of linguistics in two ways: as familial analogy (*ager* and its offspring) and as a rabbit-hole to Rome's ideal rustic roots, a time when all citizens worked the land and gained definition from its limits—as Cato's second-century BCE cliché has it, the best compliment for a Roman is to call him a farmer, and Varro himself picked up on this in his earlier survey of what makes Roman agriculture tick.[7] This passage also proposes an intimate connection between place, space (a place which enables or is defined by movement), and human identity. *Ager* signifies both 'field' and 'territory'. As such it is a key example of how Varro's second triad revolves around a vocabulary-set tying (Roman) movement, spatial organization, social convention, and linguistic expression tightly together.

---

[4] Books 1–4 (and 11–25) are not extant. Based on Varro's comments, Book 1 probably introduced the 25-book study; Books 2–4 (on etymological theory; addressed to P. Septumius, specified as having been Varro's quaestor during a tour of duty under Pompey's command) and 5–7 (the specifics of how topographic and temporal terminology develop; addressed to Cicero) form the first hexad. Books 8–10 focus on how words generate new words, and the role of language in use. Books 11–25 are likely to have concluded Varro's discussion of analogy and irregularity, before moving on to deal with syntax.

[5] 'quemadmodum uocabula rebus essent imposita', *Ling*. 8.1. All quotations from Varro use Roland G. Kent's 1951 (Loeb) edition. Translations are my own. Important words are underlined and emboldened.

[6] *Ling*. 5.13.

[7] Cato *Agr. Praef.* 2; compare e.g. Varro *Rust*. 2.1–4. See Martin (1995) for a succinct introduction.

We can expect some programmatic force to attach to this new beginning, so Varro's decision to start into the 'natural' complexity of semiotics using an agricultural analogy is intriguing. Varro sets the direct, purposeful action of *peruenero* (arrive at), against the more labyrinthine meanderings of *aberraro* (go astray), hinting that to think deeply about words is analogous to specified kinds of movement. The associative corollary, from a cognitive linguistic perspective, is that walking in particular ways (embedded in his choice of verbs) is conducive to exploring structures of communication. Varro's two first-person singular verbs oppose the guided and controlled movement of a man walking the paths or avenues of his estate or porticoes, or perhaps moving through a well-known urban layout (*peruenio* suggests an informed arrival at an anticipated point), to the ill-informed and poorly guided wanderings of one who has missed his way in a maze of blind alleys and wrong turns—an uncivilized, even irrational space.[8]

The efforts of the right people, studying language appropriately and for the right reasons, lodge words and their conjunction as meaningful discourse in two key areas of Roman interest—family history, and property and landownership.[9] Moreover, the agricultural metaphor of the tree and its life course suggests that language—and in particular, communication—is invested with an active and ethical principle.[10] Varro's investigation fashions Roman intellectual practice as goal-oriented and structured, but also emphasizes how even the basic toolkit for self-expression is always changing and therefore requires ongoing surveying if it is to remain fit-for-purpose. Discourse, whether conversational, literary, or rhetorical, takes time. It is through words (that is to say, by being understood as discourse) that movement is enriched with Roman meaning, and as Varro observes just before, temporality only exists where there is movement, and movement depends on the existence of place and body, and generates action.[11]

The significance of farming in the formation of elite identity and its nostalgic ethos (landholding was the acceptable basis for senatorial wealth and maintaining one's position on the censorial roll) helps to make an active and creative engagement with topography central to being Roman. On a day-to-day basis, Rome's hinterland still bustled with family smallholdings,

---

[8] *OLD* s.v. *peruenio*: 1, 2, 6; *aberro*: 1–3, 5. There is a hint, here, of the Epicurean saying '[we should] try to make the later stretch of the road more important than the earlier one, as long as we are on the road; and when we get to the end [of the road], [we should] feel a smooth contentment'. *Sent. Vat.* 48, in Inwood and Gerson (1994: 38).

[9] Despite lacking Varro's overall introduction, looking at his earlier study of agriculture (e.g. *Rust.* 1.8–10) gives us a flavour of how the practice of farming could mainline Roman 'virtue'.

[10] See Varro *Ling.* 8.4: kinships between words can—as for people—be primarily agnate or gentile. On rhetoric as an engine of culture and social cohesion: Narducci (1997). In Peluso (1996), boundary issues, agricultural labour, and identity in Indonesia offer a useful comparison.

[11] *Ling.* 5.12.

market-gardening, and small-scale agriculture, in addition to larger estates; much of the unremarkable white noise of movement in and out of the city is likely to have been generated by traffic associated with this kind of sourcing of farm-produce from family fields, large and small. Thus the 'field', fixed and immobile at first glance, grounded by definition in the soil, and delimited from the wider landscape through being given boundary markers (whether 'rooted' but constantly in motion, like the living tree, or fixed but repositionable stones), gains kinship through language with the human agents responsible for transforming and managing the environment (*agrarii*), and allusively ties in the urban world to which (as Varro commented disapprovingly in his study of agriculture) wealthy 'farmers' increasingly migrated (the *agricolae*).[12] There is particular piquancy in this material when we recall that Varro might be classed as a man intimately involved in land management (*agrarius*): he served on Caesar's land commission (59 BCE), set up to assist in the foundation of military veteran colonies in Campania. In the wake of Sulla's earlier land confiscation and redistribution amongst his supporters, this commission was emblematic of the increasingly personal manipulation of land rights, use, and demarcation for political ends.[13]

Varro later observes that inflection (that is, words with stems from which other related forms develop) makes a flourishing semiotic system possible.[14] The study of language (*langue*) is possible because there is a core vocabulary that generates the limitless profusion of discourse (*parole*). Varro is talking to his politically compromised but still high-profile and influential addressee, Cicero, about how best to understand 'our' language's constitution as a particular vocabulary set with specific real and epistemological relationships to the stuff of Roman life.[15] It is inevitable that Varro's relationship with Cicero colours our reading of his treatise on Latin: both authors were acutely aware of the potentially dangerous implications of inapposite nuances and turns of phrase in the years preceding Caesar's assassination, and during the wave of proscriptions that followed.[16]

---

[12] Cf. *Ling* 7.4. Varro specifically castigates the *paterfamilias* (head of household) who sneaks away from his country duties to live within the walls (*Rust.* 2.3). Lefebvre (1991: 234–5) analyses the life-and-death qualities of how city and countryside connect. Campbell (1996: 91, 92–3) discusses trees as boundaries, and notes the significance of how boundaries work in an era when land confiscations are in progress; see also Campbell (2000: 79–81, 93–5 [Hyg.]; 111–13 [Sic. Flacc.]) for Roman land surveyors on tree boundaries.

[13] Varro *Rust.* 1.2.10. Cf. Cic. *Leg. agr.*, and Williamson (2005: 63–77).

[14] *Ling.* 8.3.

[15] Varro observes that he's writing to Cicero about why certain matters (*res*, a resonant term in Latin, suggesting both the concrete and the conceptual—e.g. *res publica*, the Commonwealth, or State; public affairs) necessitated or gave rise to the Latin language (*Ling.* 5.1). On the relationship between language, politics, and identity more generally, Beccaria (2008: 127–45).

[16] Wiseman (2009: 107–30) elegantly sets out how these two self-consciously elder statesmen orbited (and perhaps teased) each other. Cicero, of course, lost his life; Varro escaped. See also Leach (1999).

When he turns to analogy in Book 8, Varro starts with words where inflection, and thus also generation of new and evolving meaning, are possible. He draws on Dio of Alexandria for the notion that grammatical case and time are the key factors for defining how words function in context.[17] Here, two key types exist: nouns and verbs, with their corollaries, adjectives, and adverbs. The clause runs: 'ut homo et equus, et legit et currit' (such as 'person' [/man] and 'horse', and 'gathers' and 'runs'). Exemplifying his two primary parts of speech in this context, therefore, are person (*homo*), horse (*equus*), plus the third-person singular verbs s/he gathers (or discerns, selects, peruses, learns about, or picks out—a path; *legit*), and s/he runs, or hurries (*currit*). *Ut* (such as) and polysyndeton show the separate qualities of the terms by making an exemplary list of them; at the same time, the terms are linked through repetition (*et*, and) and semantics. Running (enforced or unthinking speed) is what horses do; a person (and the resonances of the verb *lego* encourages reading *homo* as 'man' and 'citizen') moves carefully, in an unhurried fashion, to no one's order, and for a purpose that is in part fulfilled through the movement and in part through its outcome.

We can think about this in de Certeau's terms:

> ... the functionalist organization, by privileging progress (i.e., time), causes the condition of its own possibility—space itself—to be forgotten.[18]

Here, Varro's second noun-and-verb pair (horse/runs) exemplifies de Certeau's 'functionalist organization' by unthinkingly (or reactively) galloping. For the running horse or (implicitly, unfree or thoughtless) person, meaning is generated by the time taken from start to finish rather than from any considered interaction between subject and act and environment. Human (male, citizen) movement, defined by *legit*, is by contrast all about the thoughtful and discriminating relationship between progress and space, and the broader assumption of higher cognitive goals. In this way, one agenda item for Varro's study might be defined as the enrichment and redefinition of the relationship between citizen and the complex structured social space of the city. This hypothetical goal is achieved by mapping individual identity onto Latin's vocabulary of movement, and thus defining what it means to move. If one defines activity as hurrying or speeding, what gets lost? A sense of place (a runner concentrates on the track ahead, not the off-course vistas and environment) and therefore a sense of the Roman and civilized self.[19]

---

[17] *Ling.* 8.11. Dio, an Academic philosopher, came to Rome as an ambassador from Egypt, and died under suspicious circumstances (57/56 BCE). It is tempting to interpret Varro's decision to make Dio the primary source for this (Aristotelian) division of parts of speech as an allusion to Cicero's interest in the affair (Cic. *Acad.* 2.12; *Cael.*).

[18] de Certeau (1984: 95).

[19] Famously, Suetonius attributes 'σπεῦδε βραδέως' (make haste slowly) to Augustus as a favoured maxim (*Aug.* 25.4). Whilst the issue also exercised Greek thinkers, tensions in Roman

After this section, 'running' will not reappear on the agenda until 8.53. Here, having just redeployed 'human being' and 'horse' to explain proper nouns he moves on to describe nouns that form declensions in their own right, yet which in terms of sense and semantics have their roots in other words. His example of a noun deriving thus from a verb is *cursor* (runner) from *currere* (running; to run). Using *currere* and its offspring *cursor* to exemplify a recurring family trait in language (the 'stem') implies that the action of the parent (verb) always looms over the child's achievements. The 'runner', above all else, is one who runs; the action takes priority over the identity of the subject.[20]

That this is more than a coincidence is strongly suggested by his earlier inclusion of *cursor* in a set of 'professional' or special interest activities associated with highly specific and for the most part menial types of movement (for example, juggler, swimmer, boxer).[21] We also see this at work in Varro's return to *currere* a few sections later in Book 8, where the discussion homes in on intransitive active verbs, and Varro's two key examples are *currere* and *ambulare* (to walk, stroll), which are remarkable, he claims, for their formation of only two participles.[22] Again, Varro is interested in making the nature of how one moves define who one is. Someone who runs can be dismissed (or pigeonholed) as a 'runner', significant only for his speed; a running man, in other words, is a type or product of his activity, rather than an individual, someone who moves heedlessly (running is never analogous with intellectual achievement or reasoned activity) and is presented as an ideal example of how this morphology works in practice. This idea gains nuance when read in the context of Varro's earlier comment that a street 'runs'.[23]

---

identity associated with the lack of a myth of autochthony may have made the space/place/identity nexus particularly juicy in the late Republic when traditional sociocultural norms were subject to fierce testing; see Farrell (2001: 19–21).

[20] Cf. *Ling.* 5.11: the relationship between body, movement, place, time, and action is illustrated by equating the body with a runner racing on a track (*stadium*). This emphasizes running as a goal-oriented action with a specified duration. It also highlights that running is not part of elite identity. For appropriate elite activity, e.g. Leach (2003: 164–5, appendices A–C) on Pliny's prescriptions; O'Sullivan (2006) outlines the symbolic and emblematic connotations of walking.

[21] *Ling.* 5.94 wraps up with more action-terms—grape-picker, tracker, and hunter—but editorializing comment separates them from the cut-and-dried, 'service' activities. Game-hunting could be cast as heroic elite activity (as Varro seems to suggest here), but in context, 'hunter' (as professional noun) hints at the arena. The uncommon term 'swimmer' (*natator*) also features; the 'swimmer', without bathhouse context, is similarly cut off from citizen activity and civilization. Cf. Sall. *Cat.* 4: hunting and farming are classed as 'servile' and activities with no higher intellectual function. Unpicking agriculture from this: Martin (1995: 83). Holleran, in this volume, discusses menial movement in the streets of Rome.

[22] *Ling.* 8.59. Though of course this diktat excluding a perfect passive participle is probably for effect since such a participle can exist. Later, *curso* ('run back and forth') and *cursito* (ditto) figure as examples of words where spoken use can change the form of a word (*Ling.* 10.25).

[23] *Ling.* 5.48.

Despite linking the two words' 'root' verbs at 8.59, Varro's discussions of *cursor* never mention *ambulator* (someone who is strolling along).[24] Also unmentioned in the text but lurking in the peripheral semantic field is what we might think of as the 'slow' version of *cursor*: *pedes* (pedestrian). Taken in its Roman context—a context which is developed allusively in Varro's discussion of *pes*, 'foot'—this term has military overtones (foot soldiers; those without a public horse; harking back to the military formula *equites peditesque*, cavalry and infantry) and thus has close connections with ideas of citizenship and participation in public affairs.[25] Context adds another layer: Varro was writing at a time when swiftness of operation was increasingly presented as a feature of Roman military success, and Caesar was keen to develop on Pompey's reputation for speed. Speed was fast becoming a trope for characterizing a successful military general.[26] This emphasis adds a new dimension to a military forced march. Highlighting *pes*, we might argue, emphasizes a mode of moving rather than the movement itself. The missing *pedes* is one who moves on foot at a given speed—the noun *pes* (foot) functions metonymically, and generates the action (walking)—just as linguistically, the poetic 'foot' is the engine of the speed and rhythm of verse. Moreover, using one's feet emphasizes that the person is in charge of the movement, whereas being a runner (where the noun derives in this scheme from the verb) suggests that the action is the driving force. Upfront, *pedes* would apparently have more linguistic density than *cursor*. One explanation for its omission is that *pedes* and its associations blur the line between the different modes of citizenship in play (the *pedes* may be a solid infantryman, but this is not in practice the kind of Roman for whom Varro is writing). We might even speculate that evoking the foot soldier might have brought with it an unwelcome reminder of the militarization of contemporary politics, and the politicization of military recruitment. Nevertheless, the pedestrian still forms an associative link with the similarly absent stroller.

Already Varro has intimated that the three exemplary activities that define Roman citizen existence in time and space are sitting, strolling, and talking.[27] In this trio, elite Roman political, deliberative, and social practice are summed up. One sits (politically, perhaps at Senate meetings; contemplatively in one's

[24] *Ambulator* is absent from the extant text; it does not seem to have been a term in common use, but Varro's linguistic remit here should make it an obvious example. On this term used for tradesmen, see Holleran, in this volume.
[25] On *pes*: *Ling.* 5.95 (e.g. someone who has established himself in business—*negotium*—is 'on his feet'). The formulaic nature of the conjunction of riding and walking as the poles of Roman citizenship is evident at e.g. Cic. *Leg.* 3.7.3. See also e.g. Hor. *Ars P.* 113. This formula might also be lurking in the shadows when Varro links the terms man, horse, and run.
[26] See Cic. *Leg. Man.* 30 on Pompey's Mediterranean clean-up. On Caesar and speed, e.g. Caes. *B Afr.* 1–2; Murphey (1977: 240) analyses *celeritas* (speed) in Caes. *B Gall.* The topos recurs later: Vell. Pat. 2.41.1, 42.3, 50.4; Suet. *Iul.* 7; Plut. *Vit. Caes.* 11.
[27] *Ling.* 6.1; we return to this later.

library, or at one's villa); one walks conversationally rather than to get from A to B;[28] and one speaks (*loquor*)—whether in the Forum, in politics, in the courts, or in conversation. This rosy-hued version of a citizen-day excludes the increasingly dangerous implications of achieving and maintaining such a lifestyle: the traditional progress along the *cursus honorum* (the 'course' of political offices through which an aspirant senator advanced—ideally by making timely progress rather than charging headlong to the finishing line of the consulship, then a wealthy province to govern, and possibly immense political influence). Growing concerns about excessive personal ambition (*ambitus*) driving Roman politics in the first century BCE bring the metaphorical racetrack signalled by *cursus* (cognate with *currere* and *cursor*) more clearly into focus than a sober translation of *cursus honorum* would suggest, and repeated legislation to curb the rapid progress of determined individuals points up a general concern to slow politics down.[29] By speeding up the *cursus* to gain early consulships, Pompey, and then Octavian, were in these terms professionalizing the political career, making it a job of work rather than public service. With public office becoming a race, citizen movement too is subject to stresses.

Varro tackles *cursus* only once in the text as we have it, and it is an apparently throwaway reference: he tells us that some word-forms, all connected in some way to verbs, clearly have an inbuilt temporal quality.[30] Varro offers a perfect indicative active verb (*legisti*, you read, collated), a noun derived from a verb (*cursus*, the act of running; a race; a charge; a timely progression), and a present participle (*ludens*, playing). The nexus is suggestive, juxtaposing elite concerns (reading, processing information; a political career) with the bread and circuses of late Republican street politics (races and—often gladiatorial—games), and it is in this context that we might look again at (the absent) *ambulator*. Another noun closely linked to *ambulo* via the prefix *ambi-* (around) is *ambitus*. In contrast to his omission of 'stroller', Varro plays considerable attention to this word.[31] Varro's interest in forms prefixed with *ambi-* is flagged up early in Book 5, and is pertinent here because its first appearance is in an investigation of how *ambitus* signifies 'side-road' or 'ring road', or 'go-around'.[32] His etymological rationale makes *amb-itus* derive

---

[28] See Macaulay-Lewis, in this volume.

[29] *Ling.* 5.80–2 sets out the public offices. Minimum ages were set for each office by 180 BCE (*Lex Villia Annalis*), but this was an ongoing issue, followed up by Sulla's *Lex Cornelia de Magistratibus* (82 BCE), and Caesar's *Lex Iulia Municipalis* (45 BCE); see Beck (2005). On politics and youthful ambition, Eyben (1993: 51–66); Harlow and Laurence (2002: 111–12) summarize on Octavian.

[30] *Ling.* 6.35.

[31] Meaning 'round', or 'about', *ambi-* (also, *am-*) is cognate with the Greek ἀμφί, and other Italic forms. *Ambitus* derives in the first instance from *ambeo*, 'to do the rounds', or 'to canvass'.

[32] *Ling.* 5.22.

from *iter* (route), because it (sounds as if it) is worn away (*teritur*) through heavy, repeated use (the 'around' makes the 'rounds' of potential voters, or supporters, a kind of backstreet subterfuge). *Iter* is also a term associated with land surveying and military forced marches, encouraging the reading in of land confiscations, veteran resettlement, and a civil war context (as discussed above), but joining them up with Caesar's topographically organized survey of the Roman citizen body.[33] Emphasizing that *ambitus* (despite its chiaroscuro) is a Roman way of operating derives from an urban point of reference. Varro explains that the word has historical weight in the context of urban morphology because it features in the definitive and traditional core of Roman legislation—the Twelve Tables—as representing the extent of a building's walls (*parietes*) when expressed as a circuit (and this sense is, as we shall see, important later).

Additional capital is made out of *ambitus* when Varro proposes that it also has a role to play in generating environmental discourse: *amnis* is specified as a river that 'goes around' (*circumeo*) something—here, the example is the Tiber which loops around the Campus Martius and the city—just as a candidate for office 'goes around' the people when canvassing (*ambit*); both thus are what we might call 'going around the houses'.[34] By means of the detailed attention paid to forms using *am(bi)-*, Varro defines electioneering as a particular kind of movement. It circumscribes both the candidate and those he canvasses, containing them, forcing them into particular directions, and implies a lack of directness that sits uneasily with traditional aristocratic self-fashioning. This is developed by Varro when he quotes Lucilius, inventor of Roman satire, making *ambages* (meanderings; beating about the bush; confusion) part of a lit.-crit. joke.[35] Varro runs with this by deriving the term from *ambe* (not otherwise known). He takes no chances then by spelling out how the word *ambe* is itself embedded in *ambitus* and *ambitiosus* (twisting and turning; populist; showy; over-ambitious). This repeated harking back to the problems associated with terms depending on *ambi-* emphasizes its negative connotations when it defines a person's identity. Suddenly, the absence of *ambulator* makes more sense. It is hugely important that for men such as Varro, 'pedestrian speech acts'—power-strolling through the *allées* and avenues of villas, urban porticoes, or promenades—need to be kept just as that: an activity that makes up one part of the citizen day, not a quality that defines a (right-thinking) citizen; hence, *ambulo* works effectively for Varro's audience, and *cursor-currere* works for everyone outside the thinking-citizen circle.[36]

---

[33] Campbell (2000) evidences *iter* as a surveying term; *OLD s.v. Iter*: 2, samples military usage from this era. Suet. *Iul.* 41.3 gives us Caesar's radical new-style survey, on which, Wallace-Hadrill (2008: 290–2).
[34] *Ling.* 5.28.
[35] *Ling.* 7.30.
[36] de Certeau (1984: 97–102).

Nevertheless, *ambulare* (how citizens move) is, in Varro, just a step from *ambitus* (peregrination; ramble; going the long way round; taking the backroads), which came to signify corrupt electioneering or more generally political malpractice.[37] Even strolling in citizen fashion is therefore susceptible to redefinition, and Roman voting practice too was all about orderly movement: to vote, citizens had to walk across raised gangways or platforms known as *pontes*, bridges.[38] The gaps in Varro's archaeology of language hint at such available subtexts, but there is nothing implicit in his emphasis on the unbreakable bond between citizen speech and movement through space in time: changes to one make for shifts in meaning for both. This chapter's next section turns to a defining feature of Varro's quest for an ideal paradigm-set demonstrating the interplay between human action, human speech acts, and the placial quality of the world he inhabits.[39]

FRAMING MOVEMENT AND SEEING THE SITES

Ideally, citizen movement means moving at a measured and unhurried pace—the kind of pace that a toga and the heat of the Roman summer make necessary for those who have the status and time. This idealized 'walking rhetoric' encourages contemplation and reflection on the self and one's role in the wider community.[40] Much of the day-to-day movement in the city of Rome must, of course, have been far less leisured or considered—we can imagine the striding gait of the burdened litter-carriers, the wheeled traffic, noise, and machinery associated with major building projects,[41] the panniered mules offloading cargo and ferrying produce to market, the slaves dashing between tasks—but Varro's vision simplifies the rhythms of the city and translates the polyglot qualities of urban movement so that the city becomes an exclusive space intelligible to a specific community. Barthes observes: 'the city is a discourse and this discourse is truly a language: the city speaks to its inhabitants, we speak our city'; for Varro, the implicit 'we' signifies a group happy to identify with his ageing and politically compromised protagonists, 'Varro' and 'Cicero', and to read and rescript new discourses of the city in the light of Varro's urban tableaux and itineraries, leaving the pragmatic experience of urban dirt, sprawl, and hurly-burly in the shadows.[42]

---

[37] e.g. Cic. *De Or.* 2.105; *Fam.* 11.17.1; Caes. *B Civ.* 3.1.4.
[38] Taylor (1966: 39–56).
[39] Olwig (1996).
[40] de Certeau (1984: 100–2).
[41] See Favro, in this volume.
[42] de Certeau (1984: 118–22). See also (e.g.) Campbell and Kean (1997: 162–4).

This chapter's focus on Varro has shown us a series of key moments when he addresses running, strolling, and politicized movement. Focusing now on the first two, for the runner or the man taking a stroll, through the city in particular, tracing and narrativizing a route is paramount. This means being able to read the signs in the 'verbal' maze, and to derive meaning from the perceptual field that refreshes and personalizes each individual route or path in a productive way:

> Video a **uisu**, <id a ui>: qui<n>que enim sensuum maximus in **oculis**: nam cum sensus nullus quod abest mille passus sentire possit, **oculorum** sensus uis usque peruenit ad stellas... **Cerno idem ualet**... Dictum **cerno** a **cereo**, id est a **creando**; dictum ab eo quod cum quid **creatum est**, tunc denique **uidetur**.[43]
>
> Tueri duo significat, unum ab **aspectu** ut dixi, ... Alterum a **curando** ac **tutela**, ...[44]
>
> '**I see**' is from '**sight**', <which is from 'strength'>; for the greatest of the five senses is in the **eyes**, so that whilst no one sense can sense that which is a mile off the sense of the **eyes** reaches as far as the stars... '**I perceive**' **has the same qualities** [or '**power**']... '**I perceive**' is said to derive from *cereo*, that is to say *creo*, '**I create**', from the **act of creating** (*creandum*); on this account it's said that when something **has been created**, then at last it **is seen**.
>
> *Tueri* has two meanings, one, the sense of **vision**, as I have said, ... the other being '**caring for**' and '**guardianship**', ...

As this shows, what one sees is the most potent way of interacting with and gaining a controlling perspective on the environment. Varro equates 'seeing' to 'understanding' in a way that elides distinction between linguistic origin and functional value as a basis for knowledge.[45] In this way, he makes 'creation' the connective principle for a set of acts that define the way humans understand the world and produce meaningful space. The act of seeing is triggered by there being something to see; making something implies an understanding of its basic principles, and having created (or perhaps 'defined') a thing, it is then available to be looked at. The moment of creation (when a thing is complete) is the moment it is seen as itself, and defined as being in its final form. To see something is therefore on one level to make it complete (the city fails to exist without participating inhabitants defining themselves as citizens), and on another, to undertake a protective and nurturing role towards it (*tueri* in its second sense).

Varro's definitions suggest that seeing the city clearly, as a totality and as a series of almost infinitely varying tracks and routes, means taking some

---

[43] *Ling.* 6.80, 81.
[44] *Ling.* 7.12.
[45] Varro here seems to be playing with an idea developed more fully in Merleau-Ponty (1962: 3–83), whereby the physical and epistemological nature of 'seeing' alters the viewer and the thing seen.

responsibility for its meaning. The city-as-discourse is thus produced by its inhabitants, and these inhabitants—that is to say, the subset of citizens to whom Varro implicitly speaks—therefore have a duty (in Varro's world-view) to control and deploy language as fully and effectively as possible. This is how they can make the city (whole) and enrich and develop its semiotics (*cerno*, *creo*) whilst at the same time restoring and reclaiming any inappositely forgotten *quartiers* and their ghosts (*tueri*). These semiotic game-hunters are, as we shall see, being set on the track of specific, worthy, quarry.

Exploring Varro's linguistic topography makes it very clear how 'history and speech are inscribed into a spatial matrix to which a community or an orator returns'.[46] What we also see clearly is the relationship between time and knowledge (and knowledge gains cultural value by persisting), and between memory and communication.[47] Reading Varro's Rome as a site of violent and radical change presents challenges, but also opportunities for productive interpretation. Varro's Rome (and the audience he addresses) was convulsed with civil struggle on a grand scale, but also on the most local and day-to-day level; physical confrontation as a manifestation of political opinion was increasingly becoming a feature of the city and the radicalization and personalization of politics was most keenly felt by those who had most to lose from a changing status quo. Civil war, missing family members, proscribed properties, the dislocation of the regularizing pattern of elections and the attendant disruption of the usual ebb and flow of ex-consuls and their entourages to and from the provinces, join a sense of the shattering of social norms and values, bubbling through Varro's study. Varro's response to the ways power was ebbing and flowing is to make the unusual and the analogous ripples in language into a cornerstone of how a people develops and thrives.

> Vetustas pauca non deprauat, multa tollit. Quem puerum uidisti formosum, hunc uides deformem in senecta. Tertium seculum non uidet eum hominem quem uidit primum. Quare illa quae iam maioribus nostris ademit obliuio, fugitiua secuta sedulitas Muci et Bruti retrahere nequit. Non, si non potuero indagare, eo ero tardior, sed uelocior ideo, si quiuero. Non mediocres enim tenebrae in **silua** ubi haec captanda neque eo quo peruenire uolumus **semitae** tritae, neque non in **tramitibus** quaedam ob*i*ecta quae euntem retinere possent.[48]
>
> There are few things which the passage of time does not distort, and many which it eliminates. The one you once saw as a beautiful boy, you now see twisted by old age. The third generation does not see a person in the same way as the first saw him. Therefore those things which oblivion has taken even from our ancestors, these escapees not even the assiduous pursuit mounted by Mucius and Brutus

---

[46] Niebisch (2009: 331).
[47] Feeney (2007) details the cultural resonances of calendar and chronology. Compare the impact of the increasing specificity in measuring and fixing time in the early 20th century; Kern (2003: 10–35).
[48] *Ling.* 5.5.

could recover.⁴⁹ Even if I myself am not able to hunt down this quarry, I shall not on this account be slower in the chase, indeed I'll even be swifter if I'm able. For there is no trifling darkness in the **wood** where these are to be captured and they leave no well-trodden **paths** to take us to where we want to go. Nor indeed do the **tracks** lack obstacles which can delay the hunter.

Varro's wood is one where Heideggerian *Holzwege* predominate: the world he is mapping is one populated by ghosts and confusing shadows, deceptive echoes of communicative practices that are vanishing both as a result of the passage of time and due to the inability of earlier generations to capture them securely and put them back into productive use.⁵⁰ The wood has 'paths' and 'tracks', conjuring up a different image to the language of high-speed highways and paved streets (*uiae*; as noted above, 'streets' can 'run'), and Varro's conceptual framework encourages readers to rethink familiar itineraries and patterns of movement.⁵¹ Varro's multilayered conceit: 'on the ways of frequenting or dwelling in a place'—here, a 'Rome' defamiliarized as a wood—draws on the topographic structures of cognition that underpinned the ancient Art of Memory by emphasizing how speaking without moving is impossible, and how movement to be meaningful necessitates an epistemological frame.⁵² Varro's linguistic Snark Hunt shows how the goalposts for any study or investigation inevitably change because human time passes, on an individual and societal level, as the 'hunt' progresses.⁵³

This passage is interesting for another reason: it shows again how Varro conceptualizes intellectual endeavours as analogous to specific kinds of movement and momentum. This is in part defined by the real or imagined *locus* of the movement (here, a dark and confusing wood) and in part by the relationship between the imagined topography and the realities of the act (here:

---

⁴⁹ The names dropped by Varro are tantalizing. 'Mucius' may be Q. Mucius (referenced later: *Ling.* 6.30) Scaevola (cos. 117 BCE). This Mucius features as a key player (and font of legal wisdom) in Cicero's dialogues (e.g. *De Or.*, *Rep.*). If this is the man, then Varro undermines his authority. Alternatively, Q. Mucius Scaevola (cos. 95 BCE) also features in Cicero (*De Or.* 3.10, *Brut.*), and had a name for expertise in jurisprudence too (assassinated by L. Iunius Brutus Damasippus in 82 BCE). The other possible candidate is P. Mucius Scaevola (cos. 133 BCE); a jurisconsult (Cic. *De Or.* 1.212, 2.222-4), his name is later linked with 'Brutus' as one of the founders of Roman jurisprudence, though Q. Mucius outdoes him (Cic. *De Or.* 2.142; *Dig.* 1.2.2.39-42). A pairing of a 'Mucius' and a 'Brutus' may have become proverbial (see e.g. Cic. *Brut.* 130). It remains possible that Varro was deliberately vague to encourage the widest range of comparisons.

⁵⁰ *Holzwege* are unmapped woodland tracks that sometimes lead nowhere; see Heidegger (1950).

⁵¹ In this sense Varro's approach to reclaiming language as a process of moving through space in an acculturated fashion is one that prefigures Bachelard's poetics of space (1994: 56-66, 183-210). Laurence (1999: 58-9) defines 'road'.

⁵² de Certeau (1984: p. xxii); Rossi (2006: 8-10). In general: Yates (1966); Small (1997).

⁵³ '... the impossible voyage of an improbable crew to find an inconceivable creature' (Sidney Williams and Falconer Madan, cited at Carroll 1974: 21).

marking up the real-time process of undertaking linguistic scholarship with the memory or fantasy of a hunting expedition and its rhythms).[54] These implications might be read in terms of 'rhythmanalysis', whereby the search for language and the generation of epistemological authority is in its own right a weightily temporal undertaking which is characterized by complex ebbs and flows analogous to real-world action.[55] Varro presents the scholar's task as one which has all the urgency, aristocratic allure, manliness, and authenticity of the hunt, but the scholar is not 'professionally' tagged as a *uenator* (huntsman).[56] We might suspect that this comes too close to the servile overtones we observed in connection with *cursor*. Similarly, we sense that Varro is not eager to have his audience assume that his goal is purely intellectual: Varro's scholarly rather than professional pursuit of the fleetest and most prized aspects of a language in flux is designed, we might speculate, to show how a Roman can serve the *res publica* not just by fighting, or by deliberating in the Senate, but also by investigating how the place-world and the characteristic ways of moving through it are governed by lines of communication to the past that need to be kept open.

> Nunc singulorum uerborum origines expediam, quorum quattuor explanandi gradus ... Tertius gradus, quo philosophia ascendens peruenit atque ea quae in consuetudine communi essent aperire coepit, ut a quo dictum esset **oppidum**, **uicus**, **uia**.[57]
>
> Now I shall clear up the origins of individual words, of which there are four classes of explanation ... the third class is that which philosophy, aspiring, achieved, and which began to reveal what characterizes words in common use, such as the reason for the terminology '**town**', '**neighbourhood**' [/'block'], '**thoroughfare**' [/'street'].

Philosophical endeavour opens up the true sense of everyday vocabulary, and the terms Varro cites to make this clear are emblematically urban and features of civilization: 'town', 'neighbourhood', and 'thoroughfare' or 'street'. For Varro, *uicus* means a street with buildings on either side, or a group of buildings along a street, and so it mediates between the two concepts.[58] More generally, we are again seeing how urban movement (patterns created by and endlessly reinventing the morphology of urban space) takes on exemplary status for studying the nature of the Latin language. Scholarly enquiry— in Varro's scheme—shows how the city-and-forest is a product and generator of discourse, but without memory the city and citizen discourse disintegrates

---

[54] Compare de Certeau (1984: 173), 'to read is to be elsewhere ... it is to constitute a secret scene ... Marguerite Duras has noted: "... darkness gathers around the book"'.
[55] Lefebvre and Régulier-Lefebvre (1985).
[56] Cf. Varro *Ling.* 5.93.
[57] Varro *Ling.* 5.7–8.
[58] Varro *Ling.* 5.145, 160. See now Wallace-Hadrill (2008: 264–9).

(into, we could imagine, trackless forests). All the ways of being a citizen and living in the city depend on a shared, agreed, intelligible, and editable memory, operating on a range of private, public, individual, collective, and phenomenological levels. Varro's responses may reflect a wider sense of how the traditional direct connection between citizen identity and the city of Rome was being tested by imperial expansion.[59]

> Meminisse a memoria, cum <in> id quod remansit in mente **rursus mouetur**; quae a manendo ut manimoria potest esse dicta...... Ab eodem monere, quod is qui monet, proinde sit ac memoria; sic **monimenta** quae in sepulcris, et ideo secundum uiam, quo praetereuntis admo*n*eant et se fuisse et illos esse mortalis. Ab eo cetera quae scripta ac facta memoriae causa **monimenta** dicta.[60]
>
> 'To remember' derives from 'memory', since in this there is **movement** which is a **return** to what 'has remained' in the 'mind'. This may connect to 'remain' via a form *manimoria*...... From the same is 'to remind', because he who 'reminds' takes on the role of memory. Thus also the '**memorials**' which are on tombs and indeed along the highways, in order that they may 'admonish' the passers-by that they themselves [memorials and readers/viewers] are mortal too. From this, other things which are written and done for the sake of 'memory' are called '**monuments**'.

Memory necessitates a 'return' to what persists in the mind itself—one's personal mental library, and the complex mnemonic catalogues that permit access to its contents—and makes a direct connection between the empirical, the phenomenological, and the personal worlds of perception and recall. The person who enables this 'return' to the stored images and ideas is transformed by Varro here into a physical embodiment of memory itself, and juxtaposed with the marble roadside sepulchral monuments that perform the same function. Memory, in this way, gains real-world dimensionality, and thus partakes in the discourse of the city.[61] Each person who activates one of these mnemonic hyperlinks gains an avatar: a kind of speaking sepulchre, containing and articulating ancestral memories that need constant recall and re-examination to survive.[62] More prosaically, Varro is also making this about travelling the roads in and out of town—any town, but most lavishly, Rome itself. Touring the eternal homes of ancestors means going out of or into the city, along roads that emphasize Rome's historical power over the landscape.

---

[59] David Newsome rightly urged me to make this explicit; see Zehnacker (2008: 432). Cf. Cic. *Acad.* 1.9 (discussed below).

[60] *Ling.* 6.49. This idea had some currency, cf. Cic. *Fr. Epistle* 3.7 (to Caesar)—the word '*mon*ument' ad*mon*ishes as to its meaning; its focus should be exemplary, a future memory.

[61] Troilo (2005: 89–95) discusses how tombs featured in Italian unification. Rome's municipal cemetery (Campo di Verano) saw a wave of significant monuments during this era; see Henneberg (2004).

[62] See Jaeger (2008: 32–47).

Mapping the tombs, stellar and unremarkable, draws the passer-by back (or forwards) into the city where the most noteworthy ancestors made their name.

This passage's use of funeral monuments to connect up real-time, chronology, social memory, and individuals in motion draws on a number of different kinds of time and movement, developing themes laid out early on in this three-book group.[63] Speaking or recalling a word, just like seeing, sensing, or other forms of information-processing, means taking a trip back into the mental reference library for consultation or to make a withdrawal.[64] With each word triggering such a journey (time and action), buildings, monuments, and citizens alike are all contributing to a system that teeters on the brink of entropy unless an organizing principle intervenes. This chapter moves towards a conclusion, therefore, by exploring Varro's whistle-stop infrastructural tour of the city of Rome as a sophisticated essay on how Roman movement interacts with 'the semantic charge given to the city by its history' to create epistemological order with real world implications.[65]

We have already noted that *ager* is significant early on in Book 5. It soon crops up again. The *ager Romanus*, we learn, defined the first phase of the city: its three (*tris*) parts generated three tribes (*tribus*), making landscape the engine of early citizen identity. The tribes were, Varro learnedly proposes, Titienses (from Tatius, the king who helped Romulus merge Sabine and Roman peoples), Ramnenses (from Romulus), and Luceres (from Lucumo: Tarquinius Priscus).[66] These three ethno-topographic 'tribes' subsequently somehow—Varro offers no explanation—meshed with a set of toponyms to produce four urban tribes named for areas of the city: Subura, Palatine, Esquiline, and Colline, plus one other—Romilia—nestling under Rome's walls. This scheme, however unbelievable, shows how closely language and identity are tied to place, and especially to the notion of Rome as an up-and-down city of more than one hill in which moving from one part to another is likely to involve a cultural realignment. *Ur*-Roman field-land had three parts, and so the people in each were tagged as one-third of a territorial whole, each marked up with one political figurehead. These primal and amorphous 'thirds' lacked a rationale for how or why they should be plotted meaningfully against urban space, so instead Varro's explanation makes three hills, a valley, and the peri-urban plain become markers for how the city develops and positions its inhabitants. The city in this way seems almost to pre-date its formal habitation, with high-profile power figures only subsequently giving way to topography as the organizing principle.[67]

---

[63] *Ling.* 5.12.
[64] In detail: Farrell (1997).
[65] Barthes (1997: 160). Varro's historical survey of Rome, *Ling.* 5.145–68.
[66] *Ling.* 5.55. There is no particular logic to these explanations.
[67] Varro *Ant. fr.* 5 makes 'the city' an 'author', equivalent to a painter or builder: it creates its own unique organizing principle.

Places and people thus act as markers for each other in this keynote topography: the resulting primitive city, taken at face value, is one where walking is not just a way of moving through space but an actualization of the different characters of space on the three distinct hills and the low ground of the valley (bounded by hills) and the plain. That we are not intended to see this as a group of wholly separate settlements (potentially with little day-to-day interaction) is evident not just in how Varro constitutes the tribes as 'thirds' and marks them up using connected political figures, but also through his specification that the group of hills and the valley they tower above are themselves unified by city walls which themselves reached out to encircle and define the new city. The city walls, as Varro has already hinted, make this a place where crossing boundaries (hill to valley to hill; through the walls) is a necessity for citizen identity.

> Ubi nunc est Roma, Septimontium nominatum ab tot montibus quos postea urbs muris comprehendit; . . . Hunc antea montem Saturnium appellatum prodiderunt et ab eo Lati\<um\> Saturniam terram, ut etiam Ennius appellat. Antiquum oppidum in hoc fuisse Saturnia\<m\> scribitur. Eius uestigia etiam nunc manent tria, quod Saturni fanum in faucibus, quod Saturnia porta quam Iunius scribit ibi, quam nunc uocant Pandanam, quod post aedem Saturni in aedificiorum legibus priuatis parietes postici 'muri \<Saturnii\>' sunt scripti.[68]

> Where Rome now is, was called 'Sevenhills' from the same number of hills which later the city embraced with walls; . . . This [the Capitoline] was previously called the Saturnian Hill according to the authorities, and from this Latium is the Saturnian Land—as indeed Ennius calls it. It is recorded that an ancient town, Saturnia, was on this hill. There are now still three vestiges of it remaining: that there is a shrine of Saturn at the entranceway; that there is a Saturnian Gate which Junius writes of there, which they now call Pandana; that for private buildings behind the temple of Saturn the legislation terms the rear walls 'Saturnian Walls'.

> Quartae regionis Palatium, quod Pallantes cum Euandro uenerunt, qui et Palatini; \<alii quod Palatini\>, aborigines ex agro Reatino, qui appellatur Palatium, ibi conse\<de\>runt; sed hoc alii a Palanto uxore Latini putarunt. Eundem hunc locum a pecore dictum putant quidam; itaque Naeuius Balatium appellat.[69]

> In the fourth region sits the Palatine, so called because the Pallantes—or Palatines—came there with Evander; others derive Palatine from the original inhabitants of the Reatine territory which was called Palatium, and who settled there; still others think that it derives from Palanto, the wife of Latinus. And it's this very same place that certain others think was named for the 'flock'—thus Naevius calls it Baaa-latine.

When tackling the topography and social organization of the city in its more developed phase, Varro recommences with the hills—now rearticulated as Sevenhills Ville, a place name and etymology that works well with the other

---

[68] *Ling.* 5.41, 42.   [69] *Ling.* 5.53. Varro, of course, is a Reatine.

'obvious' topographic etymologies he deploys here, but on which he never fully delivers the goods once the tour starts.[70] First up comes the Capitoline, with its prehistoric tags.[71] This roll call prioritizes Jupiter (whose temple foundations threw up the etymologically resonant head—*caput*—from which the hill's contemporary name is derived), but also shows how vestigial topographic signs from prehistory continue to impact upon the present, and to change the way one might trace a path in which the Capitol was a feature. The Capitol is the *head*land, a place whose current height is greater than its origins (as shown by the excavation down into its earlier phase of occupation): contemporary Romans get a greater crick in their necks (or more vigorous exercise in the climbing) than those first settlers. This first bundle of key sites, linked to the Capitoline, includes the Aventine and Velabrum—again, we see that movement is built in. In the first place, two (or more) hills, as noted above, make for up and down movement and opposing points of view; secondly, emphasizing the Velabrum focuses readers on the relationship between the city and the Tiber, the movement of traffic, goods, and people the river made possible, and also points up the blurry boundaries of a floodplain where reclaimed land is always in danger from the rising river.

At this point Varro abandons the Hills as an organizing principle, and shifts to an itinerary based around the ancient Shrines of the Argei (divided into four urban regions). Of the four Regions treated here by Varro, we might want to pause first at the Palatine because it provides a usefully compact example for how Varro's surveying technique works.[72] Varro commences by deriving the name from the ethonym Pallantes; aka the Palatines who, he adds, arrived with Evander. Next, we meet some less favoured etymologies, but Varro highlights his interest in the derivation from wandering Greek pastoral folk (Evander and his Arcadians) by deploying 'Baaa' as a bookend: the aetiology concludes with a jokey etymological suggestion that 'Palatine' is more obviously a pun on the baaa-ing sheep—Baaalatine.[73] What all the explanations have in common is an emphasis on the role of immigrants. This raises an important issue—it presents a paradigm for reading hills as attractive forces,

---

[70] Holland (1953: 21–3, 31–4) discusses the oddness of Varro's abandonment of the hills as markup, and wonders if there had been some slippage between the notion of an early fortified hill-town (*saepti montes*), and the similar sounding *septimontium* (Seven Hills—only missing the dipthong '*ae*', a drift in *parole*, noted elsewhere, e.g. *Ling.* 7.96) which Varro is either drawing on for his own purposes or (as Holland thinks) unaware of. Gelsomino (1975) follows up in detail, but see more recently *LTUR s.v. septimontium*. Rodríguez-Almeida (2002: 13–21) suggests that Varro's use of the hills hints that he had a map to hand.

[71] See Zehnacker (2008: 426).

[72] Region IV (primarily, the Palatine) takes up 5.53–4; the Suburan region, 5.46–8, the Esquiline, 5.49–50, and Region III, 5.51–2.

[73] This kind of pastoral markup was clearly in the air. For (typically, vague) presentation of Rome's aborigines as shepherds, see e.g. Livy 1.7.8–9, Ov. *Fast.* 2.271–9, Prop. 4.9.3, Verg. *Aen* 8.359–61.

drawing people to them, and movement is thus a defining factor in their semiotic viability.[74] Varro moves briskly on, then, to the other core sites he has chosen to represent Region IV: the Cermalus (south-west face of the Palatine) and Velia (just north-east of the Palatine).

The <u>C</u>ermalus is marked up as part of the story of Romulus and Remus because it highlights how old Latin did not always distinguish between 'c' and 'g', and thus by association of spelling, draws in the twin *germani* (brothers). This rationale calls to Varro's mind the brothers' finding at the base of the Fig Tree (the Ficus Ruminalis, near the Lupercal on the Cermalus side). So we have another, archetypal moment where apparently informal and undirected movement results in a moment of intense significance for the proto-city, if we take the 'pastoral' hint as the passage as a whole encourages. Varro does not in fact comment on the version that they were found by shepherds. Instead he reintroduces shepherds immediately afterwards when explaining the Velia: the Palatine shepherds used to pluck (*uellere*) the wool from the sheep before they had worked out how to shear them.[75] The Palatine, therefore, becomes a node where significant individuals wander into history.

Varro's choice of entries in the complete mini-handbook of sights to see (5.41–54) characterizes a very particular view of what constitutes the core features of the city and its Regions, and how they should be connected up. We are then given a succession of manmade and Roman construction projects. First come some edited high- and lowlights, namely (associated with the Capitoline) the Temple of Jupiter Optimus Maximus, the Temple of Saturn, a narrow passage (*fauces*) possibly part of the Clivus Capitolinus, the Porta Saturnia 'now' known as Pandana, some unnamed houses backing onto the Temple of Saturn, and (moving to the Aventine) the Temple of Diana, the Velabrum and Via Nova, linked by the Ferry Chapel (commemorating the Velabrum as a transit site). Next comes a regionary tour, organized loosely around the twenty-seven shrines of the Argei (or 'Argives'). Region I provides an Argei shrine on the Caelian, the Vicus Tuscus and its famous statue of Vertumnus, then linking the Caelian and Carinae is the fourth Argei shrine (on a street by the Temple of Minerva); next the Sacra Via (topped and tailed by the Chapel of Strenia—a goddess of physical health—and the Arx), leading into the Clivus Capitolinus as it exits the Forum. Adjacent to the Carinae is the sixth Argei shrine.

Shifting into Region II we 'pass' some ghostly sites (lookouts and a former artificial grove from the regal period) before reaching a cluster of contemporary sacred groves amidst which is the Chapel of the Oak-Grove Lares, and

---

[74] See also e.g. Varro *Ling.* 5.43 on the 'Adventine'; cf. Ov. *Fast.* 4.63–84. Farrell (2001: 37–9) proposes a link between language and empire; Narducci (2004: 29–54) discusses villas and 'Greekness' in Cicero.
[75] *Ling.* 5.54; see 5.165.

then some nameless streets (*uiae*); the tour here ends with five further Argei shrines glossing our experience of the Esquiline so far.[76] Region III presents as a primarily religious topography—its five hills are named for shrines. Two of these hills, and their famous sites, have however overwritten the other three; thus we see distinctly only the Altar of Jupiter Viminius, and Quirinus' sanctuary. In fact Varro informs us that 'Quirinal' had such resonance that other less distinctive hills were subsumed under its name; evidence comes from a further allusion to the documentary evidence Varro favours here (alluding to what seems to be some sort of venerable record: *Sacrifices of the Argei*). The list of four Argei shrines adds in a few more temples as landmarks: of Quirinus, of Apollo, of Salus, of Deus Fidius; plus a Vicus (Insteianus) distinguished solely by its Argei shrine.[77]

Varro's study is, of course, massively incomplete; this means that definitive assertions are hard to support. Nevertheless we can see from extant complete sections that he must have quite carefully avoided drawing contemporary figures, monuments, and politics into the picture, whilst at the same time giving a unique historical character to Rome as a city defined by controlled and managed change and movement.[78] We might read this as a retreat to an '*anthropological* stage of social reality', where directional 'indicators...were objects invested with affective significance...Egregious aspects of the terrain were associated perhaps with a memory, perhaps with particular actions which they facilitated...The networks of paths and roads made up a space just as concrete as that of the body'.[79] Rome becomes a place shaped by the Shrines of the Argei (notionally, migrants from Greece), and given coherence by the associated records that explain their relationship to a network of temples, roads, gates, and dwellings. We might wonder whether these Shrines too are part of a lost world, marking up a contemporary stroll with echoes of an almost incomprehensible phase in urban life.[80] The only explicitly embodied figure in the tour is Vertumnus, chief of the Etruscan pantheon, and a shape-changing god who watched over the turning seasons, and stood his ground (*sto*) on the road linking the Forum Romanum (politics) to the Forum Boarium (trade) via the imagined archaic wetland of the Velabrum.[81] Otherwise, this is a ghost town.

---

[76] Four associated with groves on the Esquiline (the use of directional terminology gives an unverifiable aura of precision); one with Juno Lucina (where a grove seems to have been part of a temple complex).

[77] As David Newsome pointed out to me, Varro—for no obvious reason—does not locate all twenty-seven Argei shrines (he only mentions fourteen; Region IV has two).

[78] Zehnacker (2008: 429–30) contrasts this with Cic. *Rep.* 2 and Strabo 5.

[79] Lefebvre (1991: 192).

[80] Wallace-Hadrill (2008: 260–4) goes into detail.

[81] The idea that the Velabrum once comprised permanent wetland and areas of swampy ground, rather than just being a flood-plain, is economically sketched by Varro (cf. *Ling.* 5.156), but it also features e.g. in Tibullus 2.5.33 and Prop. 4.9.5; cf. Ov. *Fast.* 6.405–7. Tiber floods must

We see this exclusion of contemporaneity even more clearly when Varro returns to topography later in Book 5.[82] On the Forum Romanum, for example, his major excursus focuses on the Lacus Curtius, a strange echo of Rome's mysterious, swampy historical identity, formalized in gleaming marble, ideally showcasing his antiquarian expertise.[83] The Forum itself might seem an ideal testing ground for Varro's understanding of Rome as a vortex, pulling all and sundry in, but subsequently Varro only revisits the Forum as a pendant to his discussion of the Circus Maximus and Circus Flaminius.[84]

One prosaic and immediately visible reason why the Forum Romanum might have seemed too delicate a site for semiotic archaeology is that the Forum and Curia were in a state of flux, with the Forum increasingly a place where radical political upheaval in public affairs was manifest. In addition to the construction work on the Basilica Aemilia/Paulli and Basilica Iulia, the old Curia (Hostilia), famously, was temporarily replaced by a floating venue after a succession of changes (and destruction by fire) encouraged Caesar to rebuild it, a project still incomplete on his death. Caesar's new forum was at the same time jostling with the Forum Romanum as the place where cutting-edge ideology was monumentalized and laid out for public inspection. Tackling the Forum Romanum's major sites at greater length may have presented the wrong kinds of challenge to an author interested in laying bare layers of meaning in scholarly fashion—both too straightforward, if done blandly, and too tricky if worked up in depth. Cicero's Catilinarian speeches provide an excellent comparison for how writing up the Forum became a political act.[85] As discussed above, Varro was alert to the way that an effective monument could (indeed, was required to) speak for itself. The odd, archaic Lacus Curtius, unlike more contemporary sites, is both a satisfying puzzle and a monument without direct recent political resonance or intelligibility. Varro speaks authoritatively on its behalf, its own language, we might assume, having been safely forgotten.

Opening Book 6, Varro notes that this section of his study deals with activities characterized by time taken: the three examples he gives are sitting, walking and talking ('ut sedetur, ambulatur, loquontur' 6.1). This trio represents, as I suggest above, the core and defining activities for Varro's ideal audience. Placing these three activities right at the beginning of Book 6 shapes the governing principle for Book 6 as a whole. Varro goes on later to link speech

---

have given a sense of some impermanence to this area, but clearly it was fully integrated in urban use. The 'memory' of a more watery state may help to point up how urban Rome has imposed new order; see Aldrete (2007: 167–9).

[82] Varro *Ling.* 5.145–64. See Zehnacker (2008: 424–5) for a detailed list.
[83] *Ling.* 5.148–50; see Spencer (2007) on Livy's Lacus Curtius (so noteworthy for Livy that he does it twice).
[84] *Ling.* 5.155–8.
[85] Vasaly (1993: 60–75) is excellent on this.

to the suitably aristocratic practice of hunting (comparing *loquor* and *uenor* semantically in their conjugation of participles; 8.59), and this ties in nicely with his earlier use of a hunting analogy, as we saw at 5.5. Thus we find that the philosopher-etymologist—which is how Varro classes himself at one point—can aspire to be able to reveal the true nature of key, common vocabulary (*oppidum, uicus, uia*), but this goal depends on activities that start to seem more appropriate to the suburban villa (*rus*) than to Varro's framing of city life.[86]

THE SCHOLAR AND THE CITY, 40s BCE

Varro's conceptual dialogue—whereby, for the right reader, moving through urban space can become like strolling the elegantly laid out avenues of a villa—must have seemed ever more removed from real-world experience of the *urbs* (city) in the mid-40s and later, in the years after Caesar's assassination. For public-access venues where citizens might stroll expansively and converse in an atmosphere conducive to intellectual activity one has to look outside the walls, to the portico-garden attached to Pompey's theatre on the Campus, and to the Campus itself—which retained a parkland quality that might just about evoke the archetypes of *amoenitas* that glance to Plato's *Phaedrus*. Finally of course there were the immediately peri-urban gardens—by invitation only, for the most part—that hugged the city walls and overlooked Rome from the Pincian and Esquiline Hills. These private luxury villas, clustering closely around the city, were becoming an increasingly significant feature of citizen *habitus*. Such villas became in effect the powerhouses of culture and a driving force behind political and sociocultural realignment.[87]

In the light of this it is difficult not to speculate about where Varro is undertaking his reading, research, and composition for this massive linguistic study. The villa is one traditional site for being an intellectual, but for Varro we have an additional option that plays well with his emphasis on the utilitarian qualities of the villa in this study.[88] We might perhaps visualize Varro at work in Rome itself, amidst the historical scrolls of the Censors' library, housed in the Atrium Libertatis (Hall of Liberty).[89] This is an attractive proposition, particularly since we might suppose that the public library which was founded there by C. Asinius Pollio—using the spoils of his Parthian Triumph (39 BCE)

---

[86] *Ling.* 5.8. See Wallace-Hadrill (2008: 269–75) on streets and civilization.
[87] Wallace-Hadrill (1998b).
[88] Villas are named as the destination for the highways (*uiae*) along which produce (*fructus*) was transported (*conueho*), Varro *Ling.* 5.35.
[89] On censorial records: e.g. Livy 43.16.13, 45.15.5.

to restore the building and create a new space for bibliophiles—was probably in the air during the 40s BCE when it was, according to Suetonius, one of Caesar's pet projects.[90] Suetonius states that Varro himself was the man entrusted by Caesar with the task of curating this planned collection, which presumably was intended to rival that at Alexandria. A good story, which we have no reason to disbelieve. The comprehensive library that Varro may already have been beginning to collate and catalogue presents a charming (if imaginary) background for a man trying to define the hermeneutics of language, and the nature of communication, in a pan-Mediterranean empire.

The layout and groundplan of the Atrium Libertatis remains a mystery, but one linguistic connection encourages me in seeing a role for the library in Varro's long-awaited completion of *De Lingua Latina*. We have seen how city and text are inextricably linked in Varro's linguistic field; we have also seen how strolling becomes an ideal citizen pursuit when devoted to meaningful conversation. Rome's porticoes present a venue for strolling that echoes Athens' Stoa of Zeus Eleutherios (Liberator), the site of a key dialogue between Isomachus and Socrates concerning appropriate behaviour, in Xenophon's *Oeconomicus*.[91] 'Eleutherios' was Zeus' title in thanksgiving for his help in delivering Athens from the Persians, thus preserving its 'liberty'. Rome's rumbling conflict with Parthia (Persia's successor) during this era—one of Caesar's unfinished plans was to avenge Rome's overwhelming defeat at Carrhae—is built into Pollio's foundation (with hindsight at least) in the Hall of Liberty, and it makes for a tantalizing nexus of ideas. Stoas, of course, and in particular the 'Painted' Stoa in Athens, were the home of Stoic philosophy.

A stoa was a portico within which the kind of perambulation that generated academic and philosophical developments and advances in understanding was most at home, and Xenophon cites Socrates' explanation of why farming is the best activity for a citizen in that same Stoa of Zeus the Liberator.[92] Xenophon's Socrates makes farming the only activity which allows citizens time to combine personal and public duties, which takes us back to Varro's field archaeology metaphor for tracing the development of communication.[93] Varro's role in planning a Public Library for Rome, for Caesar, and finally for Pollio, in a place marked up as a (Latin) Stoa of the Liberator, must have sensitized him and his readers even further to the spatial as well as linguistically translational qualities of reading the city of Rome as ongoing discourse, and as an ideationally acute markup of their own 'walking rhetorics' through urban space.

---

[90] Suet. *Iul.* 44.2.
[91] Xen. *Oec.* 7.1 (set in the late 5th century BCE).
[92] *Oec.* 6.8–11.
[93] Kronenberg (2009) unpacks the agri-textual links from Xenophon to Rome.

In conclusion, we can turn briefly to Cicero. His revised version of his *Academica* (set in the mid-40s BCE) famously pays a tribute to Varro (still working on *De Lingua Latina*) that has significance in this context:

> tum ego, 'sunt,' inquam, 'ista Varro. nam nos in nostra urbe **peregrinantis errantis**que tamquam **hospites** tui libri quasi **domum deduxerunt**, ut possemus aliquando qui et ubi essemus **agnoscere**.'
>
> Then I commented: 'Yes, that's the case, Varro. For when we were in our very own city yet still **wandering** and **straying** as if **strangers**, it was your books, so to speak, that **led us home**, so that we were at last able to **recognize** who and where we were.'[94]

Here Cicero, speaking in character and writing some time in 45 BCE, makes Varro's books agents of change. It may be Varro's (now only fragmentary) study *Antiquities* that Cicero has in mind, but his rhetoric in this subsequent work continues to carry the torch. Moreover, by identifying how Varro's antiquarian and scholarly turn of mind has direct force for the kinds of identity crisis that Cicero imagines, we can see a rationale for Varro's continuing interest. Citizens trying to orientate themselves within a rapidly changing urban and political landscape need a guide who knows the highways and byways in order to survive. Varro's interest in the relationship between what things mean and what they represent, and between theory and practice of language, makes him in Cicero's account into such a way-finder. His antiquarian scholarship and wide-ranging interests, combined with his public success in finding his feet repeatedly with Caesar despite his close ties with Pompey, made him a voice for a coterie whose ability to move through the city in a manner that reflected their self-conceptualization had suffered. Far more than Cicero himself, Varro the thinking man's action hero uses memory, politics, military service, and scholarship in an unparalleled way—as suggested here by Cicero, he undertakes to write out a guide leading his fellow citizens back to being Roman.

---

[94] Cic. *Acad.* 1.9.

2

Literature and the Spatial Turn

Movement and Space in Martial's *Epigrams*

Ray Laurence

The city of Rome is often introduced to students via Juvenal's third *Satire*, which provides an iconic view of a dystopian metropolis which became relevant for the twentieth century's discussion of the city in history.[1] However, now in the twenty-first century, there is increased interest in Martial's first twelve books of *Epigrams* as creating an image of Flavian Rome and the transition of the city from Domitian through Nerva to Trajan.[2] These are books of poems that have much in common with the postmodern approach to the city as text, which were ultimately embedded into Juvenal's third *Satire* in a longer less disjointed text.[3] Both Martial and Juvenal create an image of a city that is different from all others and perhaps best represented through fragmentation and juxtaposition.[4] However, as we shall see via an investigation of spatial signifiers of place within Rome, by which I mean named monuments, houses, and parts of the city of Rome, it is almost impossible to pin down the vision of the city in Martial that moves from one position to another. Choosing to investigate the *Epigrams* via its spatial signifiers allows us to step beyond

---

[1] Laurence (1997); Larmour (2007).
[2] Rimmell (2008); Fitzgerald (2007: 177–86) focuses on the parade of urban inhabitants and their variation. Nisbet (2003) for epigram under Nero. Compare Miller (2007) on the presentation of Rome in satire. See also Roman (2010), which appeared after this chapter was written.
[3] Colton (1991: 85–144) for analysis of relation of content of Martial to that of Juvenal. However, in terms of genre epigram is not congruent with satire and we need to bear in mind Miller's (2007: 139–40) argument that movement associated with 'desire' in satire is 'essentially sadistic; the desire of the reader is led from one scene of grotesque degradation or uncrowning to the next by an enchainment of scenes of debasement': a structure that is only in part located within Martial's *Epigrams*, which sit between satire and the 'carnivalesque' of elegy, and can juxtapose elements found in both.
[4] Fitzgerald (2007: 4–7, 106–38) on the role of juxtaposition in the *Epigrams* and the difficulty of reading social contradictions.

the well-worn paths of historical research that focus on patronage, *amicitia*, gift exchange, urban–rural antithesis, and the interpretation of Martial's relationship to Domitian.[5] Underpinning the discussion in this paper is a perspective drawn from postmodern geography that places a stress on both the representation of the spatial and temporal contexts or what has been characterized as the 'spatial turn' within the social sciences in the 1990s.[6]

The poems in the first twelve books of the *Epigrams* written by Martial exist in the context of Domitian's Rome between 86 and 101 CE, and hence are located in a clear dated context—the rebuilding of the city after its destruction by fire in 80 CE.[7] Yet, Martial had lived in the city since 64 CE and experienced the rebuilding of the city after the fire in that year.[8] Martial's experiencing and reading Rome in other texts is also incorporated into his *Epigrams* from the 80s and 90s CE, a process by which texts from earlier periods intrude from the city's past under Augustus.[9] Martial, like Juvenal, writes both himself and the city of Rome into his text and through their chosen genres these authors 'construct themselves as subjects in a particular social and historical context'.[10] This creates a realism that is not found in others, but as Barbara Gold stresses this is not necessarily believable: the *Epigrams* are not and cannot be used as a straightforward source for 'social history', but can instead be investigated to reveal the role of the city in Martial's 'self-presentation and constructed subjectivity'.[11] This urban subjectivity sits in a defined temporal context that was itself subject to change as was the author's viewpoint of it: to seek

---

[5] Spisak (2007) provides an overview of these themes and how they can be conceptualized in the 21st century, but perhaps is undermined by Fitzgerald's (2007: 107-38) observations on contradictory juxtapositions as a literary device for the construction of the social world embedded in the *Epigrams*. Interestingly, the use of place is not defined by Fitzgerald (2007: 68-105) as a characteristic of a book of epigrams, but recognizes the importance of them (2007: 185-90), linking these to the earlier works of Ovid and Catullus, rather than the Flavian city. However, the use of earlier authors does not preclude a contemporary Flavian urban landscape—see Geyssen (1999) on Mart. 1.70 and Ovid.

[6] I do not re-rehearse the development of this methodology for understanding the Roman city and its representations here, but refer readers to earlier publications: Laurence (1994a, 1997a, 1997b, 2007) and also to these publications for relevant bibliography.

[7] Swann (1994: 10-31) on the need to contextualize poetry to a specific Rome, rather a generic timeless city of Rome. See Fearnley (2003) on the shift in nature of Martial's work under Nerva and Trajan. Darwall-Smith (1996: 101-252) provides the fullest account of the building programme under Domitian.

[8] Mart. 5.7.

[9] Swann (1994) sets out the place of Catullus in Martial's Rome. Rimmell (2008: 10) on contexts of Augustan and Domitianic Romes in poetry. Sullivan (1991: 153).

[10] Gold (2003: 595). See Pailler (1982) for a brief treatment of urban space in Martial.

[11] Gold (2003: 596-7). See papers in Morello and Gibson (2003) for readings of Pliny's *Epistles*.

consistency from Book 1 circulated in 86 CE to other books written even a couple of years later might be a mistake.[12]

Embedded within the *Epigrams* is a vision of the city and the sequencing of movement through urban space in Rome. It should be noted from the outset that these are journeys that appear real, but are as much journeys made in the realm of the imagination made vivid by reading Martial's works.[13] This causes the position of the author to resist definition, seen most clearly in the representation of his home, initially in Book 1, in a *cenaculum* overlooking the Porticus Vipsania (see below).[14] At no point in the first two books does he actually say where he lives apart from this view and a location *ad Pirum* ('at the pear tree').[15] It is a location (*regione*) that he says he has inhabited for a long time into his old age—he would have been in his late forties. Most modern authors place his apartment on the Quirinal, but that seemingly definite location replaces a view and a place name in the text.[16] His location resists such spatial definition, unlike that of his books for sale in the Argiletum at specific *tabernae* located with reference to their position in the city or places within the city. The latter are the focus of this chapter, and I leave aside the representations of the urban population, discussed for example by Victoria Rimmell, that are not located with respect to specific spatial locales within the city.[17]

Movement is fundamental to Martial and his works, which he sees travelling from Rome to the limits of the Empire.[18] However, there is another journey in his works—the crossing of Rome, a unique urban experience, not achieved in other smaller cities, for the obvious reason that within fifteen minutes of walking in Pompeii, or Bilbilis (Martial's home town in Spain), you will have left the city. Rome, in contrast, was a metropolis and journeys of several miles were made to visit friends and patrons.[19] This feature made the urban experience of Rome quite distinctive from that of elsewhere, and is a feature worthy of consideration in its own right. There are no shortage of references to this phenomenon, but to date these are seen to give realism to the

---

[12] Coleman (1998: 339) suggests that Book 1 was established in 86 CE, Book 3 in 87, Book 4 in 88, Book 5 in 89, Book 7 in 92, Books 8 and 9 at some point between Dec. 92 and Dec. 95 (probably Book 8 in 93), Book 10 (reissued) in 104, Book 11 in Dec. 96, and Book 12 in perhaps 101. However, as both she and Henriksén (1998: 21-2) point out there is a marked increase in reference to Domitian in Books 8 and 9, which is matched by a marked decrease in the number of spatial signifiers appearing in these two books of the *Epigrams*. In short, these books are quite different to those that have gone before.
[13] Rimmell (2008: 181).
[14] Mart. 1.108, see Rodríguez-Almeida (2003: 19).
[15] Mart. 1.117.
[16] Howell (1980: 329-33, 348-52); Rimmell (2008: 35).
[17] Rimmell (2008) provides an admirable guide to the *turba* in Rome found in Martial's works.
[18] Rimmell (2008: 181-9).
[19] Gold (2003).

Epigrams and to provide options for the praise of the emperor, yet at the same time there is a recognition that these are not haphazard listings but are a key to the 'architectural semiotics' of Domitianic Rome that was understood in antiquity as a phase of fundamental transformation of the city.[20] What is examined in this paper is the representation of the city of Rome in Martial's *Epigrams*; the illusion of realism via an internal structure that is underpinned by both a spatial and a temporal dynamic, both associated with movement.[21]

LOCATING MARTIAL IN BOOKS 1 TO 3: DISCOVERING DOMITIAN'S AUGUSTAN ROME (86–87 CE)

Martial locates himself in Rome, first in Book 1, in a *cenaculum* (not a 'garret' by any means) overlooking the laurels of Vipsania—an apartment in which he has lived for many years.[22] It is not a *domus*, but is considerably better in terms of accommodation than the *cella* or single room that others inhabit.[23] Later, in 1.117, he informs the reader that he lives *ad Pirum*, up three flights of steep stairs. This is taken by many to be on the Quirinal Hill overlooking the Porticus Vipsania and other monuments on the Campus Martius.[24] Also within sight, perhaps, is the Aqua Virgo adjacent to the Porticus Vipsania.[25] It should perhaps be remembered that the Porticus Vipsania contained Agrippa's map of the world, and Martial's view of the portico may be an allusion to him looking onto the whole world, or as Pliny puts it: 'the world as spectacle for the city'.[26] After all, Martial's books of *Epigrams* are seen by him to go out to the whole world and include the whole world within them.[27] His *cenaculum* is a place from which to gaze upon the city.[28]

The locale of the *Epigrams* expands from this *cenaculum* across urban space to the Campus Martius. At 2.14, a Selius is seen hurrying around the public

---

[20] Sullivan (1991: 147–8). Social semiotics are also deployed via the inclusion of nearly 800 names of persons across the books, see Vallat (2008) for discussion. On the transformation of the city under Domitian in 4th-century writers, see Anderson (1983).

[21] For reading of space-time dynamics from literature, see Laurence (2007: 154–66); Riggsby (2003).

[22] Mart. 1.108; Hermansen (1982: 17–53) on meaning of *cenaculum*; DeLaine (2004) for a full spatial investigation of these types of properties in Ostia Antica.

[23] In e.g. Mart. 3.25.

[24] Howell (1980: 348–50).

[25] Mart. 4.18.

[26] Plin. *HN* 3.17, 'orbis terrarum urbi spectandus'. Boyle (2003a: 37); Evans (2003); Connors (2000: 227–8) for exploration of the theme and Sellius' perambulation of the Campus Martius in Mart. 2.14.

[27] Rimmell (2008) for analysis of the texts for this aspect.

[28] Perhaps similar to Patrick Geddes's Outlook Tower in Edwardian Edinburgh, see Meller (1990) for details, compare to Fredrick (2003) on the gaze and the city of Rome.

spaces of the region in search of a dinner invitation beginning at the Porticus Europae, then onto the Saepta, then the Temple of Isis, then the Porticus Argonautarum, onto the Hecatostylon and the Horti Pompeiani.[29] It is a logical route around a series of public monuments that can also be found, in which Martial, located at Forum Cornelii, imagines the writer Canius Rufus idling in the Schola Poetarum, stepping into a porticus of a temple (perhaps of Isis), idly strolling in the Porticus Argonautarum and in the afternoon in the Porticus Europae in the sun amid the box trees.[30] Both Selius and Canius Rufus are expected by Martial to bathe after these activities with the former bathing again and again, desperate for an invitation to dine, in the four main *balnea* of Rome, those of Fortunatus, Faustus, Gryllus, and Lupus and the three *thermae*; whereas the latter is faced with a choice of which of *thermae* to use: Titus', Agrippa's, or Tigellinus' (= Nero's).[31] These monuments associated by Martial with his colleagues create a locale in the southern Campus Martius and an association with idle leisure, prior to bathing and then logically onto dinner, if an invitation were forthcoming. The sequence is confirmed at 3.44.[32] The only journey out of the region takes the reader to the *thermae* of Titus adjacent to the Colosseum, which Martial prefers to those of Agrippa.[33] This is a locale of *otium*: defined temporally as lying in the part of the day that is after the midday, and defined spatially in the *campus*, the *porticus*, shade, water from the Aqua Virgo, and the *thermae*.[34]

The locations mentioned in the *Epigrams* map onto the region of Rome most seriously affected by the fire of 80 CE, that Dio suggests destroyed the following monuments: the temples of Serapis and of Isis, the Saepta, the Temple of Neptune, the *thermae* of Agrippa, the Pantheon, the Diribitorium, the Theatre of Balbus, the *scaena* of the Theatre of Pompey, the Porticus Octaviae, and the Temple of Jupiter Optimus Maximus on the Capitol.[35] Not all buildings on this list appear in Martial, but clearly a number were functioning within ten years of this major fire. Martial's locale is a rebuilt landscape of monuments originally constructed in the Campus Martius by Agrippa and his relatives, but all readers would realize that it was Domitian who had

---

[29] Prior (1996) for further discussion of Mart. 2.14. Also Rodríguez-Almeida (2003: 45–63) for discussion of the topography of this and other fictional progress through the monuments of the Campus Martius.

[30] e.g. Mart. 3.20. On Martial's location at Forum Cornelii, see Mart. 3.4. Connors (2000: 227–8) suggests the route of Selius has a parallel in myth, but the itinerary of Canius disrupts this interpretation. Notice also some convergence in the itinerary with Ov. *Ars Am.* 1.61–71, but after this Ovid takes the reader on a tour of other parts of Rome (Miller 2007: 152–5). See Macaulay Lewis, in this volume, on walking in porticoes.

[31] See also Mart. 5.70 with Rodríguez-Almeida (1989).

[32] Compare Mart. 3.68; 4.8.

[33] Mart. 3.36.

[34] Mart. 4.8, 5.20.

[35] Cass. Dio 66.24; Darwall-Smith (1996: 96–7, 239–40) for discussion.

provided for their restoration and that these monuments attributed to Agrippa and his relatives were, in reality, Domitian's. This factor causes the subjective realism of Martial's locale of home and *otium* within Rome to become a subtle means of praising the emperor for the restoration of this region of Rome, noted also by John Geyssen in connection with the topographical journey of Martial's book from the Argiletum to the Palatine in 1.70.[36] The realism of the passages in the *Epigrams* has caused modern scholars to overlook this area of Rome as a key area of Flavian rebuilding.[37] Unlike the sources that post-date the *damnatio memoriae* of Domitian, there is no mention of that emperor's name in connection with the buildings. Hence, the restoration of this region of Rome cannot be fitted into what David Fredrick sees as measurement by an 'Augustan yardstick' of 'good' Vespasian and 'bad' Domitian.[38] In the context of the reign of Domitian (in Martial's *Epigrams*), Domitian acted in the manner of Augustus in the *Res Gestae* and did not place his own name on the buildings he had restored. A quite different image of Domitian is that found in Suetonius' text, in which we find the buildings restored 'under his *titulus* and without any *memoria* of the first builder'.[39] Within the *Epigrams*, the region of the southern Campus Martius contrasts with the Palatine, where there were many statues of Domitian on the Clivus Palatinus and a sense of being close to the gaze of the current emperor.[40] Martial creates a distinction here between the two locales that is lost in the literature on Domitian's building activities after *damnatio memoriae*. Those texts extend the pattern of the Palatine to the whole city: creating not an Augustus restoring buildings in the name of the original builder, but an anti-Augustus (a Domitian) filling the city with his statues and triumphal arches.[41]

LOCATING MARTIAL'S BOOKS: DEFINING MOBILITY AS A COMMODITY

In contrast to Martial's own locale of the southern Campus Martius found in Books 1 and 2, his books themselves are located in the *tabernae* of the Argiletum: a location that places them adjacent to the public buildings of

---

[36] Geyssen (1999) for discussion.
[37] Darwall-Smith (1996) devotes little space to it; whereas Packer (2003) places emphasis solely on new projects of Domitian.
[38] Fredrick (2003: 201–2); Saller (2000) on writing of biographies of Domitian.
[39] Suet. *Dom.* 5, 'sed omnia sub titulo tantum suo ac sine ulla pristini auctoris memoria'; Fredrick (2003: 222–3).
[40] Mart. 1.70. For comparison, see Thomas (2004) on role of statues of Domitian in the articulation of urban space in the Forum Romanum.
[41] Suet. Dom.13; Cass. Dio 67.8. Fredrick (2003) for full analysis of post-*damnatio memoriae* literature, see also Saller (2000).

the 'Forum of Pallas' (or Forum Nervae), and the Templum Pacis.[42] The Argiletum in Martial is (loosely) defined as a place that is between here and the *fauces* of the Subura, a locale where the tortured remains of bodies hang and at which sits a female barber or *tonstrix*.[43] It is a place to which people go or pass through and can conveniently purchase a book of epigrams.[44] The Argiletum is created via epigram as a place that is associated with monuments, but lies outside the monumental spaces in close proximity to a very different region—the Subura. It is from here that his books of epigrams travel to the houses of the wealthy on the Palatine passing first the Temple of Castor, then the Temple of Vesta, past the House of the Vestals and on up the Sacra Via to the Clivus Palatinus, passing the palace of Domitian and the temples of Bacchus and Cybele to their final destination the *atrium* and *penates* of a house.[45] It is, of course, a literary journey that can be found in the works of Ovid in exile.[46] Yet, variants can be found in Martial with destinations in specific houses of named individuals and their associated household gods specified variously as *lares* or *penates*.[47] It is from the bookshops of the Argiletum that Martial's poems travel across Rome and become integrated into the sounds of the city as they are read or sung.[48]

Like his books, Martial does not just reside in his preferred locale close to his *cenaculum* near to the southern Campus Martius. From here, he has to travel to the houses of the wealthy. His locale is separated from that of these men. Lupercus lives a long way off.[49] A distance of 2 miles separates Martial from Decianus, or a four-mile round trip, and he was not at home.[50] Attendance at the *salutatio* is a key activity that creates a situation of movement across the city at the beginning of the day.[51] As seen in Martial, there is a commodification of gift exchange between patrons and clients in his city of Rome under the Flavians, in which the presence of the client via a journey at a

---

[42] Mart. 1.117. Boyle (2003b: 33) places the shops of the Argiletum in the Forum Transitorium, but this is an over-interpretation of the lines 'Libertum docti Lucensis quaere Secundum, limina post Pacis Palladiumque forum' (1.2.7–8). Compare Anderson (1984: 119–29); Rodríguez-Almeida (1988a with plan); Roman (2001: 126–9).
[43] Mart. 2.17.
[44] Mart. 1.117.
[45] Mart. 1.70. The route is discussed by Coarelli (1983: 40–1), critiqued by Ziółkowski (1989: 229–31) and Geyssen (1999: 727 n. 20).
[46] Ov. *Pont.* 4.5; *Tr.* 3.1. Howell (1980: 265–71); Sullivan (1991: 150–1); Geyssen (1999); Hinds (2007) on the literary relationship between Martial and Ovid. However, Roman (2001: 126–9) sees a connection with Hor. *Epist.* 1.20.
[47] Mart. 3.5, also 3.31; see below for further discussion of *lares* and *penates*.
[48] Mart. 5.16, compare 6.60, 9.97. Fitzgerald (2007: 139–66) discusses the role of epigram in society at Rome.
[49] Mart. 1.117.
[50] Mart. 2.5.
[51] Mart. 4.78, 4.8; note also that a book might replace its author at the *salutatio*, e.g. Mart. 1.108.

patron's house becomes a commodity.[52] The clients, in the *Epigrams*, are often seen to have been purchased to attend on their patron with the twist that the patron often had borrowed the money to make the purchase of them, his splendid toga, his litter, and his long-haired slaves.[53] The exchange between services (*opera togata*) provided by Martial via his attendance could be withheld or denied on numerous grounds, but equally access to a patron could be denied to Martial, as can be found in numerous epigrams. For example, after an arduous journey uphill to a house on the Esquiline via the Clivus Suburanus, Martial arrives at the house wet and covered in mud to find the *ianitor* announcing that the patron is not at home.[54] Movement, in itself, becomes a commodity measured in miles, and associated with a specific time of day: the first two hours of the day were Martial's time of the *salutatio*; the third and fifth hours place client and patron in the forum and associated with legal cases; whereas the period after this is associated with rest, exercise, dinner, and reading his poems.[55] The journey of patron and client leads from the house of the patron at the *salutatio* to the location of legal cases and rhetoric in the three fora of the city, the Forum Romanum, Forum of Caesar, and Forum of Augustus.[56] The first six hours of the day are associated with these two locales and what Martial defines as *negotium*; whereas activities in *Epigram* 4.8 associated with the period after the sixth hour were seen as *otium* and placed back into the locale of the southern Campus Martius discussed above. However, this does not imply that *opera togata* or the duties of the toga clad client did not spill over into the space-time of *otium*.[57] Processions of patrons with toga-clad clients are located by Martial in the Saepta and onto the *thermae* of Agrippa right up to the tenth hour of the day, but the normative pattern for bathing would seem to be from the eighth hour.[58] The Campus Martius with its monuments was also a place for the exchange of services and the emblematic display of this commodity with individuals seeking dinner invitations in the monuments associated with *otium* and the emperor's rebuilt Augustan Rome.[59] This patterning of movement located in Martial's first three books of the *Epigrams* converges with Henri Lefebvre's concept of 'rhythmanalysis' of crowds in Mediterranean cities. The rhythm of the crowd in the city is hard to

---

[52] Gold (2003) for recent discussion of patronage and gift exchange in Martial; see also Rimmell (2008). Spisak (2007) provides a lengthy discussion of the underpinning of ancient societies via gift exchange.
[53] e.g. Mart. 2.57, 2.74.
[54] Mart. 3.46, 5.22. For possible topography of this route defined in 5.22, see Palmer 1976: 5 (see also Mart. 7.73.4).
[55] Mart. 2.5, 4.8, also 8.67.
[56] Mart. 2.64, 3.38, 5.20, 7.28, 7.65; Rodríguez-Almeida (2003: 19).
[57] *Otium* is defined at Mart. 5.20.
[58] Mart. 2.57, 3.36, 10.48.
[59] Mart. 2.11, 2.14; Prior (1996).

grasp in an objective manner, but through its representation by Martial in epigram it can be revealed as an underlying structure of the city of Rome.[60]

LOCATING JULIUS MARTIALIS AND MARTIAL'S SUBURBS (88 CE)

Martial's gaze across Rome changes perspective in Book 4 of the *Epigrams*, by imagining his place in the city through the eyes of Julius Martialis, a friend who was at least 60 years of age to whom he 'sends' his books, and has known since about 67 CE.[61] From this villa he can not just gaze but measure Rome.[62] He occupies a *domus* in the Via Tecta from which he, like Martial in his *cenaculum* overlooking the Porticus Vipsania, can engage in *otium* within the Campus Martius—monuments mentioned include the Campus, the porticus, shade, the Aqua Virgo, and the *thermae*.[63] However, in Book 4 Martial states that his friend now inhabits a few *iugera* and a villa on the Janiculum, that is so close to the city that he suggests it is more a *domus* (urban house) than a *rus* (country place).[64] His view takes in all seven of Rome's hills and *tota Roma* and can look beyond in the distance to the Alban Hills and Tusculum and looking in the other direction is the *sub urbe* (suburb) of Fidenae, and Rubrae and the grove of Anna Perena to the north of the city. He can see the travellers on the Via Flaminia and Via Salaria, as well as the boats and barges at the Mulvian bridge. The villa is both closer to Rome than those found at Praeneste, Tibur, or Setia and on a smaller scale than these villas. He can look and see all of Rome and its suburbs, and those coming to the city. This view of the adjacent city is also revealed to Julius Martialis in his library in this rural property through reading Martial's poems.[65]

The sixth book of the *Epigrams* is dedicated to Julius Martialis and is sent to him in its opening poem.[66] In this book, Martial announces that he has purchased a *praedia rustica* (rural property).[67] This shifts the subject matter

---

[60] Lefebvre (1996: 228–40; 2004). David Newsome kindly drew my attention to the convergence of Martial with rhythmanalysis.
[61] Mart. 1.15, 12.34; Howell (1995: 99–101).
[62] Mart. 4.64; Vout (2007) for discussion.
[63] Mart. 3.5, 5.20.
[64] Mart. 4.64. Liverani (2005: 83) suggests, in line with earlier views, that the expression *longum Ianiculi iugum* places the *horti* and villa not on the modern Gianicolo, but in the area of Monte Mario as part of range of hills ('la lunga catena del Gianicolo'), a view followed by Soldevila (2006: 436–7).
[65] Mart. 7.17, see Vioque (2002: 136–43).
[66] Mart. 6.1.
[67] Mart. 6.5.

of a number of epigrams in the book from the city to the countryside. The property at Ficiliae (Ficulea) complements his urban property, close to the Temple of Flora.[68] This is not just a shift from the city to the countryside, but a shift from the city and long journeys to Baiae to a shorter journey from Rome to quiet in Nomentum and his small house and few *iugera* of fields.[69] This refers back to the previous poem, in which it is stated that the *thermulae* of Etruscus in Rome were unrivalled, not even by the baths of Baiae, and excuses the need to travel to Campania.[70] In previous books, Baiae had made a regular appearance, but from this point onwards the pleasure resort disappears from Martial's view that had previously included life in Rome and journeys to this resort on the bay of Naples.[71] Martial is at pains to stress that his *rus* depends on the *urbs* for the supply of typical Italian produce: figs, goats, olives, and cabbages; whereas Regulus had produce sent to him from estates in Umbria, in Etruria, at Tusculum, and at his *rus* at the third milestone.[72] Martial's is a property in need of repair and Lucius Arruntius Stella provided the tiles to re-roof the villa.[73] However, by 105 CE, fifteen years on from his purchase of the rural property at Nomentum, guests come at the eighth hour to visit for a dinner to which the *horti* contribute and there is a *vilica* to tend it.[74] Yet, his orchards produce poor fruit and gifts of apples continue to be purchased in the Subura.[75] Even, if hunting was successful, Martial's cook insisted on cooking a boar with the additional luxuries of Falernian wine and pepper.[76] It would seem though that the ownership of this property and mules to draw his carriage from the city marked Martial's status and his success.[77] These were things others envied. It is seen as a small property and complements his *domus* in the *urbs* with its emblematic *lares* and *penates*, but without piped water from the nearby Aqua Marcia—a move up from the *cenaculum* in Book 1.[78]

---

[68] Mart. 6.27. Mari (2004) on location and relationship to Nomentum and Via Ficulensis—the old name for the Via Nomentana (Livy 3.52). Quilici and Quilici Gigli (1993: 25–9) on sources and *passim* on topography. Howell (1995: 145–6) sees the purchase occurring in Mart. 1.105 unconvincingly and perhaps also in 5.62 when Martial says he has purchased unlocated *horti*.

[69] Mart. 6.43.
[70] Mart. 6.42.
[71] Mart. 3.20, 4.57.
[72] Mart. 7.31.
[73] Mart. 7.36.
[74] Mart. 10.48, also at 7.49 and the produce of eggs and fruit from his *suburbani horti*.
[75] Mart. 10.94.
[76] Mart. 7.27.
[77] Mart. 8.61, 9.97. On status and travel in the city and from the city see Poehler, in this volume.
[78] Mart. 9.18.

DOMITIAN'S NEW ROME (92–95 CE)

There is a shift in content of the *Epigrams* in Books 8 and 9, where the city and its sequences found in earlier books vanishes. Martial is located in a *domus* in the city and in a *rus sub urbe* outside the city.[79] What increases in these two books, written in the mid-90s, are references to the emperor. Domitian takes centre stage.[80] After all, Book 8 opens with a dedication to the emperor and closes with a statement of adoration of the *princeps*.[81] Key to understanding the book is the sense of the emperor's return from campaign that is directly alluded to via the monuments of a temple of Fortuna Redux and a triumphal arch, as an entry point into the city from the north.[82] Yet, Martial does rather more than fawn to his *princeps*. He sets out a sense of urban renewal, by which the wonders (*miracula*) of the past understood as the age of their grandparents are restored (such as the monuments on the Campus Martius so central to the early books of the *Epigrams*, whilst at the same time the new (*nova*) monuments are constructed.[83] This statement is a prelude to what follows in Book 9, a focus on the new and restored temples set up by Domitian in Rome that are listed.[84] These monumental emperor-focused epigrams are startlingly juxtaposed with misogynist epigrams, discussing Galla or Paula, prior to returning to the subject of the emperor as conqueror of the Rhine, and so on.[85]

Subsequent epigrams conjure up themes of meals and gifts, but the sense of the spatial is absent. Martial would seem to reserve the spatial in Book 9 for the emperor's monuments: we find the Temple of the Gens Flavia on the Quirinal, on the site of the *domus* in which Domitian had been born and a new architectural form—part Mausoleum and part temple.[86] Later, the new temple to Hercules at the eighth milestone on the Via Appia provides Martial with the means to compare Domitian with the mythical hero Hercules, and to recount the deeds of his emperor from saving Rome from Vitellius in 69 CE down to his more recent exploits against the Dacians.[87] Significantly, the god Hercules in this new temple had the facial features of the *princeps*. Even though the spatial aspect of Martial's work is centred on the emperor, there is still some space for other aspects of the city: Galla seated in the middle of the Subura, or the wealth of goods on sale in the Saepta, or the movement of a client following

---

[79] Mart. 8.61, 9.18, 9.91.
[80] Coleman (1998); Henriksén (1998).
[81] Mart. 8.82.
[82] Mart. 8.65; Darwall-Smith (1996: 130–3).
[83] Mart. 8.80.
[84] Mart. 9.3.
[85] Mart. 9.4, 9.5, 9.6, 9.8.
[86] Mart. 9.20, 9.34; Darwall-Smith (1996: 159–65).
[87] Mart. 9.64, 9.65, 9.101.

a patron.[88] These are images that cause a separation between the emperor's monuments and the life of the inhabitant, but tend towards the generic, rather than having a specific spatial signifier or any implications of the movement associated with them.

NERVA'S AND TRAJAN'S NEW ROME (POST-96 CE)

The next books of *Epigrams* to be published first in Nerva's (Book 11) reign and then in Trajan's (Book 10) move away from the subject matter of Books 8 and 9, with the emperor and his new buildings looming over the lives of the urban population. Instead, although Nerva is present in Book 11 with numerous references in the opening epigrams to a new regime, his presence is similar to that of Domitian in Martial's earlier works—there but not dominant.[89] We also find in the opening epigram instructions to this new book to go to a sequence of places that Martial had inhabited in an earlier book: the Porticus of the Temple of Quirinus, the Porticus Europae, the Porticus Pompeii, and the Porticus Argonautarum.[90] All are locations close to his *domus* on the Quirinal and his hours of *otium* and poetic leisure in the Campus Martius. Interestingly, whereas before his new book was sent to a shop in the Argiletum; it now circulates in the public spaces of the southern Campus Martius. The sequencing of structures from this region appears in 11.47: *balnea* (baths), the shade of the Porticus Pompeii, the entrance to the Temple of Isis, the plunge pool of the Aqua Virgo in the *thermae* of Agrippa. There is a step back here from the content of Books 8 and 9 with new temples built or rebuilt by Domitian to a different locale that is intrinsically Augustan (or Agrippan) in origin that had been restored in the earlier part of Domitian's reign. An interesting appearance in the sequencing is the word *balnea* rather than *thermae*, which might be utilized to refer to the smaller luxurious baths associated with individual proprietors, from which Martial might direct his guests for dinner that was close to his house on the Quirinal.[91] It is as though the new freedom associated with Nerva causes the social world of epigram to revert to Martial and some friends, rather than the focus under Domitian with the emperor taking centre stage rebuilding the temples of Rome.[92] As a return to normality or establishment of an ideal form of government under Nerva, Martial can link himself inter-textually to Ovid and Catullus most clearly in

---

[88] Mart. 9.37, 9.59, 9.100.
[89] Coleman (2000: 33–9) on shifts in emphasis in Martial after 96 CE.
[90] Mart. 11.1, compare earlier at 2.17.
[91] Mart. 11.52.
[92] Rimmell (2008: 165–6).

this book and to establish a pedigree for his poetry to what he might have regarded as his grandfather's age and that of Augustus.[93]

The revised version of Book 10 of the *Epigrams*, issued after Book 11, features a similar return to earlier matters.[94] This book of poems travels, presumably from the Argiletum, across the Subura and up the Clivus Suburanus passing the large fountain of Orpheus to the house of eloquence revealed by Pliny to be his (it was Pliny who would fund Martial's journey from Rome to Bilbilis and a life away from Rome located in *Epigrams* 12).[95] The book also marks the return of the client greeting his patron at dawn and walking in the mud as he follows his patron's sedan chair through the city.[96] This might be seen as a response to the passing of a law freeing a client or freedman from giving evidence against a patron.[97] Again, as a temporal and spatial phenomenon, normality has returned with clients moving round the city before dawn to the *salutatio*, following the patron, bathing at the eighth hour and not being invited by the patron to dinner. Clients are free to interact and know intimate secrets of their patrons, but are also denied any rewards for their duties at dawn and through the day. There is one difference though, now Rome has four rather than three forums—presumably here the fourth is the Forum Transitorium.[98]

MARTIAL AND THE SPATIAL TURN: MAPPING SPACE AND TIME

Rimmell has recently argued that Martial's books of *Epigrams* are not just about the representation of the city of Rome, but *create* the city and, for her, provide 'a crash course in living in Rome'.[99] However, what seems unclear is: for whom is Martial's Rome written? He locates himself being in Rome and his books as being sung in Rome, but read right across the Roman world to the very fringes of the empire. Examples of migrants coming to Rome are found in the *Epigrams*: Tuccius comes from Spain, but once he hears reports of the reality of the client's *sportula* (dole) turns back at the Mulvian Bridge without entering the city. Another, Sextus, plans to perform in legal cases in the *fora*.[100] His books are dedicated to members of the elite and duly sent to their houses

---

[93] Hinds (2007); Rimmell (2008: 165–80).
[94] Mart. 10.2 for revisions to the book; see Coleman (2000: 36).
[95] Mart. 10.19; Plin. *Ep.* 3.21; see Merli (2006) on Martial's departure from Rome.
[96] Mart. 10.10, 10.18, 10.56, 10.58, 10.70.
[97] Mart. 10.34; Plin. *Pan.* 42.
[98] Mart. 10.28; Rodríguez-Almeida (2003: 19).
[99] Rimmell (2008: 9).
[100] Mart. 3.14, 3.38.

whether in the city or in the suburbs. Both migrants and the elite are situated in the books of *Epigrams* as outsiders, who have little experience of Martial's situation as a client in Rome—walking through the wet streets at dawn to a *salutatio*. The small size of the books of *Epigrams* allow for their portability and ability to cross-space, with the content of the books creating the specificity of the city of Rome with a very economical use of specific monuments or the naming of specific places within the *urbs*.

Much has been written about the relationship of Martial to the emperor Domitian.[101] Although Martial recognized the new emperor Nerva as heralding a new age of *libertas*, he had previously described the rule of Domitian as the age of *libertas* at its greatest.[102] The two statements under different rulers lead to the conclusion that the use of such phraseology is emblematic of the rule of any emperor. It is perhaps too easy to see in Martial's Rome the oppression of Suetonius' *Domitian* the tyrant. A foil to this viewpoint comes from the monuments included and omitted from the *Epigrams*. On opening the books circulated under Domitian, we expect to find the building projects of that emperor placed at the very centre of the Rome that is established by Martial. This does not occur. The monuments of Rome are enumerated: two theatres—Marcellus' and Pompey's but not Balbus'; three *thermae*—Agrippa's, Nero's/Tigellinus', and Titus'; three fora with the occasional mention of a fourth (completed under Nerva), and four *balneae* as well as the small *thermae* (*thermulae*) of Etruscus.[103] What is absent is any mention of the circuses or Domitian's arenas on the Campus Martius—the Odeon and the Stadium. As we have seen, the Campus Martius of the *Epigrams* is an Augustan landscape of porticoes, the Aqua Virgo, the Saepta, and the Temple of Isis. The only place in the *Epigrams* in which Domitian's building projects appear, apart from in Books 8 and 9, is in connection with a discussion of the critics of the new temple of Jupiter on the Capitol.[104] The only new Domitianic buildings to be mentioned are the temples found in Books 8 and 9. This causes the writing of the city to focus both on the monumental and also to provide a gaze onto the juxtaposition of these temples set up by a devoted emperor to the gods with, for example, the aged prostitutes *in media subura* (in the middle of the Subura).[105]

The regions of Rome are delineated through the city's geography, by the names of the hills and also by some low lying regions of the city (the Subura

---

[101] Howell (2009: 63–72) points out how the relationship and flattery shapes our modern reaction to the *Epigrams* and its author; Coleman (2000) explores the difference in the *Epigrams* with the replacement of Domitian with Nerva and Trajan; Spisak (2007: 61–7) sees the relationship within a framework of *amicitia*.
[102] Mart. 9.2, 5.19.
[103] Mart. 6.42.
[104] Mart. 5.10, see Howell (1995: 86–8) for discussion.
[105] Mart. 9.37.

and Transtiberim). The former in the *Epigrams* are places of destination leading to the houses of Martial's social superiors; whereas the latter are the location for his inferiors: prostitutes, hawkers, sellers of fruit and vegetables.[106] His own position is that of a client in-between what Tacitus would describe as the *plebs sordida* and the elite.[107] His books reside also in this position of in-between, in the shops of the Argiletum between the Subura and the entrances to the Temple of Peace, the Forum of Pallas (Transitorium), or the Forum of Julius Caesar from whence the book can move to the houses of the wealthy on the hills of the city. Intriguingly, when Martial has left Rome and lives in Bilbilis, he sends his book of *Epigrams* to libraries on the Palatine and to the consul Stella, whose house stood in the Subura.[108] However, Martial like Juvenal in *Epigrams* 12.18 moves around the Subura and across the city to the entrances (*limina*) of the houses owned by the wealthy on the hills of the city. The spatial in all of these cases is rather generic and made specific through Martial as the author/subject joining together these two urban worlds, whether as a client walking through the mud behind a litter-borne patron, or in purchasing fruit and vegetables in the Subura that will be a gift for a friend.

The measurement and counting of things is a feature of the books of *Epigrams*. There are twelve books, including monuments: two theatres, three *thermae* (imperial baths), four famous *balneae* (bathhouses), first three and then later four fora.[109] The enumeration of the city includes the presentation of time. A client's duty to their various patrons is a form of measured time that lasts from the first hour to the tenth.[110] The first two hours begin with the *salutatio*, and are followed by three hours attending speeches located in the fora, to be followed by rest, prior to bathing at the eighth hour prior to dinner at the ninth and it is the tenth hour that is the time for Martial's poems—probably recitation.[111] This positioning of the client as a dependant moving across the city, not as a free citizen but as a poorly treated friend, causes space and time to intersect, in ways that are familiar and are essential for understanding the spatial turn in geography—the movement of a client becomes a commodity for purchase by a patron.[112] However, what is being presented to the reader of the *Epigrams* living in the *urbs* or *sub-urbs*? The distribution of archaeological finds of sundials in the ancient world points to the universality of dividing the day into hours, but what might be distinctive in the *Epigrams* are the activities associated with temporal sequencing; emblematic statements

---

[106] See Holleran, in this volume, for discussion of marginality and mobility of these occupations. See also Howell (1980: 193).
[107] Tac. *Hist.* 1.4.
[108] Mart.12.3.
[109] Rimmell (2008: 94–139).
[110] Mart. 1.108, 3.36, 4.8, 7.51, 8.67, 10.48, 10.70, 11.52.
[111] Mart. 4.8, 7.51.
[112] Laurence (2007: 182–91) for implications for reading of Pompeii.

about the life of the client in Rome, as significant as his threadbare toga or the food he is given by his patron. Yet, there is a sense of urban transformation within the time-geography specifically of Rome—at the eighth hour, when the shifts of the praetorian guard change, the Temple of Isis shuts and *thermae* of Nero have cooled to a perfect temperature.[113] People generally go to bathe in preparation for dinner at this hour in the *Epigrams*—perhaps a not dissimilar temporal activity of those readers outside Rome, but Martial always names the baths attended and evokes the specificity of Rome rather than cities in general. Equally important in terms of time-geography is the tenth hour, since this is when all duties are performed and, coincidentally (but not included in the *Epigrams*), the hour at which vehicles are allowed back into the city.[114] There are two hours of the day left that are empty of duties, of bathing, and of dining; there is an unstated contrast with life outside Rome away from not just the duties associated with patrons, but also the noise of the city beginning before dawn, something else that the patron was stated as being shielded from.[115] Martial inhabits a time-geography that is situated in Rome, in which the use of time and the mention of specific bathing establishments evokes a life spent in Rome—that of the client who wishes to write, but cannot do so due to the time commitments demanded by a patron or number of patrons.[116] It allows for the representation of a life in which the *opera* (work) of a client prevents him from acting as a free citizen and denies him the opportunity to write additional poems.

Within the format of the *Epigrams*, there is frequent reference to a spatial texture of the city. Martial's urban residence, in contrast to his small property at Nomentum, a *rus*, is presented as a *domus* (house), but can also be presented as his urban *lares* (referring to the household gods), or represented by the term *penates* (the gods of the store cupboard).[117] The words associated with household deities often replace the word *domus* in the *Epigrams*. However, the *lares* are not seen to be associated with the present owner of the house. A rich man's property portfolio (*praedia*) could include many *lares*, and Martial's 'humble' friend, Julius Martialis, inhabits the *lares* of Daphnis.[118] There is a contrast with the use of the word *penates*, which Martial associates with the wealth or well-being of a household and associated with the owner.[119]

---

[113] Mart. 10.48.
[114] *Tabula Heracleensis* 56–61; Crawford (1996: 365) for text and 374 for translation; Laurence (forthcoming).
[115] Mart. 12.57, 2.38 for escape to Nomentum from those he does not wish to see.
[116] This paragraph owes much to Andrew Riggsby, whose paper (2003) on space-time in Pliny's *Epistles* has somewhat adjusted my view of the representation of this societal aspect in Martial's *Epigrams* as set out with reference to Pompeii, Laurence (2007: 154–65).
[117] Mart. 9.18.
[118] Mart. 3.5, for other references to owning *lares* see 3.31, 10.58, 11.82.
[119] Mart. 1.70, 7.27, 8.75.

As could the word *atrium*, usually in the plural form *atria*.[120] Returning to the *lares* of Daphnis, this is given as the place where Julius Martialis resided and is located by Martial on the Via Tecta. This is as close as we have to an address to which a book of *Epigrams* is sent. Julius Martialis, unlike other 'recipients' of Martial's books, is not a member of the wealthy elite and hence does not own a house that was an imposing palace or well known as a landmark associated with a famous man.[121] This description of the position of Julius Martialis' house can be understood more fully in connection with the fragments of marble maps of Rome that are annotated with names associated with each property division.[122] As we saw in the case of Julius Martialis, his household *lares* are not named after him but another person (Daphnis) and form the equivalent to the name of the house. In consequence, it is logical that the names on the annotated fragments of the maps of ancient Rome are unlikely to have referred to the current owners, but to the names of the *lares* associated with each property.[123] This system of nomenclature would also avoid the confusion of the rich man owning many properties in Rome, instead he owns many *lares* that are differentiated. Before closing this discussion of spatial terms found in Martial, mention needs to be made of the term *limen* or entrance; this is a word associated with structures from the most humble (a prostitutes' *cella* with its inscription to identify the trade inside) to the houses of the wealthy and the public monuments of the city (the Temple of Mars Ultor or the Temple of Peace).[124] Implicit in its deployment in the *Epigrams* is a sense of position before and the possibility of movement into the structure or the denial of access.

The terminology discussed in the previous paragraph allows for the relationship between space as represented in the text of the *Epigrams* to be related to space as represented in the surviving fragments of maps incised into marble that we possess from ancient Rome. These maps divide space with attention to three elements: property boundaries, entranceways to those properties, and the streets. We have seen in the discussion above that the first two of these features can be located in Martial's *Epigrams* and are associated with the deployment of the words *lares* and *limen*. I would suggest we also can locate the third spatial concept of the ancient maps of Rome—the streets.

---

[120] Mart. 4.40.
[121] Compare Mart. 1.70 with 3.5.
[122] Wallace-Hadrill (2008: 301–12) for concise and precise discussion.
[123] This is also the case in the opening of Plautus' play the *Aulularia*, in which the *lar* is associated with the house from its point of construction. The association of the name of the builder of a house and the *lares* may have obviated the need to constantly update the map of Rome, so that the ownership of property is reflected. However, it needs to be stated that this would apply to private rather than imperial property (denoted as *Caesaris*) or public property on the fragments of the *Forma Urbis Romae*.
[124] Mart. 11.45, 1.2, 7.51, 12.18; see Larmour (2007: 181) for use of *limen* in Juvenal.

Epigrams 7.61 refers to the disruption of the relationship between these three spatial elements, fundamental to the cartographic representation of Rome—the *popinae* (bars) and *tabernae* (shops) had encroached from their entrances and boundaries to cause the streets (*vici*) to be reduced to narrow muddy footpaths (*semitae*). This disruption of the map or city space (*totam urbem* in the text) had been regulated by the emperor Domitian to ensure that the barber, wine-seller, cook, and butcher were kept within the limits of their property and that the *vici* (streets) had expanded into *viae* (thoroughfares). Domitian had restored order to the spatial division of Rome or, in other words, had placed human activity back to where it should occur on the maps of Rome.

Very few street names appear in Martial (a feature also of the Severan Marble Plan); when they do, they are either within Martial's locale of his house—notably the Via Tecta, or are symbolic of the centre of Rome—the Sacra Via, or the street associated via its name with the patricians and the greatest *Romanitas*—the Vicus Patricius.[125] There is one final spatial expression that needs commenting on: that of the crossroads (*compita*) that was used by the Flavian writer Pliny the Elder as a means to define the city of Rome with its 265 *compita larum*.[126] Martial imagines Book 7 of the *Epigrams* being read throughout the city, and gives the locations as at dinners, in the forum, in houses (*aedes*), in the porticoes, in the shops, and at the crossroads (*compita*).[127] It is not streets but their intersection that creates position, something that suggests that these marked space to a greater degree than the *vici* (meaning both streets and neighbourhoods), with which *compita* were associated and with which magistrates had an association; inscribing their names upon the compital altar.[128] Echoing Pliny, Martial creates movement through the city as movement through the *compita* of the city.[129] These crossroads had become adorned with altars and could be associated with fountains that created the crossroad as a landmark within the city, a point at which a person or in Martial's case his books, could go the right or the wrong way in their travel to their destination in the city.[130] The *compita* had become the landmarks that informed a mental map for travel through Rome, which is a distinctive element found in the *Epigrams* as Martial or his books move through space. However, this is a conception of urban space that is derived

---

[125] Mart. 8.75, 12.2, 7.73, 10.68.
[126] Plin. *HN* 3.66.
[127] Mart. 7.97.
[128] Mart. 6.64, 10.79 for appearance of *compita* and a *magister vici* in definition of social difference between leading men (*proceres*) and other inhabitants of Rome. Wallace-Hadrill (2008: 259–312) on the relationship between the spatial organization of the *vici* and social control of the city.
[129] Mart. 6.64 'lanius per compita portat'.
[130] Mart. 10.19.

from maps. This is made clear when Martial looks onto the city in his own *Epigrams*. Whether, in Book 1 with a self-conscious stare from an apartment (*cenaculum*) at the location of the map of Rome in the Porticus Vipsania,[131] or a view of Rome from the Janiculum, which allows Martial to measure the city—meaning perhaps to create a mental map of landmarks.[132] These views inform the combination of spatial elements: movement *through* the city, the delineation of space, and the location of the occasional landmark re-create an image of life *in* Rome. In so doing, the spatial authenticates the text of the *Epigrams* (perhaps not Books 8 and 9) as being in Rome without resort to listing monuments, and locates Martial as living there and as a person who could represent the city in text. Fundamental for that representation is a sense of movement through space.

[131] Plin. *HN* 3.17.   [132] Mart. 4.64.11–12; Vout (2007) for discussion.

3

Measuring Spatial Visibility, Adjacency, Permeability and Degrees of Street Life in Pompeii

Akkelies van Nes

At present, research on the urban environment by means of space syntax theory and similar methodologies tends to focus on macro-scale spatial conditions—the study of large-scale urban networks. However, micro-scale analysis—at the level of connection between buildings and streets—should not be neglected. Buildings are spatially connected to streets through their entrances. It is the place where private life intersects with public life. Pompeii presents an ideal case study through which it is possible to analyse this private–public relationship. With this in mind, spatial methods were developed and tested to analyse the topological relationship between private and public space. Issues such as various degrees of intervisibility between windows and doors, density of entrances, the topological depth between private and public space, and degree of *constitutedness* were taken into account. These forms of analysis offer insights into the nature of spatial conditions that underpin a number of different issues including the vitality of street life, urban safety, social interactions, and the interdependence of these issues of urban life. Fundamental for our understanding of spaces within cities is that all would seem to depend on degrees of adjacency, permeability, and intervisibility that can be analysed at a variety of different levels in terms of scale from micro- to macro-. This chapter outlines some of the ways in which spatial analyses can contribute towards an understanding of the relationships between movement and the levels of relative interaction within a street network. It helps provide further context for space syntax analyses, a detailed application of which can be seen in Stöger's chapter, in this volume.[1]

---

[1] For a previous examination of Pompeii's macro-scale using space syntax methods, see Fridell-Anter and Weilguni (2003).

When applying methods of spatial analysis at excavated sites, we can form frameworks around which to interpret our data, based on empirical knowledge from a present urban context. This is necessary for establishing the link between spatial properties and its impact on street life in a built environment. Only then is it possible to derive from the spatial analyses degrees of street life, poverty, and social control at excavated sites. In so doing, this paper looks at the example of Pompeii in the light of recent developments of spatial analysis of this type undertaken in present-day environments. Pompeii is one of the best preserved towns from the Roman period and has a high level of surviving spatial information on which information spatial analyses may be undertaken.

METHODOLOGIES FOR MOVEMENT

Human beings are social. Even though they have their private spaces inside buildings, they also seek spaces to interact socialy or economically with others. Activities in society always take place in some form of physical space. How a society organizes its activities, privately as well as publicly, causes an impact on its built environment's spatial set-up. Conversely, a built environment's spatial structure affects how individuals behave in urban space in terms of the possibilities for social control, opportunity for economic activities, and social interaction. Such themes are evident throughout this volume.

The space syntax method, developed by Bill Hillier and his colleagues at University College London, is able to calculate how a street relates to all others in a town or city. The recent versions of the Depthmap software are able to calculate *topological distance* (how integrated a street is in relationship to all others in terms of the number of direction changes), *geometrical distance* (how integrated a street is in relationship to all others in terms of the angular relationship between them), and *metrical distance*. Moreover, the software is able to both simulate and trace movement routes of computer-generated agents within the built environment. This form of agent-based modelling is based on research observations of the modern city.[2] Researchers have applied the space syntax method in a variety of urban contexts. Hence, a substantial database of modern space syntax exists from which some general conclusions on the relationship between space and society for a *present* urban context can be made. These general conclusions provide us with some generalizations from which we can develop interpretations of the spatial analyses of excavated towns from the Roman world. For example, as we will see, Steven J. R. Ellis's discussion of probable ambulatory movement in Pompeii—based on the

[2] Turner (2004).

location and orientation of shop counters—should correspond with space syntax measures for the busiest streets and those where economic activities should be seen to cluster.[3] Similarly, one can plot the overlap between the streets which space syntax analyses indicates should have been the most integrated and most heavily used, with measures of wheel ruts that indicate actual usage in antiquity.[4] In the absence of such data from Pompeii or elsewhere in the Roman world, we might usefully turn to modern contexts for an understanding of spatial behaviour (e.g. principles of the 'movement economy', discussed elsewhere in this volume). This is not to deny the importance of localized contexts but to emphasize the universal properties of space, movement, and spatial relationships. Of course, while movement in space is a constant of human behaviour, it is culturally dependent, as other chapters in this volume discuss with regard to the Roman city.

These methods of analysis need to be briefly commented upon prior to evaluating the built environment of Pompeii with these techniques. Fig. 3.1 shows how global integration is calculated for a hypothetical small settlement. As the analyses show, the main street is the most spatially integrated street, since compared to all the other streets, it is associated with the fewest direction changes by a person travelling down it in either direction. In contrast, the back street is the most spatially segregated street. Every time one changes direction from one street to another, one takes a *topological step*. One has to change direction many times in order to reach all other streets from the back street. Hence, the back street is topologically *deep*, whereas the main street is topologically *shallow* and associated with far fewer *topological steps* to reach all the other streets.

This form of analysis of a simple settlement can easily be calculated on paper by hand. However, when analysing larger settlements, computer software has recently been developed to analyse how each street relates to all others in settlements with a much larger number of streets and capable of dealing with more than a hundred thousand street segments. The software for spatial analysis uses colour codes for the various integration values. In this way, one can see at one glance the spatial structure of settlements. The black colour shows the most integrated streets, while the light grey colour shows the most segregated streets. Since the space syntax method analyses spatial relationships quantitatively, the results can be correlated with other statistical data in the modern city: pedestrian flow rates, crime dispersal, land use values, and so on. What modern studies of this nature have found is that there is a correlation between the degree of a street network's spatial integration and measures of the degree of street life and the location of economic activities. In excavated towns, we cannot recover a similar range of data, but the results from the spatial analyses can be correlated with urban functions that have

---

[3] Ellis (2004).   [4] Tsujimura (1991); Poehler (2006).

Measuring Spatial Visibility in Pompeii

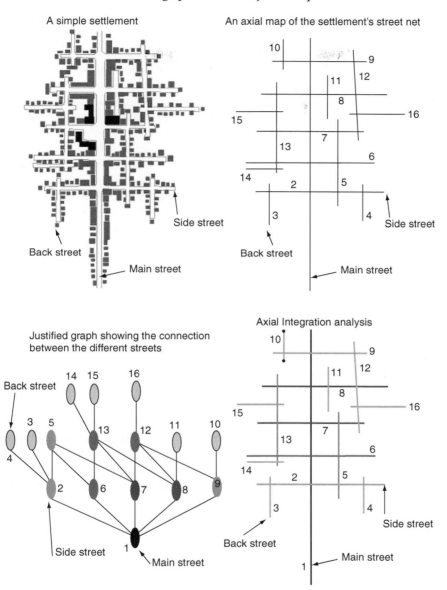

Fig. 3.1. How global integration is calculated (image by Akkelies van Nes).

been derived from archaeological data. While we lack the kind of data which helps to inform modern space syntax studies, we can nevertheless use the results of analyses as frameworks around which we can correlate other spatial data, for example: the depth of wheel ruts (comparing the map of 'traffic intensity' based on wheel ruts to a map of the most integrated streets according to a space syntax analysis), the distribution of public amenities such as streetside benches throughout the town, or the location of particular types of buildings and land uses.[5]

As research has shown at a macro-level, there are correlations between a built environment's spatial layout of its street and road network, and the location of economic activities, crime dispersal, land use along streets, and variation in property values.[6] At present, at a micro-level, few research projects have taken into account the spatial relationship between buildings and streets. These micro-spatial conditions affect the intensity of street usage, and thus also the safety of individuals in urban areas. Crime gravitates away from areas in which there is intensive usage of street frontages. This is an important factor and reflects the degree of social interaction of streets. In short, the more entrances that are directly connected to streets, the higher the degree of social interaction between inhabitants and visitors; whereas in streets with less interaction, the visitor and inhabitant are more vulnerable to crime.[7]

SPACE AND POMPEII

These observations inform my approach to the private–public relationships in Pompeii. Pompeii is suitable for this form of analysis, since there are sufficient numbers of street segments with a full mapping of the entrances and perimeter walls of properties along the streets of the city. In addition, as noted, there are numerous studies which have concerned themselves with spatial data and which can be correlated with the results of such analyses. Pompeii's inhabitants had their houses connected to public streets in different ways. Since the town was suddenly buried through an unexpected eruption of Vesuvius in 79 CE, its spatial layout is preserved. This makes Pompeii an ideal site for the use of space syntax analyses, and the interpretation of the results arising from the spatial analyses on its socially and economically related street life. This means that the spatial analysis at both a micro- and macro-level can be robust and may be correlated with the functions assigned to land use across the city by archaeologists.

When applying space syntax analyses to Pompeii or other excavated sites, it is important to decide from which time period the spatial analyses should be taken. Often, changes to the street network and buildings through history are

---

[5] On wheel ruts and the relative traffic intensity within the Roman city, see the comments by Newsome in the Introduction to this volume. On street-side benches at Pompeii, see Hartnett (2008). Stöger, in this volume, demonstrates the potential of linking space syntax analyses of street networks to the distribution of particular activities throughout the city, in this case the location of guild buildings (*scholae*).
[6] Hillier et al. (1998). [7] López and van Nes (2007).

layered upon each other. David J. Newsome's detailed survey of the changes to the street network around the Casa del Marinaio is an important example, and one in which archaeological and chronological change is tied to multi-phase space syntax analyses.[8] This enables us to measure the impact of spatial change on the movement network around the property, and goes some way to mitigating the fact that space syntax is a diachronic method: by using it over reconstructed phases, we can put chronology back into our analyses. Eric Poehler's survey of wheel ruts also points to a city in the continual process of change, whether that be repaving (and some streets closed for such projects), or the reversal of major traffic arteries because of other urban priorities.[9] Some layers might be lost while others are highly visible.

There exist several maps of Pompeii's street network. Although only two-thirds of the town's surface is excavated, it is possible to reasonably reconstruct its entire street network. This is by no means without caveats. Prior to the last decade, we may have assumed that we could simply 'extend' the street network of Ostia without many problems. However, geophysics outside the excavated areas shows that the street network does not do anything like what we would expect a Roman grid to do. Hanna Stöger describes this research in her chapter, and uses the geophysical survey to provide a larger 'control' sample against which her analyses of the excavated areas can be compared. Back in Pompeii, Hans Eschebach and Liselotte Eschebach mapped every excavated building and drew their walls and openings carefully on maps.[10] Through the items found in buildings, they also attributed functions or land use to the buildings.[11] These are recognized as bakeries, public baths, temples, taverns, wool workshops, smiths, inns, drinking places, brothels, and so on. Shops, however, are difficult to identify since items found inside buildings could be used for private use as well as for exchange. In contrast to this approach, Ray Laurence used the length of the streets in metres and divided it by the number of doorways.[12] According to him, it indicates a high number of comings and goings on these streets through those doors which is a condition for micro-scale economic activities. These two data sets can be combined. The spatial variables can be related to the socio-economic or land-use variables.

The analyses of spatial integration of Pompeii's street network indicate the location of the most 'vital' streets, or those associated with the greatest integration with the rest of the street network. Fig. 3.2 shows a global integration analysis with the location pattern of shops. It shows how accessible each street is in relation to all others. The black colour shows the highest integrated streets, and the light-grey the most segregated ones. As research has shown in present-day cities, the most integrated streets have the highest flows of

---

[8] Newsome (2009a). [9] Poehler (2006).
[10] See H. Eschebach (1970); L. Eschebach (1993).
[11] Eschebach (1993). [12] Laurence (1994: 89).

Fig. 3.2. Global integration analyses of Pompeii with the location pattern of shops (image by Akkelies van Nes).

pedestrians, and shops locate themselves in the most integrated streets.[13] Steven Ellis has discussed similar land use at Pompeii, plotting the location of *tabernae* according to likely routes of pedestrian footfall.[14] Shops are sensitive to these flows and tend to be located in positions that optimize their retail potential by reaching a large number of customers. If this optimal location is altered through changes to the spatial structure of the street network, the location of shops will also change.[15] In Pompeii, the two main cross streets, Via Stabiana and Via di Nola, have the highest integration values, followed by the Via dell'Abbondanza. These high values result from their position and size within the orthogonal street grid.

The forum is typically considered to have been the most significant meeting place for public life in a Roman town. At the forum were located the most important urban public buildings with their containing functions, such as temples, judicial centre, basilica, and a *macellum*.[16] These urban functions

---

[13] Hillier et al. (1998); van Nes (2005).   [14] Ellis (2004).
[15] van Nes (2002). See Newsome (2009a) for details of changes at Pompeii, with references to the changes to the Insula of the Menander as a result of changing traffic patterns (Ling 2005).
[16] Laurence (1994: 20).

developed as attractors for human movement (cf. Newsome, in this volume), and were associated with economic activities as well. However, the forum does not have the highest spatial integration values in Pompeii.[17] It is located one topological step from the highly integrated Via di Nola. When analysing the oldest part of Pompeii (Reg. VII and VIII), the forum had the highest value of spatial integration. At some point prior to the development of the final format of Pompeii, it was thus its economic, political, and religious centre. At the time of its destruction in 79 CE, the economic activities had spread out to the most integrated main streets as an outcome of the expansion of the density of urbanism. Looking at this expansion, we can analyse the pattern further by a 'two-steps analysis', which defines the area of a city that can be reached when changing direction twice from any particular street. This allows for the definition of the expansion of urbanism into this area. In Pompeii, we find that nearly the entire area of the city is 'two-steps' or two directional changes away from the main thoroughfares of the city, which also are associated with the greatest density of side-streets. As seen above, it was on these integrated streets that shops were concentrated, due to their topological and metrical centrality (a pattern found also in modern cities).[18] The black shading in Fig. 3.2 shows the locations of shops. These correspond with the places in Pompeii with greatest global, as well as local, integration values from the macro-scale analyses. Shops were located in the places with the greatest spatially defined levels of integration at both the local and global level in Pompeii: in other words, in places with the highest potential for passing, moving, and trade.

A recently developed measurement in the space syntax method, the angular analysis, takes into account the joining of streets together, in particular the angle at which streets intersect, known as the angular relationship between streets in spatial analysis. Most main routes through and between urban areas consist of long streets ending up at another long street with an angle close to 90 degrees. As expected, the *decumanus* and *cardus* are highlighted more in the angular analyses than in the global integration analyses. The correlation between the location pattern of shops and the occurrence of graffiti (or messages) on the walls[19] tends to be in streets with a high local angular integration value. One can presume that these streets were more frequented than others.

In his book, *Roman Pompeii: Space and Society*, Laurence identified the location of the various local neighbourhoods' centres on the basis of the location of public water fountains and provision of high-quality drinking water (water fountains). As he argues, these fountains would have been used by people living in close proximity to them and they probably functioned as a natural contact point between neighbours.[20] These water fountains were

---

[17] See Newsome (2009a; 2010) on the centrality of the forum in Pompeii.
[18] van Nes (2005).
[19] Laurence (1994: 98).   [20] Laurence (1994: 43).

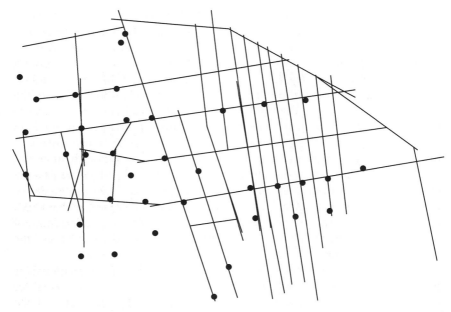

Fig. 3.3. Topological distances with a radius of 80 m with the location pattern of fountains (image by Akkelies van Nes).

nearly always located at a street junction. If the water quality was equally distributed around the town, one can presume that the locals chose the water fountain in closest proximity to their residence. These locations can be subjected to topological analyses to relate the shortest metrical radiuses of the integration values discussed above. What we find is that the water fountains were located in places with high local integration values and associated with short metrical distances. Fig. 3.3 shows the location of the various local centres for the various neighbourhoods from the spatial topological analyses with a radius of 80 m. We may also infer that there would be two or three more local centres in the unexcavated areas north of Via dell'Abbondanza.

MICRO-SCALE ANALYSES OF POMPEII

The focus of macro-scale analyses is how one street relates to all other streets in a city. In contrast, the focus of micro-scale analyses is the spatial relationships between spaces inside buildings and the streets. More precisely, it aims at defining the interrelationship of buildings or private spaces and adjacent street segments. The focus is on how dwellings relate to the street network, the way entrances to/from buildings constitute streets, the degree of topological depth from private space to public space, and intervisibility of doors and houses across streets.[21] As Jane Jacobs and Jan Gehl argue, many entrances and windows facing a street is one formula to ensure urban vitality.[22] The challenge is to *quantify* these spatial relationships. It is about measuring various degrees of urban active frontages or the relationship between buildings and streets. Only then will it be possible to gain a genuine understanding of the spatial conditions for vital street life and urban safety. There is another dimension to this. In a research project on space and crime in Alkmaar and Gouda in the Netherlands, an opportunity was provided to register various spatial relationships between private and public spaces and compare the results with numerical social and economic data in a modern context.[23] 1,168 street segments were analysed and correlated against, for example, rates of residential burglaries or thefts from vehicles. In short, there was a correlation between a street network's spatial structure, the degree of intervisibility of entrances, and the incidence of crime. To generalize, we can say that where integration values were high and the intervisibility of entrances was high, crime was low, and vice versa.

There are several ways of analysing spatial configurative relationships between building entrances and the street network. An easy way is to register the *topological depth* between private and public space. One counts the number of semi-private and semi-public spaces one has to walk through to get from a private space to adjacent public street. If an entrance is directly connected to a public street, it has no spaces between private and public space; the depth is equivalent to zero. If there is a small front garden between the entrance and the public street the depth value is one, since there is one space between the closed private space and the street. If the entrance is located on the side of the house and it has a front garden or it is covered behind hedges or fences, then the topological depth of the entrance has a value of two.

Fig. 3.4 (top) illustrates various types of relationship between private and public spaces. The black dots represent the private spaces, while the white dots represent semi-private spaces. The Casa di Octavius Quartio and the Casa della Venere in Conchiglia are typical Pompeii houses, directly connected to the highly integrated Via dell'Abbondanza. Both houses' main entrances are between two spaces where probably shops were located in the past. Even though most of Pompeii's homes housed several families and the garden/courtyard probably functioned as a place for social interaction, they all had one main entrance connected to a public street.

A street's degree of *constitutedness* depends on various degrees of adjacency and permeability from buildings to public space. When a building is directly accessible to a street, then it *constitutes* the street. Conversely, when all

---

[21] See Stöger, in this volume, on the guilds of Ostia.
[22] Jacobs (2000); Gehl (1996).
[23] López and van Nes (2007).

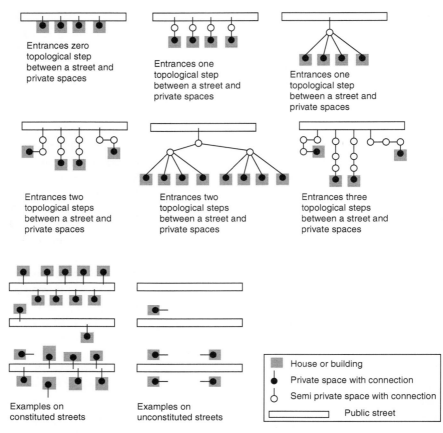

Fig. 3.4. Diagrammatic principles on the topological depth between private and public space (above) and the difference between constituted and unconstituted streets (below) (image by Akkelies van Nes).

buildings are adjacent to a street, but the entrances are not directly accessible, then the street is *unconstituted*.[24] A street segment is constituted when only one entrance is directly connected to the street. If the entrance is hidden behind high fences or hedges, has a large front garden, or is located on the side of the building, then the street is defined to be unconstituted. Fig. 3.4 (bottom) illustrates the differences between constituted and unconstituted streets. The degree of constitutedness is about the number of entrances connected to a street divided by the number of buildings located along that street.

As the results from the spatial analyses in Gouda and Alkmaar show, both micro- and macro-spatial variables are highly interdependent for

[24] Hillier and Hanson (1984: 94).

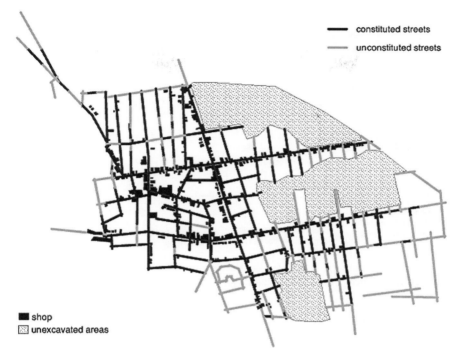

Fig. 3.5. Constituted and unconstituted streets in Pompeii (image by Akkelies van Nes).

describing an area's character—its urban vitality and the propensity for social interaction.[25] When a street segment is on a spatially integrated main route, or close to a main route, more buildings are directly connected to streets and the entrances on each side of the street are intervisible to one another. Conversely, the further away a street segment is from a main route, the more entrances to buildings are hidden away from streets and the degree of intervisibility between them is lower. These kinds of combination of macro- and micro-spatial conditions influence the degree of street life. This variable identifies the main routes through cities and shows strong correlations with the micro-scale variables.

Fig. 3.5 depicts the difference between constituted and unconstituted streets in Pompeii. Unconstituted street segments are marked with a grey colour, while the constituted ones are in black. All the shops are located along the constituted streets. The more entrances connected to a street, the higher the probability that someone comes out from a private space into public space. However, the high density of entrances connected to a street does not always imply high intervisibility. There is a distinction in the way entrances *constitute*

[25] López and van Nes (2007).

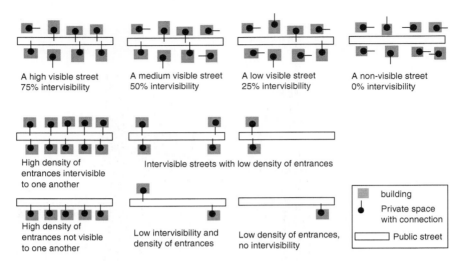

Fig. 3.6. Diagrammatic principles on the relationship between intervisibility and density of entrances (image by Akkelies van Nes).

streets and in the way they are *intervisible* to each other. The way entrances and windows are positioned to each other influences the probabilities for social control and street life.

The density of entrances and degree of intervisibility were registered separately. Thus, two buildings with two entrances facing towards each other indicate 100 percent intervisibility of doors. Conversely, a street segment with entrances on only one side of the street segment is defined to be 0 percent intervisible. In Pompeii the percentages of entrances' degree of intervisibility were grouped in 100 percent, 50 percent, and 0 percent intervisibility for each street segment. Fig. 3.6 shows some diagrammatic principles on the relationship between intervisibility and density of entrances.

The distinctiveness of this feature can be clarified further with reference to Fig. 3.7, which shows an example of a street with low intervisibility and low entrance density (left), and an example of a main street with high intervisibility and high density of entrances (right). The example on the left is a typical back street with only dwelling entrances (Via del Fuggiaschi), while the example on the right is Via dell'Abbondanza—a typical main shopping street in Pompeii.

Fig. 3.8 shows the density (above) and degree of intervisibility (below) of entrances in Pompeii. The most integrated streets have the highest intervisibility and density of entrances. When revealing the degree of constitutedness, it shows that Pompeii also had 'silent' side and back streets with no entrances connected to these public spaces. Examples on silent unconstituted streets are Vicolo del

Fig. 3.7. Example of a side street with low intervisibility and density of entrance (left) and a main street with high intervisibility and density of entrances (right) in Pompeii (photos by Akkelies van Nes with permission of the Soprintendenza Archeologica di Pompei).

Fuggiaschi and Via di Nocera, and examples on lively and constituted streets are Via dell'Abbondanza, Via Stabiana, Via Fortuna, and Via di Nola.

A combination of various micro- and macro-spatial measurements makes it possible to capture data that relate the spatial data to socio-economic circumstances and to provide an understanding of the spatial conditions for safe and vital urban areas. Through the use of Depthmap software, various macro-scale spatial variables can be calculated and visualized. For example, a street with few connections to other streets (macro-scale analyses) can still be full of social activities, provided a high density of entrances constitutes the street and when there is high visibility between public and private spaces (micro-scale analyses). The reverse can be seen in unconstituted streets with a low number of entrances and low intervisibility, but where the connections to other streets are high.

Drawing on a study of a modern urban context, we find that the further away a street segment is from the main network of routes, the greater the topological depth between private and public space. Along the main routes through urban areas, most entrances are directly connected to the street. When changing direction two times from the main routes, the average topological depth for entrances is two, while it is three in all street segments that are located more than six topological steps from the main routes. When comparing this data to that derived from Pompeii, we find numerous differences. Pompeii's orthogonal street grid is topologically shallow. By changing direction once from the main routes, which are defined as *cardus* and *decumanus*, most of the town's streets are covered. Research has shown that in a modern urban context, the more segregated a street segment is, the more monofunctional the adjacent buildings tend to be. Topological deeply located street segments usually only have a residential function, since offices, shops, and

Fig. 3.8. Entrance density (top) and the degree of intervisibility between entrances (bottom) (image by Akkelies van Nes).

public buildings tend to locate themselves along the main routes. The semi-private segments are among the topological deepest and segregated streets.[26] In Pompeii entrances to dwellings tend to dominate the side streets, whereas there is a large mixture of urban functions along the *cardus* and *decumanus*. Interestingly, the further a street segment is away from the main routes, the lower the values of spatial integration and constitutedness. The unconstituted back alleys tend to be the most segregated street segments. As the study of street segments in a present urban context clearly shows, the micro-spatial conditions of the street segment are interrelated to the macro-spatial conditions of the cities' street network.[27]

CONCLUSIONS: CONTRIBUTIONS?

How do the spatial analyses of Pompeii contribute to an understanding of the social and economic life in past built environments? The following can be said about space and society in Pompeii following such analyses:

a) Shops and bakeries locate themselves in the most integrated streets with a high number of connections to other streets in a short metrical distance. In addition, the entrances along these streets were directly connected to the street, have a high density, and a high degree of intervisibility.

b) Religious buildings (such as temples) and political institutions tend to be located one topological step (or one direction change) away from the most integrated streets. The Temple of Fortuna Augusta is perhaps an exception, since it is located on one of Pompeii's busiest junctions, of Via del Foro and Via della Fortuna. The temple's entrance is oriented towards Via del Foro, which is the second most integrated street. The entrances around these buildings have low density, but they still constitute the streets.

c) Brothels locate themselves in constituted side streets metrically close to the integrated main streets, implying that brothels were not located on the busiest streets with high pedestrian flow, but were close to streets with high levels of pedestrian flow and social activity.

d) In contrast, workshops, taverns, and drinking places are located along the main streets and in side streets topologically close to the main streets.

e) Conversely, the public baths, theatres, inns or hotels, and sport and leisure facilities are spread throughout the town's street network.

[26] López and van Nes (2007). [27] López and van Nes (2007).

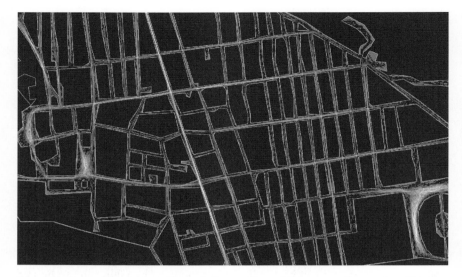

Fig. 3.9. Agent-based modelling of Pompeii's public spaces (image by Akkelies van Nes).

There are implications of this pattern for movement within Pompeii. The largest flow of pedestrians, pack animals, and equids or oxen-hauling vehicles was associated with the most integrated streets. Fig. 3.9 shows the traces of 1,500 people moving inside Pompeii's public spaces from the agent-based modelling in Depthmap. When comparing Fig. 3.9 with the spatial analyses, the highest locally and globally integrated streets (from the macro-scale analyses) combined with the highest density of entrances (from the micro-scale analyses) are the most frequented spaces in the city of Pompeii.

In general, it can be said that the way a society organizes its functions spatially and the way its spatial structure affects human behaviour, in terms of the location pattern of its activities, has not changed much in two thousand years. The same tendencies of space and human behaviour can be seen in urban centres in present cities. Micro- and macro-spatial relationships play a crucial role in the socio-economic life of human beings in built environments. All these activities depend on the spatial configuration of the street 'plinth', that is defined as the point or points at which buildings meets the street. It is at the street plinth where the micro-scale and macro-scale analyses are interconnected, and which determine the degree of liveliness in streets. The micro-spatial structure of urban street plinth affects the direct interface of public and private life of a built environment's inhabitants and visitors in an informal way. It has always been like this in built environments. Even though Pompeii's street network was planned according to ancient requirements with its strict orthogonal street grid with two main cross streets intersecting in the middle, the location process of buildings along streets occurred organically. The most integrated and crowded streets had the highest attraction for the location of an adjacent building's entrance. Hence, a street network's macro-spatial structure affects the degree of attraction to economic and social activities. In Pompeii, most entrances are packed together at the shortest sides of their insulae due to high spatial integration values on the streets on that side. These micro- and macro-spatial conditions made these locations optimal for the exchange of goods, shopping activities, social interaction, and the vitality of street life in Pompeii's main streets.

The application of space syntax and the micro-scale tools contributes to understanding how ancient cities functioned and the relationship between spatial layout and socio-economic activities. The spatial analyses of the street network give indications of possible functions in adjacent buildings where identifiable artefacts are lacking on archaeological sites. Finally the statistical data from the micro- and macro-scale spatial analyses and the agent-based modelling can give indications on the degree of vitality in shopping streets, and where the largest flow of human movement took place in the past. The results set out here in this chapter provide an additional dimension to recent work on the streets of Pompeii and set out a framework for comparison with more traditional (and less mathematical analyses) of the spatial aspect of the city, in which movement is asserted as meaningful. As noted, such analyses can be correlated with other archaeological studies of movement and land use, the better to understand the relationship between a street's position within the urban network and the variety of activities that took place there.

4

Towards a Multisensory Experience of Movement in the City of Rome*

Eleanor Betts

'Now imagine to yourself every kind of sound that can make one weary of one's ears'.[1]

Imagined journeys through the city of Rome have until now focused on the visual (monuments and buildings coming into and out of view) and the kinaesthetic (the physical movement of walking) senses.[2] None of these journeys have considered the relationship between vision and movement with the sounds, smells, or tastes someone making them would have experienced and navigated by, nor developed discussion of the kinaesthetic (or haptic) experience beyond that of walking (how would a blind person find their way around the ancient city?).[3] Nor have they considered the emotional impact of the stages of the journey: whether the person feels a sense of arriving home or leaving the familiar, or perhaps excitement and trepidation as they enter the city for the first time or venture to an unfamiliar part of it.[4] These accounts, despite noble attempts to bring the city to life, have allowed the visual to dominate and even where they claim to consider a range of senses, this 'has been filtered through a visualist framework'.[5] This visualist bias is also

---

* Thanks to Ray Laurence for inviting me to present an early version of this paper at the Classical Association conference (Birmingham 2007) and to the participants for their enthusiastic discussion; the Tavoliere-Gargano Prehistory Project funded by the British Academy, University College London, and the National University of Galway, Ireland.

[1] Sen. *Ep.* 56.1, 'Propone nunc tibi omnia genera vocum, quae in odium possunt aures adducere'. See also Ihde (2005: 61–6).

[2] Purcell (1987); Favro (1996).

[3] For a factitious account of sounds and smells in Pompeii cf. Hopkins (1999: 10–13).

[4] Chapters in this volume begin to address the second aspect. Cf. also Hopkins (1999: 10–11); Mayer (2007).

[5] Bull and Back (2005b: 1–2). This shortcoming does not belong only to the realm of Roman historians, but also to the humanities and social sciences more widely.

identifiable in the archaeological and written records of Rome: Augustan Rome was a highly visible and visualized city, particularly viewed through the eyes of the elite.[6]

So, it is perhaps unsurprising that there has been a move in recent years to develop a deeper understanding of how the populace negotiated their way around Rome, with much work focused on intervisibility, considering views in and across the city and its parts.[7] Digital modelling of the city has taken this further, enabling visual questions about specific parts of Rome to be answered, and more focused questions to be asked. There is certainly a place for these more detailed and rigorous visual analyses of the city, constructed around the archaeological evidence and against which the narratives of our sources, with all their biases and artistic devices, can be tested. The problem is not that this picture is, inevitably, fragmentary, but that it is a *picture* (albeit a three-dimensional one). Fortunately, there is some recognition that the multisensory experience of the city is more valuable than this sanitized visualization. Favro wrote recently that 'It is time to break the tyranny of sight and explore all experiential aspects of past cities', suggesting that 'Digital re-creation models facilitate examination of acoustics, climate and temperature, lighting and the sensation of time'.[8]

In wider circles, the endemic dependence on the visual has been recognized, voiced, and in some cases acted upon.[9] The *Sensory Formations* series, which takes each sense separately, presenting methodologies and case studies grounded in phenomenology, through its refusal to privilege one sense above the others and recognition of the interconnectedness of the senses, develops multisensory approaches for the social sciences and humanities.[10] Each book in the series begins to develop syntax for the individual senses, aiming to move away from visualist language.[11] Human geography's theoretical approaches to space and place have influenced research into the ancient city, and as new methodologies for the study of 'sensual culture' emerge, these will further enhance our understanding of the ancient world. Sensory experiments in the built environment can assist us in formulating sensory-oriented questions, such as, how far would particular voices have carried? How would these have helped someone navigate their way around the city? Questions of

---

[6] Plin. *HN* 11.139; Suet. *Aug.* 28.
[7] Favro (1996, 2006); Vout (2007).
[8] Favro (2006: 333).
[9] Ingold (2000: 251); Schmidt (2005: 43); Smith (1999); Mills (2005); Hamilton and Whitehouse (2006: 166).
[10] See e.g. Howes (2005), esp. for further bibliography on sensory research; Bull and Back (2005a); Drobnick (2006a).
[11] e.g., Schafer's 'soundscapes' (2005: 37) and 'clairaudience' (2005: 25), Drobnick's 'olfactocentrism' (2006b: 1), and Porteous' 'smellscape' (2006: 91–2); Bull and Back (2005b: 1) on 'scopic metaphors'.

inside/outside, private/public, female/male use of space interest classical scholars; a multisensory approach enables further exploration of these questions.[12]

Multisensory maps enable definition of locales within the city and movement between them identifiable by the distances and directions travelled by sounds, smells, and tastes plus changes in or continuity of texture and the intervisibility/invisibility of sights. Each sense has its own range and sphere of influence, so each creates its own map of the city. Locales within the city are defined by the senses as much as they are by the architecture and spaces that provide the (visual and haptic) structure. Senses function in very specific ways, each dynamically contributing to our perception and definition of space and place, the relationship between *'here*ness' and *'there*ness' establishing what is inner/private, outer/public, and where the boundary between these lies. The senses define individual and collective, cultural identities ('I', 'We', and 'Being-in-the-world') as well as how 'I' and 'We' define places and move around and between locales. Tracing locales and journeys in the city via isolated senses gives only one-fifth (or -sixth) of the map; to be multidimensional these maps need to be overlaid, creating a multisensory map of the city.[13]

This chapter will introduce how a framework for multisensory mapping of the ancient city may be developed from sensory data recovered from classical literature, epigraphy, and the archaeological record for ancient Rome, utilizing methodologies from phenomenological fieldwork (empirical visual, auditory, and olfactory data collection) and theoretical approaches from sensual research. Perhaps unsurprisingly, the senses are not treated equally by our sources; sounds and sights are privileged, but a range of smells, tastes, and haptic or kinaesthetic experiences can be inferred, such as the smell and taste of sausages and the wines of the *popinae*, or the luxurious feel of silks in the Vicus Tuscus.[14] This bias highlights the importance of using a range of specific

---

[12] Relph (1976); Bull and Back (2005b: 5). These dichotomies apply better to the study of locales within the city than movement across it, so are largely beyond the scope of this chapter. Current projects include the Tavoliere-Gargano Prehistory Project and the Applying Auditory Archaeology to Historic Landscape Characterisation Project (http://www.cardiff.ac.uk/hisar/people/sm/aa_hlc/index.html), whilst Smith's 'acoustical archaeologist' approach to his study of early modern England is more firmly based in documentary sources (1999, 2005). Smith's methodology, 'un-airing' the sounds of early modern England via documentary, architectural, and material cultural evidence may also be effectively applied to the multisensory mapping of ancient Rome.

[13] Bull and Back note that 'The reduction of knowledge to the visual has placed serious limitations on our ability to grasp the meanings attached to much social behaviour, be it contemporary, historical or comparative' (2005b: 2); also 'if we listen to it the landscape is not so much a static topography that can be mapped and drawn, as a fluid and changing surface that is transformed as it is enveloped by different sounds' (2005b: 11), which can be extrapolated to include olfactory and tactile maps. See also Schmidt (2005).

[14] Sen. *Ep.* 56.2; Mart. 1.41.9–10; 11.27. For this reason, methodologies for studying soundscapes and smellscapes are the main focus of this chapter (Bull and Back 2005a; Drobnick 2006a).

questions to interrogate the sources.[15] As an introduction to new approaches, this chapter is designed to prompt more questions than provide detailed answers.

THE PANOPLY OF SENSES IN ANCIENT ROME

A multisensory map of Rome is a topographical map of the experienced city, which combines the 'soft' (experiential) and 'hard' (empirical) data of phenomenology with the 'hard data' of known structures, urban form and fabric, within specific chronological periods.[16] It cannot map the entire city, as we do not have a complete topographic plan of Rome's built environment (in any period), but it can extend the plan of specific areas because it uses empirical evidence to map their spheres of influence, and it can produce a multidimensional map. It can also map elements of the city described in the literature, but for which the archaeological record is incomplete, such as the sounds, smells, tastes, and sensations experienced in the Subura. Smells, tastes, and sounds may enrich the character of a place, but sound and odour are no respecters of visually or kinaesthetically discrete places. Acknowledging this means that archaeologically visible places in the city can provide the starting point for analysis, but potentially extended or contracted via exploration of interaudibility/inaudibility and the distances and directions travelled by odours. In this way it is possible to build up a more complex and complete picture of the ways in which specific aspects of Rome's urban environment were experienced and perceived, as well as addressing (at least in part) difficulties arising from the multilayered and fragmentary nature of the city's topography.[17]

Multisensory maps do not merely orientate in space but also in time: different sounds, smells, tastes, and sensations occur according to the time of day or year. Just as digital models of Rome can mimic climatic conditions, a multisensory map can be drawn for a hot summer afternoon, when sounds of game-playing (voices, thrown dice, movement of counters), the stench of the Cloaca Maxima, the salt taste of sweat, and the hum of insects locate the wanderer turning a corner of the Basilica Iulia and heading into the Vicus Tuscus.[18] Time is a crucial factor in conditioning our experiences of and our responses to the city.

---

[15] See Smith (2005: 133); Mills (2005: 2–3).
[16] The tension between a desirable but impossible objective study of the past and the inevitable subjectivity of interpreting the past from a different cultural perspective is widely recognized (Brück 1998; Hopkins 1999: 2–3; Favro 2006: 332).
[17] Smellscapes will also be 'non-continuous, fragmentary in space and episodic in time', as will soundscapes (Porteous 2006: 91).
[18] Claridge (1998: fig. 112, 242); Dalby (2000: 213).

Streets and monuments are comparatively static markers which fix the visual and kinaesthetic map in time for a period. External to the person, they serve to guide a stranger around the city, with public monuments in particular pinpointing locations.[19] Odours work with the architecture, as well as against, to define both places and spaces in the city, the distinction being that smellscapes are more transient.[20] Each sensual element of the multisensory map has its own temporal rhythm. Porteous describes smellscapes as cyclical, recurring 'daily, weekly, seasonally or annually', also varying over the course of a day and night, stronger at dawn and dusk.[21] Tastescapes operate in a similar way, certain food and drink being seasonally available, or consumed at particular times of day. The visual and kinaesthetic maps created by the form and fabric of the city have the longest duration, but the details of these have their own, quicker, rhythm, such as the way light moves around the forum, and the very act of walking through the city. Soundscapes overlap with these static and evanescent landscapes, ambient sound punctuated by specific, directional noises, some of which describe the immediate locale (a conversation, a crying baby), others fixing a locale (street traders, the roar of a crowd) or an incident (a dog barks, someone shouts).

The most private sensory sphere is that of internal experience of sensations: taste is registered via the mouth, smell via the nose, heard sounds and the sounds an individual produces reverberate in the body, touch is the most intimate (the skin is the largest organ of the human body), sight the furthest removed from self, since 'Looking is centrifugal; it separates you from the world'.[22] Home is the next, an intimate locale, where familiar, habituated, smells and sounds in particular, but also the repeated experiences of other senses, interweave to create a sense of place. Smells are immediate, evocative, 'earthy and animalistic', linking biological constants with cultural variables and so intrinsic to the construction of place.[23] Smells and sounds, alongside the sights, tastes, and textures of the fabric of streets, architecture, and natural features, predominate in construction of locales within the city, the sensory sphere of social interaction. The most public sensory sphere, where all senses are in play, each one coming acutely into consciousness at different times and in different spaces, depending on the volume or unexpectedness of the sound, the rankness or intensity of the smell, the brightness and heat of the sunlight, the unevenness of the street, is the mapping of the city. This sphere is not the whole city, rather, the richest sensory tapestry of movement within the city.[24]

---

[19] Plaut. *Curc.* 462–86; Ter. *Ad.* 572–84; Ov. *Tr.* 3.1.21–34; Mart. 1.70. Cf. Ling (1990a and b); Moore (1991).
[20] Margolies (2006: 110).
[21] Porteous (2006: 99).
[22] Handel (1989: p. xi).
[23] Drobnick (2006b: 1).
[24] We know that Rome could not be perceived in its entirety (see Laurence, in this volume; Frischer et al. 2006; Vout 2007).

As these sensory spheres grow larger and more public, the senses become increasingly alert, although not necessarily all with the same force. In the most private spheres, the familiar is physiologically apprehended by the senses, but is not necessarily consciously experienced, particularly smell, a result of habituation.[25] When moving around the city, the panoply of senses is used, consciously and subconsciously, to navigate, creating a multisensory map of the city which changes according to the individual (gender, age, status) and period (time of day, season, year, government).

Different senses operate in different directions: with the body as the axis, the experiential field of sound, smell, and touch is spherical, whilst vision is restricted to a broadly forward-facing plane and taste has the most restricted field. Variety in the sensitivity, direction, and distances of perception by the different senses is what gives multisensory maps of the city a richness beyond that of visual and kinaesthetic maps. Just as a panorama or vast building occupies the entire field of vision, loud (unwanted) sounds become noise that assails the ears, deafening any inner dialogue, and pungent (unpleasant) smells overpower familiar, personal scents.[26] The suddenness with which these strong sensual stimuli come upon us affects our emotional response to the environment, but can also facilitate orientation and movement within the city. Being multidirectional, sounds and smells do not respect space and are able to permeate through and between locales, connecting them where vision may be obscured. They cross the boundary between private and public places, such as the smell of cooking meat wafting from windows and doors onto the street, or a dog barking from indoors at a passer-by.[27] By moving through space, sounds and smells can help create places; the orator speaking from the Rostra, whose voice is carried into the forum, or the smell of smoked cheese in the Velabrum.[28] The shouts of hawkers and shopkeepers may help the wanderer to orientate himself and make his journey in the city's streets. The light scent of oil and perfume emanating from a passer-by (as well as a clean appearance) hints at the direction of the nearest public baths.[29] Not only does interaudibility operate quite differently from intervisibility, incorporating smell, taste, touch, and sound into mapping social experience of the city extends places into spaces (the baths into the street) or contracts architecturally defined

---

[25] e.g., 'Hearing is a physiological constant; listening is a psychological variable' (Smith 1999: 7). On the 'habituation effect' and its relationship to the insider/outsider dichotomy, see Porteous (2006: 90). For an example of a Roman's habituation of sound, see Sen. *Ep.* 56.3–5.

[26] Sen. *Ep.* 56.1–2; Mart. 4.4.6–11, 6.93, 9.62.1–4, 12.57; Bull and Back (2005a: 1); Ihde (2005: 62).

[27] The smell of meat cooking in an enclosed outdoor area travels a minimum distance of 17 m and a maximum of 122 m (Hamilton and Whitehouse 2006: 178), indicating that cooking smells could easily have travelled from apartments and bars to the streets outside, or from temples to the *area* beyond (Suet. *Claud.* 33).

[28] Mart. 11.52.10, 13.32; Plin. *HN* 11.240–2.

[29] See Mart. 13.101; Juv. 8.159; Stat. *Silv.* 1.5.

places into more discrete locales (a conversation with a bookseller). Tastes, smells, and sounds are evocative and embedded with cultural as well as individually personal meanings, so to gain an understanding of those social experiences we must turn to the literary and epigraphic sources. Tracing the recorded distances and directions travelled by particular scents or sounds within specific locales of the city, where walls inhibit or streets conduct, but where other, stronger smells and sounds mask the one(s) being analysed, enables the facts and fictions in the vignettes painted by sources such as Juvenal, Martial, and Seneca to be considered.

THE SENSORY EXPERIENCE OF MOVEMENT THROUGH SPACE

Fuelled by a commendable desire to learn more about the lower echelons of Roman society and the city's backstreets, recent publications, with their 'thick descriptions', provide opportunities to take a multisensory approach to the interrogation of daily life in the city.[30] These works have succeeded in repopulating Rome and, more importantly, specific locales within the city, providing a foundation on which to build multisensory maps of the city at given periods of the Republic and Empire. For this foundation to be firm, it is essential to use the archaeological evidence for the city, alongside available digital models where appropriate, to identify the distances and directions that different sounds and smells would have travelled, the locales where particular tastes would have been experienced, the textures of the city, the visibility of certain coloured fabrics, or reflective materials, and so on. Tracing these details onto the physical remains of the ancient city connects locales via senses other than sight, which contributes to the mapping of movement around the city. Different architectural fabrics have different textures and sensations, so a haptic map of the movement from the Subura to the Forum Augustum in the first century CE traces a route from many-textured insulae of timber, concrete, and travertine doorframes to the rough *peperino* tufa and travertine of the boundary wall, culminating with the cool, smooth Luna marble of the Temple of Mars Ultor; this is movement from uneven ground, muddy and littered with debris, to smooth and slippery marble steps and floors.[31]

---

[30] See e.g. Gowers (1995); Dalby (2000); Larmour (2007); Holleran, in this volume.

[31] Mart. 7.61.6; *FUR* Stanford 10Aab, 10g, 11a,11c; Claridge (1998: 158). Favro hints at this when she describes the experience of walking from the 'warrenlike' Subura into the Forum Augustum, although she focuses on the visual experience of the fabrics (1996: 175–6).

Methodologies have already been developed for recording soundscapes and visual landscapes, though the other senses are as yet poorly represented.[32] Fully recording sensual experiences involves practice and also an ability to detach from the over-stimulation of the modern world.[33] For example, in the Roman period, the loudest sound heard would have been a thunderclap (120 decibels), with very few sounds reaching an intensity above that of a human shout (75 decibels at 1 m distance). The most constant sound would have been crowd noise.[34] Measurements of cities in the early twenty-first century show that constant, average, traffic noise reaches 80 decibels.[35]

There are several ways by which to establish a phenomenological framework for Rome. The standing remains and *Forma Urbis Romae* provide a topographical map, albeit incomplete, whilst further archaeological evidence (particularly reliefs) and descriptions of specific places in the city in the literature add further topographic information, even if they cannot be pinpointed exactly on the ground. As noted, literature, epigraphy, and sculpture provide sensory information, some direct (the sound of sawing or a carriage clattering by) and some indirect (the extension and contraction of muscles as weights are lifted, the sensation of water on skin in the pool).[36]

To what extent would the smells, tastes, sounds, and sensations associated with particular localities have contributed to wayfinding in the city? Certain dominant smellscapes and soundscapes, such as the stench of tanneries across the Tiber, would become landmarks by which to locate and navigate, pinpointing districts rather than fixed points.[37]

Daily life is not concerned only with the mechanics of mapping and wayfinding in the city, nor with purely utilitarian experiences of the city. Experiencing Rome, like experiencing any other environment, would have elicited emotional responses, dependent upon who you were and what you were doing within the city, along with a whole set of cultural conditions.[38] The daily routine of visiting a local bakery would provoke a minimal emotional response, unless it was 57 BCE and the grain shortage was continuing, the bakery had been closed for weeks, but this morning you smelt freshly baked bread again, its aroma intensified in the early morning air, and rushed there by the quickest route.[39] Clearly, it is impossible to know exactly how a Roman

---

[32] Smith (1999); Bull and Back (2005b: 4); Mills (2005); Hamilton and Whitehouse (2006). On the poor representation of smellscapes cf. Drobnick (2006b: 3).

[33] Bull and Back (2005b: 6–7); Schafer (2005).

[34] Stat. *Silv.* 4.5.49, 'Est et frementi vox hilaris foro.'

[35] Smith (1999: 49–50). Fortunately, the technology to screen out modern elements of the soundscape, such as traffic noise, exists, enabling valid audibility surveys to be carried out on the site of ancient Rome. See Mills (2005: 26–9).

[36] Seneca (*Ep.* 56.1–5) describes 17 different sounds which, when extrapolated, also give information about smell, taste, and touch.

[37] Mart. 6.93.4; Juv. 14.202–4. Cf. Margolies (2006: 107–10).

[38] See Favro (1996: 227–8).   [39] Cic. *Dom.* 11; Dio 39.9.2; Porteous (2006: 98–9).

citizen was feeling one September morning, but we can use recognized patterns in emotional responses to aid interpretation of the city. Encountering Rome for the first time surely inspired awe in most visitors; entering the hubbub of the city, its sights, sounds, smells, might have provoked excitement, but also confusion and some trepidation as the visitor sought to orientate himself and find the way from the Porta Maggiore to the Forum Romanum. Senses are heightened by the unfamiliar; the antithesis of habituation is the most public sensory sphere, the unknown cityscape. First impressions count.[40]

Perhaps the most significant achievement that a multisensory approach to mapping the city of Rome can make is to increase our understanding of dichotomies between city and countryside, or inside and outside, and between public and private areas and places within the city, their use, and where the boundaries between public and private were blurred. The emotional reaction to passing through the Porta Maggiore and being inside or outside the city can be considered, responses that have traceable patterns in the body of human knowledge, such as vulnerability at being beyond the city limit. But how vulnerable does one feel, when one can still see, hear, smell, and feel the proximity of the city? At what distance does the sense of security dissipate when various amenities, stalls, and traders are located outside the gate?[41] Or what was the effect of being inside or outside the wall screening the Forum Augustum from the Subura? Did the sense of being inside or outside depend on your perspective? The shape of the wall suggests that the Subura was *outside* (i.e. outside the elite, political, physical centre of Rome) and the *area* and Temple of Mars Ultor *inside*. But what was the effect of the wall? Did it deaden the noise of the Subura and stop it disturbing the tranquillity of the temple, and so, the sacrifices taking place there? It certainly made the Subura invisible. Did it make the inhabitants of the Subura feel more enclosed and so more secure in their environment, or excluded further from the elite and Augustan politics? There is scope for considerable work to be done on the question of cultural and emotional meanings of being inside or outside a building or area of Rome, such as the fora, temple precincts, the Curia, palaces, theatres, amphitheatre, baths, and the ways that spaces *within* these structures were further delineated and divided.

A preliminary analysis of how someone would have moved through the Forum Romanum follows.[42] It should be noted that the route taken, places paused at, and emotions experienced would depend on the purpose of the visit and activities taking place on a given day, such as a sacrifice at one of the

---

[40] Ling (1990a); Mayer (2007).
[41] See Malmberg and Bjur, in this volume.
[42] This example incorporates data collected as part of the Tavoliere-Gargano Prehistory Project and the author's preliminary experiments in Rome. Visual, sound, and smell data have been collected; taste and haptic experiments have yet to be conducted.

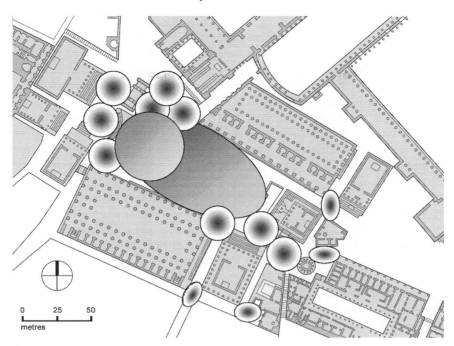

Fig. 4.1. Soundscapes in the Forum Romanum (image by courtesy of Marie Saldaña).

temples, a *contio*, or elite funeral. On a hot, dry, still late morning in early August, slightly raised male and female voices carry across the open space of the Forum Romanum, but not so easily along it, resounding off (the facades of) the Basilica Aemilia and Basilica Iulia (a distance of 52 m) (Fig. 4.1).[43] These voices stand out from small groups of people conversing, whose speech is clearly audible up to 25 m, the maximum distance between buildings in the monumentalized areas of the Forum.[44] Normal speech helps create discrete, private locales in the more enclosed spaces of the Forum Romanum, some of which are intervisible, but none of which are interaudible without the voice being raised. This is one example of the sensory sphere of social interaction in practice. The missing data from this analysis are whether or not voices carried

---

[43] When data was collected in the Forum Romanum it was noted that the smellscape was comparatively sterile, body odour and perfume predominating, and these only in close proximity; the interior of the Curia smelt musty (during this period the doors were kept closed overnight). Traffic noise was unobtrusive/less intrusive than anticipated, a combination of habituation and the height of the Via dei Fori Imperiali above the ground level of the Forum Romanum. See http://dlib.etc.ucla.edu/projects/Forum/ for a digital model of the forum.

[44] Opportunistic survey of a small settlement (75% enclosed) on a hot, dry, still late morning in early July recorded a range of 5–27 m for normal conversation amongst small groups of adults, where intimate conversation could be heard at a distance of up to 5 m and animated conversation up to 27 m (Tavoliere survey, in preparation).

between the exterior and interior of the buildings and the effect of the acoustics of interiors on voice amplification within the structure and from inside to outside.[45] Speeches, waiting, and people coming and going characterize the Basilica Iulia. The building, the steps and street outside it define the place; the soundscape and touchscape dominate, identifiable by a tension between movement and stillness, raised voices, low conversations, and silences.[46]

Meanwhile, from the Rostra, with the conditions as they are today, an orator's voice will project a little over halfway along the length of the forum, unless he shouts, whereas on a cool, damp day he could be heard from the Temple of Divus Iulius.[47] However, this does not take into consideration the voices of the crowd, and both the orator's and the crowd's voices would reverberate between the basilicas, limiting the distance and clarity of audibility.[48] The orator could be seen further than he could be heard, visible throughout the Forum Romanum, particularly when gesticulating, the arrangement of the raised podium of the Rostra and open forum below facilitating this.[49]

Analysis of the clarity of speech and visibility of a speaker in the forum (considering gestures and dress), with attendant variables such as a large audience haranguing the speaker, or a small, near silent, audience on a cold day, facilitates more detailed and accurate evaluation of sources such as Cicero and Quintilian. This in turn enables formulation of a better understanding of the relationship between speaker and audience—elite and populace.[50] For example, Cicero delivered his speech in defence of Milo on 4 April 52 BCE, in unusual circumstances. The Curia having been burnt down, Milo's trial was held in the forum before an audience of Senators. The makeshift court was demarcated by a military guard stationed in front of the temples and Cicero complains that 'the usual circle of listeners is missing; the habitual crowds are nowhere to be seen'.[51] Movement across the forum has been prevented and

---

[45] Information which architectural acoustics studies can provide (Smith 2005: 130).

[46] Plin. *Ep.* 5.9; the gaming boards etched into the steps provide good evidence for people waiting outside the Basilica Iulia and a prevalent touchscape. On game boards, see Trifilò, in this volume.

[47] In an open space, in warm, dry conditions, a man's raised voice reaches 50–70 m, whereas his shout reaches 67–118 m; in cool, damp conditions his raised voice can reach 120 m (Hamilton and Whitehouse 2006: 176). Despite the difference in climatic conditions, higher pitched voices evidently travel further, but sound conducts better in cool, damp conditions.

[48] Mart. 6.38.5–6 for the 'hubbub...and the dense encircling crowd [*clamor...densum corona vulgus*]' of the Forum Romanum. For an example of the crowd drowning out the speaker cf. Plin. *Ep.* 9.13.19.

[49] On level or raised ground, with no obstructions, in clear, sunny conditions, small hand gestures are visible to a distance of 160–85 m, whilst large, sweeping gestures can be seen up to 250–320 m (Hamilton and Whitehouse 2006: 176; Tavoliere Prehistory Project, in preparation). For sweeping gestures cf. Cic. *Brut.* 224; Quint. *Inst.* 11.3.118; Mart. 6.19.9. Cf. Bell (1997: 2) and Corbeill (2003: 127–32) for the physical relationship between speaker and audience.

[50] See Bell (1997).

[51] Cic. *Mil.* 1, 'non enim corona consessus vester cinctus est, ut solebat; non usitata frequentia stipati sumus'.

the populace has been deliberately distanced from proceedings, to ensure that the trial can be conducted appropriately, but Cicero also informs us that 'from any and every point overlooking the forum you can see crowds gazing this way'.[52] Clearly the people were never far away from the action, but how direct was their interaction? It seems that the circle of guards at Milo's trial was placed so as to prevent the people not only from interrupting, but also from hearing the speaker. The cool, possibly damp, conditions would have allowed Cicero's well-trained voice to carry throughout the forum, almost as far as the steps of the Temple of Castor and Pollux, but the vast crowd was outside this area (Fig. 4.1). A proportion of the crowd would have been able to see him, his careful hand gestures and larger arm movements, also his white toga, visible throughout the Forum Romanum, except where buildings and monuments, members of the crowd, and the military cordon obscured his figure.[53] Consequently, discussion of the trial amongst the populace would be based predominantly on interpretation of Cicero's body language, rather than on what was said, presenting both Milo's and Clodius' supporters and detractors with an opportunity to perpetuate their own accounts of proceedings.[54]

SITES, STREETS, AND SENSES IN ANCIENT ROME

Descriptions of food and drink, produce and refuse enable smellscapes and tastescapes to be mapped. The physiological effects of these on the individual can be assessed, their role in social interaction, and their spheres of influence (pervading the street, entering and exiting buildings, or highly localized to the shop, vendor, or consumer). Noise dominates Juvenal's and Seneca's accounts; characters loom large in these as well as Martial's and Petronius' satires. Soundscapes are therefore dominated by people going about their trade, by animals and traffic; also by visits to Rome's public buildings (whether for leisure or out of obligation). Some of the sounds described can be extrapolated to enhance the smell- and taste-maps. References to materials, archaeologically identifiable remains, objects from daily life, people, enable haptic and sensuous elements of the map to be constructed.

Rome's often crowded, narrow, uneven, and less than straight streets suggest that visibility along them was often restricted to short sections, with

---

[52] Cic. *Mil.* 3.1, 'neque eorum quisquam, quos undique intuentis, unde aliqua fori pars aspici potest'. See also Cic. *Cat.* 4.14, 'plenum est forum; plena templa circum forum, plena omnes aditus huius templi ac loci'.

[53] White and beige linen can be seen clearly up to a distance of 320 m in bright sunlight (Tavoliere Prehistory Project, in preparation). Hamilton and Whitehouse (2006: 176); Corbeill (2003: 130–3, 138).

[54] See Laurence (1994b) on political knowledge being gained through hearsay and rumour.

aural, olfactory, and haptic cues more relevant to inhabiting and navigating them.[55] The commercial districts of the Forum Boarium, Forum Holitorum, Velabrum, and Vicus Tuscus provided ample opportunity to navigate by smell, taste, sound, and touch. Each of these locales had its distinctive characteristics, beyond the visual landmarks of monuments in the locality, such as the sweet smell of cattle mixed with their pungent manure in the Forum Boarium.[56] Specifically named by the sources are butchers, bakers, oil merchants, and soothsayers in the Velabrum; fish, fruit, poultry, perfume, silk, and rent-boys are available in the Vicus Tuscus.[57] The spread of literary sources for the items on sale suggests that during the Republic and early Empire this was a diverse market area. The restricted soundscape of soothsayers in intimate conversation with their customers and the smellscape of livestock, dung, blood, meat, flour, bread, and oil (Velabrum), of fish, poultry, fruit, and perfume (Vicus Tuscus) pertain to c.250–30 BCE, perhaps later.[58] Which of these odours predominated would vary according to the time of day or year and their degree of habituation. Did the perfume stand out (especially in the evening) because it was exotic and designed to mask? Did the narrow street and summer afternoon heat cause an overwhelming smell of fish? Writing a little later than the other sources, Martial introduces the luxurious feel of silk, contributing to the mapping of haptic experience.[59] It is Martial who describes a bookshop opposite the Forum Iulium, books *rasum pumice* ('smoothed with pumice'), presenting fragments of a tactile map for shopping in the areas surrounding the Forum Romanum.[60]

What makes the multisensory map of the Subura distinct from that of the commercial areas around the Forum Romanum is that commercial elements are interwoven with residential. Everything needed in daily life could be found here, this district a microcosm for the city as a whole. Descriptions of the Subura as noisy, dirty, sordid, and dangerous may be exaggerated—even if they were not, we can assume a degree of habituation by the district's inhabitants.[61] Martial draws our attention to the calls of barbers (male and female), identifying the entrance to the Subura.[62] The sounds and smells of *popinae* and butchers interwove with these, the streets bustling as the various *tabernae* spilled out into the streets.[63] Whilst we are led to believe that

---

[55] Cic. *De Lex Ag.* 2.96. For a multisensory example, see Mart. 7.61.
[56] See Cohen (2006: 120–1); Porteous (2006: 98–100).
[57] Plaut. *Curc.* 482–3; *Capt.* 494; Hor. *Sat.* 2.3.226–30; Mart. 11.27.
[58] Intimate conversation between men can be heard up to a distance of 5 m; in hot weather a flock of sheep can be smelled downwind at 70 m, in open air the smell of cattle dung reaches 50 m (Hamilton and Whitehouse 2006: 177–8; Tavoliere survey, in preparation).
[59] Mart. 11.27.11.
[60] Mart. 1.117.16; compare also 1.2–4, the latter's opening word *contigeris*, from *contingo* (to touch).
[61] See Cohen (2006: 120–3); Porteous (2006: 90).
[62] Mart. 2.17.1.
[63] Mart. 7.61.

prostitutes were to be found throughout the Subura, Martial emphasizes the middle of the Subura as a favoured location for prostitutes and brothels, doors and curtains screening the *cellae*, these businesses representing a discrete part of the multisensory map.[64] The airy colonnades of the Porticus Liviae, providing cool and shade from the summer sun, where Martial may have sometimes recited his *Epigrams* amongst the sounds of fountains and scents of the gardens, provide a contrasting sensescape with the rest of the Subura (even allowing for Martial's exaggerated image of sordidness).[65] Further up the Clivus Suburanus, *c*.220 m past the Porticus Liviae, taking the next major right turn (southwards), leads us to a *balneum* located on the left, past a row of five shops.[66] The soundscape and smellscape of this compact set of baths, with its own portico and possible brothel or apartments behind, located in a commerical and residential area on the hillside towards the Esquiline Gate, can be re-created from Seneca's account.[67] Claridge notes that 'The character of the modern street is not unlike its ancient ancestor, lined with craftsmen's workshops and tall tenements, narrow, noisy, hot, and dusty in the summer'.[68] A fitting locale for conducting sensory experiments.

DIRECTIONS FOR MULTISENSORY MAPPING OF MOVEMENT

The examples in this chapter have demonstrated how available sensescape data can be applied to movement through Rome. Future studies will be more localized, detailed, and focused on specific time periods, which will limit the literary and archaeological evidence being used, such as a walk from the Argiletum to the Subura in 14 CE.[69] Pauses and interactions between people need to be considered, to detail the full array of sensory experience; this should not be just another walk through Rome. As each sense is brought in, new questions are raised, some of which will remain unanswerable owing to the fragmentary nature of the archaeological, epigraphic, and literary records. It is nonetheless important to ask the questions, since those that can be answered will give deeper insights into aspects of life in the city of Rome and its people.

Contemporary accounts of daily life in Rome and other Roman cities, such as Seneca's account of public baths, have often been combined with the rich

---

[64] See Mart. 9.37.1; 11.61.3–4. See also Holleran, in this volume.
[65] Mart. 7.97.12; Plin. *HN* 14.11; Claridge (1998: 303–4); See also Macaulay-Lewis, in this volume.
[66] *FUR* Stanford 10g, 10opqr.
[67] Najbjerg and Trimble (2003); Sen. *Ep.* 56.1–4.
[68] Claridge (1998: 304).
[69] By 14 CE Augustus had transformed Rome, his forum dominating this route.

architectural remains to bring the ancient city vividly to life.[70] Add a multi-sensory approach to mapping daily life in the city and we can draw a more accurate and detailed topographical map of republican and imperial Rome. It is noteworthy that some authors privilege particular senses, or at least give them more attention: Martial's descriptions have a tendency to be tactile, Seneca favours sounds and smells (if we extrapolate a little), Juvenal's focus is on noise and sordid sensations.[71] If current empirical phenomenological research into the distances that sounds, smells, shapes, and colours travel and have an effect is integrated with the physical and literary remains of the city, these data will enable more accurate interpretation of the impact daily life had on the definition and use of Rome's spaces and places. They will allow exploration of the sensory-spatial impact of particular events (Cicero's address of a *contio* on 2 January 63 BCE)[72] and activities (a visit to the *balneum* near the Porticus Liviae), which can only enrich our experience and perceptions of the city, encouraging new questions to be considered regarding the use of space and movement between places in Rome.

[70] Sen. *Ep.* 56; Favro (1996); See also Holleran, in this volume.
[71] Larmour (2007).   [72] Bell (1997: 1).

Part II

Movement in the Roman City: Infrastructure and Organization

5

The Power of Nuisances on the Roman Street*

Jeremy Hartnett

'Hey, I'm walkin' here!'

I take the epigraph of this essay from the 1969 John Schlesinger film *Midnight Cowboy*. In one well-known and much-cited scene, the streetwise Ratso Rizzo, played by Dustin Hoffman, limps along the sidewalk while dispensing wisdom to the naïve proto-gigolo Joe Buck, played by Jon Voigt. As they cross an intersection, a taxi screeches to a halt, nearly running them over. Ratso slaps the car's hood, twice yells, 'Hey, I'm walkin' here!', and then exchanges vulgarities and obscene gestures with the cabbie. I draw on this outburst and exchange for two reasons. First, in a volume dedicated to urban space and movement, this example makes the obvious point that the kinetic experience of passing through cities—whether by car or bus, on foot or horseback, in a rickshaw or litter—is rarely smooth. It recalls a famous passage of Juvenal's Third *Satire*, where the narrator Umbricius, a sort of poetic and Roman Ratso Rizzo, ponders leaving Rome for Cumae. Amidst a litany of complaints about the *caput mundi*, he bemoans the experience of walking along Rome's streets.[1] One crowd blocks his path while another crushes him from behind; an elbow digs into his side, and a hard pole, a wooden beam, and a wine jug, in turn, hit his head. Mud splatters Umbricius' legs, and a soldier's boot tramples his toe. Though exaggerated for satirical effect, Juvenal's words nonetheless remind us that numerous impediments could make passage along Roman streets chaotic, staccato, and sensorily overwhelming.

---

* I am grateful to the Soprintendenza Archeologica di Pompei e Napoli for permission to undertake this research, and to the editors of this volume for their help. Coss faculty development funds at Wabash College provided financial support.

[1] Juv. 3.232–67.

The scene also underscores that disruptions in city movement are not mere obstacles, but result from specific people, affect streetgoers in various ways, and sometimes lead to or reflect confrontations. Juvenal makes a similar point: though Umbricius is blocked and mauled by a crowd, the mass makes way for a rich man (*dives*), who sleeps or reads as he is borne on a litter. In other words, as much as he is bemoaning traffic conditions, Umbricius also points out their differentiated impact across the social spectrum. Along the thoroughfares of a Roman city, virtually every urban inhabitant—from slave to emperor—came into spontaneous, face-to-face contact, on a daily basis. Such contact across the entire population must have had a profound impact on how individual Romans conceived of the social hierarchy and their place within it. Moreover, this visibility also rendered the street a stage for people like the *dives*, on which to craft and display their public image through what they did, wore, and built, and also through their ability to create a nuisance or to move unobstructed through the street.

This chapter focuses on the connection between urban society and obstructions to the flow of traffic. In what ways was passage on urban streets obstructed and, more importantly, to what ends and with what implications? My goal is not merely to produce a list of traffic impediments, but also to explore what light the broad category of 'nuisance' can shed on various aspects of urban dynamics, including movement, law, inscribed social privilege, and visibility. My case studies for this investigation are the cities of Pompeii and Herculaneum, while I draw on legal opinions and charters, primarily from Rome, to offer a textual counterpoint and a degree of context to the material evidence.

Before getting started, a definition. I use the term 'nuisance' quite broadly to describe urban phenomena—whether elements of the built environment or the people who inhabited the city—that slowed or interfered with smooth, uninterrupted passage along its streets. Though 'hindrance' might provide a more appropriately neutral term, to my mind 'nuisance' importantly dips into the social realm and personalizes matters by implying both a person whose passage is hampered and an agent who creates the encumbrance. The more metaphorical 'nuisance', moreover, circumscribes a broader range of potential means and media that might affect streetgoers. I readily admit that this term also implies a degree of annoyance or irritation on the part of whoever was slowed. Indeed, like Juvenal's Umbricius, many other voices grumble about traffic and its causes.[2] Yet the sentiments of streetgoers towards nuisances need not have run so monolithically negative, and, as I hope to show, were likely quite mixed.

---

[2] Sen. *Clem.* 1.6.1; Hor. *Epist.* 2.2.72–80; Mart. 12.57.

THE MYTH OF EFFICIENT PASSAGE

In today's Western world, we tend to think of traffic primarily as a vehicular phenomenon, and of streets primarily as corridors for movement that is planned, efficient, and wheeled. To this mindset, slower passage is worse, and ought to be remedied by urban planners. Scholars—who know (and often read quite literally) their Juvenal, Martial, and Horace—make note of the relatively greater use of the street by pedestrians. Yet scholarly approaches to traffic and congestion in the Roman city have largely imported this modern paradigm, merely recalibrating the composition of the traffic by introducing comparatively more feet into the street.[3] A related current of thought presumes both that a rational order underlies many urban phenomena, including traffic, and that the city, having evolved over time (or, better yet, lucidly planned *de novo*), should have few inefficiencies. As such, the reasoning goes, it ought to be possible with the correct approach to read the urban form and thereby recognize the factors that shaped it. Not surprisingly, assessments of movement within the city have often correlated traffic with measurable phenomena, specifically street width.

A street-level view of the city, however, upends such connections. While the amount of space available to those making their way along a street, whether they be walking or riding, was inevitably related to the ease or difficulty of passage, a whole host of other factors impacted how, and how well, traffic flowed. Most important is the recognition that streets in Roman cities, beyond fostering movement, hosted a much different range of activities, architectural forms, and 'street furniture'.[4] Many of these fostered, hindered, or obstructed the passage of vehicles and pedestrians alike. In other words, to get an accurate picture of Roman streets and traffic, reconsidering the composition of the traffic and weighing the volume of the channel may be easiest to measure, but is not sufficient. The character of the space itself was fundamentally different. My goal in this section, then, is to shift the form of analysis by considering Roman uses of the street in a largely qualitative assessment of potential nuisances.

What follows is a quick and non-encyclopaedic list of encumbrances to traffic as illustrated by evidence from Pompeii and Herculaneum. Donkeys and other animals were apparently hitched to the sidewalk, their reins or leads passing through cuts made in the border stones. Drivers had to steer vehicles around them, while the animals perhaps also nipped at people passing along narrow sidewalks.[5] In the *Digesta* of Justinian, dangerous animals are

---

[3] Newsome (2008) offers an important critique of such a set of assumptions. Cf. Laurence (2007: 12–19).
[4] Kellum (1999); see also Holleran, in this volume.
[5] Saliou (1999: 203). It has also been suggested, not entirely convincingly, that these cuttings were used to tie ropes attached to vertical awnings in front of houses and shops: Spinazzola (1953: 60–1, fig. 67).

Fig. 5.1. The bench outside a Pompeian tavern (VI.1.2) blocked the sidewalk while offering a place to sit (image by permission of the Ministero per i Beni e le Attività Culturali—Soprintendenza Speciale per i Beni Archeologici di Napoli e Pompei).

forbidden from being tied up along public thoroughfares (*quo vulgo iter fiet*) because they might injure someone or cause other damage. Dogs, hogs, wild boars, wolves, bears, panthers, and lions are mentioned specifically, which implies that others were permitted.[6] At least 100 streetside benches jutted into the sidewalk in front of bars, houses, shops, baths, and bakeries at Pompeii, threatening to bruise pedestrians' shins even as they offered a place for the weary to rest[7] (Fig. 5.1). Similarly, though more often at Herculaneum than Pompeii, obstacles were presented by the columns or piers positioned at the edge of the sidewalk to support balconies or jetties that projected from facades.[8] Even the stepping stones that aided pedestrians likely annoyed and caused an inconvenience for the drivers of single-animal carts, such as on Pompeii's narrow and busy Via degli Augustali (Fig. 5.2).

Merchandise spilled from shop doorways into the sidewalk and street to the point, Martial seems to report, that Domitian took action against the practice.[9] Legal provisions generally outlawed anything appearing outside workshops,

---

[6] *Dig.* 21.1.40–2.
[7] Hartnett (2008); see also Holleran, in this volume.
[8] Spinazzola (1953: 113–28); Andrews (2006).
[9] Mart. 7.61; see also Laurence, in this volume.

Fig. 5.2. The narrow sidewalks and roadbed of the Via degli Augustali at Pompeii hosted much traffic (image by permission of the Ministero per i Beni e le Attività Culturali—Soprintendenza Speciale per i Beni Archeologici di Napoli e Pompei).

but made exceptions for some craftsmen, namely fullers and carpenters, who were permitted to dry cloth and store wheels outdoors, respectively, so long as they did not prevent a vehicle from passing.[10] Ambulatory street traders trudged along the sidewalk burdened with goods or may have erected temporary wooden seats or sales counters to hawk their wares along the street.[11] Wall paintings from Pompeii give some suggestion of these activities around the city's amphitheatre, in its forum, and elsewhere, while graffiti testify to street pedlars reserving certain parts of the sidewalk, especially near the Basilica along the Via Marina.[12] Even a cloth spread on the sidewalk for the

---

[10] *Dig.* 43.10.1.4, which also forbids fighting, dung-flinging, and the throwing of dead animals or skins in the street, all of which would have caused nuisances of various stripes. Cf. *CIL* IV.10488 for the decree of an aedile at Herculaneum against *stercus*-throwing.

[11] Saliou (1999: 203, esp. nn. 113–15); Holleran (in this volume and 2012).

[12] The famous painting of the riot in Pompeii's amphitheatre, from the eponymous Casa della Rissa nell'Anfiteatro (I.3.23), shows temporary stalls in front of the arena (Fröhlich 1991: 41–7), while inscriptions found around the building's perimeter (*CIL* IV.1096, 1115, 1129, 1130, 2485) testify to temporary stalls within its arched openings. Multiple ambulatory sellers crowd the open spaces of the forum in a frieze of paintings from the Praedia Iuliae Felicis (II.4.4) at Pompeii: Nappo (1989). Whether the figures in a painting from house VII.3.30 at Pompeii are purchasing

Fig. 5.3. A procession in honour of Cybele graced the doorway of a Pompeian shop (IX.7.1), and may have made its way along the city's streets (image by courtesy of the Istituto Poligrafico e Zecca dello Stato—Libreria dello Stato, Rome).

display of goods could hamper traffic, and carts halting at doorways to unload cargo, as appears to have occurred at the Casa del Marinaio (VII.15.1–2), certainly blocked vehicular movement and slowed pedestrians.[13]

Processions also occupied streetspace, temporarily paralysing traffic and turning attention to a moving spectacle.[14] Parades made their way from Pompeii's forum to its amphitheatre on the occasion of games, and may have covered the distance between the forum and theatre district as well.[15] Funerals, weddings, and other household ceremonies saw families take to the street en masse.[16] Processions of guilds or *collegia*, like that of carpenters depicted on a shop facade at Pompeii (VI.7.8–11), or of religious devotees, such as worshippers of Magna Mater who are shown mid-procession on another frescoed streetfront (IX.7.1), took over the street during their festivities[17] (Fig. 5.3). Lastly and less dramatically, in cities that offered little public green space for the population at large to escape from rooms made hot and

---

bread, as has long been held, or are part of an euergetistic distribution (Fröhlich 1991: 236–41), they certainly belong to a scene at pains to demonstrate that it was public. Vendors reserving places on the Via Marina: *CIL* IV.1768–9. Possible others on the eastern portion of the Via dell'Abbondanza: *CIL* IV.8432–3.

[13] Newsome (2009a: 128–9).

[14] The most prominent parade was, of course, the triumphal procession in Rome: Favro (1994); Beard (2007).

[15] Processions: *CIL* X.1074 (from the forum presumably to the amphitheatre; cf. the so-called Pompeii Gladiators' Relief: La Regina 2001: no. 74); Wallace-Hadrill (1995: esp. 49–50, between the forum and theatres); Coarelli (2001, within the theatre district and adjacent Foro Triangolare).

[16] Funeral processions: Bodel (1999) with texts, images, and bibliography. Wedding processions: Treggiari (1991: 161–82). Sampaolo (1997) discusses the wedding procession, albeit a mythical one, painted in the Casa delle Nozze di Ercole (VII.9.47).

[17] Carpenters: Fröhlich (1991: 320–1). Magna Mater: Fröhlich (1991: 332–3); Potts (2009).

stuffy by the Mediterranean climate, many people simply used the street as a space for lingering, where they simultaneously caught a breath of fresh air and the latest neighbourhood gossip.[18]

Considering this close-range evidence *in toto*, my initial point is that obstacles to smooth passage through the city were more numerous and more various than we might appreciate as today we dodge the occasional pram or tour group and wander through what now, absent of any vehicular traffic, seems a ghost town. Secondly, though there is a tendency to think of congestion as a vehicular phenomenon, both wheeled and pedestrian traffic encountered nuisances. The sidewalk was apparently reserved for pedestrians, but they may have been pushed into, or have chosen to walk in, the roadbed because of what they encountered along its sides.[19] Finally, the sheer prevalence of nuisances was such that interference with passage must have been expected by those making their way through the urban fabric. This point was true enough for those walking along, but, for anyone driving or leading a vehicle through the city, it must have been still more obvious that other priorities outranked efficient traffic flow.

Much has been written in the last decade about where wheeled traffic was allowed (or, just as importantly, not permitted) in Pompeii. A series of barriers, steps, drop-offs, and building projects rendered the forum and the adjoining broad boulevards, especially the Via dell'Abbondanza and the Via Marina, largely free from wheeled traffic[20] (Fig. 5.4). Scholarly attention has also been drawn to how traffic circulated within the city, specifically which way wheeled traffic ran along various streets in the north-west portion of the city.[21] But what has been less discussed is the substantial degree to which the blockages around the forum lengthened the journeys of vehicles making their way into, out of, or through the city. A cart entering Pompeii through the Porta Marina and carrying amphorae to the eastern half of the city, for instance, could not approach the forum directly along the Via Marina because the north-east corner of the sanctuary wall around the Temple of Venus narrowed the street to the point where it was impassable by wheeled traffic.[22] Consequently, drivers had to turn north along the Vicolo del Gigante, make a sharp right-hand turn on the Vico dei Soprastanti, continue on the Via degli Augustali until they turned right on the busy Via Stabiana, and then follow that street until its intersection with the Via dell'Abbondanza. In all, the alternative route added several hundred metres, at least four extra turns, and

---

[18] Among others, Franklin (1986); *CIL* IV.813, 8258–9.
[19] On pedestrians' potentially ceding parts of the sidewalk to one another: Hor. *Sat.* 2.5.15–18; Plut. *Rom.* 20.3. I thank David Newsome for these references.
[20] Tsujimura (1991: 65–8); Wallace-Hadrill (1995: 46–50); van Tilburg (2007: 136–44); Laurence (2008: 89–92); Poehler (2009).
[21] Poehler (2006).
[22] Newsome (2009a: 125).

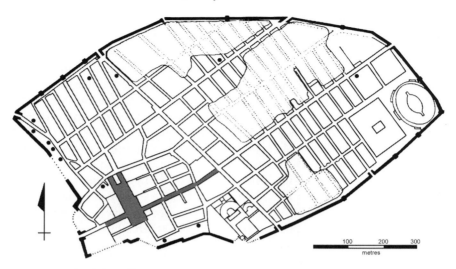

Fig. 5.4. Wheeled traffic was restricted in and around Pompeii's forum (shaded), while ramped sidewalks fronted several properties throughout the city, which are marked by dots (image by Jeremy Hartnett).

several more encounters with cross-traffic to the driver's intra-urban course.[23] Comparable inconveniences marked the passage of wheeled vehicles moving north to south (or vice versa) through much of the western half of the city. Several arguments have been advanced about why wheeled traffic was prevented from reaching or passing through the forum and its vicinity: among the possible factors are enhancing marketplace commerce, avoiding the taint of low-status activities within a ceremonial space, facilitating the reconstruction of the city's monumental spaces after the earthquake of 62 CE, and merely passively reflecting a loss of the forum's centrality.[24]

To the strain of scholarship about Roman cities that has long imagined their spaces and societies operating with a distinct precision—in which laws are followed, the city's form follows a rational and almost machine-like order, and economic optimization prevails—the lens of nuisance begins to respond with a different picture. From pedestrian to wheeled traffic, and from micro- to macro-level viewpoints on the city, a cursory survey of disturbances to traffic

---

[23] Van Tilburg (2007: 161–2) hypothesizes that a 'ring road' may have encircled Pompeii at an early date and thus minimized the intra-urban traffic of this kind, but he is sceptical of its existence in the last decades of the city's ancient life. Those portions of Herculaneum that have been excavated exhibit only one quite circuitous route for wheeled traffic to pass through the city, though it is likely that through-streets remain buried: Pagano (1996).

[24] Cf. above, n. 20. On a similar phenomenon in the city of Rome, see Newsome, in this volume.

shows that obstacles large (the forum) and small (benches, animals, etc.) were many and ubiquitous. In light of this, our understanding of Juvenal's famous passage might be reconsidered. Though many readings of Umbricius' rant spring from a modern frustration with the traffic jam, the narrator's consternation concentrates instead on the danger, physical humiliation, and social inequity of his situation, probably because smooth and direct urban passage was not attainable or even expected by Romans. In streets teeming with people, animals, merchandise, and possibly sewage, the situation was literally and metaphorically messier than has been understood. To borrow the phrase of a scholar of Roman law, in these cities we might well be encountering urban dynamics that are more reminiscent of 'old Sicily than modern Zurich'.[25]

RULES ARE RULES?

What makes nuisances and obstructions particularly fascinating is that they appear to run against the spirit and letter of Roman law. Legal sources of several types emphasize that streets must be accessible and passable. The *Tabula Heracleensis*, found in southern Italy, is apparently the final of a series of bronze plaques recording legislation, most likely of Caesarian date, regarding a host of urban issues.[26] Among other topics, the inscription details who is responsible for constructing, maintaining, and cleaning streets in and near Rome, and it assigns oversight of these responsibilities to specific city officials. In summary, all owners must maintain the street and sidewalk in front of their property (lines 20–3, 53–5); aediles are entrusted both with enforcement of this code (lines 20–8) and with ensuring that nothing is constructed in public spaces (lines 68–72). The legislation interestingly justifies these regulations by framing them in terms of urban dwellers' ability to pass through city spaces. For instance, water may not stand in the street since it would impede the populace's passage (line 23), and virtually identical language about access explains the prohibition against constructions in open spaces (lines 73–4). The inscription also details the very public process that would be set in motion when aediles encountered negligent frontagers. The transgressor's name and location were affixed to the ruling magistrate's tribunal; the aedile served notice to that person, probably by having some official visit the property; and the owner's name was again announced and inscribed when the contract for repair was let (lines 32–46). Other

---

[25] Meyer (2004: 3).
[26] *CIL* XII.593; *FIRA* 13. In general on this inscription: Crawford (1996: 355–91). On the street-related provisions: Robinson (1992: 58–79); Martin (2000); Saliou (2008).

municipal charters from elsewhere in the empire, such as the *Lex Coloniae Genetivae Iuliae* from Spain, accord with the anti-nuisance sentiment of the *Tabula Heracleensis*.[27]

The opinions of several jurists collected in the *Digesta* of Justinian echo the emphasis on eliminating nuisances. In general, jurists approach the issue in two ways: some, such as Pomponius (second century CE), frame the issue positively by emphasizing the street's accessibility by everyone, while others state a prohibition in general and negative terms, as when Ulpian (early third century CE) quotes the praetor's edict, 'You are not to do anything in a public place, or introduce anything into it, which could cause damage to such a one, except for what has been permitted to you by statute, *senatus consultum*, or edict, or decree of the emperor.'[28] At times, jurists also provide a more finely grained view. Papinian (late second to early third century CE), for example, assigns magistrates a number of responsibilities, including ensuring that nothing is left outside workshops and that 'no one digs holes in the streets, encumbers them, or builds anything on them'.[29] He also requires owners to maintain their street frontage such that vehicles are not prevented from gaining access.[30] Paul (early third century CE) sees no issue with aediles removing a potential nuisance when they come across some beds which had been purchased and then left outside a house after delivery.[31] And Ulpian, in a long treatment of interdicts related to public streets and places, offers some revealing points. He cites the praetor's edict forbidding 'doing or introducing anything in a public road or way by which that road or way is or shall be made worse'.[32] Later, he follows up with a definition: 'Making a road worse is to be understood to mean impairing its usefulness for traffic, that is, for walking or driving, as when it was level and is made steep, or when it is turned from smooth to rough, from broader to narrower, or from dry to muddy.'[33] In sum, many of the physical constructions listed above would fall within the general legal restrictions and thus be, strictly speaking, illegal. There is some question of whether the specific laws outlined in the *Tabula Heracleensis* and the exact legal provisions in the *Digesta* were in effect for Pompeii

---

[27] *Lex Coloniae Genetivae Iuliae* 104 (*FIRA* 21). Cf. *Lex Iulia Agraria* 4 (*FIRA* 12).

[28] Pompon. *Dig.* 43.7.1; Ulp. *Dig.* 43.8.2.1: 'Ne quid in loco publico facias inve eum locum immittas, qua ex re quid illi damni detur, praeterquam quod lege senatus consulto edicto decretove principum tibi concessum est.' Cf. *Dig.* 43.7.2, 43.8.1. This and the following translations are reproduced from Mommsen, Krueger, and Watson (1985).

[29] *Dig.* 43.10.1.2: μηδεὶς ὀρύσσῃ τὰς ὁδοὺς μηδὲ χωννύῃ μηδὲ κτίσῃ εἰς τὰς ὁδοὺς μηδέν.

[30] *Dig.* 43.10.3.

[31] *Dig.* 18.6.13(12)–14(13).

[32] *Dig.* 43.8.2.20: 'In via publica itinereve publico facere immittere quid, quo ea via idve iter deterius sit fiat, veto.'

[33] *Dig.* 43.8.2.32: 'Deteriorem autem viam fieri sic accipiendum est, si usus eius ad commeandum corrumpatur, hoc est ad eundum vel agendum, ut, cum plane fuerit, clivosa fiat vel ex molli aspera aut angustior ex latiore aut palustris ex sicca.' Cf. *Dig.* 43.8.2.25.

and Herculaneum. For the former, what evidence we have does point to similar conceptions of sidewalks. Namely, in both cities the sidewalk decoration and kerbstone material frequently correspond with facade decoration and presumed property lines, which implies that frontagers were responsible for the space between their property and the street.[34] For the latter, an inscription recording the public purchase of the right to block the light of an adjoining property speaks to the applicability of Roman law codes.[35]

What are we to make, then, of the apparent disjuncture between legal principle and on-the-ground actions? On the one hand, it is possible to consider the constant tension between ideology and practice in Roman society, and to see this disconnect as a situation where, despite our deep trust in clearly spelled-out laws, rules were outrightly broken.[36] By this reasoning, the laws on-the-ground were weak or ineffectual, merely spelling out what people may do, not what they in fact do. It is also tempting, on the other hand, to read these contraventions as assertions of individual power against the political and social establishment. The laws, in this vision, hold power, yet are wilfully disobeyed by those who see in the laws a representation of their elite legislators and seek to resist and thus marginalize those people's influence by acting illegally. To take an example relevant to the public space of the Roman city, Livy writes that, in 195 BCE after the conclusion of the Second Punic War, women in Rome rallied for the repeal of the *Lex Oppia* (originally passed in 215 BCE). They took en masse to the street—the very space where their behaviour and public appearance had been subject to legal restriction—and besieged the houses of the officials who threatened to block the law's abrogation.[37]

Of course, all obstructions cannot be dismissed simply as anti-authority rule-breaking, neither can they be shrugged off as unfollowed rules. A law, once made, does not exist merely in stagnant relationship to practice (i.e., is it followed or not?), but is 'a living institution that is continuously being tested, negotiated, created, and recreated within a given society'.[38] Because of this, no single explanation can reasonably account for nuisances amidst an urban population whose individual members likely experienced the law in much different ways and were driven by diverse motives. That is, just as the contents of the laws themselves were the result of debate, negotiation, and other social and rhetorical processes, we might also view their enforcement, particularly with regard to nuisances, in a similar light.[39] Both of the above explanations,

---

[34] Saliou (1999: esp. 174–85, 198–200); Hartnett (2008: 99–100, 109).
[35] *CIL* X.787; cf. *Dig.* 8.2.2, 8.2.4.
[36] On the question of the applicability of Roman law: Meyer (2004: esp. 1–11).
[37] Livy 34.1–8.
[38] Milnor (2007: 7).
[39] On law as a 'constitutive rhetoric': White (1985: 28–48). Law as discourse: Humphreys (1985).

then, might very well be valid in individual cases, but they were just two of many possibilities behind the creation and continued existence of nuisances.

The relationship between the populace and the law was highly differentiated and dependent upon many factors, such as access to legal knowledge, the enforceability of the legal code, and, not least, personal connections. We can illustrate this point with reference to a relatively simple hypothetical case involving a nuisance in the street. Let us assume that Gaius sees Quintus starting to build something that runs contrary to building regulations or that threatens Gaius' use of the street or his property in some way. Gaius' first move might not be to engage the legal system, but to seek recourse with someone more powerful and better connected than he, such as a relative or his patron, who could work informally to resolve the matter. Though our sources are vague about the obligation of urban magistrates in charge of streets to intervene in circumstances like Quintus' questionable construction, Gaius or an associate could work through such official channels by notifying the aediles.

But what if Gaius wanted to take the law into his own hands? In contrast to many of today's legal systems, which grant the right to sue to people subjected by their neighbours to intolerable inconvenience (e.g. loud music), Roman law offered no unified legal protection from the actions of neighbours, though the jurists do spell out scattered legal remedies for Romans who faced nuisances of the type I have been describing.[40] Gaius might seek a resolution by asking Quintus to cease work through a procedure known as *operis novi nuntiatio*. After Gaius went to the place where construction was occurring and served Quintus with a request to stop work, the latter could either comply or give a promise (*cautio*) to destroy the project if it was determined that Gaius would suffer harm as a result of Quintus' construction. In a different scenario, if Quintus has already completed his construction, but did so either despite Gaius' objection or while knowing that Gaius would but was not able to object (e.g. because he was out of town), Gaius might seek an interdict named 'what by force or stealth . . .' (*quod vi aut clam*). Determination would then be made of whether Gaius' claims to have been harmed were valid.[41]

Regardless of the letter of the law, a great number of factors must have influenced whether Gaius or Quintus succeeded in blocking or constructing. To what degree would Gaius have been knowledgeable about the law and his various legal options, especially since there was no one designated legal path to follow when trying to stop nuisances? How readily could Gaius gain and maintain the attention of authorities who would issue an interdict or apply pressure to stop Quintus' construction? How much could Gaius claim to be

---

[40] On law and relationships with neighbours: Rainer (1987); Palma (1988); Saliou (1994); Johnston (1999: 68–76).

[41] *Dig.* 43.24.

harmed by Quintus' actions? What would happen if there was a significant gap in wealth, personal connections, or social standing between Quintus and Gaius? The questions could continue (and we have not even considered Quintus' motivations), but my point is that even in a case such as this one, where juristic opinion appears to open at least two avenues for Gaius to pursue, matters well beyond the law could have influenced whether or to what degree he would be successful. Ultimately, the creation of nuisances, the tolerance of them, and any potential legal consequences of their presence might be considered as the manifestation of a complex contest along numerous fronts besides legal principle, not least being political clout, social standing, and practical feasibility.

THE TELLING EXCEPTIONS

If we understand Roman law not just as a static thing that exists solely to create or reflect lived reality, but also as a rhetorical act that 'encodes a highly interested representation of social life', then exceptions to rules potentially take on greater significance precisely because they privilege certain categories of behaviour by exempting them from the normal legal directives.[42] In this section, in order to draw out some of the social priorities of Roman urban dynamics, I examine the traffic impediments that were legally sanctioned in the *Tabula Heracleensis* and those that most seriously contravened legal principle at Pompeii. Immediately after the section of the *Tabula Heracleensis* assigning responsibility for road cleaning and maintenance, the text turns its attention to traffic. *Plostra*, heavy-wheeled cargo vehicles also called *plaustra* in the sources, were barred from entering the city from sunrise until the tenth hour of the day, which fell in the late afternoon (lines 66–7).[43] Such a prohibition made practical sense, since such large vehicles moved slowly and likely had trouble making tight turns, all while occupying a fair degree of the roadbed and thus leaving little room for other travellers to pass. Stacked high with materials, they threatened to crush streetgoers if something went wrong (a concern voiced in poetry).[44] Exceptions to the *plostra* prohibition fell into three main categories. First, the *Tabula Heracleensis* allowed *plostra* to pass through the city when they carried materials for the construction of temples (*aedium sacrarum deorum immortalium*) or public works, or when they took away detritus from demolitions governed by public contracts (lines 56–61).

---

[42] Quotation from Milnor (2007: 8).
[43] On *plostra/plaustra*: van Tilburg (2007: 73–4); Laurence (2008: 87–9, esp. n. 13); see also Kaiser, in this volume.
[44] e.g., Juv. 3.254–60.

Plostra were also permitted to take to the street by day when used in public rites, such as religious festivals (when the text allows for their use in carrying sacred personnel such as the Vestals, the *rex sacrorum*, and the *flamines*), triumphs, and games (specifically the *pompa* for the *ludi circenses*) (lines 62–5). Finally and somewhat less dramatically, *plostra* that had entered the city by night were permitted to leave the city, apparently at any time, while empty or while carrying away *stercus*, which here likely means faecal matter, human or animal (lines 66–7).

The *Tabula Heracleensis*, in a passage I cited above, next shifts focus to prohibitions against the blocking or occupying of public porticoes and open spaces (which presumably includes piazzas and the like), then catalogues exceptions to these provisions. Specifically listed are four circumstances: such spaces may be occupied when they are leased for the enrichment of the public purse (lines 73–6); games-givers may erect stages, platforms, or other necessary equipment (lines 77–9); magistrates may allow their attendant clerks and scribes to occupy these spaces (lines 80–1); and censors may designate them as quarters for public slaves (lines 81–2). In general, the legally enshrined exceptions to heavy vehicles and the blockage of public spaces share a striking similarity. Though somewhat diverse, in that they govern a range of participants from Vestals to scribes and a range of activities from triumphs to sleeping slaves, the exceptions allowed nuisances where the public good—by way of sacred activities, entertainment, administration, or even hygiene (in the case of human waste)—was foremost.

When attention shifts from the legal provisions to the material evidence at Pompeii and Herculaneum, we see the same priorities reflected broadly in the remains of the cities—the most serious blockers of passage along the cities' sidewalks and streets also provided public services. At Herculaneum, for instance, stone *cippi* situated at the south-east end of the so-called Decumanus Maximus rendered it impassable by wheeled traffic (Fig. 5.5). Measuring twice the width of other streets and terminating at both ends in grand architectural displays of public buildings, the Decumanus Maximus also enjoyed a monumental civic presence and likely lay along the route of a procession honouring Herculaneum's patron, M. Nonius Balbus.[45] The *cippi* also protected a bustling open-air marketplace, which was shaded by awnings stretching from the adjacent buildings to wooden posts set in the roadbed.[46] Meanwhile, where wheeled traffic was impeded at Pompeii—namely the forum, the Via Marina, and the western half of the Via dell'Abbondanza—the provision of shade may not have been as important, but these spaces certainly had a monumental

---

[45] Procession and its route: Schumacher (1976); Hartnett (2009).
[46] Maiuri (1958: 40–3, 85–7).

Fig. 5.5. An archival photograph of Herculaneum's Decumanus Maximus shows the stone *cippi* that protected wooden posts in the street from wheeled traffic (image by courtesy of the Istituto Poligrafico e Zecca dello Stato—Libreria dello Stato, Rome).

aspect (the streets are among the widest in the city), led to temples, and played host to commerce and processions[47] (Fig. 5.4).

Where wheeled traffic was not restricted, smaller-scale but still serious obstructions dotted the urban fabric. During the reign of Augustus, both Pompeii and Herculaneum benefited when an aqueduct started bringing water into the region from a mountain spring at Serino.[48] Above ground, two features of the water supply system had to be accommodated along the existing network of the cities' urban thoroughfares: fountains, usually rectangular stone basins decorated with low-relief sculpture and fed by a spout; and pressure-control towers, essentially square-based shafts of masonry whose vertical grooves protected lead pipes that led water up to and down from open-topped containers. The placement of the fountains and towers demonstrates a reluctance to block either the sidewalk or the roadway, for the fountains often span both spaces, thus compromising by interfering somewhat

---

[47] Cf. above, n. 12, for evidence of commerce on the Via Marina and in the forum. Along the Via dell'Abbondanza between its intersection with the Via Stabiana and its termination at the forum, no fewer than 45 shop-style doorways open onto the street.

[48] Aqueduct: Sgobbo (1938); Potenza (1996). Water system: de Haan and Jansen (1994).

Fig. 5.6. Fountains at Pompeii often straddle the edge of the sidewalk and are protected from wheeled traffic by large stones (image by permission of the Ministero per i Beni e le Attività Culturali—Soprintendenza Speciale per i Beni Archeologici di Napoli e Pompei).

in both realms without obstructing either entirely. Along the busiest streets, such as the eastern segment of the Via dell'Abbondanza, however, they still got in the way. As a typical example, at the south-west corner of insula IX.11, the streetward portion of a fountain was bolstered with several large stones to prevent damage or displacement by passing vehicles[49] (Fig. 5.6). The nuisance presented by fountains appears to have been great enough for owners of several properties at Pompeii to have moved their facades, 'stepping' them back from the street to make space for fountains and thus relieving the streetward impact of the installations (while also potentially luring customers to any nearby commercial enterprises).[50] On the western part of the Via dell'Abbondanza, the exception seems to prove the rule; here, where wheeled

---

[49] Wear marks from the axles of passing vehicles are evident on the side of the fountain at the south-eastern corner of insula I.13.

[50] Ling and Ling (2005: 173–4). Likely candidates include: I.10.1, VII.1.32–4, and I.12.1–2. Cf. Ling (2005). It is somewhat unclear, however, whether some properties were adjusted (or land expropriated) in response to a fountain's positioning, or whether fountains were sited in particular locations because there was adequate space. In either case, the desire to limit obstruction is nevertheless evident.

traffic was limited, the fountains at the south-east corners of insulae VII.9 and VII.14 both rest entirely in the roadbed and leave the sidewalks free for pedestrians. The towers were positioned similarly, either located behind fountains or embedded in street-facing walls, so they would obstruct traffic minimally. Across from the fountain at IX.11.1, a pressure-control tower abuts the east face of insula I.6 rather than jut into the sidewalk of the Via dell'Abbondanza, yet in this position it further constricts the already narrow and unpaved sidestreet.

A similar pattern of public obstructions continues at Pompeii's three main bath complexes, each of which seriously encroached upon sidewalks. Along the north side of the Forum Baths and the east side of the Stabian Baths, for instance, previously usable sidewalks were rendered impassable when the baths were renovated. At the latter location, moreover, the construction necessitated that the roadbed of the Via Stabiana, a primary artery of the city, be narrowed slightly[51] (Fig. 5.7). On the east side of the Central Baths, still under construction in 79, the roadway was narrowed and thus changed from a street that had hosted wheeled traffic to one that barely left room on its sidewalks for pedestrians. This final step, though inconvenient, may in fact have been intended to aid the delivery of firewood to the baths by donkeys, who could still pass through the narrowed alleyway.[52]

In sum, both the legal provisions of the *Tabula Heracleensis* and the material evidence from Campania reveal a remarkably consistent social bargain. Contrary to the legal principles prohibiting infringement on someone's use and enjoyment of the street, compromises were made in both cases. In a broad cultural phenomenon reflecting the collective nature of Roman urban life, freedom and ease of individual movement were limited or, put another way, significant infringements were tolerated in exchange for the common benefit of the populace.[53] Running water, games and a show, a spot to shop, a cleansing plunge, and more were considered worth the sacrifice. Yet we also should consider that there were specific beneficiaries to this social bargain. The legally protected obstructions of the *Tabula Heracleensis* were not addressed directly to the public at large. Instead, in virtually all of the exceptions,

---

[51] At the Forum Baths, a small entrance lobby to the women's bathing suite blocks the sidewalk along the Via delle Terme. The autonomous female baths appear to have been built in the Augustan period or later, while the lobby protrusion likely dates to the last decades of the city's ancient life: Richardson (1988: 151–2); Fagan (1999: 59). Within the walls of the Stabian Baths, at the point where they protrude into the sidewalk and ultimately the roadbed, is the complex's principal furnace room (accessible from the street, presumably for the loading of fuel, at VII.1.15). Though the phasing of this building is disputed, the complex was likely expanded during the age of Augustus: Richardson (1988: 101–5).

[52] Laurence (2008: 90). Herculaneum's one intra-urban bath complex also restricts a roadway, albeit the lesser-used Cardo 4.

[53] On a similar note, several jurists consider the beauty of the city to be significant enough to tolerate obstructions in public places (Ulp. *Dig.* 43.8.2.17, 43.8.7; Paul. *Dig.* 43.9.2).

Fig. 5.7. The eastern side of the Stabian Baths at Pompeii blocks the sidewalk and narrows a main road of the city (image by permission of the Ministero per i Beni e le Attività Culturali—Soprintendenza Speciale per i Beni Archeologici di Napoli e Pompei).

individual members of the highest ranks of society were on special display in the streets in a variety of roles. They populated the processing priesthoods, rode atop vehicles normally outlawed during prime daytime hours, and attracted the attention of all as *triumphatores*, or, less directly, as the givers of games, underwriters of public construction, and supervisors of staff and slaves. Overall, the provisions of the *Tabula Heracleensis* enhanced the visibility of the rule of law and of the social order, but in different ways. Whereas the law's enforcement repeatedly made a display of property owners who were delinquent in maintaining the sidewalk and street along their frontage, the occasions when streets' and public spaces' usefulness was compromised likewise produced an official, presumably still grander spectacle. It was one, however, that highlighted status distinctions by granting elite members of society the ability to transgress the normal rules and by ritually presenting their sanctioned contravention to the public at large. In Campania, though it is not possible to know to what extent these stipulations of the *Tabula Heracleensis* were operative, the prevailing patterns of evidence also suggest: (1) that the highest social classes not only commissioned the type of elements of the built environment, such as fountains and baths, that most obstructed traffic; and, moreover, (2) that the same people were also prominent in activities that caused nuisances in streets and public spaces.[54] A blockage to passage of the sort described, then, likely conjured a complicated set of reactions in most urban inhabitants: they were probably thankful for the amenity that blocked them, may have quietly begrudged the nuisance it caused, and likely thought on some level about the person responsible for both. Meanwhile, the elite—with their ability to skirt legal principle and their enshrined privileged position—were simultaneously conspicuous, inconvenient, and often euergetistic.

NUISANCES AND CONTROL

The previous sections of this chapter outlined various obstructions to traffic, considered how to make sense of apparent transgressions of legal principle, and spoke of the social implications of exceptions to the law. I now seek to tie related threads together by revisiting small-scale obstructions, entertaining possible motivations for their creation, and focusing specifically on the symbolic control inherent in some nuisances.

I have written of the tension among various factors that resulted in the creation of nuisances, but my discussion has largely concentrated on what or who stood in the way of nuisances. What, in fact, motivated some to obstruct the passage of their fellow urban inhabitants? One explanation is simply facility; it was probably easiest to tie one's donkey in front of where one stopped. But the example of one bench just inside Pompeii's Porta Ercolano suggests that a host of other, overlapping reasons were at play as well. The bench, running along the facade of a bar and then jutting deep into the sidewalk, likely did triple duty: it freed up the bar's interior, offered the tavern's patrons a place to sit and sip al fresco, and also drew the attention of potential customers, who, in avoiding the obstruction, were forced to notice the bar's presence[55] (Fig. 5.1).

Clear-cut cases like this are rare. In fact, one irony involved in studying nuisances is that, while it is possible to point out all manner of potential nuisances (as I have done, above), it is generally hard to get a strong grasp on where exactly in the city and to what extent there was interference. We cannot watch crowds unexpectedly bunch up, and much of what would have blocked

---

[54] e.g., donations at bath buildings at Pompeii: *CIL* X.817–18, 928, 8071; processions at Pompeii: *CIL* X.1074d. Unfortunately, little is known about who paid for the creation or maintenance of the water system in either Pompeii or Herculaneum.

[55] Hartnett (2008: 102 and 114–15).

Fig. 5.8. The ramp fronting the Caserma dei Gladiatori at Pompeii (V.5.3) shows marks left by the axles of passing vehicles (image by permission of the Ministero per i Beni e le Attività Culturali—Soprintendenza Speciale per i Beni Archeologici di Napoli e Pompei).

them is nearly untraceable in the archaeological record. Sidewalk heights, however, do provide one indication of where wheeled traffic encountered potential obstructions. Along most frontages, drivers of carts and wagons did not have to worry about passing close to the street's edge, since the sidewalk border stones were low enough for vehicle axles—which projected beyond the vehicle's body and spun at heights varying between 0.45 m and 0.75 m above the roadbed (depending on the size of the vehicle)—to glide above without incident.[56] Long horizontal gashes left in border stones by grinding axles demonstrate, however, that high sidewalks posed a potential nuisance that at times hindered vehicle passage[57] (Fig. 5.8). A city-wide survey of Pompeii's streets reveals at least forty-five places where the sidewalk was elevated more than 0.50 m above the roadbed and thus threatened to make

---

[56] Axle heights: Poehler (2009: 72–4). Carts found in Campania: Maiuri (1933: 191); Miniero (1987).

[57] Cf. Poehler (2006: 57–8) for a discussion of how marks were left on other elements of the roadway at Pompeii.

contact with axles.⁵⁸ The high sidewalks appear on all manner of streets; and some result primarily from the local lie of the land, such as those on both sides of the Via di Nocera, where the roadbed slants steeply as it exits the city through the gate of the same name.⁵⁹ Overall, it is difficult to generalize about the characteristics of these locations, particularly whether their builders intended for them to interfere with traffic or not.

Intentionality is much less in question at an intriguing subset of twelve locations, where we might observe the goals of nuisance creators. Here, in addition to having elevated sidewalks, owners did not maintain the slope of the sidewalk with adjacent properties, but exacerbated it or introduced stairs, thus creating a ramp-like effect on one or both sides of their structure's entrance.⁶⁰ The group is perhaps best exemplified by the so-called Caserma dei Gladiatori (V.5.3), a structure that was intermittently a house and apparent gladiator training ground throughout its history.⁶¹ Along the caserma's approximately 25-m-long frontage, the sidewalk rises from the eastern and western property boundaries towards the centre, reaching a height of 1.30 m above the roadbed (Fig. 5.8). Similarly, along the facade of the Casa di L. Caecilius Phoebus (VIII.2.36–7), extra courses were added to the original sidewalk border stones to build up a pedestrian zone that towers 1.20 m over the roadbed (Fig. 5.9). In front of several structures along the Via Consolare on the city's western edge, the ramping effect was also present, but was often more intense because it was limited to the area immediately surrounding the main doorway. Along this street and elsewhere, the local lie of the land provides no strong reason for the ramping phenomenon—in many cases the 'ground' floor was artificially supported by a lower level.⁶²

If, as the material evidence of these cases suggests, such ramping was deliberate, then what would motivate owners to build such constructions?

---

⁵⁸ Though not all border stones at these locations show concrete signs of where wheels left their marks, that does not imply that no nuisance was presented, for drivers may have seen a potential obstruction and deviated their route on account of it.

⁵⁹ The gigantic Casa dei Capitelli Colorati at Pompeii (VII.4.31/51) appears to have been elevated intentionally above its neighbours along the Via degli Augustali atop a coherent line of hefty border stones. Moreover, the sidewalk contributed to the likelihood of obstructed passage, since three stepping stones in the roadbed forced vehicles to stay close to the house's sidewalk, which rose 0.60 m above the road's surface. Vehicles then had to make a sharp turn to avoid the semicircular fountain at the house's and city block's south-eastern corner. By contrast, though the so-called Fullery of Ululutremulus (IX.13.4–6) had a striking streetward presentation and elevated sidewalk, its extra course of sidewalk masonry was set back from the edge of the roadbed, apparently to prevent interference with axles: Spinazzola (1953: 248–9); Hartnett (2008: 114).

⁶⁰ Related to this category of nuisance is another obstruction situated in the sidewalk: the diagonally sloping humps of masonry that outright cover the limited 'walkable' space in front of the Casa dei Vettii (VI.15.1) and on both sides of the Vicolo Storto Nuovo between VII.15 and VII.7. A confounding series of street-bordering structures on the north side of VII.15 at Pompeii might have also blocked traffic: Newsome (2009a).

⁶¹ Tommasino (2004: 28–9); Esposito (2005).

⁶² Aoyagi and Pappalardo (2006).

Fig. 5.9. The sidewalk in front of the Casa di L. Caecilius Phoebus, Pompeii (VIII.2.36–7), which was raised on extra courses of blocks, compelled pedestrians to climb up to the house's doorway (image by permission of the Ministero per i Beni e le Attività Culturali—Soprintendenza Speciale per i Beni Archeologici di Napoli e Pompei).

I would suggest that at least a pair of concerns were in play. First, the owners of these properties desired to lift their thresholds above those of the surrounding structures. Martial and Seneca both equate elevated house entrances with wealth and power, even implying that a degree of fear attended visitors as they approached.[63] At several properties, such as the Casa di Diana I (VI.17.10), elegant travertine stairs raised the threshold still higher above the ramp and were also visible to streetgoers. Though not a subtle form of image-crafting, having a higher and larger structure was nonetheless a powerful way for Roman builders to forge claims about themselves. Second, ramp builders, I contend, also wanted to attract the attention of and compel a degree of action on the part of those passing through the street. Their height above the roadbed distinguished the properties for those looking from afar, while for those passing by, the physical exertion required to walk along the sidewalk—trudging up and then descending a not insignificant incline—added a phenomenological charge and extended owners' impact out past their facades. I will

[63] Mart. 1.70; Sen. *Ep.* 84.11–12.

Nuisances on the Roman Street 157

Table 5.1 Primary uses and area of properties fronted by ramped sidewalks

Address	Name of Property	Primary Use	Ground Floor Area (in m²)
II.4.2–8,10–12	Praedia Iuliae Felicis	Semipublic	2240ª
V.5.3	Caserma dei Gladiatori	Domestic	950
VI.1.9–10	Casa del Chirurgo	Domestic	480
VI.11.18	Unnamed	Domestic	280
VI.16.15–17	Casa dell'Ara Massima	Domestic	175
VI.17.10	Casa di Diana I	Domestic	>455
VI.17.25	Casa del Leone	Domestic	>220
VI.17.27–30	Casa dei Cadaveri di Gesso	Domestic	>625
VI.17.32–6	Casa di Diana II	Domestic	1240
VII.7.19	Unnamed	Unclear	145
VIII.2.23	Sarno Baths	Semipublic	>525
VIII.2.36–7	Casa di L. Caecilius Phoebus	Domestic	>1010

ª This figure represents the enclosed floor area of the Praedia Iuliae Felicis and thus excludes the southeastern portion of insula II.4.

return to this point shortly. For now, it is enough to recognize that those who created sloping sidewalks in front of their properties may have been operating within the letter of the law, but their acts were hardly within its spirit.

What characteristics do these locations share and what might those qualities suggest about those who sought to create such nuisances and why they did so? Of the twelve ramped properties, nine can be positively identified as primarily residential structures, two are semi-public complexes, and the remains of the final one is insufficient to determine its use[64] (Table 5.1). Though it is not possible to know, in most cases, who owned or was the original commissioner of these properties, their large size provides one proxy for their owners' wealth, if not status. On average, these parcels occupy 695 m² of ground area, well more than the average house size in Wallace-Hadrill's samples from Pompeii: 266 m² for Regio I and 289 m² for Regio VI.[65] Even with the sprawling semi-public complexes eliminated and only counting the ground floor of several properties that we know spilled down the hillsides at the city's edge, the remaining properties cover an average ground floor area of approximately 558 m², which places the group comfortably in the top quartile of Wallace-Hadrill's sample of house sizes at Pompeii and Herculaneum. Additionally, six

---

[64] Determining the purpose of any structure is a notoriously difficult and perhaps inherently flawed undertaking at Pompeii (Hartnett [2008a: 102 n. 34]). The domestic spaces identified here boast substantial atria, however, and, as the data below make clear, were often quite large, which makes the identification somewhat less problematic. That said, one ramp stands in front of doorway VI.11.18, which leads to a short staircase hosting stairs to an upper floor. The area underneath the stairs is connected to a larger house, entered at VI.11.19, suggesting that this is a dependency of the sizeable property.

[65] Wallace-Hadrill (1994: 72–87, esp. table 4.2).

of the twelve properties occupy what was likely the most valued real estate in Pompeii: plots on the edge of the city that enjoyed panoramas of, and breezes from, the mountains and bay along the city's southern and western edges (Fig. 5.4). Possessing large properties in prized locations, the owners were likely among the city's elite, which may provide one explanation for how they were able to create constructions that interfered with both wheeled vehicles (via the elevated sidewalks) and pedestrians (via ramping or steps). The connection between these owners' likely standing and their nuisances potentially offers an illustration of the extra-legal process behind nuisance-creation (see above): that is, it is quite possible that these owners were able to create and maintain such nuisances precisely because of their informal power or official position within the city's sociopolitical fabric.[66] Moreover, the streetside constructions, by running contrary to the spirit of the legal principles against obstructions, may have been an assertion of the owners' ability to exist above the law.

Besides speaking to its own set of circumstances, the group of ramped properties also sheds light on my primary line of argument. In these cases, the paired nuisances for street and sidewalk suggest that having an impact on others clearly mattered. I would contend that such an influence was all the more important because of the setting of the street, which was a relatively 'open field' that was used by (and in some senses belonged to) all, but that also provided a stage where Romans vied for status and attention. Amidst other claims—articulated through one's dress, actions, retinue, etc.—that sought to make an impression on streetgoers, creating a substantial nuisance was a way of symbolically asserting one's control over the space and its denizens. It compelled actions on the part of pedestrians or drivers, whose passage was interrupted, whose attention was drawn, and who were forced to vary their course.[67] By dictating the movements of those passing through the street, nuisances created a momentary sense of order and set up a hierarchy of sorts, casting nuisance creators as the dictators of action and streetgoers as those obliged to respond. Pedestrians may have chosen to take the initiative and crossed the street or walked in the roadbed to forgo ramps, benches, and the like. But, in so doing, they nevertheless capitulated in some small way by recognizing the nuisance's potential power and responding to it. Whatever the result, whether compliance or resistance on the part of a streetgoer, it is nonetheless clear that a low-stakes dialogue about control and recognition was initiated when nuisances to movement came into play.

[66] Newsome (2009a: 132–5) offers another possible example of manipulation of street space by a powerful individual at Pompeii.
[67] Both Plutarch (*Publ.* 20) and Pliny (*HN* 36.112) mention that, by way of an extraordinary honour, the doors on the house of P. Valerius Publicola were allowed to open outwards. Plutarch suggests that making way for the doors was a sign of perpetual respect paid by the public.

CONCLUSIONS

What does a focus on nuisance suggest about urban dynamics? In a way, my final focus on ramped frontages draws together many of the themes of this chapter. First, nuisances both more and less intrusive than these were ubiquitous in cities like Pompeii and Herculaneum, and they constantly bumped wheels, forced pedestrians to step up or to the side, and caused backups on sidewalks and roadbeds alike. Our standard view of Roman cities from a 'bird's eye' view may productively encourage us, when we study urban movement, to trace routes through the urban fabric and to consider the viability of the street network as a whole. But a street-level view suggests that, if we have been expecting cities to be optimized and efficient, some rethinking is necessary. Second, working from the ground up, as was key for thinking about the causes and effects of ramp building, puts us face-to-face with those who passed through the street or those whose constructions gave shape to, and thus affected passage through, this space. From this perspective, we are encouraged to entertain the motivations, the irritations, and the *realia* of Roman life, which might be considerably more complicated than has typically been assumed. Many nuisances' apparent contravention of prevailing legal principles, for example, raises questions about how extra-legal forces such as personal relationships, political 'pull', or social status operated in dialogue with the law; what degree of access ordinary Romans had to legal recourse in the first place; and whether, given the legal steps involved, it would even be worthwhile for someone to seek an interdict against a nuisance.

Third, against the background of such negotiation, the ramp-fronted properties stand out all the more for their success in presenting an impediment to passage by wheel and foot. In a way, the obstruction they presented was not far removed from the most impactful physical nuisances in Pompeii, those presented by public amenities such as baths and the urban water system. We have no way of knowing whether the latter were enshrined through a document similar to the *Tabula Heracleensis*, but that inscription's provisions elevated certain classes of activities (largely religious and civic) and individuals (predominantly the elite who were most visible in those activities) above the push and pull of legal wrangling. Fourth and finally, while nuisances certainly presented physical hindrances to intra-urban movement, both the ramped properties and the exceptions of the *Tabula Heracleensis* suggest they could be as important on a symbolic level as well. Being above the legal fray or inconveniencing your fellow city dwellers provided one way to present one's self and to assert a degree of control over those attempting to pass through the street. Ultimately, this final note on nuisances returns attention to my first point about *Midnight Cowboy*: for studies of movement in the city, it is not just the disjointed movement of Ratso Rizzo or Umbricius that matters, but also the presence, intention, and benefits of those responsible for the disjunctures.

6

Pes Dexter

Superstition and the State in the Shaping of Shopfronts and Street Activity in the Roman World*

Steven J. R. Ellis

To move through a Roman city, whether along its streets from one neighbourhood to the next, or across the street from one building to another, brought the Roman urbanite into a kaleidoscopic array of signs and spaces that served to enable, interrupt, or block pedestrian passage.[1] The most powerful of these were also the most common: the doorway and its threshold. The Roman urban streetscape was dominated as much by doorways as by the buildings that each represented. When James Packer categorized Roman street-side doorways, he arrived at two different types: those that opened onto a shop, and those that did not.[2] Even if no precise number of shops will likely be recovered for Ostia or Pompeii, primarily because of difficulties in identification and definition, around 800 can be estimated for the excavated area alone at Ostia and around 600 for Pompeii. These figures should indicate the ubiquity of shopfront doorways in the Roman streetscape.[3] Although variations exist among the

---

* The field research for this chapter was carried out thanks to the kind permission of Pietro Giovanni Guzzo (Soprintendenza Archeologica di Pompei) and Anna Galina Zevi (Soprintendenza per i Beni Archeologici di Ostia). I thank, especially, Kyle Egerer (my research assistant) who energetically cleared, cleaned, and recorded all of the Ostian thresholds that were buried under modern soil and vegetation. Generous financial support for this research was granted initially by the Department of Classical Studies at the University of Michigan, and more recently by the Department of Classics at the University of Cincinnati. My appreciation is extended to the audience members who offered insightful and corrective feedback on the presented version of this paper at the Roman Archaeology Conference in Ann Arbor, 2009.

[1] See Hartnett, in this volume, on blockages to pedestrian movement.
[2] Packer (1971: 21).
[3] DeLaine (2005: 33) counts over 800 shops in the excavated area of Ostia, which she compares only to Rome based on the Severan marble plan. Gassner (1986: p. v) counts precisely 577 shops in the excavated area of Pompeii.

Shopfronts in the Roman World 161

Fig. 6.1. A fairly 'typical' street-front along the eastern side of the Via Stabiana at Pompeii. The narrow entrance at left is to the house (I.3.3), the wide entrance to its shop (I.3.2) (photo by Steven Ellis, with permission of the Soprintendenza Archeologica di Pompei).

types, the doorways to a shop could be distinguished in an instant from those to a house by the width of their thresholds (Fig. 6.1).[4] Born from this potent dichotomy is the now familiar metaphor: the narrower domestic doorway was dignified, privileged, and reserved, while the much-less refined shop doorway threw itself wide open to unashamedly display its insides to all who passed.[5] The street-side doorways were therefore laden with immediate significance, a veritable epithet of the function, type, and status of the property and of its occupants behind. But they also served as some of the most highly charged components in the organization of structured space, coming under the watch of the god Janus who presided over this liminal barrier between entrances and exits, beginnings and endings.[6] Whether within or into a shop, house, or

---

[4] On Roman and Pompeian doorways, see Ivanoff (1859); Wallace-Hadrill (1988: 45–6; 1994: 4–5, 118); Laurence (1995: 67; 2007: 102); MacMahon (2003a; 2003b: 91–9). A recent analysis of Ostian doorways is Stöger (2007). For a useful treatment of Hellenistic doorways see Kyllingstad and Sjöqvist (1965). The 'wide' retail entrances at Pompeii are typically around 3 m, while those to houses are typically between 1 m and 1.5 m.

[5] e.g., Wallace-Hadrill (1994: 118); MacMahon (2003b: 98–9).

[6] On Janus and the threshold, see Holland (1961). For the role of thresholds in the establishment of ritual, see Plummer (1993: 369).

temple, the threshold dictated the rules of access by activating movement from one space to the next, filtering that traffic, or forbidding it altogether. To step over a threshold was no given; each had its own set of rules, be they explicit, implicit, or even unclear. Any significant development or change to the configuration of a doorway and its threshold should therefore represent a transformation in its use and thus demand some kind of explanation for the change, along with a consideration of its implications.

The aim of this chapter is to demonstrate how changes over time in the configuration of Roman shopfronts—particularly their doorways and thresholds—can inform us about ambulatory traffic and the Roman urbanite's experience of the street. To that end, I will move beyond the Vesuvian cities to Ostia, Rome, and elsewhere, as well as from the first into the second century CE in a bid to relate the developments in the arrangement of Roman shopfronts into a broader cultural, political, and historical framework. The second half of the first century CE heralded significant changes to the shape of the Roman city. The independent and piecemeal construction of houses and shops gave way, more conspicuously than ever before, to massive building complexes that combined multiple living and retail units in previously unimagined numbers. This study of Roman shopfronts reveals how their construction and organization, formerly configured to reflect city-wide patterns in ambulatory traffic and retail competition, was wholly transformed at this pivotal moment for the emergence of a new social order. Shops were now *entered* almost exclusively on the right-hand side, causing the rearrangement of their internal and external spatial dynamics and, importantly, demonstrating a new set of urban priorities; the emphasis here will be on the entrance of space, which appears to have caused more superstitious anxiety than exit. This chapter offers an explanation for this cultural phenomenon, its origins and motivations, and considers a range of consequences that these developments had on the changing shape of the Roman city: from the rise and diffusion of the Roman 'cult of the right', to the homogeneity of a Roman retail culture, and the role that doorways and their thresholds played in the visual vocabulary of structured social space and in the rules for moving through it.

A STEP TO THE RIGHT

Some years ago, I suggested that the frontages of Pompeian bars—those with fixed masonry counters—were configured to reflect the busiest confluences of urban activity, as well as some patterns in the movement of ambulatory traffic.[7]

---

[7] For further discussion on the definition, typology, and distribution of Pompeian bars, see Ellis (2004; 2008).

My study found that apart from accumulating at the principal city gates and gravitating near monumental public spaces, a striking 85 per cent of Pompeii's 160 bars were located on a main road and/or an intersection. Expressed in another way, every one of the city's main roads were lined with bars and no fewer than sixty-three (or about two-thirds) of the ninety-six known intersections that connected the Pompeian street network were bordered by at least one bar. This is a distribution pattern that clearly reflects the close relationship between prime retail points of sale and the common confluence of people and urban activity.[8] Moreover, of the seventy-two bars that were located directly at an intersection, more than half of these placed their counters on the optimum side of the facade to allow for an unobstructed viewshed with potential customers on the busy street corner.

This tendency to align counters with areas of greatest activity and movement is demonstrated not only near intersections, but outside the main entrance and exit points of the city's public amenities: for example, opposite all of the major public bathing complexes and the so-called theatre district. Moreover, that counters could be configured to directional flows in ambulatory traffic is especially apparent away from the intersections and major monuments, such as along thoroughfares. For example, of the thirteen counters that lined the southern end of the Via Stabiana, twelve were located on the northern (left) side of the threshold to more fully occupy the viewshed of those heading north from the Porta Stabia into the city (and other such examples abound).[9] This is not to say, of course, that pedestrian traffic flowed exclusively in one direction or the other, but that along certain routes it very likely had its ebbs, flows, and pulls to which shopfronts responded. Viewshed analysis of the Pompeian bar frontages thus offers a potentially useful index of a patently important component of the urban landscape, ambulatory movement, for which we might otherwise have little understanding of, or evidence for.[10]

The configuration of bars to areas of potentially high profit and activity presents us with a very synchronic pattern. It was also one that was much localized, and although—I once imagined—the motivations were doubtless universal, the results are not necessarily consonant with other cities of the Roman world. While several of the bar counters at Herculaneum were clearly configured towards traffic heading up the Cardo V before turning left onto the Decumanus Maximus, too little of the street network and city infrastructure is

---

[8] See van Nes, in this volume.
[9] Ellis (2004: 381).
[10] Moreover, discussions of viewsheds in the Roman world have generally been confined to the more monumental architectural record than for the configuration of innocuous shops and retail entrances, for example temples, monumental arches, villas, and large houses. For monumental arches, see Marlowe (2006); for temples, see MacDonald (1982: 133–42); for temples and villas, see Kaiser (2001); for temples and houses see Kaiser (2003). Cf. DeLaine (2005: 42), who demonstrates how viewsheds operated with the *horrea* and shops in Ostia.

available to draw inferences from this pattern in the movement of people to destinations other than the forum.[11] More of the city is available for analysis at Ostia, and it is here where significant differences in the arrangement of the urban retail fabric can be registered.[12] Not only were just eight of the thirty-eight food and drink outlets located on an intersection (about 20 per cent), but, significantly for the purposes of the following argument, only two of the thirty-eight bar counters were located on the right side of the threshold.[13] This statistic represents a considerable departure from the patterns recognized at Pompeii where, because of the tendency to configure the counters to ambulatory traffic we are left with a more mixed ratio of right- and left-side counters (right: 66; left: 87).[14] It is also a statistic that warrants some explanation. Gustav Hermansen, who pioneered the study of the Ostian bar counters, naturally recognized the strong proclivity there for the counters to be placed on the left side of the threshold, but without explanation relegated the 'intriguing detail' to a footnote and instead asked: 'Was this the force of habit?'[15] This was a fair question, of course, but what kind of 'habit' was the cause and why?

To arrive at an explanation as to why the Ostian counters were so commonly located on the left side of the threshold, and to test whether these were merely random or idiosyncratic outcomes for each city, it is necessary to broaden the scope of enquiry. Given that so few bar counters survive at Ostia, a survey was made on all of the wide retail doorways that retain evidence for the location of the so-called 'night door', the smaller door that was used to enter a property when the larger 'main' door was shut.[16] These were located at one end or the other of the shuttered doorway, thus dictating that any shop furniture—be it a (no longer surviving) masonry or wooden counter, for example—could only have been placed on the opposite side to the swinging 'night door' (Figs. 6.2–6.3). Even if no such fixture—temporary or permanent—was installed here, the type of threshold itself could demonstrate at least the decision that was made between the right and left side for crossing the threshold. This could also go some way to determining why the shopfronts of one city appear to respond to ambulatory traffic patterns more so than in the other.

[11] On the distribution of bars at Herculaneum, see the brief account in De Carolis (1996: 36–7).
[12] On the distribution of retail spaces at Ostia, see DeLaine (2005). On the Ostian bars and inns more specifically, see Hermansen (1974) and esp. Hermansen (1982). On the distribution of *collegia*, see Stöger, in this volume.
[13] For a catalogue of these bars, see Hermansen (1982: 125–83).
[14] Two counters were located in the centre of the room, while five counters remain unknown because of a lack of surviving evidence.
[15] Hermansen (1982: 189, 204 n. 9). See also MacMahon (2003b: 92).
[16] A considerable but unknown number of Ostian counters were systematically removed during the excavations of the first half of the 20th century (pers. comm., Janet DeLaine). The swinging door is often called the 'night door'; e.g. Packer (1971: 22).

Shopfronts in the Roman World 165

Fig. 6.2. Threshold with shuttered groove and 'night door' into the shop at VIII.7.4, as seen from within the shop (photo by courtesy of the Pompeii Archaeological Research Project: Porta Stabia).

Fig. 6.3. The plaster cast of the shopfront doorway at IX.7.10, Pompeii. Note the 'night door' on the right-hand side, while the shutters close off the rest of the threshold (photo by Steven Ellis, with permission of the Soprintendenza Archeologica di Pompei).

The survey of Ostia reveals that the overwhelming majority of thresholds had their 'night door' located on the right-hand side—93 per cent of the total (right: 502; left: 37), reflecting the result already anticipated from the location of the counters. The same survey undertaken at Pompeii, however, returns a much different result. The numbers again echo more closely the results of the counters there, with the sum registering an approximate split at 63–37 per cent, in favour of the right side entry (right: 381; left: 226). Statistically, the results from Ostia represent a dramatic departure from the recognizable system at Pompeii. But is the difference meaningful? What should have caused the configuration of shopfronts of one city to differ so markedly from the other, and what was (or became) so special about the right side of a threshold?

THE ROMAN 'CULT OF THE RIGHT'

Explicit examples of the Roman predilection for the right over the left are as multifarious as they are countless, and can be found throughout the literary, art-historic, and archaeological records. From Pythagoras to the Elder Pliny, and many others besides, we can gauge the extent to which the Greeks and Romans associated goodness, strength, fortune, and luck with all things right-sided.[17] In this they were not alone, as most cultures, both modern and ancient, demonstrate a pronounced proclivity for duality and dichotomy in the organization of their societies.[18] The categories for which societies distinguish duality are many, but polarizing: biological (men and women), racial (western and eastern hemispheres), political (right- and left-wing philosophies), social (right and wrong behaviour), religious (good and evil), temporal (night and day), and sidedness (right and left). For the Romans this was more than a simple distinction between one thing and its opposite. The duality between right and left was above all structured by preferential choice, with the right always being favoured over the left. Only in augury do we see the left side being favoured over the right, if only because Roman augurs faced south during ritual and thus related the flights of birds to the rising of the sun in the east, which was to their left. The Greeks, in contrast, faced north during divination and thus duly preferred the right sky for divination and augury.[19] The cultural implications for left-sided elements were therefore grim: it was thought that thieves stole things with their left hands; that sexual self-stimulus was a lowly left-handed activity; and that journeys would disappoint when set out upon with the left foot first.[20]

[17] For some bibliography on ideas associated with right and left in antiquity, see English (1906); Wagener (1912); Hertz (1960); Lévêque and Vidal-Naquet (1960: 294–308); Lloyd (1962: 56–66); Poehler (2003); Wirth (2010); MacKinnon (forthcoming).

[18] Corballis and Beale (1976: 192); Needham (1973); Palka (2002: 426). Structural anthropologists point to the lateralization of the human brain as the cause of binary dualism, and the fact that around 90% of us are right-handed; the dominance of the left cerebral hemisphere causes voluntary movement and articulate speech, with the result that our actions are mostly better accomplished by our right-side limbs and muscles: Hertz (1960: 90; 1973); Lévi-Strauss (1963); van Zantwijk et al. (1990); MacKinnon (forthcoming).

[19] Cic. Div. 2.82; See also Varro Ling. 7.7; Xen. (An. 6.1.23, 6.5.2); Plin. HN 2.142. For the augural system see Linderski (1986: 2257–66, 2280–2).

[20] On thievery with the left hand: Ovid (Met. 13.110). On masturbation with the left hand: Ovid (Ars. Am. 2.705–8). On setting off with the left foot first: Apuleius (Met. 1.5.27).

Indoctrination in this 'cult of the right' was key, and from a young age children were schooled in the associated fortunes of all things right sided. Plutarch questioned whether 'in general, is it not absurd for people to accustom children to take their food with the right hand, and, if one puts out his left, to rebuke him'.[21] As a philosophical programme, none were more emphatic on the matter than the Pythagoreans (e.g., the Pythagorean 'Table of Opposites').[22] Iamblichus, for example, reported how Pythagoras had taught that it was necessary to put the shoe on the right foot first: 'When stretching forth your feet to have your sandals put on, first extend your right foot'; this was a practice that symbolized that man's first duty was in reverence to the gods.[23] Augustus, we are told, followed this superstition intently,[24] and probably the many other similar examples that Pliny detailed with an almost anthropological inquisitiveness in his *Naturalis Historia*.[25] The Roman temple, Vitruvius tells us, was to have an uneven number of steps so that the first foot to land on the bottom step, that being the right foot, was to be the same foot to step onto the podium itself.[26] Beyond philosophers and emperors and temple priests, however, these were superstitions that had resonance across the full range of the Roman social structure. So when Petronius had Encolpius and his companions enter Trimalchio's dining room, 'Right foot first! [*dextro pede*]' was the cry from the slave entrusted to the very task of ensuring that the rule of crossing the threshold was not broken by a left-footed faux pas.[27]

Little wonder, therefore, that so many allusions to the predilection of the right side over the left can be found beyond the literary realm. The art-historic record equally illustrates the right-sided phenomenon with much tradition and profusion. From gods and emperors to statesmen and greengrocers, we see the right hand being used, almost exclusively, as the active agent in greeting and in speech, as well as in gesture and contact (Fig. 6.4).[28] At a more divine level, it is almost always the right hand of an immortal that casts his will, as when Jupiter throws lightning with his right hand.[29] Fewer studies

---

[21] Plutarch (*Lib. Ed.* 5).
[22] Aristotle (*Metaph.* 1.5.986a).
[23] Iamblichus (*Protrepticus*, symbol xii).
[24] Suet. *Aug.* 92.1.
[25] The references from Pliny have been extensively documented elsewhere, e.g.: Eitrem (1915: 29) and Wolters (1935: 77–9). Some notable examples from Plin. *HN* include 17.153, 9.50, 11.245, 11.250, 24.172, 28.5, 7.15, 7. 77, 2.24 (cf. Suet. *Aug.* 92).
[26] Vitr. *De Arch.* 3.4.4.
[27] Petr. *Sat.* 30.6 makes the point clear that it was the guests who had committed the embarrassing folly of so carelessly abusing the custom, for which they were 'naturally nervous', rather than using the episode to highlight yet another quirk of their eccentric host. On the gullibility of the guests, see also Leach (2004: 82).
[28] Of the countless examples, see those from Greek vase painting as early as the 4th century BCE that can be found in, especially, Green (1999: 33–63). For this convention in Roman art, see Richter (2003).
[29] For the symbolic significance of the right hand, see Groß (1985).

Fig. 6.4. This shopkeeper on a relief at Ostia is shown standing behind a wooden trestle counter raising the right hand to communicate the price of one or more of the vegetables that are detailed on the counter (Museo Ostiense, Ostia: inv. 198; photo by Steven Ellis, with permission of the Soprintendenza per i Beni Archeologici di Ostia).

from within the archaeological record, by contrast, have engaged with the same phenomenon. Even so, some excellent work has been produced more recently on a vast array of data types. Michael MacKinnon, for example, has adroitly recognized that in Greek and Roman animal sacrifices the offering of parts of the right side of the animal represented the strong majority.[30] The implication is that the right parts of the animal represent the godly, pure, or heavenly components. From the work of Eric Poehler we now have a clearer understanding of the organization of cart traffic in Pompeii, which itself was based on the Roman systematization of driving on the right side of the street rather than the left.[31] Such a broadly integrated traffic system could only have been born from custom and habit, and the inveterate tendency to choose the right over the left in all matters, even in movement.

REGULATING THE RIGHT: REBUILDING AND REPLANNING

That a preference had developed—or had long existed—in the Roman mindset for entering a shop on the right-hand side of its threshold is now more clear, but why do we see the archaeological traces of this habit more fully formed in Ostia than in Pompeii, and expressed in such overwhelming numbers? The key here is the thirty-year gap, more or less, between the destruction of all of the shops in one city and the construction of those in the other, along with an awareness of the legal response to certain events that played out around this time in the capital itself.[32] The catastrophic fire that devastated Rome in 64 CE caused a watershed in the Roman building industry, issuing a new opportunity to apply greater urban building regulations across the full range of the urban landscape; just four of Rome's fourteen districts were untouched by the fire.[33] For while building codes had long been a challenge for the jurists—with the more successfully enacted codes being a product of the Augustan urban reform—only now with large swathes of the urban landscape destroyed could they be effectively applied to a more complete set of urban amenities that included shops and street-fronts.[34] The actual detail of the Neronian

---

[30] My utmost appreciation is extended to Michael MacKinnon for kindly sharing his forthcoming published work with me: see MacKinnon (forthcoming).
[31] Poehler (2006: 53–74).
[32] On the changing morphology of the bar counters during this time—which saw the introduction of the (water?) basin in the front of the Ostian counters, something which is never seen at Pompeii—as relating to the Roman 'sumptuary laws', see Hermansen (1974: 167).
[33] Tac. *Ann.* 15.40. Newbold (1974); Martin (1989: 19–72); Lancaster (1998: 305).
[34] For Roman building codes under Augustus, Mommsen (1887) remains the classic. See also Homo (1951) and Martin (1989).

building codes (*Lex Neronis de modo aedificiorum urbis*) are obscure, but from Tacitus we know that the city (re)claimed jurisdiction over the parts of buildings that faced onto the street or public spaces, that roads were to be widened and flanked by porticoes, that height restrictions were to be in place, that fire-retardant brick was to replace wood, and that a fire-break space was to separate neighbouring buildings.[35] These were important and widely significant regulations: Rome had to rebuild, and do so quickly given that at least 200,000 residents were now without shelter.[36] The effects of the new statute outside Rome were anything but immediate as, after all, buildings constructed prior to the issuing of the codes were exempt, while any new buildings were dependent upon local economic stimulus to be conceived of in the first place. The fires that razed parts of Rome in 80 CE must also have underlined the need for building reform, as well as providing yet another opportunity to enact them.[37]

It is difficult to chart the changes caused by the Neronian building codes in Rome itself; the surviving evidence is much too patchy. Ostia, by contrast, which naturally came under the same jurisdiction as Rome, proves a better test case for measuring the impact of the revitalized building codes.[38] Quite apart from the fact that much more of the urban fabric is available for analysis at Ostia, the key to recognizing the possible adoption of a more centralized building regulation at Ostia comes not from the outbreak of fire or natural catastrophe, but from the economic revival of the early second century CE. Trajan's establishment of Portus as Rome's principal port fuelled an economic surge in building. Almost all that survives at Ostia today dates back to this age of intense prosperity, when the piecemeal building projects of the past gave way to the systematization of massive urban complexes containing multiple living units.[39] The newly unified building projects, such as those that lined the northern *cardo* and the eastern *decumanus*, might even have been centrally conceived as rental property.[40] That a centralized building regulation was (re-)established to control this outburst of construction projects—Russell Meiggs's 'revolution in Ostian architecture'—has long been recognized, and for as early as the period under Domitian.[41] The civic observance of a building code is most palpably seen in the many Ostian buildings and insulae that were uniformly raised by about a metre in the late first or early second century CE. F. H. Wilson argued that this was the result of a deliberate policy, one that was 'controlled by statute' or some kind of local bye-law.[42] Although Meiggs

---

[35] Tac. *Ann.* 15.43 and Suet. *Ner.* 16. See also the landmark essay on these building laws in Voigt (1903); Yavetz (1958: 513); Newbold (1974).
[36] Newbold (1974: 858).
[37] Cass. Dio 66.24.
[38] Hermansen (1982: 91–6).
[39] For the changes caused to the many *horrea* at Ostia during this economic revival, see Rickman (1971: 77–8).
[40] DeLaine (2005: 33). [41] Meiggs (1960: 65). [42] Wilson (1935: 53).

pointed to a centralized regulation, he saw the changes as having been inspired by rebuilding and replanning in Rome—including the introduction of wider streets and systematized apartment blocks—created after the fire of 64 CE.[43] This was a state code rather than a civic code, and one that was possibly enforced upon new constructions directly by the curator of public works (*cura operum publicorum*).[44]

While we might never know the specific details of the Neronian (or later) building codes, it is at least clear that the systematized 'architectural revolution' at Ostia was centrally regulated and that the transformation was close to wholesale. We also cannot know if one of the regulations was to ensure that thresholds had their 'night door' placed on the right, but the suggestion is all the more likely given that the vast majority of shopfronts from this period adopted precisely the same form of right-sided threshold. The measures were also likely to have been motivated by more than the pressing need to dampen any potential city-wide outbreak of fire. Regulating the appearance of Roman shopfronts, which now uniformly lined the Ostian streets—one after the other, and in numbers never seen before—would have gone some way to ensuring a more standardized streetscape. This process of homogenization was most ostentatiously expressed by the *en masse* arrival of porticoes that now framed many of the shops and lined most of the streets.[45] The selection of one side of the threshold for the door over the other—and thus, by implication, the configuration of the counter or other shopfront furniture—satisfied this process of normalizing the street-front, and at the same time appealed to any superstitious notions or anxieties for the 'cult of the right'. Even at the most practical level, the uniformity of the shopfront threshold must have simplified mass production, a surely vital, *au courant*, factor during the early second century CE construction boom, at least at Ostia. A telling correlate is the recognizable spike at this time in the mass production of fired bricks as well as in the widespread production of the less labour-intensive and cheaper *opus caementicium*.[46]

ROME AND BEYOND

If this constellation of superstition, catastrophe, jurisdiction, and economic prosperity conspired to reshape the Roman shopfront, then the results from Ostia can hardly be seen as idiosyncratic. We should therefore expect to see the

---

[43] Meiggs (1960: 67).
[44] See Kolb (1993: 53–7); cf. Bruun (1991: 200–6); and Eck (1992: 242–3). For a good introduction to the *cura operum publicorum*, see Robinson (1992: 54–6).
[45] Equally attributed to Nero: Suet. *Ner.* 16.
[46] On the brick, see Steinby (1983: 219). Cf. Bloch (1947: 334–7); Helen (1975: 18–20); DeLaine (2002). On the *opus caementicium*, see Torelli (1980: 139–59); Rakob (1976: 366–78).

same trend in Rome itself, naturally enough, as well as among the many other towns under Roman administration, and from the late first century CE. While no other Roman site offers as much shopfront data to compare to both Pompeii and Ostia, much can still be gained from the several small handfuls of shopfronts that exist throughout the Roman world. Especially pertinent are the shopfronts that date from precisely the same period as the Ostian examples, even if they did not necessarily experience a similar construction boom as at Ostia.

In Rome itself there could hardly be a more inviting subject to begin than the Markets of Trajan, a massive complex of close to 170 mostly single-room cells that date to the early second century CE. Even if the identification of these cells as shops has rightly been called to question, and even if only three of the original thresholds survive among the fascist reconstructions, the 'night door' on these was strictly on the right-hand side.[47] It is notable that all of the reconstructed thresholds in this structure are equally right-sided entry; this modern statistic might reflect a willingness of the *muratore* to copy the type of original thresholds that were encountered here. Few other original thresholds from shopfronts survive in the capital. The one preserved threshold from the ground floor shop of the Insula Ara Coeli below the Capitoline, however, was a right-sided entry, and equally so an example from the Nova Via below the Palatine.

Better still, at the site of Lucus Feroniae, north of Rome, the threshold stones from nine shopfronts are preserved from the early second century CE, eight of which fronted onto the forum. Every one of these examples constituted a right-sided threshold. Further south, at Ordona, the thresholds of sixteen shops are preserved in the civic *macellum*, all of them right-side entry.[48] Still in the early second century CE but yet further afield at Baelo, in Spain, another *macellum* displays the same statistic with all fourteen of its shops retaining the thresholds of the right-sided entry type.[49] The six shops that survive along the northern side of the Baelo's Decumanus Maximus, just inside the city's western gate, are also exclusively right-sided entry. And while patently more shopfronts exist throughout the Roman world, it is very difficult to account for these as so few have been published, and even fewer of these publications illustrate the threshold stones in even the most basic detail. Even so, the many miscellaneous shops that are available for analysis demonstrate the pronounced bias towards a right-sided entry for the late first and early second century CE.

---

[47] Although unattested to in antiquity, the term Trajan's 'market' was applied to identify the structure from as early as the 1920s. Lugli (1929–30: 527–51) soon questioned the commercial nature of the complex, instead suggesting it could have been a distribution centre. For a good summary, see Frayn (1993: 16–17). My thanks to Lynne Lancaster (pers. comm.) for pointing out to me the locations of the few surviving original thresholds.

[48] On Ordona, see de Ruyt (1983: 82).

[49] On Baelo, see de Ruyt (1983: 44–6).

CONCLUSIONS

A return to the shopfronts at Pompeii, which have been largely excluded from this survey of late first and early second century CE urban landscapes for all obvious reasons can, I believe, underscore these developments while equally illustrating how close Pompeii came to experiencing the same transformation that swept the retail landscape of slightly later Roman cities. Two of the very latest constructions at Pompeii were massive complexes that in concept, as much as execution, differed markedly from the piecemeal constructions that typify the Vesuvian cities; they were thus a prelude to what would dominate Roman architecture from, especially, the early second century CE. One was the Central Baths, which we know were still under construction in 79 CE, while the other was the row of fourteen structurally identical shops that lined up behind a (rare, for Pompeii) portico attached to the so-called Villa della Colonne a Mosaico on the eastern side of the Via dei Sepolcri and Via Superiore outside the Porta Ercolano.[50] Of the identifiable shop thresholds that front the Central Baths (eight in total), all were right-side entry. So too were the fourteen porticoed shops outside the Porta Ercolano.

These two final Pompeian examples may thus reflect the emergence of a new-look city streetscape that was recalibrating the Roman urbanite's experience of the street. The facades of Roman shops—formerly configured on a piecemeal basis that could reflect city-wide patterns in ambulatory traffic and retail competition—were transformed into a more standardized urban retail environment, consonant with Nero's urban whitewash that lent rhythmic order to the Roman street-front of which the portico became a principal and still recognizable protagonist.[51] Directional movement need never to have changed during this period, but our (or my) willingness to use the configuration of shopfronts as a gauge for it must. For while the pull of the Roman pedestrian through a town remained particular to the urban armatures of place, the once idiosyncratic response to it was universally modified to abide by the *Lex Neronis de modo aedificiorum urbis* of the later first century CE. That the Roman 'cult of the right' was brought to the fore in this process demonstrates the quite intriguing phenomenon that old traditions could be applied to the built environment and sweep aside other rationalities, including the economic.

---

[50] On the Central Baths, see Fagan (1999). On the so-called 'Strip Building' (a term applied by Peña and McCallum 2009: 72) outside the Porta Ercolano, see Kockel and Weber (1983).

[51] On the portico being a Neronian measure in the *Lex Neronis de modo aedificiorum urbis*, see Tac. *Ann.* 15.43.

7

Cart Traffic Flow in Pompeii and Rome

Alan Kaiser

According to a popular but inaccurate story, Julius Caesar decreed that all Roman carts were to have the same axle width. As a result, it is suggested, all ruts in Roman streets and roads were the same size. Similarly, this false argument suggests that during the Middle Ages people wanting to use the Roman roads had to build their axles to match those of the Romans so that they could fit into the ancient ruts. Then, when they were building the first predecessors of modern railroads, the cart makers who were employed used the axle width to which they were accustomed to build carts. The result is that British engineers who designed the modern United States standard railroads used a gauge of 1.435 m, the distance between Roman cart wheels.[1] There is much wrong with the details of this story, not least the fact that no ancient source actually discusses a law regulating the size of Roman cart axles. The entire story is built on the claim that the width of Roman axles was uniform, a claim that is demonstrably false. The distance between ruts, a proxy measure for axle size, differs between, and even within, Roman sites by more than 0.5 m.[2]

That the story is a myth is easy to accept since it is perpetuated by people who do not have any expertise in the field, but another story has more credibility as experts have passed it on, although admittedly in publications aimed at a popular audience. According to this myth cart traffic was banned in Rome and probably other cities during the daylight hours. This story arises from a simple misreading of the *Tabula Heracleensis*, an inscription from the southern Italian city of Heraclea, which probably repeats passages from *Lex Iulia Municipalis*, a law code Caesar promulgated for Rome.[3] According to the text all *plostra* were banned from Roman streets until ten hours after sunrise, essentially the entire day.[4] The word *plostrum* is a variant of *plaustrum*, which

---

[1] Neuman, McKnight, and Soloman (1999: 97).
[2] Grenier (1934: 375–6); Pike (1967), cf. Table 7.1 below.
[3] Crawford (1996: 358–9); Nicolet (1987).
[4] *Tab. Hera.* 56–61. Crawford (1996: 355 ff.); Robinson (1992: 81).

was a slow-moving, utilitarian oxcart intended for transporting heavy loads.[5] The *lex* did make exceptions for *plaustra* hauling waste or materials associated with municipal or religious construction projects and those used in ceremonial processions. The only Latin word for cart in the text is *plostrum*. No ancient author uses either this word or its variant to refer to all vehicles in general. Thus the law was aimed specifically at functional hauling vehicles and not at passenger vehicles that also plied Rome's streets, such as the light chariot (*currus*), ostentatious carriage (*carruca* and *raeda*), or women's carriage (*carpentum* and *pilentum*).[6] Unfortunately, scholars have tended to miss this distinction and have perpetuated the idea that the law applied to all vehicles.[7] However, we should not envision the streets of Rome and other Roman cities as devoid of cart traffic in the daytime. Rather, we should envision them devoid of some oxcarts hauling goods, but dotted with carts for personal transport as well as some heavy wagons hauling construction materials, debris, or religious paraphernalia.

While these stories and their debunking appear a trivial matter, they are illustrative of an assumption of much work on traffic in Roman cities, particularly among those specializing in the evidence from Pompeii and Rome. The assumption is that Roman civic officials were concerned with facilitating both pedestrian and vehicle traffic by regulating the movement of carts. Matthews has argued that one of the duties of the *cohortes vigilum* in Rome was to arrest traffic offenders.[8] Salama has suggested that civic officials planned traffic flow in the Roman cities of North Africa by designating one-way streets and assigning civic officials to regulate the use of those streets.[9] Independently, Ciprotti and Tsujimura made the same suggestion about Pompeii, leading Tsujimura to claim that the flow of all carts in the city was 'completely systematized'.[10] Van Tilburg sees in the evidence from Pompeii a planned series of cart routes through the city regulated by civic officials.[11] Poelher agrees there were 'rules' in place and 'regulators' to enforce them.[12]

In this chapter, I challenge the assumption that civic officials in Pompeii, Rome, and other cities promulgated traffic laws to regulate and facilitate the movement of carts through the city. Instead, I outline a more complex

---

[5] cf. Auson. *Ep.* 6.21–31; Caes. *B Afr.* 9; Cato *R.R.* 62; Cic. *Div.* 1.27.57; id. *Verr.* 2.1.20.54; Hor. *Ars P.* 275–7; Isid. *Orig.* 20.12; Livy 4.41.8; Ov. *Met.* 2.176–7, 12.280; Plaut. *Aululuria* 504–21; Prop. 3.5.35; Varro *Ling.* 5.140; Verg. *G.* 2.206; Vitr. *De Arch.* 10.01.5.

[6] For a discussion of the legal restrictions on these passenger vehicles, see below.

[7] e.g., Adkins and Adkins (1994: 186); Carcopino (1940: 49); Hyland (1990: 233); Johnston (1957: 91); Matthews (1960: 23); Staccioli (2003: 21). In the original Italian version of his work, also published in 2003, Staccioli does not make this mistake.

[8] Matthews (1960: 25).

[9] Salama (1951: 88; 1994: 356).

[10] Ciprotti (1962: 34); Tsujimura (1991: 63).

[11] Van Tilburg (2007: 137 ff.). This work was originally published in Dutch in 2005.

[12] Poehler (2006: 73).

situation in which residents at Pompeii and elsewhere had the legal right to block their street to cart traffic if they so chose. Based on the evidence from Pompeii, it seems that some property owners exercised this right, excluding carts from their streets in order to rid themselves of a perceived nuisance.[13] This left cart drivers to traverse the city only along the routes from which they had not been excluded. Despite the fact that the cart routes created by exclusion were quite laborious to travel, civic officials and residents had little interest in improving the plight of cart drivers.

THE CHALLENGES OF DRIVING IN POMPEII

At first glance, the archaeological evidence from Pompeii would seem to argue in favour of a civic concern for facilitating the movement of carts within the city. The grid layout of the city is ideally designed to ease traffic flow. The street network at Pompeii grew over time and by 79 CE had at least three elements on slightly different orientations (see Fig. 7.1). The oldest part of the city in the south-western corner contains the most irregular street layout, but one can still discern a grid plan even if that plan is not standardized.[14] With the expansion of the city to the north a century later and then another expansion to the east of the Via Stabiana, the pre-Roman Pompeians utilized a standardized grid although the orientation of the two new grids differed from that of the original settlement.[15] After the Social War (91–88 BCE), Roman officials had the opportunity to shape the city's street network and impose their ideas about traffic flow over the next century and a half. The city may not have been originally laid out by Roman officials, but because of the length of time they were in charge we can take Pompeii's street network as an example of Roman urban traffic management.[16]

If civic officials intend a city to facilitate the flow of vehicles, according to modern urban planner K. Leibbrand, they must ensure the street network has 'an economic, logical, and comprehensible form'.[17] Three factors in particular make any street grid economic, comprehensible, and logical. First, the multiple routes within a grid provide users with multiple ways of reaching the same destination so that if one route becomes temporarily blocked or is crowded, one has other options for proceeding.[18] A second feature of grids is a

---

[13] For more on nuisances, see Hartnett, in this volume.
[14] Gesemann (1996: 52). See Newsome (2009a) for the deformation of this grid around insula VII.15.
[15] For the early history of Pompeii, see Zanker (1998: 31 ff.); Owens (1991: 100–1); von Gerkan (1940).
[16] Westfall (2007: 129).
[17] Leibbrand (1970: 91).   [18] Brown and Sherrard (1969: 51).

street hierarchy. Modern gridded street networks facilitate the movement of vehicles most efficiently if civic officials create a hierarchy of streets in which some streets serve as main arteries, allowing someone to cross town quickly and easily, while others serve as minor streets giving local neighbourhood access. This division is economical as civic officials can pave the main arteries with more durable, and expensive, materials than the local streets. The grids found in many Roman cities had a built-in hierarchy due to the walls that encircled them. Streets leading to city gates must have served as arterial thoroughfares while those that ended in blank city walls must have been local side streets. The arterial nature of streets leading to city gates of many Roman cities is confirmed by the frequency with which we find one or two streets linking the forum, a central destination, and a city gate in nearly all Roman orthogonal cities. The third factor that facilitates traffic flow is standardization. City blocks of an identical size and streets of a similar width as well as corners that meet at right angles achieve the goals of being comprehensible and logical. A gridded street network like that at Pompeii, to which Leibbrand draws the reader's attention, accomplished these goals and would appear to demonstrate civic interest in keeping vehicles moving. However, a problem arises in many cities, including Pompeii, when several grids laid out at different times have more than one orientation; the advantage of standardization is diminished at the intersection of the grids.[19] In the case of Pompeii, however, the similar orientations of the grids created only a few awkward corners with sharp or difficult turns. The slow speed of cart traffic would have mitigated this problem.

When we look at the street plan of a gridded Roman city, we tend to assume that carts could have moved down every street, utilizing the advantages of multiple routes, hierarchy, and standardization that a grid presents. The streets look very different on the ground, however, as there is much in Pompeii's streets to hinder the progress of a cart.[20] Indeed, a driver would have needed a mental map of Pompeii with some key pieces of information in addition to knowing the layout of the streets in order to successfully reach his destination. A cart driver would have needed to know whether a street was wide enough for the passage of only one cart or two going in opposite directions, or was so narrow that it was closed to carts altogether. Other information a cart driver would have needed would have been the direction of ruts at corners. At some Pompeian corners ruts turn in only one direction. While the depth of these ruts is usually not great, allowing one to turn in the opposite direction, it would have been easier to follow the ruts. Finally cart drivers would have needed to know the location of impediments to traffic. Impediments came in three classes: blocking stones; kerbs and sidewalks,

---

[19] Leibbrand (1970: 93).   [20] See Hartnett, in this volume.

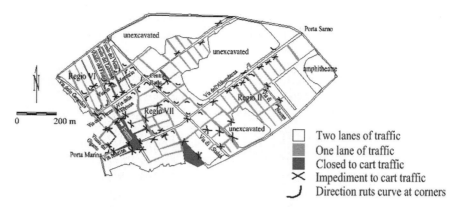

Fig. 7.1. Information of interest to Pompeian cart drivers (after Tsujimura 1991 and Poehler 2006 with additional observations by Alan Kaiser).

extended to block the entrance to a street; and building or fountain construction extending into a street and so blocking cart access.[21] Combining the work of Tsujimura and Poehler on ruts and impediments with my own observations, I have created a map of Pompeii with the information a cart driver in the city might have needed (Fig. 7.1).[22]

It is clear that the Pompeians, apparently with the knowledge and perhaps collaboration of civic officials, destroyed the advantages of multiple routes, hierarchy, and standardization presented by a gridded street network. Fig. 7.1 shows a number of impediments to cart traffic spread across Pompeii that would have made it very difficult to drive across the city because of the seemingly unsystematic nature of their distribution. If one were driving a cart west along the Via dell'Abbondanza and wanted to turn south into the neighbourhood of Regio II, one would have to have known that the majority of streets were blocked and only a few would allow for such a turn. The neighbourhood of Regio VII would have been even more difficult to drive through. While nearly all of the streets were open to cart traffic, several were only open at one end, forcing a cart driver to make several turns to reach a particular destination and, in some cases, even turn the cart around to leave. The distribution of ruts proves that carts did traverse most neighbourhoods in Pompeii, including Regio II and Regio VII, even if it was difficult to navigate them and required advanced planning.

[21] Gesemann (1996: 65 ff.).   [22] Tsujimura (1991); Poehler (2006).

While unsystematically placed impediments made it difficult to drive through much of Pompeii, a few areas were blocked to vehicles completely, complicating movement even further. By 79 CE, the forum was closed to carts. From a cart driver's point of view, this must have been extremely inconvenient; the forum is the central meeting point of many streets in the western half of the city and would have made an obvious place through which carts could go to arrive in different parts of the city. A cart driver entering the city along the Via Marina west of the forum was faced with a daunting navigational challenge to get to anywhere within the city.[23] Had this driver had the opportunity to steer through the forum, his life would have been made much easier.

The neighbourhood where impediments did the most to remove the advantages a grid presents in terms of multiple routes, hierarchy, and standardization is that surrounding the amphitheatre—Regio II—in the south-east quarter of the city. Extended kerbs and sidewalks blocked the entrance to nearly every street leading south from the Via dell'Abbondanza and east from the Via di Nocera into Regio II. Koga interpreted these types of barriers as intended to help direct water run-off.[24] They must also have been intended to block cart traffic, however, as Pompeians had the option of raising the pavement in such a way as to block water but not cart traffic.[25] These barriers would have had the effect of cutting most of Regio II off from daily traffic. However, the surface of the first side-street west of the Porta Sarno, leading from Via dell'Abbondanza to the amphitheatre, is currently buried beneath modern material (in Fig. 7.1, it is assumed that it was not impeded, if it was impeded the area of the amphitheatre was closed to vehicular traffic). The exclusion or severe restriction of vehicles within Regio II must have made life for the pot seller located at II.3.8 or the people running commercial market gardens (II.3.7 and II.9.7) very difficult if using vehicles, and might suggest that businesses of this nature may have been more likely to use pack animals for the delivery of materials to, and for the distribution of products and produce from, their premises.[26] The same must be true for those managing the amphitheatre, who had to transport animals, scenery, and other equipment necessary for a good show.

The placement of impediments in a haphazard or unsystematic manner, and the closing of certain areas to carts, would not have facilitated the movement of vehicles. The fact that the people of Pompeii chose to periodically change those impediments added one more element of

---

[23] Newsome (2009a) has a thorough discussion of the evolution of traffic routes west of the forum.
[24] Koga's drainage Type C (1992: 59).
[25] Jansen (2007: 264).
[26] On II.3.8, see Peña and McCallum (2009: 76).

unpredictability that subverted the grid. It is clear that Pompeians erected impediments to cart traffic along some streets that at one time had been open to carts. One example comes from the Central Baths, which were under construction in 79 CE. The east side of the new baths reached into the adjoining street, turning a street through which carts could once pass into one in which they could not. The ruts along this street attest to its importance to cart drivers, nonetheless the newly built narrow sidewalk and the base of the eastern wall cover up one side of these ruts. Whatever the reason for this expansion, the builders and civic officials seem to have had little concern for the impact on traffic circulation.

Further examples of the ever-changing nature of cart routes come from two separate studies in different parts of the city. In Regio VII, Newsome documented the response of the owner of the Casa del Marinaio (VII.15.1–2) to the blockage of vehicular access to the forum from the Porta Marina and Via Marina.[27] This change in traffic patterns forced cart drivers to turn north onto the Vicolo del Gigante from the Via Marina and drive past the Casa del Marinaio. Among other changes, the house owner placed a fountain in the street south of the house, leaving cart drivers only one route for getting through Regio VII. Poehler has conducted the most thorough inventory of the signs of cart passage in Regio VI and found numerous examples of changes to the streets that would have disrupted cart travel. At the time of the eruption, one section of the Vicolo del Labirinto was closed for repaving.[28] A blocking stone had recently been erected on the Vicolo del Fauno at the Via della Fortuna, blocking cart traffic. Along the east side of the Vicolo dei Vettii an enlarged shop spilled out over the sidewalk forcing someone to reconstruct the sidewalk, pushing it into the street, enveloping a stepping stone and narrowing the street.[29] Poehler also concludes that traffic along the one-lane Vicolo di Mercurio originally was predominantly eastbound but later switched to become predominantly westbound.[30] The dynamic changes which Newsome and Poehler describe further served to destroy the standardization of the grid network.

While the random placement of impediments to cart traffic and the ease with which Pompeians changed access to carts seem to have negated the multiple routes and standardization of a grid system, Fig. 7.1 presents some evidence that Pompeians did create a street hierarchy, although to a limited extent. Of the streets leading to the seven gates that we can observe, six were wide enough to accommodate two lanes of traffic moving in opposite directions. Only one street, the Via Marina, was so narrow that it allowed for one lane of traffic. In addition, five of the streets that lead to city gates had no impediments to cart traffic. The two that did, the Via Marina and the Via

---

[27] Newsome (2009a). [28] Poehler (2006: 63).
[29] Poehler (2006: 70). [30] Poehler (2006: 72).

dell'Abbondanza, impeded traffic only in the vicinity of the forum and thus these impediments appear to be part of a larger strategy to block the forum to cart traffic. Yet this is the only evidence that Pompeians took advantage of the street hierarchy. If civic officials at Pompeii had wanted to use funds for maintaining infrastructure in the most efficient way, they could have paved main arteries differently from local neighbourhood streets as the mayors and town councils of modern cities do. At Pompeii high quality basalt blocks are used uniformly across most of the city to pave arterial and side streets, a use of resources that was not the most economical.

We see other missed opportunities for utilizing the street hierarchy to its fullest extent. The Via di Nocera, which connected a city gate to the Via dell'Abbondanza, one of the city's main arteries, narrows to one lane in the block south of the Via dell'Abbondanza. This narrowing would have created a bottleneck at a crucial point in the street network. The area where the hierarchy is most subverted, however, is in Regio VI, the north-western area of the city. There the Via Consolare led to the Porta Ercolano. The street leading south into the city from the gate hugs the west side of the city, forcing cart drivers to take a left-hand turn onto either the Vicolo di Mercurio or the Via delle Terme/Via della Fortuna in order to reach any other part of the city. From the depth and number of ruts, however, it would appear that the northern route, the Vicolo di Mercurio, was the preferred route of cart drivers. This is a very odd choice as this street is only wide enough for one lane of traffic as opposed to its southern parallel neighbour which could accommodate two lanes.

While all drivers had to cope with the difficulties described so far, those driving the largest oxcarts faced an additional problem as they were funnelled into a smaller number of streets along the edges of the city. Table 7.1 presents a sample of rut measurements from across Pompeii. The vast majority of ruts were about 1.40 m apart with only a few of a larger size. Pike suggested that the difference in rut size corresponds to a difference in the animal providing traction.[31] She studied the carts still in use around the Mediterranean, particularly in Italy; the donkeys, mules, ponies, and oxen used to pull these carts are probably not so different in size from their ancient counterparts.[32] Measuring from the centre of metal-clad wooden tyres, she found fairly consistent sizes that fell into four categories. Her narrowest category (1.15–1.20 m) belonged to carts drawn by hand or by donkeys. The next widest category (1.35–1.40 m) consisted of carts drawn by a single horse. Pike found that oxen in the modern Mediterranean region are normally yoked in pairs to provide traction and so only donkeys and horses were associated with vehicles that required a single

---

[31] Pike (1967: 593–605).
[32] For a discussion of the role of these animals in transportation, see Vigneron (1968: 139ff.); Laurence (1999: 123ff.); Raepsaet (2002: 217ff.).

animal. A pair of either horses or oxen pulled carts that were in her next category (1.45–1.50 m). Finally a pair of oxen pulled the largest vehicles (1.55–1.70 m). Pike's categories for the distance between wheels is nearly identical to the categories Grenier proposed for the distance between ancient Roman wheel ruts, suggesting that the distance between ruts can be a proxy indicator for the type of animal that provided traction for the vehicle.[33]

Returning to the evidence from Pompeii, horse carts with passengers or light loads travelled most of the city's streets, while oxcarts with heavy loads had much more circumscribed routes. Pairs of ruts with a distance of 1.35–1.40 m between them, and thus falling into Pike's category for horse carts, are ubiquitous across the site (Table 7.1). In contrast, pairs of ruts with a width of 1.55 m or larger, and therefore falling into Pike's category of ox-drawn carts, are found only in the Via Consolare, the Vicolo di Mercurio, and the Via di Nocera (see Fig. 7.1). This does not mean that oxcarts were restricted from the rest of the streets in the city. The evidence from the Via Stabiana was inconclusive because the great confusion of ruts made it difficult to match pairs in order to obtain a measurement between them. As for the rest of the streets in the city, oxcarts may have traversed the streets in such small numbers that they left no ruts while their smaller counterparts used the same streets in much greater numbers leaving ruts. Oxen pulling a *plaustrum* with a heavy load of goods, raw materials, or waste would have been a common sight primarily on the edge of Pompeii. In the interior of the city donkeys or horses pulled passenger vehicles such as the *currus*, *carpentum*, *pilentum*, or *raeda*, the latter of which was also used for hauling light loads of goods or raw materials.[34]

If it is possible to document so many places where impediments and the inefficient use of street hierarchy disrupted cart traffic flow, the obvious question is: who was responsible? To answer this question it may be useful to examine who benefited from disrupting cart traffic. In some cases the answer may be collective; some or all of the residents of a particular street

---

[33] Grenier (1934: 375–6). The archaeological evidence of a few excavated carts helps to confirm that we can use Pike's data to identify the type of animal providing traction for an ancient cart. According to Röring's reconstruction of the large, covered, four-wheeled Wardertal wagon, dating to c.200 CE, the distance between wheels is about 1.60 m; the design of the wagon requires it to be pulled by two animals just as Pike would predict (Röring 1983: 50). A two-wheeled vehicle from Patuscha, Bulgaria, required two animals and also had a width of 1.55 m between the wheels (Röring 1983: 70–1). A four-wheeled wagon designed to be drawn by two animals found in the Villa della Ariana in Stabiae has about 1.48 m between the wheels (Tsujimura 1990: 61). The two-wheeled gig from the Casa del Menandro at Pompeii had a width of 1.42 m between its wheels, which falls between Pike's categories of vehicles drawn by a pair of animals or a single horse (Tsujimura 1990: 61). This vehicle was designed to be pulled by a single animal.

[34] For the use of a *raeda* in the movement of goods, see *Cod. Theod.* 8.5.8.1; Juv. 3.10–20; Mart. 3.47; SHA *Max.* 6.9.

Table 7.1 Sample of rut widths from within Pompeii

Distance between ruts (m)	Part of rut measured	Type of Traction (Pike)	Regio	Location
1.35	centre	single horse	VI	Vico dei Vetti
1.40	centre	single horse	I	I.10.14 (House of Menander)
1.40	centre	single horse	II	Via di Nocera near intersection with Via dell'Abbondanza
1.40	centre	single horse	II	Porta Sarno
1.40	centre	single horse	II	Between blocks and near intersection with Via di Castricio
1.40	centre	single horse	VI	Vico dei Vetti
1.40	centre	single horse	VI	Via di Vesuvio at VI.16.21
1.40	centre	single horse	VI	Via delle Consulare
1.40	centre	single horse	VI	Vicolo di Mercurio
1.40	centre	single horse	VI	Via delle Terme outside VI.6.1 (House of Pansa)
1.40	centre	single horse	VI	Vicolo delle Fullonica between the Vicolo di Mercurio and the Via delle Terme
1.40	centre	single horse	VII	Vicolo Storto north and south of intersection with Via degli Augustali
1.40	centre	single horse	VII	Via degli Augustali at the entrance to the forum
1.40	centre	single horse	VIII	Vicolo del Gigante at intersection with Via Marina
1.40	centre	single horse	VIII	Intersection of Via della Regina and Via dei Teatri
1.40	centre	single horse	VIII	Via del Tempio d'Iside
1.40	centre	single horse	IX	Via degli Augustali, east of the Via Stabiana
1.40	centre	single horse	IX	Vico di Tesmo near intersection with Via dell'Abbondanza
1.40	centre	single horse	IX	Via Stabiana at IX.3.10
1.43	centre	—	VI	Via delle Consulare
1.44	centre	—	VI	Vicolo di Mercurio
1.50	centre	pair of horses or oxen	VI	Via de Mercurio between the Vicolo di Mercurio and the Via delle Terme
1.56	centre	pair of oxen	VI	Vicolo di Mercurio
1.60	centre	pair of oxen	II	Via di Nocera near the city gate
1.60	centre	pair of oxen	VI	Via delle Consulare

Note: All measurements made by the author in June 2003.

may not have wanted carts passing their properties and so agreed to erect an impediment. One example of collective decision making is the blocking of the forum; indeed we see the hand of the civic government in this decision. In other cases we see evidence for individual property owners altering traffic patterns. Prior to the eruption of Vesuvius, the owner of the Casa del Fauno, a house that takes up an entire city block, appears to have chosen to repave the streets surrounding his house. As part of this project, a blocking stone was added to the Vicolo del Labirinto. The ruts in the street testify to its earlier use by cart drivers, but this house owner seems to have decided it was better to block them from passing one side of his property.[35] In another example the owner of the Casa di Vettii also set a stepping stone leading to his front door over a rut.[36] This did not block the street to carts, but it forced cart drivers to swing wide of his door.[37]

It is difficult to see all of these obstacles as part of some civic plan to facilitate cart traffic as they did more to disrupt it. Instead, individual or groups were deciding whether to allow carts to pass along their streets or to exclude them. While Pompeians were less likely to tamper with a few through-routes, particularly those that led to city gates, they seem to have had no problem impeding cart traffic in the city's side streets. This observation raises two more questions; did residents block traffic in contravention of the law and why did they make it so difficult for carts to traverse the city? While we cannot answer these questions from the archaeological evidence from Pompeii, the legal and literary evidence from Rome may suggest answers.

ROMAN TRAFFIC LAWS

Assuming laws regarding traffic in Pompeii were not so different from those in Rome, it seems likely that the actions of individual Pompeians to exclude carts from certain streets did not contravene the law. Rome, in fact, had very few traffic-related laws. As will be seen below the laws in the surviving legal corpus and references in literary texts are concerned with maintaining streets and controlling access to the status symbols that highly decorated carts could be. These laws have virtually no intent to facilitate cart traffic flow. Indeed, some laws gave Rome's residents the right to control the traffic that passed along their streets, taking the possibility of creating a coherent city-wide traffic policy away from civic officials.

Laws regarding Rome's urban infrastructure would have done more to hinder than to facilitate traffic flow. Rome's earliest law code, the Twelve Tables, established the width of a *via*, or street, at 8 Roman feet or about 2.4 m.[38] Later Roman law established the width of the space between two buildings,

---

[35] Poehler (2006: 63).
[36] Poehler (2006: 70).
[37] See also Hartnett, in this volume, for other examples.
[38] Varro *Ling.* 7.15; Adam (1994: 41) establishes the Roman foot at 0.2957 m; 8 Roman feet is 2.3656 m. See now Tuppi (2010).

a space intended to accommodate a city street, to be 10 Roman feet, or about 2.9 m.[39] A street built to the letter of the law in Rome would have been wide enough for the passage of only one cart at a time. Varro claims that the common type of street called a *semita* was half the width of a *via*,[40] which would literally place the width of a *semita* at about 1.2 m, a width that would preclude the passage of nearly all cart traffic since, as Table 7.1 indicates, most urban ruts and the carts that made them were wider than this. Even if one does not take Varro literally, particularly since the Justinian Code states that only *viae* had a fixed width, but accepts that the term *semita* was used for streets with a width narrower than that of a *via*, a street would not need to be too much narrower in order to bar cart traffic.[41] Varro's claim that it was difficult or impossible to drive through an *angiportus*, a Latin term also used for roads more narrow than a *via*, shows further that not all urban streets were designed to accommodate cart traffic.[42] Of course many Roman cities had very wide show streets, *plateae*, which would have been wide enough to accommodate not only vehicles moving in opposite directions, but pedestrians as well.[43] *Plateae* were an urban luxury, however, and were not common in Roman cities.[44] Most streets would have been narrower than a *platea*.[45]

Surviving Roman legal codes make it clear that building owners were responsible for keeping the streets in front of their property free of obstacles and in good repair, which would have aided the passage of traffic.[46] No doubt the civic government undertook to pave streets periodically, but Saliou concluded from her examination of the uniformity, or lack thereof, of styles and materials used in paving sidewalks and streets at Pompeii that individuals could also choose to repave the streets around their buildings or contiguous property owners could work together to pave the street along their entire block.[47] The result of individuals or groups paving sections of streets at separate times would have resulted in uneven patterns of wear in the paving. At Pompeii one stretch of the Via della Fortuna had to have been paved not long before the eruption. As one walks towards the Via Stabiana along the Via della Fortuna the deep ruts suddenly disappear and the road surface shows no sign of wear. The lack of uniformity in a street's maintenance must have been a factor a cart driver considered when planning a route through any Roman city.

---

[39] *Cod. Theod.* 4.24.
[40] Varro *Ling.* 5.35.
[41] Javolenus, *Dig.* 8.3.13.2.
[42] Varro *Ling.* 6.41.
[43] Bejor (1999).
[44] For a more thorough discussion of these Latin terms for streets see Kaiser (2011a) and (2011b: 25–45) and Wallace-Hadrill (2008: 269–70).
[45] See Macaulay-Lewis, in this volume, for a discussion of widths of streets on the Severan *Forma Urbis Romae*.
[46] e.g. Ulp. *Dig.* 43.8.2.32, 43.8.2.45; Papin. *Dig.* 43.10.3.
[47] Saliou (1999: 194ff.).

With the responsibility to maintain a street came a right to control it. City residents could seek to limit cart access to a specific street or part of a street. The legal presumption for public thoroughfares was that they should be open to the passage of all, but interdicts could be placed on certain roads.[48] In order to indicate a generic right of way for all modes of transportation, the Justinian Code uses the phrase '*via iter actus*'.[49] Ulpian explains that a right of *via* allowed one to walk or drive a vehicle on a particular road, an *iter* limited passage to those on foot, an *actus* was a more liberal right allowing one to walk and drive cattle or a vehicle.[50] While these rights often appear intended for country roads, the rights of *via* and *iter* were specifically extended to the public and private streets of cities.[51] In addition to limiting types of traffic, the application of *iter* or *via* could be very specific, regulating the time of day a street could be used or the type of vehicle that could traverse it.[52] We cannot know the motivation for those landowners seeking these interdicts, but the jurist Julian finds it reasonable to allow pedestrians to pass large urban homes only during the day, perhaps to control the level of noise.[53] Even if only a few urban landowners utilized their right to impose these interdicts, it must have greatly complicated traffic flow in the city.

One way modern governments keep traffic moving is by controlling the behaviour of drivers. The ubiquitous laws of modern governments standardizing the construction of vehicles and streets as well as regulating the behaviour of drivers have very few counterparts in Roman law. The Theodosian Code has the most extensive statement of driving requirements, limiting the size of loads and threatening harsh penalties for wagon makers who produced carts capable of carrying loads larger than the prescribed amounts.[54] These later laws applied only to vehicles engaged in the *cursus publicus*, the inter-urban imperial postal system, however, not to all vehicles throughout the empire or used in an urban environment. Kendal interprets the purpose of these laws to be for limiting wear on road surfaces while van Tilburg argues they were for controlling the cost of the service, and for limiting the burden on the requisitioned animals employed to pull these carts.[55] Either way, facilitating traffic flow on the *cursus publicus* does not seem to have been a major factor in their promulgation.

---

[48] Ulp. *Dig.* 43.8.2.25, 43.8.2.32; Papin. *Dig.* 43.10.3.
[49] e.g. Ulp. *Dig.* 8.3.1; Pompon. *Dig.* 18.1.66; Paul. *Dig.* 20.1.12, 33.2.1, 39.3.17; Pompon. *Dig.* 43.7.1; Ulp. *Dig.* 43.8.2.23–5, 43.19.1.7, 45.1.72.
[50] Ulp. *Dig.* 8.3.1. Other jurists confirm and elaborate these distinctions, cf. Pompon. *Dig.* 8.1.13; Paul. *Dig.* 8.3.7; Modestinus *Dig.* 8.3.12, as does Varro in his discussions of these same terms, cf. Varro *Ling.* 5.22, 5.34, 5.35.
[51] e.g. Julian *Dig.* 8.4.14; Pompon. *Dig.* 18.1.66.
[52] Julian *Dig.* 8.4.14; Marcellus, *Dig.* 8.6.11.
[53] Julian *Dig.* 8.4.14.
[54] *Cod. Theod.* 8.5.
[55] Kendal (1996: 142); van Tilburg (2007: 56 ff.).

While we know of laws that dictated urban driver behaviour, none of these seem oriented towards improving traffic flow but rather towards limiting the inconvenience carts caused city residents. Suetonius and Cassius Dio report that Claudius required drivers and those riding horses to dismount and walk through cities in Italy, a mandate later renewed and extended to cities across the empire.[56] While such a law must have slowed traffic, making it safer for pedestrians, it certainly would have done more to hinder cart traffic than to help, it required drivers to take more time to reach their destination. Dismounting was also a sign of respect, so this law may have been intended to force drivers, men at the bottom of the social ladder, to show proper etiquette towards urban residents.[57] The only imperial edict that would have positively affected traffic flow in the city of Rome is Hadrian's prohibition on 'carts with enormous loads' from entering the city, which appears to be a renewal of Caesar's ban on *plaustra* in Rome as well as an extension of it to all hours, not just the daylight hours.[58] This law would have helped keep traffic moving and limited the damage to the road surface. Considering the vast scale of construction projects Hadrian sponsored in Rome, however, he must have provided exemptions for carts hauling his building materials.

Some urban laws did control the right to own or operate a vehicle. The point of these laws was to control who had access to symbols of status rather than to ensure that drivers adhered to regularized and safe behaviour. Alexander Severus, for instance, gave senators the right to ride in the parade carts, *carrucae* and *raedae*, that had been ornamented with silver. The Roman senate allowed a blind senator the very unusual right of riding to Senate meetings in a chariot or *currus*, compensating for his blindness.[59] Later civic and military officials were granted the right to ride in carts inside towns.[60] The senate granted matrons the right to ride in the processional vehicles, *carpenta* and *pilenta* while the Oppian Law, a late-third-century BCE sumptuary law, limited not only women's rights to own and wear expensive jewellery and clothing, but also their right to ride in vehicles while in Rome.[61]

Originally only priests were allowed to ride up the Capitoline Hill in a *carpentum*, but later Claudius' wife Agrippina gained the same right, earning her a great deal of prestige.[62] Elagabulus was criticized for making laws that regulated not only which classes of women could ride in a *carpentum* or

---

[56] Cass. Dio 61.7b; Suet. *Claud.* 25.2; SHA *Marc.* 23.4.8; van Tilburg (2007: 132–4).
[57] cf. Sen. *Ep.* 64.10.
[58] SHA *Alex. Sev.* 42.1, 'vehicula cum ingentibus sarcinis'. See Newsome, in the Introduction to this volume, on the importance of semantic variation in legal texts pertaining to movement and traffic.
[59] Plin. *HN* 7.43.141.
[60] *Cod. Theod.* 14.12.1.
[61] Ov. *Fast.* 1.617–28; Livy 5.25.9. On the Oppian Law: Livy 34.1.3–4.
[62] Tac. *Ann.* 12.42.

pilentum, but also by which animal these women's carts could be drawn.[63] This criticism does not question the principle that allowed the emperor to dictate who could ride in specific types of carts, but seems to suggest that the emperor went too far by dictating the trivial details. These are the types of laws that controlled the ownership and operation of vehicles. They reflect the Roman obsession with displaying social status and controlling access to symbols of power. None of these laws would have functioned to control or improve traffic flow; that was not their intent.

The laws we know of regarding traffic flow touched on the issue obliquely, not directly. The Roman world had no equivalent to modern traffic laws. Street signs, backed by the power of the law, did not exist; no law prohibited the direction a driver could go on a street, the direction they could turn, or when they had to yield. Roman property laws required one to keep the street in front of one's property in repair, but also gave one the ability to block specific types of traffic at specific times of the day. Legal restrictions concerning the use of vehicles were related to conspicuous consumption and social status, not to traffic flow. Pompeians who chose to erect an impediment to cart traffic appear to have been within their legal rights.

ROMAN ATTITUDES TOWARDS CARTS

The law did not prevent the residents of Rome, and apparently Pompeii, from restricting cart access to their streets. In fact these residents had a strong incentive for taking advantage of their legal right. Roman authors express many reasons why city dwellers would not have wanted a cart passing along their streets. Our authors had nothing but complaints about cart traffic, especially the *plaustrum*. Ancient authors deemed the presence of these heavy wagons and the oxen pulling them a nuisance and even an outright danger. Horace and Seneca lament the noise and dust of wagons.[64] Ausonius and Juvenal deplore the shouting and unsavoury vocabulary of drivers.[65] Pliny complains that buildings shook violently when wagons passed.[66] In Aurelian's time, 'the use of a carriage in a city was odious', although this passage seems to suggest that by the fourth century CE such attitudes were relaxing.[67] The social stigma of riding in a wagon was so great that long before he was crowned emperor, Aurelian chose to switch from riding in a wagon to riding a horse

---

[63] SHA *Heliog.* 4.4.
[64] Hor. *Epist.* 1.17.6–8; id. *Sat.* 1.6.41–4; Sen. *Ep.* 56.4.
[65] Auson. *Ep.* 6.21–31; Juv. 3.235.
[66] Pliny, *Pan.* 51.1.
[67] SHA *Aurel.* 5.4, 'quia invidiosum tunc erat vehiculis in civitate uti'.

in order to enter Antioch, despite the fact that he was recovering from a serious wound.

Wagons hauling materials were not just a nuisance, they also had malevolent connotations.[68] One sewer contractor feared that the weight of columns a man named Scaurus was transporting to the Palatine Hill would damage his work under the streets of Rome and demanded Scaurus give him a security deposit.[69] Pliny does not mention any damage to the sewers, so the man's fears were probably overblown. Juvenal worries that if an axle should break and a wagon pours a heavy load on a passer-by, 'the poor man's crushed body disappears completely, as does his soul'.[70] Plautus recorded Roman fears of the malevolent nature of wagons by relating what appears to be a common expression during his day literally meaning 'the wagon is overturned', but figuratively meaning to suffer some piece of bad luck.[71] Caesar's law banning *plaustra* from the streets of Rome during the day is probably an attempt to protect residents from the perceived dangers of *plaustra* by limiting the large wagons' passage to a time when most Romans were not using the streets. These fears were not completely poetic hyperbole; things did fall out of carts causing property damage and bodily harm.[72] Libanius describes an incident in which he was nearly thrown from his horse in Antioch when unexpectedly confronted by a muleteer and presumably his cart.[73] Libanius' misadventures notwithstanding, we read of people fearing carts more than we read of actual accidents within cities.

The literary evidence shows us a deeply engrained suspicion of, and disdain for, cart traffic, especially heavy wagons. The residents of Pompeii seem to have shared these feelings. Some individual property owners and neighbours took charge of their streets and chose to exclude cart traffic or make the passage of a cart difficult.[74] In the case of the forum, members of the civic government may have chosen to close this part of the city to carts in order to preserve the dignity of the space and keep out the noise. Along residential streets, the exclusion of carts would have made the streets safer, quieter, and more prestigious. The great distrust Roman authors express for *plaustra* in particular seems to have been shared by the residents of Pompeii. The widest ruts in the city, those made by these large oxcarts, appear only on the Via

---

[68] See Favro, in this volume.
[69] Plin. *HN* 36.2.6.
[70] Juv. 3.260–1, 'obtritum vulgi perit omne cadaver more animae'.
[71] Plaut. *Epid.* 591, 'plaustrum perculi'.
[72] Ulp. *Dig.* 9.2.27.33; Alfenius *Dig.* 9.2.52.2.
[73] Libanius *Autobiography*, 216–17. Libanius had another dangerous encounter with horses in the streets of Antioch; cf. *Autobiography*, 259.
[74] Cf. *CIL* IV.8.13: 'otiosis locus hic non est, discede morator' ('this is no place for idlers, move along loiterer'). However, it is equally probable that this sign was intended for pedestrian loiterers rather than cart drivers.

Consolare, the Via di Nocera, and the Vicolo di Mercurio, three streets at the edges of the city. This last street is particularly isolated from daily activities in the city. The street is heavily rutted indicating that it was one of the most heavily used. Yet few of the buildings that line the street have their primary entrance onto it, instead having backdoors. The street has the same feel as a back alley in a modern city, a place usually ignored by those who live along it.

We must not forget, however, the mixed nature of Pompeian neighbourhoods. Beside residences it was not uncommon to find shops and workshops. Manufacturers and shop owners may have suffered great inconvenience from the vehicle impediments placed in their streets. Fortunately, there were ways to overcome some of the obstacles to cart movement within the city.

HELPING THE CART DRIVERS

Although the Pompeian cart driver would have faced serious challenges navigating the city, a few features of urban life would have helped to ameliorate the inconvenience to an extent. Some residents of the city actually had some incentive to ease the passage of some vehicles. Manufacturers and retailers could certainly have used a *plaustrum* or *raeda* to move raw materials or finished goods for sale. The elite also had incentive to allow the passage of some carts because not all were as disliked as the large wagons hauling goods. The Latin writers held vehicles for personal transportation in high regard and they often describe them as serving as a status symbol.[75] The *currus* was among the most prestigious urban vehicle; we read of its use in religious processions, triumphal processions, and races, as well as its elaborate decoration in gold, silver, and ivory.[76] We also read of a *carpentum*, *raeda*, and *carruca* with silver decorations;[77] a *raeda* and *covinnus* being given as gifts and a gilded *carruca* equal in value to a farm.[78] Pliny complains of the moral corruption of the Roman character indicated by the use of gold and silver statuettes to decorate carriages.[79] Thus even if a wealthy Pompeian had no desire to allow a *plaustrum* to drive down his street, he may have left the street open for the passage of his own *currus* or *carpentum*.

Although kerbs and blocking stones could prevent cart entry into a street, these methods of exclusion were reversible. One could have temporarily removed blocking stones or bridged kerbs with wooden ramps. Unfortunately,

---

[75] Amm. Marc. 14.6.9; Juv. 7.178-81; Vigneron (1968: 170).
[76] Gold: Flor. 1.1.5.6; Prop. 1.16.3; Silver: SHA *Aurel.* 33.2; Ivory: SHA *Aurel.* 33.2; Ov. *Pont.* 3.4.35; id. *Tr.* 4.2.63.
[77] *Carpentum*: Flor. 1.37.5-6.; *raeda* and *carruca*: SHA *Alex. Sev.* 43.1; SHA *Aurel.* 46.3-4.
[78] *Raeda*: SHA *Heliog.* 21.7; *covinnus*: Mart. 12.24; gilded *carruca*: Mart. 62.
[79] Plin. *HN* 34.48.163.

neither technique would have left an archaeological signature. While the occasional entrepreneur may have found ways to move a load-bearing cart into a restricted part of the city, there would have been other vehicles given temporary access on festival days. *Currus* and *carpenta* played an important role in religious and civic processions. Such processions must have taken place at Pompeii. No doubt any procession associated with the imperial cult would have gone through the forum while one preceding gladiatorial games would have entered Regio II. For these special occasions we must envision exceptions being made. Again, because these were infrequent events, these carts would probably not have left traces in the pavement.

Another way to overcome some of the difficulties of driving a cart around Pompeii would have been to enlist the aid of helpers. Pack animals and human porters (*saccarii*) could bypass the many impediments to cart traffic with relative ease.[80] Carts could stop on the most easily accessible streets and be unloaded. Some carts were preceded by a 'runner' (*cursor*) with a whip who would clear the way and who could make sure a one-lane street was clear of oncoming traffic.[81] While it is unclear how frequently such assistants were employed, the figural representation of one in a mosaic in Ostia in the Terme dei Cisiarii and literary reference to *cursores* in Rome suggests they were seen in at least these two cities.[82]

Ultimately, the movement of carts through a city may not have been of great importance simply because the volume of cart traffic must have been small in comparison to pedestrian traffic.[83] Carts and the animals for traction were expensive, putting them out of the reach of most urbanites.[84] The number of storage facilities with ramps over the kerb that would have allowed carts to enter is small at Pompeii and other sites.[85] When someone living in Rome needed a cart or pack animal for some reason, there appear to have been organizations ready to rent them.[86] Rome appears to have had a plaza, the Area Carruces, a name derived from the word for the luxury wagon, the *carruca*.[87] Because the Area Carruces was probably near the Porta Capena and inscriptions indicate it was the seat of the Schola Carrucarum, a guild that dealt in matters of transportation, some have suggested that people entering

---

[80] Van Tilburg (2007: 132); *Cod. Theod.* 14.22.
[81] Matthews (1960: 24); van Tilburg (2007: 124 and 139); for examples from ancient texts, see Sen. *Ep.* 87.9, which refers to an inter-urban context and *Ep.* 123.7, and Suet. *Nero* 30, both of which seem to refer to both an inter- and intra-urban context.
[82] Van Tilburg (2007: 124–5); Mart. 3.47.14.
[83] Van Tilburg (2007: 55).
[84] Van Tilburg (2007: 77 and 84).
[85] See Poehler, in this volume, for a detailed examination of the 36 ramped properties in Pompeii.
[86] Van Tilburg (2007: 47).
[87] Hor. *Carm.* 1.9.18.

Rome from outside the city would park their carts there and utilize the litters available in the *area*.[88] This would have reduced the amount of cart traffic in Rome. Finally, the scale of most ancient Roman cities precluded the need for a personal transportation cart. Walking across Pompeii between the gates placed the furthest distance from one another, the Porta Marina and the Porta Nola, takes just fifteen minutes. Going from any point in the city to any other would certainly have required less time.

CONCLUSIONS

Residents of Pompeii made driving a cart in the city a challenging task. The streets were laid out for the most part in a grid fashion that utilized a hierarchy to an extent. Residents and/or civic officials made the majority of the streets leading to and from city gates two lanes wide and inserted few impediments to vehicular passage, while secondary streets that did not lead to city gates were generally one lane wide and had many impediments. Cart drivers who could remain on the main thoroughfares would not have had too much difficulty navigating the city, provided they did not need to go west of the forum. The cart driver who had to leave these main streets would have had a difficult time navigating the city, as impediments were not placed in a systematic way, making passage through the city unpredictable. Residents complicated things further for cart drivers by changing which streets were passable and which were not. Because of the danger and noise presented by carts, as well as the low class associations with the presence of large carts, some Pompeian property owners sought to exclude them from their neighbourhoods. Regulation of traffic flow seems to have been in the hands of individuals who made decisions about the movement of carts through their neighbourhoods for their own reasons without taking into consideration the necessity of traffic flow through the entire city. Indeed, the primary guiding principle of many of these decisions was simply 'not on my street', although other urban residents seem to have had other competing ideas about the best use of street space that went beyond the presence or absence of carts.[89]

Julius Caesar did not regulate the width of axles or ban all carts from the city of Rome during the daytime. Once we set aside these myths and look at the remaining evidence for the concern Roman officials had for cart traffic flow, it becomes increasingly clear that it was not a high priority. Officials seem to have been more concerned about limiting displays of wealth and prestige

---

[88] Platner and Ashby (1929: 50); Richardson (1992: 32).
[89] Cf. Kaiser (2000: 60 ff.) and Newsome (2008: 443).

through the use of specific types of vehicles than with where those vehicles went. Instead traffic management was handled on a block by block basis by the people who lived along those blocks. No doubt moving anything by cart through Pompeii, or any other Roman city, must have been difficult. However, considering the low regard with which those hauling goods appear to have been held, it is not clear that anyone other than the cart drivers concerned themselves with their troubles.

8

Where to Park?

Carts, Stables, and the Economics of Transport in Pompeii

Eric E. Poehler

A critical factor in the growth of preindustrial economies has been the capacity of those economies to develop effective methods of transportation.[1]

Roman methods of transport are normally thought to have been so backward that they severely hindered economic development... It is primarily the high cost of land transport that makes historians pessimistic about the capabilities of the Roman transport system.[2]

How can scholars reconcile the complex commercial landscape of a city like Pompeii with the commonly held belief in the inefficiency of land-based transportation? Are we too pessimistic about the costs of land transport or do we misunderstand the economic sophistication of the urban landscape? On the one hand, it is hard to ignore the hundreds of (apparently successful) commercial properties at Pompeii and the complex economic relationships they reveal in their physical associations with other buildings, especially houses. Similarly, the enormous reconstruction efforts following the earthquake(s) of 62 CE demonstrate that vast quantities of materials were transported in and out of the city in a short amount of time.[3] On the other hand, the idea of the inefficiency of land transport has been undermined by Laurence, who has shown the comparison with waterborne transportation to be misleading.[4] Recent research by the author has also demonstrated that wheeled traffic in Pompeii was organized into an efficient system governed by the rule of right-side driving, the imposition of congestion-relieving detours, and by

---

[1] Martin (2002: 151).   [2] Harris (1993: 27).
[3] Dobbins (2007: 173–5; 1994).   [4] Laurence (1998b: 126–32).

alternating series of one-way streets.[5] The confluence of these facts—that a large amount of goods were continually transported into and out of the city and that the municipal authorities acted to facilitate that supply—argues in general for transport to have been more efficient economically than is generally accepted. Without stepping too strongly into the modernist/primitivist debates, it now seems more certain that the answer to this apparent paradox is the need for new commonly held beliefs. What remains to be understood, however, is how this efficiency was accomplished and how it was enacted within the urban environment and beyond.

This chapter takes the first step towards addressing this question by considering one aspect of Pompeii's transportation economy: the storage of wheeled vehicles.[6] Rather than attempting to define the transport economy outright, it is better to ask, as Glenn Storey has done for the Roman economy more broadly, what are the 'kinds of [economic] institutions and how did they work?'[7] The traffic system can be seen as one such institution, but less formal mechanisms—the economic reasons that carts were travelling in the first place—were certainly more important underlying motivators. This kind of segmented approach is necessitated by an absence of evidence. While mules and human porters made up the greatest proportion of the means of transport, these methods of conveyance simply don't make the same physical impact (e.g., ruts and worn street features).[8] Additionally, mules and porters are far less archaeologically recognizable in their secondary contexts. Thus, how does one determine if a manger in a bakery represents the storage of animals for transport, traction, or both?[9]

Carts, however, are singular in purpose, traceable through their impacts across the urban environment, and identifiable through secure architectural proxies. Although carts made up only a portion of the total means of transport at Pompeii, the evidence for their circulation and storage make them an excellent sample from which to draw conclusions about the economic rationalities governing transportation choices in and around the ancient city. The following discussion therefore first focuses on those locations where carts can be observed to leave the space of the street—ramps—and the spaces to which

---

[5] Poehler (2006: 59–74; 2009: 38–121). See also the chapters by Hartnett and Kaiser, in this volume.

[6] The term 'transport economy' is used here to mean all economic relations of the act of transporting goods and people as well as those operations that support such transportation. As there is no body of water within Pompeii, all transport is necessarily land-based even if significant quantities of materials were carried by boat prior to their arrival in Campania.

[7] Storey (2004: 107).

[8] These groups are represented in epigraphic (mules, CIL IV.113, 97, 134; porters, CIL IV.497) and artistic (Nappo 1989: 80–1, figs. 1–2, 4; Ling 1990b: 56, fig. 4.14; Guasti 2007: 140, fig. 3) evidence at Pompeii.

[9] e.g., draught animal skeletons were found in a stable (no ramp is known) at the bakery at IX.12.6–7. See Varone (2008) for the most recent work and bibliography.

they led—stables. An examination of the spatial and social distributions of these features then reveals important clues about who owned carts and/or the means to serve those travelling and how these distributions reflect different economic strategies within the transport economy.

The specificity of our conclusions about the transport economy is constrained by the specificity of the evidence that can be brought to bear upon the question. Thus, even if the evidence for the traffic system's organization can demonstrate the direction that a cart could travel on each excavated street in the city, without knowing where a vehicle began its journey and its intended destination, the restrictions on direction offer an abstraction of how traffic did circulate generally rather than any historical description of the particular and purposeful paths across the city. To assess the reason for travel one must establish a destination.

The most obvious destinations for travel in Pompeii are monumental buildings and the city gates. For cart drivers and pedestrians alike, monumental buildings were important landmarks that required a large volume of supply. For example, the *macellum*, the baths, and the amphitheatre all would have required vast quantities of animals, fuel, and equipment appropriate to their activities.[10] The monumental building as a destination, however, is only different from the humble shop in scale; although we know what must have been delivered, we cannot observe it archaeologically. Conversely, it is known that all traffic that entered and exited the city passed through the seven city gates. Compared to monumental buildings, gates offer a specific point at which the passage of vehicles can, necessarily, be established. Because *every* entering cart passed under the gates, however, there is no differentiation in purpose; gates suffer from the opposite problem, a lack of specificity.

RAMPS AND STABLES

The identification of a ramp leading out of the street combines the means to definitively establish the presence of vehicles with the ability to limit the purposes of their travel by the types of properties they entered. Moreover, as markers of destination, ramps are important because their different forms offer evidence for the economics of cart driving generally as well as the investment choices of specific properties. When a ramp is compared to other treatments of sidewalk space—such as its elaboration through decorative pavements or its truncation by a building to a line of kerbstones or even less—

---

[10] On evidence for animal processing at the *macellum*, cf. Maiuri (1947: 166); also McKinnon (2004: 184–94).

it becomes obvious that the choice to construct a ramp indicates the high importance of wheeled vehicles and the material supply they carried to that property.[11] Just as an oven or a millstone attests to the kinds of economic activities that took place within a property, a ramp signals the specific need for a large volume of material to be brought into and/or out of the property or for a high number of vehicles to be staged within the property.

Of the thirty-six identified ramps, three ramp types can be identified at Pompeii: Paved Ramps, Rut Ramps, and Side Ramps. Paved Ramps are openings in the kerbstones where an inclined section of paving permits access for wheeled vehicles between the street and a stable area (Fig. 8.1). Rut Ramps are identified by rutting worn into an uninterrupted line of kerbstones, demonstrating the movement of wheeled vehicles between the street and a stable area (Fig. 8.2). In some cases, the kerbstones have been cut or worn down completely to create a smooth, but very short incline.[12] Side Ramps are sections of a street where the kerbstones have been removed and replaced by an area of paving stones, usually inclined, in order to provide a place for wheeled vehicles to park out of the space of the street (Fig. 8.3). Some ramps are identified only by the existence of guard stones at the corner of the ramp—either at the street or at the door to the stable—or by an opening in the kerbstones. These ramps are identified, but categorized as Unknown type. Except for Side Ramps, all ramps lead to a stable area. As part of the property's economic infrastructure, stables were even more costly than ramps because stables were far larger in area and competed for space with other activities that might have been more productive. Economically, the equation is simple: one could profitably use the space to operate an inn with stabling services or one could reduce their own transport costs by owning the means of transportation.

Properties with ramps and stables (hereafter transport properties) have two important distributions: across the types of properties that make up the urban fabric and across the space of the city.[13] The first distribution shows that while the variety of transport properties was relatively restricted, the kind of ramp employed was strongly correlated with a type of property to which it led. Table 8.1 expresses these associations. Thus, nearly half of all Paved Ramps led to inns, which is almost four times more than any other identified transport

---

[11] On the elaboration and truncation of sidewalks see e.g. that on the west side the Casa dei Vettii, which was already only the width of the kerbstones, was covered by decorative masonry.
[12] These are I.4.28, VII.13.17, and IX.1.28.
[13] The identification of these transport properties (listed in Table 8.3) was made through on-site survey conducted between 2007 and 2009 combined with research into the publication record of each property. The main sources consulted were Eschebach (1993), Fiorelli (1875), Jashemski (1973, 1977, 1979), Mau (1899), Mayeske (1972), Overbeck (1884), Packer (1978), van der Poel (1986), and the publications *Pompeianarum Antiquitatum Historia*; *Giornale degli Scavi di Pompei*; *Notizie degli Scavi di Antichità*.

Fig. 8.1. Paved Ramp at I.1.4 (photo by Eric E. Poehler with permission of the Soprintendenza Archeologica di Pompei).

Fig. 8.2. Rut Ramp at II.5.4 (photo by Eric E. Poehler with permission of the Soprintendenza Archeologica di Pompei).

Fig. 8.3. Side Ramp at VIII.6.10 (photo by Eric E. Poehler with permission of the Soprintendenza Archeologica di Pompei).

Table 8.1 Distribution of ramps across property types

Property	Paved Ramp	% of Group	Rut Ramp	% of Group	Side Ramp	% of Group	Un-identified	% of Group
Commercial—Inn	8	47.1	2	22.2	0	0.0	1	20.0
Commercial—Bakery	1	5.9	0	0.0	1	20.0	0	0.0
Commerical—Agricultural	1	5.9	2	22.2	0	0.0	0	0.0
Commercial	2	11.8	1	11.1	0	0.0	1	20.0
Theatre	0	0.0	0	0.0	1	20.0	0	0.0
Residential—Workshop	0	0.0	0	0.0	3	60.0	1	20.0
Residential	2	11.8	3	33.3	0	0.0	2	40.0
Unknown	3	17.6	1	11.1	0	0.0	0	0.0
Totals	17	47.2	9	25.0	5	13.9	5	13.9

property. Rut Ramps are much more evenly distributed across all transport properties with large houses having a slightly larger proportion (33 per cent). Side Ramps are more variable in their forms and therefore their potential costs. The 36-m-long Side Ramp on Via Stabiana at the *teatrum tectum* is the type form for these ramps, though the only example on this scale. Other Side

Table 8.2 Spatial Distribution of transport properties across property types

Entrance	Ramp Type	% of Property Area used for Stable	Property Type Identification	At Gate?
1.2.22	Unknown Guard Stones	20.2	Commercial	Yes
VI.16.23	Paved Ramp	13.9	Commercial—Inn	Yes
VI.17.1	Paved Ramp	17.0	Commercial—Inn	Yes
VIII.7.1	Paved Ramp	33.1	Commercial—Inn	Yes
I.1.3	Paved Ramp	34.7	Commercial—Inn	Yes
I.1.8	Paved Ramp	39.7	Commercial—Inn	Yes
III.11.a	Paved Ramp	Unknown (unexcavated)	Commercial—Inn	Yes
IV.5.c	Paved Ramp	Unknown (unexcavated)	Commercial—Inn	Yes
VI.1.1	Rut Ramp	36.3	Commercial—Inn	Yes
VI.15.18	Rut Ramp	38.2	Commercial—Inn	Yes
VI.1.4	Unknown	49.7	Commercial—Inn	Yes
I.20.5	Paved Ramp	6.1	Commercial—Agricultural	Yes
II.5.4	Rut Ramp	Unknown (unexcavated)	Commercial—Agricultural	Yes
III.7.6	Rut Ramp	Unknown (unexcavated)	Commercial—Agricultural	Yes
VI.1.24	Paved Ramp	1.6	Residential (A/P)	Yes
Vl.1.22	Paved Ramp	27.7	Unknown	Yes
IV.3.h	Paved Ramp	Unknown (unexcavated)	Unknown (unexcavated)	Yes
IX.2.24	Paved Ramp	27.0	Commercial	No
1.8.12	Paved Ramp	65.2	Commercial	No
IX.1.28	Rut Ramp	31.3	Commercial	No
1.3.27	Paved Ramp	32.2	Commercial—Bakery	No
Vl.6.17	Side Ramp	n/a	Commercial—Bakery	No
II.4.4	Paved Ramp	0.9	Commercial—Inn	No
IX.9.12	Unknown Guard Stones	14.0	Residential (A)	No
VII.1.43	Paved Ramp	2.9	Residential (A/P)	No
I.4.28	Rut Ramp	1.6	Residential (A/P)	No
VI.6.13	Rut Ramp	2.0	Residential (A/P)	No
Vll.13.17	Rut Ramp	2.7	Residential (A/P)	No
I.10.14	Unknown	8.3	Residential (A/P)	No
Vl.3.28	Side Ramp	n/a	Residential—Workshop (Bakery)	No
VII.2.22	Side Ramp	n/a	Residential—Workshop (Bakery)	No
VIII.6.10	Side Ramp	n/a	Residential—Workshop (Bakery)	No
1.13.14	Unknown	4.9	Residential—Workshop (Wine)	No
VIII.7.17–21	Side Ramp	n/a	Theatres	No
III.7.2	Paved Ramp	Unknown (unexcavated)	Unknown (unexcavated)	No
III.6.2	Rut Ramp	Unknown (unexcavated)	Unknown (unexcavated)	No

Ramps are identified by large sections of sidewalk that have been replaced by street paving,[14] while still others demonstrate parking occurred on the sidewalk area itself.[15] There is very little variation, however, in the property association of Side Ramps: 80 per cent are found at bakeries, either within a residence or as a stand-alone workshop.

The typology of ramps and their associated properties (Table 8.2) indicates that there were several choices available to those who were interested to invest in vehicular access to their property. Cost, durability, and visibility were the most important factors influencing this choice. The paved Side Ramps were the most costly in terms of materials, but this may have been offset by their size, which allowed multiple deliveries to occur simultaneously. Moreover, the absence of a stable meant that more of the internal area could be used for productive activities. Thus, the association of Side Ramps with bakeries fits well with an expected investment strategy in which the use of space and animal power would have been primarily for baking and/or milling. Similarly, the association of Paved Ramps with the hospitality industry demonstrates that Paved Ramps not only provided a durable, even surface for vehicles to traverse, they also advertised stabling services to those entering the city.[16] Conversely, Rut Ramps were not intended to be as obtrusive, an observation which helps to explain their strongest correlations with the secondary entrances of houses and with the large vineyards near the Porta di Sarno.[17]

The second distribution of transport properties is spatial; they are either clustered very near the city gates or spread out more widely across the rest of the city (Fig. 8.4). Drawing a 100-m arc around each of the city gates circumscribes seventeen of the thirty-six transport properties in the city.[18] While the number of transport properties found in proximity and away from gates is equivalent, the area of the city near the gates is only 19.9 per cent of the total area.[19] Of the ramps near the gates, fourteen open directly onto the main, two-way streets leading to a gate with the remaining three found on one-lane streets that intersect or parallel that main street.[20] Four of the seven gates (Ercolano, Stabia, Sarno, and Nola) have transport properties located on the street immediately inside the gate. In fact, at each of these gates the very first doorway inside the city is accessed by a ramp.[21] As a whole, the concentration

---

[14] These are VII.6.10 and VI.3.28.
[15] These are VII.2.22 and VI.6.17. Parking on the sidewalk area would have inconvenienced pedestrians, on which more broadly see Hartnett, in this volume.
[16] Bar counters also were oriented to advertise to potential customers. Ellis (2004: 381–4).
[17] These are II.5.4 and III.7.6.
[18] 100 m was chosen arbitrarily. Expanding or contracting this arc might alter the percentages, but the clustering of transport properties very near the gates would still show as significant.
[19] The total area of the city is approximated to be 628,048.22 m^2 and the total area within 100 m of a city gate is 125,187.43 m^2.
[20] These are I.2.22, VI.1.22, and VI.1.24.
[21] These are II.5.4, III.11.a, VI.1.1, VI.17.1, and VIII.7.1.

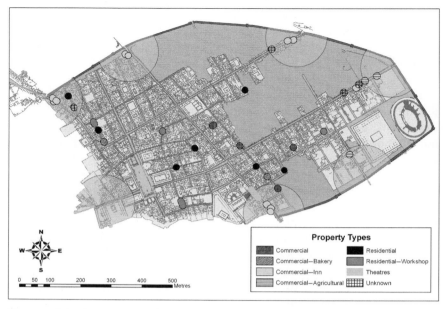

Fig. 8.4. Distribution map of transport properties (image by Eric E. Poehler).

of so many ramps and stables near the gates elides with the expected high volume of traffic here and demonstrates the economic importance of traffic for the operation and development of the city. In fact, this traffic establishes the gates as important peripheral generators of urban space.[22]

More specifically, the kinds of traffic-related space generated at the gates as well as away from them both reveal particular economic rationales. Table 8.3 shows the spatial distribution of transport properties in relation to their distribution across property types. Property types are listed along the left side of the table while the number and percentages of these types are shown in two groups of three columns. The first column lists the number of each property type near to or away from the gates, while the second column (% of Category) expresses this difference in number between the two groups as a percentage. The third column (% of Group) shows the percentage that a particular property type makes up within each group, that is, either at or not at the gates.

Two trends are remarkably clear from this tabulation. The first is the considerable concentration of inns at the city gates: over 90 per cent (90.9) of all inns with a ramp are located here. Of these, 70 per cent have Paved Ramps. Such a concentration of inns shows this service to have been a very profitable use of space, especially when compared to residential properties

---

[22] On gates as generators of space, see Malmberg and Bjur in this volume, for examples in Rome.

The Economics of Transport in Pompeii

Table 8.3 List of all transport properties

Property Type	At Gates	% of Category	% of Group	Not At Gates	% of Category	% of Group	Total
Commercial—Inn	10	90.9	58.8	1	9.1	5.3	11
Commercial—Bakery	0	0.0	0.0	2	100.0	10.5	2
Commercial—Agricultural	3	100.0	17.6	0	0.0	0.0	3
Commercial	1	25.0	5.9	3	75.0	15.8	4
Residential—Workshop	0	0.0	0.0	4	100.0	21.1	4
Residential	1	14.3	5.9	6	85.7	31.6	7
Theatre	0	0.0	0.0	1	100.0	5.3	1
Unknown	2	50.0	11.8	2	50.0	10.5	4
Totals	17		100.0	19		100.0	36

with a ramp (14.3 per cent). In fact, the one house near the gates is the only non-commercial property to have wheeled access.[23] The second trend revealed by Table 8.3 is the disproportionate number of residential properties with ramps spread throughout the rest of the city. Fully 85.7 per cent of all houses with stables are found away from the gates. These six houses make up 31.6 per cent of the group away from the gates, a percentage that is approached only by houses that have workshops within them (21.1 per cent). The increase in the number of houses with stables in this group equals the combined total of commercial properties (31.6 per cent), with the dual nature of Residential—Workshops nicely balanced in between. The strictly residential properties are remarkable not only for their having vehicular access, but also for the presence of status architecture—all have an atrium and six of seven also have a peristyle—and their very large size. The average size of these houses is 1344.7 m², more than three times the average area (404.7 m²) of all houses in the city.

Such a large average area suggests that there was a threshold in residential size, likely related to a larger need for supply, where it became both economically viable and socially desirable to dedicate an area to the storage of a vehicle and draught animals. Averages, however, have outliers and the equation is not that simple. The 1344.7 m² average is made up of two houses over 2000 m², two houses below 600 m², and three houses spaced evenly between these.[24] The decision to invest in transport infrastructures is not simply proportional to property size. If it were, one would expect more than four of the twenty-five houses larger than 1000 m² to be equipped with a ramp and stable. Instead, the

---

[23] This excludes the two unidentified properties, one of which (IV.3.h) was probably an inn.
[24] These are: I.4.28 at 2,308.11 m², VI.6.13 at 2,263.39 m², I.10.14 at 1,835.95 m², VI.1.24 at 1,124.69 m², VII.1.43 at 789.17 m², I.10.17 at 590.71 m², and IX.9.12 at 500.86 m².

purpose of a ramp and stable is intertwined in the specifics of a particular property. Well-studied houses, such as the Casa del Menandro (see below), offer strong evidence for the urban house being one node in a larger economic unit. Houses without a ramp and stable, however, rarely reveal the reason for a choice not made. Although we should not underestimate the volume of their supply, without a ramp and stable it is safe to assume that most of that supply, by mass and/or weight, did not require a cart. At the same time, those few residential properties with transport infrastructure therefore represent the highest level of investment in the transport economy. This division, based on ramps and stables, leads to two conclusions about the supply of houses in Pompeii. First, the majority of houses were supplied by mules and porters, a fact that gives still greater importance to these methods of conveyance. Second, as important as they were, these mules and porters can only be represented by their archaeologically observable proxies, ramps and stables.

In combination, these two intersecting trends—the property distribution of ramps and the spatial distribution of transport properties—reveal two distinct economic rationales for constructing an access to and specific area for the storage of wheeled vehicles. The clustering of inns with ramps near the gates demonstrates that property owners saw the compression of traffic at the gates as an economic opportunity and invested in these particular commercial infrastructures to offer a specific service. The number of inns and the competition between them to provide ramps at the very first doorways inside the gate indicate the success of these enterprises. Conversely, away from the gates, ramps most often led to large private residences, illustrating a very different economic rationale. Rather than advertising themselves with well-paved ramps on main thoroughfares, these ramps and stables were constructed because the supply to the house, for whatever purpose, was great enough to outstrip the abilities of donkeys and porters. Moreover, this supply was both so large and so constant that it was economical for the property owner to also own (a significant portion of) the means of supply. Of course, the total volume of traffic at a particular property or for the city as a whole is far beyond what the evidence from stables can demonstrate. Still, these two divisions of transport properties operate as proxies for what is archaeologically invisible (i.e., donkeys and porters) and represent two modes of the transport economy, the Household Mode and Commercial Mode, which at times complemented and competed with each other.

THE HOUSEHOLD MODE OF TRANSPORT

The Household Mode of transport was driven by the constant demand for supplies required to maintain the people and activities of an elite house as well as its associated, if not always architecturally interconnected, dependencies. Although this claim about the elite house seems obvious, packed within it are three assumptions that need to be outlined. The first is that the elite house was often filled with large numbers of people, including many beyond the family and slaves of the paterfamilias.[25] The second assumption is that Roman elites did not abhor involvement in trade, even if they chose to distance themselves from it physically through architecture and administratively through middlemen.[26] Finally, the elite urban townhouse might be best understood as one part of a larger economic portfolio that often included urban rental and/or production properties as well as rural *villae* whose relationship to the townhouse is more reciprocal than the Consumer City model admits.[27]

How the Household Mode actually worked, that is, how it gave economic advantage is revealed by an overlooked aspect of the well-known episode in Cato's *De Agricultura* (22.3). Factoring into his decision about the economics of purchasing a mill from Suessa or Pompeii were not only price differences, but also the transport costs.[28] At 25 miles distant from his villa, Cato reports the transportation cost for bringing the mill from Suessa was 72 sesterces using six men for six days, while the cost for delivery from Pompeii, 75 miles away, was nearly four times that amount, 280 sesterces. These figures seem to show transportation costs growing dramatically with distance. There is however, an important difference. While the trip to Suessa was a round trip as Cato's own carts were used, making the trip a total of 50 miles, the mill from Pompeii was being delivered.[29] Comparing these costs per mile, the trip to and from Suessa was 1.44 sesterces per mile while the trip from Pompeii was 3.73 sesterces per mile.[30] The difference in price is striking; transportation costs are two and a half times less when one owns the means of transport. Such variance in cost is the economic advantage behind the Household Mode of transport.

---

[25] Wallace-Hadrill (1994: 95).

[26] Aubert (1993: 172–8); Wallace-Hadrill (1991).

[27] Or, to put it another way, as Purcell (1995a: 172) has done, the townhouse is more like a villa: 'the topography and institutions are those of a town but the social and cultural fabric is that of a cluster of *villae* and their dependencies'.

[28] Laurence's (1998b) conclusions and methodological approach are adopted in the following calculations.

[29] Laurence (1998b: 131–6); Yeo (1946: 221–4).

[30] These figures assume the costs of the return trip from Pompeii are not included in the Commercial Mode, since these drivers could contract to carry other materials for the return trip. The rationale for this assumption is that in the competitive market of the Commercial Mode, businesses that routinely 'deadheaded' (i.e., returned empty) would quickly lose out to those who did not because the later businesses could therefore offer prices that were lower by 50% or more. There is no reason to assume that 'deadheading' was any more common in the Household Mode. Since the vehicle necessarily had to go and return, however, there is no change in the cost of the travel per mile even if, by percentage, the profitability of the trip could be increased. In the Household Mode, costs per mile are relatively fixed; in the Commercial Mode, costs per mile are price-dependent.

On the other hand, this same example also demonstrates that the Household Mode was restricted by distance. If Cato had sent his own vehicles to Pompeii, the return trip of 150 miles would have cost 216 sesterces by his figures, or 77 per cent of the cost to have it delivered. More importantly, the time necessary for the trip would have tripled as well, taking away six men and six oxen from other work for eighteen days and delaying the arrival of the mill by nine days. The value of lost production from men and machinery, though not discussed by Cato, would have made the Household Mode of transport less efficient than the Commercial Mode at this distance.[31]

The Household Mode was therefore economically limited to a relatively local circulation, within which the urban residence was one of several nodes repeatedly visited. Any rural properties associated with the house would certainly have been common destinations, as the transport needs of the villa were greater than those of the urban residence.[32] Varro is explicit about the link between availability of the infrastructure of transport—especially roads and rivers—and the success of a villa.[33] Moreover, if Strabo's description of the area of the Bay of Naples as having been given over completely to villas and agricultural production by Augustan times is accurate (and archaeological evidence seems to support it), then the success of that production is in part related to the strong integration of the economic networks of urban and rural production and consumption.[34] The means of that integration was the efficiency of the local transportation of raw materials and finished products, an efficiency made possible by effectiveness of the Household Mode of transport.

The Casa del Menandro (Fig. 8.5) exemplifies the complex economic interconnections of an urban household in the service of its owner's social, political, and commercial ambitions and the extra-urban sources of its supply and wealth. At 1,835 m^2, the Casa del Menandro is the ninth largest house in the city and includes an atrium, peristyle, suite of dining rooms, private bath complex, and large service quarter. Thus, it is easy to imagine that the day-to-day operation of the house would have required an enormous amount of food and drink appropriate to master, visitor, and slave.[35] The existence of a large (169.53 m^2) stable area in the south-east corner of the house, the remains of a

---

[31] If the cost of travelling to Suessa was 72 sesterces or 2 sesterces per day/per person (including animals), it is not unreasonable to assume that the same figure is close to the minimum value of the labour that these same persons and animals would produce should they have not gone on the trip. Therefore, at least this amount should be calculated as loss in productive value for the trip to Pompeii.

[32] Whittaker (1985) argues from epigraphic and archaeological data on amphorae that most transportation occurred between private properties whether owned by one or multiple individuals. More recently, Pena (2007: 61–118) has argued for commercial resale and reuse of amphorae.

[33] Varro *Rust*. 1.2.23; Laurence (1998b: 134–5).

[34] Strabo 5.4.8; Moorman (2007: 436, fig. 28.1).

[35] Artefactual information, including nineteen human skeletons, from Allison (n.d.).

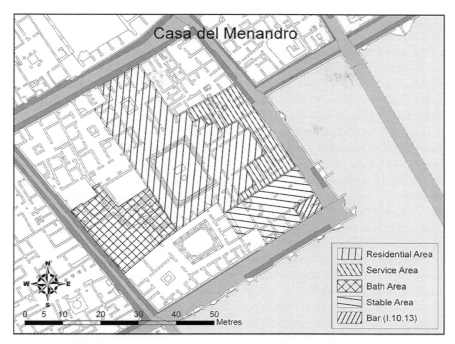

Fig. 8.5. Plan of the Casa del Menandro (image by Eric E. Poehler).

cart and the other artefacts found there physically demonstrate the means to transport the volume of materials suggested by the size, architecture, and amenities of the house.[36]

The finds in the stable area further illustrate the Household Mode of transport by associating real objects with the means of their conveyance. First, over forty amphorae were found against the south wall of the stable area, indicating either the delivery of (likely) wine for consumption within the house, sale in the associated bar property (I.10.13), or the transportation of empty vessels for filling elsewhere.[37] A handmill for grinding grain was also found here, one of five discovered in the house, along with two piles of lime and various ceramic and metal finds. Throughout the rest of the house, the excavators found at least fifty-nine lamps, lanterns, lamp bases or other

---

[36] On the cart, stable, and finds cf. Allison (2004: 106–7, 110–12, fig. 5.18); Ling (1997: 108–14, 139); Maiuri (1933: fig. 179). The large stable yard and its manger area was divided in a manner suggesting four animals were kept here, although no skeletons were found. This absence led Maiuri (1933: 193–4) to believe the animals were in the fields at the time of the eruption, while Allison (n.d.) follows Ling (1997: 114) in noting a number of possible reasons for their absence.

[37] Cf. Ellis (2005: 268–70, figs. 48b–48c) on this property. The later interpretation is preferred by Ward-Perkins and Claridge (1980: n. 88).

lighting fixtures as well as five masonry ovens for cooking and heating, all of which needed a constant supply of oil and wood for fuel. In the service quarter of the house a great supply of agricultural tools were recovered that exceeded the needs of the peristyle garden.[38] A plough was even reportedly discovered within the house.[39] These objects, in association with the structural arrangement of service quarters in the house's final phase help to identify the Casa del Menandro's role in supporting 'an agricultural operation outside the city'.[40]

The combined evidence of architecture and artefacts as well as literature (though not discussed here) describing the rich material culture of Roman households shows the variety of objects that were constantly moving into and out of large urban houses. Moreover, the kinds of objects—the agricultural implements, building materials, and large quantities of storage amphorae—demonstrate the interconnection of many houses and rural production. The house was not an economic dead-end for production, the parasite of the consumer city model, but rather was an integral part of the movement of goods between the urban and rural environments. Indeed, Cato's mill was transported from the city into the countryside. The Household Mode of transport was the most cost-effective means by which this integration occurred, as the owners of urban and rural properties moved materials to, from, and between these locations for their own internal supply, the supply of their dependants, or to deliver raw materials or finished goods to urban, local, or foreign markets.

THE COMMERCIAL MODE OF TRANSPORT

The Commercial Mode of transport operated alongside the Household Mode and complemented it by offering transportation services to those who could not afford or did not need to own their own means of transport. Three main factors informed the decision to rent transportation or to own it (i.e., the Household Mode), not least of which was the high cost of an animal and its maintenance.[41] Beyond the cost of owning the means of transport, the frequency of the need to travel and distance to be covered greatly impacted the profitability of the endeavour and therefore the choice of the mode of transport. As the discussion of delivering Cato's mill demonstrated, in a short trip to a local destination (perhaps a frequent trip from Cato's villa) owning the means of transport was far more cost-effective. On the other hand, a single trip

---

[38] These tools were: eighteen pruning knives, four hand mills, ten hoes, three picks, a mattock, a shovel, a rake and a set of shears (Allison n.d.).
[39] The plough is recorded by Kolendo (1985) and cited in Purcell (1995a: 165 n. 56).
[40] Ling (1997: 105).
[41] Martin (2002: 164 n. 59).

to a distant town offered no advantage and further excursions would have even operated at a loss.

In the Commercial Mode, the means of transport were hired from a range of contractors who might have been an individual with a single cart and donkey for hire to a business with a large number and variety of vehicles, draught animals, and manpower. Thus, the proprietor of a *taberna* might hire a cart and draught animal to pick up a load of stones and the *dolia* to build a masonry counter in order to offer and advertise the sale of prepared foods. For the owner of a large private property who is charged with the responsibility of repaving the adjacent street, a number of carts, drivers, draught animals, and slaves would need to be contracted at once so that the large amount of lava needed for paving stones could be brought to the street in quantity and with efficiency. In these examples, individuals enter into two different kinds of contracts. The first is through *locatio conductio rei*, a contract whereby property is rented, but no services are part of the agreement. Conversely, in the second example, the contract falls under *locatio operis* in which the contractor is liable for the entire project, including the objects being delivered.[42] An example of the latter, preserved in a second-century CE papyrus, is instructive:

> 'And I will weigh and give to your cameleer another twenty talents for loading up for the road inland to Coptus, and I will convey [sc. the goods] inland through the desert under guard and under security to the public warehouse for receiving revenues at Coptus, and I will place [them] under your ownership and seal, or of your agents or whoever of them is present, until loading [them] aboard at the river, and I will load [them] aboard at the required time on the river on a boat that is sound, and I will convey [them] downstream to the warehouse that receives the duty of one-fourth—the charges for conveyance through the desert and the charges of the boatmen and for my part of the other expenses'.[43]

The language of this contract gives us a glimpse of the complexities of the Commercial Mode of transport and the detailed manner in which those complexities—the costs including duties, the route, timing, security, storage arrangements, and communications with agents—were set out and agreed upon.

Pompeii was typical of Roman towns concerning the spatial organization of the Commercial Mode of transport. If one wished to hire transportation services he went to the city gates to obtain it.[44] The clustering of transport properties near the city gates demonstrates the density of wheeled vehicles expected to be and (since they were the only way in or out of the city) necessarily found at the gates.[45] Literary and epigraphic evidence for the

---

[42] Martin (2002: 156–7).
[43] Translation by Casson (1990: 195). Cf. also Meyer and van Nijf (1992: 127–8).
[44] Casson (1994: 179–80); MacMullen (1974: 70, 72).
[45] Kleberg (1957: 50) noticed this concentration but with inns generally, not ramps specifically. Cf. also DeFelice (2007: 478).

concentration of transport services at the city gates support the archaeological data. Characters in Apuleius and Plautus, for example, find lodging very near to the gates amongst a number of other commercial properties.[46] Inscriptions dedicated by members of the mule-drivers' association (*collegium iumentaio*) also place cart drivers near the gates of Roman cities.[47]

These properties, however, operated mainly as part of an economy geared towards providing for those who were already travelling. Therefore, we must distinguish between the services *of* transportation and the services *for* transportation even if these were undoubtedly intertwined in the business of running an inn. Paradoxically, it is in this distinction and also in this place (and at bakeries) that the Household and Commercial Modes of transport can be seen to intersect. For example, had Cato's men come to Pompeii to pick up the desired mill (Household Mode), they might have taken lodgings in one of the inns near the city gate, not only for their comfort but also for the security of their animals, cart, and cargo.[48] Or, if Cato had chosen to have the mill delivered from Pompeii (Commercial Mode), the vehicles, animals, and slaves might have been contracted though the manager of the same inn. Finally, when not in use for the Household Mode of transport, the cart and mules from the Casa del Menandro (for example) might have been posted near the Porta Stabia for their profitable use in the Commercial Mode of transport.[49]

The economic opportunity to cater to travellers as they entered and left the city not only determined where inns with ramps would be located at Pompeii, but also influenced the form of the building and kinds of services offered. The inn at I.1.4–6 provides a typical example (Fig. 8.6).[50] A broad, Paved Ramp (Fig. 8.1) leads from the street into a wide corridor (A) before opening into a courtyard for vehicle storage (B). As the carts moved through the corridor, their hubs hit the masonry piers, leaving a characteristic rounding as evidence of their passage (Fig. 8.7).[51] These brick piers supported an upper storey accessed by an internal staircase, but in several examples an external stairway is found leading from the street, a fact that suggests a separate apartment might have been operated by the inn.[52] Such an interpretation is bolstered by the location of the external stairway. It is often placed between the ramp entrance of the inn and the entrance to a bar (L) and/or shop (K) connected to

---

[46] Plaut. *Pseud.* 658; Apul. *Met.* 1.21.   [47] *CIL* V.5872 (Mediolanum).
[48] On the liabilities in transportation, cf. Martin (2002: 159 nn. 39 and 40).
[49] Martin (2002: 154–8). This is an alternative version of Maiuri's explanation for the absence of the draught animals in the Casa del Menandro's stable area.
[50] DeFelice (2007: 478) describes the inn at VI.1.2–4 for its similar typicality, while Packer (1978: 6–9) does the same for I.1.6–9.
[51] The rounding begins at approximately 50 cm, which matches the axle height of the carts found in the Casa del Menandro (see above) and the Villa di Arianna at Stabiae. On the later cart see Miniero (1987). For another example of the rounding from wheel hubs as proxy evidence for particular types of movement, see Newsome (2009a) on the Casa del Marinaio (VII.15.1–2).
[52] These are I.1.7, VI.1.3, and VIII.1.2.

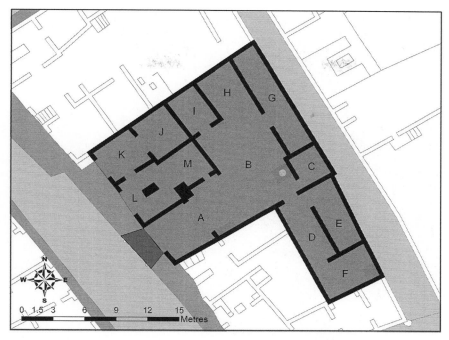

Fig. 8.6. Plan of the inn at I.1.3–5 (image by Eric E. Poehler).

the inn. These businesses not only communicated internally with the inn, but also opened wide doorways on the street to attract customers not accommodated by the inn.[53] The additional commercial enterprises associated with inns demonstrate a sophisticated and diversified business strategy to attract customers whether on wheels or on foot.

CONCLUSIONS

The Household and Commercial Modes of transport identified in the archaeological remains at Pompeii add an intention to the circulation of traffic that is absent in the articulation of the traffic system itself. By adding and defining destinations (established through the presence of ramps and stable areas) it becomes possible to distinguish the internal supply of large households from the flow of traffic on commercial errands. Conversely, the complexity of individual properties (e.g., the Casa del Menandro) and their (discernable)

---

[53] All inns with ramps except one (VI.1.1) have associated retail properties.

Fig. 8.7. Rounding on brick pier in I.1.4 (photo by Syd Evans with permission of the Soprintendenza Archeologica di Pompei).

urban investment strategies demonstrate that these were overlapping rather than opposing sectors of the transport economy. Indeed, the large inn at VI.1.2–4, immediately inside the Porta Ercolano, was once connected to the Casa delle Vestali and previously served as its stable area. Once it became more profitable to commercialize that space, the area was closed off and a new ramp was built at the rear of the house.[54] It is even tempting to link the assumed profits of this change of operation to the Commercial Mode with the lavish decor of the following phase of the house. Regardless, the wheeled access at the rear of the Casa delle Vestali maintained its connection to the Household Mode, the efficiency of which was crucial in local transportation to the internal integration of the regional market around Pompeii. Similarly, the cost-effectiveness of Commercial Mode over relatively longer distances permitted time and price difference to matter, a fact that served to tie different regional economies together. Thus, Cato's mill would have cost only 15 per cent more by having it delivered from Pompeii than fetching it himself from Suessa,[55] a margin more than made up for in the production of his men and animals.

The economics of transportation at Pompeii is perhaps best summed up in an example. The fifth doorway on Via dell'Abbondanza west of the Porta di Sarno is accessed by the broad Paved Ramp leading into the Praedia Iuliae Felicis. The ramp announced its stabling services while the rest of the frontage invited travellers with its high kerbs, projecting stairway, and monumentalizing architecture. An advertisement on the facade detailed the amenities to investors and visitors alike:

> To rent for the period of five years from the thirteenth day of next August to the thirteenth day of the sixth August thereafter, the Venus Bath fitted for the well-to-do, shops with living quarters over the shops, apartments on the second floor located in the building of Julia Felix, daughter of Spurius.[56]

The commercial investment in this building was enormous in scale and complexity: bigger than the Central Baths in area, the property joined two formerly separate insulae and completely suppressed a pre-existing street.[57] The speed of the property's construction following the earthquake(s) of 62 CE and its advertisement for rent before the eruption seventeen years later attests to the expectation of substantial profit from the traffic that would pass by this location due to its proximity to the amphitheatre and two city gates.[58] The size of the building, its well-appointed rooms, and the quality of the services it

---

[54] Jones and Robinson (2007: 398–9).
[55] Laurence (1998b: 127).
[56] *CIL* IV.1136, 'In praedis iuliae SP F felicis locantur balneum venerium et nongentum, tabernae, pergulae, cenacula ex idibus aug primis in idus aug sextas, annos continuos quinque. S Q D L E N C.' Translation by Bernstein (2007: 529).
[57] Evidence for the previous flow of traffic on this street still exists on the former corner kerbstones remaining in the kerb. The street has also been detected in excavations within the *praedia* (Parslow 1998: 203–6).
[58] Parslow (1998: 206).

offered, however, belies the similarities in location and function that the *praedia* shares with humbler hospitality properties at the other gates.

At the same time, the shop (II.4.1), baths (II.4.6), bars and restaurant (II.4.5, 7), the inn and its garden (II.4.2, 3, 8–12) all required a substantial flow of materials themselves. In fact, the stable area where visitors might have parked their vehicles was directly connected to the furnace for the baths, which was in need of fuel and ash removal. The design of the Praedia Iuliae Felicis reveals these combined concerns for the property's internal supply and for the attraction of both local (baths) and non-local (inn) traffic. The advertisement shows a similar connection, even conflation in the progression of commercial and residential activities described: from commercial baths to 'shops with living quarters' to second-floor apartments. There is also a commensurate progression in the amount of supply each would need and the amount of traffic it would receive. Thus, the bath area was equipped with a ramp and stable, the shops were given wide doorways, and private apartments were relegated to the upper floors. Within the singular design of the Praedia Iuliae Felicis, the Household and Commercial Modes of Transport intersected to the point of ambiguity. Such uncertainty, however, underscores the interdependence of these modes and the ease with which Pompeians could move between them. Moreover, the decision to invest in transport related services generally (e.g., inns and taverns) and infrastructures specifically (e.g., ramps and stables) shows the importance of this sector of the overall economy and establishes it as a particular driving force in the development of the urban landscape.

9

The Spatial Organization of the Movement Economy

The Analysis of Ostia's *Scholae**

Hanna Stöger

Ostia's built environment has attracted wide-ranging research interest in recent years; still, the city's spatial organization has remained a neglected field of study, with only limited attention given to formal methods of spatial analysis. This chapter seeks to address this imbalance, offering insights gained from a systematic spatial analysis of Ostia's street network and its guild buildings (*scholae*), applying space syntax concepts and techniques.[1] To appreciate the spatial behaviour of the guilds (*collegia*) questions about the interaction between the guild buildings and the city's streets and public spaces need to be asked and answered. In this chapter, the relationship between internal and external space at Ostian *scholae* is analysed and related to the analysis of the street network itself.[2] This provides new insights not only into

---

* Thanks to Ray Laurence for organizing the session on 'Interaction in the Roman City: Understanding Movement and Space' at the European Association of Archaeologists conference in Malta (2008), where this paper was first delivered, and to the editors for their invitation to contribute to this volume. Thanks to Janet DeLaine for her challenging questions on an earlier version, and Akkie van Nes and the space syntax community at the 2009 Stockholm symposium for their encouraging comments and suggestions. Much of this study is drawn from parts of my PhD research and I thank my supervisors, H. Kamermans and L.B. van der Meer, and above all J. Bintliff for always being constructive, supportive, and pertinently critical. Finally, my thanks go to the Soprintendenza of Ostia and their team of curators and archaeologists. This work has been possible only because the Soprintendenza has supplied unbounded support and encouragement for this research.

[1] Space syntax is a collective term for a set of theories and techniques for spatial analysis pioneered by Hillier and Hanson (1984) and further developed by the Bartlett Institute at University College London. See van Nes, in this volume.

[2] Stöger (2009).

the specific case studies of Ostian guild buildings but into the principle of the movement economy in the development of Roman urbanism.

This chapter begins with an overview of Roman guilds and the history of their study, before taking a closer look at Ostia's street network and introducing the data sets used for spatial analysis. Following this, the concept of the movement economy and the principles of space syntax are explained. This is followed by the analysis of the individual buildings and the street network, presenting the results and their interpretation in light of how they contribute to our understanding of the movement economy in Roman cities.

A HISTORIOGRAPHY OF OSTIAN *COLLEGIA* AND *SCHOLAE*

The guilds, the so-called *collegia* or *corpora*, were probably the most important private associations in Roman society. Organized on the basis of voluntary membership, the guilds pursued common goals with stated religious, social, or professional objectives, which in practice often overlapped. Their members belonged to the lower classes (*tenuiores*), below the three orders (*ordines*) of senators, knights, and municipal decurions.[3] This class distinction seemed foremost a legal one, at the same time the members must have been of good financial standing since their memberships involved considerable financial commitment.[4] The guilds could hold property and inherit legacies.[5] Their investment in urban property becomes primarily visible through their guild buildings, the so-called *scholae*.[6] These often formed part of a larger building complex, sometimes comprising entire *insulae* with diverse land use.[7]

At Ostia, around sixty different guilds and their activities have been identified on the basis of epigraphic evidence.[8] The guilds are mainly connected to port activities (for example, the guilds of the shipowners, the weight controllers, the grain measurers, and the bargemen), but also to services required by the city's inhabitants (for example, builders).[9] Concurrently, these guilds dealt with the social and religious needs of the local community. In one way or

---

[3] Bollmann (1998: 22).
[4] Ausbüttel (1982: 46–8).
[5] See Meiggs (1973: 312 n. 4).
[6] See Ausbüttel on donations and investments by guild members to construct or embellish guild buildings (1982: 43).
[7] See Hermansen (1982: 95–121) on urban property owned by Ostia's guilds; and specifically Hermansen's assessment of possible guild property based on Roman building laws; contra Bollmann (1998: 213–21).
[8] Chevallier (1986: 153–7).
[9] Hermansen (1982: 56).

another, the guilds covered almost every aspect of the town's life, involving a considerable proportion of the population. Information about the social activities of guilds comes almost entirely from dedicatory inscriptions, decrees conferring offices, and legal codes.[10] As such, comparatively little is known about daily routines or less 'celebrated' activities; of the commemorated activities, religious observance, acts of patronage, reciprocation, and conviviality are most prominent.[11] These activities were often attached to particular and identifiable locations, the so-called *scholae*. Thus the guild buildings played an important role in second-century CE Ostian society, marking locations which potentially sustained greater social interaction than many other places. Out of a larger number of Ostia's possible guild buildings, eighteen have been archaeologically identified as *scholae*.[12] Their identification is based on the combined evidence of architectural remains and inscriptions found *in situ*, often corroborated by iconography and decor of wall paintings, floor mosaics, or statuary.[13]

There is a long tradition of research on *collegia* and *corpora*, which attracted interest as early as the sixteenth and seventeenth centuries.[14] In the nineteenth century, when ancient historians were inspired by their personal experience of newly founded 'bourgeois' voluntary associations, their research interest was principally focused on the legal and political status of Greek and Roman *corpora*. With the compilation of the *Corpus Inscriptionum Latinarum* at the end of the nineteenth century, research on this subject advanced significantly. Waltzing's four-volume study included a collection of all the available relevant epigraphic and literary material, and remains a landmark on this topic.[15] The interest in Roman associations peaked for a second time in Italy in the 1930s/40s, when the corporate state ideology of Italian fascism prompted a renewed fascination with Roman associations, reflected in the resulting historical studies.[16] De Robertis, the foremost authority on the legal status of the Roman guilds, produced academic analyses that bore no trace of the

---

[10] See Bollmann (1998: 37–9) on social activities performed by guilds with references to relevant epigraphic sources. On Ostia's guilds and their activities see Bollmann's catalogue entries A27–A45 (1998: 275–345). These entries describe identified *scholae* with inscriptions attributed to them. In addition, Bollmann's catalogue C provides inscriptions referring to *scholae* which have not been identified archaeologically; catalogue entries C 29–37 are relevant to Ostia (Bollmann 1998: 470–1).
[11] Patterson (1994: 233).
[12] Bollmann (1998).
[13] Bollmann (1998: 275–345); cf. Hermansen (1982: 85–6); cf. Laird (1999, 2001) for a controversial view. I am not sure it is clear why only seventeen are considered *scholae* from 'a larger number of possible guild buildings'.
[14] Ausbüttel (1982: 11).
[15] Waltzing (1895, 1896, 1899, 1900).
[16] Ausbüttel (1982: 13).

environment in which they were created.[17] Recent scholarship has mapped the modern evolution of the ancient concept of the Roman guilds, particularly elucidating how the political and social movements of the nineteenth and twentieth centuries in Western Europe have shaped scholarly work on this ancient phenomenon.[18]

Until the 1980s, research on guilds seemed to be firmly in the hands of historians, relying exclusively on epigraphic and literary sources; the material culture of guilds was only explored through epigraphic material and its references to certain buildings and related objects. Even when large-scale excavations in Ostia (between 1938 and 1940) substantially broadened the material record, it took almost two decades before the first essays concerned with *scholae* or the buildings of the guilds appeared.[19] Such excavations gave the first indications of what these buildings looked like, when identified by epigraphy.[20] Subsequently, various properties of guilds and their architectural characteristics were published independently. The first compilation of all Ostia's presumed *scholae* appeared in 1982.[21] Other publications, although presenting combined archaeological and epigraphic evidence, concentrated on comparative studies of single groups of *collegia* and their respective type of *schola* in various Roman cities (e.g. the *scholae* of the Augustales).[22] To date, the most complete survey of the combined archaeological evidence and epigraphic sources is Bollmann's *Römische Vereinshäuser*, a study which sought to understand the *scholae* as a means of self-representation within a civic and urban context.[23]

With reference to Ostia, earlier studies have already identified the *collegia* as one of the major forces driving the development of urban social networks.[24] Earlier studies have proved important in interpreting their architectural structures, informed by Roman building laws, literary analogies, and by the careful examination of topological characteristics.[25] However, in terms of the spatial context of *scholae*, most studies are descriptive and normative.

---

[17] In 1971, after decades of research, De Robertis (1971) published a two-volume history of Roman corporations, including his earlier works produced during the crucial period of Italian fascism; see Ausbüttel (1982: 13) and Perry (2001: 205).
[18] Perry (2006).
[19] See Bollmann (1998: 17).
[20] Slater (2000: 495).
[21] Hermansen (1982: 55–89).
[22] cf. Bollmann (1998: 18). Flambard's essay provided a model for the integration of architecture and epigraphy. Cf. Slater (2000: 493); Flambard (1987).
[23] Bollmann (1998).
[24] Meiggs (1973); Kockel (1993).
[25] For the study of the architecture of *scholae*, see Bollmann (1996: 195–200); Steuernagel (2004: 176–209); on Roman building laws, see Hermansen (1982); on literary analogies see Egelhaaf-Gaiser (2000, 2002); on topological characteristics see Steuernagel (2005: 73–80).

To remedy this, we can adopt a unique, integrated spatial approach that combines aspects of space syntax analysis at the level of the individual building and at the level of the town plan. Since archaeology has to make the most of limited evidence, the incorporation of space syntax into any study of wider perspectives offers insights that reach beyond observations made from building plans and visible structural remains. Space syntax's methods not only provide evidence for the intricate organization of urban space, but also investigate the active role of space in the constitution of society, through considering the ways in which social processes map into built form.[26] To this end, the remainder of this chapter investigates whether the spatial organization of Ostia's *scholae* matches their presumed integrative role in society; a role suggested by previous investigations based on ancient literary sources and intuitive approaches to urban space and movement.

OSTIA'S PUBLIC DOMAIN: THE STREET NETWORK AND PUBLIC PLACES

To appreciate how Ostia's guilds organized and negotiated space, the locations and distribution of their *scholae* need to be studied within the wider context of the city's street network (Fig. 9.1). We can begin with a brief overview of previous research on that network, before proceeding to more detailed analyses that integrate data from the buildings with data from the city's streets.

As noted, Ostia's street network has been relatively neglected by researchers in the twentieth century. Yet, a small number of studies have made significant contributions to the better understanding of the city's streets and their underlying formation processes. Valuable insights on the earliest road system come from van Essen's work of the 1950s.[27] He identified several stretches of oblique streets cutting diagonally through the urban fabric, applying approaches similar to Conzen's method of town plan analysis.[28] Van Essen focused on those features which run against the grain of the general street network and recognized in these oblique courses the imprint of the old road system, existing long before the foundation of Ostia's *castrum* (of the military colony). Important work on the wider road system, linking Ostia with its rural hinterland, is provided by surveys carried out in the Pianabella area. Located southeast of Ostia, an extensive orthogonal street grid can be identified, dividing large plots of land, predominantly used for agriculture.[29]

---

[26] Anderson (2005); cf. Newsome (2009a: 26).   [27] Van Essen (1957: 509–13).
[28] Conzen (1960).
[29] Heinzelmann (1998), see also Bradford for an earlier survey of the Pianabella area (1957: 242–3).

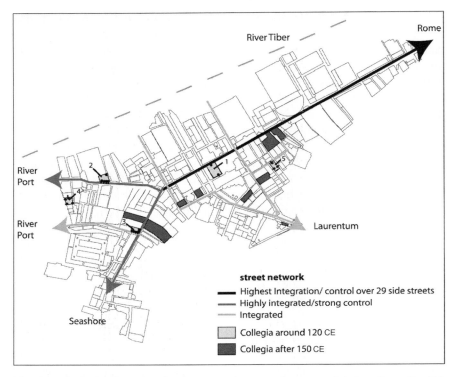

Fig. 9.1. Ostia in the second century CE (excavated areas only), location of guild buildings along the main roads (Guild buildings 1–5 have been selected for spatial analysis) (image by Hanna Stöger).

Ricardo Mar has contributed to our understanding of the historical development of the city and territory of Ostia.[30] In his earliest article, he used three levels of interrelated analyses to demonstrate how the city progressively reconstituted itself: the course of the streets, the system of urban land division, and the typology of buildings.[31] From Mar's study it becomes clear that Ostia's street network reflects long-term processes, including the need to mediate between territory, city, and the movement associated with the arrival of goods and peoples to Italy and then onward, through Ostia, towards Rome, or elsewhere. Mar demonstrates how the origins of Ostia's road system reach back to a communication system that long preceded the foundation of the so-called *castrum*. The foundation of the *castrum* provoked a noticeable change and resulted in an adjustment of the road system based on new factors. Subsequently a coherent system of streets grew around the *castrum*, developing into a network

---

[30] Mar (1991: 81–109) and, more generally, Mar (2008: 125–44).
[31] Mar (1991: 84 n.16). On the 4th century streets, see Gering (2004).

according to the most frequented directions around the settlement, responding to major access roads (the Via Ostiensis to and from Rome, and the Via Laurentina, leading towards Laurentum), and linking up with gates and city walls. In this new balance between the demands of the territory and those of the town, the urban road system that can still be seen today developed.

Naturally, the long-term occupation of Ostia and the transformations that took place over time left a mark on the physical nature of the street network. The consecutive street levels layered on top of each other provide a preliminary dating sequence, as well as information on the materials employed for the paving of streets.[32] After all, the large basalt blocks which pave Ostia's streets, as they are visible today within the excavated areas, are a statement of a conscious choice made by a city that had learnt how to keep its roads dry, and had the financial resources to invest into such durable materials.[33] Surely not much of a concern in antiquity; however, anyone of today's visitors who has experienced a day of walking Ostia's streets can confirm how unsympathetic the basalt paving is to pedestrian movement.

Turning to wheeled traffic on Ostia's roads, whilst ruts caused by wheeled traffic are visible on the basalt paving of the eastern *decumanus*, it is not clear whether Ostia followed Rome in imposing temporal controls on vehicular traffic.[34] Also, Ostia appears to have responded differently to movement and traffic than Pompeii, which had developed a structured approach including one-way systems and restrictions on vehicles.[35] For Ostia, it is difficult to establish whether the accommodation of wheeled traffic was ever taken so far as to completely, or partially, restrict certain roads for the sake of traffic. However, there are some almost defining events, most notably, when the course of the Cardo Maximus was disrupted by the placing of the temple to Augustus and Roma (beginning of the first century CE) and later the Capitolium (about 120 CE).[36] These interventions had the effect of closing the north–south axis for wheeled traffic and thus isolating the northern and southern part of the forum area from through traffic (Fig. 9.2).[37] No interventions seem to

---

[32] Heinzelmann (1999: 84–9).

[33] Basalt does not have capillary properties; hence it does not pull water (verbal communication by Enrico Rinaldi, conservation expert at Ostia). On the supply of basalt to Ostia see preliminary results in Black et al. (2009: 705–30).

[34] Laurence (2008: 87–8).

[35] Laurence (2007, 2008a), Poehler (2006); see Kaiser, Poehler, and Hartnett, in this volume for discussions of these issues in relation to Pompeii.

[36] The construction dates for the Capitolium are *c.*120 CE, based on brick stamps (Calza 1953: 215); while the dating of the temple to Roma and Augustus has not been firmly established. Construction dates have been suggested for the period of Tiberius, linked to the spread of the cult of Augustus. The cult had not been introduced to Rome during the Emperor's lifetime, whereas it was allowed in other cities. More recent work suggests mid-to-late Augustan dates for Ostia's temple to Roma and Augustus (Pavolini 2006: 106). For an accessible overview, see Mar (2008).

[37] See Newsome, in this volume, on this broader trend in the city of Rome.

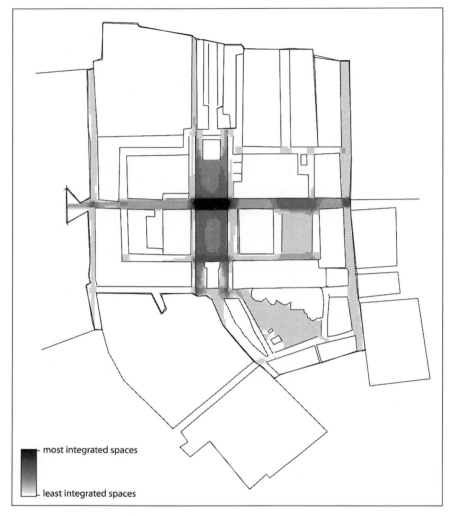

Fig. 9.2. The Forum in the Antonine period: the visually most integrated spaces are concentrated along the *decumanus*, while the north–south axis remains less integrated (Visual Graph Analysis using Depthmap (UCL) produced by Hanna Stöger).

have affected the east–west/west–east movement through Ostia along the eastern *decumanus*—and its continuation, the Via della Foce—leading to the mouth of the river. While the *decumanus* stands out as the lifeline that runs through Ostia, the Tiber undoubtedly played an equally important role within the city's system of movement; all communication with the areas north of the river

The Analysis of Ostia's Scholae 223

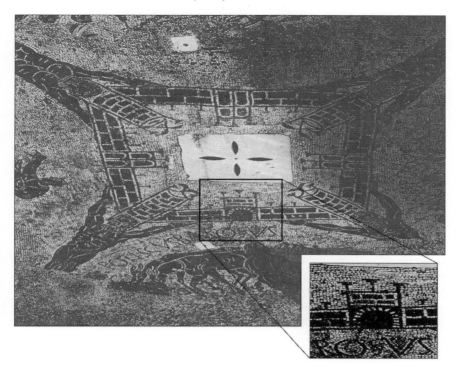

Fig. 9.3. Terme dei Cisiarii (II.ii.3): detail from the mosaics in frigidarium C (photo by Hanna Stöger).

required some form of crossing.[38] In addition, river transport could have potentially accounted for the movement of bulky goods within the confines of Ostia.

An interesting clue to Ostia's streets is detailed by the mosaics from the Terme dei Cisiarii (Baths of the Coachmen) located close to the Porta Romana. The mosaics display a highly stylized city wall as a border ornament; within the centre of the mosaic is a second city wall, presumably symbolizing Ostia's earlier city walls associated with the *castrum*. There are four gates within the stylized Ostian walls. Notwithstanding their symbolic nature, the gates are rendered in some detail, visibly distinguishing between three gates with two narrow doors placed next to each other, and one gate with a single wide door opening (Fig. 9.3). It is striking that a clear distinction had been made between gates which seem to be narrow enough to restrict passage only

---

[38] References to guilds active in the ferry business suggest plenty of interaction via the river, e.g. *corpus scaphariorum et lenunculariorum traiectus Luculli* (the guild of the operators of skiffs and ferryboats at Lucullus' crossing); *corpus traiectus togatensium* (the guild of the civilians' crossing); *corpus traiectus marmorariorum* (the guild of the marblemen's crossing). See Hermansen for a detailed survey of Ostia's guilds (1982: 56–9, 239–41).

to pedestrians, while only one gate seems wide enough to allow for wheeled traffic. This may indicate that Ostia's city gates regulated wheeled traffic, possibly leaving the *decumanus* as the only street with a road clearance wide enough to accommodate two-way wheeled traffic.

INTEGRATED DATA SETS: BUILDINGS AND STREETS

The integrated approach developed here, combining aspects of space syntax at the micro-scale level of individual buildings and at town plan level, is a promising way to capture the spatial properties of Ostia's *scholae*. By exploring different ways of formal spatial assessment, a better understanding of the spatial organization and its associated social activities can be achieved. The term 'exploring' needs to be stressed since it is impossible to establish in advance which spatial dimensions are likely to be the most relevant.[39] Space syntax offers the theories and the tools, but it is the researcher's task to discover which representations and which measures capture the spatial logic of a particular system.[40]

From a larger group of *scholae*, dating to the second century CE, five were assessed in more detail (Table 9.1). These form a small but coherent sample. All were built during the first half of the second century CE, within a period of around twenty years. Ostia's street network provides the second field of data for analysis. This includes the streets and public spaces which make up the street network of the excavated area, which amounts to about one-third of the original expansion of Ostia, as well as the extended street network based on the preliminary results of the geophysical surveys, tentatively assessed for control purposes only.[41]

Table 9.1 Sample for spatial analysis: five selected guild buildings, Scavi di Ostia

Names of the Guild Buildings	Site Reference	Date	Location
Casa dei Triclini	I.xii.1	*c*.CE 120	*decumanus*/forum
Aula e Tempio dei Mensores	I.xix.1–3	*c*.CE 112	Via della Foce
Domus di Marte	III.ii.5	*c*.CE 127	*decumanus* (west)
Domus accanto al Serapeo	III.xvii.3	CE 123–6	Via del Serapide
Caseggiato dei Lottatori	V.iii.1	*c*.CE 120	Via della Fortuna Annonaria

---

[39] Hillier and Hanson (1984: 122–3).
[40] Thaler (2005: 326).
[41] Only a tentative first assessment of the street network in the areas has been made, based on preliminary results which were communicated by Michael Heinzelmann at the 105th Annual Meeting of the Archaeological Institute of America San Francisco, California, 3 Jan. 2004; the final publication of the DAI Ostia project, conducted between 1996 and 2001, is expected shortly; see also Heinzelmann (2002, 2005).

Ostian *scholae* and Roman *scholae* in general, are characterized by varied layouts and a lack of formal architectural language. The ultimate confirmation comes from epigraphy, which alone provides certainty that a building was used by a *collegium*.[42] While they display architectural diversity, their functional role seems to be consistent. Above all, these buildings had to offer appropriate premises to accommodate a range of activities performed by the *collegia* (banquets, religious and cult practice, as well as formal and informal encounters and gatherings).

Although the small sample size of five *scholae* does not offer sufficient statistical material to allow for a strictly quantitative assessment, still some general characteristics can be observed and compared. Space syntax is well equipped to compare different ground plans, since it permits the assessment of architectural structures of very different spatial configurations.[43] Space syntax does not attach functional labels to separate parts of the buildings; instead it understands buildings as structured configurations of space, which form patterns of movement and encounter.[44] Given the fragmentary nature of archaeological data, such a value-free characterization seems most welcome in archaeological research, the more so since 'labelled spaces' with evident land-use properties and clearly defined functions are often only found in exceptional sites such as Pompeii, where spaces can be identified through well-preserved finds and detailed architectural records. This is rarely the case for Ostia.

Here we can briefly introduce the *scholae* selected for analysis, in order to clarify their specific urban settings.

The **Casa dei Triclini** (I.xii.1): The *schola* is located on the southern side of the eastern *decumanus*, bounded by the Via della Forica in the south and separated by a colonnaded passage from the forum proper in the west. The area east of the *schola* was, in the fourth century, occupied by a colonnaded space, the so-called Foro della Statua Eroica, creating an extension to the open areas of the Terme del Foro. Earlier structures, contemporaneous with the Casa dei Triclini belonged to another bath, dating to the Hadrianic period, which was built over by the Foro della Statua Eroica.[45] The Casa dei Triclini is securely identified as the meeting place of the *corpus fabriorum tignuariorum* (guild of the builders) by an inscribed statue base, dedicated to Septimius Severus.[46]

The **Aula e Tempio dei Mensores** (I.xix.1-3): These buildings are located within a trapezoid enclosed space, situated at the northern side of the Via della Foce. The area further incorporates a courtyard and a range of rooms east of the temple.[47] The *schola* complex seems structurally and functionally linked to

---

[42] Bollman (1998); Slater (2000).   [43] Lawrence (1990: 75); cf. DeLaine (2004: 161-3).
[44] Grahame (2000: 40).
[45] Pavolini (1986: 108); Cicerchia and Marinucci (1992: 20-2); Calza (1953: 128, fig. 32).
[46] *CIL* II.54569. Calza (1927: 380); Pavolini (1986: 137); Hermansen (1982: 62).
[47] Bollmann (1998: 291-5); Hermansen (1982: 65-6); Calza (1953: 125).

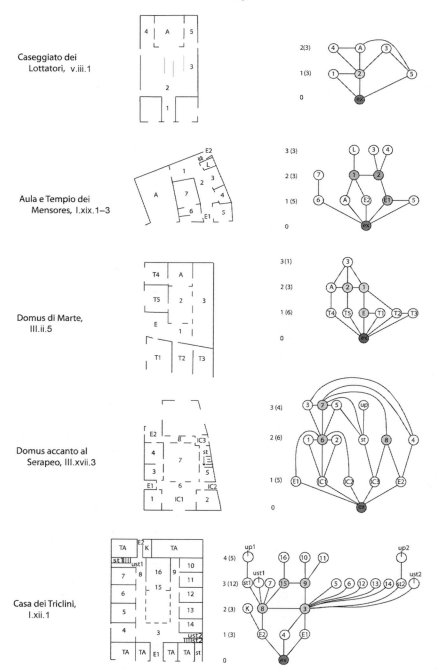

Fig. 9.4. Ground plans of guild buildings display diversity of layout (image by Hanna Stöger).

warehouses (Horrea dei Mensores, I.xix.4) and occupies the south-eastern corner of an insula otherwise composed exclusively of warehouses. Along the eastern boundary of the insula, a street leads from the Via della Foce to the Tiber, separating this insula from the Terme del Mithra. It has been identified as the seat of the *collegia* of the *mensores* (the grain measurers), on the basis of both epigraphy and iconography.[48]

The **Domus di Marte** (III.ii.5): Located right at the corner where the western *decumanus* and the Cardo degli Aurighi intersect, the so-called Domus di Marte enjoyed a prominent location. Bounded by a commercial building (part of the Domus sul Decumano (III.ii.3,4) on the *decumanus* to the north, and the Trajanic *horrea* (III.ii.6) on the Cardo degli Aurighi to the west, the guild building occupies space within an *insula* of largely commercial use. The original structures of the Domus di Marte date to *c.*127 CE, with successive alterations taking place over time.[49] Central to the argument identifying the building as a *schola* is the marble altar with the inscription 'Marti/Aug/ Sacrum'.[50]

The **Domus accanto al Serapeo** (III.xvii.3): This so-called *domus* forms part of a larger set of buildings dedicated to the cult of Serapis. The complex (three interlinked buildings III.xvii.3–5) is located within a triangular area on the southern side of the Via della Foce, extending from the Caseggiato di Bacco e Arianna (III.xvii.5) towards the warehouse or *horrea* (III.xvii.1) on the Cardo degli Aurighi. The central building hosted the temple dedicated to Serapis (III.xvii.4), while the two other buildings, found on either side of the temple, acted as spaces for banquets and meetings. These buildings were linked through a system of doorways and service corridors, running along the rear of the buildings. The complex was built between the years 123–6 and inaugurated in 127 CE.[51] Later interventions blocked the original interconnections between the buildings, and new entrances were created on the so-called Via del Serapide.

The **Caseggiato dei Lottatori** (V.iii.1): This *schola*, originally classified as *domus*, was located at the northern end of an insula bordered by Via della Fortuna Annonaria, Via delle Ermette, and Via della Casa del Pozzo.[52] To the south the *schola* is built against the northern wall of the earlier neighbouring building. That this building was built earlier is evidenced by the two walled-up doors, which originally led to a plot later occupied by the *schola* itself. The *schola*'s original structures date to the Hadrianic period with subsequent interventions taking place in several phases. Being surrounded on three sides

---

[48] Bollmann (1998: 290).
[49] Calza (1953: 222).
[50] Bollmann (1998: 308–9); Hermansen (1982: 75–6).
[51] Bloch (1959: 226).
[52] For the identification as a *domus*, see Calza (1953: 236). For the identification as a *schola* see Hermansen (1982: 76–7).

by streets, the building had doors on all of them, maximizing the accessibility of its location.

We can now consider Ostia's street network and its spatial analysis. Any form of analysis requires a coherent data set; Ostia's street network is difficult to sample. Owing to the long-term occupation of the city, it is not easy to identify a consistent street network in any single period through secure archaeological evidence. Thus we need to clarify many central issues for the analysis: we need to know under which criteria can streets be considered to be part of the public space shared by the community. Which streets were public, semi-public, or private? How can we deal with large open spaces like the forum area with four borders? And above all, we need to be aware of the fact that there is a subjective judgement to be made when decisions are taken to identify units of space.[53] In practical terms, nothing beats field work and survey data. Ostia's *Pianta delle regione e degli isolate* (1954) serves as a good indicator for the street network present within the excavated areas, marking all those streets and squares which had paved surfaces and therefore where public use can be assumed.[54]

If we want to understand city traffic, we need to understand the architecture of the street and the flow of movement through it.[55] While this might be possible for any modern street network, which allows for pedestrian counts and traffic surveys, ancient street networks need to be studied differently. Space syntax offers suitable concepts and techniques and above all adds the principle of 'movement economies' to the discussion. The concepts of the movement economy postulate that the configuration of the urban street network (the urban grid itself) is the key determinant of movement flows and co-presence in space, and hence, the urban grid prioritizes certain locations.[56] This can be best understood by looking at any town or city. There people carried out their activities, involving numerous journeys which have their origin and destination distributed more or less everywhere.[57] Consequently, every journey in an urban system has three elements: an origin, a destination, and the series of intervening spaces that are passed through on the way from one to the other. The passage between origin and destination is considered to be the by-product of movement. Streets that are easily accessible and better connected to other streets are more likely to be selected as passage routes between other pairs of streets; thus well-integrated streets attract more passing movement.[58] For this reason, most journeys from side-street origins to

---

[53] Grahame (2000: 29).
[54] Calza (1953) provides site plans of Ostia covering the excavated areas (scale 1:500) and a general overview of Ostia's division into five arbitrary regions.
[55] Laurence (2008: 88).
[56] Hillier and Vaughan (2007).
[57] See Hillier (1996b) on the theoretical underpinnings of the 'movement economy'.
[58] See Hillier (1996b: 53).

side-street destinations are likely to pass through one or more segments of the main street, thus making the main street a better location for movement seeking urban activities or land use. Conversely, other types of activity, like residential use, might have sought a location away from the main streets to minimize the possible interference of through movement.[59]

Ostia's *scholae* form part of the movement economy within the city. Their preferred location is on the major streets of the city, and there would seem to have been a conscious initiative to seek these locations, which can be described as 'movement-seeking'. These issues are further elucidated via detailed analysis of the street network presented in the following sections. This will allow us to consider whether the concept of the movement economy offers a suitable model for the explanation of the citywide distribution of *scholae* in Ostia (see Fig 9.1).

SPACE SYNTAX ANALYSIS OF THE *SCHOLAE* IN RELATION TO THE STREET NETWORK

The basic principles of space syntax need to be outlined before we can proceed to analysing the spatial location of *scholae*.[60] Since the techniques of analysis used for the assessment of the built space have been thoroughly explained in studies applied to Pompeian houses and to Ostia's *medianum* apartments, they do not require further detailed comment.[61] General trends and problems in the archaeological application of space syntax have also been discussed elsewhere.[62]

However, we do need to be aware of the general principles of space syntax theory and methods. These are based on two formal ideas, reflecting both the objectivity of space and our intuitive engagement with it. First, space constitutes an intrinsic aspect of all activities that human beings carry out. Second, human space is not about the properties of individual space (like size, material, and decor), but about the configuration of spatial units, i.e. the interrelation between the many spaces that constitute the layout of buildings or cities. The properties of space are expressed through its configuration; configuration is the means by which space acquires social meaning and has social consequences.[63] Space syntax claims that all human activities have a necessary spatial geometry: movement is linear, interaction requires a convex space, in which all points can see each other, and from any point in space we see a

---

[59] Hillier (1996b).
[60] See Hillier and Vaughan (2007) for an introduction to space syntax's concepts and tools.
[61] Grahame (2000); DeLaine (2004). For a diachronic application of space syntax to the development of a Pompeian neighbourhood, see Newsome (2009a). See also van Nes, in this volume.
[62] Thaler (2005); Cutting (2003).
[63] Hillier and Vaughan (2007).

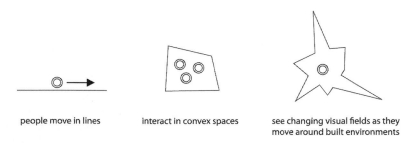

Fig. 9.5. The spatial geometry of human activity (redrawn from Hillier and Vaughan 2007 with permission from Bill Hillier).

shaped visual field, called an isovist, that can vary (Fig. 9.5).[64] Accordingly, the core tools of space syntax include axial, segment, convex, and isovist analysis, with Depthmap software (UCL) strengthening the analytical toolbox through Visual Graph Analysis (VGA). The latter correlates the result of the analysis of space syntax with visual analysis.

To examine the spatial configuration of Ostia's *scholae*, two space syntax tools are employed: access diagrams (justified graphs) and spatial values.[65] Access diagrams transform a standard architectural plan into a graph, which allows qualitative and quantitative assessment of the topological properties of space (Fig. 9.6). The spatial values applied comprise two independent space syntax measures: control values and real relative asymmetry (RRA).[66] These measures respond to the buildings' local and global spatial properties. While control values concern a space and its immediate neighbouring spaces (local), RRA deals with a certain space and its relationship to all other spaces within the system (global). These measures help in assessing the potential of different building layouts for interaction between the different groups who used the building: the inhabitants (the guild's members) and those visiting the buildings. Hence, access data offer indications about those spaces which are potentially destined for interaction and those which were more likely to have provided privacy.

By correlating the controlling spaces in *scholae*, identified by control values in excess of 1, with those spaces with very low RRA values, interesting insights have been gained on those locales where most interaction took place. As

---

[64] Hillier and Vaughan (2007).
[65] For a more detailed discussion of the analysis and the results see Stöger (2009: 108:4–7); the difference factor, a spatial value used to gauge the interior/exterior relationship of buildings, did not add significant information to the spatial values (RRA and control) used in this study. For difference factors see Hanson (1998: 28–36).
[66] See Grahame (2000: 29–36).

Casa dei Triclini, I.xii.1

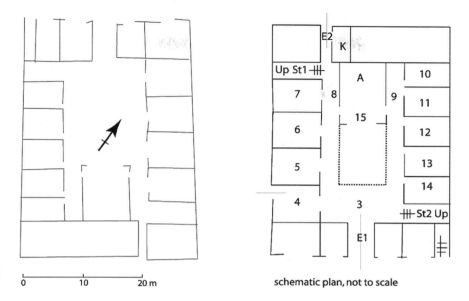

j-graph - Casa dei Triclini, I.xii.1

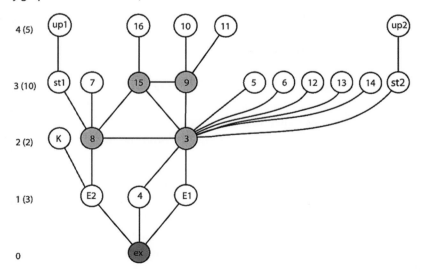

Fig. 9.6. Casa dei Triclini, I.xii.1; access analysis (j-graph) (image by Hanna Stöger).

Table 9.2 Access values for spaces with high interaction potential

Name of guild building and site-reference	Potential function	No.	RRA[a]	C.V.[b]	Neigh-bours[c]	L.I.P.[d]	G.I.P.[e]	Pres. Av.[f]
Casa dei Triclini, I.xii.1								
	Foyer	3	0.27	7.7	11	high	high	high
	Passage	15	0.46	1.55	4	high	high	high
	Portico	8	0.49	2.25	5	high	high	high
Aulae e Tempio dei Mensores, I.xix.1–3								
	Outside	ex	0.571	2.33	5	high	high	high
	Courtyard	1	0.701	2.25	4	high	high	high
	Passage	2	0.631	2.66	4	high	high	high
Domus di Marte, III.ii.5								
	Outside	ex	0.377	2.666	6	high	high	high
	Courtyard	2	0.527	1.416	4	high	high	high
	Foyer	1	0.527	1.083	3	high	high	high
Domus accanto al Serapeo, III.xvii.3								
	Stairs	st	0.382	1.83	4	high	high	high
	Outside	ex	0.594	1.41	4	high	high	high
	Triclinium	A	0.382	1.712	6	high	high	high
Caseggiato dei Lottatori, V.iii.1								
	Courtyard	2	0.196	2.166	5	high	high	high
	Outside	ex	0.588	1.033	3	med	high	m/h
	Undefined	5	0.588	1.166	3	high	med	m/h

[a] real relative asymmetry;
[b] control value;
[c] number of neighbouring spaces;
[d] local integration potential;
[e] global integration potential;
[f] presence availability (qualitative indications of interaction potential: low–medium–high)

DeLaine's study of Ostia's *medianum* apartments has already shown that we can expect the highest consistency between local and global integration in those rooms which functionally define a building.[67] Within the *scholae* these spaces are movement-related spaces, such as passages, foyers, courtyards and corridors, and above all the exterior carrier space, the city's street network (Table 9.2).

The data derived from access analysis revealed that four of the *scholae* attribute high local and high global integration potential to the outside carrier space; these buildings are to a large extent defined by their relationship with the exterior.[68] These *scholae* maximize their street frontage to take full advantage of their location by making their buildings as permeable as possible to promote social encounter at the interface between *scholae* and public domain. This brings us to questions on how well the *scholae* were integrated within the

[67] DeLaine (2004: 158). [68] Stöger (2009).

The Analysis of Ostia's Scholae 233

Table 9.3 Overall size of ground floor space, including *tabernae*

Guild Buildings	Street front on major roads (m)	Ground floor area (m^2)	*Tabernae* in total, and on major roads (m^2)
Casa dei Triclini, I.xii.1	28.3	1,239.00	275.00 / 159.00[b]
A/Temp. Mens, I.xix.1–3	24.8	540.70	30.10
Domus di Marte, III.ii.5	37.7[a]	356.00	145.00
D.ac.al Serapeo, III.xvii.3	—	407.60	—
Casa dei Lottatori, V.iii.1	15.9	455.80	67.40

[a] Total 37.7 m divided into 14.2 m on the western *decumanus* and 23.5 m on the Cardo degli Aurighi.
[b] Total of *tabernae* space 275.00 m^2; 159.00 m^2 of *tabernae* are located along the eastern *decumanus*/forum.

street network. Before formal analysis, the most straightforward aspect of the buildings—their overall size—needs to be considered.

Table 9.3 provides an overview of the total ground floor area and the extent of the street frontage of the *scholae* under discussion. It needs to be remembered that the small sample size does not allow for a strictly comparative approach or quantitative assessment, yet some observations can be made. As Table 9.3 indicates, the extent of street frontage does not simply increase in proportion to the overall plot size; instead this spatial aspect reflects the inherent property divisions, and above all situational responses to the specific setting of the guild buildings, as the corner position of the Domus di Marte demonstrates. *Scholae* dedicate their street frontage largely to *tabernae* (commercial outlets—shops and bars). Table 9.3 lists the unit sizes inclusive of *tabernae* as far as they are part of the bounded space of the *scholae*, even if these are not accessed directly from within the *schola*. Wherever *tabernae* remain strictly independent of the internal spaces of guild buildings, these were not included in the space syntax analysis when access values were calculated. Nevertheless, in terms of the total overall size of the unit, these independent *tabernae* constitute a considerable part, therefore they should help to throw light on how *scholae* relate to the street space in general. Again, Table 9.3 shows that there is no proportional relationship between the total unit size and the area covered by *tabernae*. Instead, the presence or absence of *tabernae* reflects a direct response to how well the street is integrated on which the *scholae* are located. In other words, *scholae* do not seem inclined to dedicate space to *tabernae*, unless these can be located along busy streets. Similar to any movement seeking business, the guilds studied tend to locate those spaces destined for transactions along the most accessible streets, while back streets are less likely to be attractive locations for *tabernae*, for obvious reasons.

COLLEGIA AND THE MOVEMENT ECONOMY OF OSTIA

As discussed, Ostia's *collegia* and the location of their buildings are well documented.[69] The city-wide distribution pattern of *scholae* makes it apparent that there was no clustering (see Fig. 9.1). This might have been expected given the comparable functions these buildings fulfilled. One only needs to think of today's banking districts, where functionally similar buildings are clustered to reinforce each other in an additive way, while single landmarks are likely to be weak references by themselves.[70] Their preferred location along Ostia's major thoroughfares and access roads has been interpreted as alluding to status and striving for association with the public buildings of the forum area.[71] Following Lynch's concept of place legibility, the image strength of a building rises when it coincides with a concentration of associations. Hence creating public associations could have been a powerful motif for several *collegia*, successfully put into effect by those *scholae* seeking the vicinity of the forum. Still, the location of choice might have been moderated by the realities of available urban space, as well as the guild's financial standing. In other cases the locality of certain guild seats appears dictated by proximity to their professional field, for example the Aula e Tempio dei Mensores, located next to storage facilities possibly used to store grain. Others again opted for closeness to their particular temple of worship, for example the Domus accanto al Serapeo.

Despite their distribution across the city, certain areas were almost devoid of *scholae*, in particular the area to the north of the *decumanus*, where disintegration into smaller plot size was apparently prevented by the large-scale development of public buildings and warehouses.[72] In fact, those few *scholae* found north of the *decumanus* and its continuation the Via della Foce, never reached the depth of the urban plot that conventionally characterizes the *domus*. Instead their layout is shallow, often not extending further into the insulae than the front row of *tabernae* (e.g. Aula del Gruppo dei Marte e Venere, Aula e Tempio dei Mensores, Mitreo Sacello, close to the Porta Romana). A different picture is presented by those *scholae* located south of the *decumanus*. Here the guild buildings were found to conform to the ideal of the plot size laid down by the original property divisions,

---

[69] Meiggs (1973: 324–7); Hermansen (1983); Zanker (1994: 273); Bollmann (1998); Steuernagel (2005: 79).
[70] Lynch (1960).
[71] Bollmann (1998: 195–9).
[72] Bollmann (1998: 196).

The Analysis of Ostia's Scholae

when the land outside Ostia's *castrum* was divided into fairly regular land parcels.[73] In exceptional cases, property parcels were joined back to back and alongside each other, allowing the creation of large *scholae* like the Schola del Traiano, along the western *decumanus*. This in itself is an unconcealed statement of good financial standing. Similar assumptions can be made for other *scholae* which managed to secure a prime location along Ostia's major access routes.

To take these observations further, a full analysis of the street network was undertaken. Ostia's street network comprises a total of 150 street units; these have been analysed using UCL's Depthmap software,[74] producing axial graphs, calculated for two forms of integration: radius-n and radius-3.[75] Radius-n is a global measure showing the degree of accessibility a street has to all other streets in the city, taking into account the relation between all streets and all other streets within the system,[76] whereas radius-3 is a local measure based on the analysis of all streets accessible within a certain topological radius, here a radius of two other streets. The graphs produced are visually rendered along a colour range from red to blue, with the most integrated streets marked red to orange and the least integrated ones marked dark blue.[77] To overcome the impartial evidence, we can consider the street network within the excavated area as a delimited subset, confined by the river as a natural boundary and the city gates as additional boundary markers. Within the street configuration analysed, the main access roads, the eastern and western *decumanus*, as well as the Via della Foce, leading from the forum to the river harbour, clearly emerge as the most integrated streets, serving the east–west/west–east movement within the city (Fig. 9.7). These results are confirmed by the preliminary analysis of the complete street network, using 476 street units (Fig. 9.8). The eastern and western *decumanus* are the most integrated streets within that larger street network. However an interesting shift of gravity can be observed, giving more weight to movement towards the areas south-east of the Porta Laurentina, where fairly dense urban development appears to have occurred. This issue needs to be further explored, once the final results of the geophysical survey of Ostia have been published.[78]

---

[73] Mar (1991) on land division in Ostia during the Republican period.
[74] Depthmap (UCL-Version) 7.12.00d.
[75] For theoretical assumptions underlying space syntax values see Hillier and Hanson (1984); integration as defined in Hillier and Hanson (1984: 108–9).
[76] Hillier and Hanson (1984: 108–9).
[77] Figs. 9.7 and 9.8 are in grey-scale only.
[78] See Heinzelmann (1999b) for preliminary results of the geophysical surveys in the unexcavated areas of Ostia.

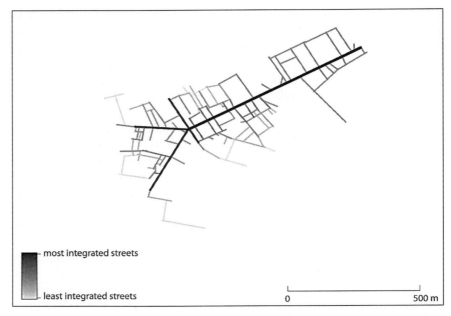

Fig. 9.7. Street network, excavated areas only; axial analysis, integration (HH, n-streets, 150) (image by Hanna Stöger).

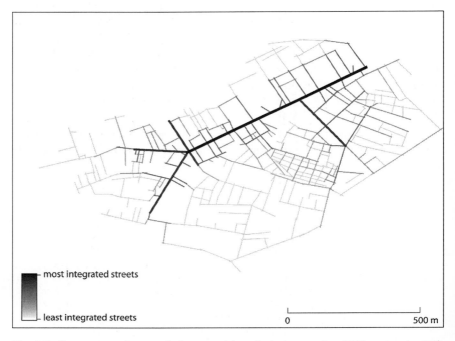

Fig. 9.8. Street network, extended area; axial analysis, integration (HH, n-streets, 476) (image by Hanna Stöger).

The Analysis of Ostia's Scholae 237

Within the scope of this chapter it is sufficient to establish that the eastern and western *decumanus* and the Via della Foce remain the most accessible streets, even within the larger area of the city revealed by geophysical survey. These streets are the most integrated elements within the network of streets and were directly linked to each other. They form a close network and can be interpreted as streets that would have facilitated the intelligible movement through a system that follows 'globalizing rules'.[79] These are the streets which were most likely the ones to be used by everybody (visitors and transient population as well as by the local population) since they direct movement to the centre from possible places of arrival and departure. In addition, as stated above, movement from origins in a side street to destinations in a side street were most likely to pass through the main streets, thus creating more movement along the main streets and higher potential for chance interaction. Hence the street grid itself contributes to making the main streets the busiest in the city, as it has been indicated.

Table 9.4 Global integration values for all streets [n-150] (2nd column) within the excavated areas only

value	Integration	Selection[a]				
	all streets (n=150)	Decumanus/ Forum	Via della Foce	Decumanus (west)	Via del Serapide	Via Fortuna Annonaria
		Casa dei Triclini	Aula e Tempio dei Mensores	Domus di Marte	Domus accanto al Serapeo	Caseggiato dei Lottatori
Average	1.42907	2.98062	2.44057	2.46635	1.6894	1.4324
Minimum	0.8107	2.98062	2.44057	2.46635	1.6894	1.4324
Maximum	2.98062	2.98062	2.44057	2.46635	1.6894	1.4324
Stand. Dev.	0.374281	0	0	0	0	0
Count	150	1	1	1	1	1

[a] Selection of streets where guild buildings have been located. These streets show integration values higher than average

[79] 'Globalizing rules' ensure a proportional relationship between the streets and the blocks of a city. With regard to movement, globalizing rules have the effect of maintaining the coherence of the growing city from the point of view of the individual (a stranger), moving around the system; see Hillier (1993: 63). A good example comes from the pre-modern City of London: whichever city gate was entered, the 'centre' could be reached in three axial steps, provided only that, at every point of choice, the longest available line of sight was followed. In this way the street network preserves a limited depth of access ('shallowness' cf. Hillier 1993: 63–4), which is much needed for intelligible movement within the city.

Table 9.5 Global integration values for all streets [n-467] (2nd column) including the unexcavated areas

value	Integration	Selection[a]				
	all streets (n=467)	Decumanus/ Forum	Via della Foce	Decumanus (west)	Via del Serapide	Via Fortuna Annonaria
		Casa dei Triclini	Aula e Tempio dei Mensores	Domus di Marte	Domus accanto al Serapeo	Caseggiato dei Lottatori
Average	1.01881	1.84646	1.63297	1.67106	1.12384	1.24956
Minimum	0.532099	1.84646	1.63297	1.67106	1.12384	1.24956
Maximum	1.84646	1.84646	1.63297	1.67106	1.12384	1.24956
Stand. Dev.	0.227051	0	0	0	0	0
Count	467	1	1	1	1	1

[a] Selection of streets where guild buildings have been located. These streets show integration values higher than average

The most integrated streets emerge as the ones on which the *collegia* chose to locate their buildings (Tables 9.4 and 9.5). While some *scholae* were situated along less integrated streets, still they were never more than two streets away from the main road. In this sense the *collegia* were movement-seeking, some more than others. Foremost of these was the Casa dei Triclini, the guild seat of the builders, the *fabri tignuari*. This *schola* was located along the eastern *decumanus* in close proximity to the forum. It was fully embedded into the public space of the wider forum area, and actively, through the way the building's entrances were structured, incorporated large portions of public space within its own external circulation space (Figs. 9.9 and 9.10). Again, other guild buildings enjoyed corner locations at the intersections between main and side streets (e.g. Domus di Marte). This brings to mind Hillier and Hanson's traditional corner-shop model,[80] which exploits the basic potential of its location and adapts its structure to maximize chance encounter at the interface with public space. Apart from the traditional corner-shop model, to which some of Ostia's *scholae* seem to conform, there are also cognitive reasons for such choice locations: pedestrians are known to heighten attention at corners. There, decisions are prompted about which direction to take, thus associations are likely made with buildings at corners.[81]

[80] Hillier and Hanson (1984: 176–7). [81] Lynch (1960).

Fig. 9.9. Isovist areas visible from the entrances to the Casa dei Triclini (Isovist 1, 2, and 3) (image by Hanna Stöger).

CONCLUSIONS

A spatial analysis of Ostia's *scholae*, using space syntax methods, has allowed us to examine them at the micro-scale of the individual buildings and their specific location, as well as at the scale of the entire town plan, through their distribution among Ostia's streets. The two scales of assessment and

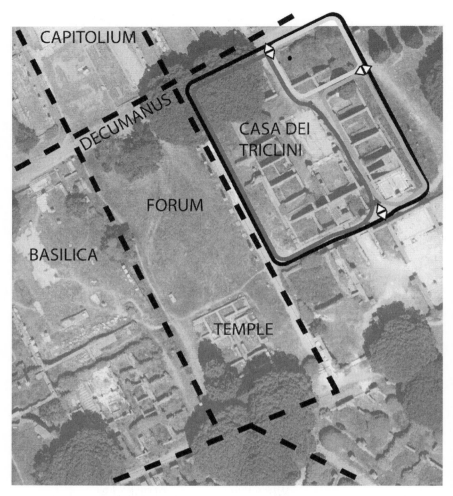

Fig. 9.10. The Casa dei Triclini and its external circulation space (image by Hanna Stöger).

their results must be considered together, since they represent interdependent spatial factors which inevitably influence each other.

First, the study of the spatial logic of the individual ground plans (access analysis) of the *scholae* was able to capture the spatial organization of the buildings and recognized them as largely defined by the outside space, the public domain of Ostia's streets. Their outward focus suggests that the *scholae* had a high potential for promoting contact and communication at the interface with public space. Second, from the assessment of the overall size of the buildings and the way they related to their local settings, several key observations were made. It emerged that the buildings strove to exploit their street frontage to utilize and optimize the spatial advantages of their specific location. This affected the way the entrances of the buildings were structured and defined street frontage best suited for the location of the *tabernae* associated with these buildings used by guilds. These *tabernae* were preferably placed along busy main streets and well-integrated side streets, responding to peak pedestrian flows ('movement-seeking' behaviour). Third, the pronounced outward orientation of the *scholae* is complemented by their location of choice: the *collegia* preferred to locate their buildings along the most easily accessible ('integrated') streets within Ostia's street network. Their exposed location not only gave the *scholae* a high public profile, but also enhanced their capacity to benefit from the concentration of movement that occurred along the main streets. While some *scholae* are found on side-street locations, these are still well-integrated streets when considered within the whole system and at the same time always in close proximity to the main streets. Street corners were preferred locations for *scholae*, exploiting both the movement economy and the heightened cognitive awareness of the urban user at street junctions.

One of the stated intentions of the analyses presented here was to test whether the concept of the movement economy could offer a suitable model for the explanation of the citywide distribution of *scholae* in Ostia. Other Roman cities including Pompeii and Empúries, show similar concentrations of movement-seeking land use (commercial, public, and religious) located along streets that were easily accessible from outside and inside the city.[82] For Pompeii factors such as ambulatory traffic and the forces of economic rationality have been identified as accounting for distribution patterns of retail outlets in areas of greatest social activity.[83] One might rightly ask what makes Ostia and its *scholae* different, or more to the point, what can a space syntax analysis of these selected buildings, and their settings, add to our understanding of the spatial organization of the Roman city that we do not already know?

The approach offered here adds a different theoretical perspective to the current discussion on the spatial organization of Roman towns, and above all it produces the statistical values to place these spatial relationships on a quantitative foundation. The shift from intuition to theory-led analysis is justified through the importance of recognizing the fundamental relationship between the urban grid (Ostia's street network) and human movement, as defined by the principles of the movement economy. This allows a broader understanding of movement, away from attractor-driven, directed movement with emphasis on direction and location. In fact, when we look at cities as movement economies, our attention shifts to the so-called movement by-

---

[82] Cf. Raper (1977); see Laurence (2007); Kaiser (2000); and Ellis (2004).
[83] Ellis (2004).

product that occurs, along the passage between origin and destination, from more or less everywhere to everywhere within the city.

Critically, space syntax analysis of Ostia's total street network created integration values for each street to all others; from which the most integrated streets have been identified as the ones which would have encouraged the greatest amount of circulation. However, to attain unbiased explanatory strength, it is crucial that the analysis of the street network is carried out completely independent of any specific land use or activities associated with buildings along a street. In the study of space in Ostia, we must prioritize purely spatial measures, in this case through the calculation of the significance of each street within the hierarchical access network of the entire urban network. Once the relative integration of each street has been established, the next step is the evaluation of its associated land use. At this point we can objectively match the *scholae* to the movement value of their associated streets.[84] In other words, the presence of the *scholae* can in itself attract people, but it cannot counteract the influence of the street's relative position in defining high and low flows of people. In this way we may avoid the circularity of explaining the position of a building by describing only its specific location in the city without further analysis and explanation of space across the whole city.

Within the framework of Ostia as a movement economy the following scenario can be presented concerning *scholae*. Placed along the most integrated streets, they were easily accessible to people directly seeking them as destinations; these buildings attracted direct, purposive movement. However by virtue of their highly advantageous location within the urban grid, the *collegia* benefited from general circulation, which included accidental interactions between people heading to different destinations. In this way, the *scholae* contributed to the 'urban buzz' that occurs where a larger number of different activities coincide, involving people going about their business in different ways, but still prioritizing the same space. Ostia's *decumanus* and immediately adjacent streets would have been such an area of 'urban buzz', where movement was channelled and activities converged. Space syntax tools have helped us to identify spatial configurations at Ostia, which would not have been visible to merely qualitative or intuitive glance at street plans; the better to understand *collegia*—and the decisions behind their locations—within the city's movement economy.

---

[84] Cf. Kaiser (2000: 48–56) for selected space syntax approaches applied to Roman Empúries, Spain.

Part III

Movement and the Metropolis

10

The Street Life of Ancient Rome*

Claire Holleran

Reconstructing the physical environment of Rome has long been a concern of those who study the ancient city.[1] Mapping Rome has been popular since the Renaissance, but in the twentieth century, there was a move to 3D reconstructions of the ancient city. The initial approach to such reconstructions was to build physical models, such as that currently on display in the Museo della Civiltà Romana. This large plaster-of-Paris model depicts Rome in the time of Constantine, and was constructed between 1935 and 1971 under the supervision of the architect Italo Gismondi.[2] More recent attempts to reconstruct Rome have made use of increasingly sophisticated techniques of computer modelling. The Rome Reborn initiative, for example, is an ongoing project to create a series of 3D digital models illustrating the development of Rome from the Bronze Age to the early medieval period.[3] The project's first task is to recreate the city as it would have appeared on 21 June 320 CE and uses the older plaster-of-Paris model as the basis for much of this digital reconstruction. These new computer models are remarkable tools for visualizing the topography of Rome; they enable a much fuller understanding of movement through the city, and the ways in which different districts, streets, and

---

* I would like to thank Eleanor Betts, Fred Jones, Elizabeth Macaulay-Lewis, John Patterson, John Pearce, and the editors for their valuable comments on a previous draft of this paper. Any remaining errors or omissions are my own responsibility.

[1] Attempts have been made since the Renaissance to reconstruct the ancient city: Ammerman (2006: 298). For discussions on reconstructions of Rome in general, see various contributions to Haselberger and Humphrey (2006), esp. Ammerman; Favro; Packer. See also contributions to Hinard and Royo (1991), particularly those of Ciancio Rossetto; Hinard; Pisani Sartorio; Royo. Also in Fleury and Desbordes (2008), see esp. Lecocq; Madeline; Poulle.

[2] Packer (2006: 310–11); Ciancio Rosetto (1991); Pisani Sartorio (1991). See also Bigot's plaster model of Rome at the Université de Caen, discussed in Hinard and Royo (1991).

[3] http://www.romereborn.virginia.edu—See also the modelling projects being undertaken at the UCLA Experiential Technologies Center: http://www.etc.ucla.edu. Similarly, a team based at the Université de Caen is developing an interactive computer model of ancient Rome: http://www.unicaen.fr/services/cireve/rome/index.php.

buildings communicate with each other, as well as the visual impact of particular monuments. Nonetheless, even leaving aside issues such as the fragmentary nature of the evidence for the topography of many areas, or the limitations of illustrating a single point in time, these models are not entirely satisfactory since they present a rather sanitized version of ancient Rome. The streets in particular are little more than empty spaces between buildings; they are devoid of trees, plants, moveable objects, animals, and above all, people.

These are not only depopulated cities but they are also extremely clean cities, without rubbish, dirt, mud, or graffiti. The multisensory experience of moving through the streets of ancient Rome, that is the crowds, the sights, the sounds, the smells, and the changing of all these factors at different times of the day and in different seasons, are thus difficult to comprehend in these topographical models.[4] Yet as has been argued in the case of London, 'streets without people constitute a void'.[5] The aim of this paper is to repopulate the streets of Rome and to consider the varied activities that took place in these spaces. The latter part of the paper focuses in particular on the commercial life of the Roman street, highlighting the central role that the streets played in the functioning of the urban economy.

THE STREETS OF THE METROPOLIS

Before we turn to people and activities, however, we should first consider the physical environment of the streets themselves. Pliny argues that if the streets of Rome were laid out end to end, they would have covered a distance of almost 60 miles (90 km).[6] Unlike the archetypal Roman city of the western provinces, the city of Rome itself was not laid out according to an organized grid system, but evolved much more organically, developing a more haphazard system of streets and alleyways of varying widths. The irregularity of streets in Rome can be seen on various fragments of the *Forma Urbis Romae*, particularly those slabs depicting the notorious area of the Subura; fragment 10g, for example, shows the relatively wide street of the Clivus Suburanus, alongside which are a chaotic jumble of alleyways, courtyards,

---

[4] Hartnett (2008: 91) makes a similar point about architectural ground plans. For the seasonality of pre-industrial cities, see Stenton (2007: 65). Artistic or filmic reconstructions of ancient Rome can be more successful in recreating the actual experience of living in the city (see e.g. the recent HBO/BBC production *Rome*, or Ridley-Scott's *Gladiator*) although these representations of Rome are subject to a different set of analytical criteria; a discussion of such reconstructions is beyond the scope of this paper. For multisensory maps of Rome see Betts, in this volume.

[5] Fishman (1979: 8).

[6] Plin. *HN* 3.66. He bases this claim on the measurements taken by Vespasian and Titus as censors in 73 CE.

arcades, and buildings of irregular shapes and sizes.[7] This irregularity is also reflected in the Latin terminology used for these streets, which Pliny describes as a series of *viae*, *vici*, and *compita*.[8] A *via* was a wide street or road, which the Twelve Tables stipulated should be at least 8 feet wide (*c*.2.5 m) on straight lengths and 16 feet (*c*.5 m) wide on the curves, enabling two carts to pass comfortably.[9] The roads radiating out from the gates of Rome are all of this type, but within the city there are only a few known *viae*; the Sacra Via, the Nova Via of Caracalla, the Via Lata (a continuation of the Via Flaminia), and the Via Tecta.[10] Other narrower roads were known as *vici*, a term also used to designate neighbourhoods; *semitae*, tracks or narrow streets; *angiportus*, alleyways; *clivi*, sloping streets; and *compita* or crossroads.[11] Some of the streets in Rome were apparently narrow enough for tenants of apartments to talk to each other and even touch through windows and from balconies on the upper floors.[12] The wide boulevards of nineteenth-century European cities, or indeed of some Roman provincial cities, such as Leptis Magna with its wide Severan road, were not a feature of ancient Rome; pressure on space, an awkward topography, and a resolute regard for individual property rights resulted in a city of narrow streets and alleyways.[13]

For those who were not brought up in Rome, the city must have appeared to be a bewildering labyrinth of streets. By no means all streets had names, and there were no house numbers or street signs to help visitors or newcomers to find their way around. Addresses in Rome were approximate and orientation was usually done by way of a particular landmark such as a temple, a colonnade, a city gate, the house of a prominent individual, or even a tree; an address on a slave collar from Rome, for example, instructs the reader that the wearer should be returned to the barber's shop near the Temple of Flora.[14] In a different context, we see a character in a play by Terence giving comic

---

[7] See also fragments 10a-h; l-m; o-r; 11a-d. For detailed descriptions, analyses, and photographs of the surviving fragments, see http:/formaurbis.stanford.edu. For a discussion of this project, see Koller et al. (2006).

[8] See Laurence (2008: 99-105) on *vici* in particular. See also Stambaugh (1988: 188); Carcopino (1940: 57-8).

[9] Varro *Ling.* 7.15; Stambaugh (1988: 188).

[10] The Via Tecta was perhaps synonymous with the Via Fornicata mentioned by Livy (22.36.8).

[11] The majority of known street names in Rome are *vici*. See *LTUR* V: 151-201.

[12] Mart. 1.86. Stambaugh (1988: 183); Carcopino (1940: 58). For Rome's irregular, winding streets ('flexi atque enormes vici'), see Tac. *Ann.* 15.38. Also note Suetonius' comment (*Aug.* 45.2), that Augustus liked to watch fighting take place in the narrow alleyways of the city. For the width of streets in Rome, see Macaulay-Lewis, in this volume (268, Table 11.1).

[13] Grand boulevards in 19th-century Europe are most famously exemplified by Haussmann's redevelopment of Paris. For the role of topography and property rights, see Wallace-Hadrill (2003: 193-4).

[14] *CIL* XV.7172; Paoli (1963: 140-1). In general, see Ling (1990a: esp. 210-12). Also Castrén (2000: esp. 15-18). For trees as a geographical marker, see Hartswick (2004: 14, 157 n. 148). See also Favro (1996: 4-11). cf. Lynch (1960: 2-6).

directions which were intended to get the listener lost; the victim was instructed to look for a portico, a shrine, a fig tree, the house of Cratinus, a temple to Diana, a gate, a pond, a bakery, and a workshop.[15]

Although visitors to Pompeii are familiar with paved streets lined with raised sidewalks and stepping stones to facilitate the crossing of these streets, this picture cannot be transferred easily to Rome. The initial dirt roads of early Rome apparently began to be paved in 238 BCE, starting with the Clivus Publicus leading from the Circus Maximus to the Aventine. Livy then informs us that in 174 BCE, the censors paved streets (*viae*) throughout the city, and Dionysius of Halicarnassus includes paved roads among the great achievements of Rome, alongside aqueducts and sewers.[16] Other literary evidence, however, points to inadequate or poor paving. Ammianus complains of uneven paving in fourth-century Rome, and the streets could also be muddy, suggesting that paving was not universal.[17] Seneca describes uneven and muddy alleyways, Martial praises Domitian for ensuring that praetors were no longer forced to walk through mud, while Juvenal's character Umbricius complains that his legs become plastered in mud while walking through the city.[18] When Vespasian was aedile, Caligula apparently ordered mud to be stuffed into his toga as a punishment for failing to clean the streets adequately.[19]

The narrowness of many of the Roman streets probably precluded them from having sidewalks; even in the age of the car and the scooter, many Roman streets remain too narrow for sidewalks, and pedestrians and vehicular traffic must negotiate the same space. The division of streets into a 'roadway' and a sidewalk is not commonly depicted on the surviving fragments of the *Forma Urbis Romae*, although there are four fragments identified by Macaulay-Lewis that may depict sidewalks;[20] other fragments depict arcades or colonnades fronting onto streets, which could of course function as sidewalks.[21] If the fragments in question do portray sidewalks, this indicates that the cartographers were interested in representing these on the map; the fact that only four sidewalks are depicted on the surviving fragments of the *Forma Urbis Romae* suggests then that these were not a common feature at Rome.

---

[15] Ter. *Ad.* 571–83; cf. Ov. *Tristia* 3.1.21–34.
[16] Livy 41.27.5; Dion. Hal. *Ant. Rom.* 3.67.5; Stambaugh (1988: 188–9); Robinson (1992: 59); Claridge (1998: 9). Although Livy claims that it was the censors who were responsible for paving the streets in 174 BCE, elsewhere he attributes this task to the aediles (Livy 10.23.12, 10.47.4).
[17] Amm. Marc. 14.6.16; cf. Petron. *Sat.* 79.
[18] Sen. *De Ira* 3.35.5; Mart. 7.61.6; Juv. 3.247–8. cf. Mart. 3.36.4, 5.22.5–6, 10.10.8, 12.29(26).8. A clear intertextuality can be detected here between Martial and Juvenal, with the latter picking up on the muddy streets described by the former: Colton (1991: 135).
[19] Suet. *Vesp.* 5; Cass. Dio 59.12.3.
[20] Fragments 341, 495de, 591, 619a. See Macaulay-Lewis, in this volume.
[21] e.g. fragments 8bde, 10a–b, 21d, 273abcd, 320ab.

Certainly there are few archaeological traces of such constructions in the city and little mention of them in the surviving literary record.[22] Perhaps there was less need for sidewalks in Rome, as from the late Republic onwards, vehicular traffic was banned from the centre of Rome during the day, clearly the busiest time for pedestrian traffic in the city.[23] That said, not all vehicular traffic was banned during the day; the notable exception to this ruling was carts involved in state-sponsored construction projects, together with those that were part of religious processions or triumphs. During times of major state-sponsored construction in the city, particular thoroughfares leading to the building sites would still have been extremely busy with vehicular traffic bringing building materials into the city, and taking waste and debris away from the site; in fact, banning other vehicular traffic may well have had the (intentional?) effect of increasing the efficiency of public building projects.[24] A division of pedestrian and vehicular traffic would surely have been welcomed by pedestrians on these particular routes, particularly given the danger of being crushed by falling material or runaway carts, but the authorities' concerns appear to have lain elsewhere.[25]

Moving about at night must have been particularly problematic, since not only were vehicles allowed back into the city, but there was little in the way of street lighting. In the early evening, when *tabernae* were still open, there will have been some light emanating from the open fronts of these units. A passage in the *Digest*, for example, describes a confrontation between a *tabernarius* and a man who stole the lamp that he had placed above the counter adjacent to the footpath.[26] The streets in general, however, must have been very dark. Indeed, our literary sources hint at the problems people faced in finding their way around cities at night and in the early morning. When Petronius' characters make their escape from Trimalchio's lengthy dinner party, for example, they get so lost that they only make it home just before dawn.[27] Umbricius complains of the danger of the dark streets for those who were unaccompanied, although the wealthy, with their large retinues of people carrying torches, must have found it much easier to move around at night.[28]

---

[22] Robinson (1992: 61). Plautus (*Curc.* 287) implies a separation between the higher sidewalk (*semita*) and the street (*via*), which suggests that the audience were at the very least familiar with the concept, even if these were not common in Rome. cf. Plaut. *Merc.* 115–16; Saliou (1999: 198).

[23] *Tabula Heracleensis* (often called the *Lex Iulia Municipalis*), ll.56–67. Vehicular traffic was allowed into the city again at the tenth hour (see also Favro, Hartnett, and Kaiser, in this volume).

[24] For a discussion of the logistical problems of transporting construction material within the city of Rome, see Favro in this volume; DeLaine (1997: 98–100).

[25] For the dangers posed by construction-related traffic in Rome, see Tib. 2.3.43–4; Hor. *Epist.* 2.2.65–86; Juv. 3.254–61; Alfenus *Dig.* 9.2.52.2 (discussed by Martin 2000). See also Favro, in this volume; Laurence (2008: 87).

[26] Alfenus *Dig.* 9.2.52.1.

[27] Petron. *Sat.* 79.   [28] Juv. 3.282–301.

The streets of Rome were heavily used and it was therefore desirable to ensure that they were kept clean and maintained. Property owners were legally responsible for repairing and maintaining the street directly alongside their property, under the supervision of the aediles. If this was not done correctly, then the aediles could arrange for any necessary work to be done and bill the property owner accordingly; if this were not paid then a fine of 50 per cent of the initial cost could also be levied.[29] Property owners were also charged with ensuring that there was no standing water in the streets. The paving and maintenance of public streets (*viae publicae*) in Rome, however, appears to have been the direct responsibility of the aediles.[30] The aediles were also traditionally responsible for ensuring that the streets of Rome were clean.[31] This may tell us something about how the authorities understood the status of streets; those which were under magisterial responsibility were perhaps viewed as autonomous public spaces, while those that were cleaned and maintained by bordering property owners had a more ambiguous status as semi-public spaces, as a public continuation of private property.[32]

Despite the clear concern of the authorities to ensure that streets, and where applicable sidewalks, were clean and well maintained, and that vehicular traffic was regulated, our literary sources suggest that the reality was rather different. The streets of Rome that emerge from ancient literature are not the sanitized versions of computer reconstructions, nor are they the ordered spaces that surviving traces of ancient legislation might imply. The literary streets were very much alive; they were dirty, dangerous, crowded, smelly, and almost incessantly noisy.[33] The streets were not merely a conduit through which people passed in order to move around the city, but were the arena for much of the social, political, and economic discourse that took place in Rome. For many the street itself was the destination, rather than part of the journey. This in itself would have affected movement through the city. What Levitas describes as the 'typical streaming behaviour of modern pedestrian

---

[29] The *Tabula Heracleensis* contains detailed rules for the cleaning and maintenance of streets: see II.20–82; Crawford (1996: 355–91). Also see Papin. *Dig.* 43.10.1. For a detailed discussion of laws related to the care of streets, see Robinson (1992: 59–69). A similar system appears to have been in place in Pompeii, where Saliou's (1999) detailed survey of the construction of the sidewalks reveals a piecemeal arrangement that reflects the fact that residents were responsible for the area of paving directly in front of their property; more homogenous pavements in some areas suggest a more collective response, either on the part of the property owners or the town council, or perhaps the paving of some sidewalks as acts of civic euergetism.

[30] *Tabula Heracleensis*, II.24–8.

[31] For aediles and cleaning, see Plaut. *Stich.* 352; Suet. *Vesp.* 5. The *Tabula Heracleensis* (II.50–2) indicates that the *quattuorviri* also played a part in street cleaning.

[32] For a discussion of the street as a public and/or private space, see Saliou (2003: 46–8; 1999: 200–4). cf. Ulp. *Dig.* 9.3.1.1–2, 43.8.2, 43.8.17; Paul. *Dig.* 43.8.1; Pompon. *Dig.* 43.7.1.

[33] For the noisy and crowded nature of ancient streets, see Hor. *Sat.* 2.6.27–31; *Epist.* 2.2.65–86; Sen. *Clem.* 1.6; Mart. 1.41; 5.22; 12.57; Juv. 3.232–314.

traffic on crowded streets', derived from the need for speed and efficiency, would not be universal, since efficient movement through the city was not always the goal.[34] Others were of course trying to get to a specific destination, such as Horace strolling along the Sacra Via on his way to visit a sick friend; others were walking solely for pleasure.[35] To understand fully the place of the streets within the urban infrastructure, we thus need to consider the ways in which these spaces were used by the inhabitants of Rome.

THE COMMERCIAL AND SOCIAL LIFE OF THE ROMAN STREET

If the streets were a destination in their own right, how do we explain this phenomenon? One explanation lies in Rome's Mediterranean climate and the culture of outdoor living that this encouraged.[36] Although winters could be harsh, they were generally short, and the climate was conducive to living outside, particularly if there were spaces that were shaded from the sun; Tacitus' comment that people complained about the lack of shade in Nero's new wider roads should be seen in this context.[37] The pre-industrial system of communication in Rome also contributed to the importance of the street. In societies without modern forms of communication, the street remains central to the transmission of information;[38] in Rome, Horace describes a rumour spreading rapidly from the Rostra through the *compita* of the city.[39]

Rome also experienced a high population density. Population estimates for the city vary, but at its most populous in the late Republic and the Principate, Rome was most likely home to around one million inhabitants.[40] The absence of transport systems prevented Rome from spreading outwards in the manner of a modern city, and the majority of inhabitants thus lived within a 2-mile

---

[34] Levitas (1986: 231). See also van Tilburg (2007) for the application of modern notions of traffic efficiency to his analysis of the streets of Rome, drawing a relationship between a street's width and its 'efficiency'; if the majority of the traffic were in fact pedestrian then different analytical criteria are required. See Newsome (2008: 443, 446) for a useful critique of van Tilburg's approach.

[35] Hor. *Sat.* 1.9.1–2. See Macaulay-Lewis, in this volume for the distinction between walking for leisure and walking for transport.

[36] For the significance of cultural differences in the use of streets, see Winter's contrast of the attitudes of 19th-century Parisians and Londoners to their respective streets; the former were proud of their streets, which they viewed as a place of recreation, whereas the latter saw them primarily as a means of transit. Winter (1993: 20).

[37] Tac. *Ann.* 15.38, 15.43.

[38] Czarnowski (1986: 209–10); Levitas (1986: 233).

[39] Hor. *Sat.* 2.6.50–1.

[40] For a detailed discussion of the evidence regarding the population of Rome, see Lo Cascio (2000a).

radius of the monumental centre.[41] Space was therefore at a premium. As early as the third century BCE, the pressure to house people resulted in multi-storey buildings in the vicinity of the Forum Boarium. Attempts may have been made to limit the height of buildings in the city, but the need to reiterate legislation suggests that the rules were repeatedly flouted.[42] Indeed, the pressure on space in Rome resulted in high prices for building plots and a tendency for tall buildings to be constructed in relatively narrow spaces; additional floors were also added to existing buildings.[43] A handful of examples of these multi-storey buildings survive in Rome, most notably the second century structure at the foot of the Capitoline. It was in such multi-storey apartment blocks, or insulae, that the majority of the inhabitants of the city lived. Conditions were particularly grim and crowded on the upper floors of apartment blocks, where the cheapest accommodation was to be found.[44] At these levels, rent was probably payable daily or weekly, reflecting the instability faced by many in Rome.[45] The facilities in apartment blocks were also very basic; archaeological evidence for latrines, hearths, and kitchens is generally lacking, and most blocks were without a direct water supply.[46] Housing in Rome may have been lavish for the wealthy few in their *domus*, but for the majority conditions were grim. In other societies, when housing is poor and cramped, the street tends to become an extension of the home, and the venue for much of the social interaction—and the entertainment—of communities; the outcome was probably much the same in Rome.[47]

Economic necessity also drove people onto the streets; it was here that they searched for employment and engaged in various informal ways of making money. The opportunities for employment in the city were restricted, dictated at least in part by the elite preference for slave and freed labour.[48] The demand for domestic labour, for example, was low, as the majority of this work was

---

[41] Van den Bergh (2003: 450) argues that 1,200,000 inhabitants lived in an area of about 8 square miles. For high population density in Roman towns in general, see MacMullen (1974: 62–3). Also Scobie (1986: 427–33).

[42] According to Livy (21.62.3), an ox escaped to the third floor of a building. Augustus set a limit of 70 feet (Strabo 5.3.7) and Nero reiterated height restrictions on buildings in the aftermath of the fire of 64 CE (Tac. *Ann.* 15.43); Trajan then lowered the height limit to 60 feet (Sext. Aur. Vict. *Epit.* 13.13). Aldrete argues that a building of 70 feet, that is, within the legal height limits of the Augustan period, would still be around seven storeys high: Aldrete (2007: 106–7). For a discussion of the height of buildings based on the *Forma Urbis Romae*, see Madeleine (2008).

[43] Vitr. *De Arch.* 2.8.17; Herodian 7.12.5–6.

[44] Mart. 2.53.8, 3.30.3, 7.20.20; Suet. *Gram.* 9; Juv. 3.198–210. Tertullian (*Adversus Valentinianos* 7) likens the heaven of the Valentiniani to a Rome apartment block, with floor piled upon floor and countless rooms and staircases.

[45] Van den Bergh (2003: 443, 452–4, 466–7); Frier (1977: 35).

[46] Stambaugh (1988: 178).

[47] Winter (1993: 68); Gutman (1986: 154–5); Levitas (1986: 233).

[48] See Holleran (2011) for a more detailed discussion of the employment market in Rome.

undertaken by slaves and freedmen from within the household; these workers also took care of the administration of households and estates. Similarly, production and commerce was undertaken by slaves and freedmen, who were very often trained and supported in a particular craft or business by their owner or patron. Thus although the majority of people in Rome had to work to support themselves, the social and institutional infrastructure of the city restricted the economic opportunities available to them. They relied above all on casual work, primarily as porters or in the construction industry. Since the demand for labour fluctuated, workers must have been hired for varying lengths of time, from a day to weeks or even months, while some people were probably hired for the completion of a particular task.

This affected the use of the street in various ways. Casual labour is inherently unstable and this led to many in Rome living correspondingly unstable lives. Competition for jobs also likely kept wages low, resulting in widespread structural poverty in the city.[49] Those who were reliant upon casual labour lived in cheap and cramped accommodation, which in itself would drive people onto the streets in the daytime. If work were not available, some of these people may also have been temporarily homeless, and were then forced to spend even more of their time on the streets. Furthermore, the street was a centre for communication and one of the major ways in which potential workers heard about employment opportunities. Workers seeking casual employment probably congregated in central areas, but employers and workers must also have been connected by the information networks found in the streets and neighbourhoods.[50] Moreover, for those who were unable or unwilling to find work as casual labourers at the docks or on building projects, the creation of informal work opportunities on the streets provided an essential survival strategy. This was particularly true for women, who were unlikely to work as labourers but needed to find some way of making enough money to survive.

The streets were thus central to the commercial life of Rome; these were likely to have been vibrant economic spaces which played a crucial role in the functioning of the urban economy. Street traders, for example, were a fundamental part of Roman street life and a ubiquitous presence in the city, enabling the distribution of goods and providing employment for many. Buying and selling was a simple way to make money in Rome; street trading required few skills, little education, and minimal initial capital. Even lacking capital may not have been a problem, as this could be circumvented by tapping into the credit networks that most likely existed on the streets of Rome. Not only did street

---

[49] Holleran (2011). For an alternative view, that the scale of immigration to Rome implies higher wages, see Scheidel (2007: 335–6).
[50] In the parable of the vineyard workers in the New Testament, potential workers are found gathered in the marketplace (Matt. 20:1–16).

trading provide a safety net for those unable to find work elsewhere, it also suited the purchasing patterns of many in Rome. Lacking storage facilities and reliant on unstable, fluctuating incomes, people tended to purchase provisions on a daily basis;[51] street sellers thus provided an essential service as they sold goods in small amounts at low prices. This is a phenomenon that is well documented in other diverse cities, such as nineteenth-century London and contemporary cities of the developing world. As at Rome, these cities do not have adequate opportunities available in the formal employment market and they lack effective social security systems; they thus demonstrate high levels of street trading and hawking, as buying and selling absorbs the jobless and the underemployed.[52]

A wide variety of goods were sold in the streets of Rome, from both fixed stalls and ambulant sellers. The Latin terminology used to denote street traders and hawkers underlines this distinction between those who sell at fixed points and those who were mobile; *ambulator* and *circitor*, for example, both derive from verbs which relate to movement (*ambulo* and *circumeo* respectively), while *circulator* suggests sales from a fixed point, stemming from the verb *circulo*, which implies the formation of circles around oneself.[53] In our literary sources, we hear of street traders and hawkers dealing in goods as diverse as ointment, gladiatorial programmes, wigs, and luxury goods.[54] However, these traders dealt primarily in essentials, such as clothing, and above all, food, including figs, pulses and grain, bread, fish, vegetables, grapes, and milk, and hot prepared food such as sausages, pastries, and pease-pudding.[55] Pictorial representations of trade also suggest the erection of fixed stalls in public spaces in Rome, although it must be noted that it is difficult to distinguish in reliefs between depictions of market stalls and street stalls. A pair of first-century CE funerary reliefs from Rome, for example, portrays the sale of cloth in a colonnade, while reliefs from Ostia show vegetables for sale on a temporary trestle table and a woman selling poultry, hares, snails, and fruit from behind a stall made up of cages containing most of her (live) stock.[56] A sarcophagus

---

[51] Tac. *Hist.* 4.38.

[52] The links between street trade and the structure of the urban economy in Rome are explored further in Holleran (2011; 2012). For the potential analogy with the credit networks that underpinned street trading in 19th-century London, and also in contemporary cities of the developing world, see Holleran (2012).

[53] Other terms used to denote street traders and hawkers include *circumforaneus*, *institor*, and *propola*.

[54] Ointment: Plin. *HN* 12.103.6. Gladiatorial programmes: Cic. *Phil.* 2.97.3. Wigs: Ov. *Ars Am.* 3.167.8. Luxury goods: Sen. *Ben.* 6.38.3.4; Man. 5.408; Jer. *Adv. Iovinian* 1.47.

[55] Clothing: Plaut. *Aul.* 512; Plin. *HN* 18.225.3; Juv. 7.221; Ulp. *Dig.* 14.3.5.4. Food: Plaut. *Capt.* 813–16; Lucil. 5.221–2; Cic. *Div.* 2.40.84; Cic. *Pis.* 67.10; Hor. *Sat.* 6.111–14; Sen. *Ep.* 56.2; Calp. *Ecl.* 5.97; Petron. *Sat.* 7; 14; Mart. 1.41.5–10; Apul. *Met.* 8.19; Ulp. *Dig.* 14.3.5.9; *CIL* VI.9683; *ILCV* 685b.

[56] A similar relief from Arlon shows the sale of fruit from a trestle table, while a relief from Bordeaux depicts the sale of fruit and grain from sacks: Liversidge (1976: 102). A famous frieze

from Narbonne depicts an ambulant fruit seller carrying his wares in a large basket on a shoulder strap.[57] For those in Rome who lacked even the smallest sum of money to invest in things to sell and who were unable to tap into any credit networks, scavenging for items to recycle and to resell was one avenue open to them. We know, for example, that some people scavenged for broken glass which was then used in the production of new glass, or collected old fabric which could be used to patch clothes or make blankets, while others dealt in second-hand goods;[58] there was little that was not reused or recycled in Rome.

All of these traders were active sellers of their wares and the cries of street traders and hawkers must have been a loud and compelling characteristic of the Roman street experience. Most famously, Seneca complained of the food vendors who competed for the attention of customers near his apartment. In this particular case, the sellers are in the bathhouse, where there was a concentration of hungry customers; similar vendors would have served the crowds attending games in the Circus, amphitheatre, and theatres, although these cries must have been commonplace across the city.[59] Indeed, disdainful elite authors liken poor oratory to the shouts of such sellers.[60] Pictorial representations of traders also indicate their engagement with potential customers; sellers are shown touching their produce as though inviting customers to do the same, while most have their arms raised as though seeking to attract the attention of passers-by. Prices were also arrived at through haggling, which adds to the social interaction inherent in Roman shopping.[61]

Other workers were to be found plying their trade in the streets of Rome. Some teachers, for example, held classes outside in the open air, teaching on the streets from the early hours and apparently disturbing local residents.[62] Barbers could be seen shaving customers, prompting jurists to discuss the legal issues arising from the possibility of injury. Ulpian thus speculates as to who would be to blame if a ball from a nearby game knocked a barber's hand, causing him to cut the throat of his customer.[63] Some of these 'street barbers' were based in *tabernae*, using the area directly in front of their units as an

---

from the *praedia* of Iulia Felix in Pompeii also depicts a wide variety of goods for sale in the forum.

[57] Magaldi (1930: 15).
[58] Broken glass: Stat. *Silv.* 1.6.73–4; Mart. 1.41.3–5, 10.3.3–4. Fabric: Vout (1996: 211–12). Second-hand dealers (*scrutarii*); Lucil. 1282; Gel. 3.14.10.6 (quoting Lucilius); Apul. *Met.* 4.8.27. Holleran (2012).
[59] Sen. *Ep.* 56.2. cf. Cic. *Div.* 2.40.84; Mart. 1.41.5–10.
[60] Petron. *Sat.* 68.6–7; Quint. *Inst.* 2.4.16, 10.1.8; Mart. 10.3.2; Plin. *Ep.* 4.7.6.4.
[61] Sen. *Ben.* 6.17.1; *Ep.* 42.8.
[62] Mart. 9.68, 12.57.4–5.
[63] Ulp. *Dig.* 9.2.11. pr. cf. Mart. 7.61.7.

additional workspace, but other barbers, who presumably charged less, could work directly from a stool in the street, providing there was a nearby fountain or basin to provide water.[64] Prostitutes of both sexes also plied their trade in the streets. The English verb 'to fornicate' famously derives from the Latin term *fornix* meaning an arch, where some ancient prostitutes worked, but others were to be found on the corners and back streets of Rome, as well as in tombs on the outskirts of the city; as McGinn points out, 'commercial sex does not always require a building'.[65] Just as the cheaper barbers worked in the streets, so the cheaper prostitutes were to be found there rather than in the brothels of Rome, a distinction that holds true even today; many of these men and women may have turned temporarily to prostitution when they were unable to find other ways of making money.[66] Women in particular, whose employment opportunities were particularly scarce, and who could unexpectedly face poverty following the death or desertion of parents or husbands, were likely to have been forced into prostitution by poverty.[67] More flamboyant skills were also on display in the streets, as entertainers sought to derive an income from loiterers and passers-by. The crowds thus mingled with jugglers, sword-swallowers, fire-eaters, snake exhibitors, and even philosophers who charged for their teachings.[68] Fortune tellers and dream interpreters also worked on the streets of the city. The fortune tellers frequented by the poorer inhabitants of Rome were generally found around the Circus or along the

---

[64] For barbers based in *tabernae*, see Vitr. *De arch.* 9.8.2.8. For these establishments spilling out into the street, see Mart. 7.61.7.

[65] On the association of *fornix* with prostitution, see Wallace-Hadrill (1990); McGinn (2004: 251). For prostitutes in alleys, see Hor. *Carm.* 1.25.10; for crossroads, see Prop. 4.7.19; McGinn (2004: 253, cf. 29). Also Evans (1991: 138). Martial (2.17) implies that a female barber (a *tonstrix*) sitting at the entrance to the Subura is in fact a prostitute, although this may reflect the prejudice of Martial as much as the reality of the situation (cf. Mart. 6.66). For the association of prostitutes with tombs, see Mart. 3.93.13–15; Juv. 6.015–16; Patterson (2000: 103).

[66] Prostitutes based in brothels were very often slaves or former slaves, whereas freeborn people who made money in this way were more likely to be forced to work on the streets, particularly if prostitution was a means of supplementing other income; working on the streets could also mean the avoidance of a pimp (*leno*) or madam (*lena*) demanding a share of a prostitute's income.

[67] Ulp. *Dig.* 23.2.43.5; *Cod. Theod.* 15.8.2; Procop. *Aed.* 1.9.4–5. McGinn (2004: 68–71) argues that poverty alone is not a sufficient explanation for prostitution, since not all poor women become prostitutes, but straitened circumstances and perhaps the potential earnings when compared to other work must have been a significant factor in the decision-making process of such women. For potential earnings, see McGinn (2004: 51–5). Also see Evans (1991: 140–1) for poverty forcing women into prostitution. cf. Ter. *And.* 70–9, 797–9; *Heauton.* 446–7.

[68] Entertainers: Sen. *Ben.* 6.11.2; Tert. *Apol.* 23.1; *De praescr. Haeret* 43.1. Jugglers: Manil. 5. 168–71. Sword-swallowers: Apul. *Met.* 1.4. Snake-charmers (or perhaps beggars who threaten passers-by with snakes, an old practice apparently still observable in modern Taipei in China, for which see Parkin [2006: 80]): Mart. 1.41.7; Celsus 5.27.3c; Paul. *Dig.* 47.11.11.pr. 1. Philosophers: Sen. *Ep.* 29.7.1.

ramparts of the Servian walls, but such services were probably available in the local neighbourhood streets also.[69]

The marginal economic activities described above were often seen as little more than glorified begging.[70] By its very nature, begging must take place in the public arena and the most popular spots were unsurprisingly some of the busiest; places where pedestrian and vehicular traffic was forced to slow down and even stop, such as at the gates of the city, or at the entrance to bridges, were favourite haunts of beggars.[71] However, desperate people were found throughout Rome, on the streets, in alleyways, on corners, and at crossroads, begging from passers-by; many of these may have been physically disabled and thus unable to support themselves in any other way.[72] Others survived through crime, particularly theft. Theft from the baths and from apartment blocks in Rome was common, but street crime was also a danger, particularly at night.[73]

Commercial activity also spilled over from the *tabernae* that lined the streets of the city; these multi-purpose units housed a wide variety of businesses, encompassing production, services, and retail.[74] The encroachment of *tabernae* into the streets was apparently severe enough in the late first century CE for Domitian to enact legislation stating that all *tabernae* should keep within their thresholds. This action was praised by Martial, who describes bars, cookshops, barbers, and butchers spilling out onto the public street; tellingly, he comments that Domitian's legislation means that praetors no longer have to walk through the mud of the street.[75] This not only hints at the state of some of the streets but points to the main concern of the authorities, which was to ensure that the thoroughfares of the city remained passable. This remained a problem in the following centuries, when Papinian recorded further legislation intended to keep the streets clear of obstructions from *tabernae*.[76] Nothing was to be left outside workshops, other than cloth

---

[69] Plaut. *Mil.* 692-94; Cic. *Div.* 1.132 (inc. passage from Ennius); Hor. *Sat.* 1.6.113-14; Juv. 6.542-7; 6.588-90; Artem. 1 pr; Parkin (2006: 78).

[70] Parkin (2006: 77-9).

[71] Gates: Juv. 3.15-16. Bridges: Sen. *De Vita Beata* 25.1; Juv. 5.8; 14.134; Mart. 10.5.3; 12.32.25; Grey and Parkin (2003: 286).

[72] Juv. 5.8; Lucian *Cyn.* 1; Dio Chrys. *Or.* 32.9; Apul. *Met.* 1.6.1-2; Grey and Parkin (2003: 286). Children taught to beg by parents: Juv. 12.57.13; Paul. *Dig.* 25.3.4; Parkin (2006: 77). Disabled beggars: Sen. *Clem.* 2.6.3; Parkin (2006: 70-1).

[73] Burglary: Plin. *HN* 19.59; Suet. *Aug.* 43.1; Juv. 3.302-5; Paul. *Dig.* 1.15.3.2; Ulp. *Dig.* 47.18.1-2. At the baths: Plaut. *Rud.* 382-5; Catul. 33; Sen. *Ep. Mor.* 56.2; Petron. *Sat.* 30.8; 93; Paul. *Dig.* 1.15.3.5; Ulp. *Dig.* 47.17.1; 47.18.1.1; 47.18.2. Street crime: Plaut. *Amph.* 153-64; Juv. 3.278-311; 10.19-22; Tib. 1.2.25-6; Prop. 3.16. Street crime perpetrated by Nero (or 'impersonators'): Tac. *Ann.* 13.25.1-3; Suet. *Nero* 26; Dio 61.8-9; cf. SHA *Verus* 4.6-7. Pickpockets (*sacularii* and *derectarii*): Ulp. *Dig.* 47.11.7; 47.18.1.2. Also Nippel (1995: 97); Robinson (1992: 206).

[74] For a detailed discussion of the function of *tabernae*, see Holleran (2012).

[75] Mart. 7.61. See Laurence, in this volume.   [76] Papin. *Dig.* 43.10.1.3-5.

left by fullers to dry on racks, and wheels put outside by carpenters. Furthermore, no rubbish was to be put out into the street, including animal skins and carcasses; this particular directive was perhaps aimed at butchers and tanners, although there are some hints in the sources that in Rome the latter were in any case restricted to Trastevere.[77] An aedile also had the right to remove couches from the street, again implying a concern with keeping the streets clear.[78]

Despite this concern, there is no evidence of any legislation in Rome intended to deal directly with street trade, unless this was to be included under the legislation described above. Certainly at Pompeii we know that aediles granted some stallholders permission to trade in sought-after locations; *dipinti* around the exterior of the amphitheatre or near temples record the names of stallholders and their official authorization to trade, but there is no surviving evidence of this practice in Rome.[79] Even if such permission to trade were required, the resources—and probably the will—to ensure that such rules were enforced were lacking, and most street traders, particularly the ambulant sellers, were likely an accepted part of life in the city, operating largely under the radar of the authorities. Indeed, although the division of Rome into *regiones* and *vici*, and the keeping of what appear to be detailed records of topography and property ownership in the city demonstrate a clear desire to map, to understand, and ultimately to control the urban environment, the fluidity and mobility of itinerant sellers made the streets difficult to regulate.[80] The concern with keeping the streets clear, however, may well have encouraged those working in the streets to be as mobile as possible; bulky fixed stalls may have attracted unwanted attention and were probably less common than easily moveable stalls composed of small tables or mats on the ground, or trays that could be hawked around neighbourhoods. Custodians stationed at the entrances to public buildings such as porticoes may also have prevented sellers from entering, thus maintaining the *dignitas* of the space.[81]

The streets of Rome then were an important part of the commercial landscape of the city.[82] Commerce added to the commotion of the streets,

---

[77] Mart. 6.93.4; Juv. 14.202.

[78] Paul. *Dig.* 18.6.13(12); Robinson (1992: 71). See also legal references in n. 32, above, which imply that the street was viewed by the jurists first and foremost as a public right of way. cf. Saliou (2003: 46).

[79] Amphitheatre: *CIL* IV.1096; see also *CIL* IV.1096a, 1097a, 1115. Temple of Venus: *CIL* IV.1768, 1769. Also see *CIL* IV.677, 2996.

[80] The Via Anicia map fragment, for example, details property ownership alongside topography: Wallace-Hadrill (2008: 304–7); Trimble (2007: 379–80). For detailed record keeping in the city of Rome, see Wallace-Hadrill (2008: 297–312); Nicolet (1991: 197).

[81] For controlled access to Rome's porticoes and portico-temples, see Macaulay-Lewis, in this volume. cf. Suet. *Aug.* 40.5.

[82] Street sellers, charlatans, and mountebanks were a ubiquitous part of street scenes in Renaissance Rome also; Welch (2005: 34–60).

which were already crowded with people both moving through them and using them as an extension of their personal living space. Hartnett has demonstrated that in Pompeii at least, benches were frequently found on sidewalks, and such structures may also have been found along some of Rome's streets, forming a central focus for socializing.[83] Neighbourhood shrines and altars, together with water basins, often placed at crossroads, must also have been local centres of gossip and social life. Indeed, for many in Rome, the streets—and their associated bars and inns—were the primary location for socializing. The danger that ball games posed to those being shaved by barbers in the streets indicates that such games were played, but other more sedentary games were commonplace.[84] Gambling for money was technically illegal in Rome; it was permitted only during the Saturnalia, although betting on the outcome of sporting events was legal all year round. However, gambling with dice remained a popular pastime, even among the emperors; books were even written on the subject, most notably by Claudius.[85] Such games needed little more than a dice-box, a board, and either knucklebones or dice.[86] Dice were also used to determine moves in certain board games. While such games were popular in bars and inns, as well as private houses, gaming also took place outside in public areas. The steps of public buildings in Rome, for example the Basilica Iulia, show traces of gaming boards carved into the stone, and games being undertaken in public could attract groups of onlookers.[87] For some, gambling may even have led to a lucky windfall and the provision of some much-needed cash.[88]

---

[83] Hartnett (2008). For socializing, see esp. Hartnett (2008: 117). Also Franklin (1986) for an example of a space in Regio VII at Pompeii which served as the social centre of the neighbourhood.

[84] For the popularity of ball games, also see Sen. *Ep. Mor.* 56.1; Mart. 10.86, 12.82, 14.45–8, 14.163; Plin. *Ep.* 3.1; Petron. *Sat.* 27; Sid. Apoll. *Epist.* 2.9.4, 5.17.6–7. Also see Galen's treatise, *De parvae pilae exercitur* (*Exercise with the Small Ball*). For more on possible danger: Alfenus *Dig.* 9.2.52.4. For ball games in Rome, see Harris (1972: 84–100, 105–11); Balsdon (1969: 163–7).

[85] Illegal: Hor. *Carm.* 3.24.58; Paul. *Dig.* 11.5.2.1; Robinson (1992: 204–5). The number of dice and knucklebones discovered by archaeologists testifies to the popularity of dice games. cf. Plaut. *Curc.* 355–8; Colum. 1.8.2; Cic. *Sen.* 58; Mart. 4.66.16, 5.84; Juv. 1.88–90; Sid. Apoll. *Epist.* 2.9.4, 5.17.6–7; Amm. Marc. 14.6.25. At the Saturnalia: Mart. 4.14, 5.30.8, 5.84, 11.6, 14.1.3; Balsdon (1969: 387 n. 96). Popularity with emperors: Sen. *Poly.* 17.4; Suet. *Aug.* 70.2, 71.2–4; *Calig.* 41.2; *Claud.* 5, 33.2; *Nero* 30.3; *Dom.* 21; Dio. 64.2.1; SHA *Verus* 4.6; *Comm.* 4.9.1. Women also: Ov. *Ars Am.* 2.203–83, 3.354–60; Prop. 2.33A.26; Balsdon (1969: 154–5, 387 nn. 101–3). Books: Ov. *Tr.* 2.472–84; Suet. *Claud.* 33.2. Balsdon (1969: 154, 387 n. 99).

[86] For more details on dice and board games, see Purcell (1995d); Balsdon (1969: 155–8).

[87] For more on game boards cut into paving stones in fora, at the sides of roads, on the steps of temples, and on the seats of theatres, see Trifilò in this volume; Bell (2007); Bell and Roueché (2007); Purcell (2007); Roueché (2007). For gamblers and drinkers spending time in the fora, the crossroads, and the streets of 4th-century Rome, see Amm. Marc. 28.4.29. cf. Sen. *Tranq.* 14.7; Balsdon (1969: 158, 389 nn.125–6).

[88] See Purcell (1995d: 22) for gambling as 'economic behaviour'.

To the groups of people passing along the streets and to those for whom the street was the destination, we should also add animals and their obvious contribution to the waste on the streets.[89] Live animals were driven through the streets to be sold at the markets and in the butchers' shops of the city; without refrigerated transport, keeping animals alive for as long as possible was the best way to ensure fresh meat. Given the general ban on carts in Rome during the day, goods that needed to be moved either had to be carried on foot by porters or packed onto transport animals such as mules or donkeys, adding to the animal traffic in the city; at Pompeii, holes cut through the border stones of sidewalks were presumably used to tether such animals.[90] Stray or wild animals were also a common sight in Rome, particularly dogs. Dogs were favourite pets of women and children, and were also kept as guard dogs in wealthy homes, but many in Rome were far from domesticated.[91] Martial creates a particularly evocative image of a dying beggar in Rome who hears dogs howling in anticipation of eating his corpse; Suetonius also casually tells a story about a dog dropping a human hand beneath Vespasian's table.[92] Occasionally wolves even found their way into Rome, causing havoc in the city.[93] Animals would also play a key role in religious processions as they were led to particular altars in preparation for their sacrifice. Indeed on certain days, religious and also funerary processions would add to the commotion of the Roman street.

CONCLUSIONS

The streets of Rome were not then the clean sanitized thoroughfares of our reconstructions, nor were they the clear ordered spaces implied by some of the legislation concerned with ancient streets. Rather they were crowded, dirty, noisy, and chaotic, with economic and social activities competing for space.[94]

---

[89] See Hor. *Epist.* 2.2.75 for a 'mad dog' and a 'muddy pig' in the street.
[90] Pack animals: Laurence (2008: 87). Holes in sidewalks: Hartnett (2008: 92); Saliou (1999: 203).
[91] As pets of women and children: Val.Max. 1.5.3; Sen. *Cons. Ad Marc.* 12.2; Mart. 1.109; Petron. *Sat.* 64; Juv. 6.654; Balsdon (1969: 151, 386 n. 82). The mosaics at Pompeii warning visitors to beware of the dog (*cave canem*) suggest that guard dogs were a common sight (for example, the mosaic at the entrance to the Casa del Poeta Tragica, IV.8.3), as does the famous plaster cast of a chained dog who perished in the eruption of Vesuvius (found in the entrance to VI.14.20). cf. Petron. *Sat.* 29. We hear much less about cats in Rome, and Balsdon (1969: 51) argues that these were not particularly popular until the 1st century CE.
[92] Mart. 10.5; Suet. *Vesp.* 5.4.
[93] Cass. Dio 53.33.5, 54.19.7, 78.1.6: in 23 BCE, 16 BCE, and 211 CE respectively. Also see Liv. 33.26.8–9.
[94] For the pre-modern street in general as both a commercial and a social space, see Stobart et al. (2007: 87, 108).

Theoretically the streets were the space in which people of all ranks, wealth, and age mixed together.[95] This interaction was, however, strictly controlled and the experience of the streets differed according to social status. The pedestrian traveller without wealth or status was at the mercy of the streets, but wealthier inhabitants of Rome could create their own pathway through the city. Travelling in a litter, for example, was one way of creating a private space within the public arena; surrounding oneself with a retinue also enabled the wealthy not only to demonstrate their status publicly, but served also to create a barrier between themselves and the turmoil of the Roman street.[96]

Yet for many the streets were much more than an arena for display or for asserting status through separation; they were central to their lives and even to the creation of their identity. If movement through the streets of Rome was as difficult as some of our sources suggest, then people likely remained largely within their local area, resulting in the relative isolation of neighbourhoods and the development of a sense of local community.[97] In recent years, scholars have increasingly recognized that ancient Rome was a fragmented city, a city of neighbourhoods, with a 'cellular structure operating on a local level';[98] living, working, and socializing in a particular area must have contributed to an individual's sense of identity and to a sense of ownership of a neighbourhood.[99] Rome may have been a city of migrants, with a fluid and changing population, but basing oneself in a specific neighbourhood must have been a good way for new migrants to integrate themselves into the city.[100] The streets of ancient Rome were thus far more than a physical network of pathways connecting the city; they were the support network for the lives of the inhabitants and it was here, against a vibrant commercial and social backdrop, of movement and interaction in space, that these lives were largely played out.

---

[95] Whyman (2007: 43). This was in fact said about the streets of 18th-century London, as depicted in John Gay's *Trivia: or, the Art of Walking the Streets of London* (1716), where we see another dirty, congested, disorientating city. The sentiment, however, holds true for ancient Rome, and Gay's readers were expected to see the poetic streets of Rome through the poetic streets of London. See Brant and Whyman (2007: 5); Braund (2007).

[96] Juv. 3.239–44, 282–301. cf. Robinson (1992: 74). Similarly, in early modern London, coaches were an increasingly popular way of enclosing the elite and creating a barrier between themselves and pedestrians: Whyman (2007: 56).

[97] For some sense of community even in the crowded and constantly changing neighbourhoods of 19th-century London, see Winter (1993: 57).

[98] Wallace-Hadrill (2003: 195). See also Wallace-Hadrill (2008); Lott (2004).

[99] For this phenomenon in modern cities, see Czarnowski (1986: 209). For the street as a space in which to carve out an identity, see Stobart et al. (2007: 109). For the social encounters and the use of space that help to construct a public culture in cities, see Lott (2004: 129).

[100] For migration to Rome, see Holleran (2011).

11

The City in Motion

Walking for Transport and Leisure in the City of Rome*

Elizabeth Macaulay-Lewis

The ancient sources suggest that there were two main types of walking that occurred within the Roman city: walking for transport and walking for leisure.¹ Walking for transport involved walking in the streets and sidewalks of Rome. People in Rome, whatever their social position, walked, although the elite often used litters. Walking was the primary way for Romans to navigate the city and to conduct their daily business, as most forms of carts were outlawed during the day.² Walking for leisure was also an important element of Roman's urban experience; these walks, which were generally, although not always, located in Rome's public porticoes and portico-temples, were a popular form of leisure activity from the time of their first construction in the late Republic.³

---

* I would like to thank Amanda Claridge, Janet DeLaine, Claire Holleran, Saskia Stevens, and the editors who read versions of this paper. Their comments have greatly improved this paper; all errors and omissions remain my own. I also owe special thanks to the editors for including my paper in this volume.
¹ O'Sullivan (2003: 70–90). Also cf. Mart. 1.70, 3.5, 3.100, 10.19, and 12.3; Juv. 3.236–48.
² O'Sullivan (2003: 73–82). On the laws regarding carts and wagons, see below and the chapters by Kaiser and Favro, in this volume.
³ O'Sullivan (2003: 82–90); Macaulay-Lewis (2008: 89–148). Porticoes were a prominent element of Rome's cityscape. They are also a problematic and little-studied architectural form. Single porticoes often lined streets and facades of buildings of Roman cities, particularly in the Near East. There were double porticoes, which generally were formed of a portico with a double aisle. There were also triple porticoes; it is unclear whether a triple portico was a ∏-shaped or a triple aisled portico. The most famous example of what may have been a triple portico, is Porticus Triplices Miliariae found in Nero's Domus Aurea, whose exact form remains debated, cf. Papi (1995: 55). Suet. *Ner.* 31, is the only ancient source to mention this portico, writing 'ut porticus triplices miliarias haberet'. *Porticus* is in the accusative plural here, suggesting that there was more than one portico, but the exact form as described in the Latin is far from clear. Quadriporticoes are formed of four porticoes that are interconnected and enclose an open space. It should be noted that the form *porticus* appears as both the singular and plural, so the context of

Thus far, archaeological evidence has played a limited role in the study of walking and movement in the Roman city; the streets, the identified context for walking for transport, are mentioned but seem a blank framework in which the colourful and varied walks of the Romans were set.[4] The ancient sources are also limited when it comes to understanding the experience of walking in Rome's monumental porticoes, the primary setting for leisured walking. Ancient authors give us a broad sense of the decor, gardens, art, and use, but little sense of the design or how space affected movement.

Architectural historians have become increasingly interested in how Rome's metropolises were constructed, experienced, and navigated. In the study of Roman urbanism, as well as Roman architecture, scholars have used fictional walks, based in part on the ancient sources, as means to recreate the urban experience of Rome and other cities.[5] Most notable is Favro's study of the urban image of Augustan Rome, which deployed three walks through Rome to chart its transformation under Augustus.[6] The archaeological evidence for sidewalks and streets played a minor role, although considerable thought was given to the monumental building programmes and changes in topography of Augustan Rome in this work. Three studies, however, have fruitfully drawn on archaeological evidence to study Roman cities and their sidewalks and are particularly relevant here. The second volume of MacDonald's well-known study on the architecture of the Roman Empire examined the urban armature of cities and explored how Roman architecture was designed to accommodate the needs of pedestrian traffic and to aggrandize the cities of the provinces.[7] His discussion and definition of thoroughfares considered the total width of these spaces and their associated porticoes, colonnades, or sidewalks where appropriate.[8] His identification of width as a critical factor in determining the purpose of a street and sidewalk is also important for understanding the experience of walking in Rome, as I will argue below.

Hartnett's doctoral thesis on the streets of Roman Italy and his work on the street-side benches in Pompeii present useful approaches to studying the nature of Roman streets.[9] Hartnett's archaeological study of the streets of

---

the word tells the reader whether it is a singular portico or not. For the purpose of this paper, the term monumental *porticoes* or monumental portico-temple refers to the public quadriporticoes found in the city of Rome that housed temples, garden, fountains, and art. For a discussion of the archaeological evidence for porticoes, cf. Gros (1996: 95–120). See also Zanker (2010: 48–61).

[4] Hartnett (2003) is a notable exception.
[5] e.g., Yegül (1994: 121–41).
[6] Favro (1996). Her itineraries, while stimulating and creative, assume a common experience that probably did not exist. The creation of a fictional itinerary, if applied in a considered and specific way, can be very useful. Purcell (1987: 187–90) used such a walk to study the transition from Rome to its suburbs.
[7] MacDonald (1986: 5–73; esp. 32–51).
[8] MacDonald (1986: 32–5).
[9] Hartnett (2003, 2008). See also his chapter in this volume.

Pompeii and Herculaneum demonstrated these streets were filled with people and served as a stage for self-presentation and social interaction.[10] His study of benches considered the width of Pompeian streets and sidewalks, arguing that the width of a sidewalk, as well as the space available for walking, shaped the experience of walking in Pompeii.[11] His study highlights the obvious but fundamental point that the study of width is essential for our understanding of Rome's streets, sidewalks, and the ancient experience of them.

Building on this previous work, this paper poses two basic, yet fundamental questions: how does the study of space inform us about the nature of walking in Rome and how does movement through a space allow us to understand the construction, purpose, and meaning of that space? This paper proposes a new approach to the study of movement in the city of Rome by establishing an 'archaeology of walking', utilizing archaeological evidence and the *Forma Urbis Romae* (henceforth the *FUR*) alongside the ancient texts, to identify different experiences of walking in the city of Rome and the nature as well as the atmosphere of Rome's streets, sidewalks, and portico complexes. First, I consider the archaeological evidence for Rome's streets and sidewalks, the spatial context for walking for transport. A detailed examination of the *FUR* follows in order to understand Rome's streets and the experience of walking there. Finally, the archaeological evidence for the nature of Rome's monumental porticoes and portico-temples, their depiction on the *FUR*, and their function as a location for leisure and leisured walking in the city of Rome are discussed.

THE *FORMA URBIS ROMAE*: EVIDENCE AND PROBLEMS

Because the survival of archaeological evidence is highly inconsistent across Rome, we are often dependent on the *FUR* for information on Rome's topography, including streets and porticoes. The *FUR* was commissioned between 203 and 211 CE under Septimius Severus and Caracalla.[12] The monumental plan, $c.13 \times 18$ m, was mounted on a wall of the south-eastern room in the southern portico in the Templum Pacis. It is one in a series of plans that depicted part or the totality of the city of ancient Rome; there was probably a Flavian predecessor that was also displayed in the Templum Pacis.[13] Sadly

---

[10] Hartnett (2008: 91–119; 2003: 19–74).
[11] Hartnett (2008: 112).
[12] Reynolds (1996).
[13] These plans include the Via Anicia fragment and the Via Labicana Plan. Other marble plans have survived that depict other Roman cities and architecture, such as the Isola Sacra Plan, the Amerino Plan, the Perugia Plan, and the Urbino Plan; a mosaic depicting a plan of a baths was also found, Reynolds (1996: 30–8) and Carettoni (1960: 207–10). There were also fragments

today, only 10–15 per cent of the *FUR* has hitherto been recovered, giving us a fragmentary vision of the city.[14] As a result of this partial survival, the purpose of the *FUR* remains enigmatic.[15]

The *FUR* was not a cadastral map, despite having been drawn to a scale of 1:240; rather, it seems to have reflected the Severan priorities for and the Severan conception of Rome as a city, even if these choices are unclear to us.[16] For example, the *FUR* does not always show Severan alterations or restorations: the Porticus Octaviae was restored under the Severans, but was depicted in its Augustan phase.[17] Sometimes buildings were inaccurately placed on the *FUR*, as in the case of the Templum Divi Claudi.[18]

While the *FUR* is problematic, it is nevertheless an essential document for understanding walking for transport and leisure in Rome, as Rome's monumental *porticoes*, streets, and sidewalks are well represented. Within the unidentified fragments, which compose the bulk of the surviving remains of the *FUR*, a surprisingly large number of streets, arcades, and sidewalks can be identified. The presence of streets and monumental porticoes suggests that these spaces were an important part of Rome. While these porticoes and streets may not have been shown in their actual form from 203–11 CE, the *FUR* presented the image of the monumental portico and Rome's streets as they were perceived of in the Severan period and are as close to a realistic image of Severan Rome that we have. The earlier Via Anicia fragment, which was more detailed than the *FUR*, also depicted an arcaded street.[19]

---

of a plan found in Orange, France, Carettoni (1960: 207). Cf. Rodríguez-Almeida (2002) for study of all of the known marble plans of Rome, including a marble fragment found on the Oppian Hill (41–3) and pre-Domitianic marble plan fragment found under the paving of the Forum Transitorium (61–6), and other Italian cities. For a Republican example of a painting of Carthage and a painting of Italy displayed in the Roman Forum, see Roth (2007: 286–300); Varro *Rust.* 1.2.1 and Plin. *HN* 35.22. Also cf. Wallace-Hadrill (2008: 301–12). On a predecessor to the *FUR*, cf. Castagnoli (1981: 263–4); Richardson (1992: pp. xix–xx); Darwall-Smith (1996: 64–5). Noreña (2003: 27 n. 10).

[14] Reynolds (1996: 1, 15).

[15] For theories, see Taub (1993) and more recently Trimble (2007) and Wallace-Hadrill (2008) on its role in visual surveillance.

[16] The scale is not always accurate and sometimes is closer to 1:250.

[17] Gorrie (2007: n. 37); Ciancio Rossetto (1996: 267–9). Likewise, the representation of the portico and theatre of Pompey probably reflects Domitian's rebuilding after the fire of 80 CE rather than Severan modifications, Sear (1993: 687–701).

[18] http://formaurbis.stanford.edu/fragment.php?record=17 (Rodríguez-Almeida [1981: 44]).

[19] The *FUR* fragment (32gh and 32i), like the Via Anicia fragment, both depict the temple of Castor and Pollux in Rome's Circus Flaminius. The Via Anicia has the same scale (1:240) as the *FUR*, but the *tabernae*, depicted here, are identified as belonging to Cornelia and her associates. Not only is the Tiber included on this plan, while it is not on the *FUR*, but the distances of the plan appear to be accurately scaled. The Tiber may have been denoted through pigmentation that is now lost (cf. Ciancio Rossetto 2006 on the use of colour on the *FUR*). For a more detailed discussion, cf. Conticello de' Spagnolis (1984); Rodríguez-Almeida (1988b: 120–31); Reynolds (1996: 33–5). It is also earlier, dating to the first half of the 2nd century CE, Tucci (1994: 123–8).

When the *FUR* is considered in conjunction with archaeological remains, we gain a more complete, although inevitably more complicated, picture of Rome's topography. Therefore, despite these issues, the *FUR* is useful in understanding Rome's streets, sidewalks, and monumental *porticoes*. The complex nature of the *FUR* also highlights the problem of chronology in the study of ancient Rome's topography and the experience of the city. The ancient sources from the first and early second centuries CE need to be used alongside the early third century CE *FUR* and archaeology from the whole imperial period to reconstruct an image of Rome's streets and sidewalks. Due to the fragmentary nature of the evidence this is understandable.

THE CROWDED STREETS OF ROME: LITERARY FICTION OR ARCHAEOLOGICAL REALITY?

The written sources suggest that Rome's streets and sidewalks were unpleasant places at best and dangerous places at worst. A key piece of evidence to challenge or support this literary topos is the width of Rome's streets and sidewalks. If we can determine the width of Rome's streets and sidewalks, this image of Rome's busy streets can be challenged or confirmed. Because the archaeological record is fragmentary, it is not possible to map with precision variation in the widths of Rome's streets. However, a brief summary of some of the widths of archaeologically known streets from Rome is helpful. Because much of this evidence is treated in other papers in this volume, I will only briefly mention a few.[20] The Via Flaminia, which was called the Via Lata inside the gates of city and lies under today's Via del Corso, was 4.2–4.3 m in certain places and had sidewalks, each of 3.1 m.[21] The basalt Via Biberatica, which ran through the Markets of Trajan, ranged from 6 m to 8 m in width and had marble sidewalks.[22] Some of the archaeologically excavated streets from Pompeii also shed light on the widths of Roman streets.[23] Hartnett's study, which considered the width of streets with benches at Pompeii, demonstrated the width of these streets fell into four quartiles: narrowest (17 per cent, width of

---

[20] Van Tilburg (2007: 31) gives the widths of several streets in Rome; however, his claim that only four streets met Varro's definition of a *via* is not supported by the archaeological evidence or the *FUR*. Cf. Laurence (2008: 99–104) on the meanings of *viae, vici*, and *compita* and on their role as spaces of social interaction; and cf. Wallace-Hadrill (2008: 264–312, esp. 269–76) on the role of *vici*, or neighbourhoods, in the Augustan reorganization of Rome and the streets of Rome.

[21] Quilici (1990: 73).

[22] See Newsome (2010: 208–11) on exclusion of vehicles from Via Biberatica (even though it had pedestrian sidewalks). Bianchini (1991: 102–21; 1992: 145–63).

[23] On Pompeii, see Saliou (1999: 161–218) in addition to Hartnett (2003, 2008); van Tilburg (2007: 31); Beard (2008: 53–80). On Ostia, see Stöger, in this volume.

2.0–2.5 m); narrower (29 per cent, 4.0–4.5 m); wider (28 per cent, 5.0–7.0 m); widest (26 per cent, 7.5+ m).[24] Although this study did not consider all of Pompeii's streets, enough were included to make its conclusions significant; in other words, 74 per cent of the sample of Pompeian streets were smaller than 7.0 m and 46 per cent were smaller than 4.5 m. At Pompeii there were narrow streets, especially in the older parts of the city, the narrowest of which seems to have been c.2.4 m (8 Roman feet). In the busier and planned parts of the town, streets such as the Via di Mercurio had a width of about 9.5 m (32 Roman feet); the Via di Nola and Via dell'Abbondanza were each c.8.3 m (28 Roman feet) wide; and the Via Stabiana (24 Roman feet).[25]

The *viae publicae*, the major thoroughfares that connected Rome to various cities in Roman Italy, were at least 4.1 m in width, but were as wide as 10–11 m when the pedestrian sidewalks were included.[26] The Via Ostiensis, which connected Rome to its most important port, was 4.5 m wide and had sidewalks of 1.8 m and 1.1 m, just before the road reached Acilia.[27] Excavations in 1982 showed that it widened to 4.5–4.8 m, about 15 Roman feet, at certain points.[28] The Via Severiana, which Septimius Severus restored and which followed the shore from Ostia to Terracina, was 4.1 m.[29] The Via Appia, the most famous road out of Rome, was as wide as 4.70 m at certain points and had sidewalks of 3.1 m on each side, making it nearly 10 m wide.[30] The Via Latina, another important artery connecting Rome to southern Italy, ranged in width from 3.8 to 4.1 m and had sidewalks with widths from 3 m to 3.5 m.[31] Paved highways outside Rome had a normal width of 4.74 m or 16 Roman feet.[32] This brief summary of certain Roman streets and roads suggests that they varied considerably in width, but that few roads, together with their sidewalks, were more than 10 m wide.

The *FUR* shows a wide range of Rome's streets from major named streets, such as Clivus Victoriae (frag. 5Abcd), Clivus Suburanus (frag. 10Aab, adjoining 10abcde), Vicus Sabuci (frag. 10abcde), and other thoroughfares, to nameless back streets and alleys. The width of the streets, indentified and unidentified, as well as the widths of porticoes, colonnades, and arcades on the

---

[24] Hartnett (2008: 111).
[25] Van Tilburg (2007: 31).
[26] Quilici (1990: 40–89).
[27] Quilici (1990: 42).
[28] *Viae* (1991: 72).
[29] Quilici (1990: 43). The width of the sidewalks of the Via Appia varied from location to location, probably due to the amount of pedestrian traffic that they were required to accommodate.
[30] Quilici (1990: 49).
[31] There are numerous other streets in Roman Italy whose widths are known; the smaller width of these streets which linked Rome with other cities in Roman Italy was at least 4.1 m, cf. Quilici (1990: 40).
[32] A. Claridge, pers. comm.

Table 11.1 Summary table of the widths of the identified and unidentified streets depicted on the FUR

Streets on the FUR			
Quartile	Width (m)	Number	%
1	0.0–5.5	97	39.9
2	6–10	116	47.7
3	11–15	21	8.6
4	15+	9	3.7

Total Number of Streets: 243

Note: Several more fragments with possible streets and arcades were photographed and displayed on the Standford Digital *FUR* website. However, as they note, the scaling is not consistent, nor are any scales presented with the fragments. These fragments have not been published yet; they include fn. 8–9; 23–4; 26–30; 32; 34; 37; and 41. A table with the complete breakdown of the widths of identified and unidentified streets and porticos was too large to include in this publication.

FUR are summarized in Table 11.1.[33] There are numerous *FUR* fragments where the piece is so poorly preserved or so small that it is not possible to confirm the presence of streets with confidence. Although this table is only a sample, it does start to give us a picture of the width of streets, sidewalks, colonnades, and arcades in the city of Rome. Of the 243 possible streets identified on the *FUR*, all but nine streets range from 3 m to 15 m in width, excluding all sidewalks or porticoes. The average width (including these outliers) is 7.04 m, but 213 (or nearly 87 per cent of the sample) are less than 10 m; only thirty streets (13 per cent of the sample) are 11 m or wider. If the median of the 243 streets is taken, the width drops to 6 m. Thirty-eight streets, some of whose widths cannot be confirmed, were bordered by arcades, colonnades, and sidewalks; of these only nine were bordered on both sides by such features.[34] Certain arcades could be identified on the *FUR* but the presence of a street that they bordered could not be confirmed due to the fragmentary nature of the *FUR*; likewise certain arcades and porticoes enclosed open spaces. The average width of these sidewalks, arcades, and porticoes was 4.83 m.

---

[33] Reynolds (1996: 173–5). On the *FUR*, three specific combinations of symbols denote colonnaded or arcaded sidewalks: *tabernae* fronted by an arcade (represented by a series of dashes [or serifs]); *tabernae* lined by a colonnade (denoted by a series of dots); or a raised colonnade (identified by a single line with dots).

[34] Of the thirty-eight streets with sidewalks, arcades, and colonnades, only four of these may have had sidewalks (frags. 341, 495de, 591, and 619a). A single line can denote a change in elevation, as in the case of the porticoes of Divorum and the other monumental porticoes. However, at the same time, a single line can also represent a wall.

What did these widths mean to the experience of walking in the city of Rome? The width of streets and sidewalks set a finite amount of space for walking and all other activities located here. However, the width of streets and sidewalks do not tell us how busy or congested they were. Congestion results from the width of the streets and the volume of traffic.[35] Thus, how congested a street was depended on how wide it was, as well as how busy it was. The widest street recorded on the *FUR* was 33 m (frag. 563a); this street, although it lacked sidewalks, presumably was never too busy; except if a special procession were passing by.[36] This exceptional street is unnamed and has not been located in the city of Rome. Here, the ancient sources, art historical evidence, and other archaeological evidence can help us to establish what the volume of traffic might have been for Rome. According to the *Tabula Heracleensis*, which recorded laws enacted from the time of Julius Caesar, many of Rome's streets were semi-pedestrianized spaces during the day when the majority of carts were not allowed in the city.[37] There were exceptions, however; building materials for temples and public works, as well as the removal of garbage, were permitted during the day provided a special licence was obtained.[38] Likewise, wagons, which were needed for official religious rites, triumphs, and public games, were also allowed.[39] While cart traffic was limited, pedestrians would have had to share the streets with horses, pack animals, small handcarts, and litters.[40] Many streets would have had two-directional traffic.[41] If this were the case, then each direction of traffic might have 3–3.5 m of space to accommodate all moving horses, small carts, litters, people, and the occasional wagon laden with building materials for one of Rome's many civic projects. Horses and mules might also be parked in the street.[42] When put in these terms, the actual amount of walking space begins to seem narrower than a median width of 6 m per street suggests; in other words, Rome's streets seem more congested.

The arcades, colonnades, and sidewalks that lined some of Rome's streets or fronted certain *tabernae* in theory also provided more walking space for

---

[35] Van Tilburg (2007: 41); cf. Laurence (2008: 87–8). Laurence suggests that Rome's population density rose from 575 people per hectare to 1,100 people per hectare in the 1st century CE, while Pompeii may have had between 181 and 303 people per hectare. Thus Rome's streets had to handle far more people than Pompeii's ever did.

[36] However, we should allow for the problem of the 'double bind', wherein the widest streets attract the largest volumes of movement and, as a consequence, are more congested than narrower streets despite their relative width. See Newsome, in this volume, 13.

[37] Robinson (1992: 79–82, vv. 56–61). Hadrian may have reinforced this ban, although to what degree is unclear, cf. Robinson (1992: 76).

[38] Robinson (1992: 73, v. 61).

[39] Robinson (1992: 74, vv. 62–5).

[40] Robinson (1992: 74–6); cf. Laurence (2008: 88).

[41] At Pompeii, e.g., we know that cart traffic flowed in both directions. Cf. Poehler (2006: 55); Tsujimura (1991: 58–86).

[42] See Poehler, in this volume.

Rome's inhabitants. An additional 4.83 m made a considerable difference to the experience of a street. Not only did the arcade or colonnade offer welcome shade in the summer months, but in the winter they also provided cover from rain and winds. However, as in the case of the streets, the entire 4.83 m of these spaces were not dedicated to walking; the measurements of these arcades include the space taken up by the colonnade. As each column was typically 0.6 m (2 Roman feet) or 0.88 m (3 Roman feet), the amount of walking space was again reduced; however, if the portico was an arcade, the piers, which tend to be smaller than columns, may have taken up less space.

Evidence from Pompeii gives us a sense of how busy the streets of an ancient city could be. Numerous Pompeian wall paintings, including one from the Casa del Panettiere (VII.3.30) and several from the *praedia* of Iulia Felix (II.4.3) among others, show shops and stalls set up in the streets and on the sidewalks, which limited the space of walkers (Figure 11.1);[43] people regularly spilled out into the streets as they shopped.[44] Presumably, clients waiting to enter their patrons' homes also milled about the porticoes, sidewalks, and streets, again taking up precious walking space. Those who had stopped at a bar for a drink and perhaps a light meal, often stood at the bar, again in the sidewalk.[45] Street-walking prostitutes also sold their bodies here.[46] Graffiti, found adorning the walls of many Pompeian houses, suggest not only that the street was an area of social interaction, but also that people stopped to deface these walls, again creating another spatial impediment to the flow of traffic.[47] Presumably, such defacing did not occur during the busiest times of day. Likewise, the benches of Pompeii, which were often positioned outside the houses of the well-to-do and important patrons, could impede traffic, particularly if they were 'flush' against building facades and thus projected in the sidewalk.[48] Of the 100-plus benches that Hartnett identified, most of these, however, located on wider and busier streets, often associated with prominent houses or with shops or bars. The 'wide and wider' streets, ranging in width from 5.0 m to 7.5+ m, contained fifty-three of the benches. With the benches, these sidewalks only had 2.1 m for walking space, which is considerably more space than the 1.71 m of sidewalk space that the 'narrowest' and 'narrower' streets of Pompeii had.[49]

---

[43] See Holleran, in this volume; Beard (2008: 72–8).

[44] On shops in the streets, cf. Livy 1.35.10; Dion. Hal. *Ant. Rom.*, 3.67.4. Martial praises Domitian for making shopkeepers keep the bounds of their shop to their threshold and not spill over into the street, cf. Mart. 7.61.

[45] On bars and their distribution in Pompeii, see Ellis (2004: 371–84). On shopfronts more generally see Ellis, in this volume.

[46] McGinn (2002: 7–46).

[47] Hartnett (2003: 19–77); Kellum (1999: 283–300); Laurence (1994a: 88–103); see Hartnett, in this volume.

[48] Hartnett (2008: 92).

[49] Hartnett (2008: 111).

Fig. 11.1. A wall painting from the so-called 'Casa del Panettiere' (VII.3.30), Pompeii; recent scholarship has suggested that this was not actually a temporary shop, but rather a distribution of a bread dole and that the baker was giving out bread in order to muster support for his run for a public office. Regardless of purpose, the stand still occupied valuable walking space in the streets. Several eighteenth-century engravings that reproduce paintings found in the *praedia* of Iulia Felix (II.4.3), record scenes of streets and Forum of Pompeii with people selling goods, handouts being given to beggars, and carts and horses taking up space (Beard 2008: 74, figs. 25–7). Courtesy of Museo Archeologico Nazionale, Naples, Italy/ Lauros / Giraudon/ The Bridgeman Art Library.

While we cannot confirm the presence of these impediments in Rome itself, their presence in Pompeii seems to suggest that congestion and similar happenings occurred throughout Rome's empire, including the capital, and affected the experience of walking. While the streets, sidewalks, and arcades of Rome were the arteries of the city, enabling it to thrive and its population to go about its daily business, leisured walking could also be found here; we know that men with free time would stroll the streets and porticoes of the Campus Martius, looking at the art in the porticoes or the shops, but it was largely confined to the monumental porticoes, which are discussed below.[50] These wandering individuals no doubt slowed walkers on the move, like tourists in major cities do today. Traffic, pedestrian and otherwise, varied throughout the day;[51] certain shops, even if open all day until dark, and areas were most likely busier in the mornings and others in the afternoon.[52] On special occasions, the streets of Rome became the staging ground for triumphs and funerary processions, which temporarily changed and dominated the flows of traffic within the city.[53]

The average width of 7.04 m (or median of 6 m) of Rome's streets was larger than Pompeii's, suggesting that there was more space for walking. Likewise, the presence of sidewalks and arcades, although limited in number, created additional walking space. Thus, there was more absolute space for walking in Rome than in Pompeii; however, with one million-plus inhabitants in the early second century CE, Rome and its streets were presumably far busier and more congested than Pompeii ever was. The volume of traffic and impediments to traffic shaped the experience of walking for transport in the city of Rome; the chaotic streets of the poets do seem to be the reality of Rome in light of the historical, archaeological, and art historical evidence for traffic.

PORTICOES AND PORTICO-TEMPLES: WALKING FOR LEISURE IN THE CITY OF ROME

Having discussed the nature of walking for transport and its spatial context, it is now time to consider the experience and setting of the leisured walking in Rome. Walking became a popular leisure activity in Rome in the mid-second century BCE.[54] Rome's monumental porticoes, a distinctive type of architecture,

---

[50] Mart. 3.20.10–14. Hor. *Sat.* 1.6.111–131; In Delos, cf. Ov. *Her.* 21.97–8.

[51] On the patterns of life and the sociotemporal dynamics of the Roman city, Laurence (1994a: 122–32; 2008).

[52] See Holleran, in this volume; Holleran (2010; 2012). Shops were generally open all day, but closed at night, and the bars were open into the evening, cf. Juv. 3.302–4; 8.158; Petron. *Sat.* 12; *Dig.* 9.2.52.1; Mart. 9.59. I owe my thanks to Claire Holleran for these references.

[53] Hartnett (2003: 36–7).

[54] O'Sullivan (2003: 37–57; esp. 53–7); Macaulay-Lewis (2008: 57–68).

were designed as public amenities with gardens, as well as sculptural and painting collections where people could go and enjoy their free time, participating in leisure activities such as walking.[55] They were also spaces of religious ceremonies and political events.[56] The ancient authors, in particular the love poets of the late Republic and Augustan era, write of men and women walking in the porticoes, looking for their lovers or seeking out a new one.[57] Jilted lovers might also wander in the porticoes, attempting to avoid their previous paramours.[58] Men also sought out dinner invitations there.[59] These complexes included the Aedes Herculis Musarum, the Porticus Liviae, the Porticus Octaviae, the Porticus Pompei, the Porticus Divorum, the Templum Pacis, and the Templum Divi Claudi.

Before discussing how these spaces were distinct from the porticoes and arcades that lined the streets of Rome, a few comments about terminology and chronology are necessary. First, the majority of these complexes are called either *porticus* or *templum*. In the case of the Divorum, it is merely called the 'Divorum' in the regionary catalogues and is only referred to as the 'Porticus Divorum' in one late antique source.[60] On the *FUR* (frag. 35cdefghi), it is called the 'DIVORUM'. Although the entire complex is not represented on the *FUR*, the name 'DIVORUM' occupies the central courtyard of the Divorum, suggesting that it might have been known as the Divorum at this time as well. It was composed of a portico and two temples, making it both a *porticus* and a *templum*. The Aedes Herculis Musarum is known by its temple rather than by the portico that enclosed it.[61] Thus, there is a range of identifications for these types of spaces.

Terminology always presents a challenge to interpretation. The difference in name—*porticus* or *templum*—suggests a different conceptualization, construction, or use of space. What is odd about these spaces is that they typically have a temple (*aedes*) or shrine of some sort as well as porticoes and are very similar in layout. The Divorum's ambiguous name, possibly implying both a *templum* and *porticus*, highlights the overlap of these types of spaces. Perhaps, the change in terminology, which seems to occur with the Flavian constructions, may reflect Vespasian's desire to project his building works in a certain light. The Flavian building projects in Rome were fundamental tools in their political campaign to establish their dynasty and its legitimacy. The Templum Divi Claudi, of course, was not begun by the Flavians, but Agrippina. By

---

[55] Macaulay-Lewis (2008: 89–148; 2009: 1–21). On porticoes in Roman Gaul, see Frakes (2009).
[56] Due to space constraints these other uses of Rome's monumental porticoes cannot be discussed here.
[57] Ov. *Am.* 2.2.3–4; Prop. 2.32.7.
[58] Ov. *Rem. am.* 627–8.
[59] Prior (1996: 121–41).
[60] Eutr. 7.23.
[61] There is debate over whether it was enclosed by the Porticus Philippi; Richardson (1976b: 57–64).

returning it to its original form as a *templum*, Vespasian linked himself to Claudius, the last good Julio-Claudian emperor and the man from whom he received his first important command.[62] Likewise, the Templum Pacis emphasized the right of the Flavians to rule by having restored peace to the empire. The porticoes of the late Republic and early Empire were associated with *otium* and pleasure by much of Rome's population. By calling a space a *templum*, the Flavians could distinguish each construction from previous porticoes and bring additional *gravitas* to these spaces, even if they were similar in design and use. Furthermore by naming them *templa*, Vespasian could emphasize the religious element of these spaces rather than their recreational nature, even if both aspects were important.

Furthermore, in the Porticus Pompei and Porticus Liviae, a temple was not included within the portico itself; the temple to Venus sat atop the *cavea* of the theatre of Pompey and a shrine to Concordia may have been included within the bounds of the Porticus Liviae, so neither space was necessarily a *templum*.[63] The Porticus Octaviae was erected after the temples it enclosed, so it was a new construction, which needed to be distinguished from the Temples of Juno Regina and Jupitor Stator. Thus, calling these spaces *porticus* emphasized that they were new types of constructions and that they had a different purpose to temples that they may have surrounded.

There is also the question of how to refer to these complexes. *Porticus* is a word with many different meanings in the ancient sources and can be used to describe a wide range of archaeologically known spaces;[64] therefore it should not be used as a technical, archaeological term with a rigid or sole meaning. *Templum* is perhaps less problematic, as scholars generally agree that *templa* were inaugurated spaces. But to call these spaces only *templa* does not acknowledge the importance of porticoes—at least in the case of the Divorum and the Templum Pacis. Thus, it may be better to refer to these two types of complexes as monumental porticoes and portico-temples, acknowledging the difference in their names but their similarity in form and at least one of their functions as Rome's public gardens.

Chronologically speaking, the construction of these complexes falls into two distinct groupings: first, the late Republic and the Augustan era and, second, the Flavian era. The building of these complexes seems to broadly mirror the public construction booms in the city of Rome. Unlike other public amenities, such as the baths, which continued to be built until the end of the Empire, we have no archaeological evidence for the erection of porticoes or portico-temples of this nature after the Flavians in Rome itself; the Library of Hadrian

---

[62] Darwall-Smith (1996: 53–5).
[63] A sacred grove, a *nemus duplex*, dedicated to Venus was present in the Porticus Pompei during one phase, cf. Prop. 2.32.13.
[64] Gros (1996: 95–120), for the archaeologically known forms of the *porticus*.

in Athens and the Hadrianic Caesareum in Cryene, North Africa, may be the latest examples of buildings of this nature. During the Severan period, which saw the last major building projects in Rome, no new porticoes or porticotemples were constructed; the Porticus Octaviae and the Templum Pacis were renovated. The *Historia Augusta* mentions the erection of numerous porticoes in the city of Rome, but none are known archaeologically. The apparent lack of construction of new porticoes and portico-temples with gardens after the Flavian era in the city of Rome may be a result of the survival of the evidence. Or it may be better explained by the growth of the baths as the major cultural institution and focus of leisure activities in the city of Rome. The imperial *thermae* included gardens in their grounds and had porticoed spaces, perhaps making the need for additional public gardens unnecessary. Likewise, land in central Rome may have become so expensive that open space was not economically feasible. Walking as a leisure activity seems to have continued in late antiquity, as porticoes and gardens were constructed in private residences, although there is no archaeological evidence for public porticoes or porticotemples being constructed for this purpose. The *Octavius*, a late antique philosophical debate over Christianity by Minucius Felix, is set during a walk in Ostia and suggests that walking for reasons other than transport was very much still a part of the urban experience even at this time.

The design, decor, and ambience of these spaces affected the experience of those walking there for leisure. The archaeological evidence, the *FUR*, and the literary sources enable us to understand the nature of the environment that the plantings, design, and art collections created in Rome's monumental porticoes and how they shaped the experience of leisured walking in Rome. These complexes were oases of *amoenitas*, the right type of setting for cultured leisure.[65]

First, they were destinations for Rome's inhabitants, but they may not have been open to all. Access to Rome's monumental porticoes and portico-temples was controlled; generally they had one monumental entrance, whose known widths range from 12.5 m to 24 m, and a limited number of exits and smaller service entrances.[66] The presence of a single monumental entrance and multiple service entrances into these porticoes suggests that these porticoes were destinations rather than spaces of passage. In the case of the Porticus Octaviae and the Porticus Liviae, one had to climb stairs to enter them. In the case of the Porticus Liviae, ten stairs made the complex inaccessible to vehicles, as did three steps for Porticus Octaviae.[67] The lack of numerous entrances and exits

---

[65] Purcell (1996: 123).
[66] Macaulay-Lewis (2008: 104–6). For comparison with access to fora in Rome, see Newsome, in this volume.
[67] These steps were present during the Severan phase, but the Severan propylon sits on the Augustan foundations, so there may have been some change in elevation; the single line on the *FUR* fragments suggest that there was at least one step up in the Augustan phase, cf. Gorrie (2007: 8, fig. 3).

suggests that one could not easily cut through the porticoes and portico-temples. The limited number of entrances also helped demarcate the space of the portico from the outside, making the porticoes inwardly focused spaces. Coupled with high walls and rich decorative programmes, the spaces became isolated units within Rome, which no doubt contributed to their pleasantness as locations of leisure. This limited access may have curtailed the number of people who could visit the portico at any given moment in time, reducing the volume of traffic, one of the two factors that affected congestion. Unlike Rome's imperial *thermae*, which had numerous entrances and exits, the routes into and out of these complexes were fewer in number. There were custodians or porters to monitor behaviour and ensure that the important works of art were not vandalized or stolen.[68]

As discussed, the ancient sources suggest that the porticoes were locations where Romans went to walk for leisure.[69] If these porticoes were the primary locations for leisured walking, what then made the experience of these spaces and the experience of leisured walking different from walking for transport in Rome's streets and sidewalks? Several features of these porticoes—the width of their porticoes, decor, plantings, and design—shaped the experience of leisured walking in the city of Rome.

The complexes were large; their scale indicates that these porticoes were a significant element of Rome's cityscape, suggesting that these spaces and the leisure activities, as well as the religious and political events that they housed were important (Table 11.2). The width of the individual porticoes and walks

Table 11.2 Dimensions of Rome's monumental porticoes and portico-temples

Complex	Dimensions (m)	Size (m)
Aedes Hercules Musarum	61 × 91	5,550
Porticus Liviae	120 × 70	8,400
Porticus Octaviae	132[a] × 119	15,710
Divorum	194 × 77	14,940
Templum Pacis	134 × 137	18,360
Porticus Pompei[b]	185 × 135	24,980
Templum Divi Claudi	180 × 200	36,000
Average Size		17,700

[a] not preserved completely, minimum estimate
[b] these measurements do not include the theatre

[68] Macaulay-Lewis (2008: 105–6). We know of *custos*, who reportedly guarded the library of Apollo on the Palatine (Ov. *Tr.* 3.1.67–8). Cf. the deposition of valuable objects in the Templum Pacis 'because it was a safe place' (Her. 1.14.2–3).
[69] O'Sullivan (2003: 82–90).

Table 11.3. Table of widths of porticoes and walks in Rome's monumental porticoes and portico-temples

Complex	No. of open-air and architectural walks	Average width (m)
Porticus Liviae	8	5
Templum Pacis (*FUR* phase)	6	8.16
Templum Pacis (archaeologically excavated phase)[a]	10	8.27
Templum Divi Claudi	20	8.43
Porticus Octaviae	7	6.64
Hercules Musarum	12	5
(Porticus) Divorum	8	8.13
Porticus Pompei	7	9.14
Average width for monumental porticos		7.58

Note: some of these measurements are based on the *FUR* and, as such, may not be accurate due to carving inaccuracies. This may explain why a number of the porticoes and walks in the same complex may vary by a few Roman feet.

[a] this excludes the central open walk, which has a width of 54.5 m

of these complexes is significant. The widths of the porticoed (or roofed) walks and open-air walks (or paths) in Rome's monumental porticoes and portico-temples are listed in Table 11.3. The porticoes and walks of Porticus Liviae had an average width of 5 m, as were those of the Aedes Herculis Musarum; the width of the porticoes and walks of the Templum Pacis *(FUR* phase) averaged 8.16 m, while the excavated phase saw an average of 8.27 m. The Porticus Octaviae's walks and porticoes averaged 6.64 m in width, the Divorum's 8.13 m, and those of the Porticus Pompei 7.58 m. Lastly, the walks of the Templum Divi Claudi averaged 8.43 m in width. The average width of these porticoed paths and open-aired ambulatories was 7.58 m. While this is only just over 0.5 m wider than the average street, it is nearly 1.6 m wider than the median street width and perhaps more significantly they are on average 3 m wider than the typical sidewalk, colonnade, or arcade that lined the streets of Rome (see Table 11.1). While certain streets were far wider than the widest portico or open-air walk found in Rome's monumental porticoes and while certain porticoes that lined the streets were spacious, these spaces were exceptional in size for connective architecture. This difference in width suggests that the experience of walking in a connective portico or arcade was different from walking in a monumental portico.

Rather than having one's path blocked by shops and people, the space within the walks and porticoes of the monumental porticoes was open and primarily dedicated to walking. The widths of the porticoes and walks of the monumental portico exclude the space taken up by colonnades and plantings. Several of Rome's monumental portico complexes, including the Porticus Liviae, the Porticus Octaviae, and Porticus Pompei, contained semicircular or rectangular *exedrae* that adjoined many of the individual porticoes. Although there is no archaeological evidence that identifies the function(s) of these spaces, it is possible that these *exedrae* could have served as sitting areas where walkers paused, ruminated on a philosophical topic, admired works of art on display, or enjoyed the pleasant atmosphere of the portico. Unlike Pompeii's benches that projected into the sidewalk and took up valuable walking space, these *exedrae* enabled walkers to step aside to rest and avoid blocking the flow of traffic.

The sculpture, choices in decoration, architectural materials, plantings, and water features of these monumental complexes also contributed to this atmosphere of *amoenitas*. The porticoes and portico-temples were built of extremely luxurious materials, as was typical of monumental public buildings in imperial Rome.[70] More importantly, the sculptural and art collections of Rome's monumental porticoes and portico-temples affected the experience of walking here. The subjects of the works of art in the monumental porticoes and portico-temples were extremely wide-ranging. In many cases, the works of art displayed in a particular portico reflected the political messages of the space's patron or sponsor and were often filled with treasures from Rome's victories.[71] Likewise, the inclusion of garden and water features also helped to create a pleasant atmosphere well suited for leisured walking in most of these complexes. The Aedes Herculis Musarum,[72] the Porticus Pompei,[73] the Porticus Liviae,[74] the Templum Pacis,[75] the Templum Divi Claudi,[76] and the Divorum all probably had some form of garden and/or water features.[77] A study of the Flavian complexes, the Templum Pacis, the Templum Divi Claudi, and the Divorum, demonstrates how the art, plantings, decor, and the overall design of each complex worked together to create a delightful

---

[70] The Severan propylon of the Porticus Octaviae was covered in white marble; the columns of the southern Severan portico were granite or cipollino marble, arranged in an alternating pattern, cf. Viscogliosi (1999: 143–4). The false colonnade of the Templum Pacis was composed of giallo antico columns, cf. La Rocca (2001: 195); Rizzo (2001: 237). Red granite columns were used for the remaining porticoes of the Templum Pacis, and the porticoes and courtyard flooring was paved with white marble in the archaeologically known phase, cf. La Rocca (2001: 195–207); Rizzo (2001: 236–7).

[71] These complexes were typically imbued with political meaning and symbolism, often aiming to convey the ideas and goals of the patron. On the political overtones of the Porticus Pompei, cf. Gleason (1994: 13–27); Kuttner (1999: 343–73). On the Templum Pacis, cf. Darwall-Smith (1996: 55–65); Noreña (2003: 25–43): Macaulay-Lewis (2009: 1–21); and the following ancient sources: Procop. *Goth*. 4.21; Joseph. *BJ* 7.5.7 (7.158–62.) Plin. *HN* 35.74, 102, 109; 36.27; 58; Juv. 9.22–4; Cass. Dio 65.15.1; Suet. *Vesp*. 9; Aur. Vict. *Caes*. 9.7.

[72] Viscogliosi (1996: 17–19).
[73] Gleason (1994: 13–27).
[74] Plin. *HN* 14.11.
[75] La Rocca (2001: 195–207); Rizzo (2001: 234–9); Lloyd (1982: 91–3).
[76] Lloyd (1982: 93–5).
[77] Lloyd (1982: 91–100); Carettoni (1960: 92), first suggested that certain dots were trees, not columns.

setting, far removed from the cramped streets of Rome, where one could stroll for leisure.

How then were these spaces different from the Forum Romanum and the imperial fora? Rome's fora included porticoes, open spaces, and works of art. However, the fora were different in numerous ways. They had no gardens, a fundamental element of the porticoes and portico-temples, which made these spaces so pleasant and different from the rest of Rome's urban fabric. They were the focus of civic and political life of Rome and hosted many religious and certain economic activities; they were designed for these purposes rather than for recreation. Generally, there was no restricted access to these spaces, unless there was a specific event being held; many of them had multiple entrances, as in the case of the Forum Romanum, that allowed people to flow in and out easily. Their open spaces were less structured and more flexible than that of the porticoes and portico-temples, allowing them to be used for everything from religious ceremonies to the burning of debt registers. They were the arteries of Rome; Domitian's forum became known as the Forum Transitorium, as it was a major link between the Subura and central Rome, as was the Forum Augustum.[78] People could and undoubtedly did walk in the fora for pleasure, but the majority of Romans were in the fora conducting some type of business.

WALKING FOR LEISURE IN FLAVIAN ROME: TEMPLUM PACIS, TEMPLUM DIVI CLAUDI, AND DIVORUM

The Flavian period saw a continued flourishing of Rome's public porticoes and portico-temples and walking as a leisure activity; the Flavians' building of public amenities seems to be in keeping with their larger construction aims within the city. It was also the period in which many of our ancient authors, such as Martial, were active. Vespasian, Titus, and Domitian were responsible for the construction or completion of the Templum Pacis, the Templum Divi Claudi, and the Divorum. The three remarkable complexes combined plantings, architecture, and art for political purposes, while also creating exceptionally pleasant environments where Romans could walk for leisure.

Known from the ancient sources, the *FUR*, and recent excavations,[79] the Templum Pacis was erected as a victory monument in celebration of

---

[78] But see Newsome, in this volume.
[79] For the recent excavations, cf. La Rocca (2001: 195–207); Rizzo (2001: 234–43); Meneghini and Santangeli Valenzani (2007: 61–70). For an interpretation of the Templum Pacis on the *FUR*, cf. Lloyd (1982). For a general overview of the complex, cf. Coarelli (1999a: 67–70).

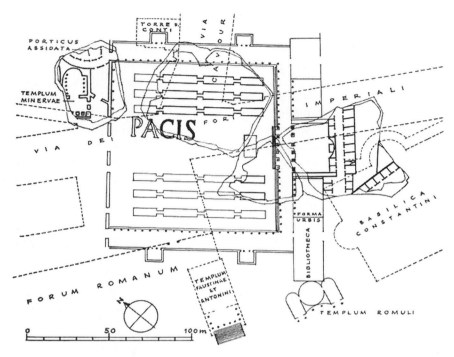

Fig. 11.2. Reconstruction plan of the Templum Pacis (Gatti 1933), FU10327 F (Fototeca Unione, American Academy in Rome).

Vespasian and Titus' victory over the rebellious land of Judaea in 75 CE. Modified by Domitian during the construction of his forum, the complex was filled with plantings, possibly water features, and an exceptional and highly politicized collection of art and objects.[80] On the *FUR* (fragments 15ab, 15c, and 16a), the open space of the Templum Pacis included two sets of three long rectangular-shaped features at one point in its history; these have been interpreted as plantings (Fig. 11.2).[81] Excavations in the late 1990s demonstrated that two sets of three rectangular features did in fact dominate the open space of the complex and that the courtyard was unpaved;[82] however, their date and relationship to the image of the Templum Pacis on the *FUR*, as

---

[80] It was constructed under Vespasian, modified by Domitian when he constructed his new forum, and renovated under Hadrian. Cf. Anderson (1982). After it had been badly damaged in a fire in 192 CE, Septimius Severus restored it again, between 208 and 111 CE, and placed the *FUR* in the room to the west of the *aedes*. An earthquake in 408 CE destroyed the building, but it was then rebuilt. In the 6th century CE another fire damaged the Templum Pacis; at this point the complex no longer functioned for its traditional purpose.

[81] Lloyd (1982: 91–3).

[82] La Rocca (2001: 195–207); Rizzo (2001: 234–43); Meneghini and Santangeli Valenzani (2007: 61–3).

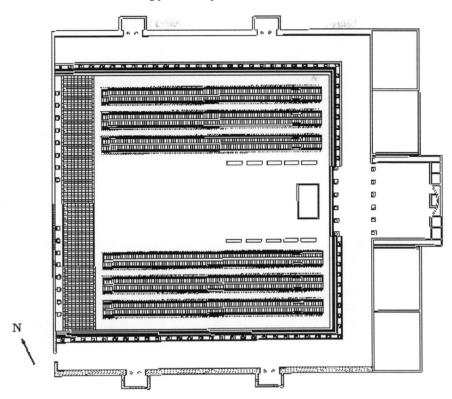

Fig. 11.3. Plan of the Templum Pacis restored on the basis of the recent excavations (Detail taken from the plan of imperial fora. Scale 1:4000. La Rocca 2001: 175, fig. 4).

well as their interpretation, remains problematic (Fig. 11.3).[83] The Italian excavators identified these rectangles as long water channels;[84] roses, which were planted in thirty-one reused amphorae, were also found here.[85] These

---

[83] The results of the excavations, to date, have only been published in a brief article format, La Rocca (2001) and Rizzo (2001), and in a summary form in Meneghini and Santangeli Valenzani (2007: 61–70); stratigraphy for the excavations has yet to be published. Furthermore, many of the claims made by the excavators, including the dating of the archaeological remains to the Vespasianic period (where a late antique date might be more probable, A. Claridge, pers. comm.) and the presence of water channels cannot be satisfactorily explained by the published reports. Hopefully, more detailed publications will resolve these problems of interpretation.

[84] R. Wilson (pers. comm.) examined the so-called water channels; he reports that he saw no archaeological evidence for a hydraulic system; A. Claridge also confirmed this (pers. comm.). Thus, the identification of these features as water features remains uncertain. Meneghini and Santangeli Valenzani (2007: 62) report that there were traces left of lead piping, but no pipes were discovered.

[85] Rizzo (2001: 239); Meneghini and Santangeli Valenzani (2007: 63). K. Gleason saw the planters in 2001 (K. Gleason, pers. comm.). The roses were identified by their carbonized plant remains, making the identification secure.

remains were dated to the original phase of Templum Pacis and were also identified as corresponding to the long rectangles depicted on the *FUR*; the reasoning for a Vespasianic date and such an interpretation is unclear.[86] Rather trying to fit the archaeological remains to the *FUR*, it is sensible to see the differences in the depiction of the courtyard of the Templum Pacis on the *FUR* and the archaeological remains as reflecting two distinct building phases in the complex's history, whose chronology has not been resolved satisfactorily. Regardless, we can say that there were probably garden features in at least two phases of the Templum Pacis.

The presence of plantings would have immediately set the Templum Pacis apart from the bustle of the neighbouring imperial fora and the loud and dirty Subura.[87] The scent of flowering roses, although a temporary, annual event, must have been a welcome relief to those plodding through Rome, likewise the spacious open-air walks would have allowed the walker much need breathing space.[88] Romans moved from the streets to these spaces, regardless of dress or station.

A collection of art also greeted one upon entering the Templum Pacis. These works included the bronze cow of Myron, which had been displayed in the Domus Aurea, and a bronze by Phidias or Lysippus,[89] as well as the menorah and other spoils taken from the Temple in Jerusalem.[90] This display of works specifically referred to Vespasian and the Flavian dynasty's rise to power by defeating the Jews and bringing peace to the Roman Empire.[91] These politicized works not only articulated Flavian ideologies and agenda, but they were also impressive and outstanding works of art. According to Pliny the Elder, paintings by the famed Hellenistic artists Timanthes, Protogenes, and Nikomachus stood here.[92] Statues of Ganymede, Scylla, Venus, the Nile, a statue of the philosopher Chrysippus, and numerous other works, were present and no doubt contributed to the pleasant and learned atmosphere of the complex.[93] A library was also included in the Templum Pacis, another

---

[86] Rizzo (2001: 238).

[87] Pliny (*HN* 36.102) characterized the Basilica Aemilia, the Forum Augustum, and the Templum Pacis as the three most beautiful buildings in the world; three spaces related by their separation by the thoroughfare of the Argiletum, which may have increased the perceived *dignitas* of the three surrounding *monumenta*. In auditory terms, Pliny laments how the Templum Pacis was so noisy that one could not properly appreciate the works of art that were displayed there (Plin. *HN* 36.27). This was probably because of the intrusive clamour from the Argiletum. The creation of the new forum in the 90s CE will have removed this intrusion. I thank David Newsome for these observations.

[88] See Betts, in this volume.

[89] Procop. *Goth.* 4.21; Plin. *HN* 35.74, 102, 109; 36.27; 58; Juv. 9.22–4.

[90] Procop. *Goth.* 4.21; Joseph. *BJ* 7.5.7 (7.158–62).

[91] Coarelli (1999a: 76); Joseph. *BJ* 7.5.7 (7.158–62).

[92] Plin. *HN* 35.74; 102; 109.

[93] On the other statues here, cf. Juv. 9.22–6; Plin. *HN* 35.74, 102, 109; 36.27; 58. Three statue bases have been found in the open courtyard of the Templum Pacis. Cf. La Rocca (2001:

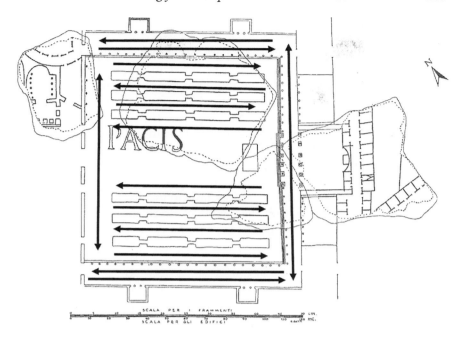

Fig. 11.4. Possible traffic patterns in the Templum Pacis (Elizabeth Macaulay-Lewis after La Rocca 2001: 194, fig. 15).

feature that encouraged literate and educated Romans to linger there. The width of these porticoes easily accommodated these works of art, which were typically placed in the intercolumniations of porticoes, and the walkers. In the *FUR* phase, the known porticoes had widths of 9 m and 15 m respectively, and the six open-air walks ranged in width from 6 to 9 m. In the excavated phase, the widths of the porticoes were 14.78 m (50 Roman feet) and the garden walks were each 5 m wide (see Table 11.3). Thus, the amount of space for walking, especially in the porticoes, was considerably more than the 4.83 m of arcades and sidewalks.

Not only did the art and plantings create an atmosphere in sharp contrast to the streets, porticoes, and sidewalks of ancient Rome, but the architectural and open space of the Templum Pacis was highly organized and linear in both the archaeologically known and *FUR* phases, particularly its open-air courtyard (Fig. 11.4). The *FUR* shows six rows of four connected rectangles that dominated the Templum Pacis' exterior space. In its two known phases, porticoes

196–201, figs. 17–19); Meneghini and Santangeli Valenzani (2007: 66, fig. 60). A statue of Septimius Severus, dressed as a philosopher, was also discovered. This reminds us that the 'collection' of art that was displayed here was not static and was added to; these works of art may not have been conceived of as a 'collection'.

enclosed the courtyard. The three true porticoes of the complex, which are known archaeologically, were situated 1.5 m above the courtyard;[94] one accessed the courtyard from the three porticoes via steps. While the elevated position of the porticoes created a clear division between the open and architectural space, one could easily walk into the open space at any point. We do not know if there were any temporary barriers that could be erected in these spaces on special occasions.[95]

More striking, however, were the possible traffic patterns that the garden and linear features created in both phases. In the archaeologically known phase, the two sets of rectilinear features ordered much of the open space of the Templum Pacis. In the *FUR* phase of the complex, these possible plantings, ordered the open space. In each phase, the linear features and the plantings also created a central ambulatory on axis with the temple of Peace and six smaller, linear ambulatories. In the archaeologically known phase, each rectangular feature was 4.7 m wide and 1–1.5 m high.[96] The width and height of these features prevented a walker from crossing over them. Presumably, the plantings (each 4 m wide)[97] had the same effect. Therefore, the ordered design of the Templum Pacis' open space and its paths created linear movement. Even if one wanted to deviate from their chosen walk, it would be difficult to do anything but turn around and backtrack, owing to the physical barrier created by the rectangular features and the plantings respectively. That said, the open space, the central ambulatory, set between the two sets of linear features was a more flexible space; thus, if a walker desired to walk in a less structured way, to meander, then this space allowed this type of walking, if not encouraged it.

Known from the *FUR* and archaeological excavations the Templum Divi Claudi also appears to have been designed with an ordered approach to its open space (Fig. 11.5). It was located atop a massive platform,[98] begun by Agrippina and completed by Vespasian.[99] The open space of the *templum* was organized around a series of rectangular and L-shaped features and a central temple. On the basis of the garden excavations at Fishbourne, Lloyd interpreted these long strips as garden plantings.[100] The presence of non-

---

[94] Rizzo (2001: 236).
[95] See Newsome, in this volume. Certain courtyards in Pompeian houses, such as the second courtyard in the Casa del Fauno (VI.12.2–5), have holes in the columns for barriers to be placed in between the columns.
[96] La Rocca (2001: 195); Meneghini and Santangeli Valenzani (2007: 61–2).
[97] The small rectangles that connected the longer rectangles were each 2 m wide.
[98] The platform shows signs of multiple phases from the 1st to the 4th centuries CE; this may suggest that the complex was also restored through antiquity although we have no record of it. Thus, the porticoes not shown on the *FUR* may simply be from another phase.
[99] Suet. *Vesp.* 9; Buzzetti (1993a: 278). Nero used the eastern side of the complex as part of a giant *nymphaeum*.
[100] Lloyd (1982: 94–5).

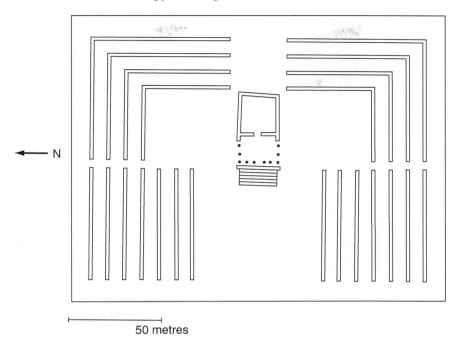

Fig. 11.5. Plan of the Templum Divi Claudi restored on the basis of the *FUR*; the temple is shown here having six columns. According to *FUR* fr. 5c, which is only reproduced in the Renaissance drawing Cod. Vat. Lat. 3439, the temple was *prostyle* and *pentastyle*. Buzzetti's suggestion that the temple was actually *hexastyle* seems more likely than *pentastyle* (1993a: 277) (Elizabeth Macaulay-Lewis after *FUR*).

architectural, rectilinear features in the open space of the Templum Pacis lends further support to the interpretation of these as plantings.

These features, each about 2 m wide, created a series of paths, 8 m wide, that defined a number of potential routes (Figure 11.6).[101] If the linear and L-shaped features were not permeable, then they shaped the traffic patterns within the Templum Divi Claudi. Within these spaces, visitors could not walk anywhere at their leisure; rather once one committed to walking on a path between two of the features, one had to stroll to the end of the path before exiting or one had to backtrack. However, like the Templum Pacis, the central open space in front of the temple was a more flexible space; its lack of plantings or features allowed visitors to walk as they chose. While walking in the Templum Divi Claudi may have been a more structured affair, there was far

---

[101] I have assumed that these features are depicted to scale; it may be that they were actually smaller than their depiction on the *FUR*, as it is very difficult to make precise incisions of objects that were 1–2 m in width when the scale of the *FUR* was 1:240 or 1:250.

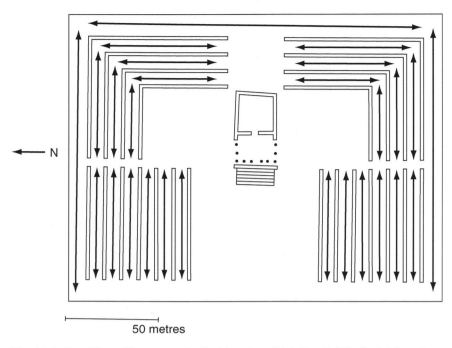

Fig. 11.6. Possible traffic patterns in the Templum Divi Claudi (Elizabeth Macaulay-Lewis after *FUR*).

more space for a walker than in Rome's sidewalks and streets; this extra space created a different experience from that of walking for transport in Rome.

The Porticus Divorum, which is only known from the *FUR* (fragments 35ab, 35cdefghi, 35cc) was designed on similar principles, although the ordering of its open space allowed more room for non-structured movement (Figure 11.7). Enclosed by porticoes only on three sides its exact purpose is uncertain; Domitian may have dedicated it to Vespasian and Titus.[102] On the *FUR*, two parallel rows of dots in the courtyard often represent trees, perhaps a sacred grove;[103] they are not perfectly aligned with the porticoes. There appears to be roughly one tree for every three columns. Considering that the Divorum may have replaced the Villa Publica, which was planted during

---

[102] Darwall-Smith (1996: 156–9); Coarelli (1995a: 19–20).
[103] Darwall-Smith (1996: 156–7); Coarelli (1995a: 20). Packer (1997; 2001) argued that two parallel rows of dots on the *FUR* (frag. 29e) in the Forum Traiani was a grove or plantings of trees. Excavations have proved that this was not the case; the entire courtyard was paved. Thus, dots do not always represent plantings but often do, as in the case of the Porticus Pompei. Carroll's excavations of the Temples of Apollo and Venus in Pompeii also seems to confirm that plantings were set in paved courtyards associated with temples, Carroll and Godden (2000: 743–54); Carroll (2008: 37–45).

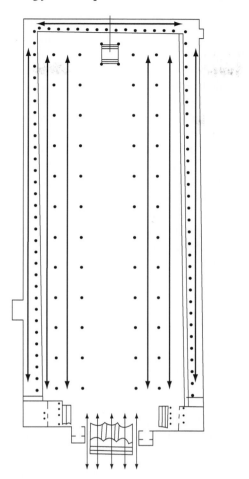

Fig. 11.7. Possible traffic patterns in the Divorum (Elizabeth Macaulay-Lewis after *FUR*).

the Republic,[104] the presence of plantings may have also continued the pleasant atmosphere established by the previous building on the site.[105]

One could walk around the three porticoes that bordered the Divorum's courtyard; these porticoes were probably at a different height, in the eastern portico's case, there may have been a single step and for the western portico, a double step. It is possible that a step or series of steps ran along the edge of all of the porticoes, not just the western and eastern porticoes, as the *FUR* indicates. Despite the different elevation between the porticoes and the

[104] Varro *Rust.* 3.2.1–6.   [105] Richardson (1976a: 159–63).

courtyard, one probably could have moved between the porticoes and the tree-lined ambulatories relatively easily and in an unstructured way, as in the Templum Pacis.

Four rows of dots, two on each side of the main axis, ordered the open space of the Divorum and created five paths for walkers. While the trees created vertical order and four linear paths, the spaces between the ranks of trees and width of the paths themselves, especially the central walk, were wide at 6–14 m, which enabled the walkers to move through the open space of the Divorum as they chose. Again, these walks were wider than Rome's sidewalks and arcades and many of Rome's streets. Thus, while the open space of the complex was ordered, it was not as rigid in its organization of space or the creation of linear walks as the Templum Pacis and the Templum Divi Claudi. Unlike Rome's streets, which were twisted to accommodate the city's topography, these complexes had a highly organized approach to open and architectural space. Such a design suggests that more ordered flows of walkers were created in the paths and colonnades of these spaces.

The exploration of the design of these porticoes and portico-temples in Rome demonstrates not only that the architectural space was highly ordered, but also that their open space was also ordered. This seems to reflect the Roman desire to impose order on their world, although reconstruction of exact flows of traffic is not possible.[106] Such an idea also seems to have extended to the Roman attitude towards the construction of leisured movement. Space was constructed in such a way to create ordered and, in certain cases, linear traffic patterns. Inevitably, walkers may have disobeyed this ordered approach; the monumental porticoes could be convenient short cuts to navigate the city and sometimes people had to reverse their intended course. While there were no doubt 'disobedient' walkers who did not follow these linear paths and the flows of traffic they created, the highly organized open spaces of the Templum Pacis and the Templum Divi Claudi demonstrate that one could not simply wander as one pleased, but had to walk, at least partially, in accordance with the design of the space. The attempt to create specific flows of traffic might have also had an important role in crowd control during specific events in such spaces. In these porticoes, there was a very different approach to movement and the experience of space as compared to the streets of Rome. The intention of Roman designers to create linear paths, walks, and traffic patterns in the public porticoes made Roman movement unique. Linear walks and traffic patterns are not obvious choices; rather, they are culturally specific.

---

[106] These routes do not appear to have had symbolic importance, although when religious ceremonies were staged, specific processional routes may have been marked off; however, from the archaeological remains and the *FUR* it is not possible to say anything else concrete.

CONCLUSIONS

In conclusion, the study of the *Forma Urbis Romae* and archaeological evidence, alongside the ancient literary texts, suggests that walking for transport was a very different experience from leisured walking, owing to the size of the streets and sidewalks, and the nature of traffic in each space. The streets and sidewalks of Rome, though wider than many of those found in Pompeii and other Roman cities, were narrower than the porticoes and open-air walks of the capital's monumental porticoes. Furthermore, while streets were often congested, Rome's monumental porticoes seem to have been less busy, making the experience of walking for transport less enjoyable than walking for leisure. Leisured walking was a more controlled, ordered, and pleasant activity, due to the construction of space in Rome's monumental porticoes and portico-temples. The limited access to these porticoes may have limited the absolute number of visitors at any given time, reducing the volume of traffic. Likewise, the width of these porticoes, their art collections and plants, as well as *exedrae* to accommodate those who had taken a rest from walking, created a very different atmosphere from the streets of Rome, which were often congested due to their width and the volume of traffic.

The archaeological record and the *FUR* demonstrate that there were different types of walking in ancient cities and that new insights into the nature of walking for transport and for leisure can be gained through the study of material culture—not merely through the study of ancient texts. The architectural and horticultural framework in which walking for transport and for leisure were set clearly shaped the nature of walking, creating two very different experiences of walking in Rome's streets or in the city's public porticoes. In Rome one could experience both the hustle and bustle of the dynamic city on the go in the streets alongside the serenity and calm of the monumental porticoes and temple porticoes. This approach, which aims to establish an 'archaeology of walking', could be applied in numerous other Roman cities to understand the nature of pedestrian traffic and how the built, urban environment shaped human walking and behaviour.

12

Movement and Fora in Rome (the Late Republic to the First Century CE)*

David J. Newsome

'Since his search had taken him out of his way along a winding path, he decided to shorten his journey on the way back, and go straight through the forum'.[1]

For Phaedrus, as for his contemporaries in Julio-Claudian Rome, the forum could be perceived and described as a short cut between other discrete places in the continuous space of the city: as used by the slave Aesop who, having been sent out to make preparations for dinner, minimized his route home by cutting through the open forum.[2] The forum offered a direct contrast to the circuitous routes of the city streets. Routes through the forum were short, while through the streets they were long. Routes through the forum were direct, minimizing the roundabout nature of paths through the city. Aside from such benefits to economies of effort, the forum was accessible, without qualification. Such is the perception, shaped by praxis, of the forum in the early first century CE.

Compare this to the early second century CE, and the younger Pliny's characterization of fora in his panegyric to Trajan. At midpoint in the text, Pliny turns his attention to the Domus Flavia on the Palatine. His description of the *domus* under Domitian casts the nature of space and accessibility as a direct reflection of the characters of its incumbent.[3] Domitian's palace

---

* With thanks to Ray Laurence, Andrew Wallace-Hadrill, Simon Esmonde-Cleary, and Gareth Sears. Translations are from the Loeb Classical Library.

[1] Phaedr. *Fab.* 3.19.5–7, 'tum circumeunti fuerat quod iter longius | effecit brevius: namque recta per forum | coepit redire'. Book 3 was written in the early 30s CE.

[2] We can consider this as one of the fables that, more closely resembling contemporary satire, 'are firmly rooted in [...] daily life in first-century Rome', Champlin (2005: 110).

[3] See Zanker (2002: 109); Wataghin Cantino (1966). For the context of the *Panegyricus*, cf. Radice (1968). On the contrast between Domitian and Trajan in the *Panegyricus*, see Noreña (2007: 247, 250).

shielded the emperor in private chambers behind a thousand doors, a common stereotype of his paranoia which, as in other texts of the period, was imagined as manifest in his organization of built space.[4] In contrast, Nerva had opened the *domus* to the public, changing the *principes arcem* into the *aedes publicae*.[5] Public accessibility is a measure of imperial magnanimity, underscored by the explicit link drawn between the text of the inscription and the habits (*mores*) of the new emperor.[6] But the choice of spaces to which Pliny compares the *domus* is worthy of note: 'no forum, no temple is so free of access'.[7] Given that fora are ostensibly public spaces, this demands explanation.

The verb Pliny uses—*reserare*—is an unusual choice, occurring only twice more in his surviving corpus: in the first instance to refer to the 'opening' of the year; in the second to refer to the 'opening' of lips that had been sealed (*obsaepio*) under Domitian's censorship.[8] In other instances, *resero* refers to the opening of doors that had previously been barred: 'unbolting' physical space.[9] Radice's Loeb translation, provided above, obscures this important point. Rather than stating that no forum or temple is so free of access, Pliny asks: 'what temples and what forum have been so reopened [*resarata*]?' The emphasis is on transition, and for this reason the verb carries more weight than the adjective (e.g. *aditus, pervium*). By referring to the forum in these terms, Pliny is drawing attention not so much to the fact that such spaces were accessible as that they had not been so before. All of this provides a curious contrast to the nature of fora a century before: open, accessible, permeable.

Taking these two opposing characterizations of fora as a starting point, this chapter examines the changes to the use and perception of fora in Rome over the first centuries BCE and CE based on movement and accessibility.[10] It begins by examining the 'through-movement potential' of the Forum Romanum and, like Phaedrus above, its role as a short cut between other urban places in the 'natural movement' of the city. However, we can also recognize a tendency for

---

[4] See Suet. *Dom.* 14.4 on Domitian's porticoes lined with reflective phengite stone (Plin. *HN* 36.163), so that he might see conspirators behind his back. On space and surveillance in Domitian's Rome, see Frederick (2003).

[5] Plin. *Pan.* 47.4. This change of name is recorded in a commemorative inscription (*ILS* 9358). Trajan is, in contrast, conspicuously accessible to petitioners (*Pan.* 48.1).

[6] Plin. *Pan.* 47.5, 'Quam bene cum titulo isto moribus tuis convenit.'

[7] Plin. *Pan.* 47.5, 'quod enim forum, quae templa tam reserata?'

[8] Plin. *Pan.* 58.3; 66.5. *Obsaepio* referred to the blockage of physical space, in both literary (e.g. Plaut. *Pseud.* 425, 'ibi nunc oppido opsaeptast via') and legal texts (e.g. *Lex Iulia Agraria* [50s BCE] KL IIII: 'ne quis eos limites decumani obsaeptos neue quid immolitum neue quid ibi obsaeptum habeto').

[9] Tib. 1.8.60; Val. Max. 2.10.2; Ov. *Tr.* 5.9.29.

[10] An early advocate of greater focus on movement was Giuliani (1985: 9): 'e più importante conoscere il rapporto in cui si poneva un monumento con le strutture adiacenti, con i suoi ingressi, i suoi vuoti, con vincolava il flusso [...] piuttosto che sapere che esso è di quel particolare anno e realizzato in quella particolare circostanza'. See more recently La Rocca (2006).

movement within the open forum to be progressively controlled. With the construction of the imperial fora, those spaces were increasingly segregated and lacked through-routes. Because the infrastructure of movement is fundamentally linked to the conception of urban space, and the 'urban disposition' which produced particular forms of organization and access, changes to movement reflect broader cultural and social changes.[11] This chapter, focusing on the late Republic through to the end of the first century CE, combines both textual and archaeological evidence to provide a more detailed and thematic interpretation of the development of movement and accessibility to Rome's central fora.[12]

The distinction between the Forum Romanum and the imperial fora can be neatly summarized as one between movement *through* and movement *to*. The distinction between the two is fundamental to how we interpret the role of urban spaces, but it has received relatively little attention in Roman urban studies. Macaulay-Lewis's examination of accessibility to portico structures indicates that their primary use and perception was as destinations, not through-routes.[13] This establishes movement *through* and movement *to* as distinct, and critical, criteria in understanding the kinetic city. However, her contrast between porticoes and fora, the latter being 'transitorial' spaces, only holds true for the Forum Romanum, and even then with historical variability. As this chapter shows, the imperial fora are akin to the destination spaces of porticoes; with fewer entrances, restricted to certain types of movement by the arrangement of space in such a way that it excluded non-pedestrian users.[14] Such architectures allow us to glimpse intention through facilitation: if routes need to mediate a change in gradient, the choice of steps rather than a ramp is a choice of the pedestrian rather than the vehicle.

Of course, controls on movement in public space need not be enforced through architecture; we must also consider other possibilities. Pliny notes that in the *aedes publicae* there were no barriers (*obicis*) and no 'grades' of entry (*gradus*) to cause humiliation.[15] Grades of entry might imply the

---

[11] On conceived space, see Lefebvre (1991). On the 'urban disposition', see Saunier (1998: 436 n. 5) and Newsome (2010: *passim*).

[12] Newsome (2010: 206–27) details such changes through the construction of the Forum Traiani and Markets of Trajan. Because of the pressures on space in this chapter, I do not go beyond the Flavian period here. See also Köb (2000: 325–40).

[13] Macaulay-Lewis (2007: 98–104 and in this volume) on restricted access to the porticoes of Livia, Octavia, and Pompey.

[14] Although my emphasis is on patterns of movement, La Rocca (2001: 186) discusses the similarities between the Porticus Metelli, the *porticus post scaenam* at the Theatrum Pompeium, and the first imperial fora.

[15] Plin. *Pan.* 47.5, 'nullae obices nulli contumeliarum gradus superatisque illa mille liminibus ultra simper aliqua dura et obstantia'. Domitian's interest in grades of privilege can be seen in his renewal of the *discrimina ordinum*, regulating seating divisions in the theatre, Suet. *Dom.* 8.3; Mart. 5.8 (see Rawson 1987).

intrusion of guards. Such roles were not unknown in the imperial house and there may have been equivalents in public spaces.[16] We should not discount the possibility of restrictions on movement that were not based on physical blockages. This would be the case for Augustus' decree that no one was allowed to linger in the Forum (Romanum), or the spaces around it, if not wearing a toga without a cloak—a decree to be enforced by the aedile.[17] Suetonius' use of *consistere* reveals a spatial concession to movement, in that people so attired could not stop there, but could continue to move through.[18] This may indicate the continued importance of the Forum Romanum as a through-route. We are not told how such a decree worked in practice, but in another occurrence relating to the control of space, Suetonius adds that this was enforced *per militem*.[19] Augustus' decree that no one were to stop in the Forum Romanum if not in the toga without a cloak requires the participation of other parties in that cultural *habitus*: the *togati*, to dress in the manner (now) appropriate for their spatial activities, or the aedile, to enforce this decree in instances of transgression.[20] Architectural control, however, requires no complicity from its users. It shapes (and constrains) possibilities of use. So, for example, to ensure that wagons do not pass through a certain space, a physical barrier making it impermeable to wagons is a more direct, effective, and (in theory) permanent manner of controlling space than would be the passing of a decree prohibiting wagons.

Through-movement potential has important implications for how we understand the configuration of built space. This links to broader spatial theory of 'natural movement', which posits that because movement in urban contexts can, within reason, be from anywhere to anywhere else, most urban movement is created not because of 'attractors' but because of the layout of urban space itself.[21] In the simplest terms, routes are made of three key elements: an origin,

---

[16] Suet. *Claud.* 35.1 on *scrutatores* checking for weapons concealed by visitors to the emperor's *salutatio*. Servile roles existed that were related to the control of access, such as the slave *ab admissione*, responsible for the admission and, by inference, for the refusal of visitors. On the emergence of this role in Augustan Rome, see Treggiari (1975: 51). See Ov. *Tr.* 3.1.67–8 on the *custos* of the Bibliotheca Apollonis on the Palatine. Evidently there existed *procuratores* for the Forum Traiani (*CIL* VI.41285a).

[17] Suet. *Aug.* 40.5, 'negotium aedilibus dedit, ne quem posthac paterentur in Foro circave nisi positis lacernis togatum consistere'.

[18] Rolfe's Loeb translation of 'to appear in the Forum' obscures the distinction between motion and motionlessness. Suetonius' other uses of *consistere* all relate to a lack of movement: *Tib.* 18.1 on stopping by the Rhine and not crossing (*non transmisit*); *Tib.* 64, on not allowing people to stop as his familial litter passed by; *Cal.* 22.2, on standing between the statues of Castor and Pollux; *Cal.* 53.1, on being unable to stand still; *Cal.* 57.3, on standing in heaven besides Jupiter.

[19] Suet. *Tib.* 64, 'prohibitis per militem obuiis ac viatoribus respicere usquam vel consistere'.

[20] On the *habitus* of the toga, see Wallace-Hadrill (2008: 38–57, esp. 41–6); Edmondson (2008: 23).

[21] Hillier (1996: 120). See also the Introduction and the chapter by Stöger, in this volume.

a destination, and the space passed through on the journey. The centrality of the spaces passed through is a by-product of the origin and the destination. This also suggests that the prime motive for through movement is distance minimization. This links centrality to accessibility. Indeed, where we talk of a centre being the most integrated space, we can also call it the space with the most through-movement potential.

THE FORUM ROMANUM AS A SHORT CUT IN THE CITY

The urban centrality of the Forum Romanum was largely related to its function as the place passed through on other routes. It was at the broad convergence of multiple streets, paths, and steps, and formed a logical minimization of both distance and effort for movement that was passing from one side of Rome to the other and which wished to avoid either circumventing or traversing the city's hills.[22] The physical centrality of the Forum Romanum is a legacy of its position relative to the hills of Rome. The Cloaca Maxima, which ran through the forum, was considered by Pliny to flow through the middle of the city.[23] Other references to the 'middle of the city' are less specific but we can still say something of the general use of the term where relevant to patterns of urban movement. Livy provides one such example, discussing the movement of troops through the *media urbis*, from the Porta Collina to the Aventine.[24] As the Porta Collina was on the north of the city, near the Castra Praetoria, and the Aventine was on the south, we might reasonably infer that the route through the 'middle of the city' was down the Vicus Longus, and thereafter through the Forum Romanum. This connected to the Vicus Tuscus which led to the Aventine by way of the Forum Boarium.[25] The Forum Romanum might also be considered the middle of the city in Livy's description of a wolf rushing into Rome in 196 BCE. This entered through the Porta Esquilina, made its way to the forum, then by way of the Vicus Tuscus and the Cermalus it exited through the Porta Capena.[26]

---

[22] I use the term 'broad convergence' because although such routes led to the area, there was no 'junction' as such. The Forum Romanum is a crossroads only in a rather loose sense. Cf. Rykwert (1988: 59) for the forum as intersection of the *decumanus* and *cardo* of early Rome. The convergence of roads at the Miliarium Aureum was an idealistic rather than an accurate representation of space (Plut. *Galb.* 24.4, εἰς ὃν αἱ τετμημέναι τῆς Ἰταλίας ὁδοὶ πᾶσαι τελευτῶσιν, on which see Newsome 2009b: 29–30).

[23] Plin. *HN* 36.94.

[24] Livy 3.51, 'Porta Collina urbem intrauere sub signis, mediaque urbe agmine in Auentinum'.

[25] Livy 27.37.15, 'inde vico Tusco Velabroque per bovarium forum in clivum Publicum'.

[26] Livy 33.26.8–9, 'lupus Esquilina porta ingressus, frequentissima parte urbis cum in forum decurrisset, Tusco uico atque inde Cermalo per portam Capenam prope intactus euaserat'.

Livy refers to the area as a *locus frequentissimus*. This is not the forum itself but the route there—the busy streets of the Esquiline. The route towards the forum from the area inside the Porta Esquilina was relatively direct. The Argiletum led to the Forum Romanum directly, while the Clivus Pullius led to the south-west of the Fagutal, the probable location of the Carinae. It was by such a route that Ovid imagined flute players from Tibur, lulled into sleep, arriving at the Forum Romanum at night. Their *plaustrum* moved from *media urbis* and ended *in medio foro*.[27] Elsewhere, Livy referred to the Carcer as being in *media urbis*, near to the forum.[28] The Carcer stood beside the Clivus Argentarius, between the *loci celeberrimi* of Piso's house further west and the Comitium and Aedes Concordiae further east.[29] The forum itself is not the *media urbis*, only a part of it. The rather loose applicability of this term to the forum area is also found in Plutarch, who describes the removal of the body of the murdered Tiberius Gracchus from the Capitoline to the Tiber—through the μέσος πόλεως.[30] Arguably, this route would be by the Clivus Capitolinus, then by the Vicus Iugarius.

In the remainder of this section, I do not wish to review the physical topography of the Forum Romanum in detail, but rather to focus on the ways in which movement to and through this space influenced its perception and its representation in contemporary texts.[31] That our sources perceived the Forum Romanum in a similar way—as a short cut on longer routes—reflects a broader cultural perception of space, based on praxis common to writers and their readers in the first century CE.[32]

Perhaps speculatively, we can begin with an earlier (second century BCE) scene in Terence's *Adelphoe*, the *locus classicus* on navigating the Roman city, including: 'This is a much nearer way, and with much less chance of missing it'.[33]

A similar portent is presented by Cassius Dio, who says that in 16 BCE a wolf entered the forum by the Sacra Via: 54.19.7, λύκος τε γὰρ διὰ τῆς ἱερᾶς ὁδοῦ ἐς τὴν ἀγορὰν ἐσπεσὼν ἀνθρώπους ἔφθειρε.

[27] Ov. F. 6.683–4, 'iamque per Esquilias Romanam intraverat urbem, et mane in medio plaustra fuere foro'. Recalling the discussion in my Introduction to this volume, it is worth noting that the *plaustrum* moves by night, perhaps reflecting the restrictions on *plaustra* throughout the day.

[28] Livy 1.33, 'media urbis foro imminens'.

[29] On Piso's house as a *locus celeberrimus*, see Tac. Ann. 3.9.3, 'et celebritate loci nihil occultum'.

[30] Plut. Tib. Gracch. 3.3.

[31] Newsome (2010: 84–173) examines the wider topographic (and movement oriented) setting of the Forum Romanum in antiquity. For brevity it is not possible to discuss all of the details here. The landmark texts on the development of the Forum Romanum remain Coarelli (1983, 1985), with excellent summaries by Purcell (1995b, 1995c) and Tagliamonte (1995).

[32] Phaedrus knew Horace (Champlin 2005: 109, 117–20); Martial was aware of Phaedrus' λόγους (Mart. 3.20).

[33] Ter. Ad. 573–84, 579: 'sane hac multo propius ibis et minor est erratio'. For the influence of the text on interpretations of urban navigation, see Ling (1990a: 211–12) and van Tilburg (2007: 49–51).

Palombi, like Frank and Gilula before him, suggested that the text was specifically (re-)written for audiences in Rome and used fictitious landmarks inspired by those that would be immediately familiar to the viewer.[34] The play was performed at the funeral of M. Aemilius Lepidus in 160 BCE, probably *in foro*. Palombi rightly noted that we cannot expect to accurately reconstruct the maze of *plataea, angiporti, vici*, and *clivi* in Terence's routes, but we might say that, if in the vicinity of the forum, further directions to the house of the millionaire Cratinus might correspond with the aristocratic houses on the Sacra Via, and the subsequent street on the left would be that between the later Temple of Romulus and Basilica of Maxentius: the 'clivus ad Carinas'—the 'much nearer way', and the street described by Dionysius of Halicarnassus as a short cut (ἐπίτομον ὁδόν) between the Forum Romanum and the Carinae.[35]

If this suggestive but unspecific text lacks topographical certainty, there are numerous other examples of the Forum Romanum characterized as a short cut in descriptions of the city of Rome. Horace describes how he was accosted by a loquacious hanger-on as he strolled along the Sacra Via.[36] As well as speeding up and slowing down at random and unnecessary junctures along the street, Horace's attempts to dissuade his new companion included the invention of a route that should have made accompanying him undesirable to even the most determined of escorts: 'I want to see a person who is unknown to you: he lives a long way off across the Tiber, near Caesar's gardens'.[37] The attendant is not dissuaded either by the fact that he does not know the person to whom Horace is walking, or by the fact that it is a long way away, near the Horti Caesaris—someway south of the Porta Portuensis. This implies a simple point relating to movement in the city: the through-movement potential of the Forum Romanum meant that moving to it and through it in order to reach a site in the far distance was entirely credible.

Horace's text begins by stating that he was on the Sacra Via 'by chance' (*forte*), but that it was customary for him to be so.[38] These two statements indicate that meeting the impertinent hanger-on at that particular time was (un-)fortuitous but that moving along the Sacra Via itself was common. Plutarch gives a similar impression, describing the movement of Tiberius Gracchus, struck by a falling roof tile as he made his way down to the forum. This portent was all the more significant because the tile struck Gracchus rather than one of the 'many people passing by'; a level of traffic

---

[34] Palombi (2005: 25–8); Frank (1936); Gilula (1991).

[35] Ter. *Ad.* 580–1, 'scin Cratini huius ditis aedes? [...] ubi eas praeterieris, ad sinistram hac recta platea'. Dion. Hal. *Ant. Rom.* 1.68.1, τὴν ἐπὶ Καρίνας φέρουσαν ὁδόν and 8.69 as a short cut.

[36] Hor. *Sat.* 1.9. Horace's reaction contrasts with Macrobius' quip that the merry talk of a companion was as good as a lift (Macrob. *Sat.* 2.7.11, 'comes facundus in via pro vehiculo est').

[37] Hor. *Sat.* 1.9.17–18, 'quendam volo visere non tibi notum; trans Tiberim longe cubat is prope Caesaris hortos'.

[38] Hor. *Sat.* 1.9.1, 'ibam forte via sacra, sicut meos et mos'.

that was natural for that street.³⁹ But while Gracchus was moving to the forum, Horace's movement, with no specific destination, is different; his route is determined not by a choice of movement from A to B, but by 'natural movement' through the city, which gravitates to and through the forum.

That Horace was not deliberately moving *to* the Forum Romanum may be further indicated by his use of the generic *ibam* ('I was going') rather than *descendere* ('to descend'); commonly used as a metonym for movement down the Sacra Via to the Forum Romanum as a destination.⁴⁰ Cicero uses *descendere* several times to describe the activities of both he and others, talking proudly of descending to the forum amid a throng of friends; the participation of the Republican elite male in this social ritual, moving through space with attendants, was all the more significant because of the destination.⁴¹ The public nature of moving through the Sacra Via to the forum naturally made it a choice location for political confrontation, with Cicero recommending that such movement be accompanied by bodyguards.⁴² Indeed, one of the most famous anecdotes of life on the Sacra Via is the attack on Cicero by Clodius' supporters, as he was making his way to the forum—'cum sacra via descenderem'.⁴³

While descending the Sacra Via had specific connotations with moving *to* the Forum Romanum, Horace's passage is more closely related to moving *through*. We can say the same of another description which, significantly, also introduces itself with *forte*. Ovid described his return from Vestalia as follows: 'I was by chance returning [home] from Vesta's feast by the way which now joins the Nova Via to the Forum Romanum'.⁴⁴ We know from other texts that Ovid's house was located somewhere in the region of the Capitoline, so the Forum Romanum formed the short cut between the Palatine and somewhere towards the far side of the Arx.⁴⁵ Again, the through-movement potential of the forum is a key determinant of the activities that occurred there (Ovid's

---

³⁹ Plut. *T. Gracch.* 16.3–17.3.
⁴⁰ The equivalent for being forcibly moved to the forum has a similar metonymic quality: *deducentes* ('to be brought down').
⁴¹ Cic. *Att.* 1.18.1, 'cum ad forum stipati gregibus amicorum descendimus'. Those not attending the forum were said to 'not be descending', e.g. Cic. *Phil.* 2.15, 'Hodie non descendit Antonius. Cur? Dat nataliciam in hortis.' To invoke a lack of social standing, see Catullus' criticism of Naso that no man would 'descend' with him to the forum, Catull. 112, 'neque tectum multos homo <est> qui descendit'.
⁴² Cic. *Phil* 8.16, 'consul se cum praesidio descensurum esse dixit'.
⁴³ Cic. *Att.* 4.3.3, 'itaque ante diem tertium Idus Novembris, cum sacra via descenderem, insecutus est me cum suis. clamor, lapides, fustes, gladii, haec improvisa omnia'.
⁴⁴ Ov. *F.* 6.395–6, 'Forte revertebar festis Vestalibus illa quae Nova Romano nunc Via iuncta foro est'. On this passage and its relationship to the topography of the area on the north-west of the Palatine see Hurst (2003, 2006) and Wiseman (2004, 2007). Cf. Knox (2009), who instead argues that Ovid was ascending to the Palatine, not heading to cross the Forum Romanum.
⁴⁵ Ov. *Tr.* 1.3.29–30, 'hanc ego suspiciens et ad hanc Capitolia cernens, quae nostro frustra iuncta fuere Lari'.

route home) and the addition of a new route that joins the forum (*iuncta foro*) has increased this potential and attracted natural movement through this space.

Through-movement potential is also implied in the route Martial described for his book as it made for the Palatine in the mid-80s CE.[46] The origin of the route is unclear, but may be from Martial's house on the Quirinal, or from a bookshop in the Argiletum.[47] He signposts the Aedes Castoris and the Aedes and Atrium Vestae as he directs his work from the forum to the Palatine by way of the Sacra Via. Coarelli had Martial's route turn north, along the street commonly called the Clivus ad Carinas, but this has been sensibly refuted, namely because a route from the Quirinal to the Velia would not pass through the Forum Romanum but would pass through the Subura.[48] In this sense, literary description, natural movement and through-movement potential combine to favour one reconstruction over others.

Those sources that describe movement through the Forum Romanum as a particular perception of space are numerous and chronologically varied, but we can highlight the concentration of such perception in the first century BCE and CE. Late in the second decade CE, the visiting Strabo described the city of Rome in terms of the spatial arrangement of structures within it; a scheme that implies movement from one monument to the next, starting in the northern Campus Martius and progressing towards the centre of the *Urbs*.[49] Strabo's route would thus pass through the Porta Fontinalis and around the Arx by way of the Clivus Argentarius, emerging in the north-east corner of the forum. Such a route, the product of the configuration of space and the natural movement of the city, bypasses the new imperial fora of Caesar and Augustus on the left of the road: 'one forum after another, ranged along the old one'.[50] They are defined according to their spatial relationship with the Forum Romanum but the route passes *by* them rather than *through* them.[51] Later in the century, Martial's route to the Palatine depicted the Forum Romanum in a similar manner but, importantly, it was written a decade before the

---

[46] Mart. 1.70. On Martial and the cityscape of Rome, see Prior (1996); Rodríguez-Almeida (2003); and Laurence, in this volume.

[47] On the location of Martial's home on the Quirinal, see 5.22.3–4 and 10.58.10. That the book originated in the Argiletum is favoured by Geyssen (1999: 723), noting the lack of reference to the imperial fora, of which the fora of Caesar and Augustus existed in the 70s CE. This explains the lack of detail between the Quirinal and the Aedes Castoris, where the description begins. On Martial's books for sale in the Argiletum see 1.3, 1.117, and perhaps 1.2.7–8.

[48] Coarelli (1983: 40–1), critiqued by Ziółkowski (1989: 229–31) and Geyssen (1999: 727 n. 20).

[49] Strabo 5.3.8. See Haselberger (2007) on the distinction between Republican *Urbs* and Augustan Campus. On Strabo's relationship with Augustan Rome, see Dueck (2000: esp. 85–94).

[50] Strabo 5.3.8.

[51] One might say something similar of the description in Ov. *Tr.* 3.1 (cf. Huskey 2006). In terms of movement through the city the next point on Strabo's itinerary, the Porticus Liviae, demands a continuation.

Argiletum was transformed by the construction of the Forum Nervae. This would remove one of the principal approaches to the area, and thus diminish its through-movement potential.

CONTROLLING MOVEMENT WITHIN THE FORUM ROMANUM: THE LATE REPUBLIC

Describing the transfer of popular assembly from the Comitium to the forum by C. Licinius Crassus, Varro referred to the 'septem iugera forensia'.[52] It has been noted that the *area* of the Forum Romanum does not fit this description, which was nearer to 1.5 *iugera*.[53] Coarelli revived the emendation of *septem* to *saepta*, implying an enclosed space, with distinct boundaries (from *saepio*, related to *obsaepio* discussed at the start of this chapter).[54] We can briefly consider the issue here as it relates to movement and access to the space of the forum. It has been suggested that small pits (*pozzetti*) around the forum at Cosa served the utilitarian purpose of anchoring ropes, used to divide the open space into smaller segments in order to manage the crowd that assembled there for voting.[55] Similar small pits have been found at various locations around the Forum Romanum, some clearly aligned, dating from the second century BCE, though it is clear that they are not all from the same period. Ropes around the Forum Romanum are known on an ad hoc basis from antiquity, more on which below, and Welch has recently interpreted them as anchoring points for ropes attached to the temporary structures erected for gladiatorial games *in foro*.[56] Taking a different interpretation, Coarelli argued that the pits represented the ritual demarcation of space, thus explaining the notional 'saepta iugera'.[57] However, the interpretations of ritual and practicality are not exclusive, since while serving to ritually demarcate the space of the forum, they also served as the basis for partitions in assemblies of the Comitia Tributa. Relevant to movement from the Comitium, a distinct line of nine *pozzetti* ran in front of the Republican Rostra, and indeed gives the impression of a separation between the north-west (Comitium) and the south-east (forum

---

[52] Varro *Rus.* 1.2.9.
[53] Summarized in Purcell (1995b), from Giuliani & Verduchi (1987: 33–9).
[54] Coarelli (1985: 130 and n. 24).
[55] Brown (1980: fig. 37).
[56] Welch (2007: 36–8). Plut. *C. Gracch.* 12.3–4 says that he allowed the games to be viewed free and had hired seats removed from round the forum, which implies that in normal circumstances there was management of access for the purpose of revenue. Indeed, Vitruvius (5.1.2) says access to the colonnades should be properly managed precisely for this purpose.
[57] Coarelli (1985: 125–31). The issue of such pits has been revived in a lively debate between Mouritsen (2004) and Coarelli (2005), the former arguing that pits across Republican fora are too heterogeneous to be interpreted for a single purpose.

area).⁵⁸ Other *pozzetti* appear in similar 'boundary' locations: beneath the street along the Basilica Iulia, by the Aedes Divus Iulius and just west of the Arcus Augusti, where formerly there had been a street, defining the eastern edge of the piazza.⁵⁹

But what, if anything, was the relevance of these pits for movement in and through the forum? It has been suggested that by the late Republic the forum was 'enclosed' by such *pozzetti* (or rather, by whatever structures were fixed within them).⁶⁰ However, it is clear that any practical function was removed when the forum was repaved, since most of the *pozzetti* were covered by travertine slabs in the Augustan period. By this time, neither games nor voting took place *in foro*, and there were no longer the same needs to partition the open space. The lack of practical necessity for the pits may explain their removal. While they perhaps served the practical function of allowing for the partitioning of space, the *pozzetti* did not form the basis for any physical enclosure of the forum and we should not view them as routine barriers to movement to or through this space. This is clear from their suppression by later roadways. If they did not limit movement in any practical sense, what other restrictions might there have been?

In 57/6 BCE, Cicero described the applause as Publius Sestius made his way from the Columna Maenia into the crowd assembled for gladiatorial games in honour of Q. Metellus Pius.⁶¹ Cicero gives specific detail on the origin of applause within or, as the case may be, at the edge of the forum—*ex fori cancellis*. The notion of a *cancellus* as a physical barrier is clear from other sources.⁶² Pliny described the construction of a latticed fence, similar to the *cancelli* of the theatre, and elsewhere he compares the defensive mechanism of the female polypus to a *cancellus*; spreading out its feelers, 'interlaced like a net'.⁶³ Perhaps the most obvious examples from Rome are the *cancelli* of the Circus Maximus, which closed the *carceres* from where the horses started races; they acted to prevent movement until they were opened.⁶⁴ Cicero is alone in referring to such structure(s) in or around the Forum Romanum.

---

⁵⁸ Van Deman (1922: fig. 1). A more detailed plan of their location has not been produced.
⁵⁹ Lugli (1946: 81–2); Coarelli (1985: 130); Carnabuci (1991: 264 n. 52, and 339, fig. 32).
⁶⁰ Platner and Ashby (1929: 233 n. 4).
⁶¹ Cic. *Sest*. 124, 'venit, ut scitis, a columna Maenia. Tantus est ex omnibus spectaculis usque a Capitolio, tantus ex fori cancellis plausus excitatus, ut numquam maior consensio aut apertior populi Romani universi fuisse in causa diceretur'.
⁶² e.g., Amm. Marc. 30.4.19 on the permeable *cancelli* of the Saepta (Iulia)—a space that by its very name was considered an enclosure ('cumque intra cancellorum venerint saepta').
⁶³ Plin. *HN* 9.164, 'polypus femina modo in ovis sedet, modo cavernam cancellato bracchiorum inplexu claudit'.
⁶⁴ Descriptions of them clearly reveal their role in closing space, e.g. *claustra*: Stat. *Theb*. 6.399; Hor. *Epist*. 1.14.9. *fauces*: Cassiodor. *Var*. 3.51. *fores carceris*: Ov. *Tr*. 5.9.29. Varro *Rus*. 3.5.4 discusses the design of particular fencing, described as being like that of the barriers between the stage and the theatre. The role of the *cancellarius*—gatekeeper—of the Campus Boarius is known from an inscription, *CIL* VI.9226.

There is no precise identification of where these *cancelli* may have been—all around the forum or at a specific location—or whether they were a permanent feature or were simply there in the context of the performance that was occurring. Permanent or not, Cicero is certainly referring to physical barriers, rather than barriers in the metaphorical sense.[65] The detail that Sestius came from the Columna Maenia provides further topographical context to a description rooted in physical space. In Cicero's description, Sestius had moved from the Columna Maenia, and the applause rose up from the Capitoline and the barriers of the forum. This would suggest that the *cancelli* were in the north-east corner of the forum, near the column and in the space between the forum and the Capitoline.

We must remember that Cicero is describing a particular event at a particular time. That this event was the provision of gladiatorial games necessarily complicates matters, since the *cancelli* may have been nothing more than temporary barriers erected for the purposes of crowd control or to separate the spectator space from the makeshift arena floor.[66] However, in *De Oratore*, written in the years before 55 BCE and therefore broadly contemporary with the events described in *Pro Sestio*, Cicero makes another reference to the *forensibus cancellis*.[67] The context is not *spectacula* but, seemingly, routine business in the forum. Indeed, later *cancelli* gave their name to legal scribes who worked within the forum, behind the *cancellus*.[68]

The stationing of guards around the forum in order to limit access is in evidence in a number of sources, such as Asconius' commentary on Cicero and the stationing of Pompey's guards 'in the forum and all the approaches around it'.[69] Cicero himself vividly described the closure of the approaches to the forum:

> First all the entrances to the forum were fenced off, with the result that, even if no armed guard stood in the way, nevertheless it was not possible in any way to enter into the forum unless the barricades were torn away; and there were indeed guards stationed around, so that, just as an enemy's access to a city is hindered by

---

[65] For which, see Cic. *Quinct*. 36.

[66] That *cancelli* could be easily added or removed is later implied in the *Digest* (*Sab.* 30.41.10), on property that could be bequeathed in order to settle a debt. The code states that things cannot be bequeathed if it means they have to be detached from the building. *Cancelli* and awnings are excluded from this, because they are easy to recover ('Sed si cancelli sint vel vela, legari poterunt'). See also *Dig.* 33.7.12.26 in which *cancelli* are considered movable *instrumenta* rather than an integral part of the structure of the house.

[67] Cic. *De Or.* 1.52.

[68] In the 6th century CE, Cassiodorus (*Var.* 11.6) provided the most detailed description of the duties and circumstances of the *cancellarius*, keeping inferiors in their proper place from behind the *cancelli* of his compartment ('latere non potest quod inter cancellos egeris'). Although the text is later than the direct concerns of this chapter, Cassiodorus implies that the role and its physical circumstances are much older ('vide quo te antiquitas voluerit collocari').

[69] Asc. *Mil.* 41, 'praesidia in foro et circa omnis fori aditus Pompeius disposuit'.

towers and fortifications, so too you would see the people and tribunes of the plebs pushed back from entry into the forum.[70]

Movement, or rather the restriction of movement, is given central importance here. Such examples include armed guards controlling movement, but given their power in numbers it was equally possible for the crowd to block access, either through the erection of barriers or by their own physical presence. When Dolabella, in 47 BCE, proposed some unfavourable laws regarding debts and property rents, the crowd are said to have 'erected barriers around the forum, setting up wooden towers at some points'.[71] When Vitellius, in 69 CE, tried to relinquish power at the Rostra and then return home, he found that every path out of the forum had been blocked (*interclusum aliud iter*), except for the Sacra Via, and so he made his way back to the Palatine and back to power.[72]

Barricading the forum through the use of a rope was not uncommon, and seems to be the quickest and easiest way in which people could take control of the space. Dionysius tells of a great gathering that occupied the forum, from which the tribunes summoned the relevant citizens after dividing them into tribes through the use of rope barriers.[73] Appian records something similar in the late Republic when the supporters of Antony had roped off the forum during the night, much to the surprise of the Senate.[74]

Octavian, we are told, stood by the rope during this time. We can identify an increasing habit of restricting space and movement in the late Republic and under Octavian/Augustus. For instance, the Lacus Curtius—*in medio foro*—was fenced off from the wider *area*.[75] Varro says that the site was struck by lightning in 102 BCE, and was enclosed by decree of the Senate.[76] Coarelli has argued that the famous relief of Curtius' horse plunging into the *lacus* would have been part of the balustrade that fenced off the area from the wider

---

[70] Cic. *Phil.* 5.4.9, 'Primum omnes fori aditus ita saepti ut, etiam si nemo obstaret armatus, tamen nisi saeptis revolsis introiri in forum nullo modo posset; sic vero erant disposita praesidia, ut, quo modo hostium aditus urbe prohiberentur castellis et operibus, ita ab ingressione fori populum tribunosque plebi propulsari videres.'

[71] Cass. Dio 42.32.3, ὡς οὖν τοῦτό τε προεπήγγελτο καὶ ὁ ὄχλος τά τε περὶ τὴν ἀγορὰν ἀποφράξας καὶ πύργους ἔστιν ᾗ ξυλίνους ἐπικαταστήσας ἕτοιμος παντὶ τῷ ἐναντιωθησομένῳ σφίσιν ἐπιχειρῆσαι ἐγένετο.

[72] Tac. *Hist.* 3.68. See also the Flavians seizing the Area Capitolina in 69 CE, and barricading themselves within by piling up statues at its entrances, Tac. *Hist.* 3.71, 'in ipso aditu vice muri obiecisset'.

[73] Dion. Hal. *Ant. Rom.* 7.59.1, χωρία τῆς ἀγορᾶς περισχοινίσαντες, ἐν οἷς ἔμελλον αἱ φυλαὶ στήσεσθαι καθ' αὑτάς.

[74] App. *Bell. Civ.* 3.30, ἐλθούσης δὲ τῆς κυρίας ἡμέρας ἡ μὲν βουλὴ τὴν λοχῖτιν ἐνόμιζεν ἐκκλησίαν συλλεγήσεσθαι, οἱ δὲ νυκτὸς ἔτι τὴν ἀγορὰν περισχοινισάμενοι τὴν φυλέτιν ἐκάλουν, ἀπὸ συνθήματος ἐληλυθυῖαν.

[75] On its location *in medio foro* see Livy 7.5, Plin. *HN* 15.20, Dion. Hal. *Ant. Rom.* 11.42.6, 14.11.20–1; Val. Max. 5.6.1.

[76] Varro *Ling.* 5.150, 'ex S.C. septum esse'.

forum.⁷⁷ Intriguingly, the outline of the Augustan *lacus*—that is, the irregular polygonal area of travertine paving (above earlier layers in cappellaccio and tufa)—occupies a space that fits precisely within the gaps between the late Republican galleries that run beneath the forum *area*. This may be because the galleries were designed so as to avoid the shrine, above, or it may more likely be because the shape of the *lacus*, and thereby its enclosure, were post-Caesarian. Varro may place the decision to enclose the site at the end of the second century BCE, but the actual enclosure is arguably late Republican or Augustan. Also dating to the late Republic are coins showing the Sacrum Cloacinae with a *cancellus*, the remains of which have been identified in the forum and which do indeed have gaps for the fitting of a metal gate.⁷⁸ These two monuments might suggest, following the evidence presented above on the enclosure of the forum, that the enclosure of individual sites within it was becoming increasingly common in the late Republic and early imperial period. Away from the Forum Romanum, Strabo notes how the Mausoleum of Augustus, on the Campus Martius, was surrounded by a circular iron fence.⁷⁹

Finally, we see changes to movement accessibility also reflected in one of the most conspicuous late Republican constructions in the forum—the Aedes Divi Iuli. This temple defined the eastern limit of the forum when it was constructed in the late first century BCE, and it is an interesting example of the modification of space and spatial practice.⁸⁰ It was constructed on the site where the people had cremated Caesar's body after his assassination.⁸¹ The temple is at an earlier location of particular significance; a *locus* of popular gathering, and is thus interesting in terms of movement and congregation in the forum. The location had accumulated numerous monuments in the late Republic: the equestrian statue of a toga-clad Q. Marcus Tremulus, who twice conquered the Samnites, was erected *ante aedem Castorum*, and Lucius

---

⁷⁷ Coarelli (1985: 126 and fig. 41).

⁷⁸ The coins depicting the shrine date to 39 BCE. See Livy 3.48 on the location near the Basilica Aemilia. Nash (1968: i.262-3) has images of both the coin and the marble remains. See also the enclosure of an *angiportum* halfway up the Clivus Capitolinus, in which refuse from the Aedes Vestae was deposited every 15 June and which was closed by a gate; Varro *Ling.* 6.32; Fest. 466L, 'Stercus ex aede Vestae XVII Kal. Iul. defertur in angiportum medium fere clivi Capitolini, qui locus clauditur porta stercoraria'.

⁷⁹ Strabo 5.3.8, κύκλῳ μὲν περικείμενον ἔχων σιδηροῦν περίφραγμα.

⁸⁰ On the *Regia* as the earlier limit, see Serv. *Aen.* 8.363, 'Regiam [. . .] in radicibus Palatii finibusque Romani fori esse'; Coarelli (1983: 68). See now Sumi (2011).

⁸¹ There is no indication that the forum was blocked off when Caesar's body was brought there for cremation, but Plut. *Caes.* 68.1 notes, in fact, that railings (κιγκλίδας) as well as benches and tables were removed from about the forum (he does not say specifically from where) and heaped onto the funeral pyre: Ἐπεὶ δὲ τῶν διαθηκῶν τῶν Καίσαρος ἀνοιχθεισῶν εὑρέθη δεδομένη Ῥωμαίων ἑκάστῳ δόσις ἀξιόλογος, καὶ τὸ σῶμα κομιζόμενον δι' ἀγορᾶς ἐθεάσαντο ταῖς πληγαῖς διαλελωβημένον, οὐκέτι κόσμον εἶχεν οὐδὲ τάξιν αὐτῶν τὸ πάθος, ἀλλὰ τῷ μὲν νεκρῷ περισωρεύσαντες ἐξ ἀγορᾶς βάθρα καὶ κιγκλίδας καὶ τραπέζας.

Antonius had a similar honour in the same space.[82] In this area too a *Senatus Consultum* was erected in 159 BCE, in an area referred to as a *locus celeberrimus*.[83] When considering the pressures on space it may be significant that initially only a column was erected in memory of Caesar.[84] It was not until the construction of the temple under Augustus that the space was fundamentally altered. This presents a key theme, to be discussed in more detail below: while the public erected a column beside the earlier road, the ruling power built a temple over the road.

From Cassius Dio, we learn that when it was first built, the temple had the right of asylum.[85] Yet less has been said about what is arguably a more intriguing aspect of this right, that it was rescinded (Dio does not give a date for this but implies that it happened relatively quickly, as soon as men began to congregate there). Dio goes on to note that after men began to congregate in that region, the 'right' of asylum was revoked, existing in name but not in any practical, physical sense: 'for it was so fenced about that no one could any longer enter it at all'.[86] Marruchi speculated that the front of the temple was separated from the forum by means of a marble barrier, and, writing in 1906, noted that 'not far from here are still to be seen some remains of this'.[87] Dio is not clear on exactly *what* was fenced off—the altar and niche before the temple, or the temple itself. In either case, it was evidently a physical intervention, with a *terminus post quem* of 29 BCE. Thus the enclosure of the space and the denial of access can be dated to the Augustan period or after. This fits with our wider chronology for the alteration of space in the forum, and is an example of restrictions on movement that are built rather than simply decreed. We see the move from regulation by cultural *habitus* to regulation by architecture.

While recognizing the continued importance of the Forum Romanum as a permeable space, we can see restrictions on space in the late Republic which were formalized under Augustus. The through movement characteristic of the Forum Romanum would mean that enclosing the space entirely would be a difficult and unpopular act. Rather we see the emergence of isolated controls aimed at reducing accessibility and increasing control over movement. The control of movement within the public space of the forum began for

---

[82] Q. Marcus Tremulus: Plin. *HN* 34.23–4, 'ante aedem Castorum'; Livy 9.43.22, 'quae ante templum Castores posita est'. Lucius Antonius: Cic. *Phil.* 6.13, 'in foro [. . .] ante Castoris'.

[83] *ILLRP* 512, 'sub aedem Castores'. On the Aedes Castoris as a *locus celeberrimus*, see Cic. *In Verr.* 2.1129, 2.5.186.

[84] The column was inscribed 'Parenti Patriae' (Suet. *Iul.* 85) but was quickly removed by Dolabella (Cic. *Phil.* 1.5). Sumi (2011: 209–1) on the movement of Caeser's body in the forum.

[85] Cass. Dio 47.19.1–2.

[86] Cass. Dio 47.19.2, οὕτω γὰρ περιεφράχθη ὥστε μηδένα ἔτι τὸ παράπαν ἐσελθεῖν ἐς αὐτὸ δυνηθῆναι.

[87] Marucchi (1906: 83).

temporary purposes in rope but later became permanent in travertine. What we see is the emergence of an urban disposition that recognized the cultural and political significance of enclosing public space and restricting movement. At precisely this period we find the imperial fora developing to the north-east of the Forum Romanum. Here we can recognize distinct differences in terms of through-movement potential and accessibility. These new fora were designed environments, rather than modifications to existing, organically developed spaces. As such, the imperial fora reflect the contemporary urban disposition and are physical manifestations of a cultural logic of space: not only reshaped movement in space but also, therefore, reshaped interactions in society.

THE IMPERIAL FORA AND LIMITED ACCESSIBILITY

The overall history of the imperial fora in relation to movement and accessibility can be summarized as follows: the Julio-Claudian fora of Julius Caesar and Augustus lie alongside existing thoroughfares, and neither form routes nor substantially alter them; the changes under Domitian remove existing thoroughfares and limit access. The archaeological data underpinning this narrative is complex and, for the sake of brevity, this section provides only the most pertinent evidence in support of the broader theme of changes to the infrastructures of movement and accessibility.[88]

The Forum Iulium was built on land bought from private ownership to the north-east of the Forum Romanum, extending 'usque ad Atrium Libertatis'.[89] The contentious location of the Atrium Libertatis need not further divert us here, but the reference reveals how Caesar's project was conceptualized in terms of *existing* spatial boundaries.[90] In terms of streets, the closest to the Forum Iulium was the Clivus Argentarius, which led from the Forum Romanum to the Campus Martius. The importance of this route around the Capitoline hill was demonstrated by the construction of a colonnade—an architectural space for movement—linking the Porta Fontinalis to the altar

---

[88] For specific archaeological details, see the synthesis in Newsome (2010: 174–249).
[89] Cic., *Ad Att.* 14.16. The most detailed survey prior to the recent excavations was Amici (1991). Ulrich (1993) provides important criticisms of earlier reconstructions, while Westall (1996) is an important discussion of the representative power of the new forum.
[90] Defining the expansion as far as a building rather than a street gives some indication of the standard topographic signifiers in Cicero's Rome. Something similar could be said of Augustus' later comment in the *Res Gestae* (20), in which the Basilica Iulia is presented as 'inter aedem Castoris et aedem Saturni', rather than the more topographically specific 'inter vicum Tuscum et vicum Iugarium'.

of Mars in the Campus Martius, constructed by M. Aemilius Lepidus and L. Aemilius Paulus in 193 BCE.[91]

Von Gerkan suggested that the Aedes Venus Genetrix was built on top of an existing gateway and a street which ran directly north from the Clivus Argentarius out of the Forum Romanum.[92] Ulrich disagreed on the details but likewise suggested that the staggered nature of the apses to the west of the podium represented the impediment posed by a pre-existing structure, and conjecturally mapped a northbound extension of the Clivus Argentarius at this point.[93] If there was not a street running at the point near the podium of the temple then the building ran exactly 'usque ad Atrium Libertatis' and fitted in accordingly. Alternatively, the complex extended as far as the line of a street that cut north from the Clivus Argentarius. In either reconstruction, the Forum Iulium goes to great lengths to mask its inability to remodel these external spaces by concealing its irregularity behind staggered apses.

Excavations in 2006 revealed that the course of the Republican Clivus Argentarius was changed during the Caesarian constructions, to follow the orientation of the new project and no doubt this was related to the terracing of the slope for the creation not only of the new forum but also of the *tabernae* on the slope of the Arx.[94] The excavators call this an eloquent testimony to the modifications and state accurately that the Caesarian project cut through the existing street. However, we can see that any such modifications did not suppress the street but reoriented it. The change is not prohibitive to movement, and if it is, it was a change that presented a remedy (the new orientation of the street). The Forum Iulium may not have fitted seamlessly into the earlier street network, but nor did it *remove* any streets in the surrounding area. In this regard, it would not have changed the through movement potential of the Forum Romanum, with both the Clivus Argentarius and Argiletum still accessible.

The same basic approach to existing infrastructures of movement can be discerned with the Forum Augustum. Ostensibly, this forum was constructed because of pressures on movement, or at least increased demands on litigation, in the Forum Romanum.[95] One of the most frequently cited passages relating to Augustan building projects states that the new emperor built his forum to a

---

[91] Livy 35.10, 'aedilitas insignis eo anno fuit M. Aemilii Lepidi et L. Aemilii Pauli: multos pecuarios damnarunt; ex ea pecunia clupea inaurata in fastigio Iouis aedis posuerunt, porticum unam extra portam Trigeminam, emporio ad Tiberim adiecto, alteram ab porta Fontinali ad Martis aram qua in Campum iter esset perduxerunt'.

[92] Von Gerkan (1940: abb.13–14).

[93] Ulrich (1993: 63, 69, fig. 1).

[94] See Meneghini & Santangeli Valenzani (2007: 32 and fig. 17). The Republican street, with a sewer lined with tufa, skirted the Capitoline in the 2nd century BCE.

[95] Suet. *Aug.* 29.1, 'Fori exstruendi causa fuit hominum et iudiciorum multitudo, quae videbatur non sufficientibus duobus etiam tertio indigere'.

smaller size than he would have liked, and that this act of constraint was responsible for the peculiar layout of the precinct. This is particularly clear at the north-east, behind the Aedes Martis Ultoris, where the perimeter of the forum was conditioned by the existing street that ran from the Quirinal towards the Argiletum. Suetonius provides the most well-known statement on Augustus' forum and its relation to existing urban space: 'he built his Forum narrower' because he was unable to dispossess local residents.[96]

Wallace-Hadrill has suggested that the spatial restraints on the Forum Augustum do not show Augustus' inability to remove individual owners, but his unwillingness to do so because this demonstrated his *civilitas* as *pater urbis*: 'we are not looking at the limits of imperial power, but its self-definition and self-representation'.[97] This involves the respect of citizen and property rights, and infrastructures of movement are similarly spared any significant alteration or replacement. Accordingly, the limits of the Forum Augustum, which must have been visible to anyone walking along the streets outside, can be read as *civilitas* inscribed on the city plan. However, in terms of how it related to movement and traffic, the Forum Augustum was not accessible to vehicles and was divorced from the streets whose course it so assiduously respected. Regarding the routes from the direction of the Quirinal and the Subura, these entered the forum either side of the Aedes Martis Ultoris. That these were designed solely for pedestrian use is evident from the deep flights of stairs. No other monumental entrance has so far been discovered, and the connections with other fora did not incorporate wider through movement from the city streets.

Cassius Dio's account of the dedication of the Aedes Martis Ultoris in the Forum Augustum is interesting in considering patterns of movement to and through the new forum, because the uses of the space imply a change in function from previous fora, which relate also to access.[98] Amongst other details, Dio includes the following: 'that [Augustus] himself and his grandsons should go there as often as they wished'.[99] Dio's statement is significant because of the implication that the free will of attendance for Augustus, Gaius, and Lucius is worthy of record. This, presumably, records a special dispensation granting the freedom of unqualified access to those named in the decree. This records the discrepant accessibility of this space and underlines the notion that going there as often as one wished was not possible to all. The restrictions on access to the Forum Augustum help us to understand why the Miliarium Aureum was not located there. Instead, it was placed in the Forum

---

[96] Suet. *Aug.* 56. 'Forum angustius fecit non ausus extorquere possessoribus proximas domos'. Here, *angustus* might better be translated as 'less spacious', since the Forum is not disproportionately narrow.

[97] Wallace-Hadrill (1982; 2003: 194). Zanker (1988: 155–6) and Geiger (2008: 57) agree.

[98] Cass. Dio 55.10.1–5.

[99] Cass. Dio 55.10.2.

Romanum, a place of through-movement potential and a space perceived as being that where the roads of Italy intersected.[100] This choice represents the recognition that those roads did not intersect the new imperial fora, neither physically nor in the perception of that space. It was a different kind of space, with a different relationship to the city and, by extension, to Roman Italy.

We have noted that Strabo described moving to the Forum Romanum from the Campus Martius, and he passed by the new fora to the side of the road. This positions them alongside movement; they were not themselves used as routes between other disparate spaces, as the Forum Romanum was, nor were they integrated into wider strategies of distance minimization or through-movement potential. In terms of the changes brought about by the new fora in the Julio-Claudian period, we can characterize this not as imposition but as negotiation and adaptation to the existing infrastructures of movement. We see properties replaced with new monumental spaces, but the major thoroughfares from the Republican period—namely the Clivus Argentarius and the Argiletum—survive any Julio-Claudian interventions. It is not until the Flavian period that we see significant changes in this area which adversely affect the through-movement potential of the surrounding areas. However, the imperial fora of Caesar and Augustus represent the first step in this process, by removing these monumental spaces and positioning them alongside—and detached from—wider urban movement. Moreover, their design corresponds to the kind of porticoes discussed by Macaulay-Lewis in her chapter in this volume: their arrangement befits their role as destinations rather than routes; they are places of movement to, not movement through.

In the Flavian era, and returning to the theme of movement in texts, we can note the way in which, in the late 90s CE, Martial contrasted the location of a quadrifrontal monument to Ianus with that of its predecessor.[101] Although the location of the monument has not been satisfactorily determined, it was likely somewhere in the location of the street of the Argiletum. Martial speaks to the new shrine: 'formerly you lived on a passage in a tiny dwelling, where Rome in her crowds trod the thoroughfare. Now your threshold is encircled by Caesar's gifts, and you number as many fora, Ianus, as you have faces'.[102] We can note the distinction between its former and latter states. Before the 90s CE, the shrine had been on a route, a particularly busy one ('plurima qua medium Roma terebat iter'). The position of the Argiletum in the natural movement of

---

[100] On the Miliarium Aureum see Mari (1996). It was erected in 20 BCE to commemorate Augustus' inauguration of the office of the *cura viarum* (Cass. Dio 54.8.4), and its location at the intersection of movement is described by Plut. *Galb.* 24.4. Newsome (2009b) discusses this in more detail than is necessary or practical here.

[101] Mart. 10.28.

[102] Mart. 10.28.3–6, 'pervius exiguos habitabas ante penates, plurima qua medium Roma terebat iter: nunc tua Caesareis cinguntur limina donis, et fora tot numeras, Iane, quot ora geris'. That the Ianus was quadrifrontal is confirmed by Serv. 7.607.

the city can be inferred from another epigram, describing how his reader would be in the habit of going there, communicated without any specified purpose and assumed without doubt because it was common praxis, recalling Horace's habitual strolling on the Sacra Via.[103] But in the mid-to-late 90s CE, the shrine had been enclosed (*cinguntur*) and, the key inference of the text, was no longer on a busy route trod by the crowds of Rome. The through movement potential of this space—one of the principal approaches to the Forum Romanum—was transformed by the construction of the Forum Nervae and the Porticus Absidata, began by Domitian and completed by Nerva. The Porticus Absidata is often ignored in studies of the imperial fora, but it is arguably the most conspicuous example of the overall logic of space, manifest in one relatively small construction.[104]

It is clear that the Porticus Absidata occupied the space where several pre-Flavian roads joined. The image of the *porticus* as the monumental node to which all streets in the local area gravitated is arguably true but this does not mean that it simply funnelled movement and traffic into the area behind it. One might envisage the Porticus Absidata to be the principal, monumental entrance to the zone of the imperial fora. On the other hand, evidence suggests that we might consider it a monumental barrier.

To Heinrich Bauer—the principal investigator of the area in the 1970s and early 1980s—movement through the Porticus Absidata would be part of a continuing crescendo of architectural ornament as one moves to the imperial fora, finally arriving in the piazza of the Forum Nervae by the Temple of Minerva. Accordingly, the form of the monument is not related to its function, and it is interpreted merely as a showpiece, signalling the entrance to the monumental zones beyond. Stated definitively: 'Si può tralasciare la questione della funzionalità dell'edificio [...] la forma di questo monumento non è determinata da funzioni pratiche'.[105] How accurate is this? We might argue, instead, that in terms of patterns of movement to and through the imperial fora, the form of the Porticus Absidata is inseparably related to its function.

Where the *porticus* met the street there is evidence for low marble steps between the two, through which access to the centre of the *porticus* was granted. Bauer's reconstruction carries this step across the entire width of the *porticus*, a measure which, if accurate, reflects the segregation of vehicle traffic from the busy streets onto which it faced. Furthermore, while unable to reconstruct a wall around the central space, Bauer noted that three holes in the marble stylobate, *c*.50 cm apart between the central columns in the north-west

---

[103] Mart. 1.117.9, 'Argi nempe soles subire Letum'.
[104] The most detailed study to date is Bauer (1983). See also Colini (1934, 1937) in the context of the excavations of the Templum Pacis.
[105] Bauer (1983: 181).

corner of the structure, indicated that provision was made for a gate of some kind—bringing to mind the depictions of *cancelli*, discussed above.[106]

An article by Pierre Gros has recently addressed the Porticus Absidata after years without inquiry, focusing on 'l'ambiguïté du mouvement'.[107] Gros follows Bauer in emphasizing the transitional nature of the Porticus Absidata, from the streets north to the new forum. However, here the similarities end. Where Bauer championed the notion that the Porticus Absidata represented an inviting, monumental entrance, collecting up the roads and filing them neatly into the imperial fora, Gros suggests the opposite: the Porticus Absidata's main function was to *prevent* movement rather than facilitate it. The physical space of the Porticus Absidata, then, was constructed not only to fit into the space left over by the various encroaching structures of the Forum Augustum, Temple of Minerva, and Templum Pacis, but for 'sélections discrètes mais efficaces pour éviter toute "invasion" incontrôlée'.[108] Accordingly, the physical constraints on the space, which control movement along clearly defined and easily controllable routes, create the imperial fora's security checkpoint at the critical juncture between the surrounding space of the *populus* and the imperial space within: 'n'est pas celle de l'accueil, mais celle du filtrage'. From reception to filtering, the role of this space is entirely transformed. Given its location, so too must be our interpretation of the spaces to which it gave access.

CONCLUSIONS: SPATIAL CHANGE AND SOCIAL CHANGE

The physical changes to fora, and patterns of movement to and through them, were the result of changes to the urban disposition and different ways of conceiving of city space. This means that we are not simply describing architectural and spatial change, we are identifying cultural change. The imperial fora allow us to consider the ways in which patterns of movement inscribe power in the space of the city, namely, the design of inaccessibility into ostensibly public spaces: the transformation of the street into the piazza, and the piazza into the precinct. This necessarily creates differential levels of access, a key sociocultural metaphor.

As has been noted, 'only when the central authority of the city can arrogate to itself the power to expropriate can broad avenues be cut through existing fabric'.[109] In this case, for 'broad avenues' we can read 'fora', but the message

---

[106] Bauer (1983: 118). [107] Gros (2001: 137).
[108] Gros (2001: 133). [109] Wallace-Hadrill (2003: 193).

remains the same: we can posit a correlation between the imperial power and the degree to which space was transformed. In short, early developments see the replacement of existing parcels of land (one type of space is replaced with another, within the same footprint), while it is only later that we see streets and wider networks being modified (one type of space is replaced with another, on a different and expanded footprint that cuts across street networks).

Pliny tells of how Cato had wished for the Forum Romanum to be paved with murex stones—which were particularly sharp, presumably in order that they might deter loitering.[110] This may betray a desire to make the space less than appealing—and influence movement within it—but, significantly, Cato the censor can only attempt to dissuade, rather than prohibit. In terms of the first imperial fora, we can refer to the two sources which are most often used to provide context for their development—Cicero's letter to Atticus, on the Forum Iulium, and Suetonius on Augustus, for the Forum Augustum.[111] Among other details harvested by topographers, Cicero notes the extraordinary cost of purchasing land for the new forum, a point he clarifies by adding that they were 'unable' to convince the private owners to sell for less. Later, as discussed above, Suetonius explains the unusual shape of Augustus' forum as a result of his reticence to dispossess local residents.

There may be some politics in Suetonius' explanation. His use of *extorquere*, with its connotations of compulsion (by physical force) and, above all, the suggestion that Augustus *chose* not to exert his influence, is more aggressive and less conciliatory than Cicero's use of *non poterat* (unable to) to describe the negotiations for the Forum Iulium. As such, one can imply that Augustus *could* have forcefully relocated the *possessores*, and changed the infrastructures of movement in the area, but made the decision not to. The use of *(non) ausus*, with its implications of daring, implies that had Augustus forcefully removed the owners of their property, he would have been inviting criticism for developing urban space in such a way that his position could not yet justify. This relates changes to patterns of movement to the emergence of imperial power. This chapter has aimed to outline how movement is meaningful in this context. Examining the history of movement and access in the fora enables us to see movement at its most metaphorical, indicative of a cultural revolution in society, manifest in built space.

---

[110] Plin. *HN* 19.24, 'quantum mutati a moribus Catonis censorii, qui sternendum quoque forum muricibus censuerat'. Isid. *Et.* 16.3.3 noted how murex stones were particularly sharp (*acutissimae*). Pliny's reference to sharp stones of Cato's project (184–3 BCE) may reflect the sensibilities of one now accustomed to walking on travertine (as the forum was so paved in 78–4 BCE, and again in 14 BCE), rather than be taken literally.

[111] Cic. *Ad Att.* 14.16: 'contempsimus sexcenties HS; cum privatis non poterat transigi minore pecunia'; Suet. *Aug.* 56: 'Forum angustius fecit non ausus extorquere possessoribus proximas domos'.

13

Movement, Gaming, and the Use of Space in the Forum

Francesco Trifilò

As a multifunctional public space, the forum was a focus for practices—such as that of erecting honorific statues—which were located in order to achieve maximum visibility and thus establish an essential connection with users of that space.[1] Therefore, the placement of many typical forum-features reflects an explicit acknowledgement of the importance of flows of traffic entering and passing through this public space. Evidence for this type of connection between material culture and forum-users, but hitherto few studies of their spatial context, includes carved game boards. These artefacts have a specific role in informing us about the use of public space in antiquity, because these inscribed boards were user-generated and therefore can reveal how the space of a forum was occupied on a daily basis. Moreover, their placement within the forum is intrinsically tied to visibility, to co-presence, and thereby to movement.

In this chapter, I argue that the location of carved game boards provides significant evidence for the way in which Roman public space was manipulated and redefined by its users. I do so through a combined approach examining both the cultural and spatial context of the games. The first concerns the interpretation of game boards as games of skill, which I differentiate from games of chance. As such, these games possessed higher status in Roman society. This is fundamental for understanding the cultural context of such artefacts. Following this, the second part of this chapter concerns the spatial attributes of recorded game boards found in the Forum Romanum and in the forum of Timgad (Numidia). By comparing their locations within the space of these fora, I demonstrate their crucial connection with

---

[1] Trifilò (2008: 115–17; forthcoming). The author is currently undertaking a field-study of the game boards located in the Forum Romanum in Rome. Regrettably, this could not be undertaken on time for presentation in the present chapter. It is anticipated, however, that many of the arguments treated here will be further developed in forthcoming work by the author.

movement and, consequently, with visibility. Drawing together the two levels of context, I outline the role of spaces for gaming within spaces of movement.

GAME BOARDS IN THE CONTEXT OF FORA

Despite a long-standing interest in the presence of games in the ancient city, only three studies have focused specifically on their locations within public areas. The first of these, published by Thédenat in 1923, includes detailed descriptions of the carved game boards and of their exact location as part of a more general work on the Forum Romanum.[2] Bendala Galán's 1973 article on the carved board games found in the town of Italica in Spain forms the second of these,[3] with the third being a report on carved and portable game boards found in the fort of Abu Sha'ar on the Red Sea coast.[4] With only three studies conducted, we may conclude that this form of evidence has not attracted the interest of archaeologists, even if there is an acknowledgement of its existence, although it needs to be recognized that a carved game board need not always survive *in situ* and might be reused for a variety of purposes—particularly in the north-west provinces. As a result, relatively few sites can provide us with direct evidence, despite the wide distribution of game boards across the Empire. For this reason, analysis will focus on evidence found in the Forum Romanum with comparisons drawn from the forum of Timgad.

Despite knowledge of their presence, particularly on the better preserved pavements of fora of Roman Africa, carved game boards are rarely mentioned, rarer still are they analysed in great detail.[5] One exception is Boeswillwald, Cagnat, and Ballu's (1905) description of the antiquities of Timgad. Here, the authors detail, within the overall description of the forum, the type and location of carved game boards within the town's civic space.[6] The description is complemented by Ballu's separate detailed drawings recording the area of Timgad's forum and the adjacent theatre.[7] However, such detail in the available records is rare. The variations present in the examples will be balanced by concentrating on comparable properties, among which the most important is the relationship to movement and access.

In the forum of Timgad there are five game boards (Fig. 13.1). The first of these is a game of the *duodecim scripta* type (see below). The game is carved

---

[2] Thédenat (1923: 216–21).
[3] Bendala Galán (1973: 263–72).
[4] Mulvin and Sidebotham (2004: 602–17).
[5] Romanelli (1970: 107). Lavan (2008: 206–7, 209) has recently outlined their presence in the public spaces of Sagalassos.
[6] Boeswillwald, Cagnat, and Ballu (1905: 19–21, 27–32).
[7] Ballu (1902).

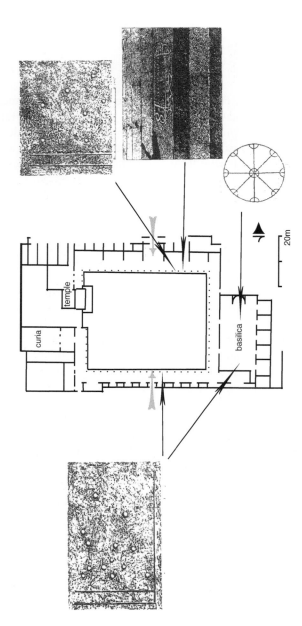

Fig. 13.1. Distribution of game boards within the forum of Timgad. The lack of an original illustration for one of the games in the basilica is compensated with an indication of its type (image/plan by Francesco Trifiló).

near the edge of the step separating the internal *porticus* from the *area* (the paved piazza of the forum), and is located to the immediate left of the forum's main entrance.[8] The game is roughly carved on the pavement and has the following three lines of words forming its core design: *venari, lavari, ludere, ridere, occest vivere.*[9] Located next to the game of *duodecim scripta* and on the same step is a game of *mancala*.[10] Two more games are located inside the basilica. The first is a game of *merels*, located along the central axis of the basilica floor, and the second is a game of *mancala*, also located along the central axis of the basilica, at a distance of 3 m from its southern end. One more game of *mancala* is located on the pavement of the internal *porticus*, in the immediate vicinity of the access route connecting the forum with the area of the theatre (and the road that separates them).

The pattern that emerges links the location of the game boards to the access routes to the forum. The most obvious quality that these access routes have was that they form zones that concentrated or channelled the flow of movement, when compared to other parts of the open space of the forum. Placing these game boards close to routes connecting the forum with other spaces, or in a 'busy' space such as the basilica, situated these games within places of social encounter, as opposed to places of privacy or segregation.

A significant parallel to the forum in Timgad is to be found in the Forum Romanum in Rome. Here, a larger number of carved game boards have been recorded and offer a further important insight into the role of gaming in fora. Thédenat's plans include the precise location of carved game boards spread throughout the surviving horizontal surfaces of the forum and of its buildings (Fig. 13.2).[11] The games consist of all the main known types: *ludus latrunculorum, merels, mancala*, and *duodecim scripta*. Given their state of preservation and greater numbers, I shall focus on three game clusters. The first is represented by the games on the surfaces of the Basilica Iulia. The second is that of the games on the pavement of the *area* of the Forum Romanum. The third is a cluster of game boards located on the side passageways of the Arch of Septimius Severus.

In the Basilica Iulia, the game boards were clearly concentrated under the building's *porticus* and on the steps connecting it to the forum's *area* and to the Vicus Tuscus. Equally interesting is the fact that if the floor of the basilica is divided into areas according to game clustering; those closer to the connection between the forum and the roads leading to it appear to contain a greater number of carved game boards. On the other side, where the basilica's *porticus* is externally enclosed by a line of rooms, there are proportionately fewer carved

---

[8] Boeswillwald, Cagnat, and Ballu (1905: 19–21).
[9] 'Hunting, washing, playing, laughing, this is living'.
[10] Boeswillwald, Cagnat, and Ballu (1905: 27–32).
[11] Thédenat (1923: 216–24).

316   Francesco Trifilò

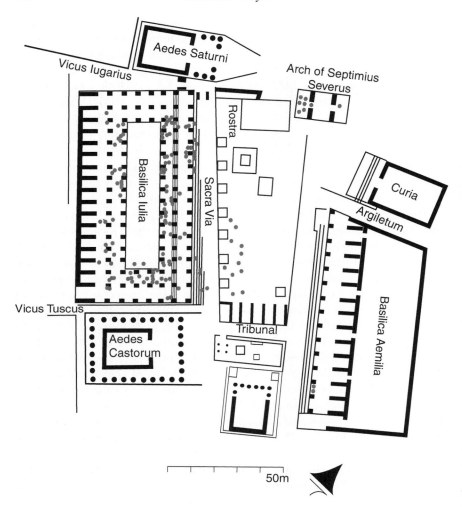

Fig. 13.2. Map of the carved game boards recorded in the space of the *Forum Romanum* (after Thédenat 1923).

game boards concentrated in four very small clusters.[12] The distribution of game boards on the Augustan travertine pavement of the *area* is very similar to the one found in the Basilica Iulia. It consists of only one cluster of games, concentrated towards the southern corner of the *area*, along the Sacra Via.[13]

---

[12] The complete absence of carved game boards on the side of the basilica, which is alongside the Vicus Iugarius, may depend on the degree of pavement preservation. I have been unable to determine this with any certainty.

[13] H. Thédenat does not address the problem of the relationship between the games and the columns placed along the Sacra Via by Diocletian. This is a problem linked to the dating of the

The third cluster, located on the surface of the side walkways of the Arch of Septimius Severus, has six board games on the pavement of the western walkway, compared to a single one found on the pavement of the eastern walkway.[14]

The similarities present in the distribution of these artefacts in the space of the two fora are crucial indicators of their role as evidence for the use of public space in antiquity. However, a full understanding of the significance of their presence requires a full understanding of their meaning as markings carved on the surfaces of buildings.

CARVED MARKINGS AND THEIR STUDY

Markings on buildings are commonly found in both public and private contexts within Roman cities. These markings are, as a rule, incorporated into studies of graffiti, a category that has become, in time, a definition for markings *tout court*.[15] Graffiti are commonly distinguished as being traces placed on surfaces not originally meant to receive them. As with painted markings, graffiti turn any surface 'into a surface for writing'.[16]

Research into the spatial properties of graffiti in the Roman world has been undertaken by Mouritsen and Laurence respectively, although neither with a specific focus on the conceptual implications of their role in the urban landscape. Mouritsen's work on the electoral graffiti in Pompeii is part of a wider project on the epigraphy of the ancient town and concentrates on the distribution and form of electoral notices located throughout the city.[17] By observing these electoral notices, Mouritsen was able to describe both the circumstances surrounding their creation and the lives of those responsible for their production. In so doing he found that candidates for magistracies posted electoral notices in areas of intense activity, which, by definition, meant increased visibility. In addition to this, he also noticed that each candidate selected a specific area of the town in which to advertise his presence in the election. Mouritsen attributes this to the existence of a constituency-like

---

game boards. In fact, if these are carved before the end of the 3rd century CE, their relationship with the columns is only consequential to the placement of the columns themselves. It is difficult to accept that the games were carved after the columns were placed in the *area*. The likelihood is that the carved game boards reflect earlier arrangements of the *area* that we can only define as Augustan or post-Augustan. This does not, in any case, substantially alter the relationship of the game boards with the Sacra Via and the southern end of the forum. See Thédenat (1923: 266).

[14] Thédenat (1923: 235).
[15] *The Shorter Oxford English Dictionary*, i (1983: 877).
[16] Plesch (2002: 168).
[17] Mouritsen (1988: 44–60).

subdivision within the town.[18] He also extends his interpretation to include the surfaces on which the electoral notices were painted. This results in his suggestion that the surfaces of private buildings were seen as public spaces and that this sort of activity was seen as an essential part of civic life.[19]

The graffiti of Pompeii were also used by Laurence to study street activity and interaction.[20] His work combined the calculation of the occurrence of doorways in streets with that of graffiti. These are related together as variables in an equation connecting them to street length. By using doorways as indicators of increased street activity, Laurence was able to produce a plan of the busiest streets in the town. By combining analyses, Laurence generated what appears to be a contradictory pattern in which the frequency of graffiti is inversely proportional to the frequency of doorways. Laurence explains this anomaly in two ways. First, the fewer doorways in a street, the more wall space there is available to write or draw upon. Second, while the greater frequency of doorways in a street relates to greater interactive activity (namely exchanges between residents), the greater presence of graffiti on streets with fewer doorways reflects movement through the streets (the main actors of which are non residents). In Laurence's words, 'Unlike the previous examination of doorway occurrences, this methodology highlights the activity of people who used a particular street, but who did not necessarily live in that street'.[21] This links such graffiti to movement *through* space.

This exploitation of the analytical potential of graffiti in the context of the urban landscape of Pompeii is a significant acknowledgement of their role in the Roman city and of their role as data for the scholar. Both of these works underline the fundamental spatial (movement) properties which this class of evidence possesses. Despite Mouritsen's attempt to normalize the space of graffiti by defining the surfaces of private buildings as perceived public space (and thus somewhat consolidating their location within accepted boundaries), Laurence's analysis proves them to possess extremely dynamic spatial properties, primarily linked to street activity and interaction.

The conceptual definition of this fundamental role of graffiti in the ancient world is not unique, but is seen also in art-historical research, and in one particular study, that of the church of San Sebastiano ad Arborio.[22] Here, the analysis of graffiti carved on the painted walls of the church has outlined the importance of this practice as well as inserting it in a conceptual framework of great relevance to the present study. The graffiti consist of phrases carved on selected parts of the church, and in particular on painted body parts of saints of local relevance. Their creation is the result of a cumulative writing process described as the product of a process of *cultural appropriation* of the church on

---

[18] Mouritsen (1988: 51–6).
[19] Also since these electoral postings are 'official looking'. See Mouritsen (1988: 58 ff.).
[20] Laurence (2007: 109–13).   [21] Laurence (2007: 111).   [22] Plesch (2002: 167–97).

the part of the community of Arborio: a process born out of the interaction of the church and its paintings with the carved inscriptions of Arborio's inhabitants.[23] This reciprocity found in this church also emerges in the conclusions reached by both Mouritsen and Laurence. Mouritsen does not limit his interpretation to the definition of marked surfaces as electoral space, but takes it further by defining the marked surfaces as defining the activity of marking itself. He demonstrates this through the suggestion, mentioned above, that the placing of electoral notices on the external walls of private dwellings appropriated these private structures as part of public space. Laurence draws a similar conclusion by defining streets as elements that form networks for movement or social interaction. Learning from the spatial process of appropriation associated with graffiti in Pompeii, we need to be aware of the role of the carving of game boards in the appropriation and redefinition of space.

Such awareness outlines the importance of activities, which are hard to detect archaeologically, in defining space such as that of fora. In this most essential space of the Roman city, the established role of architecture, statuary, and commerce in defining activity areas is relatively obvious and well studied. The same cannot be said for carved markings, which need to be reasserted in this role. In the remainder of this chapter, carved game boards will be explored in their cultural context of gambling and in their spatial context within fora.

IDENTIFYING PLAYED GAMES: GAMBLING AND REGULATION

Gaming in the Roman world took a variety of forms: some games were played on boards, some with dice, and some with knucklebones. These games would have been played by two or more opponents and would have involved varying degrees of skill and chance. Although gambling was prohibited by Roman law and generally frowned upon, gambling is a key context in which these games were often played. As such, these types of games are generally thought by scholars to have taken place in enclosed spaces such as *cauponae* (inns) and *popinae* (bars).[24] However, detailed analysis of the location of carved game boards contradicts this view, as game boards consistently appear in highly visible and movement-intense contexts such as fora, theatres, and streets. Before exploring this apparent contradiction in greater detail, it is necessary

---

[23] The difference between categories is conceptual, as Plesch and Ashley refute the preponderance of hegemony as a driving premise. In their opinion, processes of cultural appropriation take place in a context of exchange and creative response that is continuous and always of equal influence and importance (Plesch and Ashley (2002: 1–15)).

[24] Laurence (2007: 92–101).

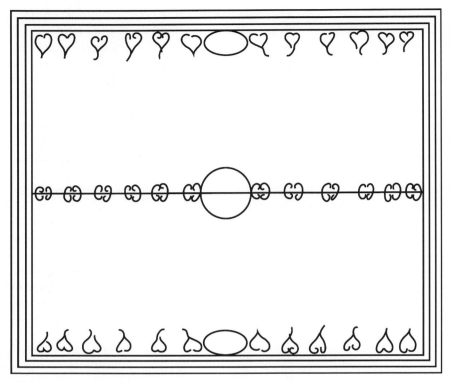

Fig. 13.3. Carved *duodecim scripta* game board (image/plan by Francesco Trifilò).

to understand the nature of the games the boards represent and the relationship these have with gambling and Roman law.

There is no single agreed system for the classification of Roman game boards. Some of the existing classifications emphasize the appearance of the carved games as their distinguishing characteristic, while others base their classification on rules of the games. This lack of uniformity is further complicated by the variety of systems used to interpret those rules. While literary sources have been extensively used for the interpretation of better known games such as the *ludus latrunculorum* (see below), the interpretation of most Roman games relies entirely on the physical evidence. As a result, some games are satisfactorily understood, while our understanding of others is limited. Where we have no literary sources to rely on, contemporary ethnological parallels form the basis of our interpretation of these games. By understanding the rules of play of each one of them, we can place game boards firmly in the landscape of Roman games at large. What follows is a list of known game types selected on the basis of their appearance. The list broadly follows the

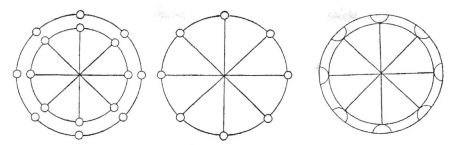

Fig. 13.4. Three variations of the *merels* game board (image/plan by Francesco Trifilò after Falkener 1892).

classification given in the *Dizionario epigrafico di antichità romane*, the basis for all subsequent interpretations of game rules.[25]

The first game type is formed of three parallel lines (Fig. 13.3). These lines are divided along their centre by a symbol or drawing. The resulting six lines are subdivided into six smaller units that may consist of letters forming words or by graphic symbols. If the lines are written, the resulting six words together normally form a motto linked to the spirit of gaming.[26] The letters also form the squares of the game board. This type of game board is widely interpreted as being the game of *duodecim scripta*, probably played in a similar way to backgammon today.[27] Another game type consists of a circle crossed through its centre by a variable (but always even) number of intersecting lines (Fig. 13.4). The game's design can be complicated by the addition of further concentric circles or by holes placed at the intersections between the circles and the intersecting lines. Another version of this game consists of a square (or rectangle) crossed at the centre by intersecting lines. This game is commonly thought to have been played as the game of *merels* is today.[28] A third type of game board consists of a variable number of holes (board cups) placed next to two parallel lines or a rectangle (Fig. 13.5). The game may have been similar to the contemporary African game of *mancala*, in which each player moves pieces in the board cups in order to capture all of the opponent's ones. This game does not seem to have involved the

---

[25] Montesano (1964–85: 2232–6, 2238) and Mulvin and Sidebotham (2004: 609–11).

[26] The presence of words has meant that these game boards are normally better recorded than the non-worded ones. This has been done in the context of epigraphy and has resulted, for example, in catalogues which largely sacrifice the spatial value of the finds in favour of the written content (Ferrua 2001).

[27] Austin (1934: 30–4); Mulvin and Sidebotham (2004: 609–11); Salza Prina Ricotti (1995: 101–2). Montesano is the only author not to make the association of this board type and the game of *duodecim scripta*. In his opinion, the latter should be identified by two known examples consisting of a game with a central cross and groups of twelve lines on its sides. The first of his examples is lost, while the second is of doubted authenticity (Montesano 1964–85: 2232–6, 2238).

[28] Mulvin and Sidebotham (2004: 608).

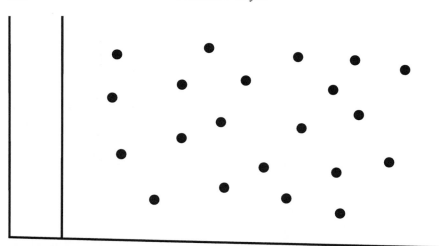

Fig. 13.5. Example of a *mancala* game board (image/plan by Francesco Trifilò).

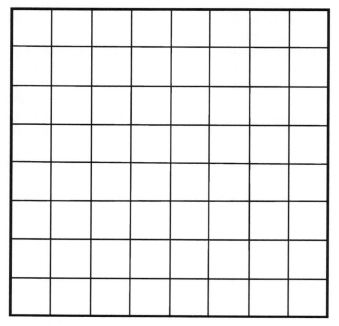

Fig. 13.6. Carved *ludus latrunculorum* game board (image/plan by Francesco Trifilò).

use of dice.²⁹ Finally, there is a game played on a square commonly divided into sixty-four further squares and spaces (Fig. 13.6). This is widely interpreted as the board used to play the game of *ludus latrunculorum*, a game of strategy apparently played without the use of dice.³⁰ A number of ancient sources provide us with descriptions of how the game was played in antiquity.³¹

The games listed above are the ones most commonly found throughout the Empire carved on horizontal surfaces or on portable game boards. As briefly shown in the description of their rules, all are interpreted as being games of skill rather than games of chance. Games of chance would have been the game of dice or that of knucklebones, which relied essentially on fate deciding winners and losers.³²

The distinction between games of skill and games of chance is fundamental to the interpretation of carved game boards, because it reflects a principle found in Roman law. What is legally differentiated is that board games could be played in the absence of gambling, whereas dice and other games of chance had little interest without the possibility of winning or losing money. Hence, board games have a specific milieu and a distinct spatiality that need not coincide with those of games of chance.

A direct connection between the game of *duodecim scripta* (as well as all other game types found carved on stone surfaces) and gambling was made by Lanciani more than a century ago.³³ His assumption almost certainly derives from the competitive phrasing found on inscribed examples of the game and from a painting from Pompeii portraying two game players arguing over its outcome.³⁴ More recently, the same specific parallel has been advanced by Purcell, who unequivocally connected gambling and board games such as *duodecim scripta* and *ludus latrunculorum*.³⁵ Purcell unites the factors of chance and of skill under the umbrella term of *alea*, which he uses to define games in general.³⁶ While Lanciani's connection lacks sufficient analysis to be deconstructed at length, the argument made by Purcell necessitates further comment, in particular regarding the choice of the umbrella term *alea*. *Alea* is a term that has no direct connection with games of skill and strategy, but refers specifically to games of chance such as *tali*, *tesserae*, or *par-impar*.³⁷ Purcell inserts 'all three of the common board-game patterns' under the definition of

---

[29] Mulvin and Sidebotham (2004: 605–8).
[30] Mulvin and Sidebotham (2004: 611–13).
[31] Varro *Ling.* 7.52; Ov. *Ars Am.* 2.208; 3.358; Bassus *Laus Pisonis* 193.
[32] See also Salza Prina Ricotti (1995: 73–108).
[33] Lanciani (1892: 97–105).
[34] *CIL* IV.3494. The author himself mentions the painting, see Lanciani (1892: 97–8).
[35] Purcell (2004: 178).
[36] Purcell (2004: 177).
[37] *Oxford Latin Dictionary*, i (1968: 94); Smith (1890: 96).

alea by noting that all three contemplate the use of dice.[38] It is therefore worth analysing to what degree dice are used in playing a Roman board game, in order to understand their real link with the word *alea*.

A game such as *merels* was probably played in turns, and might have involved the use of dice only to decide which player would have started the game. Similar to *merels* would have been the role of dice in a game of *ludus latrunculorum*, which was probably structured in a similar way to how chess is today. In both games, therefore, the use of dice was minimal and not essential. The only exception among the three most common types of board game might be *duodecim scripta*, since the game may have been played through the continuous use of dice.[39] The interpretation of the rules of the game of *duodecim scripta* is, in any case, one in which both skill and strategy play a central part. As mentioned, there are games, such as playing with the dice pot (*pyrgos, fritillus*), which rely more heavily on chance. These alone are more suited to an association with the term *alea*. Indeed, the sources mentioned by Purcell all use the term *alea* when indicating dice as objects or as a game.[40] One of these, Macrobius, goes further by making a distinction between *alea, abacus* (game board), and *latrunculi*, as separate games, each mentioned in a different section of his *Saturnalia*.[41] There is, therefore, a substantial difference between games such as *latrunculi* or *duodecim scripta* and games such as *tali* or *tesserae*. In the first group of games, a substantial use of skill and strategy was needed to reach victory. In the second, victory was decided by fate.

The particular connection between the term *alea* and games of chance is also affirmed in Roman law. This comes in the form of the section of Justinian's *Digesta* dedicated to gambling and aptly called *De Aleatoribus*. While the initial part concerns the regulation of crimes and disputes emerging in a gambling context, the remainder clearly sets out the position of gambling in the eyes of the law. The *De Aleatoribus* lists three *leges*, a *senatus consultum*, and an *edictum praetoris* all concerning gambling. Of these, the *senatus consultum* explicitly forbids playing for money (indicated by the more general expression *pecunia ludere*), 'except when one is competing at spear or javelin-throwing, running, jumping, wrestling or boxing, which are contests of strength'.[42] The connection between *alea* as games of chance and gambling is clear. Any other games played for money (including games which are contests of strength) are included under the more appropriate umbrella

---

[38] Purcell (2004: 179).
[39] Mulvin and Sidebotham (2004: 604).
[40] Naevius, 118 (in Ribbeck (1873); Suet. *Aug.* 70–1; Macrob. *Sat.* 3.16.15 (his reference is incorrect); Plaut. *Mil. Glor.* 164; *Curc.* 354–6.
[41] Macrob. *Sat.* 3.6.15; 1.5.11.
[42] *Dig.* 11.5.2 'praeterquam si quis certet hastam uel pilo iaciendo uel currendo saliendo luctando pugnando quod uirtutis causa fiat' (although the *uirtus* translated here as strength would better be translated as ability or capacity); Mommsen, Krueger, and Watson (1985).

term of *pecunia ludere*.⁴³ Furthermore, playing games of chance for money is consistently prohibited from the Republic to the age of Justinian.⁴⁴ Plautus confirms the antiquity of such a prohibition by mentioning a *lex alearia* linked to games with knucklebones: 'And furthermore, just to keep them from breaking the Dicing Act, see to it that when they give a party there is no set of *tali* amongst them'.⁴⁵ The Severan lawyers quoted in Justinian's *Digesta* confirm that there was an association between games of chance and gambling, which were prohibited by laws passed in the Republic and upheld into late antiquity. There is no parallel source confirming the same for board games, although it is worth highlighting that all literary sources indicate how games of chance were perceived as gambling games par excellence.

THE PUBLIC SPACE OF GAME BOARDS: MOVEMENT AND VISIBILITY

The introduction of laws regulating gambling provides further insight into games and their differences. As gambling was prohibited across the Empire, gamblers would have responded creatively to prohibition, by playing, for example, out of sight. This allows us to extend our focus to include the spaces where the games were played.

The spaces of gambling par excellence, *popinae* and *cauponae* (bars and inns), have been studied by a number of scholars interested in their distribution in Pompeii. Key to these is La Torre's analysis of the distribution of *popinae* and *cauponae* in Pompeii. He found that these were not evenly distributed in the urban landscape of the town, but that there were fewer *popinae* and *cauponae* around the forum and other public buildings. La Torre's explanation for this was linked to the impact of the earthquake damage suffered by the town's public buildings in 62 CE. In his opinion the damage sustained by the buildings led to a reduction in overall activity around them.⁴⁶ Wallace-Hadrill offers an entirely different interpretation to this pattern by comparing spatial patterns with the types of activity taking place in *popinae* and *cauponae*.⁴⁷ According to him, this pattern of distribution reflected a moral separation between a city of virtue (represented in particular by the forum) and a city of vice (in which gambling, drinking, and prostitution would

---

⁴³ *Dig.* 11.5.2.
⁴⁴ For details see: Berger (1953: 547, 549, 560, 623).
⁴⁵ Plaut. *Mil. Glor.* 2.2.9, 'Atque adeo ut ne legi fraudem faciant aleariae Adeuratote, ut sine talis domi agitent convivium'.
⁴⁶ La Torre (1988: 75–102).
⁴⁷ Wallace-Hadrill (1995: 43–57).

typically take place).⁴⁸ Laurence adds another dimension to this interpretation by linking the location of *popinae* and *cauponae* more specifically to three factors: the proximity of water supply, the proximity to customers (significant concentrations of *cauponae* are located close to the town's gates), and the intervention of the elite in controlling the use of property (most likely through the action of the aediles).⁴⁹ He goes a step further by examining the activities taking place in *popinae* and *cauponae* and by highlighting, in particular, the relationship between these spaces and gambling. He suggests that the section providing regulation on resolving gambling-related conflict in Justinian's *Digesta* may have had a link to *popinae* and argues that the prohibition of gambling seems to have been effectively ignored in these places.

If the prohibition of law creates a 'hidden' space for gambling, we need to understand whether games of skill (such as *duodecim scripta*) could be spatially differentiated as a consequence of their different legal status. Board games were certainly not excluded from the space of the *popinae*. One example that illustrates this is a worded *duodecim scripta* listing the food served in a Roman inn.⁵⁰ Written sources do not refer specifically to any space being commonly used to play board games, although they do tend to describe them as taking place in private contexts.⁵¹ Material evidence shows that board games were played in a large number of spaces in Roman towns.⁵²

It is worth contrasting the wide distribution of these artefacts with Cicero's observations concerning gambling in public space. He describes the activities of one Licinius Lenticula, referring to him as a protégé and fellow gamester of Mark Antony's, who had been convicted for gambling, and who 'would not hesitate to play at dice even in the forum'.⁵³ Comparing the two produces an apparent contradiction between the widespread presence of carved boards for games of skill in public spaces and the fact that in the eyes of Republican society (and later the Severan lawyers) gambling in a public space was frowned upon, if not a blatant flouting of the law. Later evidence from Aphrodisias helps our understanding of this apparent contradiction.

The site in Caria has produced a number of marble blocks on which are carved non-worded *duodecim scripta* game boards.⁵⁴ These game boards were found in key points of the town's public space and are particularly unusual, as they are dedicated as a donation to the community by benefactors. One such benefactor, Flavius Photius Scholasticus, thought to be an official of the late

---

⁴⁸ Wallace-Hadrill (1995: 55–6).   ⁴⁹ Laurence (2007: 98–100).
⁵⁰ With the apt wording: 'abemus | in cena | pullum | piscem | pernam | paonem | bena tores' 'We have (served) for dinner, chicken, fish, (pork) leg, peacock. Hunters', see Ferrua (2001).
⁵¹ Quint. *Inst. Or.* 11.2.38; Val. Max. 8.8.2; Cic. *De Or.* 1.50.217; Mart. 2.48; Ov. *Ars Am.* 3.357–80; Petr. *Sat.* 33.
⁵² Lanciani (1892: 98).
⁵³ Cic. *Phil.* 2.56, 'qui non dubitaret vel in foro alea ludere'.
⁵⁴ Roueché (2004: http://insaph.kcl.ac.uk/ala2004).

Fig. 13.7. Game of *duodecim scripta* dedicated by Photius on a reused statue base. The game is located on the northern entrance of the Hadrianic baths of the city. This is not confirmed as the original site of discovery (Roueché 2004).

fifth or sixth century CE, dedicated (probably with public money) three reused statue bases with carved *duodecim scripta* found both inside the Hadrianic baths and just outside their northern entrance.[55] Three more blocks are similarly dedicated by city officials and are themselves placed in public contexts (the theatre and the Hadrianic baths respectively) (Fig. 13.7).[56]

This evidence of game boards dedicated by town officials in late antiquity can be compared to another broadly contemporary source, Justinian's *Digesta*. In this we see the consistent application of the prohibition of gambling to the inhabitants of Aphrodisias. In addition, sources are very clear about the general aversion of Christians towards gambling and of that of Ammianus Marcellinus in the fourth century CE.[57] Therefore, Aphrodisias in the sixth century CE is a town in which the attitude towards, and with all probability the prohibition of, gambling remained unchanged. Board games, however, appear as the objects of dedications to the community by town officials, something that could not have happened had the boards been associated with gambling. These were games offered in the typical style of *beneficia*, in which a local official donates an object or building to be placed in positions of high visibility—and high movement potential—and for the benefit of the community. Furthermore, Photius' actions are repeated by other officials, thus making these more than a single or exceptional act. This does not simply constitute evidence

---

[55] Roueché (2004: inscription numbers: 68, 69, and 238; section V, 47).
[56] Roueché (2004: inscription numbers: 59, 70, 71).
[57] Pseudo-Cyprian, *Adversus Aleatores* (*Patrologia Latina* iv. 903–11); Amm. Marc. 28.4.21.

of a novel way of intending the *beneficium*, but also outlines board games as a practice understood to be of common benefit, in public space.

Clearly, evidence of games of dice or knucklebones does not have the same visibility as carved game boards as it does not require the use of carved or drawn surface markings. We do not, therefore, possess a view of the balance that may have existed between legal and illegal gaming in the urban landscape. Nonetheless, the evidence that we do possess speaks specifically of one type of game, one that did not come to the punitive attention of Roman law and did not require being hidden in a *popina*. This is important in the study of board games from a spatial perspective, as it establishes a connection between the games, movement, and visibility. In order to understand this further, we can analyse the visibility of game boards in the context of fora and suggest that games of skill were a key element in the social construction of space at the heart of the Roman city.

In particular, the distribution patterns outlined in the Forum Romanum and in the forum of Timgad show a direct relationship between game boards, visibility, and patterns of movement within the public spaces. In the former, the greater concentrations of game boards occur where the basilica meets the Sacra Via and the Vicus Tuscus, as do the game boards in the *area*. This relationship is clearly not accidental. The Sacra Via is among the oldest streets (if not itself the oldest) in Rome and was also a busy place for meetings and trade.[58] The point of connection between the Sacra Via and the Vicus Tuscus formed a key link between the Forum Romanum (and the surrounding area), the Circus Maximus, the Forum Boarium, the Pons Aemilianus, and the Pons Sublicius; this made it more than just a busy marketplace but also a key communication route in the city of Rome.[59] These factors become all the more important when the Vicus Tuscus is compared to the other road which stems off the Sacra Via, at the northern side of the Forum Romanum (in which the concentration of game boards is comparatively marginal). This road, the Vicus Iugarius, is, in fact, visibly less important than the Vicus Tuscus as a communication route.[60] The carved game boards on the surfaces of the Arch of Septimius Severus do not merely confirm the connection between games and movement; the imbalance between the eastern and the western walkway reveals the intensity of the connection between the games and access to the forum itself, rather than the slightly more peripheral space fronting the Comitium.

The nature of the contrast between busier spaces, placed next to the Sacra Via and the Vicus Tuscus, and less visible ones in the Basilica Iulia, fronted by a line of rooms, is clear. This is generated by the presence and absence of busy routes, as no other reason could satisfactorily justify such an imbalance. This is not only revealed by the distribution of the game boards, but is also made clear

---

[58] Richardson (1992: 338–9).
[59] Platner (1929: 579–80); Richardson (1992: 429); Papi (2002: 45–62).
[60] Platner (1929: 574–5).

by the small number of game boards present on the side opening towards the rooms themselves. Their presence shows that the possibility of carving game boards in this space was an option that was generally overlooked. The cluster of games found on the floor surface of the basilica provides the only real archaeological evidence of the practice of game playing in the space of the Forum Romanum. This space, though, is also used for social interaction. Viewed together, they show us how the space of social interaction escapes the rigid division operated by the architecture of the Forum Romanum. They can be used to prove how the civic space acts as a hub for the convergence of people, and how parts of its space are more efficient in this function. The comparison of this data with the findings emerging from the work of both Mouritsen and Laurence produces clear, though in some ways surprising, analogies. Carved game boards are not immediately comparable to written graffiti. Their purpose is, in fact, not to communicate, but to facilitate a specific type of interaction between two or more people.

There is clear evidence for differentiated gaming space in the Roman world: one for illegal gambling, and a largely private one, for games of skill. These contexts are ones which are more suitable to gaming as a practice, as they can be designed or adapted to enclose it tightly. In public space, gaming develops along specific patterns of distribution. These link game boards to everyday activity and movement in space, the same basic requirements for the diffusion of the graffiti in Pompeii. This element escapes more rigid definitions of gaming, a social activity but one which did not require any greater number of people than those strictly involved in the game. We have to therefore look at carved game boards in public space as possessing a role that games in dedicated spaces do not possess.

The reciprocal influence acting in processes of cultural appropriation outlined above explains the rules regulating this new role. The first part of this process regards the role played by the game boards themselves. Data show, in fact, how their distribution escapes the more rigid divisions operated by the architecture of a public space. The games are witness to a process of remodelling of the public space that concerns the games specifically. Through this process, the public space is reshaped so as to respond to needs created by emerging patterns of use. The spaces created by these new patterns of use are defined by the activity taking place within them. The quantity of carved game boards make this more obvious in the Forum Romanum, in which the space outlined by clusters of carved game boards clearly escapes the architectural divisions operated by different buildings. The identified clusters of game boards in this context act independently from architectural divisions and create a new context. In this new context, architectural space is manipulated through selective occupation. This results in the creation of a new space that is a direct witness to use outside the strict boundaries imposed by architectural division.

The clusters of game boards can also be used as a tool for the analysis of architectural space. In the case of Timgad, the relative paucity of examples does not limit interpretation of the principles of the use of space. Instead, it reinforces it by showing how, from the start, priority is given to certain locations rather than others.[61] Here, two of the spaces chosen for the location of game boards are closely connected to the forum's two main access routes. This common characteristic qualifies the 'gaming space' as one that is physically in the forum but that is, in practice, ultimately located somewhere which has a connection with the access routes to and from the 'outside'. This creates a more specific context for the games, represented by the increased presence of people. This aspect is common to both the examples observed in this chapter. The generative space of the games in Timgad is then not created solely by the forum, but also, and perhaps more importantly, by the movement of people between the adjacent theatre, the forum, and the road that separates the two. The consequences of this reflect specifically on the forum, because in the study of activity, architectural boundaries become insufficient to define it. The *duodecim scripta* game board and the game of *mancala* placed next to the main entrance to the forum repeat this pattern and confirm it in full.[62]

Further analysis demonstrates that the role of architecture is not limited to that of a passive boundary in the creation of new patterns of activity. Data show us that it has a role in defining the gaming space itself. In Timgad, the space occupied by the carved game boards does not extend outside the forum.[63] It is not, with the state of our current knowledge, possible to tell whether this was a product of independent decision making or, perhaps, impositions by local magistrates on the instruction of the decurions of local administration. This suggests nonetheless that a local perception of the forum, in addition to its patterns of use, influenced the choice of where to carve game boards. The pattern is repeated in the Forum Romanum, where we are able to observe how carved game boards remain linked to the architectural boundaries of the forum itself.

CONCLUSIONS

The sum of all these elements is a renewed definition of the public space of the forum as well as one of the space for legal gaming. The link between game

---

[61] This, regardless of whether the game boards were carved, transported on stone tablets, or marked with other means. Carving the boards represents the most immediate way of leaving a permanent mark of gaming activity. Others are not only archaeologically difficult, or impossible to detect, but would have constituted a heavy or untidy alternative to the carved examples.

[62] This pattern is also repeated by the games located in the basilica, a space of transit and encounter par excellence. The location of the games along the central axis may offer us a picture of the zoning of the space. The area around the central axis may have been easier for sedentary activity as opposed, we may tentatively propose, to the areas around the axis.

[63] Ballu (1902: tav. XVI).

boards, access routes, and, therefore, movement and encounter, enables us to expand their interpretation. Game boards aid the definition of the forum as a space for gaming, but are in turn accessories to a more general issue connected with visibility and interaction in the public space. While clusters of games show the extension of a practice that redefined space, architectural space has a role in modelling this. The role of architectural space in limiting the expansion of carved game boards represents the element of reciprocity in this process of social construction of space. Through these reciprocal influences we also obtain an archaeological redefinition of the forum as a space for interaction and visual display.

In commenting on games and on their role in society, Callois says that 'play is a luxury and implies leisure. The hungry man does not play'.[64] It is certainly tempting to place game boards in fora in a context of status and social competition, although the evidence does not allow us to speak in these terms. However, it is worth remembering the daily *negotium* of the urban elite, already described as a routine established by movement and closely connected with visibility.[65] The differing status of games of skill and games of chance may help support a specific association of games of skill with the activity of the urban elite. This differentiation is not only present in the evidence we possess for the Roman period, but is also outlined more generally by Callois himself in his classification of gaming in which the two types are distinguished as *agon* and *alea*.[66] The two are distinguished by their basic dynamics of play, linked respectively to competition and luck: with victory resulting on one side from exercise and constant commitment and on the other side by the exclusive intervention of fate. We have seen this opposition reflected in the law, as the *De Aleatoribus* tells us that money could be bet on competitive games while not on gambling. Spatial distribution reflects this in full as, in opposition to the hidden *alea*, it outlines high visual prominence as a constant factor in the distribution of carved game boards. The opposition between game boards (*agon*) and dice playing (*alea*) is paralleled by an opposition between legal and illegal and between highly visible and hidden. It is not surprising, then, that game boards would later become equivalent to statues as the objects of *beneficia*, or that a board game would be played by an emperor such as Zeno, as Agathias Scholasticus writes in late antiquity.[67]

---

[64] Caillois (1967a: p. xv).
[65] Laurence (2007: 163).
[66] Caillois (1967b: 30–4).
[67] Agathias Scholasticus *Epigrammata* 9.482. For comments on the dynamics of the described game see Austin (1934: 202–5).

14

Construction Traffic in Imperial Rome

Building the Arch of Septimius Severus*

Diane Favro

> ... up comes a huge fir-log swaying on a wagon, and then a second dray carrying a whole pine-tree; they tower aloft and threaten the people. For if that axle with its load of Ligurian marble breaks down, and pours an overturned mountain on to the crowd, what is left of their bodies?[1]

Modern pictorial reconstructions usually depict imperial Rome as clean, spacious, and tranquil (Fig. 14.1). Literary sources paint a very different story. Anyone moving through the city was jostled by raucous participants in ritual parades, solemn family members following funeral cortèges, drunken soldiers on leave, bleating sheep on the way to slaughter, litters with dozing patricians, overloaded donkeys, street musicians, slaves delivering dinners, magistrates surrounded by fawning entourages, and thousands of other pedestrians going about their daily business.[2] Then there was vehicle traffic. Carts filled with excrement, wagons loaded with vegetables, barrows with barrels sloshing wine, and huge transports with roaring animals in cages, as well as triumphal floats piled with booty and gilded carts moving statues of the gods. No wonder Julius Caesar took action.

At the end of the Republic he gathered various regulations in the composite document known today as the *Lex Iulia Municipalis*, which included a ban on

---

* Special thanks to the editors for their perceptive comments and suggestions, to Fikret Yegül for his advice and insights, William Aylward for his comments, Brian Sahotsky for his meticulous calculations and digital model making, and Marie Saldaña for her excellent work on the illustrations.

[1] Juv. 3.255–9, 'atque altera pinum plaustra vehunt; nutant alte populoque minantur. nam si procubuit qui saxa Ligustica portat axis et eversum fudit super agmina montem, quid superest de corporibus?'

[2] Van Tilburg (2007: 74–6, 120–2); see Holleran on Rome, and Hartnett on Pompeii, in this volume.

Fig. 14.1. Digital reconstruction of the Forum Romanum, third century CE (Experiential Technologies Center, UCLA; copyright Regents, University of California).

wheeled vehicles in Rome during the ten hours after sunrise.[3] The law specifically excluded, 'whatever will be proper for the transportation and the importation of material for building temples of the immortal gods, or for public works, or for removing from the city rubbish from those buildings for whose demolition public contracts have been let'.[4] The exception of state construction traffic was significant. Building in imperial Rome was both a necessity and a responsibility of all who had, or aspired to, power. Each new emperor and potential heir to the throne erected impressive public structures to promote himself, his policies, and his memory.[5] As a result, state construction traffic continuously hindered street passage in the capital during the busy daylight hours, and persisted at night for projects facing a deadline. At building sites, machinery and materials obstructed well-trod urban routes

---

[3] These provisions are also collectively known as the *Tabula Heracleensis*; Crawford (1996: 355–62). Laurence postulates that the tenth hour was selected to coincide with the dinner hour when traffic in Rome decreased; (2008b: 521–2). See also Kaiser, in this volume.

[4] Dessau (*ILS* 6085.56–67).

[5] In addition to self-promotion, emperors built to reify state policies, foster public favour, and provide jobs; Thomas (2007a: 107); Edwards (2009: 137–72).

necessitating detours that lasted for months if not years. The sheer scale of state-generated construction traffic threatened to crush Caesar's vision of freer moving traffic in Rome as readily as the load of Ligurian marble did the hapless pedestrian in Juvenal's *Satire*.

Situating the movement of construction goods in Rome's teeming context is difficult. Urban street activity is rarely documented for specific times or places. Ancient sources too often generalize the physical context and adjust the subject matter for literary or political ends. Even major events, such as processions and floods, cannot be entirely geotemporally located. The complicated transportation network of Rome with steep streets, narrow alleys, ramps, and stairs is impossible to map in full. The segments uncovered by excavations or depicted on the *Forma Urbis Romae*, provide only a partial picture[6] However, the close analysis of a single imperial project helps clarify the issues relating to building-materials' flow and its impact on ancient urban mobility.[7] The Arch of Septimius Severus erected in the Forum Romanum in the late second century CE provides an ideal subject for an experimental process analysis.[8] In contrast to the Column of Trajan and Baths of Caracalla carefully examined in other studies, the arch was relatively simple to construct.[9] However, its location deep in the heart of the city presented distinct logistical challenges and demanded solutions that, in turn, affected the broader urban environment. Comparatively rich data exists for the period: Cassius Dio and Herodian provide eyewitness accounts, while the *Forma Urbis Romae* shows the capital specifically at the time of the arch. Furthermore, invaluable recent studies of Severan Rome and ancient mobility provide rich supporting material.[10]

THE ARCH OF SEPTIMIUS SEVERUS IN CONTEXT

For generations, Roman urban displays focused on the Forum Romanum, the premier *locus* for public and private actions and ceremonies. Extant buildings weighed down with the memories of centuries were regularly restored and memorials continuously added, but the architectural form of the civic centre had been fixed in amber during the early imperial period. The erection of arches and porticoes and the Temple of Divus Iulius under Augustus clarified

---

[6] See Macaulay-Lewis, in this volume.
[7] For modern building-materials flow issues see Poortman, Norbert, and Bons (1994).
[8] Mathieu (2002).
[9] Lancaster (1999); DeLaine (1997).
[10] On Severan Rome see Gorrie (2002; 2004); Lusnia (2004); Swain, Harrison, and Elsner (2007). For ancient traffic see: Laurence (1994a, 1998, 2008a), van Tilburg (2007), Eck (2008); Martin (1999). Newsome rightly warns researchers to challenge the application of modern definitions for 'traffic', 'congestion', and 'efficiency' to ancient contexts (2008: 442–6).

a central axis through a loosely defined central open space.[11] New building was not taken lightly. As well as being fraught with meaning, any construction in the Forum Romanum faced considerable practical difficulties. By the late second century BCE not only was there little room for new projects, but access routes for construction materials were compromised by the build-up of surrounding monumental structures and complexes such as the imperial fora to the north and east, and the Temple of Venus and Roma to the south-east. In 193 CE when Septimius Severus was proclaimed emperor, the last notable addition to the forum had been over seventy years earlier with the erection of the so-called Anaglypha Traiani, a display of large sculpted panels in *l'area centrale*, rather than an actual building.[12]

Septimius Severus brought a relative degree of peace to Rome, staving off four other claimants to the throne during the infamous 'year of the five emperors'. Buying the favour of the army with money and victories, and of the general populace with games, buildings, and law, Severus ruled virtually as dictator. In 203 CE, the Senate and people of Rome dedicated a massive commemorative arch in the Forum Romanum; its bold inscription (repeated on two sides) honoured Severus and his sons for restoring the Republic and extending the empire[13] (Fig. 14.2). Though it is unclear if the emperor celebrated a formal triumph, the arch commemorated military successes.[14] The massive relief panels chronicle the Eastern campaigns and victories trumpet from the spandrels of the central opening. The monument was surmounted by bronze statues of Severus and his heirs in a *seiugis*—a six-horse chariot—flanked by trophies and soldiers.[15] Repeated depictions of Septimius Severus' sons Caracalla and Geta affirmed the establishment of a new dynasty.

The Arch of Septimius Severus is much lauded for its well-preserved dedication and complex artistic programme, but its architectural setting and location are equally notable.[16] The structure rose on the upper north-eastern corner of the Forum Romanum atop the paving of the Comitium and nearly on axis with both the Sacra Via entering the civic centre from the south-east along the Clivus Argentarius and with the Scalae Gemoniae descending from the north-west (Fig. 14.3). According to a prescient dream of Severus in 193 CE, it was when moving through the forum from the Sacra Via that a stallion

---

[11] Favro (1996: 195–9).
[12] The Anaglypha Traiani belong to the Trajanic/Hadrianic period; Brilliant (1967: 85); (Richardson 1992). The Antonine temple of 141 CE Lay just outside the Forum's centre.
[13] *CIL* VI.1033; Dessau (*ILS* 425); Brilliant (1967: 91–5); Richardson (1992: 28–9).
[14] Beard (2007: 322–3).
[15] Though the sculptures are depicted on coins, their exact original size is uncertain, but they had to be large to suit the arch's proportions; Mattingly (1975: 216 no. 320 pl. 35.5).
[16] Brilliant's (1967) comprehensive study of the arch remains the standard reference.

336 Diane Favro

Fig. 14.2. Arch of Septimius Severus in the Forum Romanum (photo by Diane Favro).

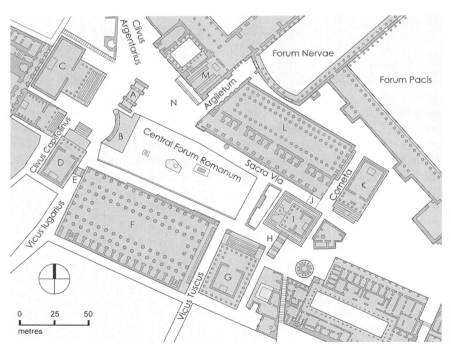

Fig. 14.3. Plan of Forum Romanum in the Severan age. The arch is marked by point A (image by Marie Saldaña and Diane Favro).

threw off his predecessor Pertinax and took the future emperor on his back.[17] The massive memorial drew upon many potent associations. The elevated position on the slopes of the Capitoline was directly in front of the Temple of Concord, goddess of social harmony. The Curia Iulia, where the Senate met, lay directly to the east. Programmatically and visually the new project linked with previous successes in Parthia including the Arch of Augustus located diagonally across the forum to the south-east, and the arch west of the Rostra associated with Tiberius who had received the Parthian standards.[18]

Like other Roman commemorative arches, that of Severus in the Forum Romanum is in essence a billboard for imperial promotion. It measures 20.88 m in height and 23.27 m in width, and rests on foundations $c.4$ m in depth.[19] The body is composed of travertine and marble blocks with brick vaulted chambers in the attic. In form, it is a triple-fornix arch with a larger central opening (6.77 m wide) and two smaller flanking ones (2.97 m wide); the side arches connect to the main opening with transverse vaulted passageways. A staircase on the south entered from a door located far above ground leads to four chambers in the attic from which another door gives on to an exterior walkway possibly used for exceptional appearances.[20] Originally stairs ran across the entire south-eastern facade to negotiate the rising ground towards the Capitoline.[21]

THE MOVEMENT OF CONSTRUCTION TRAFFIC TO THE BUILDING SITE: INFRASTRUCTURE AND OVERSIGHT

The capital city Septimius Severus inherited did not project an image equal to his ambitions (Fig. 14.4). Structures in the centre of Rome had been damaged by a fire under Commodus in 192/3 CE; other parts of the urban fabric had

---

[17] Herodian 2.9.5–6 notes that Pertinax was thrown off the stallion 'at the entrance to the forum' for one coming from the Sacra Via, suggesting a location besides the Regia. Severus was then carried to the middle of the forum where the stallion stopped. An equestrian statue was erected to mark the spot; Platner and Ashby (1929: 544); Coarelli (1980: 58–9); Richardson (1992: 145).

[18] The Tiberian arch specifically celebrated German victories, but was linked with Augustan propaganda related to Parthia; Brilliant (1967: 85–7); Favro and Johanson (2010: 27–8, 31).

[19] Two foundation courses have been uncovered, with the largest block measuring 2 m by 0.78 m; at least two more would be necessary; Brilliant (1967: 45–6).

[20] The door was elevated for security reasons, but the placement calls into question theories about Roman viewing platforms; Davies (2000: 130, 162–7).

[21] The position of the stairs remains controversial, and may have fronted only the side passages; Richardson (1992: 28); Brilliant (1967: 59–62, 89). They restricted wheeled traffic, but on special occasions could be made navigable with wooden ramps. Variations in arch stylobates on coins may indicate different degrees of permeability; Packer (1997: 86).

Fig. 14.4. Map of Severan Rome (including 3. Clivus Suburanus; 6. Forum Romanum; 9. Forum Boarium; 10. Emporium) (map by Marie Saldaña and Diane Favro).

frayed during years of neglect.[22] The Antonine plague (165–80 CE), and recent political disruptions had reduced the number of building workers, but the associated decline in new projects had kept the imperial stockpiles of building materials intact.[23] As Severus initiated an expansive building programme, designers and labourers, materials and machines gravitated towards Rome.[24] As a dynastic monument in the Forum Romanum, the arch had priority status over other Severan projects. Once plans were confirmed, contracts were issued along with orders for materials, labourers, and equipment. Immediately movement to and from the building site would have begun, continuing unbroken until completion.

---

[22] Cass. Dio 73.24.1; Gal. *De comp. med.* 1.1. On the fire and its impact on the storerooms along the Sacra Via and near the Templum Pacis, see Tucci (2008).

[23] Cass. Dio 72.14.3–4; SHA *Ver.* 8.1–2; Bruun (2003: 430–4).

[24] Most Severan projects in Rome can only be roughly dated, thus the order of creation is uncertain; Benario (1958); Lusnia (2004); Anderson (1997: 223–6).

Rome would have strained to accommodate the additional traffic. The Severan *Forma Urbis Romae* reveals a detailed knowledge of the capital's sprawling street system, but tells us nothing about traffic management. Data from various periods indicate that oversight of street maintenance and use in imperial Rome was loosely tiered under various administrative offices: traffic on Rome's smaller byways may have been overseen informally by neighbourhood officers (*vicomagistri*), the aediles supervised the care of major public streets and perhaps the licensing of large vehicles, while the *vigiles* dealt with traffic abuses, directing the more severe instances to the attention of the urban prefect (*praefectus urbis*).[25] Severus doubled the number of *vigiles*, generating speculation that they became more active in traffic management to deal with the large influx of soldiers into the capital.[26] Responsibility for the intricate planning of the delivery and removal of materials at imperial building sites involved numerous collaborating individuals or offices such as the *cura aedium sacrarum et operum publicorum* (management of sacred and public works) and the *opera Caesaris* established by Domitian. The emperor and Senate relied on individual agreements with architects, contractors, and *collegia*, including that for transporters.[27] Planning a massive building programme, Severus must have accounted for and included the involvement of transporters and of builders, which was the largest *collegium* in Rome.[28]

Writing in the first century BCE, the architect Vitruvius gave a long list of subjects to be mastered by Roman architects; a hypothetical list of hauliers of building materials in Rome would be equally lengthy.[29] Transporters and construction managers needed to know the city streets, traffic regulations, ritual and state calendars, current roadwork and building projects underway, load capabilities for wagons and animals, as well as the types and availability of lifting equipment. Preparing a shipment, an expert drover carefully plotted his route considering the weight and mode of transport, as well as the size of streets.[30] Smaller loads of building objects and tools carried by barrows, small

---

[25] State enforcement of existing laws was minimal, with action taken in response to specific legal or administrative rulings; Laurence (2008: 102–5); van Tilburg (2007: 135–6); Robinson (1992: 64, 73–6).

[26] Birley (2000: 103, 196); Robinson (1992: 106).

[27] Cf. Kolb; (1993: 53–7). On the imperial building industry see: Anderson (1997: 69, 89–92, 112–13); Lancaster (1997: 772; 2007: 18–19); Eck (1992: 242–3); Taylor (2003: 13–14); Gros (2001b).

[28] Cass. Dio 74.4; DeLaine (2000: 121–5, 132). Under Caracalla, 4–6% of Rome's workforce was in the building trade, most having trained on Severan projects; DeLaine (1997: 201); Rockwell (1993: 173).

[29] Vitr. *De arch.* 1.1.3.

[30] Burford (1960); DeLaine (1997: 99–100). As early as 450 BCE, laws stipulated that major roads (*viae*) be at least 8 Roman feet (2.37 m) wide, broadening to 16 (4.73 m) at corners; van Tilburg (2007: 128). Streets of at least 4.1 m in width could accommodate normal two-way traffic with passing on the right. Narrower streets were unidirectional or informally regulated; Lancaster (1999: 437); Poehler (2003, 2006).

carts, pack animals, or porters more easily navigated through city streets and thus required little oversight. Wagons pulled by multiple teams of oxen conveying large loads were wide, difficult to manage, and accompanied by numerous drovers, often halting traffic in both directions.[31] Tibullus described the actions of a profiteer in the time of Augustus, 'His fancy turns to foreign marbles, and through the quaking city his column is carried by 1,000 sturdy teams'.[32] Generally lengthy convoys delivered goods to the city's edges where lifting devices transferred the loads to smaller vehicles. Within Rome, expert transporters avoided steep slopes and sharp turns, directing multi-teamed conveyances to a few broad, level, and relatively direct public streets such as the Clivus Suburanus. Under Severus it is tempting to imagine that transporters consulted the marble map before beginning their deliveries; however, the enormous scale and elevated placement of the *Forma Urbis Romae* precluded detailed viewing. More likely transporters used smaller ichnographic maps owned by their individual *collegia*. The conscientious drover walked along the selected route, noting obstructions, from market goods spilling outward from shops, to braces shoring up teetering buildings.[33] He also considered vertical clearance since, as Juvenal noted, wagons were often piled with precariously high loads readily dislodged by jutting balconies or entangled in awnings.

Rome's hills and constricted crossroads continuously challenged heavy transports. Danger lurked on every steep slope. Loads readily broke loose, as documented in a court case about a mule-drawn wagon which careered down the Clivus Capitolinus killing a slave boy.[34] To minimize both deaths and damage to valuable cargo, hauliers offloaded materials to sledges or rolled on round sticks which were slowly moved on slopes using long ropes attached to pulleys, fixed bollards, or capstans[35] (Fig. 14.5). Other devices may have also been employed. According to a moralistic story, a mechanical engineer approached Vespasian with plans for a machine to move heavy stones up the Capitoline; the emperor rewarded the designer, but refused to use the machine, citing the need to employ large workforces.[36] Transporters carried long building components horizontally, and thus could not negotiate corners with right or acute angles. If no alternative route existed, the object was offloaded and lifted with cranes to negotiate the corner, halting traffic in every direction.[37]

---

[31] In long teams, almost each pair of oxen had a drover; DeLaine (1997: 99–100, 108).
[32] Tib. 2.3.-43-4, 'cui lapis externus curae est, urbisque tumultu portatur ualidis mille columna iugis'.
[33] In 92 CE, Domitian banned the blockage of streets by shop goods; Mart. 7.61; van Tilburg (2007: 38, 122); Robinson (1992); Martin (1999: 193–214). See also Laurence, in this volume.
[34] *Dig.* 9.2.52.2; Martin (2000: 204–10). Loss or damage to building materials could cause significant delays or major design alterations; Jones (2000).
[35] Adam (2003: 27); Lancaster (1999: 438).
[36] Suet. *Vesp.* 18; DeLaine (1997); Martin (2000: 212–13).
[37] Just such a scene may be described by Hor. *Epist.* 2.2.72–80.

Construction Traffic in Imperial Rome 341

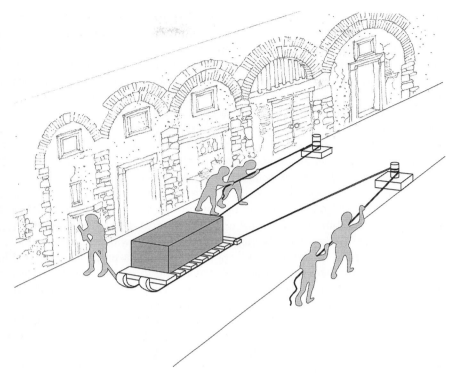

Fig. 14.5. Recreation of sledge and ropes used to lower a block on a steep urban street in Rome (image by Diane Favro).

Roman organizers and transporters of construction materials had to be familiar with legal requirements and responsibilities. Moving within the city, heavy transporters could be required to post bonds against damage to Rome's infrastructure.[38] Pliny the Elder recorded that in the late Republic, M. Aemilius Scaurus gave a security to the overseer of the sewers when moving large marble columns to his house.[39] No bond was mentioned when Scaurus later transported the same columns to a temporary public theatre in his official capacity as *curule aedile*. The story implies that action on behalf of the state provided indemnification.[40] If so, transporters carrying materials to public building projects were similarly exempt. Such protection did not extend to illegal activities. Cicero persuasively argued a case against Verres who had received state money for the purchase and transport of building materials to a

---

[38] Several ancient authors comment on tremors in the city caused by heavy transports; Sen. *Ep.* 90; Juv. 3.254; Tib. 2.3.43–4.
[39] Plin. *HN* 36.2.
[40] Possibly the route to the theatre did not cross a sewer; van Tilburg (2007: 128).

temple that were never delivered.[41] Following the widespread corruption under Commodus, Severus attempted to minimize fraud in several areas, including building. He ruled that architects and contractors, like surveyors, could be sued if they provided false measurements.[42] Regulation of licences, permits, and bans such as that regarding wheeled traffic in the city, was impossible in the bustling city centre.[43] Monitoring probably occurred at entry points, specifically toll points located at gates, the pomerial line, docks, and intersections on the major arteries entering Rome.[44] Here officials assessed the final destinations of heavy wheeled transports and approved entry of those serving imperial projects. These checkpoints became traffic bottlenecks, causing notable delays at daybreak when vehicles laden with materials clamoured to enter the capital. Severus erected a new custom gate, the Porta Septimiana, on the Tiber's right bank to monitor the importation of goods from the west.[45] In addition, he developed a large open area directly in front of his aedicular urban showpiece, the Septizodium, where those who passed inspection at the Porta Capena recuperated and negotiated with intra-urban carriers.[46]

Within the city, heavy, unwieldy construction transports blocked Rome's major arteries for hours if not days. Road closures may have been announced and posted in the Forum Romanum like the *acta diurna*, though such a paternalistic municipal act is doubtful.[47] In general blockages were conveyed by word of mouth, most likely at the neighbourhood level by the *vicomagistri* or *vigili*. During transport of lumbering loads, someone ran ahead to clear the way of pedestrians and other blockages. Minor officials could be pressed into service for this task; Plutarch tells us that at a Republican triumph numerous servitors and lictors drove back the thronging crowds who blocked the streets.[48] Of course, major state religious parades and events had priority over construction traffic. According to the *Lex Iulia Municipalis*, vehicles carrying materials to state projects could not impinge on temporary buildings or public areas in use for public games.[49]

Detailed time management analyses of Roman construction traffic cannot be reconstructed. Certainly some public projects had to meet specific deadlines. Given Septimius Severus' penchant for clustering dates, he likely established a

---

[41] Cic. *Verr.* 2.1.147.
[42] Ulp. *Dig.* 11.7.3.
[43] The Theodosian Code (compiled in 438 CE), includes edicts against the abuse of imperial transport, and describes the maximum load for each vehicle type; DeLaine (1997: 108), Burford (1960: 4).
[44] Van Tilburg (2007: 85–119), Robinson (1992: 6); Palmer (1980: 219–20).
[45] Palmer (1980: 224–6). The toll point may have been located at the crossing of a street and an aqueduct serving the new Thermae Septimianae; SHA *Sev.* 19.5.
[46] *Forma Urbis Romae* frs. 7a–b; Thomas (2007b: 330, 365).
[47] Stambaugh (1988: 140).
[48] Plut. *Vit. Aem.* 32.2.
[49] *Tabula Heracleensis* 77–9.

completion date for his arch contemporaneous with other commemorations. Vitruvius described such temporal restrictions as evil.[50] Not only could deadlines result in sloppy work, but they were unrealistic. The vagaries of sea and land transport made the precise scheduling of deliveries impossible. The seasons determined work hours, and deliveries, with the spring rising of the Tiber bringing an increased number of barges to Rome.[51] Competition for draught animals with nearby farmers also occurred seasonally, though this was probably less of a factor at Rome than elsewhere.[52] If the project fell behind schedule, additional transports may have moved to and from the Severan arch at night, joining others lumbering in the darkness.

MOVEMENT TO THE BORDERS OF THE FORUM ROMANUM

Work on the Arch of Septimius Severus project would have begun as soon as daylight illuminated the site. Labourers and transporters approached the Forum Romanum from every direction, fighting their way against the tide of private wheeled vehicles leaving the city centre. In the late second century CE, nine public ways accessed the central forum[53] (see Fig. 14.3). Four of these were unsuitable for heavy construction traffic. To the north-west, the steep Clivus Capitolinus and stepped Scalae Gemoniae led to a hilltop and thus were not used for deliveries. The Argiletum to the east was once a major entryway into the forum; however, by the Severan age the porticoes and steps of the Forum Nervae thwarted wheeled traffic.[54] The street identified as the Corneta between the Basilica Aemilia and Temple of Antoninus and Faustina was completely blocked by the construction of the Templum Pacis. Of the other five entries, at least three were spanned by arches: the Sacra Via, the street fronting the Basilica Iulia, and the Vicus Iugarius.[55] The Vicus Tuscus and Clivus Argentarius provided unobstructed access, though the latter was steep, requiring transporters to climb up and over the eastern shoulder of the Capitoline which rose approximately 10 m above the level of the forum.

---

[50] Vitr. *De arch.* 10 pref.3.
[51] DeLaine (1997: 99); Brunt (1980: 93); Maischberger (1997: 55–9).
[52] Van Tilburg (2007: 83).
[53] But see Newsome (2010) and in this volume, on the progressive restriction of those routes.
[54] See La Rocca (2006: 130–3). The restriction of traffic minimized wear and tear on the major sewer lines running under the Forum Nervae.
[55] An arch associated with Tiberius spanned the Vicus Iugarius. One to Augustus crossed the street south of the Temple of Divus Iulius, and an arch or portico spanned the Sacra Via to the north. La Rocca postulates the existence of an arch over a *diverticulum* off the Clivus Argentarius (2006: 137).

After surveyors measured the building footprint, unskilled workers cleared the area and began to dig a pit for the foundations. The first materials needed on the site were large travertine blocks quarried east of Rome at Tibur. Two primary routes could handle the transport of the stones measuring approximately 2 m in length and weighing over 3 tonnes.[56] In the first scenario, the great blocks were barged down the Anio and Tiber and unloaded either near the Milvian Bridge or on the banks of the Campus Martius. Transferred to wheeled vehicles, they entered Rome over the broad, straight Via Flaminia.[57] The rudimentary brake systems of Roman vehicles could not easily handle the last segment of the route passing up and over the steep Clivus Argentarius. Hauliers may have placed the largest stones on sledges or rollers, slowly lowering them with ropes looped around bollards inserted into the ground flanking the street (see Fig. 14.5).

The second route avoided water. For heavy loads, ancient transporters preferred the relative security of roads over the unpredictable vagaries of floods, silting, and rapids along waterways, especially tributaries.[58] Outside Rome along the wide Via Tiburtina, long teams of oxen pulled wagons laden with travertine blocks destined for the new arch. Convoys halted at staging areas where workers shifted the loads to smaller vehicles more suited to city streets.[59] The largest foundation blocks required wagons with around thirty oxen teams stretching over 75 m in length.[60] Drivers naturally selected the straightest, most direct route to the site. From a high point at the Porta Esquilina, they moved along the Clivus Suburanus which descended approximately 30 m down the valley between the Cispian and Oppian hills to the Forum Romanum. On the steeper segments sledges and ropes were called into play. In the first century Martial clearly described a scene along this route, 'I . . . surmount the steep path of the Suburran hill, and the pavement dirty with footsteps never dry; while it is scarcely possible to get clear of the long trains of mules, and the blocks of marble which you see dragged along by a multitude of ropes.'[61] The wetness may have been due to the messiness of draught animals greedily drinking at the Lacus Orpheus, or to the dousing of the taut ropes lowering the building blocks.[62]

---

[56] Brilliant (1967: 45). Travertine blocks were also used for the interior body of the arch.

[57] Maischberger (1997: 95–110). On extra-urban traffic see: Graham (2009); Black, Browning, and Laurence (2009: 713–27).

[58] DeLaine (2000: 135).

[59] Rodríguez-Almeida (1994: 215). See papers in Bjur and Santillo Frizell (2009).

[60] Estimates of load sizes and team numbers vary due to uncertainty regarding the efficiency of yokes and load capacity of animals in antiquity; cf. Wallace (1994: 51–60). The figures used derive from Xen. *Cyr.* 6.1.52, with the proposed length estimated at 2.5 m per team; DeLaine (1997: 99, 108).

[61] Mart. 5.22, 'alta Suburani vincenda est semita clivi et numquam sicco sordida saxa gradu, vixque datur longas mulorum rumpere mandras quaeque trahi multo marmora fune vides'; cf.10.19.

[62] During the raising of the Vatican obelisk in the 16th century, water was splashed on the ropes to prevent overheating; Platner and Ashby (1929: 371). Throughout antiquity liquids and grease were used to ease the movement of sledges and rollers.

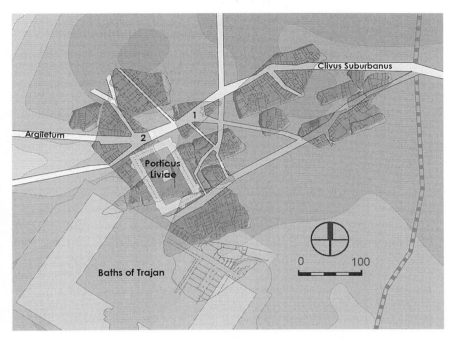

Fig. 14.6. Fragments of the *Forma Urbis Romae* marble plan showing the Clivus Suburanus and Clivus Pullius (image by Marie Saldaña and Diane Favro).

Preserved segments of the *Forma Urbis Romae* plan allow a nuanced examination of the Severan context for heavy construction traffic along this itinerary[63] (Fig. 14.6). The fragments show numerous shops along the streets. Long teams of animals with shouting drovers, tottering loads of stone, and chanting workers manning the guide ropes all affected commerce in the area, blocking access to shops or damaging goods on display. At major intersections the Clivus Suburanus widened to allow vehicles and people to pass or negotiate turns.[64] The complicated manoeuvring would have attracted attention. Rome was filled with official visitors, tourists, and of course a large population of unemployed who delighted in the unscripted drama. Imperial patrons frequently exploited the spectacular nature of transport. Hadrian and the architect Decrianus raised Nero's colossal statue to an upright position and then moved it using twenty-four elephants, a choreography obviously designed to draw a crowd.[65] Martial describes scaffolding rising in the middle of the road, and the venerable mass of the Colosseum growing before

[63] Malmberg (2009a: 40–8).   [64] Van Tilburg (2007: 27).
[65] SHA *Hadr.* 19; cf. Diod. Sic. 1.18.26–7.

observers' very eyes.[66] On tomb reliefs, representations of machinery identified the deceased's occupation, and impressed viewers with the scale and complexity of his work. The Clivus Suburanus offered an ideal 'theatre' for watching heavy transports. Onlookers gathered on the stairs of the Porticus Liviae cascading down the Oppian, some marvelling at the difficulties involved, others secretly hoping to see a bloody mishap.[67] Just to the west, the road split. Drovers turned their teams away from the northern fork, well aware that the Argiletum was impassable to heavy construction traffic. Instead, they directed their vehicles down the so-called Clivus ad Carinas behind the Templum Pacis.[68] A preserved segment of the marble plan shows a store-lined street backing the complex which, though rather narrow (about 4 m in width), could accommodate heavy transport in one direction.[69] It intersected with the Sacra Via about 250 m east of the building site for the Severan arch.[70] During days when numerous deliveries arrived from the east, organizers may have directed full wagons to enter by this route and empty ones (or those with construction debris) to exit by the Clivus Argentarius, or vice versa.

The exterior of Severus' arch glistened with marble. Standard size pieces were drawn from state stockpiles at Portus and the Emporium district.[71] For example, the Proconnesian column shafts on the arch are the same size (7 m) as those on many other imperial monuments, including in the Forum Traiani and the Arch of Constantine.[72] Unique components were made to order, with requests sent to quarries in the eastern empire and Italy long in advance of use. Pentelic and Proconnesian stones transported by ship were stored and in some instances roughly worked at Rome's harbour, Portus, before being moved by barge up the Tiber to urban docking facilities. Though unloading centred at

---

[66] Mart. *Spect.* 2.2. The scaffolding is associated with construction of the Arch of Titus; Coleman (1998: 18–19); Boethius (1952). Martial's reference to delays facing crowds exiting the amphitheatre may also allude to building work in the area as pointed out to me by David Newsome; Mart. *Spect.* 26; Coleman (1988: 193–4, 146–7). On the context of Martial writing in a city that was being rebuilt, see Laurence in this volume.

[67] Taylor (2003: 4–5); DeLaine (1997: 99–100).

[68] At roughly the same time as the arch, Severan teams repaired extensive fire damage to the Templum Pacis and put the marble map on display. During this work there may have been direct access from the Corneta or Clivus ad Carinas, allowing the central court to serve as a staging and working area; La Rocca (2006: 133–4).

[69] Dion. Hal. *Ant. Rom.* 4.39.4 notes that one of the main streets from the Forum Romanum towards the Esquiline was very narrow, and that the driver of Tullia's *carpentum* recognized that there was no room to manoeuvre around either side of her father's body in the carriageway (in this context, this is a reference to the Clivus Orbius, further away from the forum but an extension of the Clivus ad Carinas immediately beside the Templum Pacis).

[70] If this street were a *via tecta*, the height of loads would have been restricted; Fogagnolo (2006: 62–5).

[71] Fant (2001); Rockwell (1993: 145).

[72] Jones (2000: 124). Columns and relief panels cut against the grain may indicate the reworking of stones from imperial stockpiles; Claridge (1998: 76).

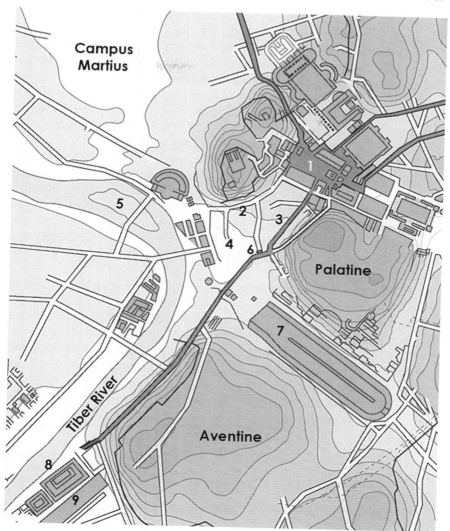

Fig. 14.7. Map of central Rome in the Severan period showing possible routes for the movement of construction traffic from the Tiber (including 2. Vicus Iugarius; 3. Vicus Tuscus; 4. Forum Boarium) (map by Marie Saldaña and Diane Favro).

the Emporium outside the old Republican wall, docking occurred all along the river below the Aventine[73] (Fig. 14.7). Planning an extensive building programme in the capital, Severus reworked the river banks early in his

---

[73] DeLaine (1997: 99); Richardson (1992: 144). Large timbers for scaffolds and machines also often entered Rome along the river.

reign.[74] Once the stones were unloaded, masons began refining the shapes to lessen the weight transported within the city and minimize work on site.[75]

Like generals provisioning a campaign, schedulers carefully regulated the supply lines, transferring materials to the location of the arch as needed. Wagons carried specific blocks northwards from the docks and warehouses of the Emporium.[76] They met heavy congestion at the Forum Boarium, a loosely defined open area where land traffic from the left and right banks converged with that from the river[77] (see Fig. 14.7). The open area was pressed into use for the turning of large timbers and stones, such as the 7-m column shafts for the Severan arch.[78] Pliny specifically mentioned masses of stone dragged through the space and expressed concern that the weight might damage the underlying sewer.[79] Recognizing the value for urban mobility, the Roman state took repeated steps to ensure the Forum Boarium remained unencumbered.[80] In the imperial period, the very name 'cattle forum' may have evoked thoughts not (as previously) of a market for meat and hides or a bull statue located there, but of bellowing ox teams straining to move construction materials.

Continuing past the Forum Boarium, transports conveying heavy building materials for the Severan arch veered north to take the Vicus Tuscus towards the Forum Romanum.[81] The shop-lined street rose a gentle 5 m following a relatively straight path through the valley between the Palatine and Capitoline. Not long after completion of Severus' arch, the *Argentarii* (bankers and merchants) of the Forum Boarium dedicated a gateway facing the Vicus Tuscus. The richly carved memorial honoured the imperial family, reflecting gratitude for the extensive business generated by construction traffic in the area.[82] Heavy transports entered the Forum Romanum between the Basilica

---

[74] *CIL* VI.31555. Carvers of the Severan marble plan used representational techniques usually reserved for monuments to depict the Porticus Aemilia warehouse at the Emporium, perhaps emphasizing Severus' promotion of trade and construction; *Forma Urbis Romae* frs. 23, 24b-c; Reynolds (1996: 74–7).

[75] Architectural lines inscribed on the plaza near Augustus' Mausoleum confirm that stones were worked near the river before transport; Haselberger (1994).

[76] Heavy loads could also have moved eastward from the docks, past the Circus Maximus, and into the Forum, though this path took more time and involved both a rise of about 10 m over the Velia, and passage through two arches. Especially large stones, such as the obelisk erected by Constantine in the Circus Maximus, were offloaded down river and entered Rome dragged over land; Amm. Marc. 17.4.14.

[77] Coarelli (1988). For the movement of stones from the Forum Boarium to the work site of Trajan's Column see Lancaster (1999: 437–8).

[78] Lancaster (1999: 437).

[79] Plin. *HN* 36.105–6.

[80] *CIL* VI.919, 31574.

[81] Livy described a procession in 207 BCE that took the same path in reverse, underscoring its directness and popularity; 27.37.14–15.

[82] This trabeated memorial served as an entry to the Forum Boarium and was erected not long after the arch in the Forum Romanum; *CIL* VI.1035; Newby (2007: 218).

Construction Traffic in Imperial Rome 349

Iulia and the Temple of Castor and Pollux, east of the triple Arch of Augustus (see Fig. 14.3). From this point they moved towards the building site, taking care not to damage the numerous statues and commemoratives, or to interfere with the workers repositioning materials using rollers, levers, or cranes.[83]

When moving less weighty or unwieldy building materials to the arch, hauliers had a greater selection of routes. Porters, barrow pushers, and pack animals conveying bricks, ropes, metals for clamps and tools, bags of mortar mixings, and wood pieces for scaffolds, formwork, and various lifting devices easily moved along most byways of Rome. For example, those carrying sacks of lime from the Emporium bypassed the congested Vicus Tuscus, instead going along the parallel Vicus Iugarius to the north and under the Tiberian arch between the Basilica Iulia and Temple of Saturn, to enter the western corner of the Forum Romanum. They had no trouble negotiating the diverse paving levels or dodging the memorials on and around the Rostra Augusti (including those added by Septimius Severus) to reach either side of the building site for the arch.[84]

TRAFFIC AROUND THE BUILDING SITE: LOGISTICS AND MOBILITY

Numerous obstructions hindered mobility around the construction site of Severus' arch in the Forum Romanum. Across the central pavement, piles of supplies rose in towering stacks (Fig. 14.8). If spread out, the 4,886 m^3 of materials used in the arch would have covered the entire central forum to a depth of almost a metre. To minimize blockage, large stones were brought as needed. Supplies in constant or repeated use were kept at hand. Forests of wood for scaffolding, formwork, and equipment along with huge coils of ropes filled the area; mounds of construction debris awaited removal. Related work activities further hampered movement as shown on a late antique mosaic from North Africa depicting carpentry, stone carving and transport, and the mixing of mortar at a building site (Fig. 14.9); a fresco from Stabiae conveys the bustle generated by labourers at various tasks including operation of a crane[85] (Fig. 14.10). The arch project also required metalworkers to forge clamps

---

[83] Wooden ramps or packed earth placed over the *crepidines* (curbstones) around the central area would have facilitated passage of wheeled vehicles; Giuliani and Verduchi (1987: 33–7, 39).

[84] Favro and Johanson (2010: 27–31).

[85] Sartorio (1988: 29); Adam (2003: 44, 163–4). On the organization, number, and seasonal occupation of the construction workers in Rome see: Cozzo (1970: 247–51); DeLaine (1997: 197–201); Brunt (1980).

Fig. 14.8. Digital reconstruction of a boom arm crane during phase 2 of constructing the Arch of Septimius Severus in the Forum Romanum (digital model by Brian Sahotsky of the Experiential Technologies Center, UCLA).

and inlay letters, painters, and bricklayers. They were joined by an army of sellers hawking food and drink to the labourers.

Accessibility around the construction site changed as work progressed. Four main building phases can be identified[86] (Fig. 14.11). The first involved site preparation and the laying of foundations. The location was not a blank slate. The Comitium's pavers occupied the space; directly to the east lay the Lapis Niger; to the west was the so-called Umbilicus Romae. The equestrian statue of the emperor rose nearby. The pavement of the expansive plaza to the southeast boasted an inlaid inscription and innumerable statues.[87] To prevent damage during construction, the sculptures were temporarily relocated and the pavement protected with a layer of sand, straw, or dirt which stirred up clouds of dust with every movement. Some features, including the Umbilicus Romae and the Rostra itself, were altered or temporarily relocated to accommodate the new arch. Carts and barrows carried away debris from the foundation pit. Reusable materials were stored nearby. Once the foundation trench was cleared, workers laid concrete or brick layers at the lowest level.[88] A chanting bucket brigade brought water for mortar from nearby fountains.

---

[86] These hypothetical phases derive from basic building requirements. Vitruvius mentioned several ancient books detailing the construction process of individual buildings, but none are preserved; Vitr. *De arch.* 7.pref.

[87] Richardson (1992: 173).

[88] Brick foundation courses have been found at the Arch of Titus; Adam (2003: 108).

Construction Traffic in Imperial Rome 351

Fig. 14.9. Late antique depiction of construction activities, from transport to the mixing of mortar and carving of stones on a mosaic from Ste-Marie-du-Zit, Sufetula; Bardo Museum Tunis (image by Diane Favro).

Fig. 14.10. Construction workers operate the pulleys of a crane while other labourers move building materials and cut blocks; first-century CE fresco from the Villa of San Marco at Stabia (image by Brian Sahotsky and Diane Favro).

Most materials were delivered to the long sides of the arch to avoid blocking the street running in front of the Curia. Bigger stones were lowered into the excavated pit from the south-east where there was ample room for simple double-arm flexible cranes. Access to the north-west side required more level changes and the obstruction of movement on the Clivus Capitolinus and Clivus Argentarius.[89]

While the foundations of the Arch of Septimius were laid, contractors oversaw the transport and preparation of materials and equipment for the second building phase: the body of the arch. Materials required included travertine for the interior of the podium base and main arched section, marble blocks for the exterior, and wood for scaffolding and formwork.[90] Large stones measuring up to 4 m in length and lifting machines clogged the broad area south-east of the arch's footprint. Carpenters fashioned formwork for the three main and two lateral arches on the higher ground to the north. As the arch rose ever higher, more and larger lifting devices were needed. The Romans employed several types of cranes, ranging from flexible single or double arm to fixed-brace, often with several simultaneously in operation.[91] Unwieldy to move through the city, such machinery was assembled from long beams at the site. Since large cranes could not pivot horizontally while lifting, they were repeatedly repositioned, adding to the flurry of motion in the forum. Crane crews varied from three to eight or more, as shown with operators of pulleys on the Campanian fresco and the famous Tomb of the Haterii relief[92] (Figs. 14.10, 14.12). Extensive guide ropes were essential; a single arm crane required a minimum of four; double arm and treadwheel cranes seven or more. For especially heavy loads the ropes were looped through several pulleys and lifted using capstans.[93] As the arch grew, engineers placed additional

---

[89] Severan events on the Capitoline cannot be firmly dated to the period of construction on the arch; Birley (2000: 90, 100, 158–9).
[90] This phase required complex calculations for the interior stairs; Brilliant (1967: 46, 64–5).
[91] Vitr. De arch. 10.2; Cozzo (1970: 255–67); Adam (2003: 43–51); Lancaster (2007: 37).
[92] Adam and Varene (1980–2: 213–38).
[93] Vitr. De arch. 10.2.7; Bingöl (2004: 44, 77).

Fig. 14.11. The four main building phases of the Arch of Septimius Severus: 1. Foundations; 2. Plinth and body; 3. Attic and roof; 4. Ornament (digital model by the Experiential Technologies Center, UCLA; copyright Regents of the University of California; image by Marie Saldaña and Diane Favro).

Fig. 14.12. Five men operate a treadwheel crane while four others deal with the pulleys and guy ropes on a relief from Tomb of the Haterii, Vatican Museum (image by Diane Favro).

cranes atop the scaffolding and on built sections of the structure with stabilizing ropes stretching to the ground.[94] Guy ropes at the Severan arch extended up to 40 m to temporary capstans or bollards, necessitating the cutting or

---

[94] For an assessment of Roman crane use see Yegül (2004: 388–9).

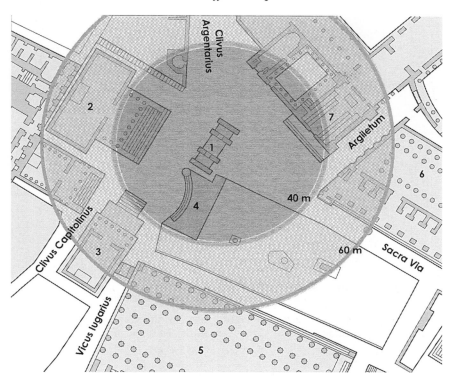

Fig. 14.13. Estimated area around the Arch of Septimius Severus encumbered by guidelines during phase 3 (drawing by Brian Sahotsky of the Experiential Technologies Center, UCLA, Marie Saldaña and Diane Favro).

removal of pavers in the central Forum Romanum, as well as the repeated relocation of workers, materials, and traffic.[95] During the construction months, the entire area around the site was criss-crossed with a network of cords that significantly compromised mobility in the forum.

In the third main phase, construction progressed above the top of the columns, over 15 m above ground on the south-east side. Cranes continued in use, with larger treadmills providing the power to raise pieces towards the sky. A single treadmill crane and its guy ropes encumbered approximately 1,500 m² at the site, hampering movement and activities in front of the Curia Iulia and Rostra Augusti and possibly along the Sacra Via (Fig. 14.13). The inclusion of vaulted rooms in the attic of the arch lessened the overall volume of

---

[95] On the final repaving of the Forum Romanum in the early-3rd century CE, see Giuliani and Verduchi (1987: 46–50, 61–6). This reorganization, as well as associated changes to the paving around the Rostra and Umbilicus Romae, may be understood as part of the refurbishment of those areas which were disturbed during the construction of the arch.

Fig. 14.14. Digital reconstruction of a lifting tower for the Arch of Septimius Severus (digital model by Brian Sahotsky of the Experiential Technologies Center, UCLA).

materials to be lifted and the size of each load; brick, dry mortar, and water for the *opus latericium* were easily raised in buckets or baskets by small cranes on the scaffolding. In contrast, big stones for the entablature and attic exterior, as well as the colossal sculptures, presented significant lifting challenges.[96] To elevate the large and unwieldy pieces the contractors employed a lifting tower on the south-east side of the arch (Fig. 14.14). Inspired in part by siege towers, the looming wooden structure rose the entire height of the arch and was fettered to the ground by extensive stays and ropes stretched to capstans.[97] Ammianus Marcellinus described such a tower used for the erection of an obelisk at the time of Constantius: '... to tall beams which were brought and raised on end (so that you would see a very grove of derricks) were fastened long and heavy ropes in the likeness of a manifold web hiding the sky with their excessive numbers ... while many thousand men turned wheels resembling millstones ... .'[98]

---

[96] At the Arch of Titus, lewis holes indicate the 16-tonne blocks were lifted up to 15 m using cranes; Rockwell (1993: 173). For the larger Severan arch a boom crane with a 40-m arm would be required, a size generally considered unfeasible.

[97] Lancaster (1999: 428–32); cf. Vitr. *De arch.* 10.13.5. Unlike cranes, lifting towers were not easily moved and offered no lateral play.

[98] Amm. Marc.17.4.15, 'digestisque ad perpendiculum altis trabibus, ut machinarum cerneres nemus, innectuntur vasti funes et longi ad speciem multiplicium liciorum caelum densitate

Blocks were rolled horizontally into the lifting tower, and then lifted using pulleys and capstans. Once at the appropriate level, they were rolled horizontally onto the large platform formed by the top of the expansive area of the partially completed arch above the entablature surmounting the columns. From this staging area smaller cranes lowered the column capitals and blocks of the projecting entablature into place; other stones were levered into position. As the arch grew vertically, workers extended the wooden tower upwards to facilitate the placement of the surmounting sculptures. Many questions arise regarding the mobility of the colossal statues. Where were they cast in Rome? How much did they weigh? Were they moved through the city in one piece as a propagandistic act or in pieces for easy lifting?[99]

The fourth and final phase of work at the Arch of Septimius Severus dealt with ornamentation and site clearance. The lifting tower was removed and the entire structure covered with wooden scaffolding extending about a metre from the surface. While sculptors did much of the carving on the reliefs and architectural pieces on the ground, they finished the more refined sculpting *in situ* working side by side on the scaffolding.[100] They were followed by teams of painters, polishers, letter carvers, and finally by workers who dismantled the wooden framework.[101] On the ground below, teams laboured to clean the surrounding section of the Forum Romanum. Smaller cranes operated to the last, situating the travertine blocks for the stairs on the eastern side, and returning statues, memorials, and pavers to their original positions.

The overall length of time required to erect the structure is not easily determined. The Romans did not regularly record specific initiation and completion dates, and were known to dedicate structures when incomplete.[102] Severus, like other Roman emperors, often clustered celebrations on the same date to create a collective impact. The following postulates a possible schedule. A fawning Senate first proposed the project either in 195 CE after Severus' first Parthian victory and the elevation of Caracalla as *Caesar*, or three years later amid the celebrations for the second Parthian victory, Caracalla's elevation to *Augustus*, and the centenary anniversary of Trajan's accession.[103] State

---

nimia subtexentes.... paulatimque in arduum per inane protentus, diu pensilis, hominum milibus multis tamquam molendinarias rotantibus metas'.

[99] The transport of colossal statues was a bravura act of engineering throughout the ancient world; Bingöl (2004: 38–41). The Propaganda value of lifting a giant statue is affirmed by the press surrounding the 2011 placement of a giant new statue of Alexander the Great in Skopje, Macedonia.

[100] Archaeological evidence indicates the reliefs, spandrels, and pedestals were carved in place; Brilliant (1967: 50, 108, 132, 151).

[101] James Packer has created a hypothetical model of the arch's colour scheme for a future publication on the Forum Romanum.

[102] Among other projects, the Basilica Iulia, Forum Augustum, and Baths of Caracalla were dedicated before fully completed; Richardson (1992: 52, 150); DeLaine (1997: 16).

[103] The emperor repeatedly associated himself with earlier 'good' rulers including Trajan; Birley (2000: 130).

architects, possibly from the army, developed the basic design and sent orders to eastern quarries for custom cut blocks. The Senate formally approved and initiated the project in 202 CE to mark Severus' return from the east. Cassius Dio described the lavish festivities for the tenth anniversary of his reign on 9 April, just two days before the emperor's birthday, and the wedding of Caracalla.[104] At this point construction had not yet advanced, since the wedding gifts were carried through the Forum Romanum in an extravagant display. The dedication (or completion) of the arch in 203 CE, a date confirmed by the titles of Septimius Severus in the inscription, coincided with formal announcements about the great *ludi Saeculares* to be celebrated the following year.[105] Such a scenario allows about one year for construction, ample time for a structure of this size and relatively simple design. The speed of completion depended, as today, on funding, availability of materials, number of labourers and machines, working space, and traffic. Delayed deliveries, obstructing events, and clogged roads all slowed progress.[106] Despite inevitable delays, the Romans did not take steps to improve urban traffic flow; congested Roman streets operated as well as they needed to, getting the materials to the site while providing ample opportunities for spectatorship.[107]

CONCLUSIONS

The building of the Severan arch significantly obstructed the most important civic stage for public events in the imperial capital. Full closure during construction was unthinkable except for short periods during complicated manoeuvres, yet the work and noise generated must have caused rescheduling and relocation. In particular, the Senate who usually convened directly east of the arch must have met elsewhere or curtailed sessions. This imposition suited Severus who considered the august body an annoyance.[108] Though the SPQR (Senate and people of Rome) commissioned the arch, he had determined its position through his dream. Situated before the Curia, the dynastic monument filled the view of anyone entering or exiting the Senate, and almost completely

---

[104] Cass. Dio 77.1.
[105] Birley (2000: 157); Gorrie (2002). Processions at the *ludi Saeculares* in 204 CE probably passed through the Severan arch; Zos. 2.4.3.
[106] The wide range of construction speeds is revealed by comparing the ratios of 'total weight of materials' to 'recorded time to complete'. That for the Colosseum was 581 tonnes/day, for the Arch of Constantine 14 tonnes/day. If applied to the Severan arch the length of building time ranges from 24 to 1,574 construction days.
[107] Newsome (2008: 443–6).
[108] SHA *Sev.* 12–13.

blocked sight of the Temple of Concordia.[109] In scale and ornamentation the Arch of Septimius Severus ostentatiously conveyed the military prowess of the emperor and his sons, with the great relief panels accentuating the huge number of soldiers under their command.[110] Erection of the arch itself reinforced the emperor's militarism already overtly evident in Rome. During construction the 'armies' of transporters and labourers melded with the hundreds of military men who crowded the capital's streets.[111] This urban reality was emulated on the arch in the register depicting a procession with numerous heavily laden wheeled vehicles moving relentlessly towards the goddess Roma.[112] The impressive lifting tower recalled military siege structures such as that memorialized on the north-west panel. In effect, the very act of creating the arch portrayed the 'visible strengths' of the emperor and his sons proclaimed on the grand dedicatory inscription.

Urbanistically, the Arch of Septimius Severus completed the work of regularizing and enclosing the central Forum Romanum initiated by Augustus. This single intervention gave clarity to the spatial form. The large arch closed off the forum's northern end where previously the space of the central area became undefined, leaking out over the Comitium and towards the north (Fig. 14.3). Standing as the fourth arched doorway of the forum, the arch led into a roughly rectangular, porticoed area with a temple on axis, an arrangement similar to the configuration of the imperial fora to the east. Like those complexes, Rome's traditional civic centre was now isolated and internalized. The overall design integrity discouraged subsequent architectural additions as did the design features that restricted mobility in the Forum Romanum. The stairs fronting the Severan arch hampered large building transports approaching from the accessible northern route, just as the level changes, porticoes, and sharp access turns compromised movement through the imperial fora. As a result, the Arch of Septimius Severus was the last major architectural addition to *l'area centrale* of the Forum Romanum in antiquity. Future monumental projects shifted to places with more space and easier access. Septimius Severus himself planned a grand entrance to the imperial residence on the south slopes of the Palatine, bypassing the complex layers of meaning, senatorial associations, and traffic challenges of the Forum Romanum.[113]

Like the man crushed by an overturned load of building materials, the impact of construction traffic has been buried. Architectural history has traditionally privileged finished structures or urban layouts as objects of

---

[109] Favro and Johanson (2010).
[110] Unlike the more narrative reliefs on Trajan's Column, those on the Severan arch did not include military construction scenes. For more on the iconography of the arch see Lusnia (2006).
[111] Cass. Dio 74.1 tactfully described the conspicuousness of soldiers in Rome during 193 CE; Birley (2000: 104, 129, 196–7).
[112] Beard (2007: 363); Brilliant (1967: 137–47).
[113] SHA Sev. 24.3.

study. The project-based construction analysis of the Arch of Septimius Severus affirms that the building process, including the traffic generated, is integral to understanding physical, political, and cultural historical contexts. In imperial Rome, the movement of materials and construction activities impacted mobility while conveying propagandistic messages. The implementation of designs that limited the accessibility of heavy transports hindered future developments. Magnificent, large-scale imperial buildings were found only where ox carts could go.

15

Movement and Urban Development at Two City Gates in Rome

The Porta Esquilina and Porta Tiburtina

Simon Malmberg and Hans Bjur

The focus of this chapter is movement through the Porta Esquilina in the Republican Wall and the Porta Tiburtina in the Aurelianic Wall, alongside an evaluation of the mutual influence of traffic and urban development from the time of Augustus to the late fifth century (Fig. 15.1).[1] Two decisive moments are identified. The first is the Augustan period, characterized by a great expansion and modernization of the infrastructure in the periphery. The second is the fifth century, with the construction of the Aurelianic Wall, resulting in the development of new urban nodes and a process of monumentalization of the Porta Tiburtina.

The interaction between movement and what are seen in urban studies as 'edge phenomena' (including city walls and customs boundaries) generated fundamental changes in the spatial configuration of the Via Tiburtina. The driving forces behind edge phenomena are high accessibility and a freedom of space and movement, which may form the basis for flexible and innovative urban development.[2] Over the course of the five centuries of the imperial period, the Via Tiburtina was subject to considerable changes, which can be seen in the emergence of new areas of habitation and production connected with the emergence of new centres of power, nodes, public spaces, and a new articulation of the street system. Fundamental to this study is the realization that the position of the Porta Esquilina and Porta Tiburtina played a decisive role not just in dictating how the city was accessed in terms of movement (and

---

[1] All dates in the text are CE unless otherwise stated.
[2] Sudjic (1992); Sieverts (2001).

Fig. 15.1. The south-eastern city edge of Rome with locations mentioned in the text (illustration by Simon Malmberg).

variations in urban flow over time), but also in determining the growth of the street network and the spatial development of the city.

The study begins by providing an overview of the Porta Esquilina at the beginning of the Augustan period, and the changes in the following decades. There follows a discussion of some of the practices concerning the nature of traffic passing through this area, touching upon the different purposes of that traffic, what means of transport were used and their demands on the infrastructure, traffic regulations and customs duties, as well as focusing on how physical structures were altered or created to accommodate the increase in traffic over time. Following this analysis of topography and movement in connection with the Esquiline in the early Empire, the discussion then shifts to the period of late antiquity, moving further out along the road to the Porta Tiburtina in the Aurelianic Wall. The focus of this part of the chapter is how the new city wall, the imperial presence, and new nodes in the form of churches and the cult of Christian martyrs affected traffic flows and urban development at the edge of the city of Rome.[3]

The area under investigation is the level ground of the Esquiline plateau, gently sloping downwards to the west and south (Fig. 15.1). The Republican

---

[3] See also Malmberg and Bjur (2009). On the Aurelian Wall see Dey (2011).

Wall, which in this part consists of the large earthworks of the Agger, more or less follows the westernmost brow of the plateau, beyond which the long ridges of the Viminal, Cispian, and Oppian hills stretch like fingers into the city and the low lying areas adjacent to the Tiber. From the Esquiline, the Argiletum and the Clivus Suburanus are the main traffic arteries leading to and from the Forum. The latter, if followed from the Forum Romanum, ascends the plateau of the Esquiline in a valley between the Cispian and Oppian hills, before reaching the crest just inside the Republican Wall. It is at this point that the Clivus Suburanus meets the wall at the southern point of the Agger, and this is the location of one of the two gates in this chapter, the Porta Esquilina. The slope between these hills constitutes the simplest route between the Esquiline plateau and the flood plain of the Tiber, and the route is presumably of a very ancient date. This is confirmed by remains unearthed just outside the present gate, probably contemporary with the original construction of the Republican Wall in the first half of the fourth century BCE.[4]

Outside the Porta Esquilina the road divided in two, with the Via Tiburtina forming a route to the east towards Tibur and the Viae Labicana–Praenestina providing a route to the south-east towards Labici and Praeneste. The ground along the Via Tiburtina gently slopes upwards towards a natural ridge, about 700 m from the Porta Esquilina, before descending towards the valley formed by the Marranella stream a further 2 km to the east. This natural ridge came to be used first for the construction of aqueducts and later for the new city wall built under Aurelian.[5]

THE AREA OF THE PORTA ESQUILINA UNDER AUGUSTUS

In the late Republic the area outside the Porta Esquilina was a dismal place on the margins of the city. At the crossroads public executions were performed, refuse was dumped in the ditch beyond the Agger, while the area south of the Via Labicana was used for the mass burial of the poor and slaves. These burials can be linked to the presence of the grove and sanctuary of Libitina, the goddess of funerals. An inscribed stone from the late second century BCE, marking the boundary of the sacred area, was found just north of the Porta Esquilina. The inscription on the stone also states that the sanctuary was part of the Pagus Montanus, marking this area as outside the city and subject to the

---

[4] Säflund (1932: 43–4).
[5] Aqua Marcia (144, 33, and 11–4 BCE), Aqua Tepula (125, 33, and 11–4 BCE), Aqua Iulia (33 and 11–4 BCE), Aqua Claudia (38–52 CE).

authority of a rural, rather than urban, district. Thus, it can be considered a republican suburb outside the Porta Esquilina.[6]

From the reign of Augustus onwards, the area underwent radical change. With the new city regions instituted by Augustus in 7 BCE, the Pagus Montanus was incorporated into the urban area of Rome. Here the words of Pliny ring true, that Rome's 'increasing spread of buildings has added a number of towns to it'.[7] This affected the whole character of the area: the location for mass burials was moved further out and the area south of the gate was taken over by the *horti* of Maecenas. Other noble families soon followed suit, transforming the area east of the city into what has become known as a 'green belt' of *horti*.[8] However, these *horti* should not be seen simply as villas and parklands. Inscriptions attest to significant production within the domains, as well as apartment blocks and burial grounds for the slaves working there.[9] During the first century, many of these *horti* shifted in ownership from private individuals to become the property of the emperors, a pattern of ownership that resulted in a barrier to urban development and exploitation into late antiquity.

Looking inside the Porta Esquilina, we can identify the development of the so-called Forum Esquilinum—mentioned by Appian in his discussion of the battle between Marius and Sulla at this gate in 88 BCE—the commercial association of which is confirmed by a group of early imperial inscriptions.[10] Interestingly, the commercial function within the walls was matched by commercial activities outside the Porta Esquilina, north of Via Tiburtina, where the Campus Esquilinus developed; associated in the late Republic with butchers and clothes-dealers, who identified their location as 'at the grove of Libitina'.[11] Similar *campi* developed also outside the Portae Viminalis and Caelimontana in the Republican Wall.[12] The roads that led through these three gates all had a regional significance, and were associated as entry points to the city for provisions and foodstuffs from the fertile eastern hinterland of Rome. As a result, informal, but nevertheless important markets developed at these gates.[13]

---

[6] Suet. *Claud.* 25; Tac. *Ann.* 2.32.3; *CIL* VI.3823; Bodel (1986: 43–51); Hinard (1987: 113–15); Wiseman (1998); Coarelli (1996, 1999b, 1999c, 1999d, 1999e). Compare the Aemiliana suburb in southern Campus Martius: Varro *Rust.* 3.2.6. See also MacDonald (1986); Patterson (2006); Goodman (2007).
[7] Plin. *HN* 3.5.67: 'exspatiantia tecta multas addidere urbes'. See also Hor. *Sat.* 1.8.14–16; Frézouls (1987); Favro (1996).
[8] Krautheimer (1980: 17). See also Wiseman (1998); Jolivet (1997). Modern use pioneered by *First Report* (1929) and Abercrombie (1945).
[9] Jolivet (1997); Coates-Stephens (2004: 33).
[10] App. *B.Civ.* 1.58; *CIL* VI.2223, 9179–80; Coarelli (1995b).
[11] Strabo 5.3.9; *CIL* VI.9974, 33870; Coarelli (1993a).
[12] Buzzetti (1993b); Coarelli (1993b).   [13] Morley (1996).

TRAFFIC THROUGH THE PORTA ESQUILINA

We can now consider how traffic through the Porta Esquilina functioned in practice: what means of transport were used and what regulations were in force to control traffic entering the city? It is important to understand the multitude of reasons for traffic and the different ways it was accomplished, in order to comprehend both the increasing requirements on infrastructure, and its impact on the physical environment around the Porta Esquilina, as well as the effect of traffic on the urban development of the area.[14]

Several conditions funnelled traffic through the Porta Esquilina. As mentioned, the hilly terrain inside the city focused traffic routes to the few convenient connections between the river's flood plain and the Esquiline plateau. The wall itself of course also acted as a barrier, even after the dismantling of the Republican Wall, because the Agger remained a linear obstacle to traffic.[15] In addition, the barrier of the imperial estates outside the walls created a corridor through which the Via Tiburtina passed, also delineated by the line of aqueduct arches, and because of the crossing points of two rivers—the Maranella and the Anio.[16] Together with the important traffic coming to the city along the Viae Labicana and Praenestina, this would have caused a large number of people and goods to be funnelled through the Porta Esquilina.

The most common traffic that passed through the Porta Esquilina was probably connected with trade. The area around Tibur was known for its fruit, figs, wine, and flowers. The eastern hinterland was also generally known for its production of vegetables and dairy products.[17] These were commodities that could not be transported over long distances, and so had to be supplied from the city's immediate surroundings. Often the burial plots outside the city wall were also used for cultivation, causing the cost of the burial plot to be offset through agricultural production.[18] Neville Morley has drawn attention to this integration between city and *suburbium*, stating that 'the countryside around the city of Rome became urbanized, not only in the density of the settlement there and the lack of a clear boundary with the city, but in its economy'.[19] In the light of this comment, we should not see the Porta Esquilina as separating the city from its eastern hinterland, but instead joining two spatially distinct parts of the urban economy together through movement.

Apart from the daily market, fairs called *nundinae* were held every eight days, which were occasions for people living around Rome to come into the city to sell their produce and buy manufactured goods, such as farm

---

[14] Laurence (2008); Mar (2008). [15] Hor. *Sat.* 1.8.14–16; Dion. Hal. *Ant.Rom.* 4.13.5.
[16] Mari (2008: 165).
[17] Hor. *Sat.* 2.4.70–1; Plin. *HN* 15.70, 17.120; Col. *Rust.* 5.10.11, 10.138; Mart. 9.60; Morley (1996: 92–5).
[18] Purcell (1987); Morley (1996: 95).   [19] Morley (1996: 92).

implements, clothes, and shoes. Often these *nundinae* were located on the outskirts of towns, close to and sometimes outside the city gates.[20] *Nundinae* probably took place at the Forum Esquilinum or the Campus Esquilinus. Annual fairs were also common, often associated with a sanctuary, and the yearly festival to the god associated with it. The regionary catalogues mention a temple to Hercules Sullanus outside the Porta Esquilina, the location of which has been confirmed by an inscription dedicated to Hercules Victor found outside the gate.[21] One may well imagine harvest fairs taking place here annually on 13 August, the festival day of Hercules Victor, which was also the day of the Vertumnalia, the festival dedicated to the gods of fruit and vegetables, and on 19 August, when the wine festival of the Vinalia was celebrated at the grove of Libitina close to the Campus Esquilinus.[22]

The Via Tiburtina had been an important transhumance route for centuries, allowing flocks of sheep summer pasture in the mountains above Tibur, while moving them through Rome to the coastal grasslands in winter.[23] Some animals, especially lambs, may have been sold off for slaughter at the city gate. As mentioned above, butchers are attested in the vicinity of the Campus Esquilinus. In the Horti Taurani, close to the Campus Esquilinus, there is a large wool manufactory attested in the first half of the first century,[24] which may have bought wool directly from passing flocks. Clothes-dealers were active at the Campus, as early as the second century BCE, and may have sold the produce of this manufactory.[25]

A lot of production was located on the outskirts of the city, in some cases because it was foul-smelling fulleries and tanneries, or slaughterhouses, or trades that involved the use of fire including potteries and glass production. Several large-scale bakeries with mills and ovens have been excavated on the Esquiline. The largest was situated at the later Porta Maggiore. Another one, with seven mills, was unearthed just outside the wall, to the north of Porta Esquilina. Both can be dated to the Augustan period, and there are also many contemporary memorials set up to the dead in the area, in which epitaphs name the deceased's profession as a baker. Apart from the fire hazard, a further reason for *pistrinae* to be located here is the possibility that water-powered mills were fed by water from the aqueducts on the Esquiline.[26] These

---

[20] Cato *Agr.* 135; Livy 7.15.13; Col. *Rust.* 11.1.23; Plin. *HN* 18.13; Plin. *Ep.* 5.4.1; Festus 176; Degrassi (1963: 300–6); Frayn (1993: 26, 34, 38–41); De Ligt (1993: 111–17); Morley (1996: 166–74); Trout (1996); Lo Cascio (2000a).

[21] *CIL* VI.330; Keaveney (1983); Palmer (1990); Frayn (1993: 133); Palombi (1996). Probably built by Sulla after victory at the gate (App. *B.Civ.* 1.58), though the cult is probably much older. There were also temples to Hercules outside Porta Capena, Porta Salaria, and close to Porta Trigemina.

[22] Ovid, *Fasti* 863–76; Livy 40.34.4; Varro *Ling.* 6.16–20; *CIL* I.1, pp. 324–6.

[23] Thompson (1988); Santillo Frizell (2009).

[24] Coates-Stephens (2004: 33).

[25] Cato *De Agr.* 135.

[26] Lanciani (1874: 36); Coates-Stephens (2004: 21–31); Cf. Wilson (2001).

facilities generated further traffic, both in raw materials to the Esquiline and in finished products to the city.

Another kind of traffic was the transport of building materials. The Tibur region was known for its access to timber in antiquity, and between the city and Rome were the famous travertine quarries.[27] It has been suggested that the increasing transport of travertine was directly related to the broadening of the Via Tiburtina into a two-lane highway in the Augustan period.[28] This indicates that the land route was favoured for the transport of stone, and possibly also timber.[29] Carts (*plostra*) may have left the quarries every four minutes during large-scale building projects such as the Colosseum.[30] This enormous level of traffic also led to an increase in the building of road stations, of which several have been excavated and dated to the Augustan period.[31] The route of the road was also important for the maintenance of the aqueducts between Tibur and Rome.[32] This involved a massive investment in infrastructure by the government, which also benefited all other forms of communication, and which may have led to an economic boom along the route.

Long-distance travel was one kind of traffic that surely benefited from the new infrastructure. The new way-stations were ideal stopping places for fast travel, either by the *cursus publicus* or by private means. It made locations along the route more accessible, and the possession of villas in the Tibur area more attractive. Tibur had by Augustan times become a magnet for aristocratic villas, whose owners could take advantage of the ample water supply offered by the aqueducts, a cooler climate, and a splendid view across the *suburbium* and sight of the city of Rome in the distance. Moreover, it was considered a healthier place to live than in the city or even the coast.[33] It needs also to be remembered that on the way to Tibur lay the sulphurous springs of Aquae Albulae, famous for their healing qualities.[34] In summary, there was probably regular traffic along the road on the part of the aristocracy and their households travelling back and forth to their villas from the city. These households had to get supplies and luxury goods from the city, leading to further traffic through the gate and out onto the main road.

Foremost among these leisurely travellers were the emperors themselves. In the first century, when most of the *horti* had come into imperial hands, the

---

[27] Mart. 7.28.
[28] Mari (1983: 366–7; 2008: 171).
[29] Juv. 3.534–9. Arguments for land transport: Laurence (1998b). The Anio presented an alternative route: Strabo 3.5.11.
[30] DeLaine (1995: 559).
[31] Klynne (2009).
[32] Mari (2008: 172).
[33] Sen. *Ben.* 4.12.3; Kolb (2001).
[34] Strabo 3.5.11. See also Peutinger Map: André and Baslez (1993: 276–7).

court often removed to these grounds outside the city wall.[35] The emperors also had villas further out from the city, of which the grandest and most famous was the Villa Adriana outside Tibur. The movement of the court was a massive undertaking. Hadrian was probably accompanied by about five thousand men on his way to Egypt, while Caesar is said to have brought two thousand men with him when he visited Cicero in Campania.[36] An imperial court located either in the *horti* or outside Tibur consisting of thousands or even tens of thousands of people involved a major logistical undertaking, leading to a massive increase in traffic through the Porta Esquilina and along the Via Tiburtina.

Like other major roads leading from Rome, the Via Tiburtina was lined by tombs, some of which were truly monumental. People regularly attended the tombs, for communal meals, for growing crops, or even to live in them.[37] The *libitinarii* (undertakers) prepared burials and handled dead bodies and had to live outside the walls of the city. They were also responsible for public torture and executions taking place at the Campus Esquilinus, which may have attracted crowds from the city. The landscape of 'hollows, garden walls, tombs and sunken lanes'[38] outside the city gate also attracted prostitutes and their customers, presumably both from inside the city and visitors who frequented the inns which concentrated around the city gates. At the inn travellers could also find a place for their animals and vehicles.[39] As demonstrated, many people lived outside the walls, while many others perhaps lived inside the city but worked outside. Much of the traffic probably consisted of local inhabitants moving between the parts of the city that were inside and those that were outside the walls.

THE DEMANDS OF TRAFFIC AND TRAFFIC REGULATION

It is necessary, now, to turn our attention to the processes of movement that we associate with such levels of traffic. This is done with a view to determining what demands were made by traffic on the gate and the road itself in the area

---

[35] Millar (1977: 18–24).
[36] Cic. *Att.* 13.52; *P.Oxyr.* 3602–5; Halfmann (1986: 84, 110). Nero allegedly travelled with 1,000 vehicles: Suet. *Nero* 30.
[37] Purcell (1987).
[38] Livy 26.10.6, 'convalles tectaque hortorum et sepulcra et cavas undique vias', trans. Wiseman (1998: 15).
[39] Mart. 1.34, 3.82. *Lupanarii* at Porta Caelimontana in Regionary Catalogues. Inns: *CIL* IX.2689; Chevallier (1988: 67–8); Laurence (1994a: 73, 81); Wallace-Hadrill (1995: 44). *Graffito* outside Porta Marina in Pompeii: *CIL* IV.1751; Hartnett (2008: 95–8). See Poehler, in this volume.

of Porta Esquilina. To do this, an overview is presented of the different vehicles, animals, and humans involved in transport. It is important to break down into categories the types of road use, the better to understand their different requirements and the accompanying traffic regulations imposed upon these different forms of transportation.

Transport of goods and persons were often made by carriage. A bewildering number of designations for different types of carriages have come down to us in literature. They can mainly be divided into light two-wheeled carts for fast travel, mainly carrying people, and heavy four-wheeled carriages, used both for goods and persons. Among the former are the covered *carpentum* and the open *cisium*, usually drawn by mules, or horses for speed. In the latter category were the *carruca*, a four-wheeled carriage mainly used for transporting persons, while the *raeda* was used to carry both humans and goods and seems to have been very versatile. The *plostrum*, which could be either two- or four-wheeled, was a low-status vehicle. It was usually associated with farming and goods transport, drawn by oxen, and proverbially dangerous in traffic.[40]

Carriages required certain widths. A restored cart in the Casa del Menandro in Pompeii is 142 cm wide (179 cm including hubs), while the four-wheeled carriage at the Villa della Arianna from Stabiae measures 148 cm (185 cm including hubs).[41] So for two-way traffic to be possible, the city gate had to be at least about 4 m wide, while a carriageway required somewhat over 3 m. Modern experiments show that two horses can pull up to 1,000 kg on a four-wheeled wagon.[42] Restrictions were placed on the *cursus publicus*, which was not allowed to transport more than 492 kg on a carriage.[43] Pack animals were probably a more common means of transport. Mules had higher speed than oxen, more stamina than horses, and needed less fodder. They could carry up to around 180 kg in panniers.[44] Mules were thus more cost-effective than wagons, or other pack animals, advantages to which we can add ease of access when compared to wheeled transport within the city. Wagons only had an advantage when it came to carrying indivisible or bulky loads. So probably most of the traffic in goods took place by pack animals, and foremost by mules. To use porters (*geruli* or *saccarii*) was another cheap and convenient way to transport goods, though only for shorter stretches.[45]

As already mentioned, much vehicular traffic was viewed as potentially dangerous. From the late Republic onwards, several attempts were made to

---

[40] Laurence (1999: 136); van Tilburg (2007: 52–3, 73); *Plostra*: Plaut. *Epid.* 4.2.22; Juv. 3.255–9; *Dig.* 9.2.52.2; Hor. *Epist.* 2.2.72–4; Auson. *Ep.* 6.26.
[41] Tsujimura (1991: 61–2).
[42] Raepsaet (2008: 590), contra Needham (1986: 312).
[43] Raepsaet (2008: 600); Leighton (1972: 72); *Cod. Theod.* 8.5.
[44] Laurence (1999: 123–6); White (1984: 133, 208); Chevallier (1988: 37); van Tilburg (2007: 72); Raepsaet (2008: 589–90).
[45] Brunt (1980: 94); Frayn (1993: 78–9); Raepsaet (2008: 589).

bring order to the chaotic situation by implementing traffic regulations.[46] This would also have a fundamental impact on the activities and urban development at gate areas such as outside Porta Esquilina. The well-known *Tabula Heracleensis* may have been part of Caesar's legislation for Rome. It is very interesting for our purposes here, because it is one of few examples of traffic regulations that have come down to us in original form, although only eleven lines long.[47] The introductory lines are worth quoting in full:

> Whatever roads lie or shall lie within the city of Rome within those areas where there shall be continuous habitation, no one...in day-time, after sunrise or before the tenth hour of the day, is to lead or drive a *plostrum* on those roads, except...[48]

There follow exceptions for building, maintaining, or demolishing temples, for public rites by Vestal virgins or public priests, for triumphs and games, and for leaving the city with an empty carriage or with refuse.

Three details in the text may be remarked upon. Carriages are only mentioned under the term *plostra*. It is possible that the legislation meant to stop only these overloaded and proverbially hazardous vehicles from entering Rome in daytime, thus reducing the danger to pedestrians from sunrise to the end of the ninth hour—the period of the day in which the greatest number of people were moving around the city.[49] Other carriages may not have been affected by the law. Secondly, the area encompassed by the law was not bounded by the city wall, but rather the 'continuous habitation'. It is clear that this term comprises areas outside the walls.[50] Regulation could in other words apply also to the extramural quarters on the Esquiline. Finally, there are a lot of exceptions to the law. Most government traffic was not affected, and the real purpose of the law may have been to facilitate that traffic, especially to bring in building materials including those transported to Rome along the Via Tiburtina.[51]

Another question is how much the law actually changed things. How practical was it to drive a wagon through Rome in the middle of the day? If you had heavy goods to transport, porters, pack animals, or light carts could probably have done the job more efficiently than heavy wagons. And if you really wanted to drive a *plostrum* inside the walls of Rome in daytime, who was

---

[46] Tsujimura (1991); Sonnabend (1992); Poehler (2006); Eck (2008); Laurence (2008); Malmberg (2009a, 2009b).

[47] *Tab. Herc.* 56–67; Crawford (1996: 355–91); Nicolet (1987); Kaiser, in this volume.

[48] *Tab. Herc.* 56–7: 'Quae viae in u(rbem) R(omam) sunt erunt intra ea loca, ubi continenti habetabetur, ne quis in ieis vieis...plostrum interdiu post solem ortum neve ante horam decimam diei ducito agito, nisi quod...'

[49] Hadrian banned heavily laden wagons from Rome: SHA *Hadr.* 22.6.

[50] *Dig.* 3.3.6; 33.9.4.2–6; 50.16.2, 154.

[51] See also Suet. *Claud.* 25; SHA *Hadr.* 22.6; SHA *Marc.* 23.8; SHA *Alex.Sev.* 43.1; *Cod. Theod.* 14.12.1.

Movement at City Gates in Rome 371

Fig. 15.2. The Via del Corso today: an example of vehicle traffic on a 'pedestrian' street within the Zone of Limited Traffic (photo by Simon Malmberg).

to stop you? Today, the Via del Corso in Rome is both within the Zona a Traffico Limitato (ZTL) and a pedestrian street, but still contains an amazing amount of traffic (Fig. 15.2).

The law probably reflected a preference for the use of carriages outside the city, and reinforced this trend through the deployment of the force of law. For goods to reach the city from daybreak to the tenth hour, they had to be reloaded onto light carts, pack animals, or porters at the city gates. In Pompeii, it seems that the vehicles normally used within the city were of the two-wheel type, since there are no separate curve traces for rear wheels.[52] The regionary catalogues mention an Area Carruces close to Porta Capena, where the Via Appia reached the Republican Wall. In the area was also the *collegium* of the coachmen.[53] A convenient way to leave the city, if you did not own your own means of transport, was either to pay for a seat on a passenger wagon, or to hire a carriage or animal.[54] These services were offered by *cisiarii*, who let wagons, and *iumentarii*, who hired out animals. These entrepreneurs

[52] Tsujimura (1991: 65).
[53] ILS 9047 (*schola carrucariorum*); Rodríguez-Almeida (1993b).
[54] Edictum Diocletiani 17.1–2.

seemed to crowd around city gates, and the Porta Esquilina was probably no exception.[55] Juvenal tells us how a family ready to leave the city loaded their belongings on to a *raeda* at the Porta Capena, while Martial pokes fun at an aristocrat for stacking a *raeda*, destined for his villa, with vegetables and animals at the same gate.[56] In the region of the Porta Capena was also situated the Area Radicaria, a place for weighing grain, and the Area Pannaria, a place of wool-trading.[57] Many of these facilities were probably present at Porta Esquilina as well. The transfer of grain and wool from the heavy highway wagons was carried out at this point, which probably also was connected to the customs levied at the gates.

THE NEW AUGUSTAN INFRASTRUCTURE

There was a great increase in traffic in the late Republic, partly because of the continued growth in urban and suburban population and partly through the large building works of Pompey, Caesar, and other Republican magnates. When peace returned under Augustus the city gates were in dire need of modernization due to new infrastructural demands. The defensive importance of the city wall was finally gone and so commercial and traffic requirements could assume priority. Without military functions, the idea to build new gates was probably motivated by propagandistic values, civic pride, and a will to define the limits of the city proper.[58] Augustus rebuilt several of the old gates in the Republican Wall, for instance the Porta Trigemina close to the Forum Boarium, and the Porta Caelimontana.[59] Scholarly opinion now also favours the Porta Esquilina as part of this programme (Fig. 15.3).[60] These gates are all located at some of the busiest entrances to the city. To get a picture of the old gates, one may use the Porta Viminalis as a comparison, since this gate was not rebuilt, and preserved its Republican form inside the Agger until excavated in the nineteenth century. It consisted of a single passageway, only 3 m wide, with room only for one wagon at a time.[61] In contrast, the new Porta Esquilina consisted of three passageways: a central one 7.16 m wide, while the side

---

[55] See also the Porta Romana at Ostia (van Tilburg 2007: 47).
[56] Juv. 3.9–14; Mart. 3.47.
[57] Rodríguez-Almeida (1993c; 1993d).
[58] Zanker (1988: 329).
[59] Trigemina: *CIL* VI.1385; Caelimontana: *CIL* VI.1384; Coarelli (1988: 42–50); Étienne (1987).
[60] Lugli (1937); Santa Maria Scriniari (1979); De Angelis Bertolotti (1983); De Maria (1988: 191, 311); Le Gall (1991); Rodríguez-Almeida (1993a); Caruso and Volpe (1995); Coarelli (1996b). Formerly identified as the Arch of Gallienus because of its inscription, but Lugli could conclude this was added in the mid-3rd century to the Augustan arch, a view shared by scholars since.
[61] Strabo 5.3.7; Säflund (1932: 65); Coarelli (1996c).

Fig. 15.3. The Porta Esquilina in 1756. Through the Augustan arch one can make out the monumental fountain outside the gate (illustration from Vasi 1756: plate 126).

arches measured 3.45 m.[62] This was ample room to have two-way (or even three-lane) traffic through the central gate, while the two smaller ones presumably were used for pedestrian traffic. Because the gate had no military use, it was equipped with neither a portcullis nor doors. The monumental gateway in marble with its three large passages shows how important this traffic route had become in the Augustan period.[63]

This widening of the gate corresponds to a similar expansion of the roadway. The pre-Augustan Via Tiburtina was between 3 m and 4 m wide between Rome and Tibur. This would make it hard to accommodate two-way traffic, except for short stretches. Augustan rebuilding widened the carriageway to between 4.10 m and 4.80 m, which made ample room for two-way traffic.

---

[62] De Maria (1988: 311–12); Rodríguez-Almeida (1993a). The side arches were demolished in 1477, while the northern side arch was excavated in 1834.

[63] Strabo 5.3.9; Patterson (1999). Compare the contemporary Porta Ercolano, Pompeii and Porta Praetoria, Aosta, or later gate at Timgad.

Moreover, Augustus also fitted the highway with sidewalks, at least as far as the bridge across the Anio, slightly less than 7 km away from the Porta Esquilina, and possibly all the way to Tibur, a distance of 35 km. The sidewalks, or rather side paths, were between 1.5 m and 6 m wide, and were presumably used by riders, pack animals, and transhumance flocks, so as not to interfere with the busy vehicle traffic. But the sidewalks also attest to a lively pedestrian traffic, reaching tombs, agricultural areas, and manufactories along the route.[64]

The Augustan rebuilding of aqueducts, the Aquae Marcia and Tepula, together with the new Aqua Iulia, changed the vista along the Via Tiburtina forever, by raising the aqueducts on high arches (Fig. 15.1). As with the Aqua Virgo on the Campus Martius, the arches of the new aqueduct probably came to form the outer limit of the built-up area in the imperial period, and travellers from the east could have had the impression of a city wall.[65] At the point where the aqueducts cross the Via Tiburtina, Augustus built a monumental arch in travertine, with an inscription commemorating his deeds. The archway is about 5.5 m wide, which could have accommodated both the new wide carriageway and a sidewalk.[66] Thus, he created a formal entrance to the built-up area, in much the same way as the Augustan arches over the Viae Flaminia and Praenestina, and the Claudian ones over the Viae Labicana–Praenestina, signalling 'a subtle change from *Suburbium* to *Urbs*'.[67]

PENT-UP MOVEMENT AT THE GATE

Often the city gates themselves are regarded as barriers to traffic. This may be true of other ancient cities, and also of Rome's Republican gates. However, through the Augustan rebuilding of the most important gates to widths exceeding 7 m, they no longer hampered traffic. But the approaches to Rome, and some gate areas, were still bottlenecks in the traffic system, caused by the implementation of a customs tax. A customs border with thirty-seven gates was installed in 73–4.[68] The border was probably the same as that delineated in 175 by stone markers, four of which have survived. Three of these were situated at the Portae Flaminia, Salaria, and Asinaria of the later Aurelianic Wall, but the fourth was found at the Porta Esquilina.[69] Thus, due to this position of the stone marker at this location, we may conclude that

[64] Mari (2008: 171–2); Quilici (2008: 562–3).
[65] See also Wiseman (1993; 2008).
[66] Richmond (1930: fig. 33).
[67] Coates-Stephens (2004: 15, 34); see also Wiseman (2008).
[68] Plin. *HN* 3.65–6; Palmer (1980) is fundamental; see also De Laet (1949: 347–9, 425–46). Customs border may have been created by Augustus: Palmer (1980: 217–18).
[69] *CIL* VI.1016a–c, 31227. Cf. known *cippi* of the extended pomerium.

the customs barrier did not extend beyond the Agger on the eastern side of the city. This probably changed with the building of the new city wall in the 270s under the emperor Aurelian, which may have also pushed out the customs barrier to the Porta Tiburtina. The Aurelianic Wall was clearly related to the customs barrier: the thirty-seven customs stations mentioned by Pliny corresponded to thirty-seven gates in the new wall. The customs rights were often sold to aristocrats, such as the privately owned customs station at the Porta Nomentana at the beginning of the fifth century.[70]

Everything, even corpses on their way to burial, passing the customs barrier was taxed, except if it was for personal use. The tax collectors had the right to inspect passing goods and any which had not been declared were confiscated.[71] The only ones exempt from customs were those of the emperor and his court, public officials, and soldiers and their families.[72] Large depots were purposely located beyond the customs barrier. At the wholesalers' storehouses the merchandise could have then passed from long-distance tradesmen to local traders with the mediation of brokers. It was up to these local traders to pay the customs when they brought their goods inside the city on their lighter, two-wheeled carriages. To improve conditions for these middlemen, public warehouses were constructed in the early third century in all urban regions so that traders without private means of storage could conveniently import goods into the city.[73] From stocks inside the city the merchandise then passed to the simple street vendor, whose costs included whatever the middleman had paid in customs.

The corruption and extortion displayed by Roman custom officials were notorious.[74] Inspection of goods, arguments, and payment must have caused long delays when passing the toll stations. A remedy in the fourth century was to display bronze plaques on wagons to signal a deal between the private proprietors of the customs stations and the owner of the wagon.[75] Presumably accounts were kept and settled at regular intervals between them. This allowed the drivers to pass through the gates without any delay, but was probably restricted to habitual travellers of some means.

So traffic was delayed at the gate, and caused this space to be associated with waiting to enter or leave the city. Consequently gate areas developed into nodes within the overall system of urban movement, places to locate inns, to find coachmen, to establish storehouses and for the sale of sex via prostitution, all of which can be defined as trades that catered for the needs of the traveller. The city gates thus became important transit areas, and we can see the gates of Rome developing as nodes to form edge phenomena.

---

[70] Lib.Pont. 1.222.
[71] Quint. Decl. 359; CJ 3.44.15; 4.61.5; CIL III.5122, 14354; van Tilburg (2007: 86–9).
[72] Dig. 39.4.4.1; 39.4.9.7; 49.14.6.1; Tac. Ann. 13.51.
[73] SHA Sev.Alex. 33.1–2; 39.3–10; Patterson (2000: 94).
[74] Plut. Mor. 518E; Lib. Or. 50.16, 29.
[75] e.g. CIL VI.32033.

EXTRAMURAL URBANIZATION AND A NEW MARKET

In the second century at the latest, high-rise buildings had spread to the area outside the Republican Wall, attested by excavations south of the Praetorian Camp, a preserved apartment building beneath Sant'Eusebio, and another insula built into the Aurelianic Wall south of the Porta Tiburtina (Fig. 15.1).[76] Significantly, housing was only located north of or close by the Via Tiburtina, an area that was in close proximity to the Praetorian Camp, which may have been a factor that attracted people to locate their homes and businesses nearby. The area to the north was probably more open to exploitation, with a road running along the outer line of the Agger, providing connections with the Portae Viminalis and Collina. It was also on the northern side that the only late antique parish church of the area, Sant'Eusebio, was located, which indicates a sizeable population in this period. The imperial estates to the south, however, remained a formidable barrier to urban development, and one may imagine the densely built-up area, hugging the Via Tiburtina, in the form of a linear town with only one main street.[77]

The market at Campus Esquilinus gradually became more formal with, in all probability, official recognition of its existence. As mentioned above, a temple to Hercules was built at the Campus by Sulla. Augustus drew a sideline from one of the aqueducts to furnish a monumental fountain, placed at the crossroads of the Viae Tiburtina and Labicana, and perfectly aligned with the arch of the Porta Esquilina (Fig. 15.4).[78] Together with the temple, the *nymphaeum* probably formed the focus of the marketplace. These early steps were followed in the early second century by the building of a *macellum* just north of the Campus. It had become too awkward to bring merchandise into the city centre, so the merchants had to move to the edge of town.

In 1872, a large rectangular court was excavated just outside the Porta Esquilina, next to the Campus Esquilinus (Fig. 15.1). It measured 80 m by 25 m and was surrounded by porticoes and shops, with its main entrance onto the Via Tiburtina to the south. Trajanic lead pipes and Hadrianic brick stamps date the building to the early second century.[79] It seems close at hand to link the construction of this new market, which can be securely identified as a *macellum*, with the destruction of the old *macellum* near the Forum Romanum only a few years earlier.[80] An important aspect is that the

---

[76] Packer (1971: 27, pl. 113); Caruso and Volpe (1992); De Spirito (1995a) Pisani Sartorio (1996d: 290); Caruso (1996); Cozza (1997); Guidobaldi (2000: 151).
[77] Carter (1983: 130–6); Kostoff (1992: 34–5).
[78] Replaced in 3rd century by larger fountain: Tedeschi Grisanti (1977, 1996).
[79] Usually identified as the Macellum Liviae. De Ruyt (1983: 163–72); Pisani Sartorio (1996b).
[80] Tortorici (1991); Pisani Sartorio (1996a).

Fig. 15.4. The monumental fountain outside the Porta Esquilina in 1606. Originally built by Augustus, it was later enlarged in the 220s (illustration from Sadeler 1606: plate 25).

new *macellum* was situated outside both the city gate and the customs barrier, which may have allowed the long-distance traders to bring their heavy wagons right up to the market and to avoid paying customs duties, which was instead levied upon the middlemen. One may compare this grouping of *macellum*, market square, city gate, and temple with similar clusters in Djemila, Timgad, or at the Porta Marina in Ostia.[81] The location of the new *macellum* might have been the reason why the customs barrier was not shifted further outwards, despite increasing urban development of the area, as similar structures had been on the Viae Flaminia, Salaria, and Asinaria by 175.

The *macellum* catered to the needs of the more well-to-do, and thus attests to the changing character of the area, providing a luxurious alternative to the *nundinae* at the Campus.[82] A few decades earlier, under Nero, the Macellum Magnum had been constructed in similar surroundings, close to the *campus* outside the Porta Caelimontana.[83] The location of these two *macella* reflected the flourishing agriculture of the eastern hinterland, as well as the separation of commercial activities from the area of the Forum Romanum itself.[84]

---

[81] MacDonald (1986), see also Forum Tauri below.
[82] Frayn (1993: 34).
[83] Cass. Dio 62.18.3; Carignani et al. (1990); Buzzetti (1993b); Pisani Sartorio (1996c).
[84] Coarelli (1986: 42–3); Morel (1987: 137–9); Patterson (2000); De Ruyt (2000).

THE LATE ANTIQUE PORTA TIBURTINA NODE

In the 270s a major restructuring of Rome's periphery took place through the building of the new city walls. In many places these caused severe upheavals for the city population and major changes to the urban fabric. This was partly the case in the Tiburtina area also, with the destruction or abandonment of houses and grave monuments in the area of the new city wall. However, in places we can establish a degree of continuity. Sections of the Aurelianic defensive circuit followed the line of aqueducts, whose arches were simply walled up to create the new fortification (Fig. 15.1). The Augustan arch across the roadway was transformed by Aurelian into a city gate, and the Porta Tiburtina was flanked by two imposing towers (Fig. 15.5).[85] The Porta

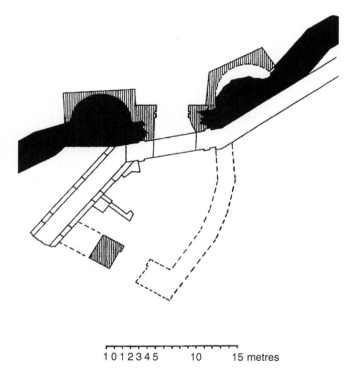

Fig. 15.5. Plan of the Porta Tiburtina. In white outline the Augustan aqueduct channels passing the road on a travertine arch; in black the first phase of the Aurelian Wall (271–9); shaded, the second phase of the Aurelian Wall (401–2), which added new towers and an inner court (illustration based on Cozza 1997: 93, fig. 109).

---

[85] Cozza (1997: 88–103); Coates Stephens (1998; 2004: 79–89), see also Stevens (1996).

Fig. 15.6. The Porta Tiburtina, inner side. Photograph taken before the removal of the inner court in 1869; the Augustan arch is visible through the court gate (anonymous photo).

Tiburtina seems not to have been an important gate in the Aurelianic phase, since it had only a single arch, compared to the four main gates of the city which in all cases had two arches. However, in the rebuilding by Honorius in 401–2, all of the gates, with the exception of the Porta Portuensis, were reduced to a single arch. In this phase of rebuilding, the Porta Tiburtina received larger towers, an inner gate, and a new decoration that comprised a travertine facade, a monumental inscription, and statues of the emperors (Figs. 15.6 and 15.7). Thus, in this second phase of the gate it received an upgrade in terms of status and counted among the most important gates, like the Porta Asinaria near the Lateran, which was also upgraded at this time.[86]

The building of this new monumental entrance to the city signals that the south-eastern part of the city had become much more important in the period between 270 and 400. In all probability, this was due to the existence of the imperial and episcopal residences in the area. The area had always been associated with places of imperial residence, the *horti* for example, and under the Severans, a new imperial residence was constructed close to the

---

[86] Pisani Sartorio (1996d; 1996e).

Fig. 15.7. The Porta Tiburtina in 1747, outer side. Both the Honorian and Augustan arches are visible; the towers are mainly from the 1450s (illustration from Vasi 1747: plate 6).

later Porta Maggiore (but was probably abandoned at the time of Aurelian). Constantine, however, seems to have extended this palace as far north as the present church of Santa Bibiana (Fig. 15.1). In the 430s the emperor again took up residence in Rome, and Valentinian's choice of location was the Constantinian palace on the Esquiline. It seems likely that this palace was renamed the Palatium Licinianum mentioned in sources from the 460s—a period in which the palace may have expanded northwards towards the Porta Tiburtina.[87]

However, there was not just palace building in the area. Inside the Porta Tiburtina a formal square was built, probably by Flavius Taurus as praetorian prefect of Italy in 355–61, and named after him Forum Tauri.[88] It is possible that the forum gave its name to the whole city district around it, known as the region of Caput Tauri, or it might derive from the *bucranium* decoration on the Augustan arch at the Porta Tiburtina (Fig. 15.6). The gate itself was also from the fifth century identified with this forum (or the decoration on the gate) since it became known as the Porta Taurina.[89] A large *nymphaeum* preserved with all its statues in good condition was excavated when part of the

---

[87] Cima (1995); Rizzo (1996); Guidobaldi (1998); De Spirito (1999).
[88] De Spirito (1995d).
[89] Pisani Sartorio (1996e).

city wall was torn down close to Porta Tiburtina in 1886. Because of the rudimentary report, it is hard to date the structure, but it may have formed part of the decoration of the Forum Tauri.[90]

The Forum Tauri and Forum Esquilinum had similar positions, immediately inside city gates which also marked the customs limit. The Forum Appiae, inside the Porta Appia, and the Forum Sallustii, inside the Porta Salaria, were similar to the Forum Tauri, and were probably also built in the fourth century.[91] In view of the spontaneous development of the Campus Esquilinus outside the Porta Esquilina the conclusion that a similar development began outside the Porta Tiburtina seems likely. With this spontaneous *campus* might be associated the church of San Gennaro, first mentioned in the 590s and situated where the Viae Tiburtina and Collatina forked only a few dozen metres beyond the gate (Fig. 15.1).[92]

The congestion around the Porta Esquilina for goods transportation to and from the *macellum* might have become overwhelming due to the increased urbanization of the area. The pressure for access to land along the Via Tiburtina was perhaps the motivation behind the parcelling up and privatization of the imperial *horti* after the construction of the Aurelianic Wall. This development was especially marked in the north and in the areas closest to the main roads.[93] This prompted the move of *macellum* facilities further to the east and out from the city's centre. The building of the new walls probably moved the customs border to the Porta Tiburtina, and may also have forced the building of a new *macellum* there. In late antiquity, parts of the *macellum* close to the Porta Esquilina had been taken over by private dwellings and a small bath building.[94] The church of San Vito in Macello may also have been constructed inside the *macellum* before 340. This indicates that this *macellum* had, at least partly, gone out of use at this time, perhaps to be supplanted by a new construction close to Porta Tiburtina.

The huge episcopal church of Liberius, built in the 350s, was presumably located between the present church of Santa Bibiana and the Porta Tiburtina, and thus probably formed the focus of the Forum Tauri. It was one way by which the bishop could become associated with this new suburban centre, and with the newly established imperial palace just to the south of the square. Liberius probably worked together with Taurus in creating this joint monumental complex. To this must be added the contemporaneous Macellum Liviae or Livianum, close to the Porta Tiburtina, because of its close association with the Basilica of Liberius. This *macellum*, as yet not archaeologically

---

[90] Gatti (1886); Caruso and Volpe (1992); Pisani Sartorio (1996d). Cf. waterworks at Porta Maggiore (Coates-Stephens 2005: 59) and Porta Esquilina.
[91] De Spirito (1995b; 1995c).
[92] Serra (2005a).
[93] Coates-Stephens (2004: 104–5).   [94] De Ruyt (1983: 166).

attested, should thus not be identified with the excavated *macellum* building close to the Porta Esquilina. It was in all probability located close to the Forum Tauri, either at the square itself or, perhaps more plausibly, beyond the gate and outside the customs limit. Its name may have derived either from its proximity to the church, or by it being commissioned by Liberius (Fig. 15.1).[95]

MARTYR CULT AND TRAFFIC

The main generator of movement and urban development around the Porta Tiburtina in late antiquity was the extra-urban sanctuary of San Lorenzo. The venerated grave of Lawrence was next to a suburban temple of Hercules (first mentioned in the third century BCE). There may be some connection between the cults, since Lawrence was celebrated on 10 August, and Hercules only two days later.[96] Constantine dedicated a large church (99m by 35m) to the saint approximately 1 km outside the Porta Tiburtina, next to the subterranean crypt of the martyr.[97] Lawrence, a deacon of Bishop Sixtus II, was executed together with his bishop during the persecutions of 258. He counts among the most important martyrs of the Roman Church, and his shrine is the most venerated after those to Peter and Paul. As early as the time of Constantine, special stairs had to be built to accommodate the steady stream of pilgrims who wanted to visit the grave. The status of Lawrence attracted many prestigious burials to the church, among those three fifth-century popes. Indeed, the pontificate of Sixtus III in the 430s began a period of unprecedented popularity of the cult. Sixtus probably undertook a large restoration of the church, and provided it with a baptistery, before being interred next to Lawrence after his death.[98] Furthermore, in the 460s Pope Hilary built an episcopal residence, a monastery, two baths, and two libraries next to the church. Later in the fifth century, basilicas to Agapitus and Hippolytus, followers of Lawrence, were also built in the vicinity, together with a hostel for paupers and pilgrims.[99] A new urban node thus developed around the martyr's grave, requiring a complex infrastructure and a close relationship with the cluster at Porta Tiburtina.

---

[95] The church of Liberius and the Macellum Liviae were located close to each other, also close to the 'Claudian aqueduct arches' and Forum Tauri. A discussion of the new findings proposed here will be presented in a separate article: Malmberg (forthcoming). See also Malmberg and Bjur (2009).

[96] Livy 26.10.3; Mari (1983: 235–8; 2005).

[97] *Lib.Pont.* 1.181; Krautheimer (1959: 94–114, 118–23).

[98] *Lib.Pont.* 1.234–5.

[99] *Lib.Pont.* 1.245, 252; Serra (2001, 2005b); Coates-Stephens (2003: 429–30); Brandt (2009).

It may be in connection with Honorius' major rebuilding of the Porta Tiburtina (Fig. 15.7) that a portico between the gate and San Lorenzo was built, although this portico is not mentioned before the early eighth century. Porticoes linking the city with the Vatican and San Paolo fuori le Mura are known to have existed in the 530s. The portico to the Vatican was probably built already in the fourth century, while that to San Paolo perhaps came about in conjunction with the new basilica which was inaugurated in around 400.[100] Honorius may thus have initiated the portico to San Paolo as well as that of San Lorenzo. An inscription by the priest Ilicius, dated to c.400, may refer to porticoes along the Via Tiburtina from the city gate to the *memoria* of Hippolytus near San Lorenzo.[101] The porticoes were probably made of wood and brick. Recent excavations for the Rome Metro have in fact unearthed fourth-century brick porticoes along the Via Labicana outside Porta Maggiore.[102] If one compares the portico to San Lorenzo with similar extramural porticoes in for instance Milan or Antioch, they were probably lined with shops and workshops.[103] The portico to the church thus provided a prolongation of the city into the extramural area. It probably entailed a formalization of the approaches to the Porta Tiburtina, and also shows the importance of the sanctuary of Lawrence and the magnetism it exerted on traffic and urban development in this area.[104] This is also shown by the changing names of the gate itself, which is attested as the Porta Sancti Laurentii from the seventh century. The Honorian archway at Porta Tiburtina was, at 4.04 m in width, much narrower than the Augustan arch, probably for defensive reasons (Fig. 15.5). An inner courtyard was added to the gate by Honorius. It probably had a military use, but could equally well have had a function in the screening of transports for toll reasons (Fig. 15.6). Despite the narrowing of the gate, there are no signs that traffic along Via Tiburtina had diminished by around 400, which is attested by the extramural porticoes, and the continued maintenance of the roadway, demonstrated by fourth-century milestones.[105]

The late fifth century marked the high point of urban and infrastructural development on the Esquiline. With the fall in population beginning in the second half of the fifth century, a contraction of Rome's inhabited areas began. After the final move of the political centre away from Rome in 476, the general expansion could not continue with the limited means at the city's disposal, and life at the eastern periphery became polarized. Although the church of San

---

[100] Reekmans (1989: 909); Quilici (1996); Serra (1998); Barclay Lloyd (2002).
[101] *ILCV* 1.2.1773; Testini (1989); Serra (1999).
[102] Rea (2009).
[103] Stephens Crawford (1990); Bejor (1999); Mango (2001); Lavan (2006); Brandt (2009); Malmberg and Bjur (2009).
[104] Valentini and Zucchetti (1940: 81, 114); Mari (2008: 172); alternative route: Serra (1998).
[105] Richmond (1930: 177); Morley (1996: 85); Laurence (2004); Mari (2008: 172); Esch (2008).

Lorenzo continued to be a focal point in the religious life, from around 500 onwards the area was overshadowed by two great religious centres of the city: the Lateran and the Vatican. With the long wars from 535 onwards the economy of the *suburbium* was severely disrupted and the maintenance of the Via Tiburtina declined. Apart from pilgrims, the road probably experienced only a trickle of traffic compared with earlier periods.

CONCLUSIONS

The two city gates, built six centuries apart, at the border to Rome's *suburbium*, reflect a universal dynamic taking place on the periphery of the city of Rome: so-called edge phenomena. As urban space, initially, was most easily accessible on the city edge, various activities could optimally be accommodated here in built structures, which grew into urban nodes with a certain capacity of transforming and reconfiguring the city's spatial system. When movement passed through this spatial system, it could be considered part of the city's total sum of movements, its movement economy. As movement has been shown to generate and attract activities where it passes by, this movement economy would represent an important urban development force.[106] As shown in this chapter, the combination of high-level infrastructure and city edge phenomena demonstrably had a fundamental impact on urban development on the Esquiline in the imperial period.

The Augustan age saw a radical change in the Porta Esquilina area, which became gentrified, but also a centre for the production of wool and bread. The existing informal markets continued to flourish. Traffic increased because of an increased urban and suburban population, through large imperial building projects and an imperial presence on the Esquiline. Building material became a prime concern, because of the travertine quarries located along Via Tiburtina. The demands on the existing infrastructure became too great, which led to the modernization of the city gate, a widening of the roadway, and increased provision of aqueduct water. New traffic regulations and a customs border turned the Porta Esquilina into a transit area, which had widespread effect on urban development outside the gate, most conspicuously through the building of a large *macellum*.

In the fifth century the effects of the building of the Aurelianic Wall can clearly be seen in the urban fabric. Most of the facilities earlier concentrated around the Porta Esquilina had by now moved to the Porta Tiburtina. An imperial and episcopal interest in the area created a node around the gate, with

---

[106] Read and Budiarto (2003); Klasander (2004); Azimzadeh and Bjur (2009).

a large church, palace, and *macellum* gathered round a formal square. A new factor in the urban development was Rome's importance as a pilgrimage centre. One of the city's foremost sanctuaries, to St Lawrence, was situated outside the gate, along the Via Tiburtina. A node with some urban characteristics was created around the church, and was linked to the gate by long porticoes, expanding the built-up area into the *suburbium*.

Endpiece

From Movement to Mobility: Future Directions

Ray Laurence

If the Greeks had the reputation of aiming most happily in the founding of cities, in that they aimed at beauty, strength of position, harbours, and productive soil, the Romans had the best foresight in those matters that the Greeks made little of, such as the construction of roads and aqueducts, and of sewers that could wash out the filth of the city into the Tiber.[1]

The papers in this volume open a new chapter in the historiography of Roman urbanism and have a wider significance for the study of both Roman archaeology and Roman history. As readers will have seen, in the previous chapters, there is a shift from the study of space to a study of movement. This alteration of focus causes the study of the Roman city to keep pace with changes in foci within the disciplines of geography and the social sciences. This new focus is the subject of exploration in this final chapter, to look for new lines of enquiry and to update thinking on space undertaken some two or three decades ago. Before embarking on this discussion, it is necessary to set out the significance of a shift from space to movement for the disciplines of Roman archaeology and Roman history.

Roman archaeologists (and to a lesser extent historians) in the late 1980s and early 1990s shifted focus from the description of the architecture and topography of the city towards a study of space and its interaction with societies situated therein.[2] This paradigm shift in the discipline (and in allied disciplines) has become known as a 'spatial turn'. This spatial turn demonstrated that societies were spatially structured, and that space was an active

---

[1] Strabo 5.3.8.
[2] Laurence (1994a); Grahame (2000); Parkins (1997).

component in the configuration of cities.³ With regard to my own work, in particular *Roman Pompeii: Space and Society* published in 1994, it can be seen how the, then new, emphasis on space tended to be at the expense of the social. This occurred also at a time when Anthony Giddens's structuration theory, with its emphasis on individual agency, was being incorporated into Roman archaeology.⁴ What the discipline was producing with this emphasis on space was a new means of describing—and writing—the Roman city. However, in so doing, we caused the city's inhabitants to be placed in space without adequately stressing the key component of movement through it.

By the end of the twentieth century, movement at a macro-level was being discussed by ancient historians in terms of both land and sea transportation, as well as in the context of migration to the city of Rome.⁵ What these studies did was to recognize that in the Roman world the level of human 'mastery' of nature was remarkably high and that massive investment in transport infrastructure was made to enable the possibility of movement, even if the opportunity to participate in the newly created movement economy was a privilege of the few.⁶ More recently, John Urry has observed that the road system of the Roman Empire allowed for the movement of objects that created an evolving and adaptive distribution. Through the use of mobile objects this enabled the integration of people who were not in face-to-face contact and who existed in what may be described as quite different societies within the empire.⁷ This observation shifts the discussion of movement at the macro-level from the realm of historians into that of the archaeologist studying the distribution of objects. In so doing, we need to understand the Roman Empire not so much as a unified space but as a mobility-system that produced material inequalities in distribution of objects and in participation in what might be labelled Roman material culture.⁸ The shift from distribution patterns towards an understanding of the mobility of objects has the potential to shift the debate over Romanization from the realm of cultural history towards an understanding of mobilities within the Roman Empire, and, in so doing, bypass the emphasis on spatial distributions associated with Roman archaeology in the 1980s and 1990s. The theoretical framework of a focus on mobility would shift the focus towards an alliance of theory and data to reinvigorate the discussion of cultural change in the Roman Empire, whilst placing a stress on difference and inequality.

---

³ Urry (2007: 34–5); Hillier and Hanson (1984) for a reading of space as an active component of society.
⁴ Giddens (1984); Barrett (1997a, 1997b).
⁵ Horden and Purcell (2000); Laurence (1999); Scheidel (2004, 2005).
⁶ Laurence (1999); Urry (2007: 51).
⁷ Urry (2007: 51).
⁸ Urry (2007: 51–2).

Moving down scale to the study of the city, the first decade of the twenty-first century has seen detailed studies of space and movement by a younger generation of scholars—most of whom are represented in this volume. However, it needs to be noted that movement has tended to be studied by archaeologists rather than by historians—the same constituency which engaged with the 'spatial turn' in the 1990s. This is a cause of concern, since there is much to recommend an emphasis of mobility for better understanding of a range of historical topics. Importantly, a schism between the theoretical frameworks in archaeology and those in ancient history is undesirable. To counteract this imbalance towards archaeological studies, I wish to set out the case for the fundamentality of movement for the historical understanding of the city of Rome, Roman society, and Roman politics.

THE *COMPITUM*

A conundrum for Roman historians has been the participation of citizens in the political process at Rome. To participate, citizens needed to move to the place of politics, either the Forum Romanum or the Campus Martius. Participation involved knowledge of politics, but the key questions are: how did citizens gain knowledge of politics and how did certain key issues mobilize a greater number of participants?[9] In the past, it has been argued that the social network associated with key elite patrons could mobilize their clients for key actions. This can be seen to be the case put forward in the *Commentariolum Petitionis*, with a focus on consular elections. Knowledge of politics can be seen to have been underpinned by rumours that moved from the Forum Romanum across the city to the *compita* or crossroads.[10] Strikingly, rumour does not spread from the forum to the houses of the elite from where the plebs gained their understanding from a patron. Rumour moves fleetingly across the city and stops or is stationary at the crossroads from where the plebs can gain access to knowledge of politics. This would suggest that there were other social networks in the city of Rome that lay beyond the formal relationship between a patron and his clients. However, we need to bear in mind that not all would participate in elections and the capacity of locations for elections was limited to some 12 per cent of the plebs.[11] Indeed, it would seem that popular politicians (such as Publius Clodius in the 50s BCE) tapped into the urban

---

[9] North (2002: 5–6) commenting on Millar (1998: 94–123) and Mouritsen (2001: 38–62).
[10] See Laurence (1994b) for references to this process; see also Mouritsen (2001: 58–62) on Clodius' mobilization of the plebs.
[11] Mouritsen (2001: 32, also 18–37) for the relationship of space, action, and level of participation.

networks that were focused on the crossroads. These very same locations would become the focus of the worship of the Lares Augusti. More importantly, for the subject of this volume, the *compita* with its shrine was an inscribed space in which the local magistrates of the *vicus* were commemorated. In some ways as well, it was an information centre with at least one such *compitum* having an inscribed calendar of the days on which major festivals would occur and a record of who had been consul from 43 BCE onwards.[12]

The crossroad or intersection of two streets has been found in Pompeii to accumulate facilities in a way that two directional street sections do not. Shrines, water fountains, bars, and so on tend to have been located at the intersection. In addition, as shown in Stöger's chapter, guilds preferred such locations for their properties. The crossroad is a point in the street network at which traffic needed to slow down to take account of the action of two or more flows of traffic joining and a need to instruct animals to turn a corner. These locations are favoured due to their position in the overall street grid, but this spatial preference becomes amplified as new facilities are added and the action of moving through a crossroad became more difficult. This causes the joins of streets designed to facilitate movement in a number of directions to become impeded through their ability to concentrate traffic and, ultimately, to become points at which traffic stops or at least has to slow down. This point of slower pace, like the Porta Esquilina or Porta Tiburtina, becomes a locale or a meeting place and a point of intersection of the locals with those passing by or deliberately seeking out people in a neighbourhood. It is at this point, the *compita*, that brokerage with outsiders might occur; this is seen most clearly in the relationship between the magistrates of the *vici* with the emperor Augustus.[13]

Within the networks of power relations presented in the *Commentariolum Petitionis*, we find the magistrates of *collegia*, those influential in *vici*, and important freedmen being identified as separate from clients or *amici*. This separation converges with Tacitus' contention that there was a respectable plebs connected to the houses of the rich—perhaps the world parodied by Martial in his *Epigrams*, and the *plebs sordida*—disconnected from the elite and seen to be interested in only bread and circuses in Juvenal's *Satires*.[14] These were people who might have been seen by Seneca as he moved through the city as being dirty and filling the narrow streets between poorly constructed insulae.[15] The important point of this discussion is that the elite were not directly connected to many sections of the plebs, and it was only

---

[12] Lott (2004) for discussion of evidence.
[13] Lott (2004) for discussion and evidence. Haselberger (2007: 224–30).
[14] Tac. *Hist.* 1.4.
[15] Sen. *De Ira* 3.35.3–5.

via brokers from the plebs that contact was made.[16] These brokers were not drawn into the patronage of a single person and acted on the edge of communication at the *compita*—a point at which they could be located. The brokers were associated with voluntary associations—the *collegia*, and/or direct involvement in a *vicus*.[17] Through involvement in these institutions, the brokers gained social capital that was increased through interaction with members of the elite. Yet, conceptually, their position was different to that of a client or a person fully dependent on a member of the elite. In effect, the broker was a member of a network who could communicate with another network—that of the elite. All parties (whether freeborn, freed, or enslaved) would have also been associated with a third network—that of a *familia*—and in most cases a kinship network. The movement across the city involved in the maintenance of these very different, yet in part overlapping, social networks may have been subject to time constraints of travel, the kind we can locate in Martial. What is unclear at present is whether the *magistri vicorum* of the city of Rome formed a network or whether they interacted in any way? Inscriptions point to their local individuality, but equally *CIL* VI.975 points to their equality in a summary listing of magistrates within each region. Given this evidence, it seems unlikely that these magistrates were isolated from each other apart from by the friction of distance. That friction, it should be stated, would not have existed in either Ostia or Pompeii due to their size; whereas in Rome—a city of hills—distance appears to have been a factor in the movement of individuals. Economies of effort thereby influence social networks.

AGENCY, STATUS, AND MOVEMENT

If we accept, as the authors in this volume advocate, that movement was a feature of Roman urbanism, this becomes more visible to us in the city of Rome, where there is a literary representation of movement and where the distances travelled were greater. Movement thus is part of the city and the structure of the world into which people were socialized as children. It is also recognized by Stöger and van Nes that the topology of the street network also provides the Roman city with a structure. If we are going to see movement in these ways, we need to relate the phenomenon of movement to Anthony Giddens's structuration theory that has been widely adopted within Roman archaeology.[18] Giddens tends to relate movement to his discussion of time and the temporality of moving from home to school to work back to school and

---

[16] *Comm. Pet.* 50. Cicero wins over brokers of the urban masses by advancing Pompey.
[17] *Comm. Pet.* 30, 32; Mouritsen (2001: 83–4, 141–2).
[18] Giddens (1984); in so doing he draws on Carlstein (1982).

home again. Certainly, there was a temporal structure to the Roman city with activities timed according to the hour in which they occurred, for example dinner at the ninth hour.[19] Meetings were held at a specific place and at a specific time, as can be seen from entries in the Sulpicii archive from near Pompeii, for example at the statue of Gaius Sentius Saturninus, in the Forum of Augustus, at the third hour of the day.[20] The latter was a preferred location for meetings, but they did vary the time. What this suggests in terms of agency is that for prearranged encounters choices needed to be made so that all parties were at the same location at the same time. Hence, a particular point was chosen or favoured to create a locale for an encounter. In contrast, chance encounters, those found represented in Martial or Ovid, were caused by habitual movement at certain times to monuments, as Newsome shows with reference to the 'natural movement' of the Forum Romanum and the propensity for chance encounters on the routes that surrounded it. It is significant that in the *Commentariolum Petitionis* there is advice that the consular candidate should move through the city at very similar times—so that people would encounter him as expected, and so that his movement became part of the urban routine, its rhythm.

The spectacle of seeing the candidate progress through the city from his house to the forum was regular and almost timetabled, and became part of the structure of the city and could be described as a *pompa* or processional show.[21] Clearly, it was the candidates' movement that was important, rather than the movement of all senators. Thus each year there was a shift in this pattern of movement according to who was a candidate and the location of their home in the city. The entry or exit of the elite from the city could also constitute an event—for example, Cicero's report of his entry into Rome via the Porta Capena (discussed in Newsome's Introduction) to the Capitol.[22] Three points of this short journey are noted—the Porta Capena with crowds watching and applauding from temples, the crowd that follows him to the Forum, and then on to the Capitol. Both these locations are marked in the letter as having large crowds, who had come to see the spectacle of an ex-consul's return from exile. As numerous chapters in the present volume have discussed, the movement of the elite should be differentiated due to their status and prominence. They were to be observed and seen, with the result that their movement reinforced their social and political position within the city. Movement made their status apparent, in a similar way that their house might also reflect their status.[23] What is set out here is only a beginning for further investigation of how movement and travel on the part of the elite was part of not just the structure of the city, but also the Roman state.

---

[19] Laurence (2007: 154–66).   [20] Camodeca (1999); Laurence (2007: 154–5).
[21] *Comm. Pet.* 34–8, 51.   [22] Cic. *Att.* 4.1.4–5.
[23] Wallace-Hadrill (1994: 4–6).

MOVEMENT AND MONUMENTS

For Strabo, the Romans were unlike the Greeks in their choice of urban priorities, as is demonstrated in the opening quotation to this chapter.[24] The Greeks, he says, focused on beauty, defence, harbours, and the fertility of the soil; whereas Romans had other priorities: roads for wagons, aqueducts, and sewers to wash dirt away from the city. The movement of goods, water, and waste defined the city with only the monuments of the Augustan Age in the Campus Martius and the fora gaining Rome a reputation for beauty. This mode of progressive description of Rome only developing beauty under Augustus and then moving across the monuments from the Campus Martius to the fora and onto the Capitol, Palatine, and Porticus of Livia is found reversed in Ovid.[25] Starting from his house, the poet in exile imagines Rome's landscape of forums, temples, marble-clad theatres, paved porticoes, grass of the Campus Martius, the *horti*, water from the Aqua Virgo, before closing with the Via Appia and wheeled transport.

These two passages, which have been influential in shaping our perception of Augustan Rome, juxtapose the role of movement with that of the monuments of Augustus and his family. Intriguingly, Ovid's Rome includes a far greater number of monuments than can be found in Martial's Rome, discussed in my earlier chapter. Both authors include a sense of movement, yet the focus is quite different—Ovid can imagine in exile spectacles of state, whereas Martial's *Epigrams* represents his duties and journeys as a client, contrasted to his *De Spectaculis* with its focus on the new landscape of the Colosseum.[26] As Don Fowler pointed out, the problem of monuments is that their meaning can change, and that there is no 'correct' interpretation of their meaning.[27] Even the monument, a seemingly static object, could become fluid and change. Given the fluidity of language found in Varro, discussed in Spencer's chapter, we cannot see meaning as static to either the physical monuments nor in attitudes to the texts that were also defined as monuments.[28] Yet, it is Varro's definition of the flow of language which stabilizes the meaning of words associated with urbanism. After all, Varro was the scholar seen by Cicero to define Roman culture: the antiquity of the city of Rome, chronology of Rome's history, religious laws and priesthoods, its regions, its places; whereas previously the Romans had wandered like foreign guests in their own city.[29] Varro gave the Romans a definition of home, which was a prelude to the creation of a

---

[24] Strabo 5.3.8. Compare Favro (1996: 252–80).
[25] Ov. *Pont.* 1.8–68. Compare Ov. *Ars Am.* 1.67–88, 3.385–98; *Tr.* 2.277–302. Boyle (2003) provides an analysis of Ovid on the monuments of Rome.
[26] Edwards (1996: 124); Boyle (2003) for examples and discussion.
[27] Fowler (2000: 193–217).
[28] Fowler (2000: 197–8) for discussion.
[29] Cic. *Acad.* 1.9; Fowler (2000: 206).

notion of home expressed via the Augustan building programme. By being the person to change the city to the greatest extent, in terms of number of monuments and over the longest period of time, Augustus created a stability of urban form—even though many of these buildings were destroyed by fire only fifty years after his death.

Most studies of the building and rebuilding of monuments in Rome by Augustus have stressed the role of imagery to reinforce the power relations of the first *princeps*.[30] However, there is another side to monuments in relation to movement. A monument is a landmark that can be utilized to navigate the city. Yet, at the same time, the monuments and the alteration of monuments associated with the building of the imperial fora altered the topology of Rome and restructured the nature of movement, as Newsome has discussed in his chapter. The urban effect was to restructure the description of the central spaces of the city for both Strabo and Ovid to create a series of monumental spaces separate or at least dominant over the rest of the *urbs*.[31] Moreover, we should perhaps see change to movement in Augustan Rome also being affected by the redefinition of the role of the *vici* and shape of the *regiones* in 7 BCE, which included the decision that Rome should be a city that did not have a defensive circuit of walls.[32] Andrew Wallace-Hadrill has recently argued that this redefinition of space in 7 BCE was combined with the cartographic representation of the city and a census that located the people within the framework of the map.[33] The fluidity of the city's expansion that blurred the boundary of urban and rural, and caused Dionysius of Halicarnassus to see Rome as an undefined city, was fixed through the imposition of the fourteen regions.[34] Mapping, though, does not explain the effect of monument building on movement. Just building a single new monument, associated with religion and duly positioned in Rome's calendar, would alter the structure of movement in the city of Rome.[35] The Augustan programme of monument building, in particular the construction of porticoes and baths, would have reshaped movement in the city or expanded existing patterns of movement, say by youths to the Campus Martius, with the accommodation of activities in new structures.

The programme of change to the city did not cease with Augustus, but commenced again after the fire of 64 CE with a similarly lengthy programme of renewal by Nero and the Flavians. This programme was different. The rubble from the ruins of the city was shipped down the Tiber and dumped in the marshes adjacent to Ostia. New streets were constructed that were wider and

---

[30] Zanker (1988) is a forceful reminder.
[31] Strabo 5.3.7; Ov. *Pont.* 1.8–68.
[32] Haselberger (2007: 230–7).
[33] Wallace-Hadrill (2008: 276–311).
[34] Haselberger (2007: 236); Dion. Hal. *Ant. Rom.* 4.13.4.
[35] Boyle (2000: 35–53) on religion and new monuments in the definition of place.

less crooked and the project was only completed by Vespasian with a new temple of Jupiter on the Capitol, his Templum Pacis, by the fora, the Temple of Claudius—spaces discussed in Macaulay-Lewis's study of leisured movement in Flavian urbanism—and the Colosseum in the middle of the city, a project seen by Suetonius to have been an idea of Augustus.[36] The city was also defined not by walls but by thirty-seven gates, was divided into fourteen regions, and had 265 *compita* with shrines of the *lares*.[37] The question is what was the nature of the movement economy of this new Rome? Certainly the gates, as Malmberg and Bjur's chapter reminds us, became central points of marketing and exchange associated with movement to and through them, whilst the *compita* continued to be places that were enumerated within the major routes that were seen to be over 70 miles in length. Yet, where was the centre? Certainly, the new monuments built by Vespasian altered not just the skyline of Rome but also navigation through the city. It is difficult for us to think of a Rome without the Colosseum as a central point (*urbe media*), let alone to imagine the change to the movement economy of the building of such an iconic monument. The issues of movement and monumental change need further investigation, but for now it is worth commenting that when Rome burns down in 64 CE, Tacitus finds space not for the monuments of the Augustan city found in Ovid or Strabo, but for those of an earlier time of Servius Tullius. Yet, his description of the scale of the fire measures it with reference to the fourteen regions created in 7 BCE; only four were not damaged. The buildings of Agrippa in the Campus Martius became a refuge, for those who had not just survived the fire but also the experience of moving through Rome's narrow winding streets that had been blocked by the terrified crowds of screaming women, helpless old people, and children, those helping others and those charging through them.[38] This must be the ultimate literary description of congestion in Rome, with the danger presented not by carts of building material, but by the raging conflagration.

STOPPING AND SPACE

When looking at movement in the Roman city, we almost automatically think of movement to the forum—the location of elite politics and limited participation in politics on the part of others. Of course, there were other activities taking place there too, as Trifilò's chapter reminds us with reference to the location of carved game boards. Yet, there are a far greater variety of journeys that need consideration. The theatres with their excellent acoustics were a

---

[36] Tac. *Ann.* 15.43; Suet. *Vesp.* 8.5–9.   [37] Plin. *HN* 3.66–7.
[38] Tac. *Ann.* 15.38–41.

particular place for the recounting and intergenerational transmission of an understanding of identity. Acoustically, all could hear the words uttered or the songs sung or the music played. Other forms of movement need consideration too, the journey to bathe and then onto dinner either at home or at another person's house would seem to have been a fundamental part of the structure of the city. The location of these functional structures within the movement economy of the city needs consideration.

There is another side to this focus on movement in the papers in this volume: stopping, or the shift from movement to a stationary state. Whether the cause of this was arrival at a destination (a house, a building site, a guild, or a public space) or was an interruption to the action of movement (a public nuisance, a sound, or traffic congestion), the result needs to be considered as a feature of Roman urbanism. The action of stopping might be best explained with reference to an event that caused many to pause in modern London. The artist Anthony Gormley created the opportunity for members of the public to occupy the empty space of the fourth plinth in Trafalgar Square (the other three are occupied by statues) for a period of twenty minutes. Reactions to this opportunity varied, but I wish to focus on one that I saw. A man, Jeremy Patterson, was brought onto the plinth wearing a toga with a folder and a public address system. A handful of people were gathered beneath the plinth, the man explained that he was going to speak in Latin with a view to promoting the learning of Latin in UK primary schools. Reactions to this spectacle were that people stopped and then wanted to discover what this strangely dressed (and to the majority, strangely speaking) man was doing. When they had gained information, they either stayed for longer or moved on (Fig. E.1). All those who had stopped had gained information about the activity taking place, something they could not have gained if they had walked or cycled past. My point is that movement alone does not, of itself, provide those that move with information. It is only if they stop that information can be acquired. On understanding this curious spectacle, most went on their way.

Another key element, drawn from viewing reactions to an address by a man in a toga from the fourth plinth, is that Jeremy Patterson needed a public address system to make his wisdom from and on Latin audible to those watching. Even with amplification, movement to a point further from the plinth caused his words to be lost and those who had stopped to be less engaged, often talking amongst themselves about the man in the toga on the plinth. This has an importance for understanding where a person stops in the city. For example, there were sound contours to all ancient spaces, as Betts sets out in her chapter; where a person stops to watch, or watch and listen, to a speech or execution or other action was a matter of choice. However, we might suggest that actual engagement with that action was reflected in the point chosen to stop, or on a crowded occasion, the point from which a person felt they could best experience events. A forum was a place of hearing and partial

Fig. E1. Movement around the Fourth Plinth in London's Trafalgar Square (photo by Ray Laurence).

hearing of action taking place; whereas a theatre was a place of hearing and audibility. Place Jeremy Patterson in Ostia's theatre and he would not have needed amplification. Yet theatres do not appear to be the spaces of rhetoric in antiquity and a key aspect of rhetoric was the possibility of not being—as well as being—heard. Technologically, space could have been designed for total audibility—but this was not a requirement acted upon to create the audible forum for clear speeches heard by a mass audience. It may be the understanding of the limits of participation in politics, based on the size of the forum, is a measure of a variable that is actually determined by the number of people to whom the human voice can travel across space.[39] This was a limitation for the size of theatres. The implications of Betts's study of movement through and between aural spaces could be applied to other archaeological contexts. One that would seem most appropriate are those places in which the function of an activity, for example manufacture, can be mapped across the archaeological context, allowing for the reconstruction of activities and their associated sounds. This could be undertaken with reference to Miko Flohr's spatial analysis of the social world of the Roman *fullonicae* to enhance our

[39] Mouritsen (2001: 18–37).

understanding of the soundscapes—as permissive networks—of the Pompeian fullery compared to those of Ostia.[40] Such an analysis might establish, for example, the degree of verbal interaction in the final phase of the production of new garments—their cleaning.

FROM SPACE SYNTAX TO SOCIAL NETWORKS

Space syntax or topology has a place in understanding the flow of human mobility, but there needs to be rather more than this to shift the meaning of these observations into the historical realm. It is here that space syntax can be allied with theories of social networks. Space syntax emerged in the 1980s and was adopted in archaeology at what might be described as the 'spatial turn' in that subject. Within ancient history, drawing on developments in the 1990s, the use of social networks has been promoted particularly by Irad Malkin in the context of a Mediterranean framework.[41] These new forms of discourse would seem to be developing separately. For example, in this volume Stöger discusses the locations of *scholae* in Ostia with a view to better understanding their relation to the streets of the city. What is seen as a separate discourse is the network of contacts that might be revealed through an examination of the names of guild members, their patrons, and so on in inscriptions. These inscriptions, it has to be said, were representations of the ordered institutional structure of the guild. However, this epigraphic evidence can be seen as a small world that has links to wider society through its patrons, family relations of its members, and so on. The patrons are particularly important for the integration of the members of a *collegium*, since through them the small world of a *collegium* is integrated with other groups in the city. This is an example of where the experience of the prosopographer using network theory and simulation could be drawn together with the archaeologist's use of space syntax. This would effectively render a more robust approach to the evidence and elucidate, even if fleetingly, the intersection between space, people, and formal, but voluntary, associations.[42] For a long time, it has been recognized that there is a sporadic clustering of guild foundations in space and time.[43] However, we may be able to analyse this problem in reverse formation. Since we now

---

[40] Flohr (2009).
[41] See papers in Malkin (2005a, 2009); Ruffini (2008: 8–40) provides a summary of the nature of network analysis in what he describes as 'A Tutorial for Ancient Historians' with a full summary of the development of the subject and its application for the elucidation of networks in antiquity.
[42] See Remus (1996) for an example of the use of network theory for the analysis of a voluntary association.
[43] Cotter (1996) traces this observation to Waltzing (1895–1900: i. 57–8)

understand the spatial location in the urban network of guild buildings, it would be possible by removal of the guild from the spatial analysis to define the nature of an Ostia without guilds at an earlier date, or after a crackdown on 'new' or 'illegal' associations.[44]

Even within a city the size of Rome, the linkages made by formal patronage, although few in number, would integrate this large population and cause strangers meeting for the first time to have a high probability of social connection. Network theory stresses that most people exist in a tight group, but that it only takes a few individuals with moderate connections to connect a group with others.[45] The perspective based on movement can redefine the role of the client in Roman society. Much has been made of the literature that articulates the client's world view, as a person insulted by his patron, and seldom in receipt of the material benefits of the relationship with his patron.[46] Yet, it is clear that people continued to be clients of patrons, it was a voluntary relationship, and we need to evaluate why this was the case. The answer need not come from the texts detailing the relationship in material terms, but in the definition of the status of the client as a person with the leisure to attend a patron from morning to night, as my chapter in this volume discussed in the context of Martial's *Epigrams*. The clients were of a different status from the *plebs sordida*.[47] Their status was defined by their role and association with their patron. Movement to the patron's house and the act of following him created a group of clients who were drawn together not necessarily from the same neighbourhoods of the city and thus like the patron had access to a greater range of information, gossip, and rumour and were importantly much more mobile than most of the population. For every patron, we can speculate that there were a number of clients, ranging from say a dozen to about a hundred in number, which in terms of the mobile sector of the city would have accounted for a high percentage of the population. Returning to that discussed above, on return to their own neighbourhoods, the clients would have an increased status through their knowledge of things happening outside the neighbourhood as well as their association with a patron. It is this knowledge that allowed Martial, as a client, to write of the city and to develop epigram as an urban genre that would be incorporated into satire by Juvenal. The client connected the patron to others outside his immediate network, and as a consequence the client occupied a role as information carrier. It is by looking at the movement of the client that we find the value of the relationship. This in turn underscores the difference between the movement of clients and the movement of, say, porters; the client had the freedom to move. Yet, the pattern

---

[44] Cotter (1996) for overview of the legislation affecting *collegia*.
[45] Urry (2007: 213–15).
[46] Seen graphically in Juv. *Sat.* 3.
[47] Tac. *Hist.* 1.4.

of movement of clients was organized with reference to their inequality with their patron and thus determined by their patron: movement is commoditized.

Patron–client relations are a means to the articulation of the movement associated with social networks in the city of Rome. However, there are other materials for the development of social networks. Shawn Grahame's analysis of the geochemistry and prosopography of brick stamps from the Tiber Valley has developed an advanced articulation of the nature of the social relations of the supply of building materials.[48] Key to the definition of such as network is the movement of a material from one point in the Tiber Valley to another. Interestingly, although bricks were imported into Ostia from the Tiber Valley, other materials were not, most notably basalt paving stone that have been shown through X-ray florescence to have come from a quite different region.[49] What this shows is that there could be quite different networks of supply according to the goods traded. Importantly, streets were paved by the city authorities at Ostia; whereas the use of brick was supplied to the privately funded builders across the city.[50] Key to the definition of supply and the network of supply is the movement of a resource for consumption elsewhere. In the case of the Tiber Valley, these materials were moved to distances of up to 30 miles and reveal the network of relationships between cities as points of consumption and the Tiber Valley as point of extraction or manufacture.

THE STATE AND TRANSPORT INFRASTRUCTURE

There is a tendency to understand the transport infrastructure (roads, aqueducts, sewers) and its stopping points (cities) through the lens of technology and the adoption of that technology across the Roman Empire (Romanization). However, to a certain extent, this misses the point—the static infrastructure that allows for the movement of people, goods, water, and waste underpinned a conception of mobility as a necessity for the expansion of Rome from a city-state into a world empire.[51] The cost of establishing this infrastructure should not be underestimated; the renewal of just 16 miles of the Via Appia under Hadrian cost 2,039,100 sesterces or, on average, 129,467 sesterces per Roman mile.[52] The use of the network of roads as an infrastructure for mobility tends not to be explained, nor does the place of mobility within Roman society.

---

[48] Grahame (2009).
[49] Stuart Black, pers. comm.
[50] Black, Browning, and Laurence (2009) on the supply of basalt for road building by the state. DeLaine (2004) on supply of brick to Ostia in the 2nd century CE.
[51] Scheidel (2005a).
[52] *CIL* IX.6075.

Underpinning that mobility is another economic cost, the time taken to travel across space, if we assume 20 to 35 miles as a reasonable journey for a single day. For a monthly meeting of the senate, individuals travelled to Rome or supported expensive lifestyles in Rome to avoid travel. Cicero, as we can see from his letters, was on the move from Rome to his country estates—as was Pliny over a longer distance to and from his home town in Como.[53] Governing a province involved movement. Yet, all this movement was for a purpose—stopping at different cities or in *villae* to meet people.[54] The network of a city-state aristocracy had in the Roman Empire expanded, but had not obviated the need for the elite to meet with each other, communicate by letter, and to meet with their social inferiors. These meetings are the points for the distribution of power and it is the mobility of the elite that delivers access for others to the power of Rome. Mobility expands the range of face-to-face contact and also the concept of the self-governing city and other cultural forms across the Roman Empire. Underpinning the Roman state was a need for meetings that would create ties between persons of different statuses, and interlink social networks. These are the occasions on which the 'weak' links that are so fundamental to our current understanding of networks were established across the space of the Roman Empire. The weak link was often articulated through the language of patronage to establish a connection over a longer time period and is seen most clearly in epigraphy.[55]

CONCLUSIONS: NEW INTRODUCTIONS

The discussion in this chapter, and the earlier chapters throughout this volume, articulates a new importance to movement for our understanding of urbanism and, for that matter, in our entire approach to the study of the Roman Empire. There is a question that has also been neatly posed by Ian Morris, asked in discussion of *The Corrupting Sea* and other work on networks in the Mediterranean:

> All history involves mobility, connectivity, and dispersed power. If these were defining characteristics of the pre-modern Mediterranean, we need ways to measure them. So far, Mediterraneanists have amassed what evidence there is but have not explained how to gauge its significance.[56]

---

[53] Cicero's letters have revealed the network of connections modelled by Alexander and Danowski (1990).
[54] Urry (2007: 230–5) on meetings and social networks.
[55] There are numerous inscriptions that utilize the language of patronage, e.g. *CIL* VI.3918.
[56] Morris (2005: 45).

In the case of mobility and movement in cities, we can turn to the analysis of space or what we might now term the infrastructure of movement. The analytical framework of space syntax provides archaeology with a powerful predictive model for how space shapes movement. However, space syntax need not be the only paradigm through which archaeology might access movement in antiquity. As Diane Favro points out in her chapter, 3D visualization provides the possibility of establishing the probable mobility constraints of movement in space. Bernie Frischer, at the University of Virginia, is looking for applications based around predictable patterns of movement to populate his virtual Rome with a view to establishing a 3D agent-based model.[57] These forms of investigation at the level of the city can be expanded to include analyses of regions and their land-based connectivity.[58] Shifts and changes to the movement economy over time are important themes for such studies. What becomes intrinsically more difficult is the examination of networks associated with sea travel, since the infrastructure of travel is the sea itself and so lacks the materiality found associated with land-based travel, associated with permanence over time. This leads to conceptual difficulties in identifying variations in the nature of movement between the networks of an emergent Greek identity in the Mediterranean posited by Irad Malkin, and those that were associated with a Roman Mediterranean, in which a Greek identity was maintained.[59] However, perhaps, the archaeologists' humble distribution maps of pottery types and other forms of material culture might establish the wider dynamic of networks of connectivity. This might be the spin-off from the University of Oxford's exploration of the Roman economy, via the re-reading of material that has been recovered.[60]

Such approaches engage what could be seen as a paradigm shift towards networks and movement with the possibility of understanding the significance of connectivity—within and beyond the city—and thus also of mobility within Roman culture. These things may be achieved in the future. For the time being, we need to be content in the knowledge that ancient historians and archaeologists have moved on from the paradigm shift associated with the spatial turn of the 1980s and now, like their colleagues in the social sciences and geography, have stepped up to the challenge of recovering the significance of movement. The next step is to begin to write cultural histories of mobility.

---

[57] Bernie Frischer, pers. comm.
[58] Graham (2006) examines the configuration of the connectivity found in the Antonine Itineraries. Similar studies of landscapes and the material connectivity (road network) associated with regions would be a possibility—e.g. southern Etruria in the Tiber Valley.
[59] Malkin (2005b).
[60] See discussion in Bowman and Wilson (2009).

Bibliography

Abercrombie, P. (1945), *Greater London Plan 1944* (London: His Majesty's Stationery Office).

Abrams, E. M., and Bolland, T. W. (1999), 'Architectural Energetics, Ancient Monuments, and Operations Management', *Journal of Archaeological Method and Theory* 6/4: 263–91.

Adam, J. P. (1994), *Roman Construction* (Bloomington: Indiana University Press).

——(2003), *Roman Building: Materials and Techniques* (London: B. T. Batsford Ltd).

Adam, J. P., and Varene, P. (1980–2), 'Une peinture romaine représentant une scène de chantier', *Revue archéologique* 2: 213–38.

Adkins, L., and Adkins, R. A. (1994), *Handbook to Life in Ancient Rome* (New York: Facts on File).

Aldrete, G. S. (2007), *Floods of the Tiber in Ancient Rome* (Baltimore: Johns Hopkins University Press).

Alexander, M., and Danowski, J.A. (1990), 'Analysis of an Ancient Network', *Social Networks* 12: 313–35.

Allison, P. (2004), *Pompeian Households: An Analysis of the Material Culture* (Los Angeles: Cotsen Institute).

——(n.d), 'Pompeian Households: An On-line Companion', http://www.stoa.org/projects/ph/index.html.

Amici, C. M. (1991), *Il Foro di Cesare* (Florence: L. S. Olschki).

Ammerman, A. (2006), 'Adding time to Rome's *imago*', in Haselberger and Humphrey (eds.), 297–308.

Anderson, J. C. (1982), 'Domitian, the Argiletum and the Temple of Peace', *American Journal of Archaeology* 86: 101–10.

——(1983), 'A Topographical Tradition in Fourth Century Chronicles: Domitian's Building Program', *Historia* 32: 93–105.

——(1984), *The Historical Topography of the Imperial Fora* (Brussels: Collection Latomus).

——(1997), *Roman Architecture and Society* (Baltimore and London: Johns Hopkins University Press).

Anderson, M. (2005), 'Houses, GIS and the Micro-Topology of Pompeian Domestic Space', in B. Croxford and D. Grigoropoulos (eds.), *TRAC 2004, Proceedings of the Fourteenth Annual Theoretical Roman Archaeology Conference, Durham 2004* (Oxford: Oxbow), 144–56.

Anderson, S. (ed.) (1986), *On Streets* (1st edn 1978; Cambridge, Mass. and London: MIT Press).

André, J.-M., and Baslez, M.-F. (1993), *Voyager dans l'antiquité* (Paris: Fayard).

Andrews, J. (2006), 'The Use and Development of Upper Floors in Houses at Herculaneum' (Unpublished PhD Dissertation, University of Reading).

Aoyagi, M., and Pappalardo, U. (2006), *Pompei: (Regiones VI–VII) Insula Occidentalis* (Naples: Valtrend Editore).

Aubert, J.-J. (1993), 'Workshop Managers', in W. V. Harris (ed.), *The Inscribed Economy: Production and Distribution in the Roman Empire in the Light of Instrumentum Domesticum* (*Journal of Roman Archaeology*, suppl. ser. 6), 171–82.
Ausbüttel, F. M. (1982), *Untersuchungen zu den Vereinen im Westen des römischen Reiches* (Kallmütz: Michael Lassleben).
Austin, R. G. (1934), 'Zeno's Game of *Tables*', *Journal of Hellenic Studies* 54: 202–5.
Azimzadeh, M., and Bjur, H. (2009), 'Discovering Space as Cultural Heritage: Hidden Properties of the Urban Palimpsest', in Bjur and Santillo Frizell (eds.), 181–91.
Bachelard, G. (1994), *The Poetics of Space*, trans. M. Jolas (Boston: Beacon Press).
Bakker, J. T. (1994), *Living and Working with the Gods: Studies of Evidence for Private Religion and Its Material Environment in the City of Ostia (100–500 AD)* (Amsterdam: J. C. Gieben).
Ballet, P., Dieudonné-Glad, P., and Saliou, C. (eds.) (2008), *La Rue dans l'antiquité: Définition, aménagement et devenir de l'orient méditerranéen à la Gaule: Actes du colloque de Poitiers, 7–9 septembre 2006* (Rennes: Presses universitaires de Rennes).
Ballu, A. (1902), *Théâtre et forum de Timgad, antique Thamugadi: État actuel et restauration* (Paris: Leroux).
Balsdon, J. P. V. D. (1969), *Life and Leisure in Ancient Rome* (London: Bodley Head).
Barclay Lloyd, J. (2002), 'Krautheimer and S. Paolo Fuori le Mura: Architectural, Urban and Liturgical Planning in Late Fourth-Century Rome', in F. Guidobaldi and A. Guiglia Guidobaldi (eds.), *Ecclesiae Urbis: Atti del Congresso internazionale di studi sulle chiese di Roma (IV–X secolo)* (Vatican City: Pontificio Istituto di Archeologia Cristiana), 11–24.
Barrett, J. C. (1997a), 'Theorising Roman Archaeology', in K. Meadows, C. Lemke, and J. Heron (eds.) *TRAC 96: Proceedings of the Sixth Annual Theoretical Roman Archaeology Conference* (Oxford: Oxbow Books), 1–7.
——(1997b), 'Romanization: A Critical Comment', in D. J. Mattingly (ed.), *Dialogues in Roman Imperialism: Power, Discourse and Discrepant Experience in the Roman Empire* (*Journal of Roman Archaeology*, suppl. ser. 23, Portsmouth RI), 51–66.
Barthes, R. (1997), 'Semiology and the Urban', in N. Leach (ed.), *Rethinking Architecture: A Reader in Cultural Theory* (London: Routledge), 166–72.
Bauer, H. (1983), 'Porticus Absidata', *Mitteilungen des Deutschen Archäologischen Instituts, Römische Abteilung* 90: 111–84.
Beard, W. M. (2007), *The Roman Triumph* (Cambridge, Mass.: Belknap Press of Harvard University Press).
——(2008), *Pompeii: The Life of a Roman Town* (London: Profile).
Beccaria, G. (2008), *Tra le pieghe delle parole: Lingua, storia, cultura* (Turin: Einaudi).
Beck, H. (2005), *Karriere und Hierarchie: Die römische Aristokratie und die Anfänge des cursus honorum in der mittleren Republik* (Berlin: Akademie).
Bejor, G. (1999), *Vie colonnate: Paesaggi urbani del mondo antico* (Rome: 'L'Erma' di Bretschneider).
Bell, A. J. E. (1997), 'Cicero and the Spectacle of Power', *Journal of Roman Studies* 87: 1–22.
Bell, R. C. (2007), 'Note on Pavement Games of Greece and Rome', in I. Finkel (ed.), 98–100.

Bell, R. C., and Roueché, C. (2007), 'Graeco-Roman Pavement Signs and Game Boards', in I. Finkel (ed.), 106–9.
Benario, H. (1958), 'Rome of the Severi', *Latomus* 17: 712–22.
Bendala Galán, M. (1973), 'Tablas de juego en Italica', *Habis* 4: 263–72.
Benjamin, W. (1985), *One-Way Street and Other Writings*, trans. E. Jephcott and K. Shorter (London: Verso).
——(1997), *Charles Baudelaire*, trans. H. Zohn (London: Verso).
——(1999), *The Arcades Project*, trans R. Tiedemann (Cambridge, Mass.: Harvard University Press).
Berger, A. (1953), 'Encyclopaedic Dictionary of Roman Law', *Transactions of the American Philosophical Society*, NS 43/2: 333–809.
Bergmann, B., and Kondoleon, C. (eds.) (1999), *The Art of Ancient Spectacle* (Washington: National Gallery of Art).
Bernstein, F. (2007), 'Pompeian Women', in Dobbins and Foss (eds.), 526–37.
Bianchini, M. (1991), 'I mercati di Traiano', *Bollettino d'arte*, 8: 102–21.
——(1992), 'Mercati traianei: La destinazione d'uso', *Bollettino d'arte*, 16–18: 145–63.
Bingöl, O. (2004), *Arkeolojik Mimaride Taş* (Istanbul: Homer Books).
Birley, A. (2000), *Septimius Severus: The African Emperor* (London: Routledge).
Bjur, H., and Santillo Frizell, B. (eds.) (2009), *Via Tiburtina: Space, Movement and Artefacts in the Urban Landscape* (Rome: Swedish Institute in Rome).
Black, S., Browning, J., and Laurence, R. (2009), 'From Quarry to Road: The Supply of Basalt for Road Paving in the Tiber Valley', in Coarelli and Patterson (eds.), 705–27.
Bloch, H. (1947), *I bolli laterizi e la storia edilizia romana* (Rome: 'L'Erma' di Bretschneider).
——(1959), 'The Serapeum of Ostia and the Brick-Stamps of 123 A.D.: A New Landmark in the History of Roman Architecture', *American Journal of Archaeology* 63: 225–40.
Bodel, J. (1986), 'Graveyards and Groves: A Study of the Lex Lucerina', *American Journal of Ancient History* 11: 1–133.
——(1999), 'Death on Display: Looking at Roman Funerals', in Bergmann and Kondoleon (eds.), 259–81.
Boeswillwald, G., Cagnat, R., and Ballu, A. (1905), *Timgad, une cité africaine sous l'empire romain* (Paris, Leroux).
Boethius A. (1952), 'Et crescunt media pegmata celsa via (Martial's De Spectaculis 2, 2)', *Eranos* 50: 129–37.
Bollmann, B. (1998), *Römische Vereinshäuser: Untersuchungen zu den Scholae der römischen Berufs-, Kult- und Augustalen-Kollegien in Italien* (Mainz: Philipp von Zabern).
Boyle, A. J. (2003a), *Ovid and the Monuments: A Poet's Rome* (Bendigo).
——(2003b), 'Introduction: Reading Flavian Rome', in Boyle and Bominik (eds.), 1–68.
Boyle, A. J., and Dominik, W. J. (eds.) (2003), *Flavian Rome: Culture, Image, Text* (Leiden: Brill).
Bradford, J. (1957), *Ancient Landscapes: Studies in Field Archaeology* (London: G. Bell and Sons).

Brand, S. (1994), *How Buildings Learn: What Happens After They're Built* (New York: Viking).
Brandt, O. (2009), 'Movement between Rome and the Sanctuary of San Lorenzo', in Bjur and Santillo Frizell (eds.), 79–93.
Brant, C., and Whyman, S. E. (eds.) (2007), *Walking the Streets of Eighteenth-Century London* (Oxford: Oxford University Press).
Braund, S. (2007), 'Gay's *Trivia*: Walking the Streets of Rome', in Brant and Whyman (eds.), 149–68.
Brilliant, R. (1967), *The Arch of Septimius Severus in the Roman Forum* (Memoirs of the American Academy in Rome, 29).
Brown, A. J., and Sherrard, H. M. (1969), *An Introduction to Town and Country Planning* (2nd edn., New York: American Elsevier Publishing Co.).
Brown, F. E. (1980), *Cosa: The Making of a Roman Town* (Ann Arbor: University of Michigan Press).
Brück, J. (1998), 'In the Footsteps of the Ancestors: A Review of Christopher Tilley's *A Phenomenology of Landscape: Places, Paths and Monuments*', *Archeological Review from Cambridge* 15/1: 23–36.
Brunt, P. A. (1980), 'Free Labour and Public Works at Rome', *Journal of Roman Studies* 70: 81–100.
Bruun, C. (1991), *The Water Supply of Ancient Rome: A Study of Roman Imperial Administration* (Commentationes Humanarum Litterarum, Helsinki: Societas Scientiarum Fennica).
——(2003), 'The Antonine Plague in Rome and Ostia', *Journal of Roman Archaeology* 16: 426–34.
Bruun, C., and Gallina-Zevi, A. (eds.) (2002), *Ostia e Portus nelle loro relazioni con Roma* (Rome: Acta Instituti romani finlandiae).
Bull, M., and Back, L. (2005), 'Introduction: Into Sound', in Bull and Back (eds.), *The Auditory Culture Reader* (Oxford: Berg), 1–18.
Burford, A. (1960), 'Heavy Transport in Classical Antiquity', *Economic History Review*, NS 13/1: 1–18.
Buzzetti, C. (1993a), 'Claudius, Divus, Templum', in *LTUR* I (A–C): 277–8.
——(1993b), 'Campus Caelimontanus', in *LTUR* I (A–C): 218.
Caillois, R. (1967a), *Les Jeux et les hommes: Le Masque et le vertige* (Paris, Gallimard), (Italian trans. (1981), *I giochi e gli uomini: La maschera e la vertigine* (Milan: Bompiani).
——(1967b), 'Jeux et sport', *Encyclopèdie de la Pléiade XXIII* (Paris, Gallimard).
Calza, G. (1927), 'Ostia—Rinvenimenti epigrafici', *Notizie degli scavi di antichità* 379–431.
Calza, G., Becatti, G., Gismondi, I., De Angelis D'Ossat, G., and Bloch, H. (eds.) (1953), *Scavi di Ostia I: Topografia generale* (Rome: La libreria dello stato).
Campbell, B. (1996), 'Shaping the Rural Environment: Surveyors in Ancient Rome', *Journal of Roman Studies* 86: 74–99.
——(2000), *The Writings of the Roman Land Surveyors: Introduction, Text, Translation, and Commentary* (Hertford: Society for the Promotion of Roman Studies).
Campbell, N., and Kean, A. (1997), *American Cultural Studies: An Introduction to American Culture* (London: Routledge).

Carcopino, J. (1940), *Daily Life in Ancient Rome: The People and the City at the Height of the Empire*, trans. E. O. Lorimer (New Haven: Yale University Press).

Carettoni, G. (ed.) (1960), *La pianta marmorea di Roma antica: Forma Urbis Romae* (Rome: Comune di Roma).

Carignani A., Gabucci A., Palazzo P., and Spinola, G. (1990), 'Nuovi dati sulla topografia del Celio: Le ricerche nell'area dell'Ospedale Militare', *Archeologia laziale* 10: 72–80.

Carlstein, T. (1982), *Time Resources, Society, and Ecology: On the Capacity for Human Interaction in Space and Time* (London: Allen & Unwin).

Carnabuci, E. (1991), *L'angolo sud-orientale del Foro Romano nel manoscritto inedito di Giacomo Boni* (Rome: Atti della Accademia nazionale dei lincei. Memorie della classe di scienze morali, storiche e filologiche 1/4).

Carroll, L. (1974), *The Hunting of the Snark*, ed. and notes M. Gardner (1st pub. 1876; Harmondsworth: Penguin).

Carroll, M. (2008), '*Nemus et Templum*: Exploring the Sacred Grove at the Temple of Venus in Pompeii', in P. G. Guzzo and M. P. Guidobaldi (eds.), *Nuove ricerche archeologiche nell'area vesuviana (scavi 2003–2006)* (Rome: 'L'Erma' di Bretschneider), 37–45.

Carroll, M., and Godden, D. (2000), 'The Sanctuary of Apollo at Pompeii: Reconsidering Chronologies and Excavation History', *American Journal of Archaeology* 104/4: 743–54.

Carter, H. (1983), *An Introduction to Urban Historical Geography* (London: Arnold).

Caruso, G. (1996), 'Le mura aureliane', in R. Volpe (ed.), *Aqua Marcia: lo scavo di un tratto urbano* (Florence: All'Insegna del Giglio), 83–6.

Caruso, G., and Volpe, R. (1992), *Colle Oppio* (Rome: Palombi).

——(1995), 'Mura serviane in Piazza Manfredo Fanti', *Archeologia laziale* 12: 185–91.

Casson, L. (1990), 'New Light on Maritime Loans: PVindob.G 40882', *Zeitschrift für Papyrologie und Epigraphik* 84: 195–206.

——(1994), *Travel in the Ancient World* (1st edn. 1974; Baltimore: Johns Hopkins University Press).

Castagnoli, F. (1985), 'Un nuovo documento per la topografia di Roma antica', *Studi romani*, 33: 205–11.

Castrén, P. (2000), '*Vici* and *Insulae*: The Homes and Addresses of the Romans', *Arctos* 34: 7–19.

Cecamore, C., Ungaro, L., and Panunzi, S. (2006), 'Il virtuale nel reale: il caso del Foro di Augusto', in Haselberger and Humphrey (eds.), 183–90.

Champlin, E. (2005), 'Phaedrus the Fabulous', *Journal of Roman Studies* 95: 97–123.

Chevallier, R. (1986), *Ostie antique: Ville et port* (Paris: Les Belle Lettres).

——(1988), *Voyages et déplacements dans l'Empire romain* (Paris: Colin).

Ciancio Rosetto, P. (1991), 'La Reconstitution de Rome antique: Du plan-relief de Bigot à celui de Gismondi', in Hinard and Royo (eds.), 237–56.

——(1996), 'Rinvenimenti e restauri al portico d'Ottavia e in piazza delle Cinque Scole', *Bullettino della Commissione archeologica comunale di Roma* 97: 267–78.

——(2006), 'Il nuovo frammento della *Forma* severiana relativo al Circo Massimo', in Meneghini and Santangeli Valenzani (eds.), 127–42.

Cicerchia, P., and Marinucci, A. (1992), *Le terme del Foro o di Gavio Massimo, Scavi di Ostia*, 11 (Rome: La libreria dello stato).
Cima, C., and La Rocca, E. (eds.) (1998), *Horti Romani* (Rome: 'L'Erma' di Bretschneider).
Cima, M. (1995), 'Gli horti liciniani e le statue dei magistrati', in M. Cima (ed.), *Restauri nei Musei Capitolini* (Venice: Arsenale), 53–69.
Ciprotti, P. (1962), *Pompei* (Rome: Editrice Studium).
Claridge, A. (1998), *Rome: An Oxford Archaeological Guide* (Oxford: Oxford University Press).
Coarelli F. (1980), *Roma* (Rome: Laterza).
——(1983), *Il Foro Romano* I: *Periodo archaico* (Rome: Quasar).
——(1985), *Il Foro Romano* II: *Periodo repubblicano e Augusteo* (Rome: Quasar).
——(1986), 'L'urbs e il suburbio', in A. Giardina (ed.), *Società romana e impero tardoantico*, vol. ii (Rome: Istituto Gramsci), 1–58.
——(1988), *Il Foro Boario: Dalle origini alla fine della Repubblica* (Rome: Quasar).
——(1993a), 'Campus Esquilinus', *LTUR* I (A–C): 218–19.
——(1993b), 'Campus Viminalis', *LTUR* I (A–C): 226.
——(1995a), 'Divorum, Porticus, Templum', in *LTUR* II (D–G): 19–20.
——(1995b), 'Forum Esquilinum', *LTUR* II (D–G): 298.
——(1996a), 'Libitina, Lucus', *LTUR* III (H–O): 189–90.
——(1996b), 'Porta Esquilina', *LTUR* III (H–O): 326–7.
——(1996c), 'Porta Viminalis', *LTUR* III (H–O): 334.
——(1999a), 'Pax, Templum', *LTUR* IV (P–S): 67–70.
——(1999b), 'Pagus Montanus', *LTUR* IV (P–S): 10.
——(1999c), 'Puticuli', *LTUR* IV (P–S): 173–4.
——(1999d), 'Pagus Montanus', *LTUR* IV (P–S): 10.
——(1999e), 'Venus Libitina, Lucus', *LTUR* V (T–Z): 117–18.
——(2001) 'Il Foro Triangolare: decorazione e funzione', in P. G. Guzzo (ed.), *Pompei: Scienza e società* (Milan: Electa), 97–107.
——(2005), 'Pits and Fora: A Reply to Henrik Mouritsen', *Papers of the British School at Rome* 73: 23–30.
Coarelli, F., and Patterson, H. (eds.) (2009), Mercator Placidissimus: *The Tiber Valley in Antiquity: New Research in the Upper and Middle Valley* (Rome: Quasar), 705–27.
Coates-Stephens, R. (1998), 'The Walls and Aqueducts of Rome in the Early Middle Ages, AD 500–1000', *Journal of Roman Studies* 88: 166–78.
——(2003), 'Gli acquedotti in epoca tardo-antica nel suburbio', in P. Pergola, R. Santangeli Valenzani, and R. Volpe (eds.), *Suburbium: Il suburbio di Roma dalla crisi del sistema delle ville a Gregorio Magno* (Rome: École française de Rome), 415–36.
——(2004), *Porta Maggiore: Monument and Landscape* (Rome: 'L'Erma' di Bretschneider).
Cohen, E. (2006), 'The Broken Cycle: Smell in a Bangkok Lane', in Drobnick (ed.) (2006a: 118–27).
Coleman, K. M. (1998), 'The *Liber Spectaculorum*: Perpetuating the Ephemeral', in F. Grewing (ed.), Toto Notus in Orbe: *Perspektiven der Martial-Interpretation* (Stuttgart: Franz Steiner Verlag), 15–36.

Coleman, K. M. (1998), 'Martial Book 8 and the Politics of AD 93', *Papers of the Leeds Latin Seminar* 10: 337–57.

——(2000), 'Latin Literature after AD 96: Change or Continuity', *American Journal of Ancient History* 15: 19–39.

Colini, A. M. (1934), 'Notizario scavi: Templum Pacis', *Bullettino della Commissione archeologica comunale di Roma*, 62: 165–6.

——(1937), 'Forum Pacis', *Bullettino della Commissione archeologica comunale di Roma*, 65: 7–40.

Colton, R. (1991), *Juvenal's Use of Martial's Epigrams: A Study of Literary Influence* (Amsterdam: Adolf M. Kakkert).

Connors, C. (2000), 'Imperial Space and Time: The Literature of Leisure', in O. Taplin (ed.), *Literature in the Roman World* (Oxford: Oxford University Press), 208–34.

Conticello de' Spagnolis, M. (1984), *Il tempio dei dioscuri nel Circo Flaminio* (Rome: De Luca).

Conticello, B. (1987), *Pio IX a Pompei: Memorie e testimonianze di un viaggio* (Naples: Bibliopolis Ed. Coll. Le Mostre).

Conzen, M. R. G. (1960), *Alnwick, Northumberland: A Study in Town-Plan Analysis* (London: Institute of British Geographers).

Corballis, M. C., and Beale, I. L. (1976), *The Psychology of Left and Right* (Hillsdale, NJ: Lawrence Erlbaum Associates).

Corbeill, A. (2002), 'Political Movement: Walking and Ideology in Republican Rome', in D. Frederick (ed.), *The Roman Gaze: Vision, Power and the Body* (Baltimore: Johns Hopkins University Press), 182–215.

——(2003), *Nature Embodied: The Power of Gesture in Ancient Rome* (Princeton: Princeton University Press).

Cornell, T. J., and Lomas, K. (eds.) (1995), *Urban Society in Roman Italy* (London: University College London Press).

Cotter, W. (1996), 'The *Collegia* and Roman Law: State Restrictions on Voluntary Associations 64 BCE–200 CE', in J. S. Kloppenborg and S. G. Wilson (eds.), *Voluntary Associations in the Graeco-Roman World* (London: Routledge), 74–89.

Cozza, L. (1997), 'Mura di Roma dalla Porta Nomentana alla Tiburtina', *Analecta romana instituti danici*, 25: 7–113.

Cozzo, G. (1970), *Ingegneria romana: Maestranze romane, strutture preromane, strutture romane, le costruzioni dell'anfiteatro Flavio, del Pantheon, dell'emissario del Fucino* (Rome: Multigrafica, reproduction of 1928 edn.).

Crawford, M. H. (1996), *Roman Statutes, Bulletin of the Institute of Classical Studies*, suppl. 64 (London: Institute of Classical Studies, University of London).

Cutting, M. (2003), 'The Use of Spatial Analysis to Study Prehistoric Settlement Architecture', *Oxford Journal of Archaeology* 22/1: 1–21.

Czarnowski, T. V. (1986), 'The Street as a Communications Artefact', in Anderson (ed.), 206–13.

Dalby, A. (2000), *Empire of Pleasures: Luxury and Indulgence in the Roman World* (London, Routledge).

Darwall-Smith, R. (1996), *Emperors and Architecture: A Study of Flavian Rome* (Brussels: Collection Latomus).

Davies, P. (2000), *Death and the Emperor: Roman Imperial Funerary Monuments from Augustus to Marcus Aurelius* (Cambridge: Cambridge University Press).

De Angelis Bertolotti, R. (1983), 'Le mure serviane nella Quinta Regione Augustea', in *Archeologia in Roma capitale tra sterro e scavo* (Venice: Marsilio), 119–29.

De Carolis, E. (1996), 'Il recupero di due impianti commerciali ercolanesi', *I beni culturali*, 4/2: 34–7.

De Certeau, M. (1984), *The Practice of Everyday Life*, trans. S. Rendall (Berkeley and Los Angeles: University of California Press).

De Felice, J. (2007), 'Inns and Taverns', in Dobbins and Foss (eds.), 474–86.

De Haan, N., and Jansen, G. (eds.) (1996), *Cura aquarum in Campania: Proceedings of the Ninth International Congress on the History of Water Management and Hydraulic Engineering in the Mediterranean Region: Pompeii, 1–8, 1994*, BABESCH, suppl. 4 (Leiden: Stichting Babesch).

De Laet, S. (1949), *Portorium: Étude sur l'organisation douanière chez les Romains* (Bruges: Rijkuniversitet te Gent).

De Ligt, L. (1993), *Fairs and Markets in the Roman Empire* (Amsterdam: Gieben).

De Maria, S. (1988), *Gli archi onorari di Roma e dell'Italia romana* (Rome: 'L'Erma' di Bretschneider).

De Robertis, F. M. (1971), *Storia delle corporazioni e del regime associativa nel mondo romano* (Bari: Adriatica editrice).

De Ruyt, C. (1983), *Macellum: Marché alimentaire des Romains* (Louvain-La-Neuve: Institut Supérieur d'Archéologie et d'Histoire de L'Art Collège Érasme).

——(2000), 'Exigences fonctionelles et varieté des interprétations dans l'architecture des macella du monde romain', in Lo Cascio (2000b: 177–86).

De Spirito, G. (1995a), 'S. Eusebius, Titulus', *LTUR* II (D–G): 239–40.

——(1995b), 'Forum Appiae', *LTUR* II (D–G): 288.

——(1995c), 'Forum Sallustii', *LTUR* II (D–G): 345–6.

——(1995d), 'Forum Tauri', *LTUR* II (DG): 347–8.

——(1999), 'Palatium Liciniani / Licinianum', *LTUR* IV (P–S): 45.

Degrassi, A. (1963), *Inscriptiones Italiae 13.3: Fasti et Elogia* (Rome: Istituto Poligrafico dello Stato).

DeLaine, J. (1995), 'The Supply of Building Materials to the City of Rome', in N. Christie (ed.), *Settlement and Economy in Italy 1500 BC to AD 1500* (Oxford: Oxbow), 555–62.

——(1997), *The Baths of Caracalla: A Study in the Design, Construction, and Economics of Large-Scale Building Projects in Imperial Rome* (*Journal of Roman Archaeology*, suppl. ser. 25, Portsmouth, RI: Journal of Roman Archaeology).

——(2000), 'Building the Eternal City: The Building Industry of Imperial Rome,' in J. Coulston and H. Dodge (eds.), *Ancient Rome: The Archaeology of the Eternal City* (Oxford: Oxbow Books), 119–41.

——(2002), 'Building Activity in Ostia in the Second Century AD', in Bruun and Gallina-Zevi (eds.), 41–102.

——(2004), 'Designing for the Market: *Medianum* Apartments at Ostia', *Journal of Roman Archaeology* 17: 146–76.

——(2005), 'The Commercial Landscape of Ostia', in MacMahon and Price (eds.), 29–47.

Dey, H. W. (2011), *The Aurelian Wall and the Refashioning of Imperial Rome, AD 271–855* (Cambridge: Cambridge University Press).

Dobbins, J. (1994), 'Problems of Chronology, Decoration, and Urban Design in the Forum at Pompeii', *American Journal of Archaeology* 98: 629–94.

——(2007), 'The Forum and Its Dependencies', in Dobbins and Foss (eds.), 150–83.

Driessen, M., Heeren, S., Hendriks, J., Kemmers, F., and Visser, R. (eds.) (2009), *TRAC 2008 Proceedings of the Eighteenth Annual Theoretical Roman Archaeology Conference, Amsterdam 2008* (Oxford: Oxbow Books).

Dobbins, J. and Foss, P. (eds.) (2007), *The World of Pompeii* (London: Routledge).

Drobnick, J. (ed.) (2006a), *The Smell Culture Reader* (Oxford: Berg).

——(2006b), 'Introduction: Olfactocentrism', in Drobnick (2006a: 1–9).

Dueck, D. (2000), *Strabo of Amaseia: A Greek Man of Letters in Augustan Rome* (London: Routledge).

Dugan, J. (2005), *Making a New Man: Ciceronian Self-Fashioning in the Rhetorical Works* (Oxford: Oxford University Press).

Eck, W. (1992), 'Cura viarum und cura operum publicorum als kollegiale Ämter im frühen Prinzipat', *Klio*, 74: 237–45.

——(2008), 'Verkehr und verkehrsregeln in einer antiken Großstadt, Das Beispiel Rom', in Mertens (ed.), 59–69.

Edmondson, J. C. (2008), 'Public Dress and Social Control in Late Republican and Early Imperial Rome', in J. C. Edmondson and A. Keith (eds.), *Roman Dress and the Fabrics of Roman Culture* (Toronto: University of Toronto Press), 21–46.

Edwards, C. (2009), *The Politics of Immorality in Ancient Rome* (Cambridge: Cambridge University Press).

Egelhaaf-Gaiser, U. (2000), *Kulträume im römischen Alltag* (Stuttgart: Franz Steiner Verlag).

——(2002), 'Religionsästhetic und Raumordnung am Beispiel der Vereinsgebäude von Ostia', in U. Egelhaaf-Gaiser and A. Schäfer (eds.), *Religiöse Vereine in der römischen Antike: Untersuchungen zu Organisation, Ritual und Raumordnung* (Tübingen: Mohr Siebeck), 123–72.

Eitrem, S. (1915), *Opferritus und voropfer der Griechen und Römer* (Kristiania: J. Dybwad).

Elden, S. (2004), *Understanding Henri Lefebvre: Theory and the Possible* (London: Continuum).

Ellis, S. J. R. (2004), 'The Distribution of Bars at Pompeii: Archaeological, Spatial and Viewshed Analyses', *Journal of Roman Archaeology* 17: 371–84.

——(2005), 'The Pompeian Bar and the City: Defining Food and Drink Outlets and Identifying their Place in the Urban Environment' (Unpublished PhD Dissertation, University of Sydney).

——(2008), 'The Use and Misuse of "Legacy Data" in Identifying a Typology of Retail Outlets at Pompeii', *Internet Archaeology* 24, http://intarch.ac.uk/journal/issue24/ellis_index.html.

English, R. B. (1906), 'The Right Hand In Roman Art And Literature' (Unpublished PhD Dissertation, University of Michigan).

Esch, A. (2008), 'Sträenzustand und Verkehr in Stadtgebiet und Umgebung Roms im Übergang von der Spätantike zum Frühmittelalter (5.-8. Jh.)', in Mertens (ed.), 213-37.
Eschebach, H. (1970), *Die Städtebauliche Entwiclung des Antiken Pompeji: Die Baugeschichte der Stabianer Thermen* (Heidelberg: F. H Kerle Verlag).
——(1978), *Pompeji* (Leipzig: VEBE. A. Seemann, Buch- und Kunstverlag).
Eschebach, L. (1993), *Gebäudverzeichnis und Stadtplan der antiken Stadt Pompeji* (Köln: Bohlau).
Esposito, D. (2005), 'Pompei, Regio V, insula 5: Relazione sulla prima campagna di scavi', *Rivista di studi pompeiani*, 16: 156-66.
Étienne, R. (1987), 'Extra Portam Trigeminam: Espace politique et espace économique à l'Emporium de Rome', in Pietri (ed.), 235-49.
Evans, J. K. (1991), *War, Women and Children in Ancient Rome* (London: Routledge).
Evans, R. (2003), 'Containment and Corruption: The Discourse of Flavian Empire', in Boyle and Dominik (eds.), 255-76.
Eyben, E. (1993), *Restless Youth in Ancient Rome*, trans. P. H. Daly (London: Routledge).
Fagan, G. G. (1999), *Bathing in Public in the Roman World* (Ann Arbor: University of Michigan Press).
Falkener, E. (1892), *Games Ancient and Oriental and How to Play Them* (London: Longmans, Green and Co.).
Fallou, A., and Guilhembet, J. P. (2008), '*Sedvm regionvm locorvm nomina* (Cicéron): La Roma antique à travers ses toponymes: les *vici*', in Fleury and Desborbes (eds.), 175-88.
Fant, C. (2001), 'Rome's Marble Yards', *Journal of Roman Archaeology* 14: 167-98.
Farrell, J. (1997), 'The Phenomenology of Memory in Roman Culture', *Classical Journal* 92/4: 373-83.
——(2001), *Latin Language and Latin Culture* (Cambridge: Cambridge University Press).
Favro, D. (1994), 'The Street Triumphant: The Urban Impact of Roman Triumphal Parades,' in Z. Çelik et al. (eds.), *Streets: Critical Perspectives on Public Space* (Berkeley and Los Angeles: University of California Press), 151-64.
——(1996), *The Urban Image of Augustan Rome* (Cambridge: Cambridge University Press).
——(2006), 'In the Eyes of the Beholder: Virtual Reality Re-creations and Academia', in Haselberger and Humphrey (eds.), 321-34.
Favro, D., and Johanson, C. (2010), 'Death in Motion: Funeral Processions in the Roman Forum', *Journal of the Society of Architectural Historians* 69/1: 12-37.
Fearnley, H. (2003), 'Reading the Imperial Revolution: Martial *Epigrams* 10', in Boyle and Dominik (eds.), 613-35.
Feeney, D. (2007), *Caesar's Calendar: Ancient Time and the Beginnings of History* (Berkeley and Los Angeles: University of California Press).
Ferrua, A. (2001), *Tavole lusorie epigrafiche: Catalogo delle schede manoscritte, introduzione e indici a cura di Maria Busia* (Vatican City: Pontificio istituto di archeologia Cristiana).

Finkel, I. (ed.) (2007), *Ancient Board Games in Perspective* (London: British Museum Press).

Fiorelli, G. ([1875] 2001), *La descrizione di Pompei*, repr. U. Pappalardo (Naples: Massa Editore).

First Report (1929), *First Report of the Greater London Regional Planning Committee* (London: Knapp).

Fishman, W. J. (1979), *The Streets of East London* (London: Duckworth),

Fitzgerald, W. (2007), *Martial: The World of the Epigram* (Chicago: University of Chicago Press).

Flambard, J.-M. (1987), 'Éléments pour une approche financière de la mort dans les classes populaires du Haut-empire', in F. Hinard (ed.), *La Mort, les morts et l'au-delà dans le monde romain, Actes du colloque de Caen 20–22 novembre 1985* (Caen: Université de Caen), 209–43.

Fleury, P., and Desbordes, O. (eds.) (2008), *Roma illustrata: Représentations de la ville* (Caen: Université de Caen).

Flohr, M. (2009), 'The Social World of Roman *Fullonicae*', in M. Driessen et al. (eds.), 173–86.

Fogagnolo, S. (2006), 'Lo scavo del *Templum Pacis*: Concordanze e novità rispetto alla *Forma Urbis*', in Meneghini and Santangeli Valenzani (eds.), 61–74.

Foucault, M. (1975), *Discipline and Punish: the Birth of the Prison* (New York: Vintage).

Fowler, D. P. (2000), *Roman Constructions: Readings in Postmodern Latin* (Oxford).

Frakes, J. F. D. (2009), *Framing Public Life: The Portico in Roman Gaul* (Vienna: Phoibos).

Frank. T. (1936), 'The Topography of Terence, *Adelphoe* 573–85', *American Journal of Philology* 57: 470–2.

Franklin, J. (1986), 'Games and a *Lupanar*: Prosopography of a Neighbourhood in Ancient Pompeii', *Classical Journal* 81: 319–28.

Franklin, J. L. (1990), *Pompeii: The 'Casa del Marinaio' and Its History* (Rome: L'Erma di Bretschneider).

Frayn, J. M. (1993), *Markets and Fairs in Roman Italy: Their Social and Economic Importance from the Second Century BC to the Third Century AD* (Oxford: Oxford University Press).

Frederick, D. (2003), 'Architecture and Surveillance in Flavian Rome', in Boyle and Dominik (eds.), 199–227.

Frézouls, E. (1987), 'Rome ville ouverte: Réflexions sur les problèmes de l'expansion urbaine d'Auguste à Aurèlien', in Pietri (ed.), 373–92.

Fridell-Anter, K., and Weilguni, M. (2003), 'Public Space in Pompeii', in G. Malm (ed.), *Toward an Archaeology of Buildings: Contexts and Concepts* (BAR International Series 1186, Oxford: Archaeopress), 31–9.

Frier, B. W. (1977), 'The Rental Market in Early Imperial Rome', *Journal of Roman Studies* 67: 27–37.

Frischer, B., Abernathy, D., Giuliani, F. C., Scott, R. T., and Ziemssen, H. (2006), 'A New Digital Model of the Roman Forum', in Haselberger and Humphrey (eds.), 163–82.

Fröhlich, T. (1991), *Lararien- und Fassadenbilder in den Vesuvstädten: Untersuchungen zur 'volkstümlichen' pompejanischen Malerei* (Mainz: P. von Zabern).
Gassner, V. (1986), 'Die Kaufläden in Pompeii' (Unpublished PhD dissertation, Universität Wien).
Gatti, G. (1886), 'Trovamenti riguardanti la topografia e la epigrafia urbana', *Bullettino della Commissione archaeologica comunale di Roma* 14: 308–18.
——(1933), 'Regione IV. Templum Pacis', *Bullettino della Commissione archeologica comunale di Roma* 61: 247–9.
Gehl, J. (1996), *The Life Between Buildings: Using Public Space* (Copenhagen: Arkitektens forlag).
Geiger, J. (2008), *The First Hall of Fame: A Study of the Statues in the Forum Augustum*, Mnemosyne, suppl. 295 (Leiden and Boston: Brill).
Gelsomino, R. (1975), *Varrone e i Sette Colli di Roma* (Rome: Herder).
Gering, A. (2004), 'Plätze und Straßensperren an Promenaden. Zum Funktionswandel Ostias in der Spätantike', *Mitteilungen des Deutschen Archäologischen Instituts, Römische Abteilung* 111. 299–382.
Gesemann, B. (1996), *Die Straßen der antiken Stadt Pompeji: Entwicklung und Gestaltung* (Frankfurt: Peter Lang).
Geyssen, J. (1999), 'Sending a Book to the Palatine: Martial 1.70 and Ovid', *Mnemosyne* 52: 718–38.
Gilula, D. (1991), 'A Walk Through Town (Ter. Ad. 573–584)', *Athenaeum* 79: 245–7.
Giuliani, C. F. (1985), 'Lettura del centro monumentale: problemi di metodo', in *Roma, archeologia nel centro*, i (Rome: De Luca Editore), 9–16.
Giuliani, C. F., and Verduchi, P. (1987), *Foro Romano: L'area centrale* (Florence: Leo S. Olschki).
Gleason, K. L. (1994), 'Porticus Pompeiana: A New Perspective on the First Public Park of Ancient Rome', *Journal of Garden History* 14/1: 13–27.
Gold, B. K. (2003), '*Accipe Divinitas et Vatum Maximus Esto:* Money Poetry, Mendicancy and Patronage in Martial', in Boyle and Dominik (eds.), 591–612.
Goodman, P. (2007), *The Roman City and Its Periphery: From Rome to Gaul* (London: Routledge).
Gorrie, C. (2002), 'The Severan Building Program and the Saecular Games', *Athenaeum* 90/2: 461–81.
——(2004), 'Julia Domna's Building Patronage, Imperial Family Roles and the Severan Revival of Moral Legislation', *Historia: Zeitschrift für Alte Geschichte* 53: 61–72.
Gorrie, C. (2007), 'The Restoration of the Porticus Octaviae and Severan Imperial Policy', *Greece and Rome* 54: 1–17.
Gould, P., and White, R. (1974), *Mental Maps* (London: Pelican).
Gowers, E. (1995), 'The Anatomy of Rome from Capitol to Cloaca', *Journal of Roman Studies* 85: 23–32.
Graham, S. (2006), 'Networks, Agent-Based Models and the Antonine Itineraries: Implications for Roman Archaeology', *Journal of Mediterranean Archaeology* 19: 45–64.
——(2009), 'The Space Between: The Geography of Social Networks in the Tiber Valley', in Coarelli and Patterson (eds.), 671–86.

Grahame, M. (2000), *Reading Space: Social Interaction and Identity in the Houses of Roman Pompeii: A Syntactical Approach to the Analysis and Interpretation of Built Space* (BAR International Series 886, Oxford: Archaeopress).

Green, J. R. (1999), 'Tragedy and the Spectacle of the Mind: Messenger Speeches, Actors, Narrative, and Audience Imagination in Fourth-Century BCE Vase-Painting', in Bergmann and Kondoleon (eds.), 37–63.

Grenier, A. (1934), *Manuel d'archéologie gallo-romaine II. L'Archéologie du sol. 1. Les Routes* (Paris: Éditions A. Picard).

Grey, C., and Parkin, A. (2003), 'Controlling the Urban Mob: The *colonatus perpetuus* of CTh 14.18.1', *Phoenix* 57: 284–99.

Griesbach, J. (2005), 'Villa e mausoleo: trasformazioni nel concetto della memoria nel suburbia romano', in B. Santillo Frizell and A. Klynne (eds.), *Roman Villas Around the* Urbs: *Interaction with Landscape and Environment* (Rome: The Swedish Institute in Rome, Projects and Seminars 2), 113–23.

Gros, P. (1996), *L'Architecture romaine: Du début du IIIe siècle av. J.-C. à la fin du Haut-Empire*, i. *Les Monuments publiques* (Paris: Picard).

——(2001a), '*Nunc tua cinguntur limina*: L'Apparence de l'accueil et la réalité du filtrage à l'entrée des forums impériaux de Rome', in J. Ch. Balty (ed.), *Rome et ses provinces: Genèse et diffusion d'une image du pouvoir* (Brussels: Le Livre Timperman), 129–40.

——(2001b), 'Les Édifices de la bureaucratie impériale: Administration, archives et services publics dans le centre monumental de Rome', *Pallas* 55: 107–26.

——(2005), 'Le Rôle du peuple de Rome dans la définition, l'organisation et le déplacement des lieux de la convergence sous l'Empire', in G. Urso (ed.), *Popolo e potere nel mondo antico* (Pisa: Edizione ETS), 191–214.

Groß, K. (1985), *Menschenhand und Gotteshand in Antike und Christentum* (Stuttgart: A. Hiersemann).

Guasti, L. (2007), 'Animali per Roma', in E. Papi (ed.), *Supplying Rome and the Empire* (*Journal of Roman Archaeology*, suppl. ser. 69), 139–52.

Guidobaldi, F. (1998), 'Il tempio di "Minerva Medica" e le strutture adiacenti: Settore private del Sessorium Costantiniano', *Rivista di archeologia cristiana*, 74: 485–518.

——(2000), 'Le abitazioni private e l'urbanistica', in A. Giardina (ed.), *Roma antica* (Rome: Laterza), 133–61.

Gutman, R. (1986), 'The Street Generation', in Anderson (ed.), 249–64.

Halfmann, H. (1986), *Itinera Principum: Geschichte und Typologie der Kaiserreisen im römischen Reich* (Stuttgart: Franz Steiner).

Hamilton, S., and Whitehouse, R. (2006), 'Three Senses of Dwelling: Beginning to Socialise the Neolithic Ditched Villages of the Tavoliere, Southeast Italy', in V. O. Jorger, J. M. Cardoso, A. M. Vale, G. L. Velho, and L. S. Pereira (eds.), *Approaching 'Prehistoric and Protohistoric Architectures' of Europe from a 'Dwelling Perspective'* (Porto: Adecap; *Journal of Iberian Archaeology* 8), 159–84.

Handel, S. (1989), *Listening: An Introduction to the Perception of Auditory Events* (Cambridge, Mass.; London: MIT Press).

Hanson, J. (1998), *Decoding Homes and Houses* (Cambridge and New York: Cambridge University Press).

Harlow, M., and Laurence, R. (2002), *Growing Up and Growing Old in Ancient Rome: A Life Course Approach* (London: Routledge).
Harris, H. A. (1972), *Sport in Greece and Rome* (London: Thames and Hudson).
Harris, J. (2007), *Pompeii Awakened: A Story of Rediscovery* (London: I. B. Tauris Ltd).
Harris, W. V. (1993), 'Between Archaic and Modern: Problems in Roman Economic History', in W. V. Harris (ed.), *The Inscribed Economy: Production and Distribution in the Roman Empire in the Light of Instrumentum Domesticum* (*Journal of Roman Archaeology*, suppl. ser. 6), 11–30.
Hartnett, J. (2003), 'Streets, Street Architecture, and Social Presentation in Roman Italy', Unpublished PhD Dissertation, University of Michigan.
——(2008), '*Si quis hic sederit*: Streetside Benches and Urban Society in Pompeii', *American Journal of Archaeology* 112: 91–119.
——(2009), 'Fountains at Herculaneum: Sacred History, Topography, and Civic Identity', *Rivista di tudi ompeiani* 19: 77–89.
Hartswick, K. J. (2004), *The Gardens of Sallust: A Changing Landscape* (Austin: University of Texas).
Haselberger, L. (1994), 'Ein Giebelriss der Vorhalle des Pantheon die Werkrisse vor dem Augustusmausoleum', *Mitteilungen des Deutschen Archäologischen Instituts, Römische Abteilung* 101: 279–307.
——(2007), Urbem Adornare: *Die Stadt Rom Und Ihre Gestaltumwandlung Unter Augustus/Rome's Urban Metamorphosis under Augustus* (*Journal of Roman Archaeology*, suppl. ser. 64, Portsmouth, RI).
Haselberger, L., and Humphrey, J. (eds.) (2006), *Imaging Ancient Rome: Documentation—Visualization—Imagination* (*Journal of Roman Archaeology*, Suppl. Ser. 61, Portsmouth, RI).
Heidegger, M. (1950), *Holzwege* (Frankfurt: Vittorio Klostermann).
Heinzelmann, M. (1998), 'Beobachtungen zur suburbanen Topographie Ostias: Ein orthogonales Straßensystem im Bereich der Pianabella', *Mitteilungen des Deutschen Archäologischen Instituts Römische Abteilung*: 175–225.
——(1999a), 'Zur Entwicklung der Gelände- und Strassenniveaus in der Nekropole vor der Porta Romana', in T. A. M. Mols and C. E. van der Laan (eds.), *Atti del II Colloquio Internazionale su Ostia Antica, Roma 1998, Mededelingen, Papers of the Netherlands Institute in Rome* 58: 84–9.
——(1999b), 'Neue Untersuchungen in den unausgegrabenen Gebieten von Ostia: Luftbildauswertung und geophysikalische Prospektionen', in T. A. M. Mols and C. E. van der Laan (eds.), *Atti del II Colloquio internazionale su Ostia Antica, Roma 1998, Mededelingen, Papers of the Netherlands Institute in Rome* 58: 24–5.
——(2002), 'Bauboom und urbanistische Defizite—zur städtebaulichen Entwicklung Ostias im 2. Jh.', in Bruun and Galina-Zevi (eds.), 103–21.
——(2005), 'Die vermietete Stadt', in P. Zanker and R. Neudecker (eds.), *Lebenswelten: Bilder und Räume in der römischen Stadt der Kaiserzeit*, Palilia 16 (Wiesbaden: Dr Ludwig Reichert Verlag), 113–28.
Helen, T. (1975), *The Organization of Roman Brick Production in the First and Second Centuries AD.: An Interpretation of Roman Brick Stamps* (Annales Academiae Scientiarum Fennicae, Helsinki: Suomalainen Tiedeakatemia).

Henneberg, K. (2004), 'Monuments, Public Space, and the Memory of Empire in Modern Italy', *History and Memory* 16/1: 37–85.
Henriksén, C. (1998), *Martial, Book IX: A Commentary* (Uppsala: Studia Latina Upsaliensia 24).
Hermansen, G. (1974), 'The Roman Inns and the Law: The Inns of Ostia', in J. A. S. Evans (ed.), *Polis And Imperium: Studies in Honour of Edward Togo Salmon* (Toronto: Hakkert), 167–81.
——(1982), *Ostia: Aspects of Roman City Life* (Edmonton: University of Alberta Press).
Hertz, R. (1960), *Death and The Right Hand* (Aberdeen: Cohen & West).
——(1973), 'The Pre-eminence of the Right Hand: A Study in Religious Polarity', in R. Needham (ed.), *Right and Left: Essays on Dual Symbolic Classification* (Chicago: University of Chicago Press), 3–31.
Hillier, B. (1996a), *Space is the Machine: A Configurational Theory of Architecture* (Cambridge: Cambridge University Press).
Hillier, B. (1996b), 'Cities as Movement Economies', *Urban Design International* 1/1: 41–60.
Hillier, B., and Hanson, J. (1984), *The Social Logic of Space* (Cambridge: Cambridge University Press).
Hillier, B., and Vaughan, L. (2007), 'The City as One Thing', *Progress in Planning* 67/3: 205–30.
Hillier, B., Penn, A., Banister, D., and Xu, J. (1998), 'Configurational Modelling of Urban Movement Networks', *Environment and Planning B: Planning and Design* 25: 25–84.
Hillier, B., Penn, A., Hanson, J., Grajewski, T., and Xu J. (1993), 'Natural Movement: Or Configuration and Attraction in Urban Pedestrian Movement', *Environment and Planning B: Planning and Design* 20: 29–66.
Hinard, F. (1987), 'Spectacle d'executions et espace urbain', in Pietri (ed.), 111–25.
——(1991), 'La Maquette comme objet scientifique', in Hinard and Royo (eds.), 281–6.
Hinard, F., and Royo, M. (eds.) (1991), *Rome: l'espace urbain et ses représentations* (Paris: Université de Paris-Sorbonne).
Hinds, S. (2007), 'Martial's Ovid/Ovid's Martial', *Journal of Roman Studies* 97: 113–54.
Hitchcock, L. (2008), *Theory For Classics: A Student's Guide* (London: Routledge).
Holland, L. A. (1961), *Janus and the Bridge* (Rome: American Academy in Rome).
Holleran, C. (2011), 'Migration and the Urban Economy of Rome', in C. Holleran and A. Pudsey (eds.), *Demographic Influences on Greco-Roman Societies and Economies*.
——(2012), *Shopping in Ancient Rome: The Retail Trade in the Late Republic and the Principate* (Oxford: Oxford University Press)
Homo, L. P. (1951), *Rome impériale et l'urbanisme dans l'antiquité* (Paris: A. Michel).
Hopkins, K. (1999), *A World Full of Gods: Pagans, Jews and Christians in the Roman Empire* (London: Weidenfeld & Nicolson).
Horden, P., and Purcell, N. (2000), *The Corrupting Sea: A Study of Mediterranean History* (London: Blackwell).
Howell, P. (1980), *A Commentary on Book One of the Epigrams of Martial* (London: Athlone).

———(1995), *Martial Epigrams* V (Warminster: Aris and Phillips).
———(2009), *Martial* (Bristol: Bristol Classical Press).
Howes, D. (2005), *Empire of the Senses: The Sensual Culture Reader* (Oxford: Berg).
Humphreys, S. (1985). 'Law as Discourse', in S. Humphreys (ed.), *The Discourse of Law, History and Anthropology* 1/2: 241–64.
Hurst, H. (2003), 'Excavation of the Pre-Neronian *Nova Via*', *Papers of the British School at Rome* 71: 17–84.
———(2006), 'The *Scalae* (ex-*Graecae*) above the *Nova Via*', *Papers of the British School at Rome* 74: 237–92.
Huskey, S. J. (2006), 'Ovid's (Mis)Guided Tour of Rome: Some Purposeful Omissions in *Tr.* 3.1', *Classical Journal* 102: 17–39.
Hyland, A. (1990), *Equus: The Horse in the Roman World* (New Haven: Yale University Press).
Ihde, D. (2005), 'Auditory Imagination', in M. Bull and L. Back (eds.), *The Auditory Culture Reader* (Oxford: Berg), 61–6.
Ingold, T. (2000), *The Perception of the Environment* (London: Routledge).
Inwood, B., and Gerson, L. P. (ed. and trans.) (1994), *The Epicurus Reader: Selected Writings and Testimonia* (Indianapolis: Hackett Publishing Company).
Ivanoff, S. (1859), 'Varie specie di soglie in Pompei ed indagine sul vero sito della fauce', *Annali dell'Istituto di corrispondenza archeologica* 31: 82–108.
Jacobs, J. (2000), *The Death and Life of Great American Cities* (London: Pimlico).
Jaeger, M. (2008), *Archimedes and the Roman Imagination* (Ann Arbor: University of Michigan Press).
Jansen, G. (2007), 'The Water System: Supply and Drainage', in Dobbins and Foss (eds.), 257–66.
Jashemski, W. F. (1973), 'The Discovery of a Large Vineyard at Pompeii: University of Maryland Excavations, 1970', *American Journal of Archaeology* 77: 27–41.
———(1977), 'The Excavation of a Shop-House Garden at Pompeii (I. XX. 5)', *American Journal of Archaeology* 81: 217–27.
———(1979), *The Gardens of Pompeii, Herculaneum and the Villas Destroyed by Vesuvius* (New Rochelle, NY: Caratzas Bros.)
Johnston, D. (1999), *Roman Law in Context* (Cambridge and New York: Cambridge University Press).
Johnston, M. (1957), *Roman Life* (Chicago: Scott, Foresman and Co.).
Jolivet, V. (1997), 'Croissance urbaine et espaces verts à Rome', in C. Virlouvet (ed.), *La Rome impériale: Démographie et logistique* (Rome: École française de Rome), 193–208.
Jones, M. W. (2000), *Principles of Roman Architecture* (New Haven: Yale University Press).
Jones, R., and Robinson, D. J. (2007), 'Intensification, Heterogeneity and Power in the Development of *Insula* VI.1', in Dobbins and Foss (eds.), 389–406.
Kaiser, A. E. (2000), *The Urban Dialogue: An Analysis of the Use of Space in the Roman City of Empúries, Spain* (British Archaeological Reports International Series 901; Oxford: Archaeopress).
———(2001), 'The Visibility of Temples and Villas in the Roman Urban Environment: A Case Study from Emporiae (Spain)', *American Journal of Archaeology* 105/2: 271.

Kaiser, A. E. (2003), 'The Application of GIS Viewshed Analysis to Roman Urban Studies: The Case-Study of Empúries, Spain', *Internet Archaeology* 14, http://intarch.ac.uk/journal/issue14/kaiser_index.html.
——(2011a), 'What Was a *Via*?: An Integrated Archaeological and Textual Approach', in E. Poehler, K. Cole, and M. Flohr (eds.), *Pompeii: Art, Industry and Infrastructure* (Oxford: Oxbow), 106–21.
——(2011b), Roman Urban Street Networks (New York: Routledge).
Kay, N. M. (1985), *Martial Book XI: A Commentary* (London: Duckworth).
Keaveney, A. (1983), 'Sulla and the Gods', in C. Deroux (ed.), *Studies in Latin Literature and Roman History*, iii (Brussels: Latomus), 4–79.
Kellum, B. (1999), 'The Spectacle of the Street', in Bergmann and Kondoleon (eds.), 283–99.
Kendal, R. (1996), 'Transport Logistics with the Building of Hadrian's Wall', *Britannia* 27: 129–52.
Kent, R. G. (ed. and trans.) (1951), *Varro: On the Latin Language*, 2 vols. (Cambridge, Mass.: Harvard University Press).
Kern, S. (2003), *The Culture of Time and Space, 1880-1918, with a New Preface* (Cambridge, Mass: Harvard University Press).
King, R. J. (2010), '*Ad Capita Babula*: the birth of Augustus and Rome's imperial centre', *Classical Quarterly* 60/2: 450–69.
Klasander, A.-J. (2004), *Suburban Navigation: Structural Coherence and Visual Appearance in Urban Design* (Gothenburg: Chalmers).
Kleberg, T. (1957), *Hôtels, restaurants et cabarets dans L'antiquité romaine* (Uppsala: Almqvist & Wiksells).
Klynne, A. (2009), 'Where Have All the Ruins Gone?', in Bjur and Santillo Frizell (eds.), 165–80.
Köb, I. (2000) *Rom—ein Stadtzentrum im Wandel: Untersuchungen zur Funktion und Nutzung des Forum Romanum und der Kaiserfora in der Kaiserzeit* (Hamburg: Kovac).
Kockel, V. (1992), 'Ostia im 2. Jahrhundert n. Chr. Beobachtungen zum Wandel eines Stadtbilds', in H.-J. Schalles, H. von Hesberg, and P. Zanker (eds.), *Die römische Stadt im 2. Jahrhundert n. Chr. Der Funktionswandel des öffentlichen Raumes, Xantener Berichte*, ii. 99–117.
Kockel, V., and Weber, B.F. (1983), 'Die villa delle colonne a mosaico in Pompeji', *Mitteilungen des Deutschen Archäologischen Instituts, Römische Abteilung* 90/1: 51–89.
Koga, M. (1992), 'The Surface Drainage System of Pompeii', *Opuscula Pompeiana* 2: 57–72.
Kolb, A. (1993), *Die kaiserliche Bauverwaltung in der Stadt Rom: Geschichte und Aufbau der cura operum pulicorum unter dem Prinzipat* (Stuttgart: F. Steiner).
——(2001), 'Transport and Communication in the Roman State: The Cursus Publicus', in C. Adams and R. Laurence (eds.), *Travel and Geography in the Roman Empire* (London: Routledge), 95–105.
Kolendo, J. (1985), 'Le attività agricole degli abitanti di Pompei et gli attrezzi agricole ritrovati all'interno della città', *Opus* 4: 111–24.

Koller, D., Trimble, J., Najbjerg, T., Gelfand, N., and Levoy, M. (2006), 'Fragments of the City: Stanford's Digital *Forma Urbis Romae* Project', in Haselberger and Humphrey (eds.), 237–52.

Kostof, S. K. (1991), *The City Shaped: Urban Patterns and Meanings Through History* (Boston: Bulfinch).

——(1992), *The City Assembled: The Elements of Urban Form through History* (London: Thames and Hudson).

Krautheimer, R. (1959), *Corpus Basilicarum Christianarum Romae* (Vatican City: Pontificio Istituto di Archeologia Cristiana), vol. ii.

——(1980), *Rome. Profile of a City, 312–1308* (Princeton: Princeton University Press).

Kronenberg, L. (2009), *Allegories of Farming from Greece and Rome: Philosophical Satire in Xenophon, Varro, and Virgil* (Cambridge: Cambridge University Press).

Kuttner, A. L. (1999), 'Culture and History at Pompey's Museum', *Transactions and Proceedings of the American Philological Association*, 129: 343–73.

Kyllingstad, R., and Sjöqvist, E. (1965), 'Hellenistic Doorways and Thresholds from Morgantina', *Acta ad Archaeologiam et Artium Historiam Pertinentia* 2: 23–34.

La Regina, A. (ed.) (2001), *Sangue e Arena* (Milan: Electa).

La Rocca, E. (2001), 'La nuova immagine dei Fori Imperiali: Appunti in margine agli scavi', *Mitteilungen des Deutschen Archäologischen Instituts, Römische Abteilung* 107: 171–213.

——(2006), 'Passeggiando intorno ai Fori Imperiali', in Haselberger and Humphrey (eds.), 120–43.

La Torre, G. F. (1988), 'Gli impianti commerciali ed artigianali nel tessuto urbano di Pompei', in A. De Simone et al. (eds.), *Pompeii: L'informatica al servizio di una città antica*, i–ii (Rome: L'Erma di Bretschneider), 75–102.

Laird, M. L. (1999), 'A Reappraisal of the "Seat of the Augustales" at Ostia', *American Journal of Archaeology* 103: 303.

Laird, M. L. (2000), 'Reconsidering the So-Called "Sede degli Augustali" at Ostia', *Memoirs of the American Academy in Rome* 45: 41–84.

Lancaster, L. C. (1998), 'Building Trajan's Markets', *American Journal of Archaeology* 102/2: 283–308.

——(1999), 'Building Trajan's Column', *American Journal of Archaeology* 103/3: 419–39.

——(2007), *Concrete Vaulted Construction in Imperial Rome* (Oxford: Oxford University Press).

Lanciani, R. (1874), 'Delle Scoperte Principali Avvenute nella Prima Zona del Nuovo Quartiere Esquilino', *Bullettino della Commissione archaeologica comunale di Roma* 2: 33–88.

——(1892), 'Gambling and Cheating in Ancient Rome', *North American Review* 155: 97–105.

Larmour, D. H. J. (2007), 'Holes in the Body: Sites of Abjection in Juvenal's Rome', in Larmour and Spencer (2007b), 168–210.

Larmour, D. H. J., and Spencer, D. J. (2007a), 'Introduction—*Roma, recepta*: A Topography of the Imagination', in Larmour and Spencer (2007b: 1–60).

Larmour, D. H. J., and Spencer, D. J. (eds.) (2007b), *The Sites of Rome: Time, Space, Memory* (Oxford: Oxford University Press).

Laurence, R. (1991), 'The Urban Vicus: The Spatial Organisation of Power in the Roman City', in E. Herring, R. Whitehouse, and J. Wilkins (eds.), *Papers of the Fourth Conference of Italian Archaeology*, i (London: Institute of Classical Studies), 145–50.
——(1994a), *Roman Pompeii: Space and Society* (London: Routledge).
——(1994b), 'Rumour and Communication in Roman Politics', *Greece and Rome* 41/1: 62–74.
——(1995), 'The Organization of Space in Pompeii', in Cornell and Lomas (eds.), 63–78.
——(1996), *Roman Pompeii: Space and Society* (New York: Routledge).
——(1997), 'Writing the Roman Metropolis', in H. M. Parkins (ed.), *Roman Urbanism: Beyond the Consumer City* (London: Routledge), 1–20.
——(1998a), 'Review of Bakker 1994', *Classical Review* 48: 444.
——(1998b), 'Land Transport in Roman Italy: Costs, Practice and the Economy', in H. Parkins and C. J. Smith (eds.), *Trade, Traders and the Ancient City* (London: Routledge), 129–48.
——(1999), *The Roads of Roman Italy: Mobility and Cultural Change* (London: Routledge).
——(2004), 'Milestones, Communications, and Political Stability', in L. Ellis and F. L. Kidner (eds.), *Travel, Communication, and Geography in Late Antiquity: Sacred and Profane* (Burlington: Ashgate), 43–58.
Laurence, R. (2005), 'Tourism and Romanità: A New Vision of Pompeii (1924–1942)', in *'Cities of Vesuvius: Pompeii & Herculaneum': A Special Issue of Ancient History: Resources for Teachers* 35: 90–110.
——(2007), *Roman Pompeii: Space and Society* (2nd edn., London: Routledge).
——(2008), 'City Traffic and the Archaeology of Roman Streets from Pompeii to Rome: The Nature of Traffic in the Ancient City', in Mertens (ed.), 87–106.
——(forthcoming), 'Traffic and Land Transportation in and near Rome', in P. Erdkamp (ed.), *The Cambridge Companion to the City of Rome* (Cambridge: Cambridge University Press).
Lavan, L. (2006), 'Street Space in Late Antiquity', in *Proceedings of the 21st International Congress of Byzantine Studies* (Aldershot: Ashgate), vol. 2, 68–9.
——(2008), 'The Monumental Streets of Sagalassos in Late Antiquity: An Interpretive Study', in Ballet, Dieudonné-Glad, and Saliou (eds.), 201–14.
Lawrence, R. J. (1990), 'Public Collective and Private Space: A Study of Urban Housing in Switzerland', in S. Kent (ed.), *Domestic Architecture and the Use of Space: An Interdisciplinary Cross-Cultural Study* (Cambridge: Cambridge University Press), 73–91.
Le Gall, J. (1991), 'La Muraille servienne sous le Haut-Empire', in Hinard and Royo (eds.), 55–63.
Leach, E. W. (1999), 'Ciceronian "Bi-Marcus": Correspondence with M. Terentius Varro and L. Papirius Paetus in 46 B.C.E.', *Transactions of the American Philological Association* 129: 139–79.
——(2004), *The Social Life of Painting in Ancient Rome and on the Bay of Naples* (Cambridge: Cambridge University Press).

Lecocq, F. (2008), 'Les Premiéres Maquettes de Rome: L'Exemple des modéles réduits en liège de Carl et Georg May dans les collections européennes aux XVIIIe-XIXe siècles', in Fleury and Desbordes (eds.), 227-59.

Lefebvre, H. (1991), *The Production of Space*, trans. D. Nicholson-Smith (Malden, Mass.: Blackwell Publishing).

——(trans. and ed. E. Kofman and E. Lebas) (1996), *Writings on Cities* (Oxford: Blackwell).

——(trans. S. Elden and G. Moore) (2004), *Rhythmanalysis: Space, Time and Everyday Life* (London: Continuum).

Lefebvre, H., and Régulier-Lefebvre, C. (1985), 'Le Projet rhythmanalytique', *Communications* 41: 191-9 [=(2003), 'The Rhythmanalytical Project', trans. I. Forster, in S. Elden, E. Lebas, and E. Kofman (eds.), *Henri Lefebvre: Key Writings* (London: Continuum), 190-8].

Leibbrand, K. (1970), *Transportation and Town Planning*, trans. N. Seymer (Cambridge, Mass.: MIT Press).

Leighton, A. (1972), *Transport and Communication in Early Medieval Europe AD 500-1100* (Newton Abbot: David & Charles).

Lévêque, P., and Vidal-Naquet, P. (1960), 'Epaminondas Pythagoricien: Ou le problème tactique de la droite et de la gauche', *Historia* 9: 294-308.

Levi, C. (2004), *Fleeting Rome: In Search of la Dolce Vita* (Oxford: Wiley).

Lévi-Strauss, C. (1963), *Structural Anthropology* (New York: Basic Books).

Levitas, G. (1986), 'Anthropology and Sociology of Streets', in Anderson (ed.), 225-40.

Linderski, J. (1986), 'The Augural Law', *Aufstieg und Niedergang der römischen Welt II*, 16/3: 2146-2312.

Ling, R. (1977), 'Studius and the Beginnings of Roman Landscape Painting', *Journal of Roman Studies* 67: 1-16.

——(1990a), 'A Stranger in Town: Finding the Way in an Ancient City', *Greece and Rome* 37: 204-14.

——(1990b), 'Street Plaques at Pompeii', in M. Henig (ed.), *Architecture and Architectural Sculpture in the Roman Empire* (Oxford: Oxbow), 51-66.

——(1997), *The Insula of the Menander at Pompeii*, i. *The Structures* (Oxford: Clarendon).

——(2005), 'Street Fountains and House Fronts at Pompeii', in *Omni pede stare: Saggi architettonici e circumvesuviani in memoriam Jos de Waele* (Naples: Electa), 271-6.

Ling, R., and Ling, L. (2005), *The Insula of the Menander at Pompeii*, ii. *The Decorations* (Oxford: Clarendon Press).

Liverani, L. (2005), 'Ianiculum', in V. Fiocchi Nicolai, M. Grazia Granino Cecere, and Z. Mari (eds.), *Lexicon Topographicum Urbis Romae Suburbium* III (Rome: Quasar), 82-3.

Liversidge, J. (1976), *Everyday Life in the Roman Empire* (London: Batsford).

Lloyd, G. E. R. (1962), 'Right and Left in Greek Philosophy', *Journal of Hellenic Studies* 82: 56-66.

Lloyd, R. B. (1982), 'Three Monumental Gardens on the Marble Plan', *American Journal of Archaeology* 86/1: 91-100.

Lo Cascio, E. (2000a), 'La popolazione', in E. Lo Cascio (ed.), *Roma imperiale: Una metropoli antica* (Rome: Carocci), 17-69.

Lo Cascio, E. (ed.) (2000b), *Mercati permanenti e mercati periodici nel mondo romano* (Bari: Edipuglia).

López, M., and van Nes, A. (2007), 'Space and Crime in Dutch Built Environments: Macro and Micro Spatial Conditions for Residential Burglaries and Thefts from Cars', in A. S. Kubat (ed.), *Proceedings Space Syntax: 6th International Symposium* (Istanbul: Istanbul Technological University).

Lott, J. B. (2004), *The Neighborhoods of Augustan Rome* (Cambridge: Cambridge University Press).

Lugli, G. (1929–30), 'I mercati Traianei', *Dedalo* 10: 527–51.

——(1937), 'L'Arco di Gallieno', *L'Urbe* 2: 16–26.

——(1946), *Roma antica: Il centro monumentale* (Rome: G. Bardi).

Lusnia, S. (2004), 'Urban Planning and Sculptural Display in Severan Rome: Reconstructing the Septizodium and Its Role in Dynastic Politics', *American Journal of Archaeology* 108: 517–44.

——(2006), 'Battle Imagery and Politics on the Severan Arch in the Forum', in S. Dillon and K. Welch (eds), *Representations of War in Ancient Rome* (Cambridge: Cambridge University Press), 272–99.

Lynch, K. (1960), *The Image of the City* (Cambridge, Mass. and London: MIT Press).

Macaulay-Lewis, E. (2008), 'The City in Motion: Movement and Space in Roman Architecture and Gardens from 100 BC to AD 150' (Unpublished PhD thesis, University of Oxford).

——(2009), 'Political Museums: Porticos, Gardens and the Public Display of Art in Ancient Rome', in S. Bracken, A. Gáldy, and A. Turpin (eds.), *Collecting and Dynastic Ambition* (Newcastle: Cambridge Scholars Publishing), 1–22.

MacDonald, W. L. (1982), *The Architecture of the Roman Empire* (New Haven: Yale University Press).

——(1986), *The Architecture of the Roman Empire II: An Urban Appraisal* (New Haven; London: Yale University Press).

McGinn, T. (2002), 'Pompeian Brothels and Social History' in *Pompeian Brothels, Pompeii's Ancient History, Mirrors and Mysteries, Art and Nature at Oplontis, the Herculaneum Basilica* (Porstmouth, RI: Journal of Roman Archaeology), 7–46.

——(2004), *The Economy of Prostitution in the Roman World* (Ann Arbor: University of Michigan).

MacKinnon, M. R. (2004), *Animal Production and Consumption in Roman Italy Integrating the Zooarchaeological and Ancient Textual Evidence* (Journal of Roman Archaeology suppl. 54).

——(forthcoming), 'When "Left" is "Right": The Symbolism behind Side Choice among Ancient Animal Sacrifices', in D. Campana, et al. (eds.), *Anthropological Approaches to Zooarchaeology: Colonialism, Complexity and Animal Transformations* (Oxford: Oxbow Books).

MacMahon, A. (2003a), 'The Realm of Janus: Doorways in the Roman World', in G. Carr, E. Swift, and J. Weekes (eds.), *TRAC 2002: Proceedings of the Twelfth Annual Theoretical Roman Archaeology Conference, Kent 2002* (Oxford: Oxbow Books), 58–73.

——(2003b), *The Taberna Structures of Roman Britain* (BAR International Series 356; Oxford: Oxbow Books).

MacMahon, A., and Price, J. (eds.) (2005), *Roman Working Lives and Urban Living* (Oxford: Oxbow Books).

MacMullen, R. (1974), *Roman Social Relations, 50 BC to AD 284* (New Haven and London: Yale University Press).

Madeleine, S. (2008), 'La Troisième Dimension des *insulae* d'après les symboles de la *Forma Urbis Romae*', in Fleury and Desbordes (eds.), 291–316.

Magaldi, E. (1930), *Il commercio ambulante a Pompei* (Naples).

Maischberger, M. (1997), *Marmor in Rom: Anlieferung, Lager- und Werkplätze in der Kaiserzeit* (Wiesbaden: L. Reichert).

Maiuri, A. (1933), *La casa del Menandro e il suo tesoro di argenteria* (Rome: Libreria dello Stato).

——(1947), *Introduzione allo studio di Pompei: Il Foro e i suoi monumenti* (Naples: R. Pironti).

——(1958), *Ercolano: I nuovi scavi (1927–1958)* (Rome: Libreria dello Stato).

Malkin, I. (2005a), *Mediterranean Paradigms and Classical Antiquity* (London; Routledge).

——(2005b), 'Networks and the Emergence of Greek Identity', in I. Malkin (ed.), *Mediterranean Paradigms and Classical Antiquity* (London: Routledge), 56–74.

——(2009) *Greek and Roman Networks in the Mediterranean* (London: Routledge).

Malmberg, S. (2009a), 'Finding Your Way in the Subura', in M. Driessen et al. (eds.), 39–51.

——(2009b), 'Navigating the Urban Via Tiburtina', in Bjur and Santillo Frizell (eds.), 61–78.

——(forthcoming), 'Monumentalizing the Porta Tiburtina Area: a Proposed New Location for the Basilica Liberii and the Macellum Liviae'.

Malmberg, S., and Bjur, H. (2009), 'Suburb as Centre', in Bjur and Santillo Frizell (eds.), 109–28.

Mango, M. (2001), 'The Porticoed Street at Constantinople', in N. Necipoğlu (ed.), *Byzantine Constantinople: Monuments, Topography and Everyday Life* (Leiden: Brill), 29–51.

Mar, R. (1991), 'La formazione dello spazio urbanao nella Citta di Ostia', *Mitteilungen des Deutschen Archäologischen Instituts Römische Abteilung* 98: 81–109.

——(2008), 'Il traffico viario a Ostia: Spazio pubblico e progetto urbano', in Mertens (ed.), 125–44.

Margolies, E. (2006), 'Vagueness Gridlocked: A Map of the Smells of New York', in J. Drobnick (ed.), *The Smell Culture Reader* (Oxford: Berg), 107–17.

Mari, Z. (1983), *Forma Italiae. Regio 1. xvii. Tibur, 3* (Rome: De Luca).

——(2004), 'Ficulensis Ager', in V. Fiocchi Nicolai, M. Grazia Granino Cecere, and Z. Mari (eds.), *Lexicon Topographicum Urbis Romae Suburbium* II (Rome: Quasar), 248–9.

——(2005), 'Herculis Templum (Via Tiburtina)', in A. La Regina (ed.), *Lexicon Topographicum Urbis Romae Suburbium* III (Rome: Quasar), 54–5.

——(2008), 'Tiburtina, Via', in A. La Regina (ed.), *Lexicon Topographicum Urbis Romae Suburbium* V (Rome: Quasar), 160–73.

Marlowe, E. (2006), 'Framing the Sun: The Arch of Constantine and the Roman Cityscape', *Art Bulletin* 88: 223–42.

Marshall, S. (2005), *Streets and Patterns* (London and New York: Spon Press).
Martin, R. (1995), 'Ars an quid aliud? La conception varronienne de l'agriculture', *Revue des études latines* 73: 80–91.
Martin, S. D. (1989), *The Roman Jurists and the Organization of Private Building in the Late Republic and Early Empire* (Brussels: Latomus).
——(2000), 'Transportation and Law in the city of Rome', in S. K. Dickison and J. P. Hallett (eds.), *Rome and Her Monuments: Essays on the City and Literature of Rome in Honor of Katherine A. Geffcken* (Wauconda, Ill.: Bolchazy-Carducci Publishers), 193–214.
——(2002), 'Roman Law and the Study of Land Transportation', in J-J. Aubert and B. Sirks (eds.), *SPECVLVM IVRIS: Roman Law as a Reflection of Social and Economic Life in Antiquity* (Ann Arbor: University of Michigan Press), 151–68.
Marucchi, O. (1906), *The Roman Forum and the Palatine according to the Latest Discoveries* (Paris and Rome: Desclée, Lefebvre & co.).
Mathieu, J. R. (ed.) (2002), *Experimental Archaeology: Replicating Past Objects, Behaviours and Processes*, (BAR International Series 1035; Oxford: Oxbow Books).
Matthews, K. D. (1960), 'The Embattled Driver in Ancient Rome', *Expedition* 3: 22–7.
Mattingly, H., Carson, R. A. G., and Hill, P. V. (1975), *Coins of the Roman Empire in the British Museum*, v, *Pertinax to Elagabalus* (London: British Museum Publications).
Mau, A. (1899), *Pompeii: Its Life and Art*, trans. F. W. Kelsey (New York: MacMillan Co.).
Mayer, R. (2007), 'Impressions of Rome', *Greece and Rome* 54: 156–77.
Mayeske, B. J. B. (1972), 'Bakeries, Bakers and Bread at Pompeii: A Study in Social and Economic History' (Unpublished PhD Dissertation, University of Maryland).
Meiggs, R. (1973), *Roman Ostia* (1st edn. 1960; Oxford: Clarendon Press).
Meller, H. (1990), *Patrick Geddes: Social Evolutionist and City Planner* (London: Routledge).
Meneghini, R., and Santangeli Valenzani, R. (2007), *I Fori Imperiali: Gli scavi del Comune di Roma (1991–2007)* (Rome: Viviani Editore).
Meneghini, R., and Santangeli Valenzani, R. (eds.) (2006), *Formae Urbis Romae: Nuovi frammenti di Piante Marmoree dallo scavo dei Fori Imperiali* (Rome: 'L'Erma' di Bretschneider).
Merleau-Ponty, M. (1962), *Phenomenology of Perception*, trans. C. Smith (London: Routledge & Kegan Paul).
Merli, E. (2006), 'Martial between Rome and Bilbilis', in R. M. Rosen and I. Sluiter (eds.), *City, Countryside and the Spatial Organization of Value in Classical Antiquity* (*Mnemosyne*, suppl. 279; Leiden: Brill), 327–48.
Mertens, D. (ed.) (2008), 'Stadtverkehr in der Antiken Welt/ Traffico Urbano nel Mondo Antico', *Palilia* 13 (Rome).
Meyer, E. (2004), *Legitimacy and Law in the Roman World: Tabulae in Roman Belief and Practice* (Cambridge: Cambridge University Press).
Meyer, F., and van Nijf, O. (eds.) (1992), *Trade, Transport and Society in the Ancient World: A Sourcebook* (London: Routledge).
Millar, F. (1977), *The Emperor in the Roman World (31 BC–AD 337)* (London: Duckworth).

——(1998), *The Crowd in Rome in the Late Republic* (Ann Arbor: University of Michigan Press).

Miller, P. A. (2007), '"I get around": Sadism, Desire, and Metonymy on the Streets of Rome with Horace, Ovid, and Juvenal', in Larmour and Spencer (2007b), 138–67.

Mills, S. (2005), *Applying Auditory archaeology to Historic Landscape Characterisation: A Report Produced for English Heritage*, http://www.cardiff.ac.uk/hisar/people/sm/aa_hlc/Text/AA_HLC_Report.pdf.

Milnor, K. (2007), 'Augustus, History, and the Landscape of the Law', *Arethusa* 40/1: 7–23.

Miniero, P. (1987), 'Studio di un carro romano dalla villa c. d. di Arianna a Stabia', *Mélanges de l'École française de Rome, Antiquité*, 99: 171–209.

Mommsen, T. (1887), *Römisches Staatsrecht* (3rd edn.; Berlin: Akademische Druck-u. Verlagsanstalt).

Mommsen, T., Krueger, P., and Watson, A. (eds.) (1985), *The Digest of Justinian*, iv (Philadelphia, University of Pennsylvania Press).

Montesano, G. (1964–85), 'Lusoria (tabula)', in E. Ruggiero (ed.), *Dizionario epigrafico di antichità romane*, iv (Rome: Pasqualucci).

Moore, T. J. (1991), 'Palliata Togata: Plautus, Curculio 462–86', *American Journal of Philology* 112: 343–62.

Moorman, E. (2007), 'Villas surrounding Pompeii and Herculaneum', in Dobbins and Foss (eds.), 435–54.

Morel, J.-P. (1987), 'La Topographie de l'artasinat et du commerce dans la Rome antique', in Pietri (ed.), 127–55.

Morello, R., and Gibson, R. K. (2003), *Re-imagining Pliny the Younger* (*Arethusa* 36) (Baltimore: Johns Hopkins University Press).

Morley, N. (1996), *Metropolis and Hinterland: The City of Rome and the Italian Economy 200 BC–AD 200* (Cambridge: Cambridge University Press).

Morris, I. (2005), 'Mediterraneanization', in I. Malkin (ed.), *Mediterranean Paradigms and Classical Antiquity* (London: Routledge), 30–55.

Mouritsen, H. (1988), *Elections, Magistrates and Municipal Elite: Studies in Pompeian Epigraphy* (Analecta Romana Instituti Danici; Rome: L'Erma di Bretschneider).

——(2001), *Plebs and Politics in the Late Roman Republic* (Cambridge: Cambridge University Press).

——(2004), 'Pits and Politics: Interpreting Colonial Fora in Republican Italy', *Papers of the British School at Rome* 72: 37–67.

Muccigrosso, J. (2006), 'Religion and Politics: Did the Romans Scruple about the Placement of Their Temples?', in C. E. Schultz and P. B. Harvey jr. (eds.), *Religion in Republican Italy* (Yale Classical Studies 33; Cambridge: Cambridge University Press), 181–206.

Mulvin, L., and Sidebotham, S. E. (2004), 'Roman Game Boards from Abu Sha'ar (Red Coast, Egypt)', *Antiquity* 78: 602–17.

Murphey, P. R. (1977), 'Themes of Caesar's *Gallic War*', *Classical Journal* 72/3: 234–43.

Najbjerg, T., and Trimble, J. (2003), 'The Severan Marble Plan of Rome (*Forma Urbis Romae*)', http://formaurbis.stanford.edu/fragment.php?record=43 (Stanford Digital *Forma Urbis Romae* Project).

Nappo, S. (1989), 'Fregio dipinto dal "praedium" di Giulia Felice con rappresentazione del foro di Pompei', *Rivista di studi pompeiani* 3: 79–96.

Narducci, E. (1997), *Cicerone e l'eloquenza romana: Retorica e progetto culturale* (Rome: Laterza).

——(2004), *Cicerone e i suoi interpreti: Studi sull'Opera e la Fortuna* (Pisa: Ets).

Nash, E. (1968), *Pictorial Dictionary of Ancient Rome*, 2 vols. (Deutsches Archaeologisches Institut; Tübingen: Verlag Ernst Wasmuth).

Needham, J. (1986), *Science and Civilization in China*: iv. *Physics and Physical Technology*, 2. *Mechanical Engineering* (Cambridge: Cambridge University Press).

Needham, R. (ed.) (1973), *Right and Left: Essays on Dual Symbolic Classification* (Chicago: Chicago University Press).

Neuman, W. R., McKnight, L., and Soloman, R. J. (1999), *Gordian Knot: Political Gridlock on the Information Highway* (Cambridge, Mass.: MIT Press).

Newbold, R. F. (1974), 'Some Social and Economic Consequences of the A.D. 64 Fire at Rome', *Latomus* 33: 858–69.

Newby, Z. (2007), 'Art at the Crossroads? Themes and Styles in Severan Art', in Swain, Harrison, and Elsner (eds.), 201–49.

Newsome, D. J. (2008), '"Traffic" and "Congestion" in Rome's Empire', *Journal of Roman Archaeology* 21: 442–6.

Newsome, D. J. (2009a), 'Traffic, Space and Legal Change around the Casa del Marinaio at Pompeii (VII.15.1–2)', *BABesch* 84: 121–42.

——(2009b), 'Centrality in Its Place: Defining Urban Space in the City of Rome', in M. Driessen et al. (eds.), 25–38.

——(2010), 'The Forum and the City: Rethinking Centrality in Rome and Pompeii (3rd century B.C.–2nd century A.D.)' (Unpublished PhD Dissertation, University of Birmingham).

——(forthcoming), 'Worn out with Walking: movement, age, and exertion in ancient Rome'.

Nicolet, C. (1987), 'La Table d'Héraclée et les origines du cadastre Romain', in Pietri (ed.), 1–25.

——(1991), *Space, Geography, and Politics in the Early Roman Empire* (Ann Arbor: University of Michigan Press).

Niebisch, A. (2009), 'Symbolic Space: Memory, Narrative, Writing', in G. Backhaus and J. Murungi (eds.), *Symbolic Landscapes* (New York: Springer), 323–37.

Nippel, W. (1995), *Public Order in Ancient Rome* (Cambridge: Cambridge University Press).

Nisbet, G. (2003), *Greek Epigram in the Roman Empire: Martial's Forgotten Rivals* (Oxford: Oxford University Press).

Noreña, C. (2003), 'Medium and Message in Vespasian's *Templum Pacis*', *Memoirs of the American Academy in Rome* 48: 25–43.

——(2007), 'The Social Economy of Pliny's Correspondence with Trajan', *American Journal of Philology* 128: 239–77.

North, J. (2002), 'Introduction: Pursuing Democracy', *Proceedings of the British Academy* 114: 1–12.

O'Sullivan, T. M. (2003), 'Mind in Motion: The Cultural Significance of Walking in the Roman World' (Unpublished PhD Thesis, Harvard University).

—— (2006), 'The Mind in Motion: Walking and Metaphorical Travel in the Roman Villa', *Classical Philology*, 101: 133–52.

—— (2007), 'Walking with Odysseus: The Portico Frame of the Odyssey Landscapes', *American Journal of Philology* 128: 497–532.

—— (2011), *Walking in Roman Culture* (Cambridge: Cambridge University Press).

Oleson, J. P. (ed.) (2008), *The Oxford Handbook of Engineering and Technology in the Classical World* (Oxford: Oxford University Press).

Olwig, K. (1996), 'Recovering the Substantive Nature of Landscape', *Annals of the Association of American Geographers* 86/4: 630–53.

Overbeck, J. (1884), *Pompeji in seinen Gebäuden, Alterthümern und Kunstwerken* (Leipzig: Englemann).

Owens, E. J. (1991), *The City in the Greek and Roman World* (New York: Routledge).

Oxford Latin Dictionary (1968), i (Oxford and London: Clarendon).

Packer, J. E. (1971), *The Insulae of Imperial Ostia* (Rome: American Academy in Rome).

—— (1978), 'Inns at Pompeii: A Short Survey', *Cronache pompeiane* 4: 5–53.

—— (1997), *The Forum of Trajan in Rome: A Study of the Monuments* (Berkeley and Los Angeles: University of California Press).

—— (2001), *The Forum of Trajan in Rome: A Study of the Monuments in Brief* (Berkeley and Los Angeles: University of California Press).

—— (2003), '*Plurima et Amplissima Opera:* Parsing Flavian Rome', in Boyle and Dominik (eds.), 167–98.

—— (2006), 'Digitizing Roman Imperial Architecture in the Early 21st Century: Purposes, Failures and Prospects', in Haselberger and Humphrey (eds.), 309–20.

Pagano, M. (1996), 'La nuova pianta della città e di alcuni edifice pubblici di Ercolano', *Cronache ercolanesi*, 26: 229–62.

Pailler, J. M. (1981), 'Martial et l'espace urbain', *Pallas* 28: 79–87.

Palka, J. W. (2002), 'Left/Right Symbolism and the Body in Ancient Maya Iconography and culture', *Latin American Antiquity* 13: 419–43.

Palma, A. (1988), *Iura Vicinitatis: Solidarietà e limitazioni nel rapporto di vicinato in diritto romano dell'età classica* (Turin: G. Giappichelli).

Palmer, R. E. A. (1976), 'Jupiter Blaze, Gods of the Hills, and the Roman Topography of *CIL* VI.377', *American Journal of Archaeology* 80: 43–56.

—— (1980), 'Customs on Market Goods Imported into the City of Rome', in J. D'Arms and E. C. Kopff (eds.), *The Seaborne Commerce of Ancient Rome* (Rome: American Academy in Rome), 217–33.

—— (1990), 'Cults of Hercules, Apollo Caelispex and Fortuna in and around the Roman Cattle Market', *Journal of Roman Archaeology* 3: 234–44.

Palombi, D. (1996), 'Hercules Sullanus', *LTUR* III (H–O): 21–2.

—— (2005), 'Paesaggio storico e paesaggio di memoria nell'area dei Fori Imperiali', in R. Neudecker and P. Zanker (eds.), 'Lebenswelten: Bilder und Räume in der Römischen Stadt der Kaiserzeit', *Palilia* 16: 21–37.

Panella, C. (1998), 'Valle del Colosseo, Area della *Meta Sudans*', in L. Drago Troccoli (ed.), *Scavi e ricerche archeologiche dell'Università di Roma 'La Sapienza'* (Rome: 'L'Erma' di Bretschneider), 43–50.

Paoli, U. E. (1963), *Rome: Its People, Life and Customs* (London: Longman).

Paoli, U. E. (1995), 'Domus Aurea: Porticus Triplices Miliariae', *LTUR* II (D–G): 55–6.

Papi, E. (2002), 'La *turba inpia*: Artigiani e commercianti del Foro Romano e dintorni (I sec. a.C.-64 d.C.), *Journal of Roman Archaeology* 15: 45–62.

Parkin, A. (2006), 'An Exploration of Pagan Almsgiving', in M. Atkins and R. Osborne (eds.), *Poverty in the Roman World* (Cambridge: Cambridge University Press), 60–82.

Parslow, C. (1998), 'Preliminary Report of the 1996 Fieldwork Project in the Praedia Iuliae Felicis (Region II.4), Pompeii', *Rivista di studi pompeiani* 9: 199–207.

Patterson, J. R. (1999), 'Via Tiburtina', *LTUR* V (T–Z): 146–7.

——(2000), 'On the Margins of the City of Rome', in V. Hope and E. Marshall (eds.), *Death and Disease in the Ancient City* (London: Routledge), 85–103.

——(2004), 'The Collegia and the Transformation of the Towns of Italy in the Second Century AD', in *L'Italie d'Auguste à Diocletien* (Trieste: École francaise de Rome), 227–38.

——(2006), *Landscapes and Cities: Rural Settlement and Civic Transformation in Early Imperial Italy* (Oxford: Oxford University Press).

Pavolini, C. (1996), *La vita quotidiana a Ostia* (Rome and Bari: Laterza).

——(2006), *Ostia* (Rome and Bari: Laterza).

Peluso, N. (1996), 'Fruit Trees and Family Trees in an Anthropegenic Forest: Ethics of Access, Property Zones, and Environmental Change in Indonesia', *Comparative Studies in Society and History* 38/3: 510–48.

Peña, J. T. (2007), *Roman Pottery in the Archaeological Record* (New York: Cambridge University Press).

Peña, J. T., and McCallum, M. (2009), 'The Production and Distribution of Pottery at Pompeii: A Review of the Evidence; Part 1, Production', *American Journal of Archaeology* 113: 57–79.

Perry, J. S. (2001), 'Ancient *Collegia*, Modern Blackshirts? The Study of Roman Corporations in Fascist Italy', *International Journal of Classical Tradition* 8/2: 205–16.

——(2006), *The Roman Collegia: The Modern Evolution of an Ancient Concept* (*Mnemosyne*, suppl. 277; Leiden: Brill).

Pietri, C. (ed.) (1987), *L'Urbs: Espace urbain et histoire* (Rome: École française de Rome).

Pike, G. (1967), 'Pre-Roman Land Transport in the Western Mediterranean Region', *Man* 2/4: 593–605.

Pisani Sartorio, G. (1991), 'Le Plan-relief d'Italo Gismondi: Méthodes, techniques de réalisation et perspectives futures', in Hinard and Royo (eds.), 257–77.

——(1996a), 'Macellum', *LTUR* III (H–O): 201–3.

——(1996b), 'Macellum Liviae', *LTUR* III (H–O): 203–4.

——(1996c), 'Macellum Magnum', *LTUR* III (H–O): 204–6.

——(1996d), 'Muri Aureliani', *LTUR* III (H–O): 290–9.

——(1996e), 'Porta Tiburtina', *LTUR* IV (P–S): 312–13.

Pitcher, R. A. (1998), 'Martial's Debt to Ovid', in F. Grewing (ed.), *Totus Notus in Orbe: Perspektiven der Martial-Interpretation* (Stuttgart: Franz Steiner), 59–76.

Platner, S. B., and Ashby, T. (1929), *A Topographical Dictionary of Ancient Rome* (London: Oxford University Press).

Plesch, V. (2002), 'Memory on the Wall: Graffiti on Religious Wall Paintings', *Journal of Medieval and Early Modern Studies* 32: 167–98.
Plesch, V., and Ashley, K. (2002), 'The Cultural Process of "Appropriation"', *Journal of Medieval and Early Modern Studies* 32: 1–15.
Plummer, H. (1993), 'Meeting Ground', in B. Farmer and H. J. Louw (eds.), *Companion to Contemporary Architectural Thought* (London: Routledge), 368–77.
Poehler, E. E. (2003), 'Romans on the Right: The Art and Archaeology of Traffic', *Athanor* 21: 8–15.
——(2006), 'The Circulation of Traffic in Pompeii's *Regio VI*', *Journal of Roman Archaeology* 19: 53–74.
——(2009), 'The Organization of Pompeii's System of Traffic: An Analysis of the Evidence and its Impact on the Infrastructure, Economy and Urbanism of the Ancient City' (Unpublished PhD Dissertation, University of Virginia).
Poortman, E. R., Norbert, H., and Bons, M. (1994), 'Information for the Management of the Building-Materials Flow', *Engineering, Construction and Architectural Management* 1/2: 139–46.
Porteous, J. D. (2006), 'Smellscape', in J. Drobnick (ed.), *The Smell Culture Reader* (Oxford: Berg), 89–106.
Potenza, U. (1996), 'Gli acquedotti romani di Serino', in De Haan and Jansen (eds.), 93–100.
Potts, C. (2009), 'The Art of Piety and Profit at Pompeii: A New Interpretation of the Painted Shop Façade at ix.7.1–2', *Greece and Rome* 56: 55–70.
Poulle, B. (2008), 'Rome vue par l'humaniste Jean-Jacques Boissard (1528–1602)', in Fleury and Desbordes (eds.), 365–75.
Prior, R. E. (1996), 'Going around Hungry: Topography and Poetics in Martial 2.14', *American Journal of Philology* 117: 121–42.
Purcell, N. (1987), 'Town in Country and Country in Town', in E. B. MacDougall (ed.) *Ancient Roman Villa Gardens, Dumbarton Oaks Colloquium on the History of Landscape Architecture* 10 (Washington DC: Dumbarton Oaks), 185–203.
——(1995a), 'The Roman Villa and the Landscape of Production', in Cornell and Lomas (eds.), 151–80.
——(1995b), 'Forum Romanum (Republican Period)', *LTUR* II (D–G): 325–36.
——(1995c), 'Forum Romanum (Imperial Period)', *LTUR* II (D–G): 336–42.
——(1995d), 'Literate Games: Roman Urban Society and the Game of *Alea*', *Past and Present* 147: 3–37.
——(1996), 'The Roman Garden as a Domestic Garden', in I. Barton (ed.), *Roman Domestic Buildings* (Exeter: University of Exeter Press), 121–52.
——(2004), 'Literate Games: Roman Urban Society and the Game of *Alea*', in R. Osborne (ed.), *Studies in Ancient Greek and Roman Society* (Cambridge: Cambridge University Press), 177–205.
——(2007), 'Inscribed Imperial Roman Gaming-Boards', in I. Finkel (ed.), 9–97.
Quilici, L. (1990), *Le strade: Viabilità tra Roma e Lazio* (Rome: Quasar).
——(1996), 'I ponti della Via Ostiense', *Atlante tematico di topografia antica* 5: 53–79.
——(2008), 'Land Transport, Part 1: Roads and Bridges', in Oleson (ed.), 580–605.
Quilici, L., and Quilici Gigli, S. (1993), *Ficulea* (*Latium Vetus* VI) (Rome: Consiglio nazionale delle ricerche).

Radice, B. (1968), *Pliny, Letters: Books 8–10 and* Panegyricus (Harvard, Mass.: Loeb Classical Library).

Raepsaet, G. (2002), *Attelages et techniques de transport dans le monde gréco-romain* (Brussels: Le livre Timperman).

——(2008), 'Land Transport, Part 2: Riding, Harnesses, and Vehicles', in Oleson (ed.), 580–605.

Rainer, J. (1987), *Bau- und nachbarrechtliche Bestimmungen im klassischen römischen Recht* (Graz: Leykam).

Rakob, F. (1976), 'Hellenismus in Mittelitalien: Bautypen und Bautechnik', in P. Zanker (ed.), *Hellenismus in Mittelitalien*, ii (Göttingen: Vandenhoeck und Ruprecht), 366–86.

Raper, R. (1977), 'The Analysis of the Urban Structure of Pompeii: A Sociological Examination of Land Use (Semi-Micro)', in D. L. Clarke (ed.), *Spatial Archaeology* (London, New York, and San Francisco: Academic Press), 189–221.

——(1979), 'Pompeii, Planning and Social Implications', in B. C. Burnham and J. Kingsbury (eds.), *Space, Hierarchy and Society: Interdisciplinary Studies in Social Area Analysis* (BAR International Series 59, Oxford: Archaeopress), 137–48.

Rawson, E. (1985), *Intellectual Life in the Late Roman Republic* (Baltimore: Johns Hopkins University Press).

——(1987), '*Discrimina Ordinum*: The *Lex Iulia Theatralis*', *Papers of the British School at Rome*, 55: 83–114.

Rea, R. (2009), 'Metropolitana di Roma Linea C, Indagini archeologiche preventive: Da via casilina vecchia al Colosseo. Risultati 2006–2009', Lecture at the British School at Rome, 6 May, 2009.

Read, S., and Budiarto, L. (2003), 'Human Scales: Understanding Places of Centring and Decentring', in *Proceedings, Fourth International Space Syntax Symposium*, vol. i (London), www.spacesyntax.org/symposia/SSS4/fullpapers/13Budiarto-Readpaper.pdf (accessed 10 Dec. 2009).

Reekmans, L. (1989), 'L'Implantation monumentale chrétienne dans le paysage urbain de Rome de 300 à 850', in N. Duval (ed.), *Actes du XIe Congrès internationale d'archéologie chrétienne* (Vatican City: Pontificio Istituto di Archeologi Cristiana), 861–915.

Relph, E. (1976), *Place and Placelessness* (London: Pion Ltd).

Remus, H. (1996), 'Voluntary Association and Networks: Aelius Aristides at Asclepeion in Pergamum', in J. S. Kloppenborg and S. G. Wilson (eds.), *Voluntary Associations in the Graeco-Roman World* (London: Routledge), 146–75.

Reynolds, D. W. (1996), '*Forma Urbis Romae*: The Severan Marble Plan and the Urban Form of Ancient Rome' (Unpublished PhD Thesis, University of Michigan).

Ribbeck, O. (ed.) (1873), *Comicorum Romanorum Praeter Plautum et Terentium. Fragmenta. Secundis Curis* (Lipsia: B. G. Teubner).

Richardson, L. (1976a), 'The Villa Publica and the Divorum', in L. Bonfante and H. von Heintze (eds.), *Memoriam Otto J. Brendel: Essays in Archaeology and the Humanities* (Mainz: Verlag Philipp von Zabern), 159–63.

——(1976b), 'The Evolution of the *Porticus Octaviae*', *American Journal of Archaeology* 80: 57–64.

——(1988), *Pompeii: An Architectural History* (Baltimore: Johns Hopkins University Press).
——(1992), *A New Topographical Dictionary of Ancient Rome* (Baltimore: Johns Hopkins University Press).
Richmond, I. (1930), *The City Wall of Imperial Rome* (Oxford: Clarendon Press).
Richter, T. (2003), *Der Zweifingergestus in der Römischen Kunst* (Möhnesee: Bibliopolis).
Rickman, G. (1971), *Roman Granaries and Store Buildings* (Cambridge: Cambridge University Press).
Riggsby, A. (2003), 'Pliny in Space (and Time)', *Arethusa* 36: 167–86.
Rimmell, V. (2008), *Martial's Rome: Empire and the Ideology of Epigram* (Cambridge: Cambridge University Press).
Rizzo, S. (1996), 'Horti Liciniani', *LTUR* III (H–O): 64–6.
——(2001), 'Indagini nei Fori Imperiali', *Mitteilungen des Deutschen Archäologischen Instituts, Römische Abteilung Römische Mitteilungen* 108: 215–44.
Robinson, O. F. (1992), *Ancient Rome: City Planning and Administration* (New York: Routledge).
Rockwell, P. (1993), *The Art of Stoneworking: A Reference Guide* (Cambridge: Cambridge University Press).
Rodríguez-Almeida, E. (1981), *Forma urbis marmorea: aggiornamento generale 1980* (Rome: Quasar).
——(1988a), 'Ancora sui supposti *Penates* della Velia in Mart., I, 70', *Bullettino della Commissione archeologica comunale di Roma*, 92: 293–8.
——(1988b), 'Un frammento di una nuova pianta marmorea di Roma', *Journal of Roman Archaeology* 1: 120–31.
——(1989), 'Due note Marziliane: i *balnea quattuor in Campo* e le *sellae Paterclianae* subcapitoline', *Mélanges de l'École Française de Rome* 101: 243–54.
——(1993a), 'Arcus Gallieni (Porta Esquilina)', *LTUR* I (A–C): 93–4.
——(1993b), 'Area Carruces', *LTUR* I (A–C): 118.
——(1993c), 'Area Pannaria', *LTUR* I (A–C): 119.
——(1993d), 'Area Radicaria', *LTUR* I (A–C): 119–20.
——(1994), 'Marziale in marmo', *Mélanges de l'École française de Rome: Antiquité* 106: 197–217.
——(2002), *Forma Urbis Antiquae: Le mappe marmoree di Roma tra la Repubblica e Settimio Severo* (Rome: L'École française de Rome).
——(2003), Terrarum dea gentiumque: *Marziale e Roma; un poeta e la sua città* (Rome: Unione Internazionale degli Istituti di Archeologia Storia e Storia dell'Arte in Roma).
Roman, L. (2001), 'The Representation of Literary Materiality in Martial's *Epigrams*', *Journal of Roman Studies* 91: 113–45.
——(2010), 'Martial and the City of Rome', *Journal of Roman Studies* 100: 88–117.
Romanelli, P. (1970), 'Topografia e archeologia dell'Africa Romana', *Enciclopedia Classica* X, vol. vii (Turin, Societa editrice internazionale).
Romano, D. G., Stapp, N. L., and Davison, M. (2006), 'Mapping Augustan Rome: Towards the Digital Successor', in Haselberger and Humphrey (eds.), 271–84.

Röring, C. W. (1983), *Untersuchungen zu römischen Reisewagen* (Koblenz: Numismatischer Verlag Gerd Martin Forneck).
Rossi, P. (2006), *Logic and the Art of Memory: The Quest for a Universal Language*, trans. S. Clucas 2nd edn. (London: Continuum).
Roth, R. (2007), 'Varro's "picta Italia" (*RR* I.2.1) and the Odology of Roman Italy', *Hermes* 135: 286–300.
Roueché, C. (2004), *Aphrodisias in Late Antiquity: The Late Roman and Byzantine Inscriptions*, rev. 2nd edn, http://insaph.kcl.ac.uk/ala2004.
——(2007), 'Late Roman and Byzantine Game Boards at Aphrodisias', in I. Finkel (ed.), 100–5.
Royo, M. (1991), 'Rome en Représentation', in Hinard and Royo (eds.), 19–28.
Rudofsky, B. (1969), *Streets for People: A Primer for Americans* (New York: Doubleday).
Ruffini, G. R. (2008), *The Social Networks in Byzantine Egypt* (Cambridge: Cambridge University Press).
Rykwert, J. (1988), *The Idea of a Town: The Anthropology of Urban Form in Rome, Italy and the Ancient World* (Cambridge, Mass.: MIT Press).
Sadeler, M. (1606), *Vestigi delle Antichità di Roma, Tivoli, Pozzuolo e Alri Luochi* (Rome: De Rossi).
Säflund, G. (1932), *Le mura di Roma repubblicana: Saggio di archeologia romana* (Lund: Gleerup).
Salama, P. (1951), *Les Voies romaines de l'Afrique du nord* (Algiers: Gouvernement Général de l'Algerie).
——(1994), 'Entrées et circulation dans Timgad (étude préliminaire)', in A. Mastino and P. Ruggeri (eds.), *L'Africa romana: Atti del X convegno di studio Oristano, 11–13 dicembre 1992* (Sassari: Editrice Archivio Fotografico Sardo), 347–52.
Sale, K. (1980), *Human Scale* (New York: Coward, McCann & Geoghegan).
Saliou, C. (1994), *Les Lois des bâtiments: Voisinage et habitat urbain dans l'empire romain: Recherches sur les rapports entre le droit et la construction privée du siècle d'Auguste au siècle de Justinien* (Beirut: Institut français d'archéologie du Proche-Orient).
——(1999), 'Les Trottoirs de Pompéi: Une première approche', *BABesch*, 74: 161–218.
——(2003), 'Le Nettoyage des rues dans l'Antiquité: Fragments de discours normatifs', in P. Ballet, P. Cordier, and N. Dieudonné-Glad (eds.), *La Ville et ses déchets dans le monde romain: Rebuts et recyclages* (Montagnac: Monique Mergoil), 37–50.
——(2008), 'La Rue dans le droit romain classique', in Ballet, Dieudonné-Glad, and Saliou (eds.), 63–8.
Saller, R. P. (2000), 'Domitian and His Successors: Methodological Traps in Assessing Emperors', *American Journal of Ancient History* 15: 4–18.
Salza Prina Ricotti, E. (1995), *Giochi e giocattoli* (Roma, Quasar).
Sampaolo, V. (1997), 'VII.9.47: Casa delle Nozze di Ercole', in G. Pugliese Carratelli (ed.), *Pompei: Pitture e mosaici*, vii (Rome: Istituto della enciclopedia italiana), 373–5.
Santa Maria Scriniari, V. (1979), 'Brevi Note sugli Scavi sotto la Chiesa di S. Vito', *Archeologia laziale* 2: 58–62.

Santillo Frizell, B. (2009), 'Changing Pastures', in Bjur and Santillo Frizell (eds.), 39–59.

Sartorio, G. P. (1988), *Mezzi di trasporto e traffico* (Rome: Quasar).

Saunier. P. (1998), 'Center and Centrality in the Nineteenth Century: Some Concepts of Urban Disposition Under the Spot of Locality', *Journal of Urban History* 24: 435–67.

Schafer, M. (2005), 'Open Ears', in M. Bull and L. Back (eds.), *The Auditory Culture Reader* (Oxford: Berg), 25–39.

Scheidel, W. (2004), 'Human Mobility in Roman Italy, I: The Free Population', *Journal of Roman Studies* 94: 1–26.

——(2005), 'Human Mobility in Roman Italy, II: The Slave Population', *Journal of Roman Studies* 95: 64–79.

——(2007), 'A Model of Real Income Growth in Roman Italy', *Historia* 56: 322–46.

Schmidt, L. E. (2005), 'Hearing Loss', in M. Bull and L. Back (eds.), *The Auditory Culture Reader* (Oxford: Berg), 41–59.

Schumacher, L. (1976), 'Das Ehrendekret für M. Nonius Balbus aus Herculaneum', *Chiron* 6: 165–84.

Scobie, A. (1986), 'Slums, Sanitation and Mortality in the Roman World', *Klio* 68: 399–433.

Sear, F. (1993), 'The Scaenae Frons of the Theater of Pompey', *American Journal of Archaeology* 97/4: 687–701.

Serra, S. (1998), 'La viabilità tardoantica e medievale dalla Porta Tiburtina a San Lorenzo Fuori le Mura: una nota', *Bullettino della Commissione archaeologica comunale di Roma*, 99: 125–44.

——(1999), 'Porticus Iici', *LTUR* IV (P–S): 124.

——(2001), 'S. Agapiti Basilica', in A. La Regina (ed.), *Lexicon Topographicum Urbis Romae Suburbium* I (Rome: Quasar), 30.

——(2005a), 'Ianuarii (Beati) Ecclesia', in A. La Regina (ed.), *Lexicon Topographicum Urbis Romae Suburbium* III (Rome: Quasar), 83–4.

——(2005b), 'S. Laurentii Basilica, Balneum, Praetorium, Monasterium, Hospitia, Bibliothecae', in A. La Regina (ed.), *Lexicon Topographicum Urbis Romae Suburbium* III (Rome: Quasar), 203–11.

Sgobbo, I. (1938), 'L'acquedotto romano della Campania, Fontis augustaei aquaeductus', *Notizie degli scavi di antichità* 14: 75–97.

Shields, R. (1999), *Lefebvre, Love and Struggle: Spatial Dialectics* (London: Routledge).

Sieverts, T. (2001), *Zwischenstadt—Zwischen Ort und Welt, Raum und Zeit, Stadt und Land* (Braunschweig: Vieweg).

Slater, W. J. (2000), 'The *Scholae* of Roman *Collegia*', *Journal of Roman Archaeology* 13: 493–7.

Small, J. P. (1997), *Wax Tablets of the Mind: Cognitive Studies of Memory and Literacy in Classical Antiquity* (London: Routledge).

Smith, B. R. (1999), *The Acoustic World of Early Modern England* (Chicago: University of Chicago Press).

——(2005), 'Tuning into London circa 1600', in M. Bull and L. Back (eds.), *The Auditory Culture Reader* (Oxford: Berg), 127–35.

Smith, W. (1890), *A Dictionary of Greek and Roman Antiquities*, i (3rd edn, London: John Murray).
Sofroniew, A. K. T. (2006), '"*Turba*": Latin', in J. T. Schnapp and M. Tiews (eds.), *Crowds* (Stanford: Stanford University Press), 30–4.
Soldevila, R. M. (2006), *Martial, Book IV: A Commentary* (Leiden: Brill).
Sonnabend, H. (1992), 'Stadtverkehr im antiken Rom: Probleme und Lösungsversuche', *Die Alte Stadt* 19: 183–94.
Spencer, D. J. (2007), 'Rome at a Gallop: Livy, on not Gazing, Jumping, or Toppling into the Void', in Larmour and Spencer (eds.), 61–101.
——(2011), *Landscape and Identity in Roman Culture (New Surveys in Classics)* (Cambridge: Cambridge University Press).
Spinazzola, V. (1953), *Pompei alla luce degli scavi nuovi di Via dell'Abbondanza (anni 1910–1923)* (Rome: Libreria dello Stato).
Spisak, A. L. (2007), *Martial: A Social Guide* (London: Duckworth).
Staccioli, R. A. (2003), *The Roads of the Romans*, trans. Stephen Sartarelli (Los Angeles: J. Paul Getty Museum).
Stambaugh, J. (1988), *The Ancient Roman City* (Baltimore: Johns Hopkins University Press).
Steinby, M. (1983), 'L'edilizia come industria publica e privata', *Città e architettura nella Roma imperiale* (Analecta Romana Instituti Danici), suppl. x: 219–22.
Stenton, A. (2007), 'Spatial Stories: Movement in the City and Cultural Geography', in Brant and Whyman (eds.), 62–73.
Stephens Crawford, J. (1990), *The Byzantine Shops at Sardis* (Cambridge, Mass.: Harvard University Press).
Steuernagel, D. (2004), *Kult und Alltag in roemischen Hafenstaedten: Soziale Prozesse in archaeologischer Perspektive* (Stuttgart: Potsdamer Altertumswissenschaftliche Beiträge, 11).
——(2005), 'Öffentliche und private Aspekte von Vereinskulten am Beispiel von Ostia', in P. Zanker and R. Neudecker (eds.), *Lebenswelten: Bilder und Räume in der römischen Stadt der Kaiserzeit* (Wiesbaden: Dr Ludwig Reichert Verlag), 73–80.
Stevens, S. T. (1996), 'Transitional Neighbourhoods and Suburban Frontiers in Late- and Post-Roman Carthage', in R. W. Mathisen and H. S. Sivan (eds.), *Shifting Frontiers in Late Antiquity* (Aldershot: Variorum), 187–200.
Stewart, P. (2003), *Statues in Roman Society: Representation and Response* (Oxford: Oxford University Press).
Stobart, J., Hann, A., and Morgan, V. (2007), *Spaces of Consumption: Leisure and Shopping in the English Town, c.1680–1830* (London and New York: Routledge).
Stöger, H. (2007), 'Monumental Entrances of Roman Ostia. Architecture with Public Associations and Spatial Meaning', *BABesch* 82: 347–63.
——(2008), 'Roman Ostia: Space Syntax and the Domestication of Space', in A. Posluschny, K. Lambers, and I. Herzog (eds.), *Layers of Perception, Proceedings of the 35th Conference on Computer Applications and Quantitative Methods in Archaeology, Berlin* (Bonn: Dr Rudolf Habelt), 322–7.
——(2009), 'Clubs and Lounges of Roman Ostia: The Spatial Organisation of a Boomtown Phenomenon (Space Syntax Applied to the Study of Second Century AD "Guild Buildings" at a Roman Port Town)', in D. Koch, L. Marcus, and J. Steen

(eds.), *Proceedings of the 7th International Space Syntax Symposium* (Stockholm: KTH), 108:1–108:12.

——(2010), 'Roman Ostia: Society and Urban Infrastructure during the 2nd century AD' (Unpublished PhD Dissertation, Universiteit Leiden).

Storey, G. (2004), 'Roman Economies. A Paradigm of Their Own', in G. M. Feinman and L. M. Nicholas (eds.), *Archaeological Perspectives on Political Economies* (Salt Lake City: University of Utah Press), 105–28.

Sudjic, D. (1992), *The 100 Mile City* (London: Deutsch).

Sullivan, J. P. (1991), *Martial: The Unexpected Classic* (Cambridge: Cambridge University Press).

Sumi, G. S. (2011), 'Topography and Ideology: Caesar's Monument and the Aedes Divi Iulii in Augustan Rome', *Classical Quarterly* 61/1, 205–29.

Swain, S., Harrison, S., and Elsner, J. (eds.) (2007), *Severan Culture* (Cambridge: Cambridge University Press).

Swann, B. W. (1994), *Martial's Catullus: The Reception of an Epigrammatic Rival* (Zurich: Olms).

Tagliamonte, G. (1995), 'Forum Romanum (fino alla prima età Repubblicana)', *LTUR* II (D–G): 313–25.

Tarpin, M. (2002), *Vici et pagi dans l'Occident romain* (Rome: École française de Rome).

Taub, L. (1993), 'The Historical Function of the *Forma Urbis Romae*', *Imago Mundi*, 45: 9–19.

Taylor, L. R. (1966), *Roman Voting Assemblies: From the Hannibalic War to the Dictatorship of Caesar* (Ann Arbor: University of Michigan Press).

Taylor, R. (2003), *Roman Builders: A Study in Architectural Process* (Cambridge: Cambridge University Press).

Tedeschi Grisanti, G. (1977), *I 'Trofei di Mario': il Ninfeo dell'Acqua Giulia sull'Esquilino* (Rome: Istituto di Studi Romani).

——(1996), 'Nymphaeum Alexandri', *LTUR* III (H–O): 351–2.

Testini, P. (1989), 'Nota di topografia romana: gli edifici del prete ilicio', in P. Pergola and F. Bisconti (eds.), *Quaeritur Inventus Colitur* (Vatican City: Pontificio Istituto di Archeologia Cristiana), ii. 781–93.

Thaler, U. (2005), 'Narrative and Syntax: New Perspectives on the Late Bronze Age Palace of Pylos, Greece', in A. van Nes (ed.), *5th International Space Syntax Symposium Proceedings* (Delft: TU Delft): 323–39.

The Shorter Oxford English Dictionary (1983), i (Oxford and London: Oxford University Press).

Thédenat, H. (1923), *Le Forum romain et les Forums impériaux* (Paris, Hachette).

Thomas, E. V. (2007a), *Monumentality and the Roman Empire: Architecture in the Antonine Age* (Oxford: Oxford University Press).

——(2007b), 'Metaphor and Identity in Severan Architecture: The Septizodium between Reality and fantasy', in Swain, Harrison, and Elsner (eds.), 327–67.

Thomas, M. L. (2004), 'Domitian's Horse of Glory: The *Equus Domitiani* and Flavian Urban Design', *Memoirs of the American Academy in Rome* 49: 21–46.

Thompson, J. (1988), 'Pastoralism and Transhumance in Roman Italy', in C. R. Whittaker (ed.), *Pastoral Economies in Classical Antiquity* (Cambridge: Cambridge Philological Society), 213–15.

Tilley, C. (1994), *A Phenomenology of Landscape: Places, Paths, and Monuments* (Oxford: Berg).

Tommasino, E. (2004), 'Oltre lo sterro: Scavi stratigrafici inediti nelle domus pompeiane', *Rivista di studi pompeiani*, 15: 15–49.

Toner, J. (2002), *Rethinking Roman History* (Cambridge: The Oleander Press).

Tonkiss, F. (2005), *Space, the City and Social Theory: Social Relations and Urban Forms* (Oxford: Polity Press).

Torelli, M. (1980), 'Innovazioni nelle tecniche edilizie romane tra il I sec. AC e Il I sec. DC', *Tecnologia, economia e società nel mondo romano: Atti del Convegno di Como 27/28/29 Settembre 1979* (Tecnologia, economia e società nel mondo romano, Como), 139–59.

Tortorici, E. (1991), *Argiletum: Commercio, speculazione edilizia e lotta politica nell'analisi topografica di un quartiere di Roma di età repubblicana* (Rome: 'L'Erma' di Bretschneider).

Treggiari, S. M. (1975), 'Jobs in the Household of Livia', *Papers of the British School at Rome* 43: 48–77.

——(1991), *Roman Marriage: Iusti Coniuges from the Time of Cicero to the time of Ulpian* (Oxford: Clarendon Press).

Trifilò, F. (2008), 'Power, Architecture and Community in the Distribution of Honorary Statues in Roman Public Space', in C. Fenwick, M. Wiggins, and D. Wythe (eds.) *TRAC 2007. Proceedings of the Seventeenth Annual Theoretical Roman Archaeology Conference, London 2007* (Oxford: Oxbow), 109–20.

——(2009), 'The Social Production and Social Construction of Roman Public Space: Creation and Change in Two Fora of Trajanic Foundation' (Unpublished PhD Dissertation, Birkbeck, University of London).

——(forthcoming), 'Public architecture and urban living in the Roman city: The example of the forum at Timgad', *BABesch* 86.

Trimble, J. (2007), 'Visibility and Viewing on the Severan Marble Plan', in Swain, Harrison, and Elsner (eds.), 368–84.

Troilo, S. (2005), *La patria e la memoria: Tutela e patrimonio culturale nell'Italia unita* (Milan: Mondadori Electa).

Trout, D. (1996), 'Town, Countryside and Christianization at Paulinus' Nola', in R. W. Mathisen and H. S. Sivan (eds.), *Shifting Frontiers in Late Antiquity* (Aldershot: Variorum), 175–86.

Tsujimura, S. (1991), 'Ruts in Pompeii: The Traffic System in the Roman City', *Opuscula Pompeiana* 2: 58–86.

Tuan, Y. F. (1977), *Space and Place: The Perspectives of Experience* (Minneapolis: University of Minnesota Press).

Tucci, P. L. (1994), 'Il tempio dei Castori in Circo Flaminio: La lastra di via Anicia', in L. Nista (ed.), *Castores: l'immagine dei Dioscuri a Roma* (Rome: De Luca), 123–8.

——(2008), 'Galen's Storeroom, Rome's Libraries, and the Fire of A.D. 192', *Journal of Roman Archaeology* 21: 133–49.

Tuppi, J. P. (2010), 'Traffic Bottlenecks in South Etraria?, Comparing the Archaic Road Cutting Widths with Ancient Vehicles', *Arctos: Acta Philologica Fennica* 44: 263–88.

Turner, A. (2004), *Depthmap 4—A Researcher's Handbook* (London: Bartlett School of Graduate Studies, UCL).

Ulrich, R. B. (1993), 'Julius Caesar and the Creation of the Forum Iulium', *American Journal of Archaeology* 97: 49–80.

Ungaro, L. (2005), 'I mercati di Traiano: aspetti funzionali e strutture', *Bullettino della Commissione archeologica comunale di Roma* 104: 205–18.

Urry, J. (2007), *Mobilities* (Oxford: Polity Press).

Valentini, R., and Zucchetti, G. (1940) (eds.), *Codice topografico della città di Roma* (Rome: Tipografia del Senato), vol. 1.

——— (1953), *Codice topografico della città di Roma* (Rome: Tipografia del Senato).

Vallat, D. (2008), *Onomastique, culture et société dans les Épigrammes de Martial* (Brussels: Collection Latomus).

Van Deman, E. B. (1922), 'The Sullan Forum', *Journal of Roman Studies* 12: 1–31.

Van den Bergh, R. (2003), 'The Plight of the Poor Urban Tenant', *Revue internationale des droits de l'antiquité* 50: 443–77.

Van der Poel, H. (1986), *Corpus Topographicum Pompeianum*, IIIA (Austin: University of Texas Press).

Van Essen, C. (1957), 'A propos du plan de la ville d'Ostie', *Hommage à W. Deonna Coll. Latomus* 28: 509–13.

Van Nes, A. (2002), 'Road Building and Urban Change: The Effect of Ring Roads on the Dispersal of Shop and Retail in Western European Towns and Cities' (Unpublished PhD Dissertation, Agricultural University of Norway).

——— (2005), 'Typology of Shopping Areas in Amsterdam', in A. van Nes (ed.), *Proceedings Space Syntax: 5th International Symposium* (Amsterdam: Techne Press).

Van Tilburg, C. R. (2007), *Traffic and Congestion in the Roman Empire* (London and New York: Routledge).

Van Zantwijk, R., de Ridder, R., and Braakhuis, E. (eds.) (1990), *Mesoamerican Dualism, Dualismo Mesoamericano* (Utrecht: R.U.U.-I.S.O.R.).

Vanderbilt, T. (2009), *Traffic: Why We Drive the Way We Do (and What It Says About Us)* (London: Penguin).

Varone, A. (2008), 'Per la storia recente, antica e antichissima del sito di Pompei', in P. G. Guzzo and M. P. Guidobaldi (eds.), *Nuove ricerche archeologiche nell'area vesuviana (Scavi 2003–2006)* (Rome: 'L'Erma' di Bretschneider), 349–61.

Vasaly, A. (1993), *Representations: Images of the World in Ciceronian Oratory* (Berkeley and Los Angeles: University of California Press).

Vasi, G. (1747), *Delle Magnificenze di Roma Antica e Moderna*, vol. i (Rome: Chracas).

——— (1756), *Delle Magnificenze di Roma Antica e Moderna*, vol. vii (Rome: Chracas).

Viae (1991), *Viae publicae romanae: Roma, Castel Sant'Angelo, 11–25 aprile 1991* (Rome: Leonardo-De Luca editori).

Vigneron, P. (1968), *Le Cheval dans l'antiquité Gréco-Romaine* (Nancy: Faculté des lettres et des sciences humanies de l'Université de Nancy).

Vioque, G. G. (2002), *Martial, Book VII: A Commentary* (Leiden: Brill).

Viscogliosi, A. (1996), 'Hercules Musarum, Aedes', *LTUR* III (H–O): 17–19.
——(1999), 'Porticus Octaviae', *LTUR* IV (P–S): 141–5.
Voigt, M. (1903), 'Die römischen Baugesetze', *Berichte über die Verhandlungen der königlich sächsischen Gesellschaft der Wissenschaften zu Leipzig* 12: 175–98.
Von Gerkan, A. (1940), *Der Stadtplan von Pomeji* (Berlin: Archäologisches Institut des Deutschen Reich).
Vout, C. (1996), 'The Myth of the Toga: Understanding the History of Roman Dress', *Greece and Rome* 43: 204–20.
——(2007), 'Sizing up Rome, or Theorizing the Overview', in Larmour and Spencer (2007b), 295–322.
Wagener, A. P. (1912), *Popular Associations of Right and Left in Roman Literature* (Baltimore: J. H. Furst company).
Wallace, W. (1994), *Michelangelo at San Lorenzo: The Genius as Entrepreneur* (Cambridge: Cambridge University Press).
Wallace-Hadrill, A. (1982), '*Civilis Princeps*: Between Citizen and King', *Journal of Roman Studies* 72: 32–48.
——(1988), 'The Social Structure of the Roman House', *Papers of the British School at Rome* 56: 43–97.
——(1990), 'Roman Arches and Greek Honours: the language of power at Rome'. *Proceedings of the Cambridge Philological Society* 31: 143–81.
——(1991), 'Elites and Trade in the Roman Town', in A. Wallace-Hadrill and J. Rich (eds.), *City and Country in the Ancient World* (London: Routledge), 241–72.
——(1994), *Houses and Society in Pompeii and Herculaneum* (Princeton: Princeton University Press).
——(1995), 'Public Honour and Private Shame: The Urban Texture of Pompeii', in Cornell and Lomas (eds.), 39–62.
——(1998a), '*Horti* and Hellenization', in M. Cima and E. La Rocca (eds.), Horti Romani: *Atti del Convegno Internazionale, Roma, 4–6 Maggio 1995* (Rome: 'L'Erma' di Bretschneider), 1–12.
——(1998b), 'The Villa as Cultural Symbol', in A. Frazer (ed.), *The Roman Villa: Villa Urbana* (Philadelphia: The University Museum, University of Pennsylvania), 43–53.
Wallace-Hadrill, A. (2003), 'The Streets of Rome as a Representation of Imperial Power', in L. de Blois, P. Erdkamp, O. Hekster, G. De Kleijn, and S. Mols (eds.), *The Representation and Perception of Roman Imperial Power: Proceedings of the Third Workshop of the International Network Impact of Empire (Roman Empire, c. 200 BC–AD 476)* (Amsterdam: J. C. Gieben), 189–206.
——(2008), *Rome's Cultural Revolution* (Cambridge: Cambridge University Press).
Waltzing, J. P. (1895–1900), *Étude historique sur les corporations professionnelles chez les Romains depuis les origines jusqu'à la chute de l'Empire d'Occident*, 4 vols. (Louvain: Charles Peeters).
Ward-Perkins, J. B., and Claridge, A. (1980), *Pompeii. AD 79* (Boston: Boston Museum of Fine Arts).
Wataghin Cantino, G. (1966), *La Domus Augustana: Personalità e problemi dell'architettura Flavia* (Turin: G. Giappichelli).
Weiss, C. F. (2010), 'Performativity of Place: Movement and Water in Second Century A.D. Ephesus', in A. J. Moore, E. Harris, P. Girdwood, G. Taylor, and L. Shipley

(eds.), *TRAC 2009: Proceedings of the 19th Theoretical Roman Archaeology Conference Southampton & Michigan 2009* (Oxford: Oxbow Books), 66–74.

Weiss, C. J. (2010), 'Determining Function of Pompeian Sidewalk Features through GIS Analysis', in B. Frischer, J. Webb Crawford, and D. Koller (eds.), *Making History Interactive: Computer Applications and Quantitative Methods in Archaeology (CAA): Proceedings of the 37th International Conference, Williamsburg, Virginia, USA March 22–26, 2009* (BAR International Series 2079, Oxford: Archaeopress), 363–72.

Welch, E. (2005), *Shopping in the Renaissance: Consumer Cultures in Italy 1400–1600* (New Haven and London: Yale University Press).

Welch, K. E. (2007), *The Roman Amphitheatre: From Its Origins to the Colosseum* (Cambridge: Cambridge University Press).

Westall, R. (1996), 'The Forum Iulium as Representation of Imperator Caesar', *Mitteilungen des Deutschen Archäologischen Instituts, Römische Abteilung* 103: 83–118.

Westfall, C. W. (2007), 'Urban Planning, Roads, Streets, and Neighborhoods', in Dobbins and Foss (eds.), 129–39.

White, J. (1985), *Heracles' Bow: Essays on the Rhetoric and Poetics of Law* (Madison: University of Wisconsin Press).

White, K. D. (1984), *Greek and Roman Technology* (London: Thames and Hudson).

Whittaker, C. R. (1985), 'Trade and Aristocracy in the Roman Empire', *Opus* 4: 49–75.

Whyman, S. E. (2007), 'Sharing Public Spaces', in Brant and Whyman (eds.), 43–61.

Whyte, W. H. (1980), *The Social Life of Small Urban Spaces* (Washington: Project for Public Spaces Inc.).

——(1981), *City Spaces, Human Places* (Boston: WGBH).

Williamson, C. (2005), *The Laws of the Roman People: Public Law in the Expansion and Decline of the Roman Republic* (Ann Arbor: University of Michigan Press).

Wilson, A. (2001), 'The Water-Mills on the Janiculum', *Memoirs of the American Academy in Rome* 45: 219–46.

Wilson, F. H. (1935), 'Studies in the Social and Economic History of Ostia, Part 1', *Papers of the British School at Rome*, 13: 41–68.

Winter, J. (1993), *London's Teeming Streets 1830–1914* (London and New York: Routledge).

Wirth, H. (2010), *Die linke Hand: Wahrnehmung und Bewertung in der griechischen und römischen Antike. HABES: Heidelberger Althistorische Beiträge und Epigraphische Studien Bd. 4* (Stuttgart).

Wiseman, T. P. (1993), 'Campus Martius', *LTUR* I (A–C): 220–4.

——(1998), 'A Stroll on the Rampart', in Cima and La Rocca (eds.), 13–22.

——(2004), 'Where was the *Nova Via*?', *Papers of the British School at Rome* 72: 167–83.

——(2007), 'Where was the *porta Romanula*?', *Papers of the British School at Rome* 75: 231–7.

——(2008), 'Rethinking the Roman Triumph', *Journal of Roman Archaeology* 21: 389–91.

——(2009), *Remembering the Roman People: Essays on Late-Republican Politics and Literature* (Oxford: Oxford University Press).

Wolters, X. F. M. G. (1935), *Notes on Antique Folklore on the Basis of Pliny's "Natural History" Bk. XXVIII*.22–29 (Amsterdam: H. J. Paris).

Yates, F. A. (1966), *The Art of Memory* (Chicago, Ill.: University of Chicago Press).

Yavetz, Z. (1958), 'The Living Conditions of the Urban Plebs in Republican Rome', *Latomus* 17: 500–17.

Yegül, F. (1994), 'Street Experience in Ancient Ephesus', in Z. Celik, D. Favro, and R. Ingersoll (eds.), *Streets of the World, Critical Perspectives on Public Space* (Berkeley and Los Angeles: University of California Press), 121–41.

——(2004), 'Review of Roman Builders, A Study in Architectural Process', *Journal for the Society of Architectural Historians* 63: 387–9.

Yeo, C. A. (1946), 'Land and Sea Transportation in Imperial Italy', *Transactions of the American Philological Association* 77: 221–44.

Zanker, P. (1988), *The Power of Images in the Age of Augustus* (Ann Arbor: University of Michigan Press).

——(1994), 'Veränderungen im öffentlichen Raum der Italischen Städte der Kaiserzeit', in *L'Italie d'Auguste à Dioclètien* (Trieste: École francaise de Rome), 259–83.

——(1998), *Pompeii. Public and Private Life*, trans. D. L. Schneider (Cambridge, Mass.: Harvard University Press).

——(2002), 'Domitian's Palace on the Palatine and the Imperial Image', in A. K. Bowman, H. M. Cotton, M. Goodman, and S. Price (eds.), *Representations of Empire: Rome and the Mediterranean World*, Proceedings of the British Academy 114: 105–30.

——(2010), 'By the emperor, for the people: "popular" architecture in Rome', in B. C. Ewald and C. F. Noreña (eds.), *The Emperor and Rome: Space, Representation, Ritual. Yale Classical Studies* 35 (Cambridge: Cambridge University Press), 45–87.

Zeggio, S. (2005), 'Roma, Valle del Colosseo, Scavo dell'area della Meta Sudans (1996–2002): Spazi urbani e storia', in P. Attema, A. Nijboer, and A. Zifferero (eds.), *Papers in Italian Archaeology*, VI (BAR International Series 1452 (I), Oxford: Archaeopress), 269–77.

Zehnacker, H. (2008), 'La Description de Rome dans le livre V du *de lingua latina* de Varron', in Fleury and Desbordes (eds.), *Roma illustrata: Représentations de la ville, actes du colloque international de Caen, 6–8 octobre 2005* (Caen: Presses universitaires de Caen), 421–32.

Ziółkowski, A. (1989), 'The *Sacra Via* and the Temple of Iuppiter Stator', *Opuscula Romana* 17: 225–40.

Index

Numbers in *italics* indicate their presence in an image, a table or a caption.
Words in *italics* are in Latin

access 6, 97, 143, 146, 180, 187, 197, 230, 275, 279, 289, 291–2, 299, 302–7, 311, 313, 328, 381
   analysis 232, 240
   network 242
   roads 221, 234, 235
   route 235, 315, 330, 335
   vehicular access 180, 201, 203
accessibility 3, 6, 21, 144, 228, 235, 290, 291, 294, 303–5, 350, 360, 361; *see also* access
agency 30, 387, 391
angiportus 185, 247, 296,
animal 11, 137, 181, 260, 369
ambulatio 46, 49
archaeology of walking 46, 264, 289
Argiletum 32, 41, 83, 86–7, 92–5, 131, 282, 295, 298–9, 306–8, 316, 343, 346, 363
Augustus 86, 293, 307, 311, 364, 372, 374, 392–3

Basilica Iulia 128, 259, 300, 315, 328, 343
boundary 8, 32, 305; *see also* separation

Campus Martius 88, 91, 94, 392
cart 141, 154, 174–93, 195–6, 207, 209, 210, 249, 260, 270, 367, 369
Capitoline 27, 73–5, 172, 187, 252, 295, 297, 301, 305–6, 337, 340, 343, 348, 352
Casa del Menandro 204–11, 369
Casa dei Triclini 225, 231, 238, 240
castrum 219–20, 223, 235
Cato 205–13, 311
   De Agricultura 205
cenaculum 83–4, 87, 89, 90
Cicero 60, 66, 80, 128–9, 297, 300–301, 311, 326, 341, 391, 400
   De Oratore 301
   Pro Sestio 301
cippi 12, 42, 148, *149*
Circus Maximus 77, 300, 328
citizen identity 36, 71–3
citizenship 57, 63
clivus 247, 296
Clivus Argentarius 305–6, 308, *338*, 343, 352
Clivus Capitolinus 75, 352

Clivus Suburanus 88, 246, 267, 340, 344–6, 363
collegium 140, 215–18, 225, 227, 234, 238, 241–2, 339–40, 389, 390
compita 98, 247, 388, 390
congestion 16, 137, 141, 272, 394
control 153, 158–9, 170, 186–7, 221, 258, 302, 304
   architectural 293
   crowd 288, 301
   of movement 6, 292, 304, 310
   social 101, 112
   of space 293
   traffic 365
Corpus Inscriptionum Latinarum 217
crowd 21, 88, 128, 391
currere 62, 64–5
cursor 62, 64–5, 70, 191

danger 188, 192, 249, 257, 340, 394
decumanus 113, 115, 170, 221–5, 227, 234–7, 242
destination 20, 59, 178, 196, 211, 228, 242, 251, 275, 292, 308
Dionysius of Halicarnassus 248, 296, 393
dirt 66, 130, 246, 282, 350, 392
distance minimization 6, 294
Divorum 273–4, 277–9, 286, 288
Domitian 81–2, 84, 86, 91–5, 98, 138, 170, 248, 257, 279, 286, 290, 291, 305, 309
domus 84, 89, 90–1, 96, 227, 234, 252, 290–1
Domus di Marte 227, 233, 238
doorway 160–62, 164–5

economic opportunity 30, 42, 204, 210, 253
economy 195, 210, 214, 384, 401
   transport 195–6, 204, 212
   urban 29, 45, 246, 253, 366
entrance 24, 95, 97, 100, 104, 108–17, 156, 163, 238, 241, 275–6, 307, 379
epigraphy 218, 225, 227, 400

foot 17, 63, 167
Forma Urbis Romae 125, 247, 248, 264–89, 334, 339, 340, 345
Forum Augustum 279, 306–7

Forum Boarium 76, 130, 252, 294, 328, 348, 372
Forum Esquilinum 364, 366, 381
Forum Iulium 130, 305–6. 311
Forum Nervae 87, 93, 263, 279, 299, 309, 343
Forum Romanum 77, 126–30, 279, 292–300, 304–6, 309–16, 328–30, 334–8, 343–4, 348–50, 359, 388, 391
Forum Transitorium; *see* Forum Nervae
fountain 98, 107–8, 149–51, 376

gambling 259, 319, 323–6; *see also* game of chance
game 312, 319, 329
   board (inscribed) 312, 326, 329, 331
   of skill 319, 323, 326, 329, 331
   of chance 319, 323, 329, 331
gaming 50, 259, 312, 315, 327, 329;
   see also game
garden 109–10, 275, 279, 284
gate 29, 177, 180, 196, 201–4, 209, 214, 223–4, 257, 364, 371–5, 379, 384
guild 44, 191, 215–41, 389; *see also schola* and *collegium*

Herculaneum 137–8, 145, 149, 157, 159, 163, 264
Horace 22, 36, 97, 137, 188, 251, 295–7, 309
hortus 24, 364, 368, 381

image 52, 91, 99, 265–6, 337
impediment 42, 177–81, 192, 272; *see also* obstruction
interaction 2, 129, 131, 261, 305
   social 100, 104, 122, 127, 230, 329
inn 30, *200*, 202, 209–14, 326

Julius Martialis 89, 96–7
Justinian 137, 144, 324–7
   Code 185–6
   Digesta 137, 144, 324–7
Juvenal 81, 82, 124, 130, 132, 135, 137, 143, 189, 248, 334, 340, 372, 389, 398
   Satire 81, 135–6, 334, 389
   Umbricius 135–6, 143, 159, 248–9

kinaesthetic 119–22

language 57–60, 66, 70, 72, 225
law 16, 144–7, 174–5, 184, 186–8, 319, 323–5, 333, 370
leisure 48, 86, 262, 265, 272–3, 275–6, 279, 289, 367
Lex Iulia Municipalis 14–15, 17–18, 34, 64, 143, 147–8, 152, 159, 174, 179, 249, 269, 332, 342, 370

Livy 145, 248, 294–5
locus celeberrimus 3, 20–26, 28, 295, 304

macellum 96, 172, 376–7, 381–2, 384
map 97, 120, 123, 130, 340, 393
   Depthmap 101, 113, 116, 230, 235
Martial 82–99, 129–32, 156–7, 260, 279, 298, 308, 344, 372, 389–92, 398
   Epigrams 82–99, 131, 389, 392, 398
memory 35, 57, 68, 70, 71
monumental building 45, 51, 196, 263
movement
   economy 26, 216, 228, 241, 384, 394
   habitus of 12, 49
   natural 6, 35, 44, 291, 293, 297
   patterns of 22, 69, 225, 307, 328, 401
   seeking 29, 44, 53, 229, 238, 241
   through 21, 45, 292, 308, 318
   through movement potential 49, 291, 293–9, 308
   to 20, 29, 44, 292, 300, 308
multisensory approach 119, 124, 132;
   see also sensory experience

neighbourhood 70, 190, 247, 261, 342, 398
Nero 94, 96, 173, 251, 345, 377, 393
Nerva 81, 92, 94, 291, 309
noise 12, 41, 60, 123, 129, 189
nuisance 42, 136, 141–7, 153, 158–9, 188

obstruction 42, 136, 143, 145, 151, 158–9, 340, 349
order 66, 137, 142, 152, 162, 288
Ostia 105, 160, 162, 164–166, 170–2, 215–42, 390, 397

Palatine 26–7, 32, 72–5, 86–7, 95, 172, 189, 276, 290, 293, 297–7, 302, 348, 359, 392
path 45, 69, 285, 288, 374; *see via* and *semita*
passage 4, 135–137, 160, 192, 275
pedestrian 63, 106, 115, 137, 140, *156*, 158, 163, 238, 241, 249, 292, 370
perception 3, 132, 291, 330, 392
   of space 22, 120, 295, 308
phenomenology 119, 121
plan 137, 169, 176, 219, 265, 315
plantings 280, 282–283, 285, 288
plaustrum/plostrum 11, 147, 175, 182–8, 295, 15, 147, 174–5, 370
Pliny the Elder 84, 93, 98, 166–7, 188–90, 246–8, 282, 294, 300, 311, 341, 348, 364, 375, 400
Pliny the Younger 290–2
Pompeii 100–17, 136–59, 160–73, 175, 178, 181, 192–3, 194–7, 209–11, 221, 241, 263–4, 267, 270, 325, 390

Index

popina 291, 325–6
population 251–2, 261, 372, 383, 398
Porta Capena 3, 19–26, 30, 52, 191, 294, 342, 366, 371–2, 391
Porta Ercolano 180, 212
Porta Esquilina 361–84
Porta Stabia 163, 210
Porta Tiburtina 52, 361–2, 375–6, 378–84, 389
porter 191, 195, 204, 369
portico 49, 79, 258, 272–9, 383
Porticus Liviae 131, 273–8
Porticus Octaviae 265, 273–5, 277
procession 42, 140, 148, 175, 187, 249
praedia of Iulia Felix 213–14, 270
public amenities 5, 104, 159, 163, 273, 279

ramp 42, 156, 159, 196–204
reconstruction 142, 195, 245, 250, 298, 306
restriction 6, 15–20, 144, 188, 300
  on movement 293, 302, 305
  on space 302
rhythmanalysis 6, 36, 40, 70, 88
Rome 68, 71–2, 79, 81–99, 118–32, 136, 162, 170, 172, 174–5, 184, 189, 245–61
Rostra 123, 128, 299, 350, 355
runner 61–3, 191
running 61, 62, 67
rus 78, 89–92, 96

Sacra Via 98, 247, 296–8, 302, 316, 328, 335, 343, 355
salutatio 87–8, 93–4
scale 100, 192
  human scale 9, 37
  macro-scale 100, 108, 113, 116, 117
  micro-scale 100, 108, 116, 117, 224
schola 216–9, 224–5, 227, 229–35, 238–42, 397
semita 98, 185, 247
Seneca 124, 129, 131, 132, 156, 188, 248, 255, 390
sensory experience 39, 124, 131
  multisensory experience 119, 131, 246
Septimius Severus 225, 264, 342, 358
  Arch of 315, 317, 329, 334–7, 359
shop 42, 106, 115, 162, 169, 173
  shopfront 159, 162–3, 171–2
  workshop 115, 144, *200*, 203, 257
short cut 6, 49, 288, 291, 294–7
Shrines of the Argei 74–6
sidewalk 28, 42, 154, 201, 248, 263–70, 278, 374
space syntax analysis 32, 44, 219, 229, 241, 401
spatial analysis 7, 101, 215, 228, 239

spatial theory 4, 293
'Spatial Turn' 2, 53, 82, 95, 386, 397
speed 13, 27, 61–3, 251
stables 30, 194, 196, 199–214
statuary 23, 217
statue 23, 282, 312, 326, *327*, 352
stepping stone 138, 184, 248
stoa 79; *see also* portico
storage 191, 195, 204, 254, 375
Strabo 298, 303, 308, 392
street
  grid 106, 114, 176, 219, 237, 389
  network 105, 176, 219–20, 228, 362
Subura 72, 87, 90–1, 93–5, 121, 124, 126, 130–2, 279, 282, 298, 307
suburbium 365, 384, 385
Suessa 205, 213
Suetonius 79, 86, 94, 187, 260, 293, 307, 311, 394

taberna 98, 106, 233, 241, 257
Tacitus 95, 170, 251, 389, 394
threshold 156, 160–7, 169, 171–3
Tabula Heracleensis;
  see Lex Iulia Municipalis
Templum Divi Claudi 265, 273, 277–9, 284–6, 288
Templum Pacis 87, 264, 273–5, 277–85, 288, 343
Terme dei Cisiarii 191, 223
Tiber 65, 74, 222, 343, 363, 393
  Valley 399
Traffic:
  ambulatory 42, 162–3
  flow 8, 174–7, 184, 192, 312
  law 175, 184, 188
  management 12, 193, 339
  pattern 180, 183, *283*, 284, 288
  regulations 3, 14, 368
  system 12, 169, 195, 211
travel 71, 196, 208, 367, 391, 401

Ulpian 144, 186, 155
urbanism 2, 107, 216, 263, 386, 390
urbs 78, 95

Varro 57–81, 185, 206, 299, 302–3, 392
  De Lingua Latina 57, 79
Vespasian 86, 260, 273–4, 278–80, 292, 294, 296, 340, 394
via 28, 45, 69, 184, 186, 247
Via Appia 267, 399
Via dell'Abbondanza 106–9, 113, 141, 148, 150, 178–9, 181, 213, 267
Via della Fortuna 115, 181, 185
Via Flaminia 266, 344

Via Ostiensis 267
Via Stabiana 106, 113, 151, 163, *183*, 199, 267
Via Tiburtina 361, 366-7, 383
vicus 38, 45, 98, 247, 393
Vicus Tuscus 10, 22, 50-1, 75, 120-1, 130, 294, 315-6, 328, 343, 347-9
Vicus Iugarius 51, 295, 315-6, 328, 343, 347, 349
villa 78, 89-90, 206, 367-8

visibility 50-1, 124, 129, 312, 325
  intervisibility 100, 109, 110-13, 120

wagon 15, 188-9, 293, 340, 370;
  see also plaustrum/plostrum
walker 46, 48-9, 278, 285, 288
wheel ruts 3, 9, 12-15, 102, 182

zoning 42, 44